Study Guide/Solutions Manual to Accompany Organic Chemistry

Fifth Edition

MAITLAND JONES, Jr.

New York University

HENRY L. GINGRICH

Princeton University

STEVEN A. FLEMING

Temple University

W · W · NORTON & COMPANY NEW YORK · LONDON

W. W. Norton & Company has been independent since its founding in 1923, when William Warder Norton and Mary D. Herter Norton first published lectures delivered at the People's Institute, the adult education division of New York City's Cooper Union. The Nortons soon expanded their program beyond the Institute, publishing books by celebrated academics from America and abroad. By mid-century, the two major pillars of Norton's publishing program—trade books and college texts—were firmly established. In the 1950s, the Norton family transferred control of the company to its employees, and today—with a staff of four hundred and a comparable number of trade, college, and professional titles published each year—W. W. Norton & Company stands as the largest and oldest publishing house owned wholly by its employees.

Composition by codeMantra
Manufacturing: RR Donnelley, Kendallville

Editor: Erik Fahlgren
Associate Editor, Digital Media: Jennifer Barnhardt
Project Editor: Carla L. Talmadge
Production Manager: Eric Pier-Hocking
Managing Editor, College: Marian Johnson
Assistant Editor: Paula Iborra

ISBN 978-0-393-93659-9

W. W. Norton & Company, Inc., 500 Fifth Avenue, New York, NY 10110
www.wwnorton.com

W. W. Norton & Company Ltd., Castle House, 75/76 Wells Street, London W1T 3QT

2 3 4 5 6 7 8 9 0

Contents

Introduction

In this Study Guide we try to go beyond straightforward "bare bones" answers to problems. Once one gets past simple questions, problem solving in organic chemistry becomes very hard to teach. The answers to many problems are intuition-intensive, and therefore it becomes difficult to explain in explicit terms how to proceed. Nonetheless, there are things that can be said that may help; there are techniques of problem solving that can be learned. In this collection of comments on solutions to all the problems that are not solved in the text itself, we try to help you to do that. As in the text, we seek in these pages to show you how to approach problems in organic chemistry in general, not just how we got from here to there in specific cases.

The exercises in these chapters become much easier when we have an idea of where we are going. What exactly are we trying to do in this problem? What tasks must be accomplished? What bonds are we trying to make or break? What rings must be closed or opened? Such questions seem simple, but it is amazing how few people really start problems with the simple question: "What happens in this reaction?" Analyze! Once a goal is in mind, the path to that goal becomes much, much easier. In a sense, a good problem solver has learned, first and foremost, to avoid "thrashing." We know that is a flip remark, but it is, nonetheless, true. A person solving a tough problem is like a bacterium swimming up a food gradient—he or she (or it) is following a pathway that "feels good." We will try to show you how to do that in this Study Guide, but there can be no denying that experience is important, and experience can be gained only by practice. Practice and more practice will teach you what feels good in terms of problem solving—of how to swim up that food gradient—but we will try to give you some hints along the way.

The problems solved in the Study Guide will recapitulate each chapter, and thus will generally start off with the easier examples and then go on to tougher stuff. Don't worry if the hard problems at the end of the sections do not come so easily; they are meant to tax you, to demand some hard work and careful thought. Some difficult problems will be dealt with best over time. If a problem resists solution, and some will, come back to it after a while; let your subconscious work on it for a while. Most research chemists carry unsolved problems around in their heads for a long time, sometimes for years, returning to them now and then. There is nothing wrong with emulating that process. People think at vastly different rates, and it is a rare situation that requires a rapid solution of a problem. (Hour exams may be an exception, unfortunately.)

Many of us who actually do organic chemistry for a living (believe it or not there are such people) typically get great pleasure from problem solving. We hope that you will be similarly stimulated. In a fundamental way that is what we humans are about. We have evolved to be curious and to turn over rocks to see what is underneath. Perhaps, thinks our ancestral hunter-gatherer, I will find something good to eat! From such imperatives we humans have become problem solvers, and it gives us pleasure to work out what's happening in unknown situations.

Here is one favorite example, which makes an important point about problem solving. This lesson is so simple as to be trivially obvious and yet at the same time so profound as to be most difficult in practice. MJ and HG had been working for some time on the synthesis of the following pterodactyl-shaped compound:

One of us (MJ) had become entrapped in devising increasingly clever "solutions" that a series of graduate students and undergraduates had not been able to make work—and for good reasons. Those clever solutions were complicated, and extraordinarily hard to carry out in a practical sense. While MJ was away one July, HG had the wit to avoid all the foolish "cleverness" and to do what we beseech our students to do—to "think simple." HG went back to basics, did the work himself, and solved the problem. MJ arrived back in Princeton and was presented with a vial containing exquisitely beautiful crystals of the long-sought compound. What's the lesson? Don't be too clever. If you ground yourself in the basics, analyze what you want to do, and then apply those basics, you will prosper.

Remember, think simple. As Ted Williams is supposed to have said: "If you don't think too good, don't think too much."

Maitland Jones, Jr.
Henry L. Gingrich
Steven A. Fleming

Atoms and Molecules; Orbitals and Bonding

1

In this chapter, you are learning a bit about atomic structure and acquiring skills that you will need throughout your study of organic chemistry. In a sense, you are learning vocabulary and grammar that will enable you to write sentences a little later and eventually to compose whole paragraphs and short stories. The problems of this chapter concentrate on tool building and require less thought and imagination than those of later chapters. That does not mean that they are unimportant. Even though much of this chapter may review what you already know, please do not skip past it until you are certain that you can write Lewis structures easily, determine the position of charges without error, and use the arrow formalism to write resonance forms with ease. These skills will be as necessary in Chapter 24 as they are now.

Problem 1.2 See Problem 1.1. If the fifth and sixth electrons in a carbon atom occupied the same orbital with parallel spins, they would have the same values for all four quantum numbers, a violation of the Pauli principle. If they occupy the same orbital, their spin quantum numbers (s) *must* be opposite (paired).

Problem 1.3 In this row, we fill the $n = 3$ energy levels, first filling the $3s$ orbital, then moving on to the higher energy $3p$ orbitals.

$_{11}$Na $1s^2\, 2s^2\, 2p_x^{\,2}\, 2p_y^{\,2}\, 2p_z^{\,2}\, 3s$
$_{13}$Al $1s^2\, 2s^2\, 2p_x^{\,2}\, 2p_y^{\,2}\, 2p_z^{\,2}\, 3s^2\, 3p_x$
$_{15}$P $1s^2\, 2s^2\, 2p_x^{\,2}\, 2p_y^{\,2}\, 2p_z^{\,2}\, 3s^2\, 3p_x\, 3p_y\, 3p_z$
$_{16}$S $1s^2\, 2s^2\, 2p_x^{\,2}\, 2p_y^{\,2}\, 2p_z^{\,2}\, 3s^2\, 3p_x^{\,2}\, 3p_y\, 3p_z$
$_{18}$Ar $1s^2\, 2s^2\, 2p_x^{\,2}\, 2p_y^{\,2}\, 2p_z^{\,2}\, 3s^2\, 3p_x^{\,2}\, 3p_y^{\,2}\, 3p_z^{\,2}$

Problem 1.4 A bond dipole will result when two atoms of different electronegativities are attached to each other. This problem requires you only to look up the electronegativities of the two atoms in the bond. The electronegativity (Table 1.8, p. 15) of each atom in the bond is shown in parentheses. As you see below, the atom of greater electronegativity will be at the negative end of the dipole. The answer uses an arrow to represent the direction of the dipole, $\delta^+ \longmapsto\!\!\!\longrightarrow \delta^-$.

(continued)

Problem 1.4 *(continued)*

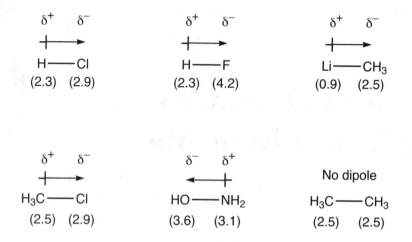

Problem 1.5   This classic problem is asked in many textbooks. It requires you to estimate the net dipole in a molecule by summing the bond dipoles. In carbon tetrachloride (CCl_4), the four dipoles cancel, and there is no net dipole. In chloroform ($CHCl_3$), the dipoles do not cancel, and a net dipole exists.

Here, the four dipoles cancel..............................but in this molecule they do not.

Problem 1.6   In each part of this problem, we will first determine the number of electrons available for bonding on each atom. For atoms in the second row of the periodic table, this will be the atomic number of the atom, less the two $1s$ electrons. Then we will see how many bonds between atoms are possible using these electrons. Finally, be careful to write the leftover nonbonding (lone pair) electrons as a pair of dots.

(b) H_2Be

Beryllium ($_4Be$) has two electrons available for bonding ($4 - 2$ $1s$ electrons), and each hydrogen contributes one electron. Two beryllium–hydrogen single bonds are formed. There are no nonbonding electrons.

H·   ·Be·   ·H   ⟶   H:Be:H   =   H——Be——H

(c) SiH_4

Silicon has four valence electrons. We know that hydrogen has only one valence electron and that hydrogen only makes one bond. So the structure for SiH_4 must be

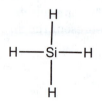

(d) CH_2Cl_2

We know that carbon has four valence electrons. Each chlorine atom has seven valence electrons and must be bonded to the carbon. If a chlorine atom were bonded to hydrogen or another chlorine, then the CH_2Cl_2 would not be a single molecule. Hydrogen and chlorine fill their valence shell by making one bond, and carbon can make four bonds. So the structure of CH_2Cl_2 must be

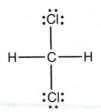

(e) $HOCH_3$

Oxygen ($_8O$) has six electrons available for bonding ($8 - 2$ $1s$ electrons). Carbon forms three covalent single bonds with the three hydrogens, which leaves one carbon electron available for covalent bond formation between carbon and oxygen. Oxygen forms the carbon–oxygen bond as well as one covalent bond with hydrogen, leaving oxygen with two pairs of nonbonding electrons.

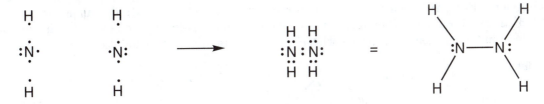

(f) H_2N—NH_2

Nitrogen has five electrons available for bonding. In this molecule, each nitrogen forms covalent bonds with two hydrogens and a third with the other nitrogen. This bonding system uses three of the available electrons, leaving a nonbonding pair of electrons on each nitrogen.

Problem 1.7

(c) Br

$_{35}$Br is in the fourth row of the periodic table, so we can ignore the 28 $1s$, $2s$, $2p$, $3s$, $3p$, and $3d$ electrons. This procedure leaves seven electrons. Like the other halogens (F, Cl, and I), Br has three nonbonding pairs and a single odd electron.

(d) OH

Oxygen ($_8O$) has six electrons available for bonding. One electron is used in forming a covalent bond with hydrogen, leaving two pairs of nonbonding electrons and a single odd electron on oxygen.

(continued)

Problem 1.7 *(continued)*

(e) NH_2

Nitrogen has five electrons available for bonding. Two covalent bonds are formed to hydrogens, leaving a nonbonding pair and a single odd electron remaining on nitrogen.

(f) $H_3C—N$

There are two possible answers to this one, and both kinds of molecule are known. Carbon has four electrons available for bonding. Three are used in forming single bonds to the three hydrogens, and the fourth is used in the single bond to nitrogen. Nitrogen has five electrons available for bonding. One covalent bond is formed to carbon, leaving four electrons. These can be used either as two pairs of nonbonding electrons or as one nonbonding pair and two odd electrons.

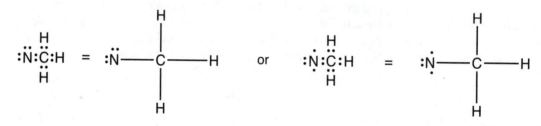

Problem 1.8 The question itself helps with the hard part—working out the connectivity of the atoms. Once again, this is an exercise in electron counting. First, determine the number of electrons available for bonding (atomic number less two 1s electrons for most atoms or a single 1s electron for hydrogen), then make single bonds. Finally, we look to see where multiple bonds can be formed.

(a) F_2CCF_2

Carbon has four electrons available for bonding and fluorine seven, including a single odd electron. As in ethylene (Fig. 1.20), each carbon forms three covalent bonds, two with fluorines and one with the other carbon atom. Thus, there are three nonbonding pairs remaining on each fluorine and a single electron left over on each carbon. These are shared in a second covalent carbon–carbon bond.

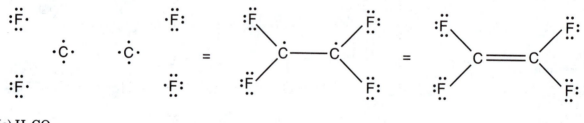

(c) H_2CO

Once again, carbon has four electrons available for bonding and oxygen six. Carbon uses two electrons to form bonds to hydrogen and one to bond to oxygen. Oxygen uses one electron in the bond to carbon, leaving one unshared electron on carbon and five on oxygen. Formation of a second carbon–oxygen bond leaves two nonbonding electron pairs on oxygen.

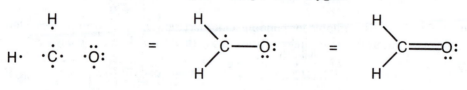

(d) H_2CCO

Each carbon has four available electrons and the single oxygen six. The left-hand carbon uses three electrons to form three covalent bonds to the two hydrogens and the adjacent carbon. The right-hand carbon uses two electrons in forming covalent bonds with the left-hand carbon and the oxygen, leaving the left-hand carbon with one electron, the right-hand carbon with two, and the oxygen with five. Formation of a second carbon–carbon bond and a second carbon–oxygen bond completes the picture, leaving the oxygen with two nonbonding pairs of electrons.

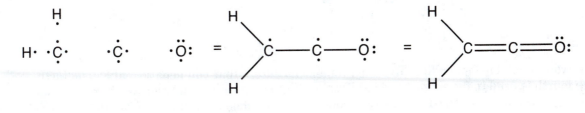

(e) $H_2CCHCHCH_2$

Each of the four carbons has four electrons available for bonding. The two terminal carbons form two covalent bonds with the two hydrogens and a third covalent bond with the adjacent carbon. Each internal carbon forms two covalent bonds with the adjacent carbons and a third to a hydrogen. There remains one electron on each carbon, allowing the formation of two additional carbon–carbon bonds.

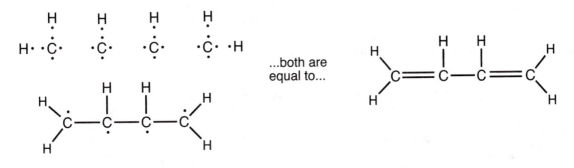

...both are equal to...

(f) $H_3C.NO$

Once again there is a methyl group ($\cdot CH_3$). Nitrogen has five available electrons, and oxygen has six ($_8O$, $8 - 2$ $1s$ electrons $= 6$). A carbon–nitrogen single bond and a nitrogen–oxygen double bond can be formed. Nitrogen is left with one pair of nonbonding electrons and oxygen with two pairs.

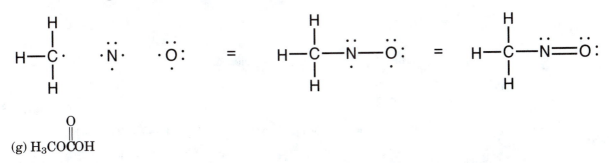

(g) H_3COCOH (with O double-bonded above the second C)

Each carbon has four available electrons for bonding, and each oxygen has six. The left-hand carbon (a) forms three bonds to hydrogen and a fourth to one oxygen (b). Oxygen (b) forms two bonds to a pair of carbons. The remaining carbon (c) forms covalent bonds with oxygen (b) and the two other oxygens (d). Each oxygen (d) forms a covalent bond with carbon (c), and one oxygen (d) forms a bond to hydrogen. This process leaves one odd electron on carbon (c), four nonbonding electrons on the oxygen (d) bound to hydrogen, and five electrons on the non-hydrogen-bound

(continued)

Problem 1.8 *(continued)*

oxygen (d). Oxygen (b) has four electrons remaining. Formation of a carbon (c) oxygen (d) bond completes the picture.

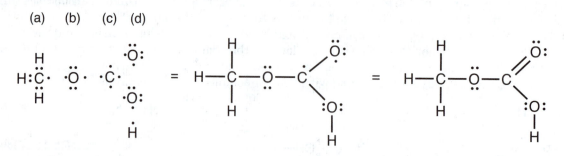

Problem 1.9   This problem is just like Problem 1.7 except that you need to make an adjustment for the charge. First, calculate the number of available electrons on the neutral atom, then add one electron for a negative charge or subtract one electron for a positive charge.

(a) $^-$OH

Neutral oxygen ($_8$O) has six electrons available for bonding ($8 - 2\ 1s = 6$). Therefore, negatively charged oxygen must have seven electrons.

One covalent bond can be made to the lone hydrogen, which supplies a single electron.

$$:\overset{..}{\underset{..}{O}}\cdot \quad \cdot H \quad = \quad \overset{..}{\underset{..}{O}}\!\!-\!\!H$$

(b) $^-$BH$_4$

Neutral boron ($_5$B) has three electrons available for bonding ($5 - 2\ 1s$). Therefore, negatively charged boron must have four electrons, allowing four covalent bonds to be made to the four hydrogens, each of which supplies a single electron. Notice that there is no pair of electrons on the negatively charged boron atom. In most negatively charged species, there is a nonbonding pair of electrons. This molecule is an exception.

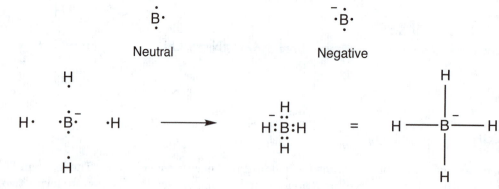

(c) $^+$NH$_4$

Neutral nitrogen ($_7$N) has five electrons available for bonding ($7 - 2\ 1s$). Positively charged nitrogen must have four electrons for bonding, allowing four single bonds to the four hydrogens, each of which supplies its single electron.

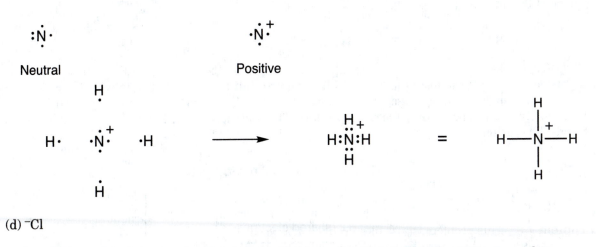

(d) ⁻Cl

Neutral chlorine ($_{17}$Cl) has seven electrons available for bonding (ignore the 10 1*s*, 2*s*, and 2*p* electrons). Therefore, negatively charged chlorine must have eight electrons.

$$:\ddot{\underset{\cdot\cdot}{C}}l\cdot \qquad\qquad :\ddot{\underset{\cdot\cdot}{C}}l:^{-}$$

Neutral   Negative

(e) ⁺CH₃

Neutral carbon ($_6$C) has four electrons available for bonding (6 – 2 1*s*). Positively charged carbon must have only three electrons for bonding, allowing three single bonds to the hydrogens, each of which supplies its single electron.

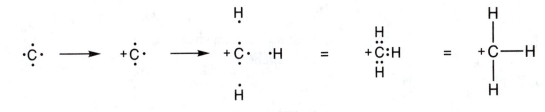

(f) ⁺OH₃

Neutral oxygen has six electrons available for bonding ($_8$O, 8 – 2 1*s* electrons), and so positively charged oxygen must have five electrons.

$$:\ddot{\underset{\cdot}{O}}\cdot \qquad\qquad :\ddot{\underset{\cdot}{O}}\cdot^{+}$$

Three single bonds can be made with the three hydrogen atoms.

$$\begin{array}{c} H \\ :\ddot{\underset{\cdot\cdot}{O}}:H \\ H \end{array} \qquad = \qquad \begin{array}{c} H \\ | \\ :\overset{+}{O}\!-\!H \\ | \\ H \end{array}$$

(g) ⁺NO₂

Nitrogen has five electrons available for bonding, so ⁺N must have four. Each oxygen has six. Two nitrogen–oxygen double bonds can be formed, leaving each oxygen with two pairs of nonbonding electrons.

(continued)

Problem 1.9 *(continued)*

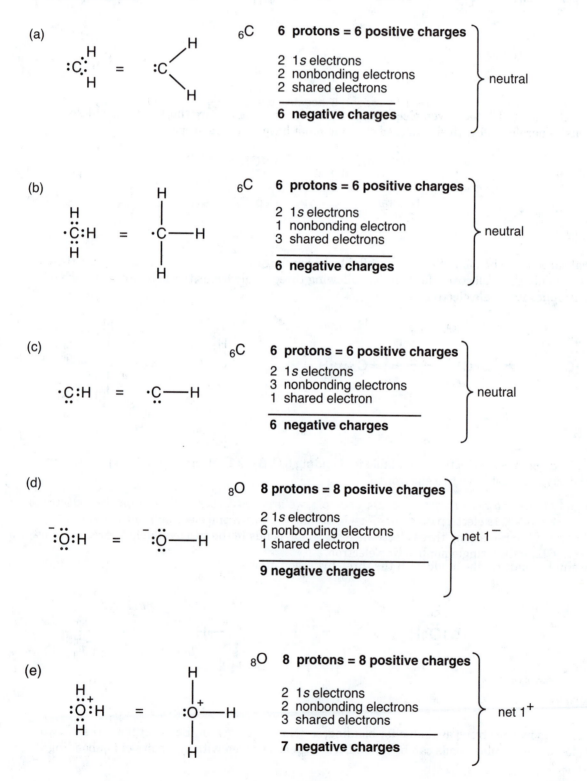

Problem 1.10   In these examples, we will first show a full Lewis structure in which each bonding electron appears as a dot, then a more schematic Lewis structure in which bonds are shown as lines and nonbonding electrons as dots. These structures will be followed by the charge calculation.

(a)

$_6$C   **6 protons = 6 positive charges**

2 1s electrons
2 nonbonding electrons   } neutral
2 shared electrons
_____
6 negative charges

(b)

$_6$C   **6 protons = 6 positive charges**

2 1s electrons
1 nonbonding electron   } neutral
3 shared electrons
_____
6 negative charges

(c)

$_6$C   **6 protons = 6 positive charges**

2 1s electrons
3 nonbonding electrons   } neutral
1 shared electron
_____
6 negative charges

(d)

$_8$O   **8 protons = 8 positive charges**

2 1s electrons
6 nonbonding electrons   } net 1$^-$
1 shared electron
_____
9 negative charges

(e)

$_8$O   **8 protons = 8 positive charges**

2 1s electrons
2 nonbonding electrons   } net 1$^+$
3 shared electrons
_____
7 negative charges

(f)

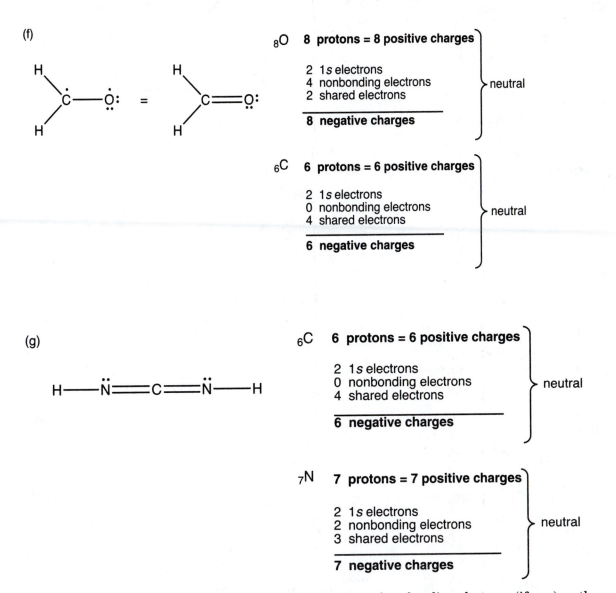

$_8$O   **8 protons = 8 positive charges**

  2 1s electrons
  4 nonbonding electrons
  2 shared electrons
_____
  8 negative charges

neutral

$_6$C   **6 protons = 6 positive charges**

  2 1s electrons
  0 nonbonding electrons
  4 shared electrons
_____
  6 negative charges

neutral

(g)

$_6$C   **6 protons = 6 positive charges**

  2 1s electrons
  0 nonbonding electrons
  4 shared electrons
_____
  6 negative charges

neutral

$_7$N   **7 protons = 7 positive charges**

  2 1s electrons
  2 nonbonding electrons
  3 shared electrons
_____
  7 negative charges

neutral

Problem 1.11   The task here is to work out the number of nonbonding electrons (if any) on the charged atom. Each answer first shows the neutral atom, then the atom with an electron added or removed to get the proper charge. Finally, electrons are used to make the bonds to the available hydrogen atoms or other groups. In (a), for example, we first see carbon with four bonding electrons ($_6$C; 6 electrons − 2 1s electrons = 4 bonding electrons), then with one electron removed to get $^+$C; finally, two of the remaining three electrons form single bonds to the two available hydrogens. Now we have $^+$CH$_2$ with a single nonbonding electron.

(a) $^+$CH$_2$

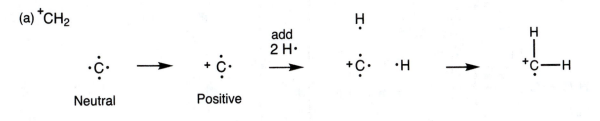

Neutral   Positive   add 2 H·

(continued)

Problem 1.11 *(continued)*

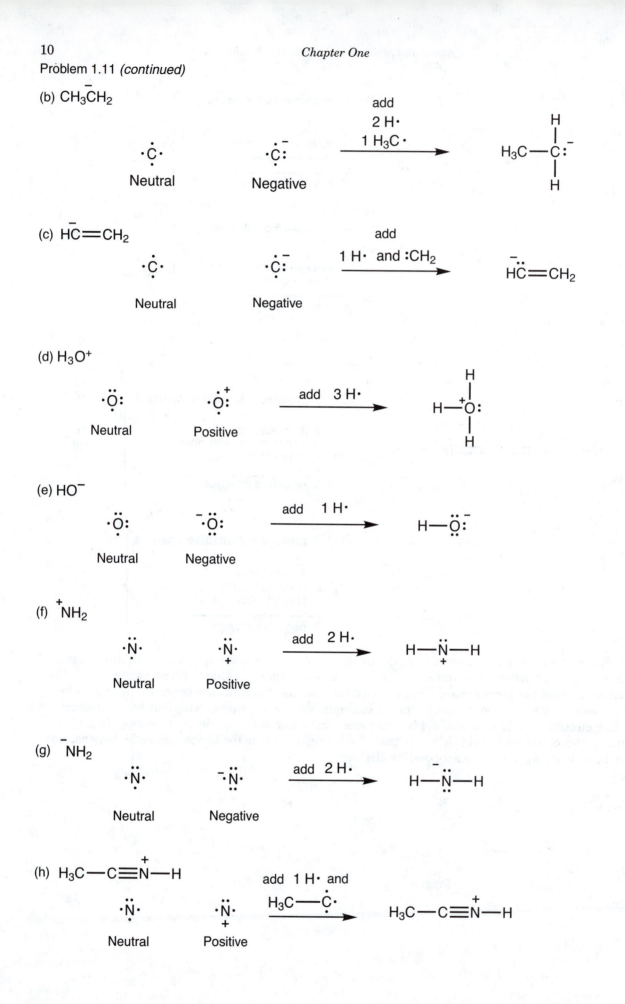

Problem 1.12   Figure 1.26 shows the structure of nitromethane, $H_3C—NO_2$. By analogy, we can write a structure for nitric acid, $HO—NO_2$.

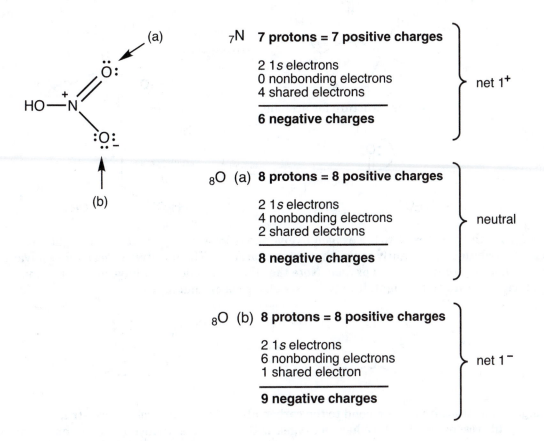

Problem 1.13   Notice that the lower arrow pushes a nonbonding pair of electrons on one oxygen to displace a bonding pair (shown only as a line in the drawing). The displaced pair winds up as a nonbonding pair on the other oxygen, and the displacing pair winds up as a new bond between the lower oxygen atom and nitrogen.

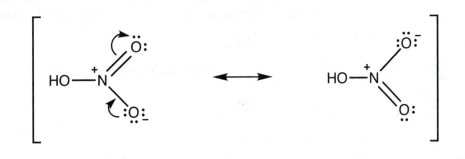

Problem 1.15

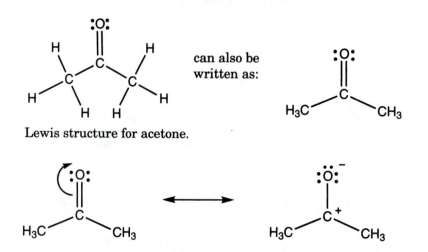

Lewis structure for acetone.

We can move the π electrons to the oxygen to give the resonance structure on the right. This species contributes significantly to the chemistry of acetone. The negative charge takes advantage of the higher electronegativity of oxygen. Note that the carbon does not have an octet in the resonance structure on the right. It only has six electrons around it.

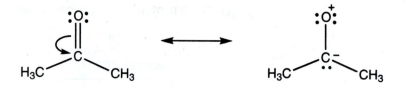

Moving the electrons from the π bond to the carbon also gives a valid resonance structure. However, this resonance structure has an oxygen without an octet (which is worse than a carbon with only six electrons, because oxygen is more electronegative). In addition, the negative charge is on the less electronegative carbon. To make matters even worse, the electronegative oxygen has a positive charge. This resonance structure does not contribute to the overall structure of acetone.

Problem 1.16   *Remember*: The double-barbed, curved arrows move pairs of electrons, and we must be careful neither to violate the rules of valence nor to move atoms. A new convention appears in the last example. When "pushing" single electrons, a single-barbed, curved arrow is used.

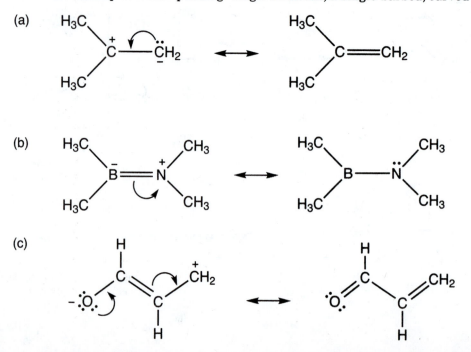

(d) Here is the tricky part to this problem. It is sooooo tempting to push the arrow as shown:

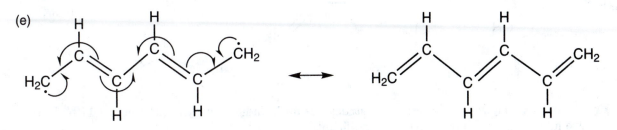

But this "pushing" makes no sense: There is no low-lying empty orbital on nitrogen to accept an electron pair and no pair of electrons on boron to push! The form on the right violates the rules of valence twice.

(e)

In this example, notice the use of *single-barbed*, or "fishhook," arrows to show the motion of single electrons!

Problem 1.17

(a)

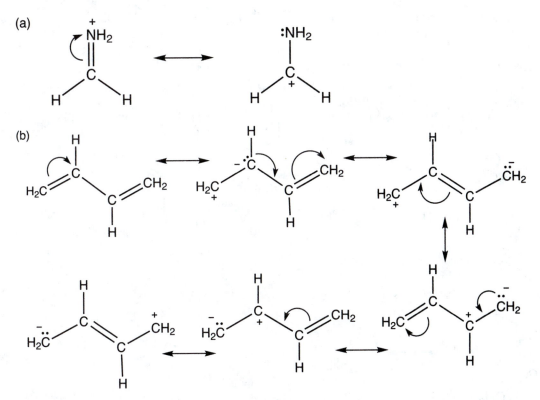

(b)

None of the resonance forms in part (b), except the first, uncharged one, is very good. Each contains one fewer bond than the first, and each requires substantial charge separation.

(continued)

Problem 1.17 *(continued)*

(d)

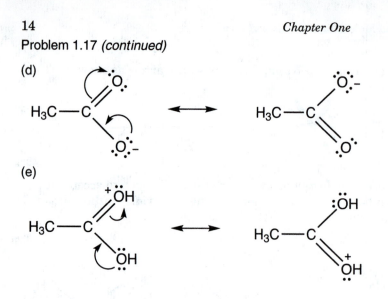

(e)

Problem 1.18   There are six resonance structures for butadiene drawn in Problem 1.17(b). Here are two more. These do not contribute significantly.

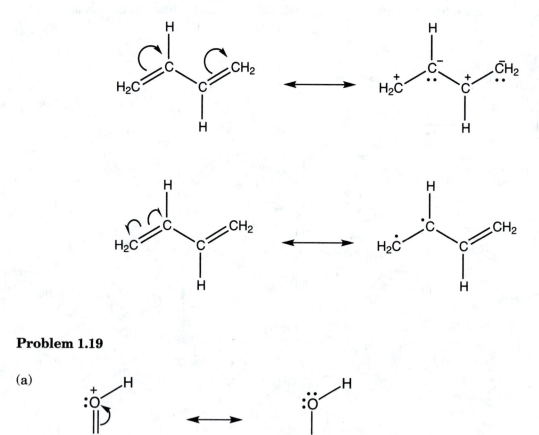

Problem 1.19

(a)

(b)

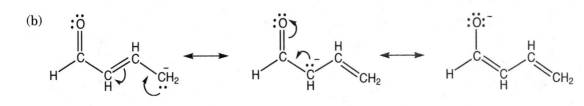

(c)

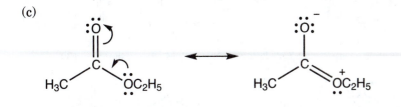

(d)

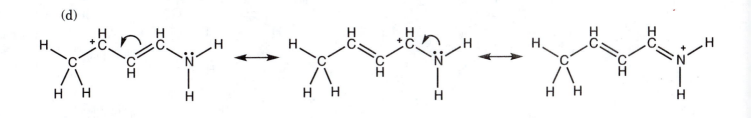

(e)

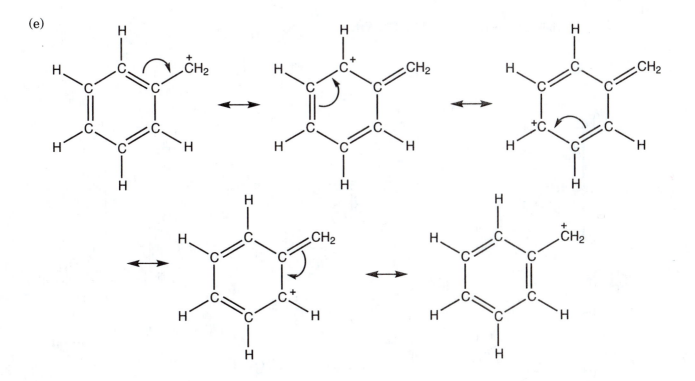

(continued)

Problem 1.19 *(continued)*

(g)

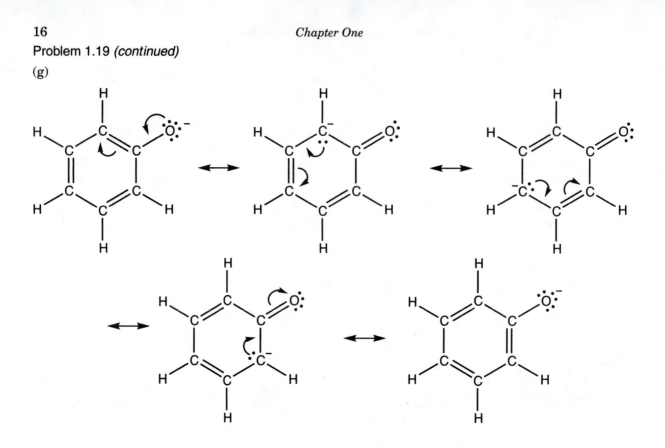

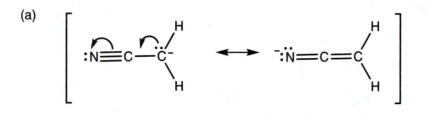

Problem 1.20

(a)

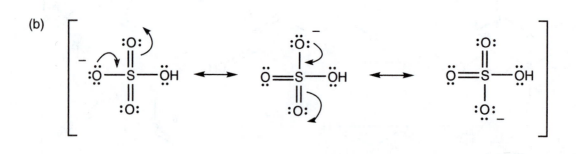

(b)

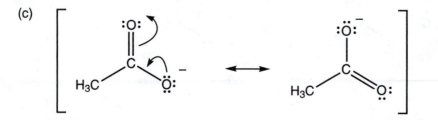

(c)

Problem 1.21

(a)

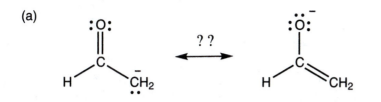

Yes, these are resonance structures. The electron pushing that allows you to go from the left structure (the aldehyde) to the right structure is shown here:

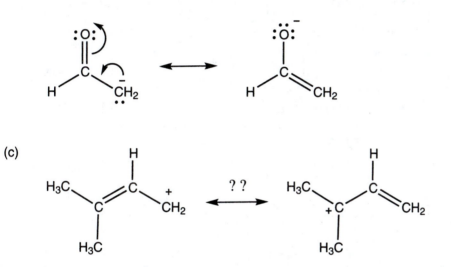

(c)

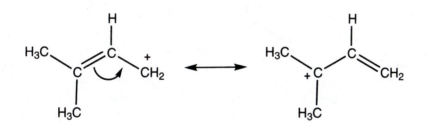

Yes, these two are resonance structures of each other. The electron pushing that allows you to go from the left structure to the right is shown here:

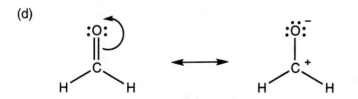

In (d), it is the traditional "carbonyl" carbon–oxygen double bond that is the better form. It contains more bonds and has no charge separation.

(d)

The carbon on the resonance structure on the right does not have an octet of electrons.

Problem 1.22

(a)

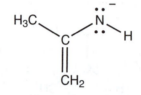

This structure contributes more
because there is one more bond.
The carbons and oxygen all have
an octet!

(b)

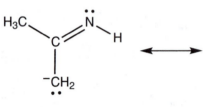

This structure contributes more
because the negative charge is on
the more electronegative atom.

(c)

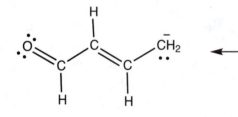   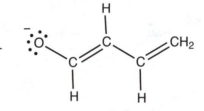

This structure contributes more
because the negative charge is on
the more electronegative atom.

(d)

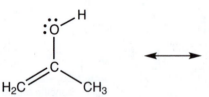

This structure contributes more
because it minimizes the charge
and because it gives more electrons
to the more electronegative oxygen.

(e)

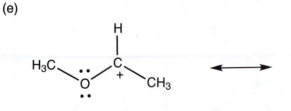   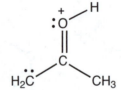   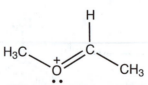

This structure contributes more
because there is one more bond. The
carbons and oxygen all have an octet!

Many of the problems in this chapter require simple combinations of orbitals to produce new orbitals, and then the placement of electrons in the orbitals. The key thing to remember is that orbitals can interact in both constructive and destructive ways; $H_{1s} + H_{1s} = \Phi_B$ and $H_{1s} - H_{1s} = \Phi_A$ are prototypal examples discussed in the chapter. The interaction of orbitals is often shown in a graphical way:

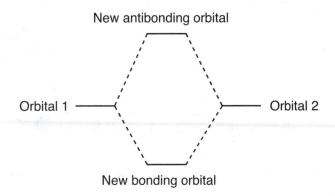

Each orbital can hold a total of two electrons. Remember that placing single electrons with parallel spins in different orbitals of equal energy gives a lower energy species than the one made by placing two electrons with paired (opposite) spins in a single orbital. We move from fairly simple examples to quite sophisticated systems in these problems, but the principles remain the same. Notice the complete absence of mathematics.

Problem 1.23   The simplest molecule is "H_2 minus something." The H_2 molecule contains only two protons and two electrons. As loss of a proton doesn't leave a molecule behind, that "something" can only be an electron. The simplest molecule must be H_2^+. Another electron cannot be lost to give something even simpler because H_2^{2+} is not a molecule. In H_2^{2+}, there would be no electrons to bind the two nuclei.

Problem 1.25

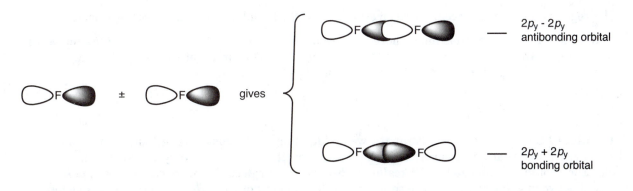

Mixing two p orbitals that are lined up head-to-head gives two molecular orbitals (MOs). These new MOs are σ and σ^*.

(continued)

Problem 1.25 *(continued)*

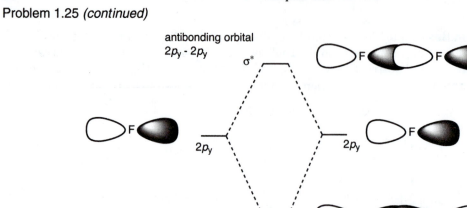

This energy diagram shows the mixing of two equal-energy $2p_y$ fluorine orbitals to obtain bonding and antibonding σ orbitals in F_2.

Problem 1.28 Fluorine is able to participate in hydrogen bonding (H-bonding). Although the strength of such interactions is theoretically calculated to be 2.38 kcal/mol, the actual observation of R—F----H—O—H is rare.

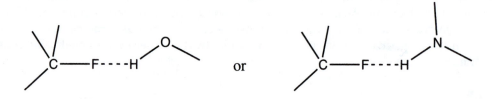

Hydrogen bonding might occur between a fluorine in R—F and an O—H or N—H hydrogen.

In addition, the C—F bond is much more polar than the C—H bond. This polarity can influence the way a fluorinated molecule interacts with other molecules. The ability of a molecule to bind with an enzyme, for example, can be altered as a result of a C—F bond.

Problem 1.29 In making this estimate, we divided the known bond strength of H_2 by 2. The supposition was that if two electrons in the bonding orbital are stabilized by 104 kcal/mol, one electron should be stabilized by 52 kcal/mol. However, we neglected to worry about the repulsive forces between two electrons in the same orbital. The energy of H_2 is *raised* by the repulsive forces between two negatively charged electrons occupying the same orbital. When an electron is removed to form H_2^+, these repulsive forces disappear!

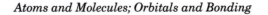

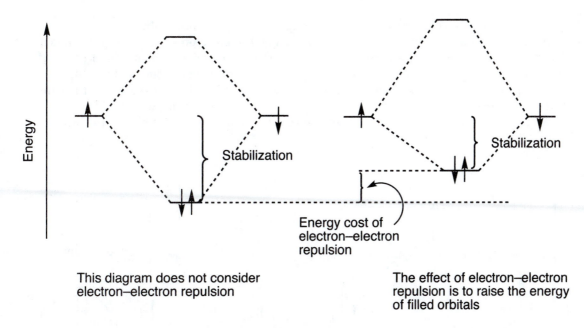

This diagram does not consider electron–electron repulsion

The effect of electron–electron repulsion is to raise the energy of filled orbitals

Look at it backward. If one electron in the bonding molecular orbital is stabilized by 64 kcal/mol, two electrons will be stabilized by 128 kcal/mol. However, these two electrons will repel each other, a somewhat destabilizing factor, and the real bond energy is only 104 kcal/mol.

Problem 1.30 The diagram for He_2^+ can easily be derived from the diagram in Figure 1.48 by removal of one electron. *Remember*: Construction of molecular orbitals from atomic orbitals does not depend on the number of electrons. The electrons are placed in the appropriate orbitals later. In this case, we first build the molecular orbitals of He_2 from two He $1s$ orbitals. In Figure 1.48, we put in four electrons to construct He_2. In this problem, you need only put in three electrons to make He_2^+.

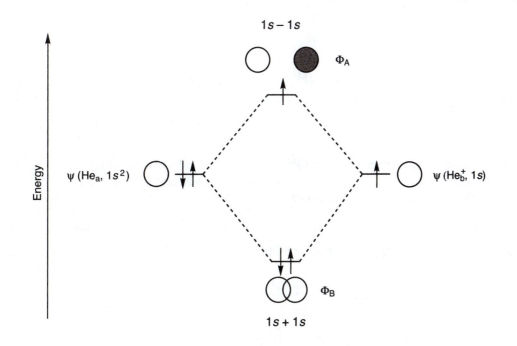

Problem 1.32   The energy of an electron in an orbital depends on both stabilizing and destabilizing forces. Whenever two electrons occupy an orbital, electron–electron repulsion is a destabilizing factor and raises the energy of the electrons in that orbital. An electron in a filled bonding orbital is moved higher in energy, closer to the energy of its constituent atomic orbitals. An electron in a filled antibonding orbital is also increased in energy but moves away from the energy of its constituent atomic orbitals. The net result is that an electron in a filled bonding molecular orbital is stabilized less than an electron in a filled antibonding molecular orbital is destabilized.

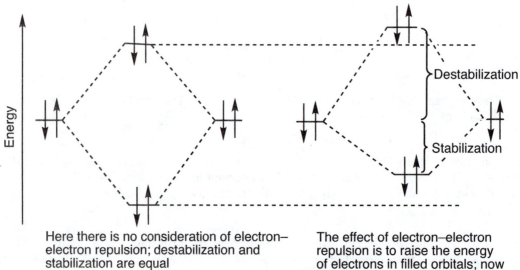

Here there is no consideration of electron–electron repulsion; destabilization and stabilization are equal

The effect of electron–electron repulsion is to raise the energy of electrons in filled orbitals; now destabilization is greater than stabilization

Problem 1.33

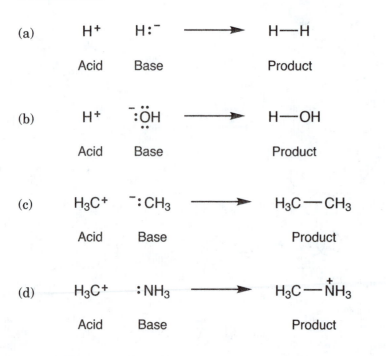

Problem 1.34 The most important aspect of electron pushing is to start where the electrons are and show where they go.

(a)

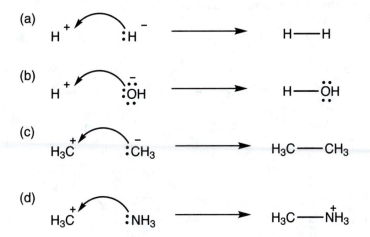

(b)

(c)

(d)

Additional Problem Answers

Problems 1.35, 1.36

(a)

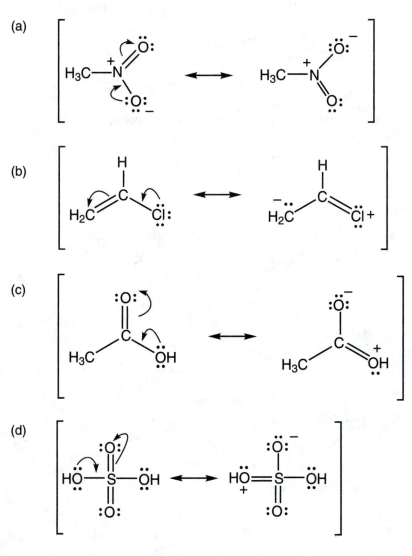

(b)

(c)

(d)

Problem 1.37   In these answers, the arrows shown will always produce the resonance form immediately to the right or below. That is, these answers are to be read left to right or top to bottom. Notice the extensive use of the double-headed resonance arrow.

(a) Carbonate

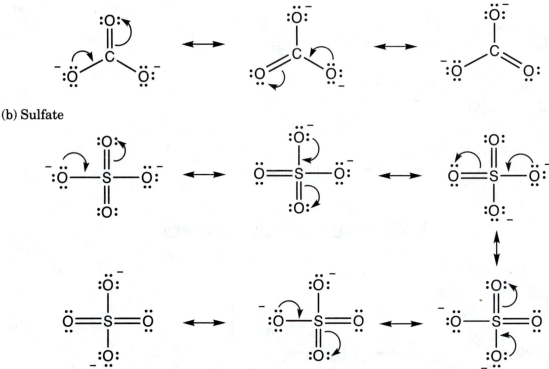

(b) Sulfate

There are also resonance forms in which sulfur bears a positive charge and three oxygen atoms are negative. Can you draw these?

(c) Nitrate

(d) Guanidinium

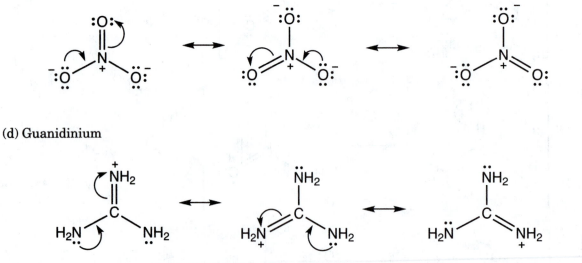

(e) Vinyl ammonium

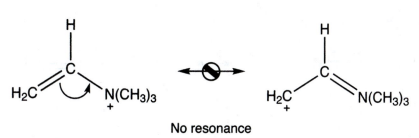

No resonance

There are no important resonance forms for this species. There is no pair of electrons on nitrogen, and five bonds cannot be formed to nitrogen. Once again, it is tempting to "push the arrow," but in this case it is best to resist that temptation. There is no empty orbital on nitrogen to receive the electrons we are trying to push.

Problem 1.38 In drawing resonance forms for these molecules, it is sometimes hard to know when to stop, as less and less stable structures are produced. We have perhaps gone too far on occasion. Can you see which forms are likely to be especially unstable and, therefore, minor contributors?

(a)

$$H_2\overset{\cdot\cdot}{C}-\overset{\cdot\cdot}{N}=\overset{+}{N}: \longleftrightarrow \overset{-}{H_2\overset{\cdot\cdot}{C}}-\overset{+}{N}\equiv N: \longleftrightarrow H_2C=\overset{+}{N}=\overset{-}{N}: \longleftrightarrow \overset{+}{H_2C}-\overset{\cdot\cdot}{N}=\overset{\cdot\cdot}{N}:$$

(b)

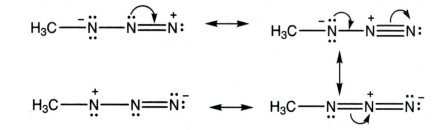

(c)

$$H_3C-\overset{+}{C}=\overset{\cdot\cdot}{N}-\overset{\cdot\cdot}{N}-CH_3 \longleftrightarrow H_3C-C\equiv\overset{+}{N}-\overset{\cdot\cdot}{N}-CH_3$$

$$H_3C-\overset{-}{\overset{\cdot\cdot}{C}}=\overset{\cdot\cdot}{N}-\overset{\cdot\cdot}{\underset{+}{N}}-CH_3 \longleftrightarrow H_3C-\overset{-}{\overset{\cdot\cdot}{C}}=\overset{+}{N}=\overset{\cdot\cdot}{N}-CH_3$$

(continued)

Problem 1.38 *(continued)*

(d)

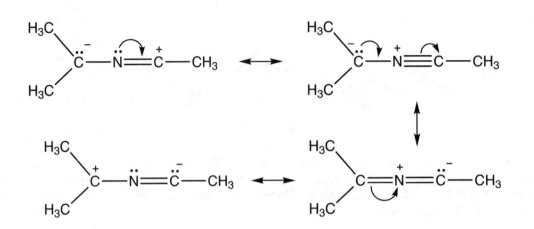

(e)

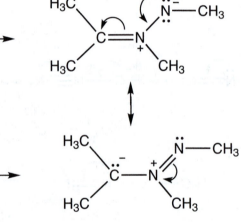

(f)

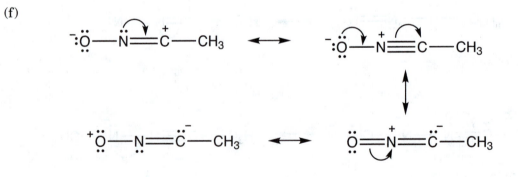

Problem 1.39

(a)

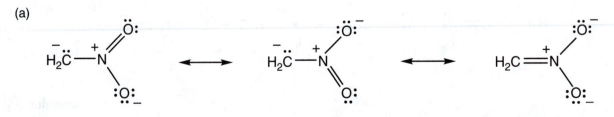

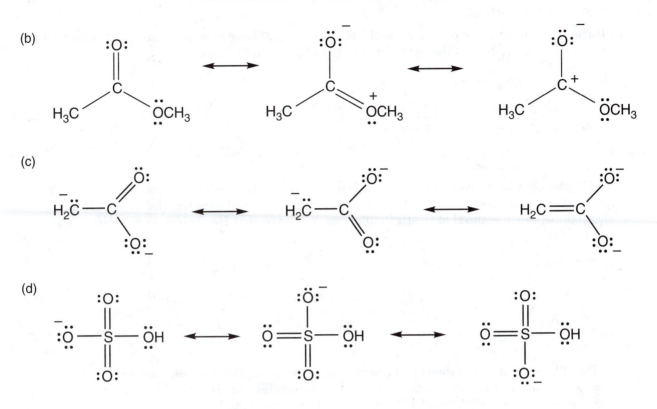

(b)

(c)

(d)

Problem 1.40

(a) The oxygen is the nucleophile and the proton of H—Cl is the electrophile in this reaction. Electrons go from the oxygen lone pair to the sigma antibond of the H—Cl, and the electrons in the sigma bond between the hydrogen and the chlorine go to the chlorine to make the relatively stable chloride ion.

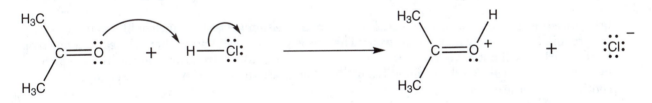

(b) The nitrogen is the nucleophile, and the proton is the electrophile. Electrons go from the nitrogen lone pair to the proton.

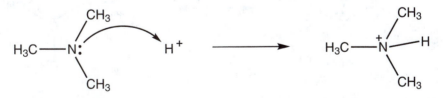

(c) The arrow needs to show the electrons going from the double bond, the nucleophile, all the way to the proton, the electrophile

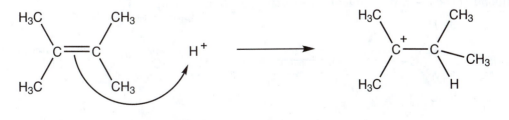

(continued)

Problem 1.40 *(continued)*

(d) The electrophile is the proton of the H—Cl. But we can't have hydrogen with two bonds. So we must show the electrons of the sigma bond in H—Cl going to the chlorine.

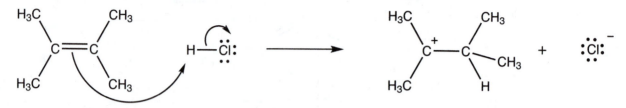

(e) The electrons in the double bond will not go to the sigma bond of the H—Br. Electrons won't make a bond with other electrons. The proper representation will have the electrons go to an electrophile, which is the H of H—Br in this case. You know that H—Br is a strong acid. It is a source of H^+.

(f) The only error in this problem is the arrow drawn from the chlorine lone pair. As the nucleophile, the nitrogen in this case, attacks the electrophile, the H of H—Cl in this case, the sigma bond of H—Cl must break. Hydrogen can't have two bonds.

(g) Notice that this process involves resonance, not a reaction between a nucleophile and an electrophile. We are only moving electrons. The arrow pushing can't stop where the carbon has five bonds. Carbon will not have five bonds because that means 10 electrons around the atom, which violates the octet rule. Carbon cannot handle five bonds. So the movement of electrons must continue up to the oxygen.

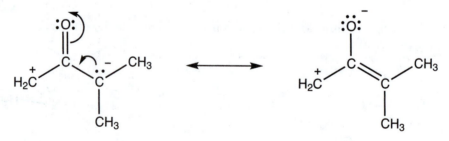

Problem 1.41

(a) Sulfide is an excellent nucleophile, and the C—I bond is weak. The nucleophile attacks the sigma antibond of the C—I. We will learn about this reaction in Chapter 7.

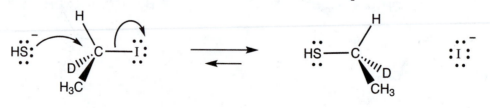

(b) Hydroxide is a nucleophile, and the carbon of a C=O bond is an electrophile. It is the antibonding orbital associated with the π bond that the nucleophile adds into. We will learn about this reaction in Chapter 16.

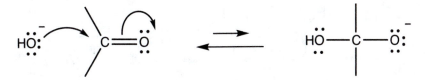

(c) The reverse reaction of the previous problem is favored in this equilibrium. The carbon–oxygen double bond is stronger than a carbon with two C–O bonds.

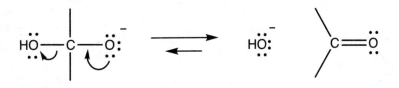

(d) This example involves resonance, not a nucleophile–electrophile reaction. These are resonance structures. Notice that no atoms are moved.

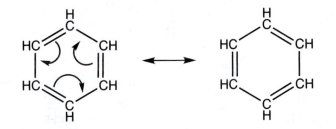

(e) This is a reaction that makes two new sigma bonds in one step. We will learn about it in Chapter 13.

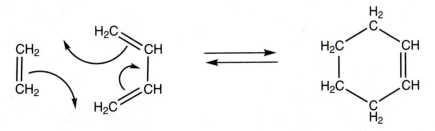

(f) Hydroxide is a nucleophile, and it deprotonates the starting molecule, as the electron pushing illustrates. We will learn about this reaction in Chapter 8.

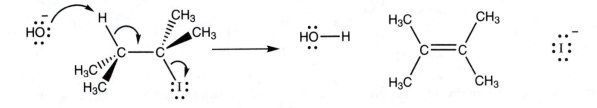

Problem 1.42

(a)

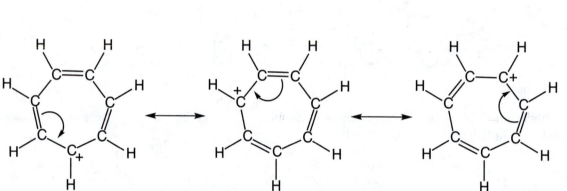

(b)

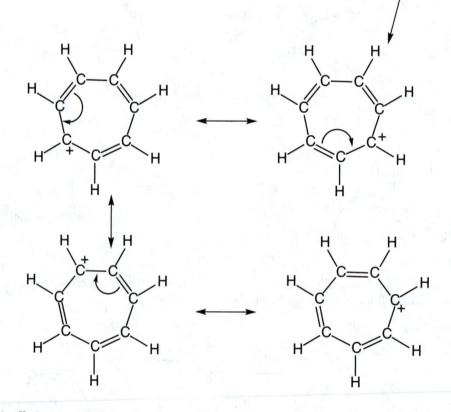

In parts (a) and (b), all ring atoms share the charge. This sets you up for part (c), in which they do not.

(c)

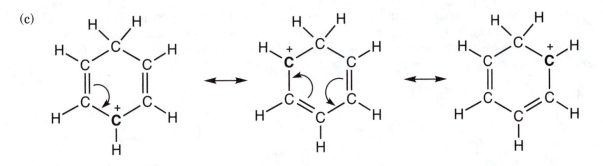

Notice the difference between (a) and (b) of this question and (c). In the first two molecules, every carbon helps share the charge. In (c), in which a CH_2 group interrupts the connectivity of $2p$ orbitals, only three carbons share the charge. We will talk more about such molecules in the future, but note now the difference between these two kinds of system. In cyclic molecules in which every ring atom has a p orbital, each atom shares the charge. In systems in which there is an insulator (an atom or atoms without a p orbital), not every ring atom will share the charge.

Problem 1.43

(a)

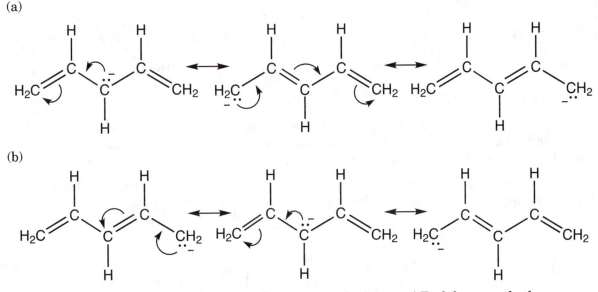

(b)

Now look carefully at these "two" species. They are exactly the same! Each has exactly the same set of three resonance contributors. Tricky, tricky, tricky.

Problem 1.44

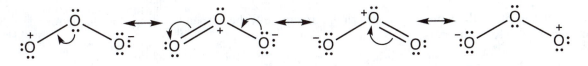

The neutral resonance form can be produced this way:

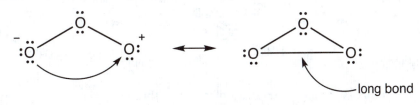

(continued)

Problem 1.44 *(continued)*

The cyclic, "no charge" resonance form *is* a resonance form, but it is not a very good one. *Remember*: No atoms may be moved in drawing resonance forms—only electrons. Resonance forms are different *electronic* structures. The long bond in the cyclic form is not a strong bond because the two oxygen atoms are very far apart.

Problem 1.45

(a)

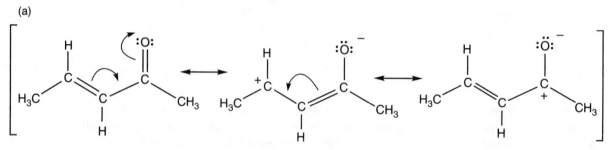

(b) This molecule is actually linear. We will learn why in Chapter 2.

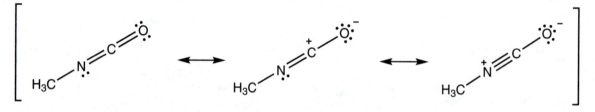

Problem 1.46

See Problem 1.45 answers that already show the electron pushing. It will always be a good idea to use the electron pushing to come up with your resonance structures. So when you are asked to draw a resonance structure, you should automatically use electron pushing to figure it out.

Often there are multiple ways to show the electron pushing. Here are equally valid options for electron pushing to get resonance structures for the molecules in Problem 1.45.

(a)

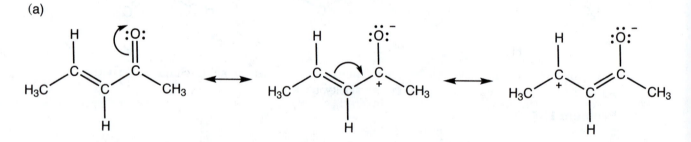

(b) This molecule will be linear because the nitrogen (and the carbon between the nitrogen and oxygen) are *sp* hybridized.

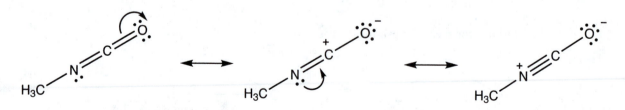

Problem 1.47   Structure (b) violates the rules of valence (there are five bonds to neutral nitrogen) and thus is not a resonance contributor.

Problem 1.48 These molecules all look alike, but a bit of electron counting shows that the charges on the noncarbon atom are different.

(a) For oxygen, the atomic number is 8, which means that there are eight positive charges in the nucleus. The oxygen atom in the three-membered ring has a total of eight electrons: two 1s electrons (never shown), four nonbonding electrons, and a share in two bonds for an additional two electrons. This oxygen is neutral. For $_8O$, 2 (1s) + 4 (nonbonding) + 2 (shared) = 8 electrons.

The other electron counts are as follows:

(b) For $_7N$, 2 (1s) + 4 (nonbonding) + 2 (shared) = 8 electrons, so N is net 1⁻.

(c) For $_{35}Br$, the calculation is tougher. Bromine is in the same column as fluorine, $_9F$, and will have the same number of valence electrons. There are 28 "core" electrons ($n = 1, 2, 3$ levels, 1s, 2s, 2p, 3s, 3p, and 3d), leaving 7 valence electrons. This bromine atom has 4 nonbonding and 2 shared electrons for a total of 6, and a grand total of 28 + 6 = 34. The Br is net 1⁺.

(d) For $_{16}S$, there are 2 (1s) + 2 (2s) + 6 (2p) + 4 (nonbonding) + 2 (shared) = 16 electrons. This S is neutral. Notice that atoms in the same column of the periodic table are similar. For example, both oxygen and sulfur are neutral.

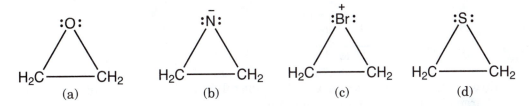

(a)   (b)   (c)   (d)

Problem 1.49 This problem reinforces the notion that atoms in the same column of the periodic table with the same number of electron dots in their Lewis structure will have the same charge. If you count for oxygen across the first three structures, you get

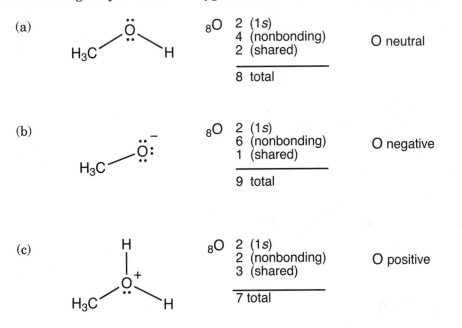

(a) $_8O$ 2 (1s)
      4 (nonbonding)   O neutral
      2 (shared)
      ———————
      8 total

(b) $_8O$ 2 (1s)
      6 (nonbonding)   O negative
      1 (shared)
      ———————
      9 total

(c) $_8O$ 2 (1s)
      2 (nonbonding)   O positive
      3 (shared)
      ———————
      7 total

(continued)

The charge determination for sulfur comes out the same:

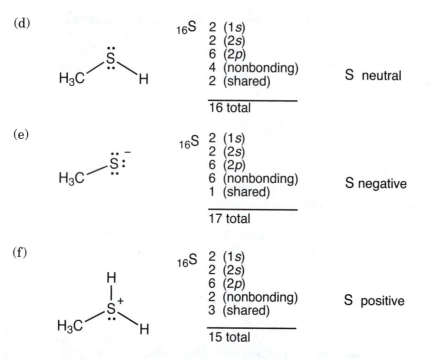

(d)

$_{16}$S   2 (1s)
     2 (2s)
     6 (2p)
     4 (nonbonding)
     2 (shared)   S neutral
     _____
     16 total

(e)

$_{16}$S   2 (1s)
     2 (2s)
     6 (2p)
     6 (nonbonding)
     1 (shared)   S negative
     _____
     17 total

(f)

$_{16}$S   2 (1s)
     2 (2s)
     6 (2p)
     2 (nonbonding)
     3 (shared)   S positive
     _____
     15 total

Similarly, nitrogen and phosphorus atoms in the same column of the periodic table are also identically charged in identical bonding situations.

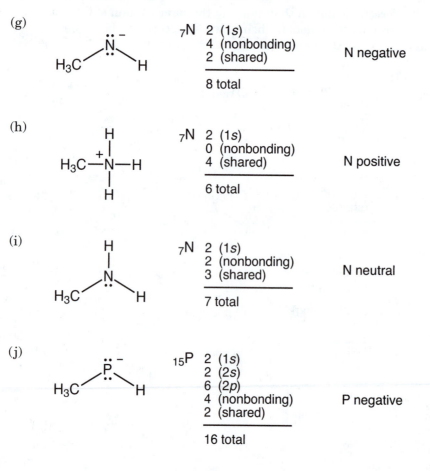

(g)

$_7$N   2 (1s)
     4 (nonbonding)
     2 (shared)   N negative
     _____
     8 total

(h)

$_7$N   2 (1s)
     0 (nonbonding)
     4 (shared)   N positive
     _____
     6 total

(i)

$_7$N   2 (1s)
     2 (nonbonding)
     3 (shared)   N neutral
     _____
     7 total

(j)

$_{15}$P   2 (1s)
     2 (2s)
     6 (2p)
     4 (nonbonding)
     2 (shared)   P negative
     _____
     16 total

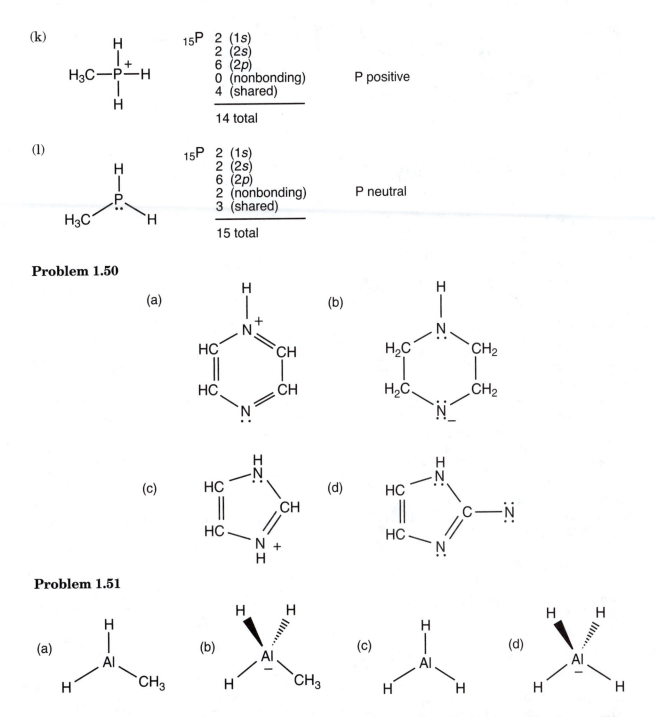

(k)

$_{15}P$ 2 (1s)
     2 (2s)
     6 (2p)
     0 (nonbonding)   P positive
     4 (shared)
     ───────
     14 total

(l)

$_{15}P$ 2 (1s)
     2 (2s)
     6 (2p)
     2 (nonbonding)   P neutral
     3 (shared)
     ───────
     15 total

Problem 1.50

(a)   (b)

(c)   (d)

Problem 1.51

(a)   (b)   (c)   (d)

Problem 1.52 Start by drawing Lewis dot structures for the atoms. For fluorine ($_9F$), there will be seven electrons available for bonding (nine less the two $1s$ electrons), and for nitrogen ($_7N$), there will be five (seven less the two $1s$ electrons). For F_2, where there is a single bond between the two atoms, there will be six electrons left over. For N_2, where there is a triple bond between the atoms, there will be only a single pair of electrons left on each nitrogen.

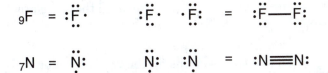

Problem 1.53 As shown in the chapter, the ground state of carbon is $1s^2\, 2s^2\, 2p_x\, 2p_y$. There are many possible excited states, including $1s^2\, 2s^2\, 2p_x^2$ and $1s^2\, 2s\, 2p_x\, 2p_y\, 2p_z$. In these species, two electrons have been brought closer together than is optimal, or one electron has been promoted to a higher energy orbital.

Problem 1.54   For these ions, we first figure out the configuration of the neutral atom. Then we add or remove electrons as necessary to accommodate the charge.

$_{11}Na = 1s^2\ 2s^2\ 2p_x^{\ 2}\ 2p_y^{\ 2}\ 2p_z^{\ 2}\ 3s$
... so $_{11}Na^+$ with one fewer electron will be

$1s^2\ 2s^2\ 2p_x^{\ 2}\ 2p_y^{\ 2}\ 2p_z^{\ 2}$

$_9F = 1s^2,\ 2s^2,\ 2p_x^{\ 2},\ 2p_y^{\ 2},\ 2p_z$
... so $_9F^-$ with one more electron will be

$1s^2,\ 2s^2,\ 2p_x^{\ 2},\ 2p_y^{\ 2},\ 2p_z^{\ 2}$

$_{20}Ca = 1s^2\ 2s^2\ 2p_x^{\ 2}\ 2p_y^{\ 2}\ 2p_z^{\ 2}\ 3s^2\ 3p_x^{\ 2}\ 3p_y^{\ 2}\ 3p_z^{\ 2}\ 4s^2$
... so $_{20}Ca^{2+}$ with two fewer electrons must be

$1s^2,\ 2s^2,\ 2p_x^{\ 2},\ 2p_y^{\ 2},\ 2p_z^{\ 2},\ 3s^2,\ 3p_x^{\ 2},\ 3p_y^{\ 2},\ 3p_z^{\ 2}$

Problem 1.55

$_{19}K \quad = 1s^2\ 2s^2\ 2p_x^{\ 2}\ 2p_y^{\ 2}\ 2p_z^{\ 2}\ 3s^2\ 3p_x^{\ 2}\ 3p_y^{\ 2}\ 3p_z^{\ 2}\ 4s = [Ar]\ 4s$
$_{20}Ca \quad = [Ar]4s^2$
$_{21}Sc \quad = [Ar]4s^2 3d$
$_{22}Ti \quad = [Ar]4s^2\ 3d^2$
$_{23}V \quad = [Ar]4s^2\ 3d^3$
$_{24}Cr \quad = [Ar]4s^2\ 3d^4$ (in fact, $_{24}Cr$ is $[Ar]4s\ 3d^5$)
$_{25}Mn = [Ar]4s^2\ 3d^5$
$_{26}Fe \quad = [Ar]4s^2\ 3d^6$
$_{27}Co \quad = [Ar]4s^2\ 3d^7$
$_{28}Ni \quad = [Ar]4s^2\ 3d^8$
$_{29}Cu \quad = [Ar]4s^2\ 3d^9$ (in fact, $_{29}Cu$ is $[Ar]4s\ 3d^{10}$)
$_{30}Zn \quad = [Ar]4s^2\ 3d^{10}$
$_{31}Ga \quad = [Ar]4s^2\ 3d^{10}\ 4p_x$
$_{32}Ge \quad = [Ar]4s^2\ 3d^{10}\ 4p_x\ 4p_y$
$_{33}As \quad = [Ar]4s^2\ 3d^{10}\ 4p_x\ 4p_y\ 4p_z$
$_{34}Se \quad = [Ar]4s^2\ 3d^{10}\ 4p_x^{\ 2}\ 4p_y\ 4p_z$
$_{35}Br \quad = [Ar]4s^2\ 3d^{10}\ 4p_x^{\ 2}\ 4p_y^{\ 2}\ 4p_z$
$_{36}Kr \quad = [Ar]4s^2\ 3d^{10}\ 4p_x^{\ 2}\ 4p_y^{\ 2}\ 4p_z^{\ 2}$

Problem 1.56

$_{14}Si \quad = 1s^2\ 2s^2\ 2p_x^{\ 2}\ 2p_y^{\ 2}\ 2p_z^{\ 2}\ 3s^2\ 3p_x^{\ \uparrow}\ 3p_y^{\ \uparrow}$
$_{15}P \quad = 1s^2\ 2s^2\ 2p_x^{\ 2}\ 2p_y^{\ 2}\ 2p_z^{\ 2}\ 3s^2\ 3p_x^{\ \uparrow}\ 3p_y^{\ \uparrow}\ 3p_z^{\ \uparrow}$
$_{16}S \quad = 1s^2\ 2s^2\ 2p_x^{\ 2}\ 2p_y^{\ 2}\ 2p_z^{\ 2}\ 3s^2\ 3p_x^{\ \uparrow\downarrow}\ 3p_y^{\ \uparrow}\ 3p_z^{\ \uparrow}$

Hund's rule states that for orbitals of equal energy, such as the three $3p$ orbitals, the electronic configuration with the greatest number of parallel spins will be the lowest in energy. Electrons with parallel spins (same spin quantum number s) cannot occupy the same orbital and are therefore kept apart, minimizing electron–electron repulsion. We faced the same problem in determining the spin state of a carbon atom. Here $_6C = 1s^2 2s^2 2p_x^{\ \uparrow} 2p_y^{\ \uparrow}$ is preferred to $1s^2 2s^2 2p_x^{\ 2}$ or $1s^2 2s^2 2p_x^{\ \downarrow} 2p_y^{\ \uparrow}$.

Problem 1.57   In oxygen, the last two electrons fill the $2p_y$ and $2p_z$ orbitals and have unpaired spins (Hund's rule). In this case, the ESR machine will find two unpaired spins.

$_8O = 1s^2\ 2s^2\ 2p_x^{\ 2}\ 2p_y^{\ \uparrow}\ 2p_z^{\ \uparrow}$

In O^+, there will be one fewer electron, and the ESR instrument will still find the three unpaired electrons in the $2p_x$, $2p_y$, and $2p_z$ orbitals.

$$_8O^+ = 1s^2\, 2s^2\, 2p_x{}^\uparrow\, 2p_y{}^\uparrow\, 2p_z{}^\uparrow$$

In O^{2-}, there will be two more electrons than in neutral O. The electronic configuration will be $_8O^{2-} = 1s^2,\, 2s^2,\, 2p_x{}^{\uparrow\downarrow},\, 2p_y{}^{\uparrow\downarrow},\, 2p_z{}^{\uparrow\downarrow}$ and the ESR instrument will see no unpaired electrons.

In neutral neon ($_{10}$Ne), there are 10 electrons, so in $_{10}$Ne$^+$, there will be only 9. The electronic configuration will be $_{10}Ne^+ = 1s^2,\, 2s^2,\, 2p_x{}^{\uparrow\downarrow},\, 2p_y{}^{\uparrow\downarrow}\, 2p_z{}^\uparrow$. Once again, the ESR instrument will find a single unpaired electron.

Fluoride, F^-, has the electronic configuration of Ne. All electrons are paired, and the ESR instrument will seek in vain for an unpaired spin.

$$_9F^- = 1s^2,\, 2s^2,\, 2p_x{}^{\uparrow\downarrow},\, 2p_y{}^{\uparrow\downarrow}\, 2p_z{}^{\uparrow\downarrow}$$

Problem 1.58 Both carbon and oxygen are neutral.

$_6$C	$_8$O
2 (1s)	2 (1s)
2 (shared)	2 (shared)
2 (nonbonding)	4 (nonbonding)
⎯⎯⎯⎯⎯	⎯⎯⎯⎯⎯
6 total	8 total
C is neutral	O is neutral

As oxygen is more electronegative than carbon, the dipole will be

$$\overset{\longrightarrow}{\underset{\delta^+\quad\ \delta^-}{C=\!\!=\!\!=O}}$$

The second Lewis structure is

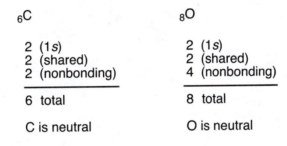

In the new Lewis structure, the carbon is negative and the oxygen positive:

$_6$C	$_8$O
2 (1s)	2 (1s)
3 (shared)	3 (shared)
2 (nonbonding)	2 (nonbonding)
⎯⎯⎯⎯⎯	⎯⎯⎯⎯⎯
7 total	7 total
C is negative	O is positive

So, to the extent that this second resonance form is important, the dipole will be in the opposite direction:

$$\overset{\delta^-\qquad \delta^+}{\underset{\longleftarrow}{:\overset{-}{C}\!\!\equiv\!\!\overset{+}{O}:}}$$

The dipoles in the two Lewis structures (two resonance forms) will tend to cancel each other out. The result is a very small observed dipole.

Problem 1.59 This molecule also has two important resonance forms, but the dipole is in the same direction in each and will reinforce. Formaldehyde will have a larger dipole moment than carbon monoxide.

Problem 1.60 Structure **B** will have no dipole moment, as the dipoles cancel. Therefore, the observation of a dipole moment for CH_2F_2 eliminates **B** as a possibility. However, in planar structure **C**, the dipoles reinforce. This molecule will have a dipole moment and cannot be distinguished from **A** on this basis.

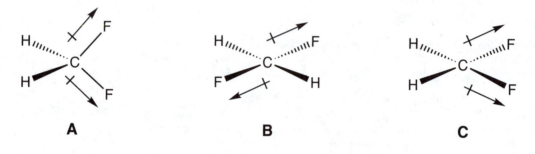

| A | B | C |

Problem 1.61

Homolytic cleavage

$$:\!\overset{..}{\underset{..}{Br}}\!-\!\overset{..}{\underset{..}{Br}}\!: \quad \longleftrightarrow \quad :\!\overset{..}{\underset{..}{Br}}\!\cdot \quad \cdot\!\overset{..}{\underset{..}{Br}}\!:$$

Heterolytic cleavage

$$:\!\overset{..}{\underset{..}{Br}}\!-\!\overset{..}{\underset{..}{Br}}\!: \quad \longleftrightarrow \quad :\!\overset{..}{\underset{..}{Br}}\!^{+} \quad :\!\overset{..}{\underset{..}{Br}}\!:^{-}$$

Problem 1.62 It all starts normally enough, so n must increase monotonically. The $1s$ shell fills with two electrons so $s = \pm 1/2$. Lithium (Li) and beryllium (Be) are normal as the $2s$ shell fills. However, there seem to be no $2p$ orbitals. So l must not be the same as in our universe. In the new universe, $_5$B is similar to $_1$H and $_3$Li, and $_6$C is similar to $_2$He and $_4$Be. The $3s$ shell must be filling with these two atoms.

$_1$H $= 1s$
$_2$He $= 1s^2$
$_3$Li $= 1s^2, 2s$
$_4$Be $= 1s^2, 2s^2$

(as no $2p$ orbitals are available in this universe, start to fill the $3s$ shell in boron)

$_5$B $= 1s^2, 2s^2, 3s$
$_6$C $= 1s^2, 2s^2, 3s^2$

Now we find six atoms filling $_7$N through $_{12}$Mg. These must be the three $3p$ orbitals.

$_7$N $= 1s^2, 2s^2, 3s^2, 3p$
$_8$O $= 1s^2, 2s^2, 3s^2, 3p^2$
$_9$F $= 1s^2, 2s^2, 3s^2, 3p^3$
$_{10}$Ne $= 1s^2, 2s^2, 3s^2, 3p^4$
$_{11}$Na $= 1s^2, 2s^2, 3s^2, 3p^5$
$_{12}$Mg $= 1s^2, 2s^2, 3s^2, 3p^6$

Apparently, l in this universe has the value $l = 0, 1, 2, 3 \ldots (n-2)$.
So, when $n = 1, l = 0, n = 2, l = 0$ (no $2p$ orbitals), $n = 3, l = 1, m_l = -1 \ldots 0 \ldots +1, s = \pm 1/2$.

Now the $4s$ shell must fill.

$_{13}$Al $= 1s^2, 2s^2, 3s^2, 3p^6, 4s$
$_{14}$Si $= 1s^2, 2s^2, 3s^2, 3p^6, 4s^2$

Now the $4p$ orbitals fill to complete the periodic table, as shown below.

$_{15}$P $= 1s^2\ 2s^2\ 3s^2\ 3p^6\ 4s^2\ 4p$
$_{16}$S $= 1s^2\ 2s^2\ 3s^2\ 3p^6\ 4s^2\ 4p^2$
$_{17}$Cl $= 1s^2\ 2s^2\ 3s^2\ 3p^6\ 4s^2\ 4p^3$
$_{18}$Ar $= 1s^2\ 2s^2\ 3s^2\ 3p^6\ 4s^2\ 4p^4$
$_{19}$K $= 1s^2\ 2s^2\ 3s^2\ 3p^6\ 4s^2\ 4p^5$
$_{20}$Ca $= 1s^2\ 2s^2\ 3s^2\ 3p^6\ 4s^2\ 4p^6$

Problem 1.63

(a)

$$H_3C\cdot \quad \cdot \ddot{\underset{\cdot\cdot}{Br}}{:} \quad \rightleftharpoons \quad H_3C\!-\!\ddot{\underset{\cdot\cdot}{Br}}{:}$$

Table 1.10 gives the bond energy of the carbon–bromine bond, 73 kcal/mol. This reaction involves pure bond making, an exothermic process. The reaction would be written

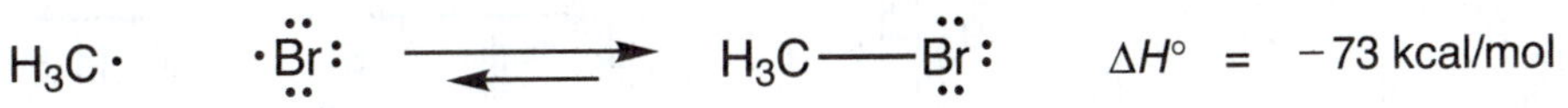

$$H_3C\cdot \quad \cdot\ddot{\underset{\cdot\cdot}{Br}}{:} \quad \rightleftharpoons \quad H_3C\!-\!\ddot{\underset{\cdot\cdot}{Br}}{:} \qquad \Delta H^\circ = -73 \text{ kcal/mol}$$

(b)

$$H_3C\!-\!\ddot{\underset{\cdot\cdot}{Cl}}{:} \quad \rightleftharpoons \quad H_3C\cdot \quad \cdot\ddot{\underset{\cdot\cdot}{Cl}}{:}$$

By contrast, this reaction involves only the breaking of the carbon–chlorine bond. This reaction is endothermic by about 84 kcal/mol. The reaction would be written

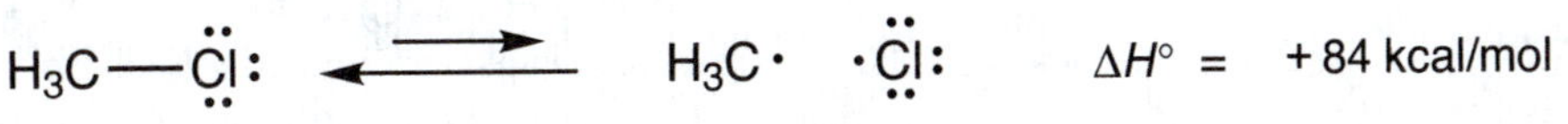

$$H_3C\!-\!\ddot{\underset{\cdot\cdot}{Cl}}{:} \quad \rightleftharpoons \quad H_3C\cdot \quad \cdot\ddot{\underset{\cdot\cdot}{Cl}}{:} \qquad \Delta H^\circ = +84 \text{ kcal/mol}$$

(c)

$$H_2C\!=\!CH_2 \ + \ H\!-\!H \ \rightleftharpoons \ \begin{matrix} H_2C\!-\!CH_2 \\ \mid \quad\ \ \mid \\ H \quad\ \ H \end{matrix}$$

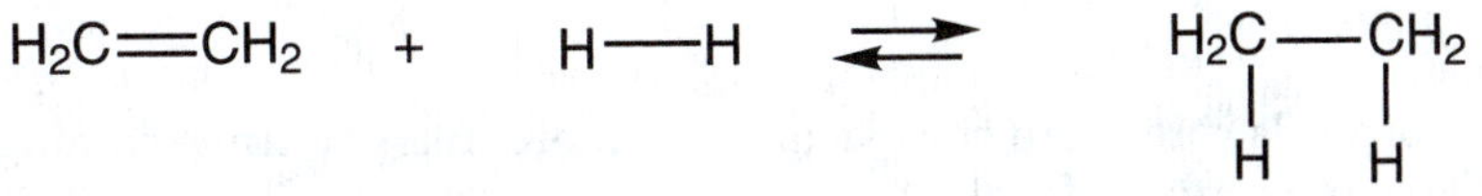

This reaction is more complicated than those of parts (a) and (b). Bonds are both made and broken. Assume that the four carbon–hydrogen bonds in the starting material are exactly balanced by the four carbon–hydrogen bonds in the product.

(continued)

Problem 1.63 *(continued)*

Bonds broken in starting material (kcal/mol)   Bonds made in product (kcal/mol)

π $H_2C{=\!=}CH_2$ 66   two   C——H   200

H——H   104

_____

170

This reaction is exothermic by 30 kcal/mol (200 – 170 = 30). The reaction is written

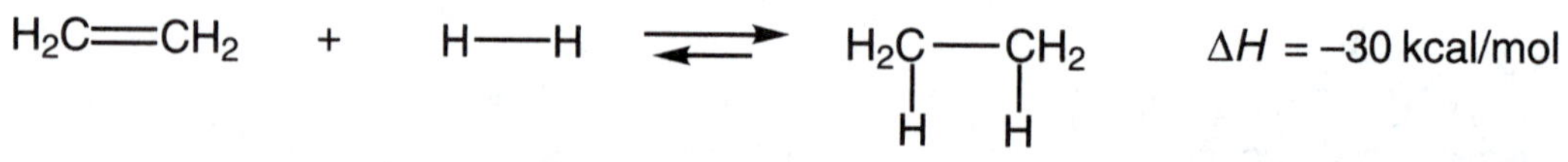

$H_2C{=\!=}CH_2$ +   H——H $\rightleftharpoons$ H_2C——CH_2 $\Delta H = -30$ kcal/mol

H   H

(d) $H_2C{=\!=}CH_2$ +   H——$\ddot{\underset{..}{C}}l$: $\rightleftharpoons$ H_2C——CH_2

H   :$\ddot{\underset{..}{C}}l$:

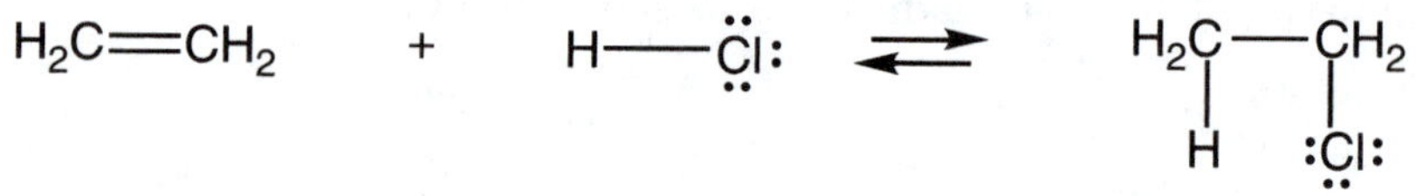

Bonds are both broken and made in this reaction.

Bonds broken in starting material (kcal/mol)   Bonds made in product (kcal/mol)

π $H_2C{=\!=}CH_2$ 66   C——H   100

H——Cl   103   C——Cl   84

_____   _____

169   184

This reaction is exothermic by about 15 kcal/mol (184 – 169 = 15). The reaction is written

$H_2C{=\!=}CH_2$ +   H——$\ddot{\underset{..}{C}}l$: $\rightleftharpoons$ H_2C——CH_2 $\Delta H = -15$ kcal/mol

H   :$\ddot{\underset{..}{C}}l$:

Problem 1.64

(a) As there are three orbitals going into our calculation, there must be three coming out. The H_3 molecule will have three molecular orbitals.

(b) First of all, remember that the problem tells us how to construct H_3. Place the new H in between the two hydrogens of H——H. The interaction of Φ_B with $1s$ will yield two new molecular orbitals, **1** and **3**.

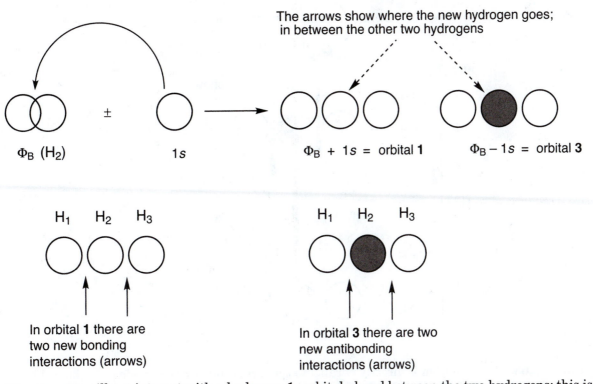

The arrows show where the new hydrogen goes; in between the other two hydrogens

Φ_B (H$_2$)   1s Φ_B + 1s = orbital **1** Φ_B − 1s = orbital **3**

H$_1$ H$_2$ H$_3$

H$_1$ H$_2$ H$_3$

In orbital **1** there are two new bonding interactions (arrows)

In orbital **3** there are two new antibonding interactions (arrows)

However, Φ_A will not interact with a hydrogen 1s orbital placed between the two hydrogens; this is a net-zero interaction, as the new bonding interaction is exactly canceled by the new antibond.

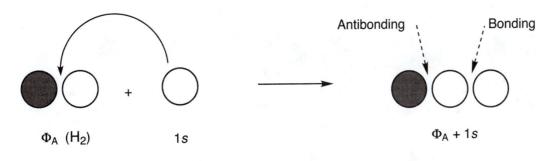

Antibonding   Bonding

Φ_A (H$_2$)   1s Φ_A + 1s

New antibonding exactly cancels new bonding— there is no net interaction between the orbitals

So, the three orbitals for HHH are **1** and **3** and the old Φ_A of H$_2$, modified only by the moving apart of the two hydrogens. Let's call this one **2**. The center dot shows the position of the middle hydrogen. The sign of the wave function at this point is zero.

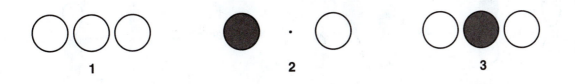

1   **2**   **3**

(continued)

Problem 1.64 *(continued)*

(c) You can easily order these new orbitals by simply counting the nodes.

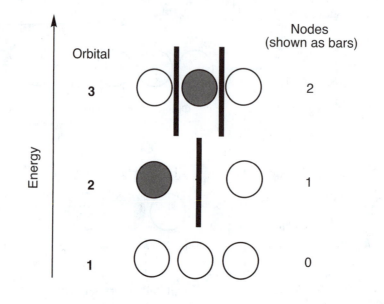

Problem 1.65   The situation is directly parallel to the formation of HHH, made in Problem 1.64. Once again, there will be three molecular orbitals. Two new orbitals, **A** and **B**, are formed by the interaction of a single $2p$ orbital with $(2p + 2p)$.

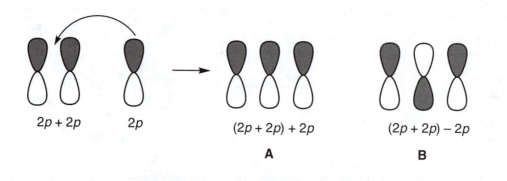

The third orbital, **C**, comes from $(2p - 2p)$, which does not interact with a $2p$ orbital placed in the middle. Once again, this is a net-zero situation as the bonding and antibonding interactions exactly cancel.

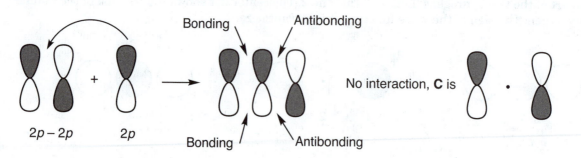

The dot in **C** simply marks the position of the central atom.

The new orbitals **A**, **B**, and **C** can be ordered by counting nodes:

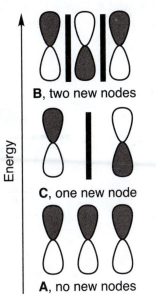

Problem 1.66

(a) All you need to do is to bend the orbitals for HHH into a triangle to generate the molecular orbitals for the triangular H_3. The direction of energy change is determined by noting whether a new bonding or antibonding interaction is created.

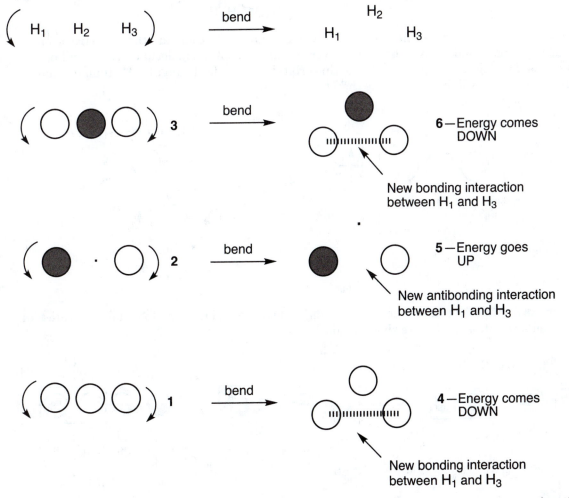

(continued)

Problem 1.66 *(continued)*

(b) Apply what the answer to (a) tells us. This answer shows how the energies of **1**, **2**, and **3** will change as bending occurs to make **4**, **5**, and **6**. Notice that orbitals **5** and **6**, each with one node (shown as a bar in the figure), are placed at the same energy.

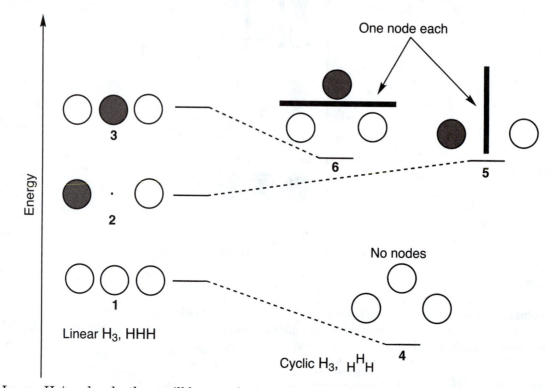

(c) In any H_3^+ molecule, there will be two electrons. In neutral H_3, there are three electrons, one from each hydrogen, and H_3^+ will have one fewer. So, only the lowest molecular orbital will be occupied. As the diagram for part (b) shows, this orbital is lower for triangular H_3 than for linear H_3. The bent species will be more stable than the linear molecule.

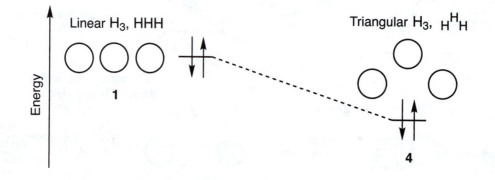

Problem 1.67 We take combinations of $\Phi_B \pm \Phi_B$ and $\Phi_A \pm \Phi_A$ placed end to end to generate four new molecular orbitals. Nodes are shown as dashed lines.

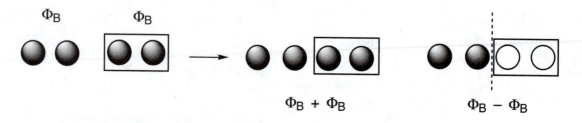

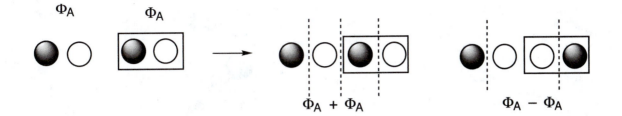

Counting nodes leads to the following order. The four electrons are placed as shown:

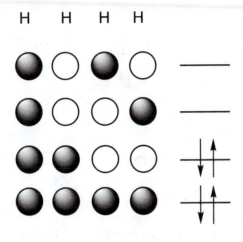

Problem 1.68 Here are the interactions:

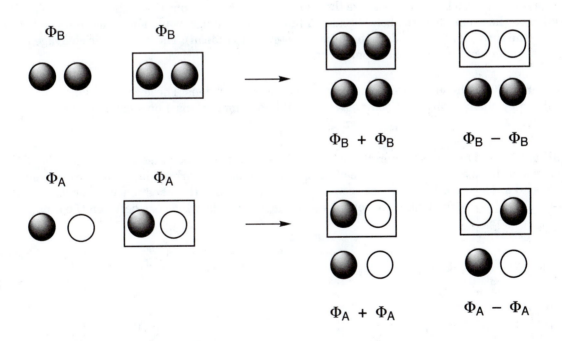

One new orbital has zero nodes, two have one node, and one has two nodes. The order will be as shown. For neutral, square H_4, there will be four electrons placed as shown. Note the parallel spins in the singly occupied orbitals. Nodes are shown as dashed lines.

(continued)

Problem 1.68 *(continued)*

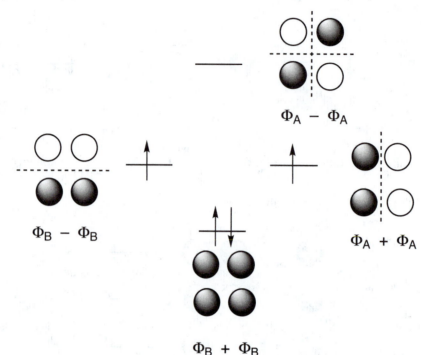

$\Phi_A - \Phi_A$

$\Phi_B - \Phi_B$

$\Phi_A + \Phi_A$

$\Phi_B + \Phi_B$

Problem 1.69   The carbon–bromine bond breaks in the first step. We can consider the two ways that bond could break: a homolytic cleavage to give a carbon radical and a bromine radical or a heterolytic cleavage to give a carbocation and a bromide ion. It is not easy to tell the difference just by seeing the bond breaking. However, there is a clue in the reaction drawn out on the right side of the screen. The product is a bromide ion. This observation tells you that the cleavage was heterolytic. It is also helpful to observe the HOMO track. Notice that the highest occupied orbital ends up on the bromine as the bond breaks. A homolytic cleavage would show electrons on both the carbon (the carbon radical) and the bromine (the bromine radical). The heterolytic cleavage has all the electron density going to the bromide, as shown in the animation.

Problem 1.70   A heterolytic cleavage gives charged species: ions. These charged ions would be stabilized by a polar solvent. A polar solvent would have more impact on such a heterolytic cleavage.

Problem 1.71   One can see from the selected picture that the calculated area for the π bond electrons is above and below the plane of the carbon atoms. This orbital is where the electrons in the π bond are located. They are not between the carbon atoms. It makes sense to conclude that π bond electrons are not as tightly held as electrons in σ bonds, which *are* between the atoms.

Alkanes

2

This set of problems provides practice in using the various hybridization schemes developed in this chapter. The major theme of the chapter is structure, and the specific focus is on the structure of alkanes. Special attention is paid to writing and naming isomeric alkanes and to using Newman projections to help us to visualize molecules in three dimensions. Practice in using the various coded two-dimensional representations of three-dimensional molecules is provided, and ring compounds are introduced.

Problem 2.1

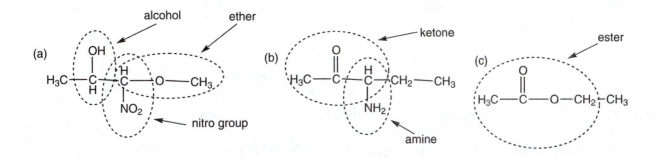

Problem 2.2   This problem asks you to consider the molecular orbital system made up of overlapping 1s, 2s, and 1s orbitals containing four electrons. Where are these four electrons? *Three* overlapping atomic orbitals will produce *three* molecular orbitals. That is all we really need to notice in this problem. There is plenty of room for four electrons. One of the three new orbitals is bonding, one nonbonding, and one antibonding. The four electrons will occupy the lowest two molecular orbitals. There is no violation of the Pauli principle here, as shown on the next page.

(continued)

Problem 2.2 *(continued)*

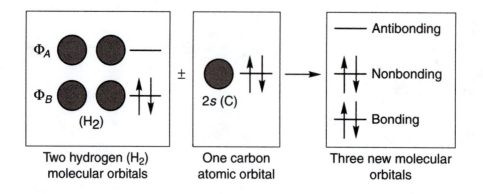

| Two hydrogen (H_2) molecular orbitals | One carbon atomic orbital | Three new molecular orbitals |

Problem 2.3 As the bonding and antibonding interactions exactly cancel, there is *no* net interaction between the orbitals. In this orientation, a $2p$ and an s orbital do not interact; they are orthogonal.

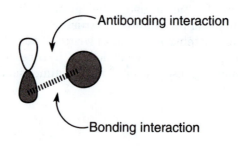

Problem 2.4 In BeH_2, the empty $2p_y$ and $2p_z$ orbitals are oriented at 90° to each other and to the pair of sp hybrid orbitals.

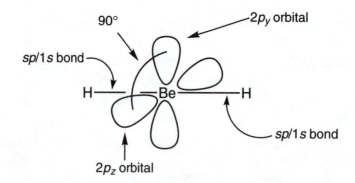

Problem 2.6 Trimethyl means that there are three methyl (CH_3) groups. Aluminum (like boron) has three valence electrons. So it is the central atom, and it can have a bond to each of the carbons of the three methyl groups. Those bonds will be at maximum distance from each other, which will give the planar arrangement with 120° angles for the C—Al—C bonds.

Lewis structure:

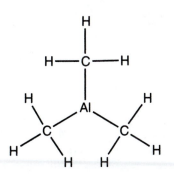

Aluminum is in the third row, and in trimethylaluminum, it only sees three electrons as its own (one electron from each bond to carbon belongs to the aluminum). So the aluminum is neutral, and it will be sp^2 hybridized. It will have an empty $3p$ orbital.

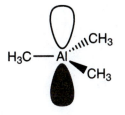

Problem 2.7 The unused empty $2p_z$ orbital on boron extends above and below the xy plane of the three boron–hydrogen bonds.

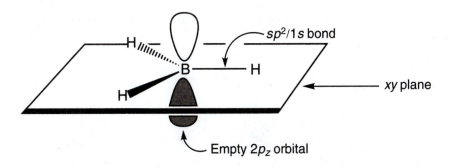

Problem 2.8 Nitrogen is attached to four atoms—the four hydrogens. Thus, we need four hybrid orbitals. These sp^3 hybrids are constructed from nitrogen's $2s$, $2p_x$, $2p_y$, and $2p_z$ atomic orbitals. The nitrogen is sp^3 hybridized and tetrahedral in shape.

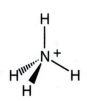

Problem 2.9

CH_2F_2	CH_2Cl_2	CH_2Br_2	CH_2I_2
CH_2ClF	CH_2BrF	CH_2FI	
CH_2BrCl	CH_2BrI		
CH_2ICl			

Problem 2.10

CH_2F_2	CH_2Cl_2	CH_2FCl		
CHF_3	$CHCl_3$	CHF_2Cl	$CHFCl_2$	
CF_4	CCl_4	CF_3Cl	CF_2Cl_2	$CFCl_3$

Problem 2.11 Exactly halfway between the two pyramids, the molecule must be flat, and therefore the carbon atom is *sp*² hybridized. One pyramidal molecule cannot pass to the other without going through a planar form, **A**.

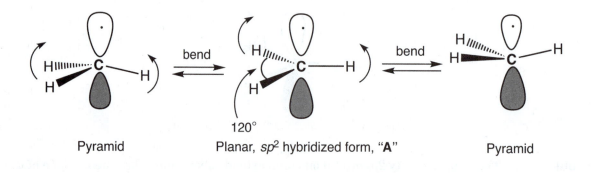

Problem 2.12 Look down the carbon–carbon bond of ethyl chloride (CH_3—CH_2—Cl) with the methyl group in front. You see three carbon–hydrogen bonds attached to the front carbon. In the rear you see the carbon represented as a circle attached to two hydrogens and a chlorine. There are three staggered forms of equal energy.

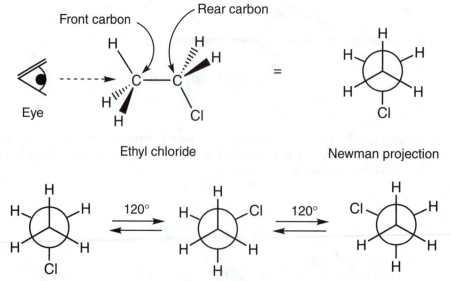

The three equivalent forms are interconverted by 120° rotations about the carbon–carbon bond.

The second compound, 1,2-dichloroethane (Cl—CH₂—CH₂—Cl), is more complicated. As you look down the central carbon–carbon bond, you see in front a carbon attached to two hydrogens and a chlorine. In back, the carbon represented as a circle is also attached to two hydrogens and a chlorine. There are three forms interconverted by rotation, **A**, **B**, and **B′**. Forms **B** and **B′** are of equal energy, but **A** is different. As **A** keeps the two relatively large chlorine groups as far apart as possible, it is lower in energy than **B** or **B′**.

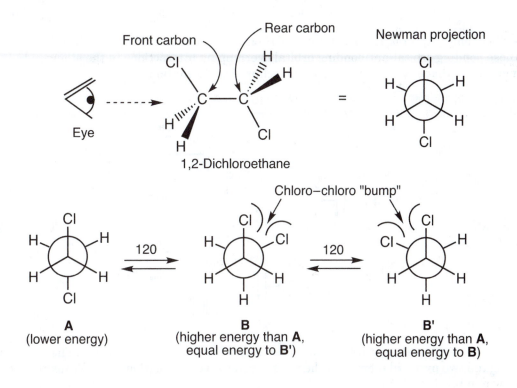

Problem 2.14

(a)

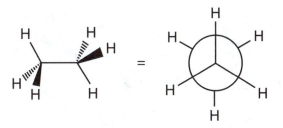

(b)

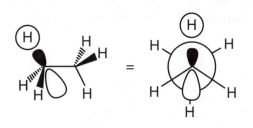

(continued)

Problem 2.14 *(continued)*

(c)

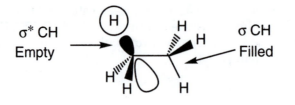

(d) The filled–empty interaction is between σ CH on one carbon and σ* CH on the other. There are six of these bonding orbital-antibonding orbital overlapping interactions in the staggered conformation of ethane. There are none in the eclipsed conformation.

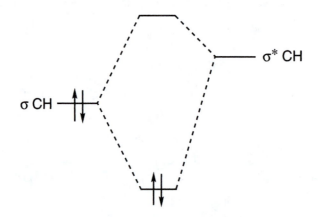

Problem 2.15 A transition state was first encountered in Problem 2.11 (p. 67). The planar form separating the two pyramidal forms of the methyl radical is a transition state. Transition states are often shown in brackets, [TS].

Problem 2.16 Start by drawing a "stick" or "sawhorse" figure for staggered ethyl alcohol.

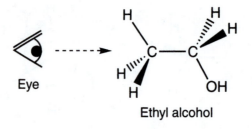

Ethyl alcohol

Next, convert this figure into a Newman projection. What do we see on the front carbon? Three hydrogens. In this projection, they appear at 120° angles.

Next, add the substituents on the back carbon, two hydrogens and an OH. Now we have our Newman projection.

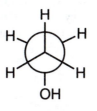

Problem 2.17 This problem can be reduced to: How many isomers of 1,2-dideuterioethane are possible? This problem is actually much tougher than it looks. Part of it is easy, but there is a hidden, and quite subtle, difficulty. By far the easiest way to attack this problem is to use Newman projections. Sight down the carbon–carbon bond and examine all possible staggered conformations.

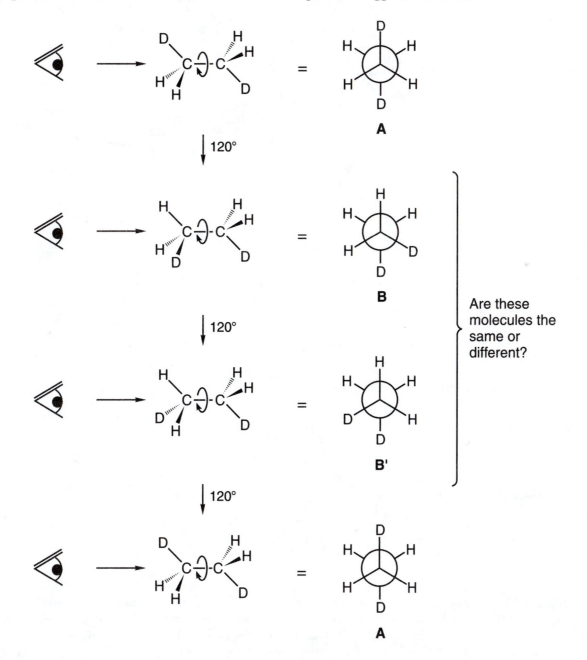

Problem 2.17 *(continued)*

The first 120° rotation takes us from **A** to **B**, and the two are clearly different. A second 120° rotation generates **B'**, and a third 120° rotation takes us back to **A** as a 360° rotation is completed. The hard part of this problem is ferreting out the relationship between **B** and **B'**. Are they the same or not? At this point, this problem is very tough to do without models. Your models will show you that **B** and **B'** are not the same. There are three different possible staggered forms of 1,2-dideuterioethane. Just how **B** and **B'** differ is an interesting question, and we will deal with it extensively in Chapter 4.

Problem 2.18

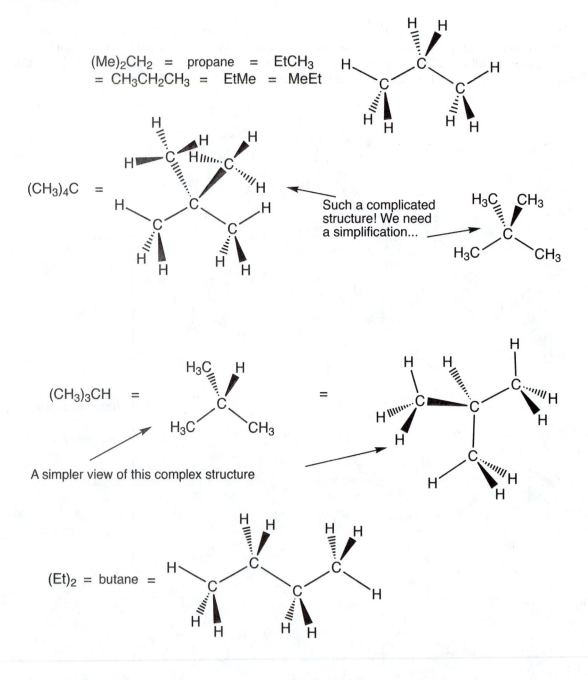

Problem 2.19   Your models will serve you better than words here. Replacement of the "end" hydrogen to give **A** and the "corner" hydrogen to give **B** leads to the same thing. The C—C—C right angle apparent in structure **B** is not real. The take-home lesson here is, *Always Remember*: Organic molecules are three-dimensional. Never trust the two-dimensional surface.

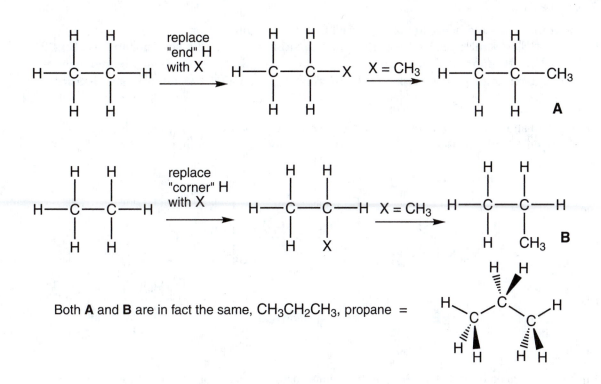

Both **A** and **B** are in fact the same, $CH_3CH_2CH_3$, propane =

Problem 2.20 The only trick here is to remember to keep all carbon–hydrogen and carbon–hydroxyl bonds staggered.

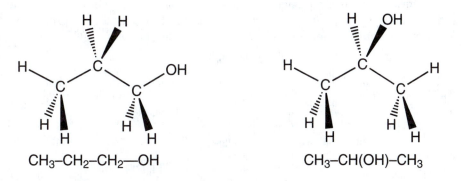

$CH_3–CH_2–CH_2—OH$ $CH_3–CH(OH)–CH_3$

Problem 2.21 In this problem you need to calculate the amount of destabilization caused by C—CH_3/C—H and C—CH_3/C—CH_3 interactions. Each C—H/C—H eclipsed interaction costs about 1.0 kcal/mol (p. 71). The transition state for the interconversion of **A** and **B** is an eclipsed form. In the figure, it is reached through a 60° clockwise rotation of the rear carbon. A second 60° clockwise rotation takes us to **B**.

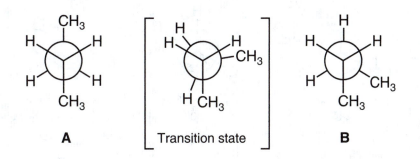

A   Transition state   **B**

(continued)

Problem 2.21 *(continued)*

In the transition state, there are one eclipsed C—H/C—H interaction and two eclipsed C—H/C—CH₃ interactions. We know that the transition state lies 3.4 kcal/mol above **A** (Fig. 2.33). The C—H/C—H interaction costs 1.0 kcal/mol, and so the two C—H/C—CH₃ interactions must produce the remaining 2.4 kcal/mol. Each one must cause about 1.2 kcal/mol destabilization.

Figure 2.33 shows that the transition state for the interconversion of **B** and **B′** lies 3.8 + 0.6 = 4.4 kcal/mol above **A**. The transition state for this conversion is shown.

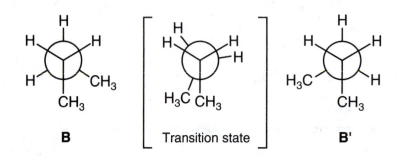

Each of the two eclipsed C—H/C—H interactions costs 1.0 kcal/mol, so the single C—CH₃/C—CH₃ eclipsing interaction must cause the remaining 2.4 kcal/mol destabilization.

Problem 2.22   For the first molecule, the Newman projection is constructed as usual. The front carbon bears two methyl groups and a hydrogen, and the rear carbon, shown as a circle, two hydrogens and one methyl group. Start at 0° with an eclipsed form and then proceed by 60° rotations of the rear carbon to generate the other Newman projections.

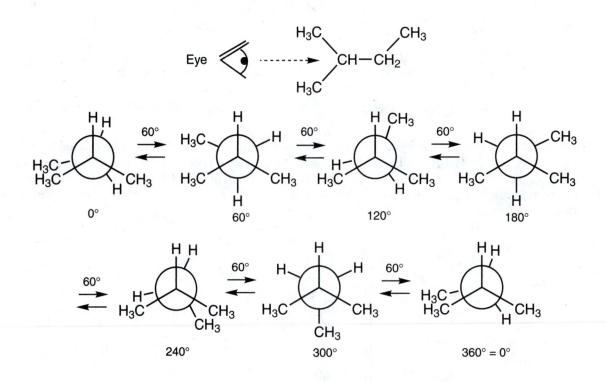

The second molecule is more symmetrical, as both carbons bear two methyl groups and a single hydrogen. The 0° and 360° forms are identical and contain two methyl–methyl eclipsed interactions. These will be the highest energy conformations. A 60° rotation leads to a staggered molecule with three methyl–methyl gauche interactions. The 120° and 240° transition states have one methyl–methyl eclipsed interaction and two methyl–hydrogen eclipsed interactions and will be lower in energy than the 0° and 360° conformations. Lowest energy of all is the 180° form, which contains only two methyl–methyl gauche interactions. The graph shows the relative energies.

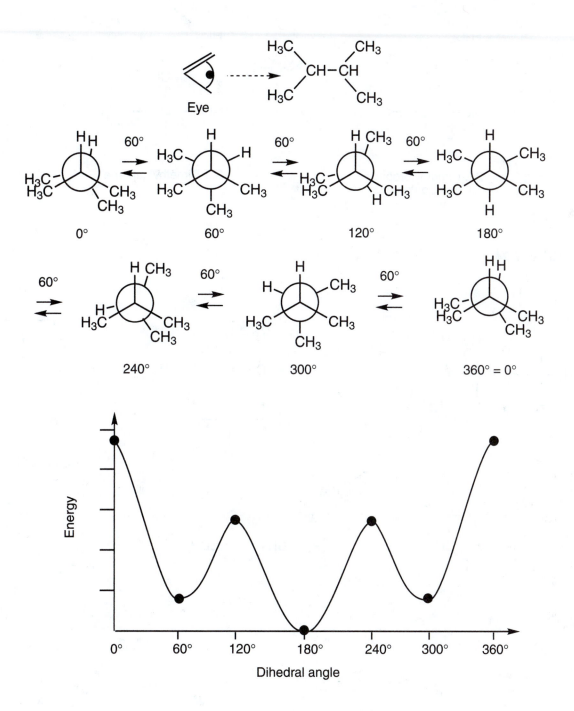

Problem 2.23 Any compound containing a carbon attached to four other carbons contains a "quaternary" carbon. The quaternary carbon is shown in boldface type in the figure.

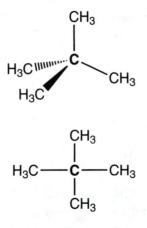

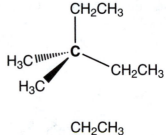

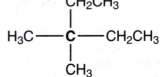

2,2-Dimethylpropane
(neopentane)

3,3-Dimethylpentane

Problem 2.24

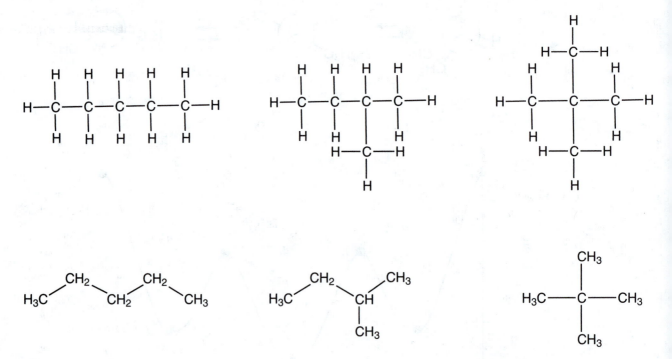

Problem 2.25

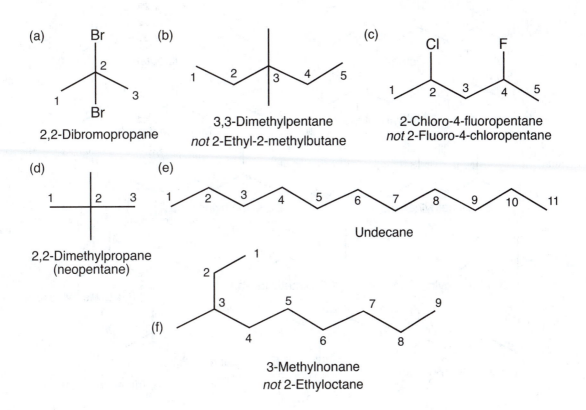

(a) 2,2-Dibromopropane

(b) 3,3-Dimethylpentane *not* 2-Ethyl-2-methylbutane

(c) 2-Chloro-4-fluoropentane *not* 2-Fluoro-4-chloropentane

(d) 2,2-Dimethylpropane (neopentane)

(e) Undecane

(f) 3-Methylnonane *not* 2-Ethyloctane

Problem 2.26

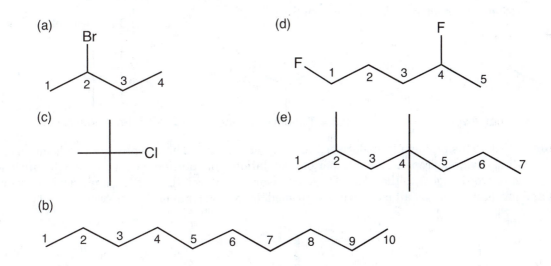

(a)

(c)

(b)

(d)

(e)

Problem 2.27   4-ethyl-5-isobutyldecane; 5-(*sec*-butyl)-4-ethyldecane.

Problem 2.28   Here we use only the most schematic representations for the molecules.

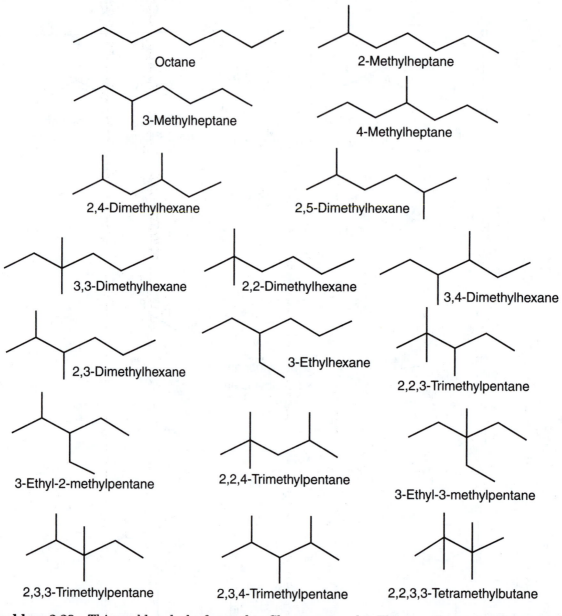

Octane

2-Methylheptane

3-Methylheptane

4-Methylheptane

2,4-Dimethylhexane

2,5-Dimethylhexane

3,3-Dimethylhexane

2,2-Dimethylhexane

3,4-Dimethylhexane

2,3-Dimethylhexane

3-Ethylhexane

2,2,3-Trimethylpentane

3-Ethyl-2-methylpentane

2,2,4-Trimethylpentane

3-Ethyl-3-methylpentane

2,3,3-Trimethylpentane

2,3,4-Trimethylpentane

2,2,3,3-Tetramethylbutane

Problem 2.29   This problem looks forward to Chapters 4 and 5. There really are two isomers of *trans*-1,2-dimethylcyclopropane. No number of translational or rotational operations will suffice to change one mirror image into the other. These two isomers are related in the same way that your right and left hands are. By all means, use your models to be certain of this answer.

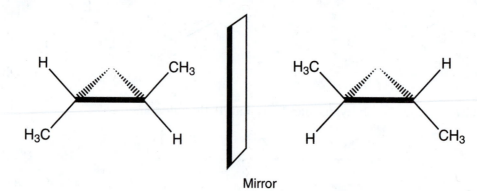

Mirror

Problem 2.32 The two rings share two carbons in the following way:

(C₈H₁₄)
The two rings share two carbons

But the problem points out that there are two of these compounds. Remember that rings have sides. There are cis and trans forms for this molecule (p. 88). The two rings can be fused together in a cis or trans fashion.

cis Form   trans Form

Problem 2.33 Finally, we come to C₇H₁₂, in which the two rings share three carbons. A "cage" structure results.

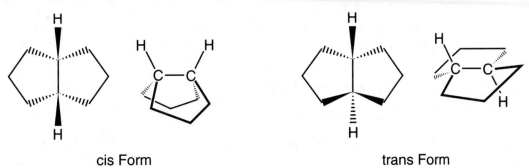

(C₇H₁₂)
Now three carbons are shared

Problem 2.34 The arrows and numbers show the different carbons. Be careful of symmetry! The only way to make a "real" mistake (as opposed to just overlooking some carbon) is to miss the symmetry of the situation and find too many "different" carbons.

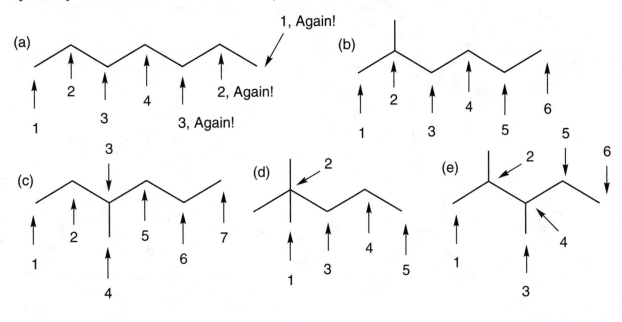

(continued)

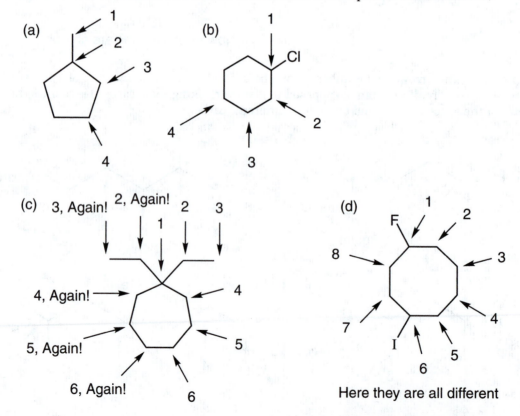

Beware of the two-dimensional page! The "repeats" are shown in the first and last examples, (a) and (i). When in doubt, make a model.

Problem 2.35   Here is more practice in seeing symmetry and not finding too many "different" carbons. In the third example (c), the repeats are pointed out. In the last example (d), all eight carbons are different. Would this still be true if the iodine were replaced with a fluorine?

Here they are all different

Problem 2.36

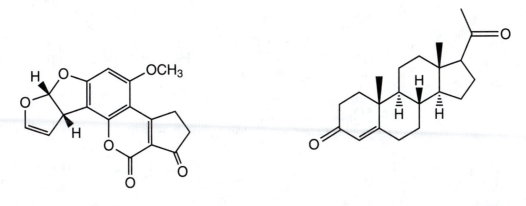

All carbons are different in both these molecules—there are 17 in the first and 21 in the second.

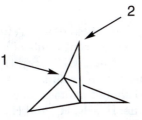

Here there is quite another situation—very high symmetry is present, and there are only two different carbons.

Problem 2.37 In propane, the methylene (CH$_2$) hydrogens are different from the six equivalent methyl hydrogens, so there will be two signals.

Tetrahydrofuran will show two signals in the ^{1}H NMR spectrum, one for the four equivalent hydrogens on the carbons adjacent to oxygen, and one for the four hydrogens on the carbons remote from the oxygen atom.

Remember that integration is relative, a ratio, not an absolute count of hydrogens.

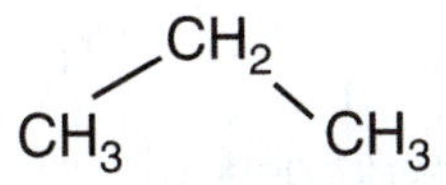

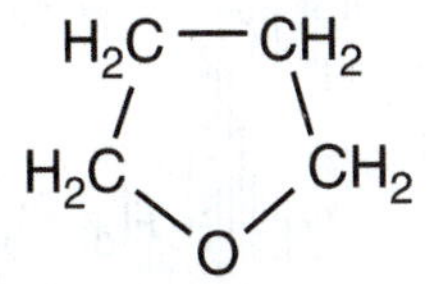

3:1   1:1

Problem 2.38 Heptane will show four signals in the ratio 6:4:4:2 = 3:2:2:1 (H_a:H_b:H_c:H_d).

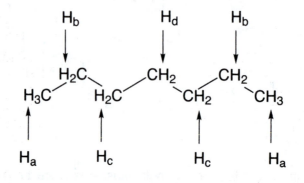

Problem 2.39 As advertised, this question takes some careful three-dimensional (3-D) viewing. It is tempting to say that there are only three different kinds of hydrogen: the six methyl hydrogens, the two methines (CH), and the methylene group (CH_2). But that analysis ignores the fact that rings have sides—the hydrogen cis to the methyl groups is not exactly the same as the hydrogen trans to the methyl groups. A look in 3-D should make it all clear. There are, in fact, four different hydrogens in *cis*-1,2-dimethylcyclopropane: H_a, H_b, H_c, and H_d.

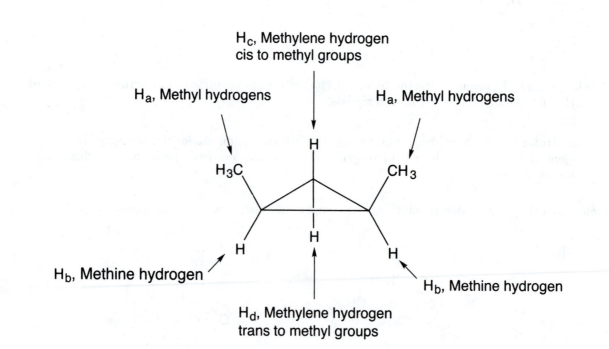

Problem 2.40

(a)

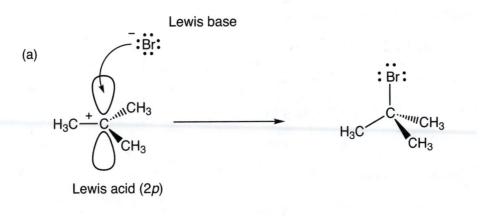

(b)

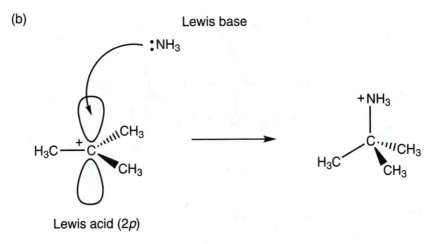

(c)

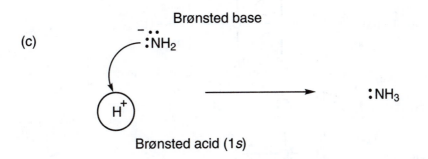

Additional Problem Answers

Problem 2.41

(a)   4-Fluoro-9-iodo-2,3-dimethylundecane

(b)   1,1,3-Trichloro-3,5-diethylcyclohexane

(c)   2,4-Dibromo-3-ethyl-2-methylpentane

(d)   5-Bromo-3-ethyl-2,2-dimethylheptane

(e)   2-Chloro-1-fluoro-3-methylcyclopentane

(f)   2,3,3-Trimethylpentane

(g)   1,1-Dichloro-1-cyclobutylbutane

(h)   4,4-Diethyloctane

(i)   2-Bromo-3-methylhexane

Problem 2.42

Et_2O is an ether:

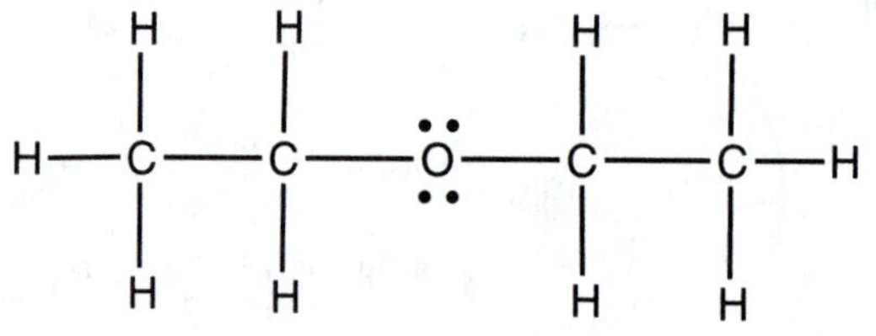

CH_3CO_2H is a carboxylic acid:

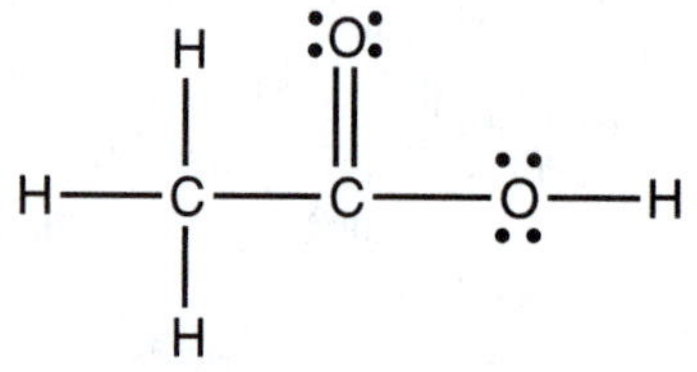

$CH_3CHOHCH_3$ is an alcohol:

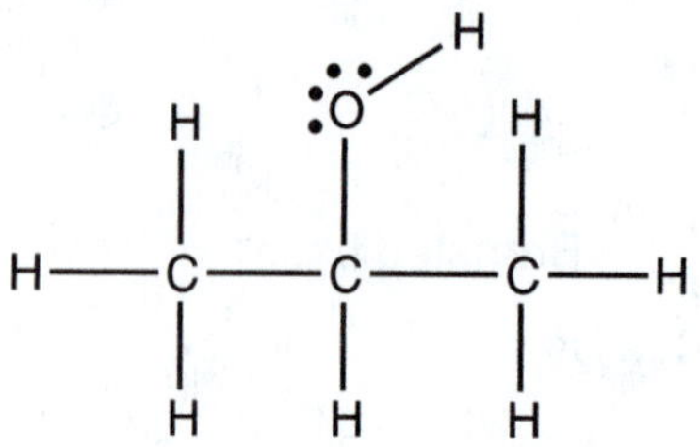

CH_2F_2 is an alkyl halide:

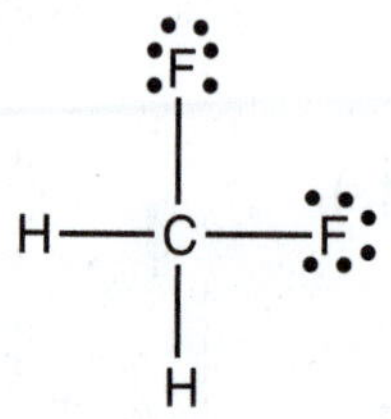

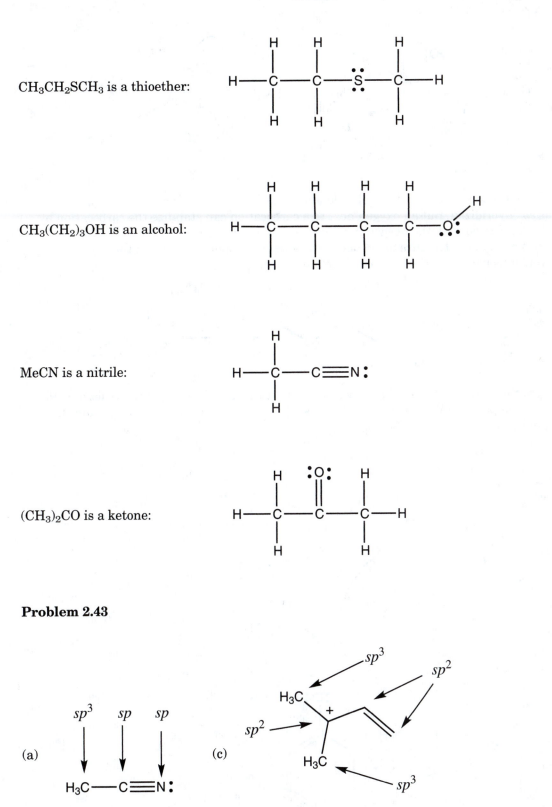

$CH_3CH_2SCH_3$ is a thioether:

$CH_3(CH_2)_3OH$ is an alcohol:

MeCN is a nitrile:

$(CH_3)_2CO$ is a ketone:

Problem 2.43

(a)

(c)

(b) This problem needs some explanation. All the carbons are sp^3. We expect the carbocation to be sp^2, but in this molecule the carbocation cannot adopt the planar geometry that is required for sp^2. Were you supposed to know that? Maybe, maybe not. But if you made a model of this structure, then you would have noticed. The structure on the right shows the empty orbital of the carbocation and the orbitals that the oxygen lone pairs occupy. Notice that the oxygen is also sp^3.

(continued)

Problem 2.43 *(continued)*

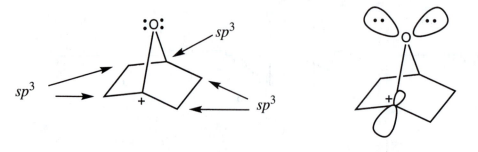

That's not all that is unique about this molecule. An oxygen with its lone pairs is normally described as sp^3 hybridized. But an oxygen next to a carbocation can stabilize the carbocation by resonance. In order to do that, the oxygen must use a p orbital. Consider the ether shown below. You can draw a resonance structure using one of the lone pairs of electrons from the oxygen. By doing this, you are saying that the lone pair is in a p orbital.

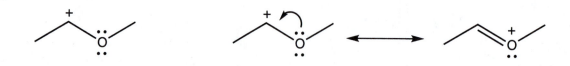

Now go back to the bicyclic ether shown above. Notice that the oxygen lone pairs are not able to line up with the empty orbital of the cation. There is no resonance between the oxygen and the carbocation in this case.

Problem 2.44

(a)

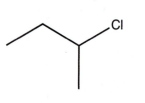

is

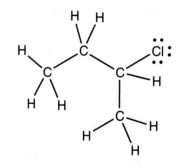

(b)

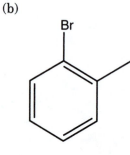

is

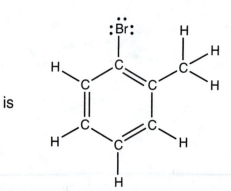

(c)

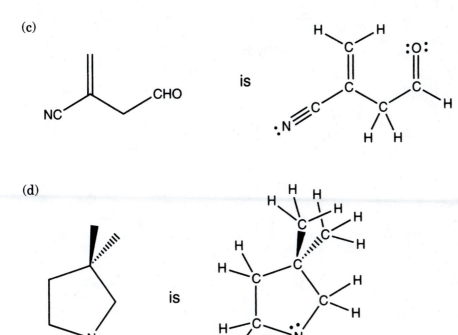

(d)

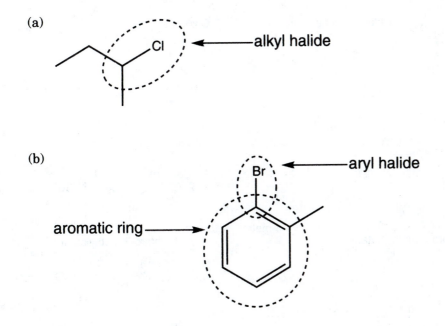

Problem 2.45

(a)

alkyl halide

(b)

aryl halide

aromatic ring

(continued)

Problem 2.45 *(continued)*

(c)

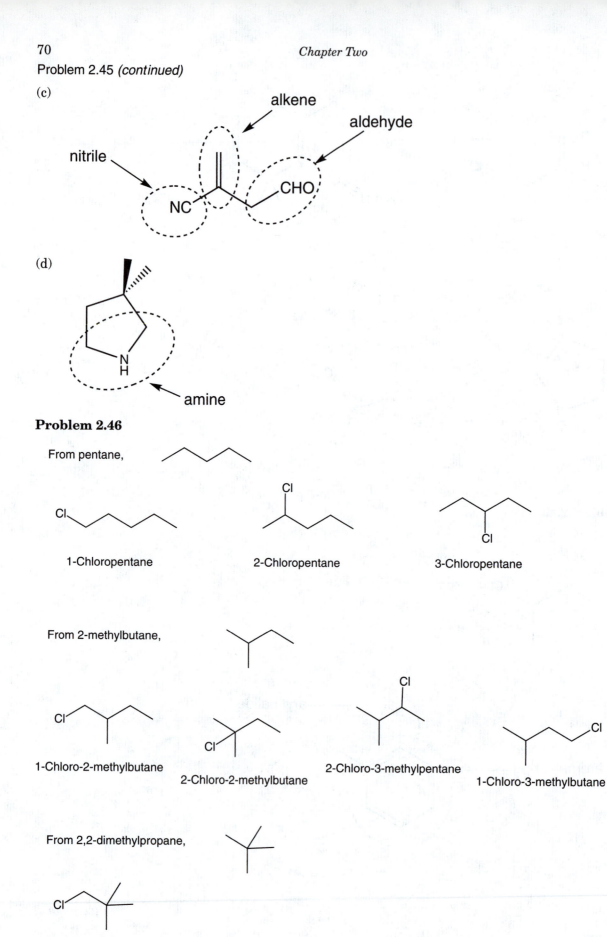

(d)

amine

Problem 2.46

From pentane,

1-Chloropentane

2-Chloropentane

3-Chloropentane

From 2-methylbutane,

1-Chloro-2-methylbutane

2-Chloro-2-methylbutane

2-Chloro-3-methylpentane

1-Chloro-3-methylbutane

From 2,2-dimethylpropane,

1-Chloro-2,2-dimethylpropane

Problem 2.47 First draw the line structure for hexane. Now find the C(3)—C(4) bond and draw in any hydrogens that are on those carbons. Use wedges and dashes to make your drawings. Now look down that bond in either direction. Represent what you see on a Newman projection.

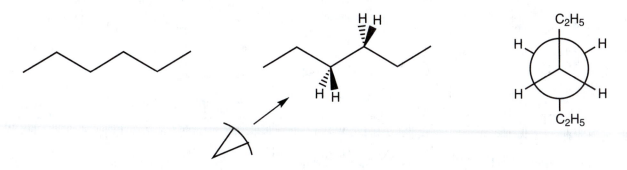

Now draw the conformational isomers (conformers) that are obtained when you rotate the front carbon in a clockwise fashion. You could rotate the back carbon if you prefer.

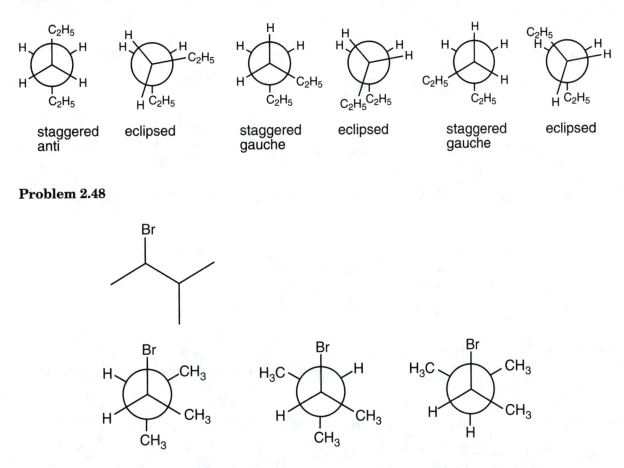

The first Newman projection has one bromo–methyl gauche and two methyl–methyl gauche interactions. The second Newman projection has one bromo–methyl gauche and one methyl–methyl gauche interaction. The third Newman projection has two bromo–methyl gauche and one methyl–methyl gauche interactions. Regardless of whether the bromo–methyl or methyl–methyl interaction causes more strain, the second projection has the lowest total strain.

The bromo group is much larger than a methyl group, but its space is made up of electrons, not atoms as in the methyl group. The bromo–methyl gauche interaction actually causes less strain than a methyl–methyl gauche interaction.

Problem 2.49

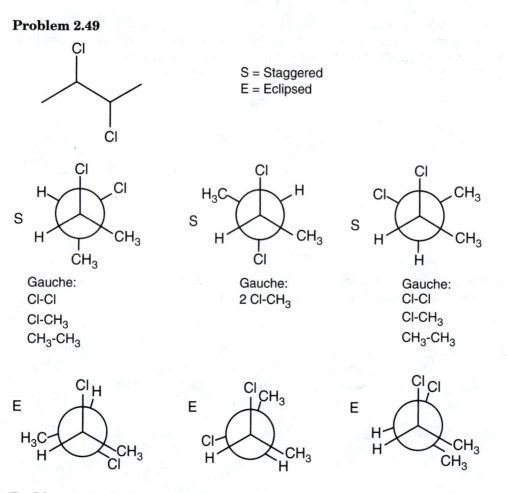

S = Staggered
E = Eclipsed

Gauche:
Cl-Cl
Cl-CH₃
CH₃-CH₃

Gauche:
2 Cl-CH₃

Gauche:
Cl-Cl
Cl-CH₃
CH₃-CH₃

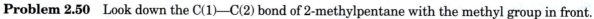

Problem 2.50   Look down the C(1)—C(2) bond of 2-methylpentane with the methyl group in front.

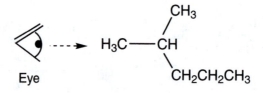

Eye

$H_3C—CH$

with CH₃ and CH₂CH₂CH₃ branches

On the front carbon you see three hydrogens. On the rear carbon you see one hydrogen, a methyl group, and a propyl (Pr) group. Start at 0°, an eclipsed form, and then proceed by 60° rotations of the rear carbon.

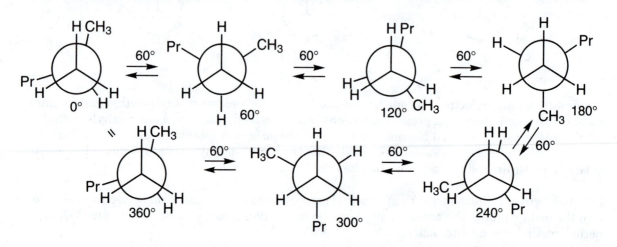

The three eclipsed conformations (0°, 120°, and 240°) are equi-energetic. These will be the highest energy conformations as they contain eclipsed methyl–hydrogen, propyl–hydrogen, and hydrogen–hydrogen interactions. The three staggered conformations (60°, 180°, and 300°) are also of equal energy. They will be lower in energy than the eclipsed conformation, as they contain only gauche, not eclipsed, interactions.

Now look down the C(2)—C(3) bond of 2-methylpentane with the methine (CH) carbon in front.

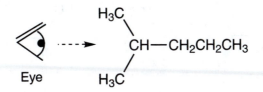

On the front carbon you see two methyl groups and a hydrogen. On the rear carbon you see two attached hydrogens and an ethyl (Et) group. Again, start at 0° with an eclipsed form and proceed by 60° rotations of the rear carbon.

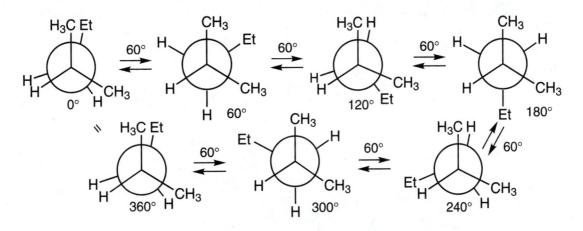

The situation is more complex from this view. The eclipsed conformations (0°, 120°, and 240°) are still energy maxima. The 0° and 120° conformations each contain a methyl–ethyl, a methyl–hydrogen, and a hydrogen–hydrogen eclipsed interaction and are equi-energetic. These are probably higher in energy than the eclipsed 240° form, which contains two methyl–hydrogen and one ethyl–hydrogen interactions. The staggered conformations (60°, 180°, and 300°) represent energy minima. The 180° and 300° conformations, which contain one methyl–ethyl gauche interaction, are equi-energetic and are lower in energy than the 60° conformation in which there are two methyl–ethyl gauche interactions.

Problem 2.51 Start by drawing the two conformations. If size were all that mattered, the two would surely be very close in energy.

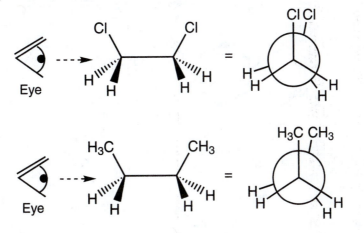

(continued)

Problem 2.51 *(continued)*

However, the carbon–chlorine bond is much more polar than a carbon–methyl bond. Note that in the eclipsed form of 1,2-dichloroethane shown, the two C—Cl dipoles are lined up. This molecule will be strongly destabilized through charge–charge opposition.

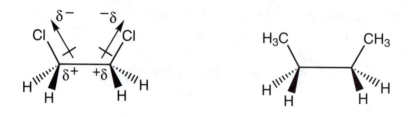

Problem 2.52   Trivalent carbon requires three bonds, so sp^2 hybridization is necessary. Divalent carbon requires only two bonds, so sp hybridization is appropriate.

(a) sp^2 (b) sp^3 (c) sp^3 (d) sp (e) sp^2 (f) sp^3

Problem 2.53

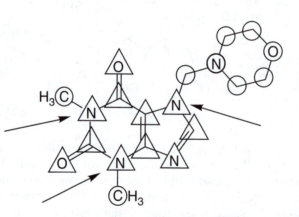

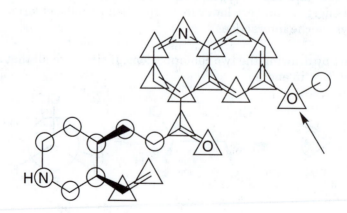

There are three nitrogens in Xanturil (the ones with arrows pointed at them) and one oxygen in Viquidil (with an arrow) that you might have decided were sp^3 rather than sp^2. Each of these atoms is sp^2 because they have a lone pair that can participate in resonance. It would be good practice to draw the resonance structures that show the reason for the sp^2 hybridization. A drawing of one of the resonance structures responsible for the sp^2 hybridization is shown. There are no sp hybridized atoms.

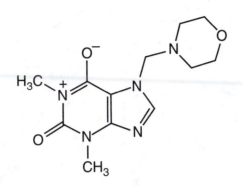

Problem 2.54 Extrapolate from the boiling points of dodecane ($C_{12}H_{26}$, bp 216.3 °C) and eicosane ($C_{20}H_{42}$, bp 343°C). So, 343 − 216.3 = 126.7. The compound C_{15} is three-eighths of the way from C_{12} to C_{20}, so we take three-eighths of 126.7 = 47.5. The boiling point of C_{15} should be the boiling point of C_{12} plus this number, 216.3 + 47.5 = 263.8 °C. This procedure works reasonably well, as the real value is found to be 270.6 °C.

Problem 2.55 As noted in Section 2.13 (p. 90), symmetry is important in determining melting points. Highly symmetrical molecules pack well into crystal lattices, and more energy is required to break up the lattice than for molecules that do not pack so well. Thus, the highly symmetrical neopentane melts 113 °C higher than pentane. However, branched-chain hydrocarbons without high symmetry tend to have lower melting points than straight-chain hydrocarbons because the branching interferes with regular packing in the crystal. Accordingly, isopentane melts 30 °C lower than pentane.

Problem 2.56

(a) Ethane yields only two compounds. You can either replace two hydrogens on one carbon or one hydrogen on each carbon. Those are the only possibilities.

$$H_3C—CH_3 \xrightarrow{\text{replace two H with X}} H_3C—CHX_2 \quad \text{and} \quad XH_2C—CH_2X$$

(b) Propane yields four compounds:

replace two H with X

(continued)

Problem 2.56 *(continued)*

(c) Butane yields six new compounds. Notice the changing code level in the representations of these compounds.

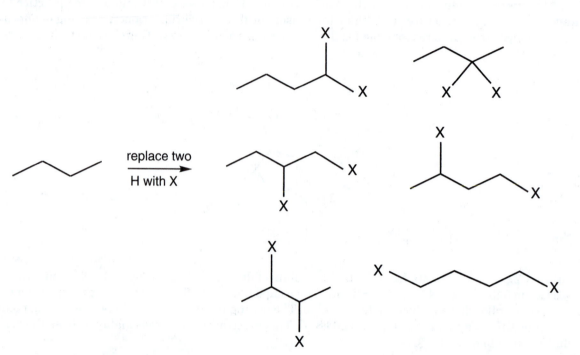

Problem 2.57 The eight-carbon molecule that has only two cyclobutanes and no ethyl or methyl groups must be the structure shown below.

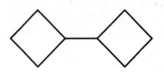

This molecule has a horizontal plane of symmetry and a vertical plane of symmetry. So there are only three different carbons in the structure, as shown.

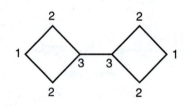

Problem 2.58

(a) There will be three signals for pentane. The methyl carbons on the ends of the molecule are identical. The methylenes on carbons 2 and 4 are identical (both have a methyl on one side and a propyl on the other). Carbon 3 is unique. It has an ethyl on both sides, so it is different from carbons 2 and 4.

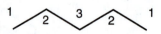

(b) There will be four signals for 2-methylbutane (isopentane). The two methyl groups on carbon 2 are equivalent.

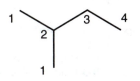

The three-dimensional drawing on the right might help you picture the identical nature of the two methyl groups on carbon 2 of 2-methylbutane.

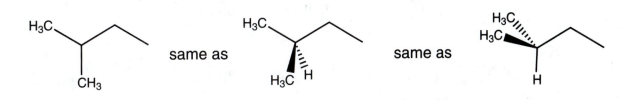

(c) There are only two different carbons in 2,2-dimethylpropane (neopentane). Each of the methyl groups is equivalent.

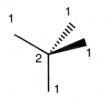

Problem 2.59

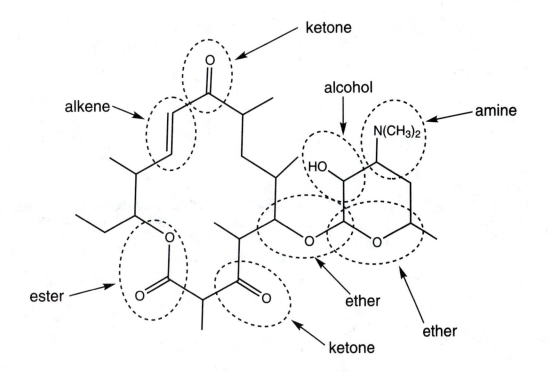

Problem 2.60

(a)

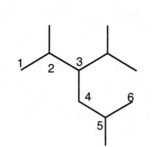

3-Isopropyl-2,5-dimethylhexane
not 3-Isobutyl-2,4-dimethylpentane
not 4-Isopropyl-2,5-dimethylhexane

(b)

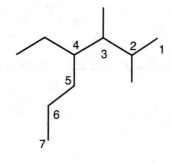

4-Ethyl-2,3-dimethylheptane
not 2,3-Dimethyl-4-propylhexane
not 4-Ethyl-5,6-dimethylheptane

(c)

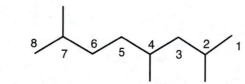

2,4,7-Trimethyloctane
not 2,5,7-Trimethyloctane

(d)

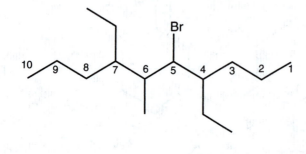

5-Bromo-4,7-diethyl-6-methyldecane
not 6-bromo-4,7-diethyl-5-methyldecane

Problem 2.61

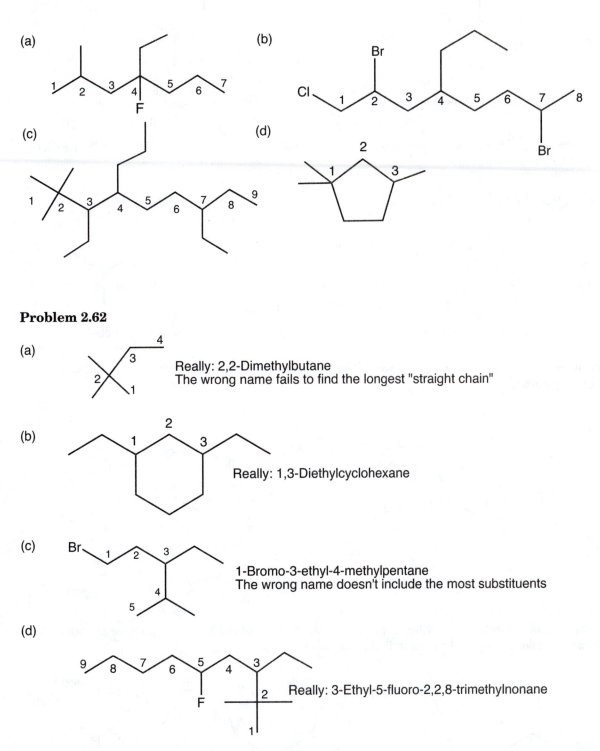

Problem 2.62

(a)

Really: 2,2-Dimethylbutane
The wrong name fails to find the longest "straight chain"

(b)

Really: 1,3-Diethylcyclohexane

(c)

1-Bromo-3-ethyl-4-methylpentane
The wrong name doesn't include the most substituents

(d)

Really: 3-Ethyl-5-fluoro-2,2,8-trimethylnonane

Problem 2.63 There are two equal possibilities for the "longest straight chain." However, there is another rule to resolve this problem. (There are always other rules!) If chains of equal length compete for selection as the main chain in a saturated branched alkane, then the choice goes to the chain that has the greatest number of side chains. See, for example, Problem 2.41(c).

(continued)

Problem 2.63 *(continued)*

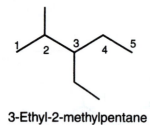

3-Ethyl-2-methylpentane   or   3-Isopropylpentane

Problem 2.64 This problem gives you a chance to apply the new rule of Problem 2.63. The correct name has two substituents.

not

2-Methyl-4-propylheptane   4-Isobutylheptane

Problem 2.65 Both 1-bromobutane and 2-bromobutane have two primary carbons and two secondary carbons.

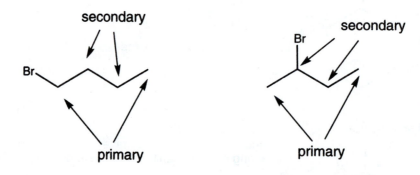

Problem 2.66 First draw 3-chlorohexane. Now draw in all of the hydrogens. There are only primary and secondary hydrogens in 3-chlorohexane.

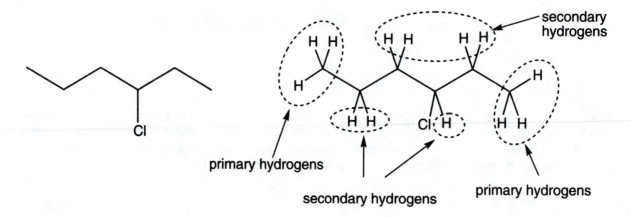

Problem 2.67

(a) This molecule is a secondary alkyl iodide. The iodo group is attached to a carbon that is bonded to two carbons.

(b) This molecule is a secondary alkyl chloride. The chloro group is attached to a carbon that is bonded to two other carbons.

(c) This molecule is a tertiary alkyl fluoride. The fluoro group is attached to a carbon that is bonded to three other carbons.

(d) Aha! This molecule is not a primary, secondary, or tertiary bromide. We only speak of primary, secondary, and tertiary carbons when they are sp^3 hybridized. The aromatic ring carbons are sp^2 hybridized. This is an aryl bromide. The aryl name refers to an aromatic ring.

Problem 2.68

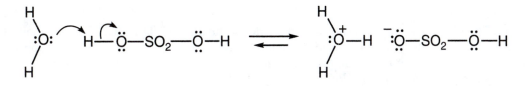

Problem 2.69

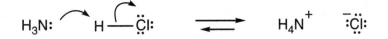

Problem 2.70

Problem 2.71   Notice that the conjugate base of H_2SO_4 (sulfuric acid) has the ability to spread the negative charge equally to three oxygens by resonance.

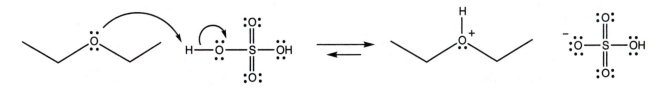

(continued)

Problem 2.71 *(continued)*

The reverse reaction is shown in this second drawing.

Problem 2.72   The animation represents the minimum energy for the species along the reaction pathway. If no rotation occurred, the product would be eclipsed. The staggered product is more stable, so in order to minimize energy, the rotation must happen as the reaction proceeds.

Problem 2.73   In the first step of the reaction, HBr reacts with the alkene. The alkene is using the loosely held π electrons to react with the acidic proton of HBr. So the alkene is the nucleophile (Lewis base) and the hydrogen of the HBr is the electrophile (Lewis acid). In the second step, a bromide ion reacts with the empty $2p$ orbital of the carbocation. So, the carbocation is the electrophile and the bromide ion is the nucleophile.

Problem 2.74   The nucleophile is the oxygen of the methoxide ion. The HOMO track shows that the electrons in the highest occupied orbital are found mostly on the oxygen. The electrophile is the carbon bonded to the bromine. The LUMO track shows that the electrons from the oxygen can mix with σ^*, the empty orbital at the backside of the C—Br bond.

Alkenes and Alkynes

3

This chapter is devoted almost entirely to structure, and the following problems reflect that emphasis. Here we explore the structural consequences of sp^2 and sp hybridization in alkenes and alkynes. There is practice in finding isomers in both cyclic and acyclic molecules. Stereochemistry becomes especially important in the alkenes and ring compounds, and there are several opportunities in the following problems for you to work on stereochemical aspects of these kinds of molecules.

Questions of energy and stability also arise. The π bonds contributing to the double and triple bonds encountered here are weaker than the σ bonds emphasized in the earlier chapters. There will be a number of chances to make assessments of relative energies in the problems that follow.

Problem 3.2   The lowest energy orbital results from mixing the four orbitals (the carbon $2s$ orbital plus the three hydrogen $1s$ orbitals) in phase. This orbital is bonding. The highest energy orbital comes from mixing the four orbitals so that there is only out-of-phase interaction (the carbon $2s$ orbital minus the three hydrogen $1s$ orbitals). This orbital is antibonding.

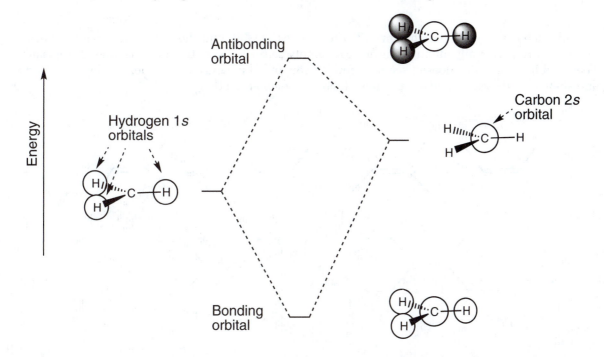

Problem 3.3   In the 90° form, the two *p* orbitals are of course also at 90°. In this arrangement, there is no overlap between the two orbitals because the bonding and antibonding interactions exactly cancel.

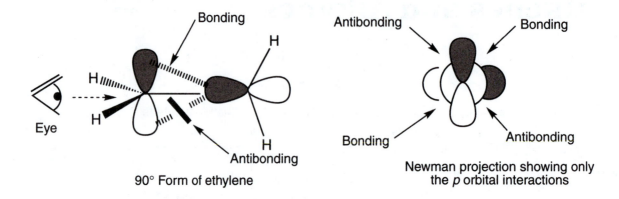

90° Form of ethylene

Newman projection showing only the *p* orbital interactions

Problem 3.4   Only two of these molecules can exist in cis/trans (*Z/E*) forms. Here they are:

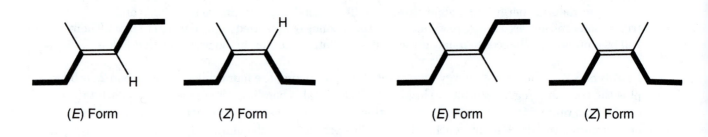

(*E*) Form   (*Z*) Form   (*E*) Form   (*Z*) Form

The others have only one possible form. These alkenes are flat, and the "different" isomers shown in the drawing are really identical, as one can simply be turned over to make the other. If this isn't clear, and if the drawing doesn't seem obvious to you, by all means make models of the "two" forms and show that they are identical by superimposing one on the other.

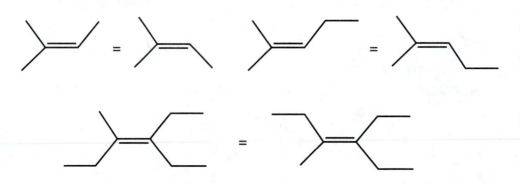

Problems 3.5 and 3.6

Pentenes

1-Pentene

(*Z*)-2-Pentene

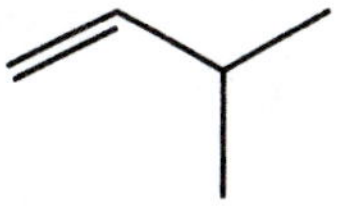

3-Methyl-1-butene

(*E*)-2-Pentene

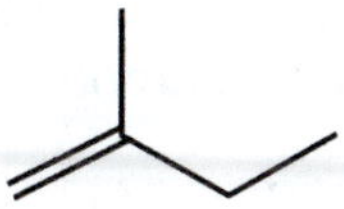

2-Methyl-1-butene

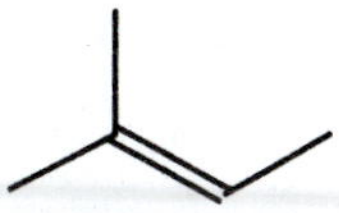

2-Methyl-2-butene

Hexenes

1-Hexene

(*E*)-2-Hexene

(*Z*)-2-Hexene

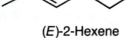

(*E*)-3-Hexene

(*Z*)-3-Hexene

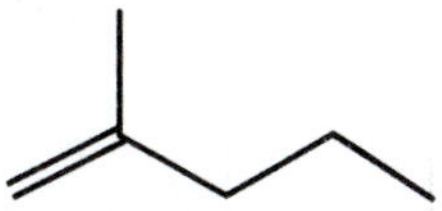

2-Methyl-1-pentene

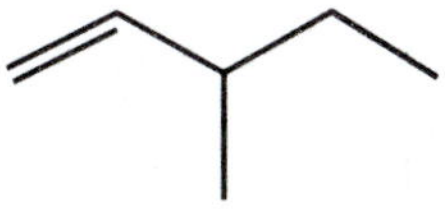

3-Methyl-1-pentene

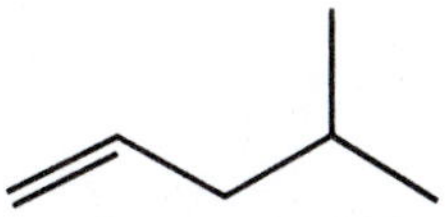

4-Methyl-1-pentene

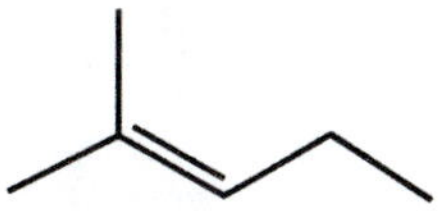

2-Methyl-2-pentene

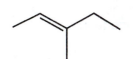

(*E*)-3-Methyl-2-pentene

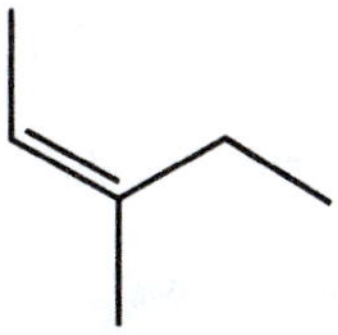

(*Z*)-3-Methyl-2-pentene

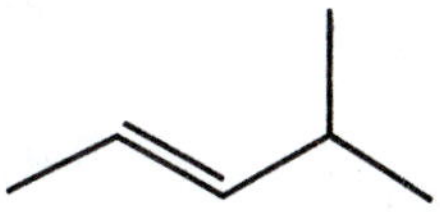

(*E*)-4-Methyl-2-pentene

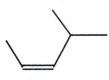

(*Z*)-4-Methyl-2-pentene

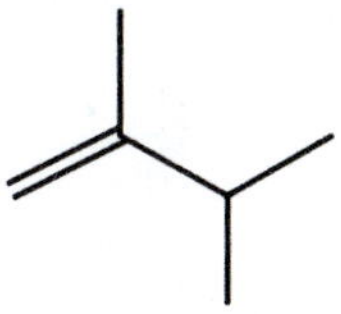

2,3-Dimethyl-1-butene

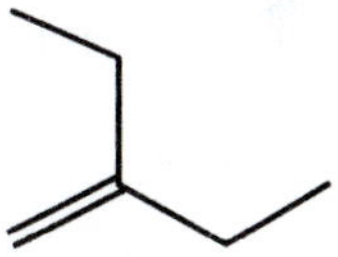

2-Ethyl-1-butene

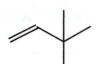

3,3-Dimethyl-1-butene 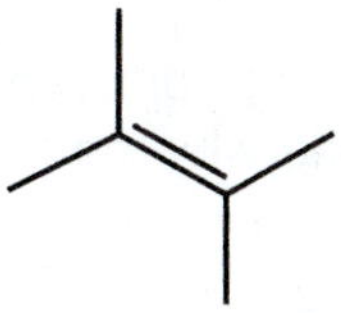   2,3-Dimethyl-2-butene

Problem 3.7

(a) $H_2C{=}CH_2$

Here the two carbons are identical. There can be only one signal

(b) $Cl{-}CH_2{-}CH{=}CH_2$

All three carbons are different. One is an sp^3-hybridized carbon attached to a chlorine, one is an sp^2-hybridized carbon attached to one hydrogen, and one is an sp^2-hybridized carbon attached to two hydrogens

(c) $H_2C{=}CH{-}CH_2CH_3$

All four carbons are different, and there will be four signals

(d)

The two methyl groups are equivalent, as are the two methine groups, and there will be two signals

(e)

Again, the two methyl groups are equivalent, as are the two methine groups, and there will be two signals

(f)

This time there is one methylene, one quaternary carbon, and two equivalent methyl carbons, and there will be three signals

Problem 3.8

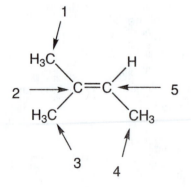

All five carbons are different, and there will be five signals. Be sure you see why the two methyls on the left (1 and 3) are different. One is cis to an H, the other is cis to another CH_3

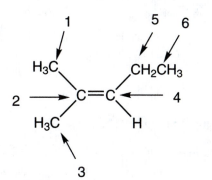

All six carbons are different. There will be six signals

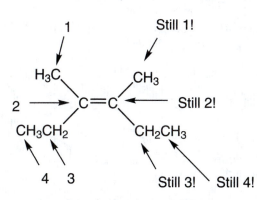

All seven carbons are different. There will be seven signals

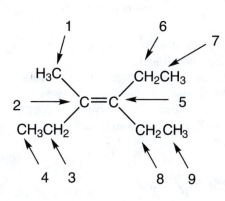

Now there is much more symmetry, and there are only four different carbons, and there will be four signals

All nine carbons are different. There will be nine signals

Problem 3.9

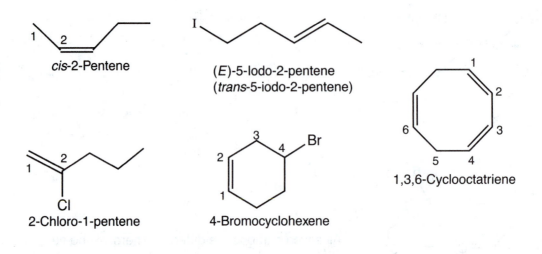

cis-2-Pentene

(*E*)-5-Iodo-2-pentene
(*trans*-5-iodo-2-pentene)

1,3,6-Cyclooctatriene

2-Chloro-1-pentene

4-Bromocyclohexene

Problem 3.11 There are two carbons in each alkene. We can refer to these two carbons as one end or the other of the alkene. Look at one end and circle the substituent with the largest atomic number attached to that carbon. Look at the other end and circle the substituent with the largest atomic number. If the circled groups are on the same side of the alkene, then the molecule is a (*Z*) isomer. If the circled groups are on opposite sides, then the alkene is the (*E*) isomer.

(a) On the left end, the C is larger than the H. On the right end, the Br is larger than the C. This isomer is (*E*)-1,2-dibromo-2-butene.

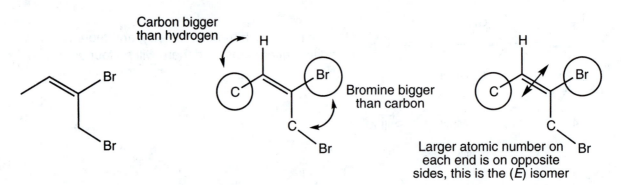

Carbon bigger
than hydrogen

Bromine bigger
than carbon

Larger atomic number on
each end is on opposite
sides, this is the (*E*) isomer

(b) On the left end, the C is larger than the H. On the right end, both attachments are carbon. So we have to go to the next attachments looking for a difference. One carbon has (C, C, H) for attachments. The other carbon has (C, H, H). The circled attachments are on opposite sides. Notice that there are two options for longest chain on this molecule. We choose numbering so that we maximize the number of substituents (see Problems 2.63 and 2.64). Which way we number does not have any bearing on the (*E/Z*) analysis. This molecule is (*E*)-3-(*sec*-butyl)-5,5-dimethyl-2-hexene.

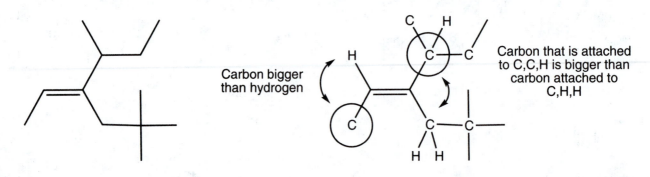

Carbon bigger
than hydrogen

Carbon that is attached
to C,C,H is bigger than
carbon attached to
C,H,H

(c) On the left end, the C is larger than the H. On the right end, both attachments are carbon. We compare the next attachments to determine which group is larger. The carbon attached to (C, H, H) is larger than the carbon attached to (H, H, H). It doesn't matter what else is attached further out on the chain. It is the first difference that is compared. The circled attachments are on the same side. This molecule is (*Z*)-3-methyl-2-pentene.

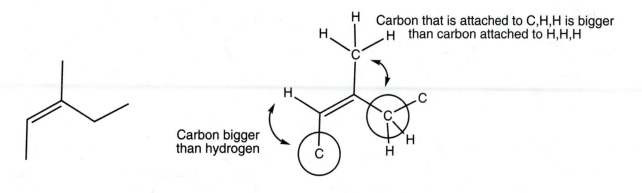

(d) On the bottom end, the C is larger than the H. On the top end, the Cl is larger than the C. It doesn't matter what else is attached further out on the chain. It is the first difference that is compared. The circled attachments are on the opposite sides. This molecule is (*E*)-3-chloro-4-iodo-2-pentene.

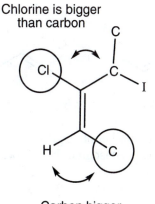

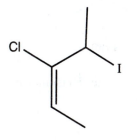

Problem 3.12   In (a), the higher priority groups are fluorine and ethyl, so the (*E*) isomer has those groups on opposite sides of the double bond. In (b), the higher priority groups are propyl and ethyl, and the (*E*) isomer must have them on opposite sides of the double bond. In (c), the higher priority groups are I and Cl. The (*Z*) isomer will have them on the same side of the double bond.

(a)   (b)   (c)

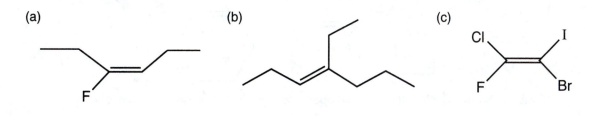

Problem 3.13 In (a), the higher priority groups are CH_3 and F. The (Z) isomer has them on the same side; the (E) isomer has them on opposite sides. In (b), the higher priority groups are ethyl (CH_3CH_2) and amino (NH_2). In (c), the higher priority groups are the methylene (CH_2) groups starting the ring. In (d), the higher priority groups are deuterium (D) and the ring carbon bearing the methyl (CH_3) group.

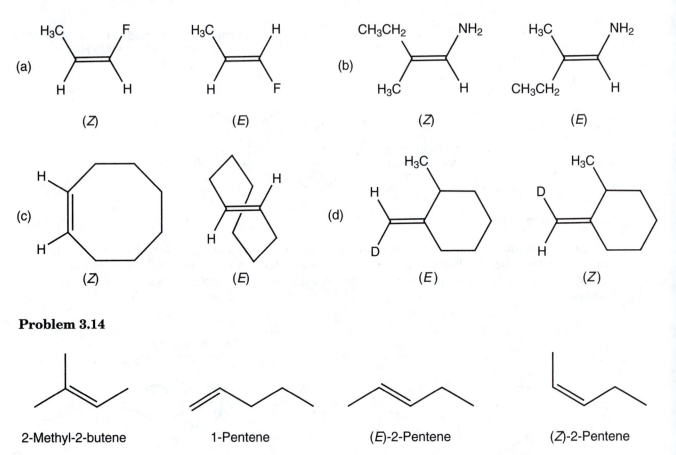

Problem 3.14

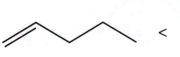

2-Methyl-2-butene   1-Pentene   (E)-2-Pentene   (Z)-2-Pentene

Based on the number of substituents on the double bond and based on the (E) isomer being more stable than the (Z) isomer, the following order is obtained:

 < < <

Problem 3.15

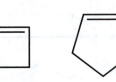

Cyclopropene   Cyclobutene   Cyclopentene   Cyclohexene   Cycloheptene

Problem 3.16

All four methine carbons are equivalent. There will be only one signal

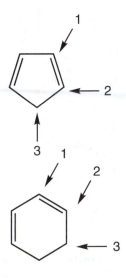

The two methine carbons are different, and there is a single methylene. There will be three signals

Three signals—two methines and one methylene. Note the symmetry

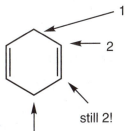

Note again the symmetry, which is probably more obvious in this molecule than the one above. There are only two kinds of carbon—and thus, two signals

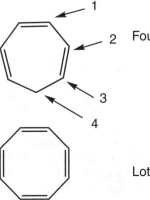

Four signals from three different methines and the one methylene

Lots of symmetry here—all eight methines are the same—one signal

Problem 3.17

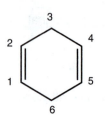

1,4-Cyclohexadiene

1,3,5,7-Cyclooctatetraene

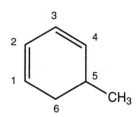

5-Methyl-1,3-cyclohexadiene

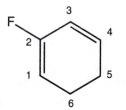

2-Fluoro-1,3-cyclohexadiene

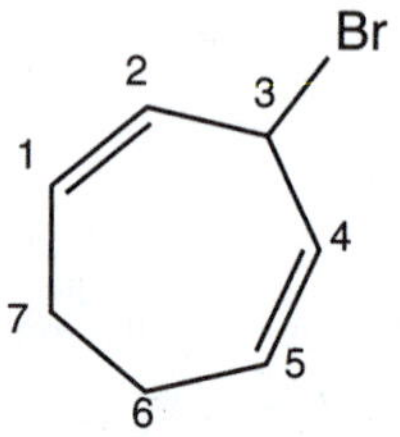

3-Bromo-1,4-cycloheptadiene

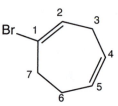

1-Bromo-1,4-cycloheptadiene

Problem 3.19

$\Delta G = -RT \ln K$. In this case,

ΔG = 11.4 kcal/mol
R = 1.986 cal/deg·mol
T = 298 K

RT = 592 cal/mol = 0.592 kcal/mol

11.4 = $-(0.592)\ln K$
$\ln K$ = -19.26
K = 4×10^{-9}

Problem 3.20   The form shown in Figure 3.48 is (*Z*). On the left-hand carbon of the double bond, the higher priority group is CH_2 and the lower priority group is H. On the right-hand side of the double bond, the higher priority group is CH_2—CHC_2 and the lower priority group is CH_2—CH_2C. As the higher priority groups are on the same side of the double bond, the compound is (*Z*).

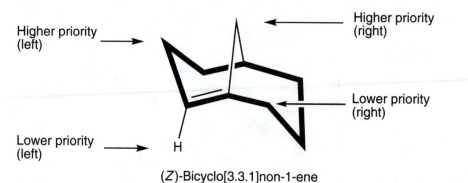

(*Z*)-Bicyclo[3.3.1]non-1-ene

Problem 3.21 The form in Figure 3.48 is (*Z*), as the higher and lower priority groups are on the same side of the double bond. On the left-hand carbon of the double bond, the higher priority group is CH_2 and the lower priority group is H. On the right-hand carbon of the double bond, the higher priority group is CH_2—CHC_2 and the lower priority group is CH_2—CH_2C. In this form, the two higher priority groups are on the same side of the double bond. Thus the correct label is (*Z*). The other form is the (*E*) structure, with the two higher priority groups on opposite sides.

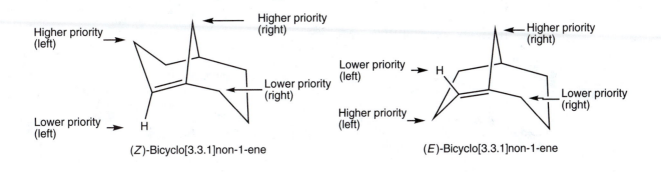

(*Z*)-Bicyclo[3.3.1]non-1-ene   (*E*)-Bicyclo[3.3.1]non-1-ene

Note that this molecule contains a double bond in two different-sized rings. The double bond of the (*Z*) form is cis in the six-membered ring and trans in the eight-membered ring. In the (*E*) form, the double bond is cis in the eight-membered ring and trans in the six-membered ring. The smaller the ring, the less stable is a trans double bond (why?), so the (*E*) form is less stable than the (*Z*) form.

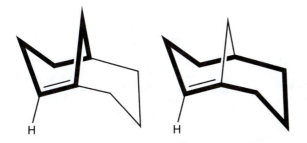

(*Z*) Form: double bond cis in six-membered ring (bold), trans in eight-membered ring (bold)   (*E*) Form: double bond cis in eight-membered ring (bold), trans in six-membered ring (bold)

Problem 3.22 The picture is exactly the same as for alkenes (Fig. 3.12, p. 108), except that there are two π bonds at 90° to each other. The diagram shows only one orbital interaction diagram and indicates both π bonds schematically.

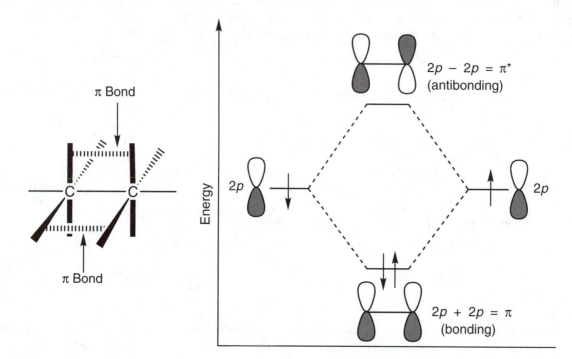

Problems 3.23 and 3.24

Pentynes

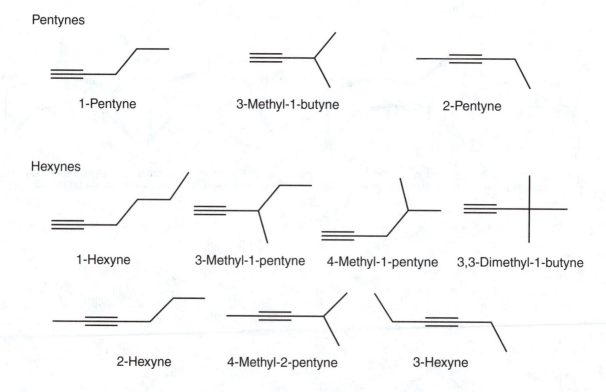

1-Pentyne   3-Methyl-1-butyne   2-Pentyne

Hexynes

1-Hexyne   3-Methyl-1-pentyne   4-Methyl-1-pentyne   3,3-Dimethyl-1-butyne

2-Hexyne   4-Methyl-2-pentyne   3-Hexyne

Problem 3.25

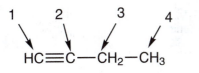

All four carbons are different—there will be four signals

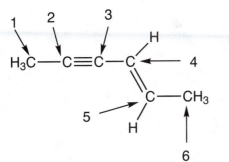

All six carbons are different—there will be six signals

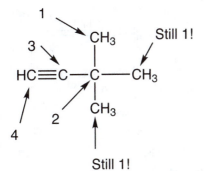

The three methyl groups are equivalent, and there are three other different carbons—four signals

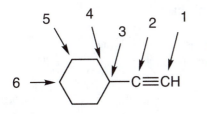

Watch the symmetry—there are only six different carbons here

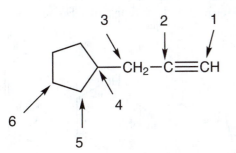

Again, there are only six different carbons

Problem 3.26

(a) C_5H_6

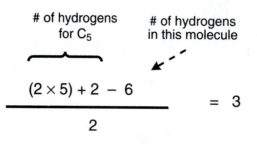

$$\frac{(2 \times 5) + 2 - 6}{2} = 3$$

# of hydrogens for C_5 — # of hydrogens in this molecule

There must be a total of three π bonds and/or rings. Some possibilities are

Three π bonds   One ring, two π bonds   Two rings, one π bond

(b) C_7H_8O

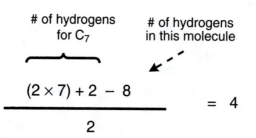

$$\frac{(2 \times 7) + 2 - 8}{2} = 4$$

# of hydrogens for C_7 — # of hydrogens in this molecule

Notice that oxygen does not get included in the calculation. A molecule of C_7H_8O must have 4 degrees of unsaturation. This means there must be a total of four π bonds and/or rings. Some possibilities are

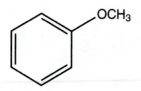 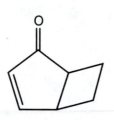 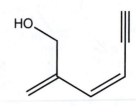

One ring, three π bonds   Two rings, two π bonds   Four π bonds

(c) C_7H_{10}

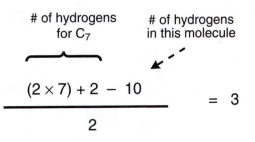

# of hydrogens
for C_7

# of hydrogens
in this molecule

$$\frac{(2 \times 7) + 2 - 10}{2} = 3$$

A molecule of C_7H_{10} must have 3 degrees of unsaturation. This means there must be a total of three π bonds and/or rings. Some possibilities are

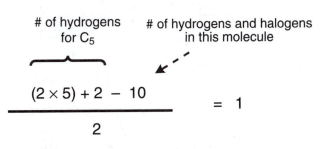

One ring, two π bonds   Two rings, one π bond   Three π bonds

(d) $C_5H_8Br_2$

# of hydrogens
for C_5

# of hydrogens and halogens
in this molecule

$$\frac{(2 \times 5) + 2 - 10}{2} = 1$$

A molecule of C_5H_8Br must have 1 degree of unsaturation. This means there must be a π bond or a ring. Some possibilities are

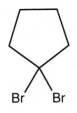

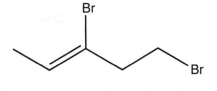

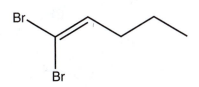

One ring   One π bond   One π bond

Problem 3.29

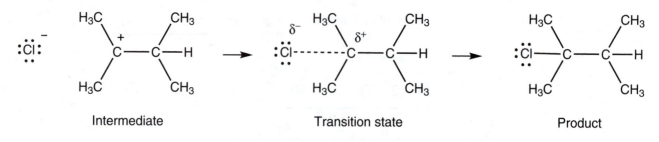

Intermediate   Transition state   Product

Problem 3.30 These reactions are "standard" additions of H—X across the double bond of 2,3-dimethyl-2-butene. Only H_2SO_4 might cause problems—you have to see it as HO—SO_2—OH, just another "H—X."

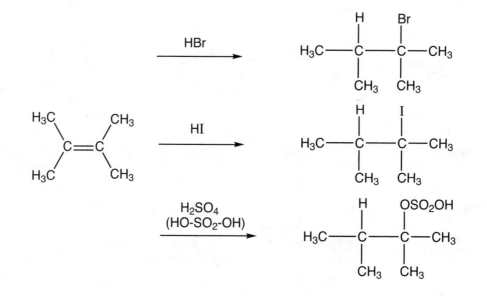

Problem 3.31

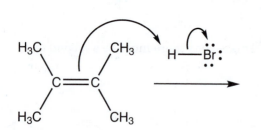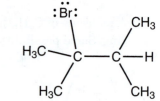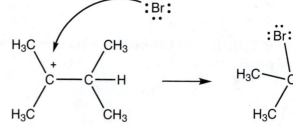

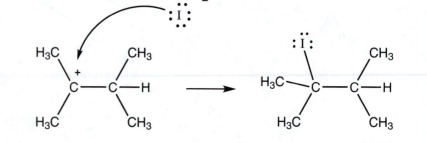

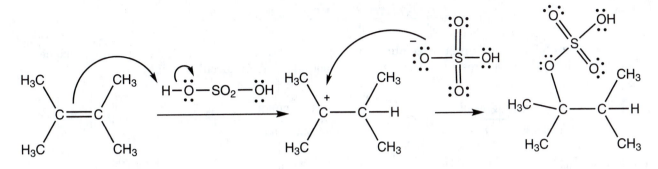

Problem 3.32 Here are two more additions of "H—X." This time, however, you have to deal with the direction of addition, the regiochemistry. In each case, the alkene will be protonated to give the *tertiary* carbocation, to which X⁻ will add. Here are the products:

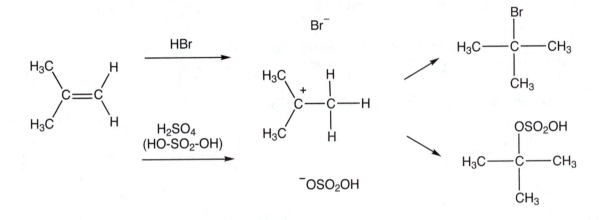

Problem 3.34 Each of these alcohols can be made through an acid-catalyzed addition of water to an alkene:

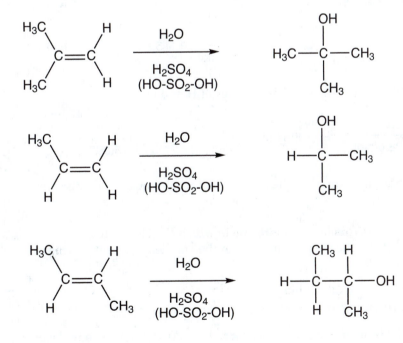

Additional Problem Answers

Problem 3.35   First, find the weight of carbon present in the carbon dioxide. This must also be the weight of carbon present in the sample compound (MW = molecular weight).

$$\text{Wt (C)} = \frac{\text{MW (C)}}{\text{MW (CO}_2)} \times \text{Wt (CO}_2) = \frac{12.011 \text{ g/mol}}{44.009 \text{ g/mol}} \times 16.90 \text{ mg} = 4.61 \text{ mg}$$

Similarly, the weight of hydrogen can be calculated from the weight of water:

$$\text{Wt (H)} = \frac{2 \times \text{MW (H)}}{\text{MW (H}_2\text{O)}} \times \text{Wt (H}_2\text{O)} = \frac{2 \times 1.008 \text{ g/mol}}{18.015 \text{ g/mol}} \times 3.46 \text{ mg} = 0.39 \text{ mg}$$

Note that the sum of the weights of carbon and hydrogen equals the weight of the sample. The sample must have contained only carbon and hydrogen.

The weight percents of carbon and hydrogen can now easily be calculated:

$$\%\text{C} = \frac{4.61 \text{ mg}}{5.00 \text{ mg}} \times 100 = 92.2\% \qquad \%\text{H} = \frac{0.39 \text{ mg}}{5.00 \text{ mg}} \times 100 = 7.8\%$$

Problem 3.36   The "missing" weight percent is oxygen, in this case, 23.50%. Now, assume a 100 g sample of the compound in question and compute the number of moles of each element present in a sample of this size. If the compound contains 70.58% carbon, 100 g of sample will contain 70.58 g, or 5.88 mol:

$$\text{C} = \frac{70.58 \text{ g}}{12.011 \text{ g/mol}} = 5.88 \text{ mol}$$

Similarly, we can determine the number of moles of H and O:

$$\text{H} = \frac{5.92 \text{ g}}{1.008 \text{ g/mol}} = 5.87 \text{ mol} \qquad \text{O} = \frac{23.50 \text{ g}}{15.999 \text{ g/mol}} = 1.47 \text{ mol}$$

Therefore, a formula that expresses the relative molar proportions of carbon, hydrogen, and oxygen is $C_{5.88}H_{5.87}O_{1.47}$.

Now you need to convert this formula into one in which the elements are present in whole number ratios. Divide through by the element present in the smallest amount, in this case, oxygen:

$$\text{C} = \frac{5.88}{1.47} = 4.00 \qquad \text{H} = \frac{5.87}{1.47} = 3.99 \qquad \text{O} = \frac{1.47}{1.47} = 1.00$$

This calculation yields an empirical formula of C_4H_4O. As C_4H_4O has a molecular weight of 68 g/mol, it can't be the molecular formula in this case because you know that the molecular weight is about 135 g/mol. However, simply multiplying by 2 gives the correct molecular formula of $C_8H_8O_2$, MW = 136 g/mol.

Problem 3.37   A molecule with the formula of C_5H_8 will have 2 degrees of unsaturation. Here are the possible acyclic isomers that you should be able to predict:

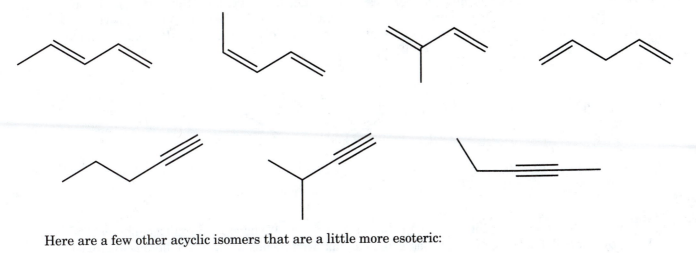

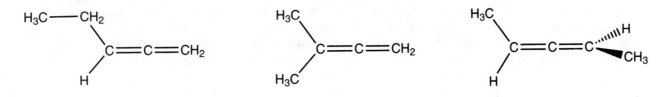

Here are a few other acyclic isomers that are a little more esoteric:

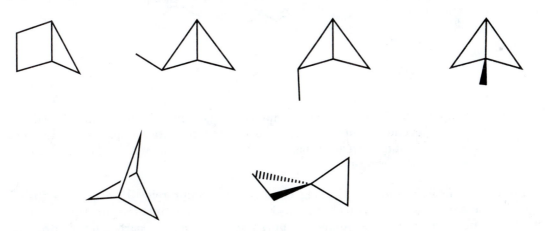

If a C_5H_8 molecule has no π bonds, it must have two rings. Here are the possible bicyclic isomers:

Problem 3.38

(a) This is an aromatic ring attached to a bromine. There are three π bonds and one ring. There are 4 degrees of unsaturation in this molecule.

(b) There are two rings and one π bond. There are 3 degrees of unsaturation in this molecule.

(c) There is one double bond and one triple bond in this molecule. There are 3 degrees of unsaturation in this molecule.

(d) There are two double bonds and no rings in this molecule. Thus there are 2 degrees of unsaturation in this molecule.

Problem 3.39

(a) This molecule is an alkene, which is the priority group. Find the longest carbon chain that contains both carbons of the alkene. There are two possible longest chains. Both are five carbons long. Pick the longest carbon chain that has the maximum groups attached. Number the chain so that the priority group gets the lowest possible number. Therefore, this molecule is a 2-pentene.

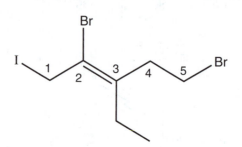

Find the substituents and indicate the location of attachment. We have 1-iodo and 2,5-dibromo and 3-ethyl. Now alphabetize the substituents in the prefix. That means we have 2,5-dibromo-3-ethyl-1-iodo-2-pentene. We aren't finished. There is the issue of (*E*) or (*Z*) isomer. On carbon number 2 the larger atomic numbered substituent is the Br. On carbon number 3 the substituents are both carbons. Each of the carbons is attached to (C, H, H). So we must look to the next attachment. Comparing the next carbons, carbon number 5 has (Br, H, H) and the second carbon of the ethyl group has (H, H, H). The circled groups are on the same side. This molecule is (*Z*)-2,5-dibromo-3-ethyl-1-iodo-2-pentene.

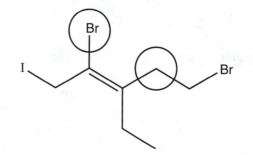

(b) The priority group in this molecule is the alkene. The larger chain is the six-membered ring. This molecule is a cyclohexene. Because the alkene is the only priority group, the cyclohexene is known to have the double bond between carbons number 1 and number 2. We don't call this 1-cyclohexene. There is no other choice.

There are two options for numbering the ring: clockwise and counterclockwise. Because the alkene carbons must be numbered consecutively, the clockwise numbering puts the substituent on carbon number 3 and the counterclockwise numbering puts the substituent on carbon number 6.

Clockwise numbering   Counterclockwise numbering

We see there is only one substituent, so there is no need to worry about the alphabetical order in the prefix. It is a 3-methyl group. We know that only the (*Z*) isomer is stable for cycloalkenes with a ring size smaller than eight carbons. So we don't need to indicate that this is the (*Z*) isomer. It is assumed to be the (*Z*) isomer. This molecule is 3-methylcyclohexene.

(c) The priority group in this molecule is an alkene. We find the longest chain with both carbons of the alkene and number the chain so that the lowest possible number is used for the priority group. We have another 2-pentene. It doesn't matter which methyl group on carbon number 2 we use in the numbering.

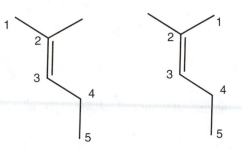

There is only one substituent, the methyl group on carbon number 2. This molecule is a 2-methyl-2-pentene. We do the (*E*) or (*Z*) analysis and we find that the two groups on carbon number 2 are the same atomic number and there is no difference in further attachment. Both carbons have attachment of (H, H, H). That means there is no (*E*) or (*Z*) for this molecule. This molecule is 2-methyl-2-pentene.

(d) The priority group in this molecule is a diene. We find the longest chain with all four carbons of the diene and number the chain so that the lowest possible numbers are used. This molecule is a 1,3-pentadiene, not a 2,4-pentadiene.

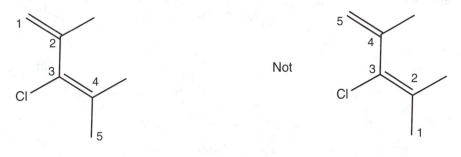

We find the substituents for this molecule. We will have 3-chloro and 2,4-dimethyl in the prefix. The attachments on carbon number 1 are both hydrogens. So there is no (*E*) or (*Z*) for the first alkene. The attachments on carbon number 4 are both methyl groups. So the second alkene is also neither (*E*) nor (*Z*). This molecule is 3-chloro-2,4-dimethyl-1,3-pentadiene.

Problem 3.40 These molecules are conformational isomers. They differ only by the rotation around a C—C single bond. You might need to build the molecule using your model set to convince yourself that they will interconvert at room temperature. The Newman projections show that the isomer (a) would be more stable. Isomer (b) is higher in energy because of steric interactions. The bulky groups are eclipsing each other, *and* the hydrogens on one of the methyls, labeled C(5) in the drawing, are competing for space with one of the C(1) vinyl hydrogens. Your model set will show this more clearly.

(continued)

Problem 3.40 *(continued)*

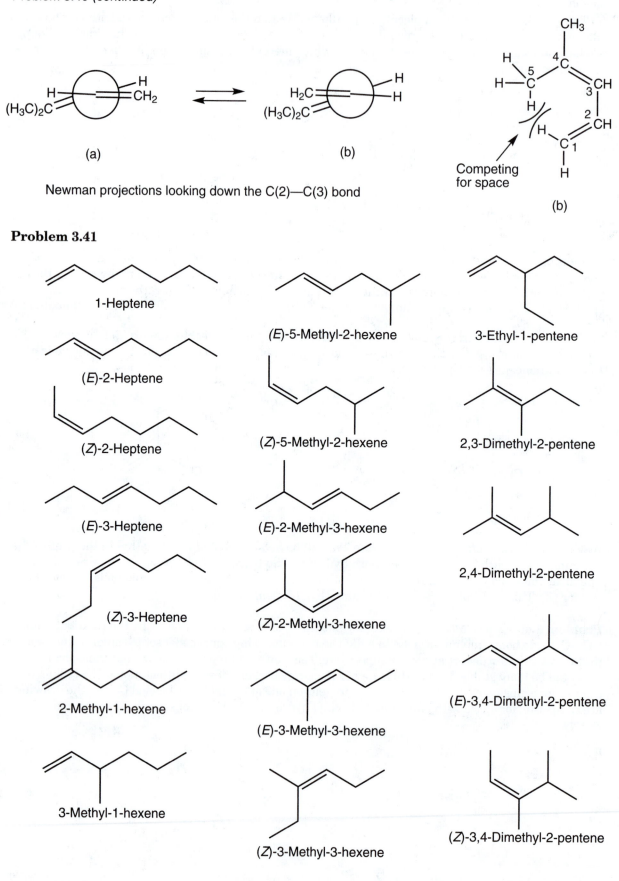

(a)   (b)

Newman projections looking down the C(2)—C(3) bond

Competing
for space

(b)

Problem 3.41

1-Heptene

(*E*)-2-Heptene

(*Z*)-2-Heptene

(*E*)-3-Heptene

(*Z*)-3-Heptene

2-Methyl-1-hexene

3-Methyl-1-hexene

(*E*)-5-Methyl-2-hexene

(*Z*)-5-Methyl-2-hexene

(*E*)-2-Methyl-3-hexene

(*Z*)-2-Methyl-3-hexene

(*E*)-3-Methyl-3-hexene

(*Z*)-3-Methyl-3-hexene

3-Ethyl-1-pentene

2,3-Dimethyl-2-pentene

2,4-Dimethyl-2-pentene

(*E*)-3,4-Dimethyl-2-pentene

(*Z*)-3,4-Dimethyl-2-pentene

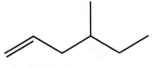

4-Methyl-1-hexene

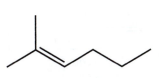

5-Methyl-1-hexene

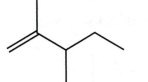

2,3-Dimethyl-1-pentene

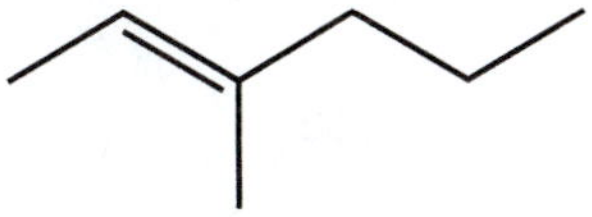

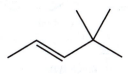

(*E*)-4,4-Dimethyl-2-pentene

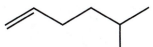

2-Methyl-2-hexene

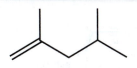

2,4-Dimethyl-1-pentene

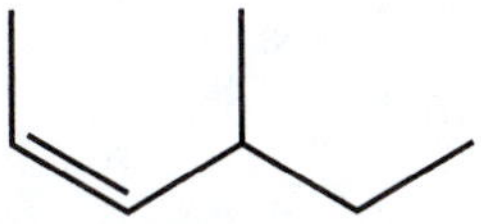

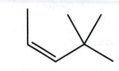

(*Z*)-4,4-Dimethyl-2-pentene

(*E*)-3-Methyl-2-hexene

3,3-Dimethyl-1-pentene

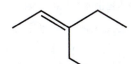

3-Ethyl-2-pentene

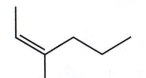

(*Z*)-3-Methyl-2-hexene

3,4-Dimethyl-1-pentene

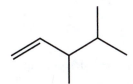

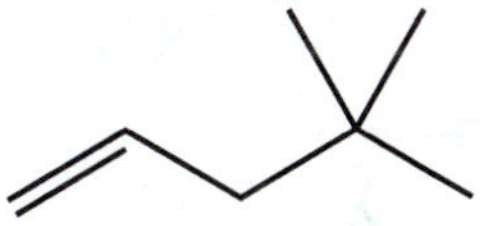

2,4,4-Trimethyl-1-butene

(*E*)-4-Methyl-2-hexene

4,4-Dimethyl-1-pentene

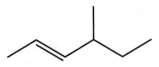

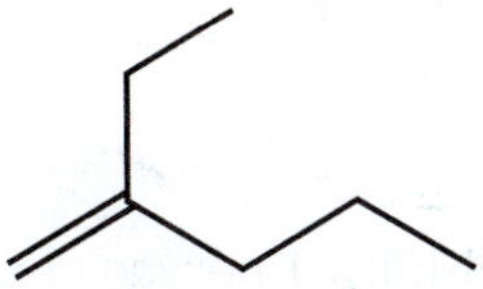

(*Z*)-4-Methyl-2-hexene

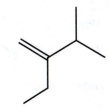

2-Ethyl-1-pentene

2-Ethyl-3-methyl-1-butene

Problem 3.42

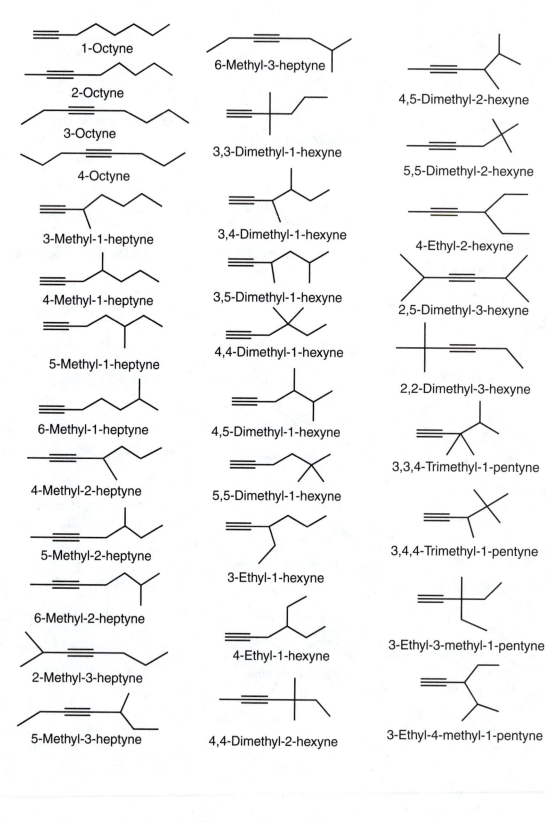

1-Octyne

2-Octyne

3-Octyne

4-Octyne

3-Methyl-1-heptyne

4-Methyl-1-heptyne

5-Methyl-1-heptyne

6-Methyl-1-heptyne

4-Methyl-2-heptyne

5-Methyl-2-heptyne

6-Methyl-2-heptyne

2-Methyl-3-heptyne

5-Methyl-3-heptyne

6-Methyl-3-heptyne

3,3-Dimethyl-1-hexyne

3,4-Dimethyl-1-hexyne

3,5-Dimethyl-1-hexyne

4,4-Dimethyl-1-hexyne

4,5-Dimethyl-1-hexyne

5,5-Dimethyl-1-hexyne

3-Ethyl-1-hexyne

4-Ethyl-1-hexyne

4,4-Dimethyl-2-hexyne

4,5-Dimethyl-2-hexyne

5,5-Dimethyl-2-hexyne

4-Ethyl-2-hexyne

2,5-Dimethyl-3-hexyne

2,2-Dimethyl-3-hexyne

3,3,4-Trimethyl-1-pentyne

3,4,4-Trimethyl-1-pentyne

3-Ethyl-3-methyl-1-pentyne

3-Ethyl-4-methyl-1-pentyne

Problem 3.43

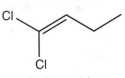
1,1-Dichloro-1-butene

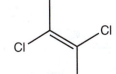
3,3-Dichloro-1-butene

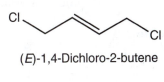
(*E*)-1,4-Dichloro-2-butene

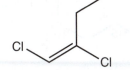
(*E*)-1,2-Dichloro-1-butene

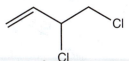
3,4-Dichloro-1-butene

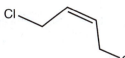
(*Z*)-1,4-Dichloro-2-butene

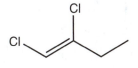
(*Z*)-1,2-Dichloro-1-butene

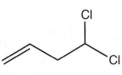
4,4-Dichloro-1-butene

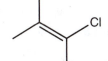
(*E*)-2,3-Dichloro-2-butene

(*Z*)-1,3-Dichloro-1-butene

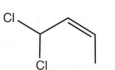
(*E*)-1,1-Dichloro-2-butene

(*Z*)-2,3-Dichloro-2-butene

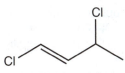
(*E*)-1,3-Dichloro-1-butene

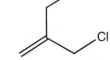
(*Z*)-1,1-Dichloro-2-butene

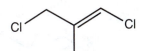
3,3-Dichloro-2-methyl-1-propene

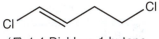
(*E*)-1,4-Dichloro-1-butene

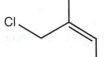
(*Z*)-1,2-Dichloro-2-butene

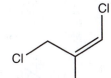
3-Chloro-2-(chloromethyl)-1-propene

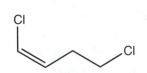
(*Z*)-1,4-Dichloro-1-butene

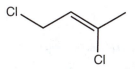
(*E*)-1,2-Dichloro-2-butene

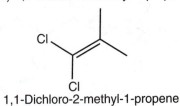
(*E*)-1,3-Dichloro-2-methyl-1-propene

2,3-Dichloro-1-butene

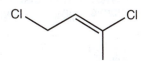
(*Z*)-1,3-Dichloro-2-butene

(*Z*)-1,3-Dichloro-2-methyl-1-propene

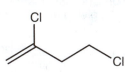
2,4-Dichloro-1-butene

(*E*)-1,3-Dichloro-2-butene

1,1-Dichloro-2-methyl-1-propene

Problem 3.44   (a) (*E*)-5-iodo-2,7-dimethyl-3-nonene (the di- is ignored, so *i*odo comes before di*m*ethyl), (b) (*Z*)-2-chloro-3-ethyl-3-hexene, (c) 4-bromo-5-isopropyl-2-octyne, (d) (*Z*)-4,4-dimethyl-2-hepten-5-yne (-ene gets priority over -yne).

Problem 3.45

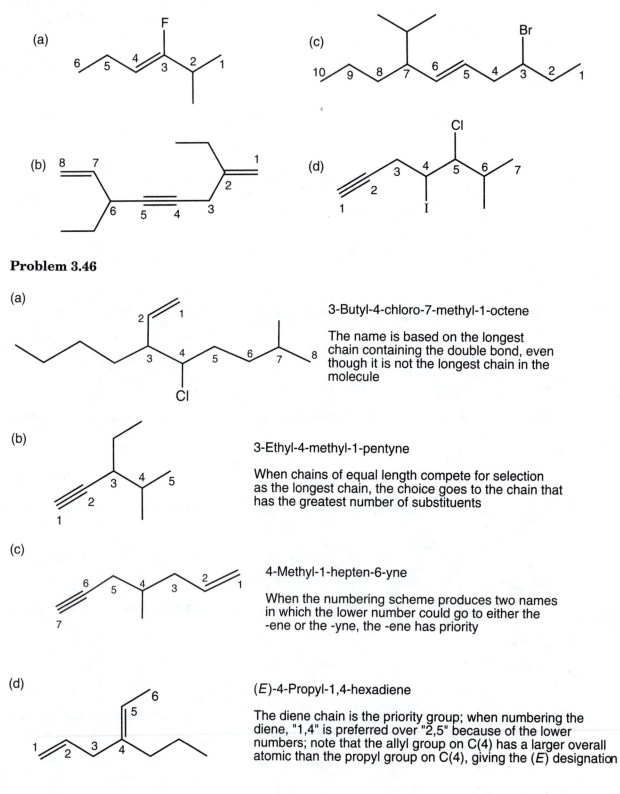

Problem 3.46

(a)

3-Butyl-4-chloro-7-methyl-1-octene

The name is based on the longest chain containing the double bond, even though it is not the longest chain in the molecule

(b)

3-Ethyl-4-methyl-1-pentyne

When chains of equal length compete for selection as the longest chain, the choice goes to the chain that has the greatest number of substituents

(c)

4-Methyl-1-hepten-6-yne

When the numbering scheme produces two names in which the lower number could go to either the -ene or the -yne, the -ene has priority

(d)

(*E*)-4-Propyl-1,4-hexadiene

The diene chain is the priority group; when numbering the diene, "1,4" is preferred over "2,5" because of the lower numbers; note that the allyl group on C(4) has a larger overall atomic than the propyl group on C(4), giving the (*E*) designation

Problem 3.47 The first thing to do is to determine the number of degrees of unsaturation in $C_4H_6Br_2$. Because we have a molecule with four carbons, the related *saturated* alkane is C_4H_{10}. The bromines in the compound $C_4H_6Br_2$ are univalent and can be treated as hydrogens for the purpose of counting. So, $10 - 8 = 2/2 = 1$ degree of unsaturation. The compounds in question must contain only one ring or one π bond. Let's take this question step by step. There are three noncyclic chains to consider: propene, 1-butene, and 2-butene. Two bromines can be arranged in the following ways:

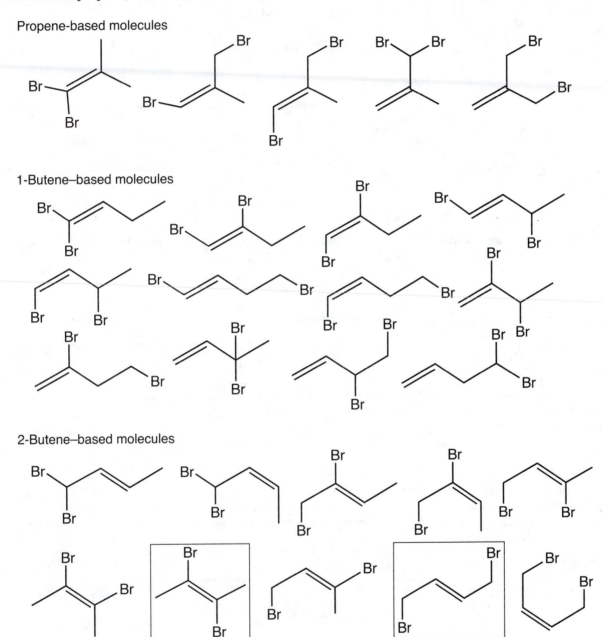

Only the molecules in the boxes have zero dipole moments.

(continued)

Problem 3.47 *(continued)*

There are five possible molecules containing a four-membered ring, but only the indicated isomer
has a zero dipole moment.

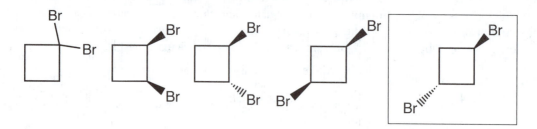

There are nine molecules containing a three-membered ring, but none has a zero dipole moment.

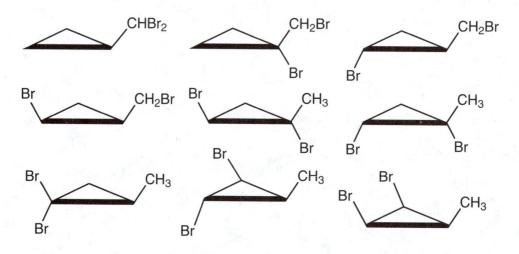

So we find 14 cyclic isomers. It has been claimed (*J. Chem. Educ.* 1992, *69*, 452—an article on finding
isomers that is well worth a look) that there are 15. Who's wrong? Well, we were. There is a 15th
isomer, and it is an easy one to find. It is the 1,2-dibromo-3-methylcyclopropane with both bromines
trans to the methyl group. How we missed it, we don't know. See how hard this "easy" stuff is?

Problem 3.48   The saturated hydrocarbon is C_4H_{10}. The formula is $(10 - 6) = 4/2 = 2$. There are 2
degrees of unsaturation in this molecule, and so the possible combinations are two rings, one ring
and one π bond, and two π bonds. The possible molecules are given below, with the ones with four
different carbon atoms in boxes.

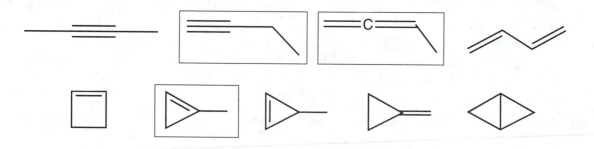

Problem 3.49 First, determine the degrees of unsaturation. The chlorine is univalent and can be treated as a hydrogen, so the related saturated alkane is C_4H_{10}, where $(10 - 6) = 4/2 = 2$ degrees of unsaturation. You must look for compounds containing two π bonds, one π bond and one ring, or two rings.

There are 10 isomers containing two π bonds:

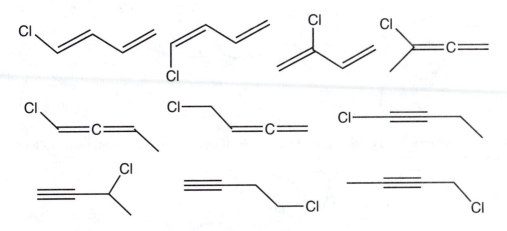

There are also 10 compounds with one ring and one π bond:

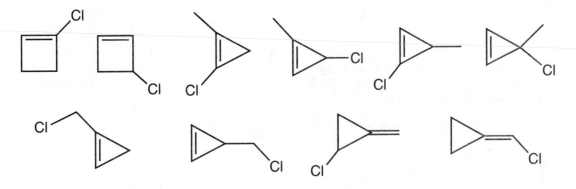

Finally, there are three structures containing two rings and no π bonds:

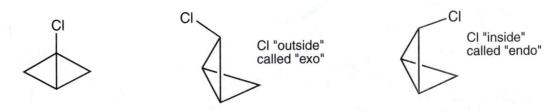

Problem 3.50 The saturated alkane would have 22 hydrogens ($C_{10}H_{22}$). The formula shows that there are 3 degrees of unsaturation in this compound $(22 - 16) = 6/2 = 3$. If there must be only two rings, and if the rings must be six membered, there must be one π bond. The possibilities are

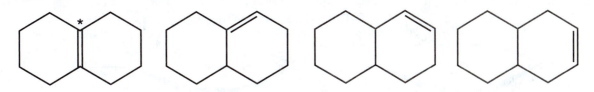

The asterisk shows the one with only three kinds of carbon: the double-bonded carbon and the two different methylenes.

Problem 3.51 An sp^2-hybridized carbon atom is shown. In the figure, two sp^2 hybrids are shown schematically, one coming toward you (solid wedge), the other retreating (dashed wedge). In the first figure, the four valence electrons of carbon are shown as dots, one in each of the three sp^2 hybrids, and one in the unhybridized $2p_z$ orbital. The C—H bonds are shown as overlapping $1s$ and sp^2 orbitals and then as schematic "line" bonds.

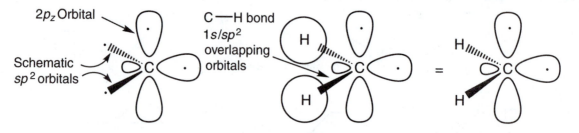

An oxygen atom has six valence electrons, and so two of the sp^2 hybrids must be doubly occupied.

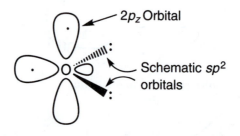

In our scheme, we will form one σ bond through sp^2/sp^2 overlap and a π bond through $2p_z/2p_z$ overlap. Don't forget that these overlapping orbitals produce both a bonding combination (σ and π) and an antibonding combination (σ* and π*). The result is a double bond, σ and π, shown as a pair of identical lines between C and O.

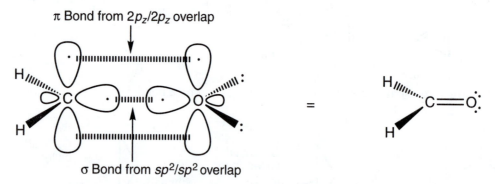

Problem 3.52

(a) We can use the sp^2 hybrid orbitals to bond to the hydrogen and the two neighboring carbons. The figure first shows the planar six-membered ring with the carbons and hydrogens and then focuses on one sp^2-hybridized carbon and the bonds to it.

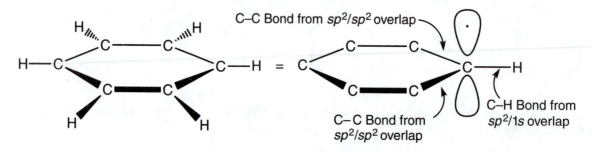

Each carbon in the ring is identical, so let's now look at the six singly occupied $2p_z$ orbitals. In the left-hand figure, the lower lobes of the $2p_z$ orbitals on the back two carbons are not shown, and the C—H bonds are not drawn in. In the right-hand figure, we draw in three π bonds made from the usual $2p_z/2p_z$ overlap.

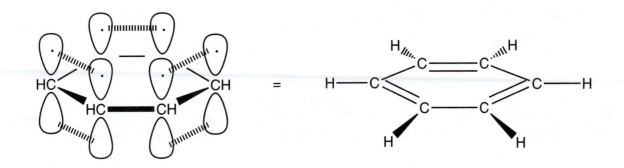

(b) Clearly, if there is only one C—C bond distance in this molecule, there cannot be simple, alternating single and double bonds, which must be different lengths. However, we made an arbitrary decision in the previous figure to let certain $2p_z$ orbitals overlap and ignore others. Why not do it another way? Our second drawing of this molecule is a resonance structure of the drawing in (a).

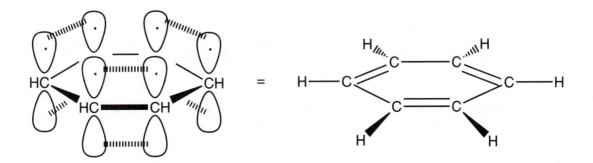

In fact, all the $2p_z$ orbitals are the same and must overlap equally with their right-hand and left-hand neighboring $2p_z$ orbitals. At this point, we can't do too much better than this—just note that there is excellent overlap among these $2p_z$ orbitals above and below the plane of the ring. This molecule is an aromatic ring, which we will discuss in detail in Chapter 14.

Problem 3.53 The two carbocations in question are

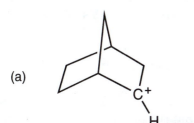

(a)   (b)

(continued)

Problem 3.53 *(continued)*

There is no problem accommodating a planar carbocation in (a):

(a)

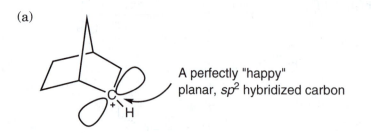

A perfectly "happy"
planar, sp^2 hybridized carbon

However, the drawing of (b) is harder to see (a model should make everything crystal clear). There is no way to flatten out this molecule at this "bridgehead" position. As carbocations are most stable when planar, this introduces a necessary instability into this cation and makes it difficult to form.

(b)

"Front view"

Eye

This carbon cannot flatten out!

"Side view"
seen from the
position
of the eye

Problem 3.54

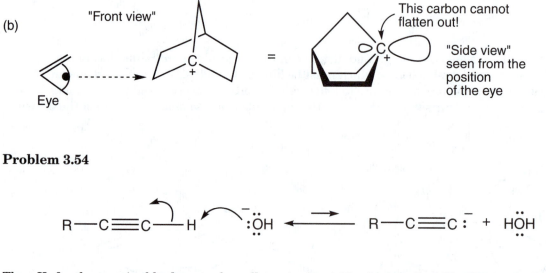

The pK_a for the terminal hydrogen of an alkyne is about 25, whereas the pK_a of water is 15.7. Water is a far stronger acid, and therefore the equilibrium lies to the left.

$$R—C\equiv C—H \quad :CH_3 \quad \rightleftharpoons \quad R—C\equiv C:^- + CH_4$$

The pK_a for the terminal hydrogen of an alkyne is about 25, whereas the pK_a of methane is at least 60. Methane is a far, far weaker acid, and therefore the equilibrium lies to the right.

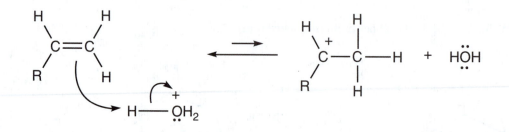

The pK_a for the hydronium ion is about –2. We will see in Chapter 7 that the pK_a of a hydrogen on a carbon adjacent to a carbocation is about –11.

Problem 3.55 Here are two HX additions. Just be careful to get the direction of addition—the regiochemistry—right. Protonation must always occur so as to give the more stable (more substituted) carbocation.

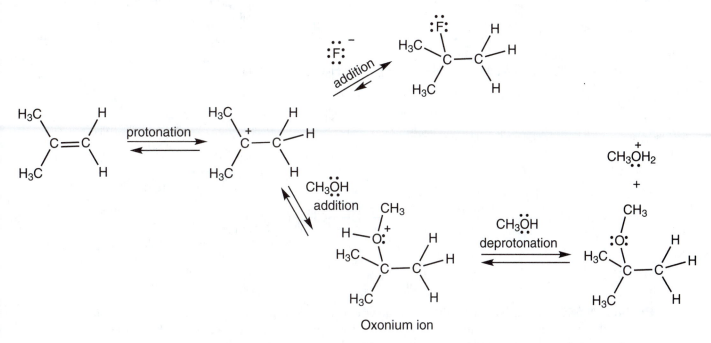

Oxonium ion

Problem 3.56 Protonation will occur so as to give the resonance-stabilized secondary carbocation, not the hideously unstable primary carbocation. The positive charge in the carbocation is shared by two carbons. The nucleophile (chloride ion here) can add at both of those positive carbons to give the two products.

This primary carbocation is less stable, and is *not* formed

Secondary carbocation is more stable, and is formed

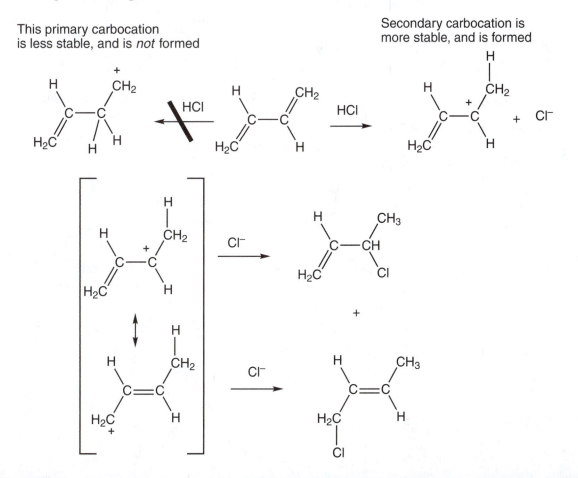

Problem 3.57   As the starting material is an alkyne, with two π bonds, the initial addition reaction leads to an alkene. Both (*Z*)-2-chloro-2-butene and (*E*)-2-chloro-2-butene would be formed. Only the (*Z*) isomer is shown. A second addition is now possible to give 2,2-dichlorobutane.

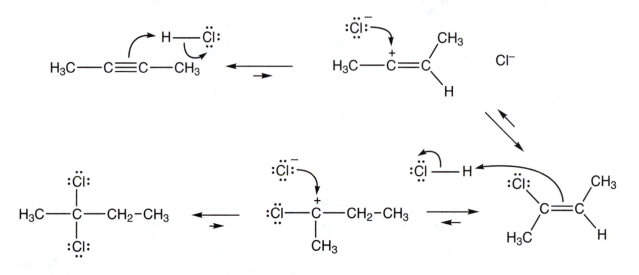

Problem 3.58   Two carbocations are possible when the initial product is protonated. One is far more stable than the other because it is stabilized by resonance. Its formation will be favored, and therefore the final product will come from addition of chloride to it, as shown.

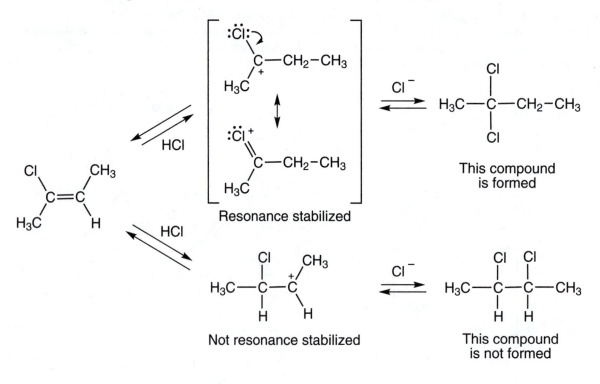

Problem 3.59 The hydrogen perpendicular to the π system is not involved. We call these nonmixing orbitals orthogonal. They don't "see" each other or "talk to" each other. The electrons can't move between the two. The electrons in the bonds to the hydrogens that are perpendicular to the π system are not able to mix with the π system because they are orthogonal to it.

Problem 3.60 The electron pushing for the reaction left to right is

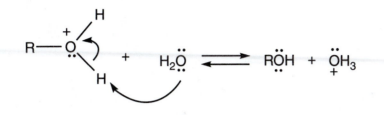

The electron pushing for the reaction from right to left is

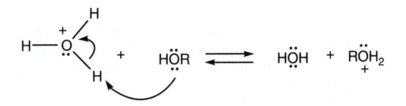

The two directions are very similar. The equilibrium constant is nearly 1. The neutral alcohol could be obtained by neutralizing the solution.

Problem 3.61 No, the π orbitals do not mix with the orbital containing the nonbonding electrons of the acetylide anion. The electrons of the carbanion are in an orbital that is orthogonal to the alkyne's π orbitals.

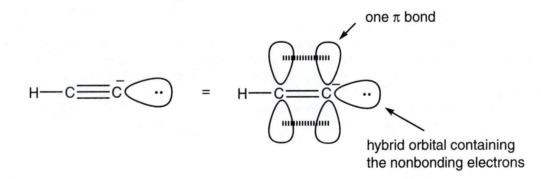

Stereochemistry

4

In this chapter, we deal with the details of molecular structure. There can be no hiding the fact that stereochemistry is difficult for many people, but it will yield to careful work and practice. Moreover, at this point we must reemphasize the idea that you cannot simply read this material and hope to get it. That approach is hopeless—repeat, *hopeless*. In this chapter, we see the consequences of the three-dimensional nature of molecules, and from now on you will have to be able to visualize molecules in three dimensions. Do not trust the two dimensions available to this book, or the blackboard, or a piece of paper.

The notion that organic chemistry is to be "read with a pencil" has been emphasized many times, and nowhere is that admonition more appropriate than here. Indeed, you should work through these problems with your models at the ready; there will be many points when they will be helpful or even essential.

Why is stereochemistry so important? Soon we are going to look at reactions in detail and try to work out how and why they proceed in the ways they do. As you will see, stereochemistry plays a huge role in this analysis. Working out the mechanistic details requires an ability to keep the stereochemical nuances of structure in mind. In turn, when we come to organic synthesis, the construction of molecules from simpler starting materials, we will always have to keep stereochemistry in mind. It is not enough to devise a synthesis that builds a molecule with its constituent parts more or less in the right place. They must be exactly right, perfectly positioned. Stereochemical control is vital, and to achieve that, you must understand the subject itself well.

Practice, practice, practice! The following problems, none of them especially difficult, give you a chance to start. Later, more complicated stereochemical analyses will accompany our discussions of many of the mechanisms of organic reactions.

Problem 4.1

(a) Carbon number 2 has four different groups (–H, –CH$_3$, –Br, and –CH$_2$CH$_3$). There is no plane of symmetry in the molecule. 2-Bromobutane is chiral.

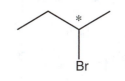

2-Bromobutane

(b) There are no stereogenic atoms. Carbon number 2 has two bromines. 2,2-Dibromobutane is achiral.

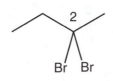

2,2-Dibromobutane

(c) There are no stereogenic atoms. Carbon number 3 has two ethyl groups attached. 3-Ethylheptane is achiral.

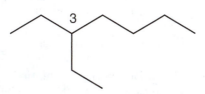

3-Ethylheptane

(d) There are no stereogenic atoms. Carbon number 4 has two propyl groups attached. 4-Ethylheptane is achiral.

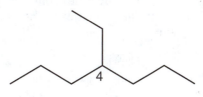

4-Ethylheptane

(e) There are no stereogenic atoms. Carbon number 2 has two methyl groups. Carbon number 5 also has two methyl groups. 2,5-Dimethylhexane is achiral.

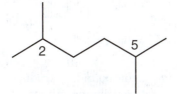

2,5-Dimethylhexane

(continued)

Problem 4.1 *(continued)*

(f) Carbon number 3 has four different groups attached (–H, –CH$_3$, –CH$_2$CH$_3$, –CH$_2$CH$_2$CH$_2$CH$_3$), and there is no plane of symmetry in the molecule. 3-Methylheptane is chiral.

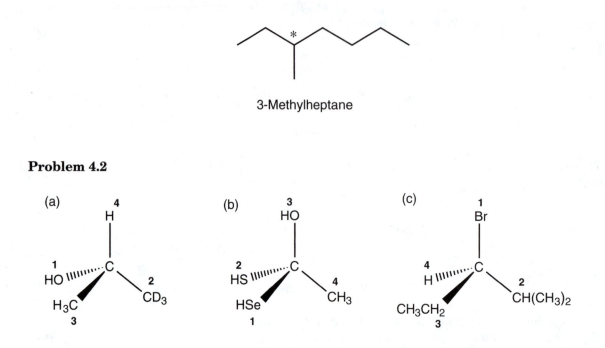

3-Methylheptane

Problem 4.2

(a) H is **4** and O is **1** on the basis of atomic number. The tie between CH$_3$ and CD$_3$ is broken by the greater atomic weight of D over H.

(b) Priorities can be assigned strictly by atomic number.

(c) Atomic number makes H priority **4** and Br priority **1**. The tie between the two carbons is broken by working out along the chain. The ethyl carbon is attached to C, H, H (**3**) and the isopropyl carbon to C, C, H (**2**).

(d) There are two stereogenic carbons in this molecule. For the rear carbon, H is priority **4**, but the other three substituents are carbon. The methyl carbon (attached to H, H, H) is priority **3**, the ring CH$_2$ group priority **2** (attached to C, H, H), and the other ring carbon **1** (attached to C, C, H). A similar tie-breaking procedure serves to assign priorities to the front carbon.

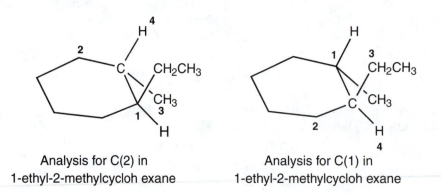

Analysis for C(2) in
1-ethyl-2-methylcycloh exane

Analysis for C(1) in
1-ethyl-2-methylcycloh exane

Problem 4.3   This problem may seem easy, even mindless. But watch out! First of all, drawing mirror images requires practice, and ring compounds introduce strange difficulties for many people. It is well worth your time to practice drawing mirror images a bit. Invent some more problems for yourself.

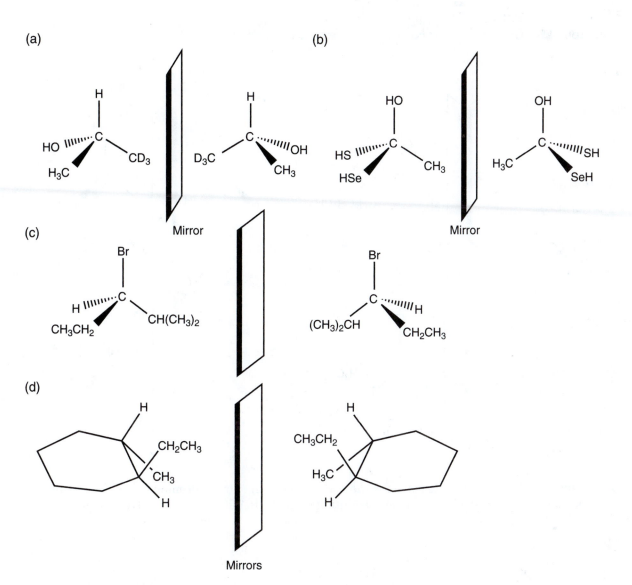

Problem 4.4 Remember, look down the C—priority **4** bond (arrow) and connect the groups **1→2→3**. The priorities were assigned in the usual way, through the Cahn–Ingold–Prelog protocol.

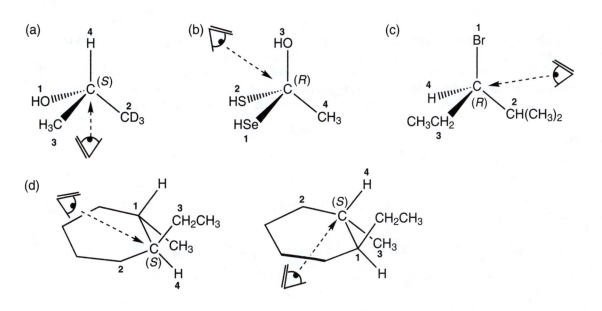

Problem 4.5   Here is more practice in drawing mirror images.

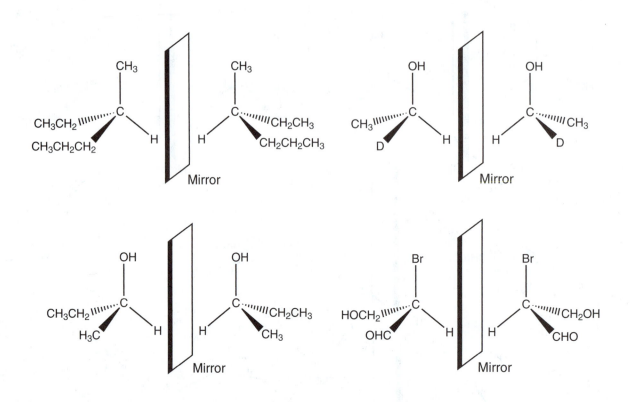

Problem 4.6   First we need to solve for the specific rotation of (*S*)-2-bromobutane. We are told that the observed rotation is + 2.31° for a solution of 10.0 g in 100 mL hexane. That means

$$[\alpha] = \frac{2.31°}{10 \text{ g}/100 \text{ mL}} = 23.1°$$

So the specific rotation of (*S*)-2-bromobutane is +23.1°. By the way, that means the specific rotation for (*R*)-2-bromobutane is –23.1°.

To determine the measured rotation for a mixture of (*S*) and (*R*) enantiomers, we first need to determine the enantiomeric excess. How much of the (*S*) enantiomer is not canceled out by the (*R*) enantiomer? There are 7.5 g of (*S*) enantiomer and 2.5 g of (*R*) enantiomer. That leaves 5.0 g of (*S*) enantiomer not canceled out by the (*R*). That amount is the enantiomeric excess in this example. Now we can use the same formula to determine the observed rotation for such a solution.

That means

$$23.1° = \frac{\alpha}{5 \text{ g}/100 \text{ mL}}$$

Solving for α, we find that the measured rotation for this solution is +1.16°.

Problem 4.8   Here is an opportunity to look for a plane of symmetry in everyday objects. A chiral object cannot have a plane of symmetry. Here are some likely answers: scissors, knives that are serrated on one side, a refrigerator that has a freezer on top or on bottom, a microwave oven, books (although a book with blank pages might be achiral), a face clock, a calendar, artwork (although there might be some art out there that has a plane of symmetry), gloves that fit one hand but not the other, shoes that fit one foot but not the other, button-up or zipper clothing (shirts, pants, sweaters, coats), a toilet that has a handle on one side, a piano (if the keys are visible), people living in the house (although I suppose it is possible, I have never met an achiral person).

Problem 4.9   Here is the interaction of an inverted shell and a pair of hands. The interactions with the enantiomeric pair of hands are still different. Turning the shell makes no difference whatsoever.

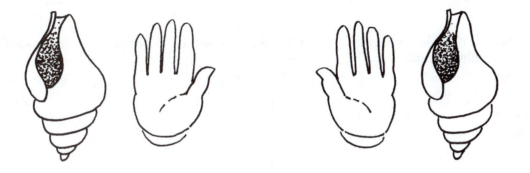

Problem 4.11   This problem is a nuts and bolts question, but, as we have said many times, cyclic systems are sometimes inexplicably difficult. So here is a chance to practice dealing with rings and priority assignment. The priorities are assigned as shown. H is (**4**), and the doubly bonded carbon is (**1**). The C—C—O gets a higher priority (**2**) than the C—C—C (**3**) because of the higher atomic number of oxygen than carbon.

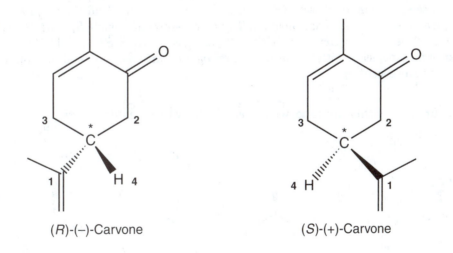

(*R*)-(−)-Carvone   (*S*)-(+)-Carvone

Problem 4.12   Ha! The mirror image of "(*R*)—(**S**)" is "(*R*)—(**S**)." Most people say, too quickly, "(*S*)—(*R*)." The boldface "(**S**)" is reflected as "(**R**)."

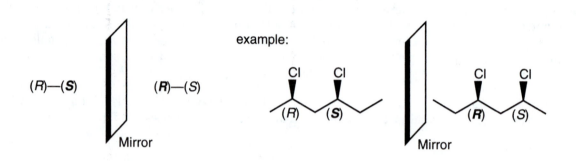

Problem 4.14   You might have drawn a very complex molecule. Or perhaps you drew something like 2,3,4-trichlorohexane.

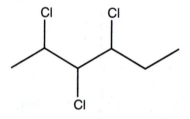

A molecule with three stereogenic carbons can have at most eight stereoisomers. Each stereogenic carbon can be either (*R*) or (*S*). So we have the following possible structures: *RRR, RRS, RSR, SRR, RSS, SRS, SSR*, and *SSS*. The enantiomers are the ones that are exactly opposite. That means that *RRR* and *SSS* are enantiomers. So are *RRS* and *SSR*; *RSR* and *SRS*; and *SRR* and *RSS*. Any other comparisons are diastereomers. That means *RRR* is a diastereomer of *RRS, RSR, SRR, RSS, SRS*, and *SSR*. The *RRS* is a diastereomer of *RRR, RRS, RSR, SRR, RSS, SRS*, and *SSS*. To make a diastereomer of any molecule with two or more stereogenic carbons, you need to change at least one but not all of the stereocenters.

If we look at our 2,3,4-trichlorohexane, we can easily draw two diastereomers.

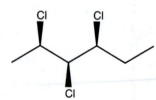   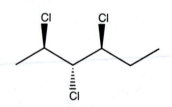

(2*R*,3*S*,4*S*)-2,3,4-Trichlorohexane   (2*R*,3*S*,4*R*)-2,3,4-Trichlorohexane   (2*R*,3*R*,4*S*)-2,3,4-Trichlorohexane

Problem 4.15

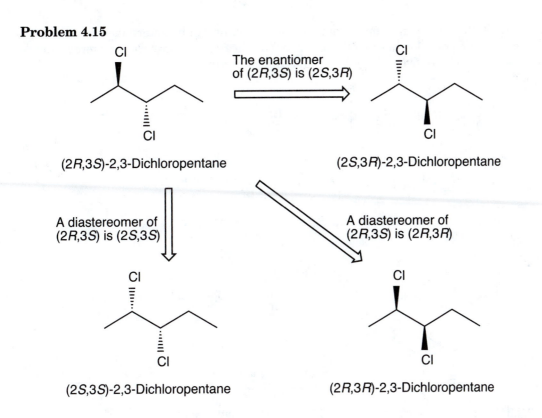

(2R,3S)-2,3-Dichloropentane

The enantiomer of (2R,3S) is (2S,3R)

(2S,3R)-2,3-Dichloropentane

A diastereomer of (2R,3S) is (2S,3S)

A diastereomer of (2R,3S) is (2R,3R)

(2S,3S)-2,3-Dichloropentane

(2R,3R)-2,3-Dichloropentane

Problem 4.16 There is no problem here. The designations "(S)" and "(R)" come from the *individual priority* assignments for the two stereogenic carbons in **A**. In the hypothetical bond-forming process we have developed, the right-hand carbon comes from (S)-2-chloro-1,1,1-trifluoropropane, in which it is the Cl that is priority **1**. In the new compound, **A**, this chlorine is no longer there. Now the priority **1** is the trifluoromethyl group, and a proper priority count shows that the right-hand carbon is "(R)."

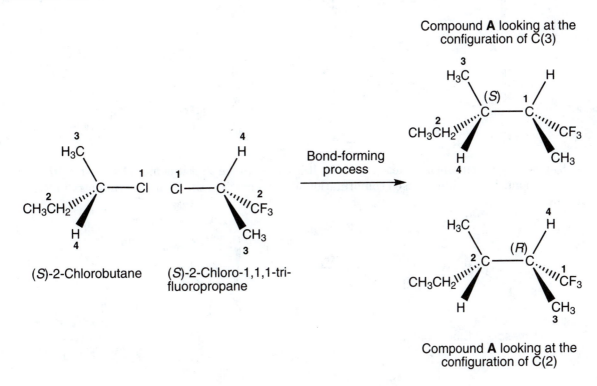

(S)-2-Chlorobutane

(S)-2-Chloro-1,1,1-tri-fluoropropane

Bond-forming process

Compound **A** looking at the configuration of C(3)

Compound **A** looking at the configuration of C(2)

Problems 4.17 and 4.19   The stereoisomers of 2,3-dibromobutane can be determined by drawing all possible structures and checking to see if there are any duplications. There are two stereogenic carbons, so we need to draw the (*R,R*), (*S,S*), (*R,S*), and (*S,R*) structures.

(a)

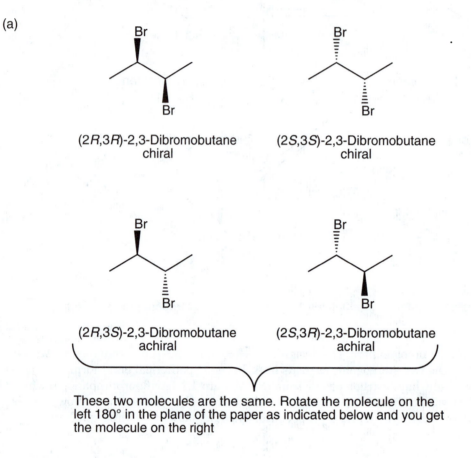

(2*R*,3*R*)-2,3-Dibromobutane
chiral

(2*S*,3*S*)-2,3-Dibromobutane
chiral

(2*R*,3*S*)-2,3-Dibromobutane
achiral

(2*S*,3*R*)-2,3-Dibromobutane
achiral

These two molecules are the same. Rotate the molecule on the left 180° in the plane of the paper as indicated below and you get the molecule on the right

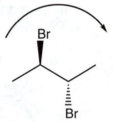

(2*R*,3*S*)-2,3-Dibromobutane is a meso molecule. It is easier to see that it is meso by rotating the C(2)—C(3) bond 180° to the eclipsed conformation. We can now see the plane of symmetry in the molecule.

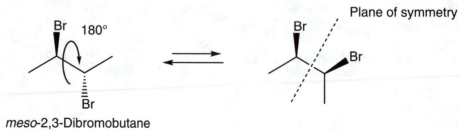

meso-2,3-Dibromobutane

(b) 2,2-Dibromo-3,3-dichlorobutane has no stereogenic carbons. There is a plane of symmetry in the molecule, which is the plane of the paper. The molecule is achiral and has no stereoisomers.

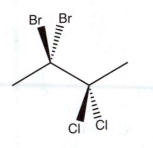

(c) 2,3-Dibromo-2,3-dichlorobutane has three stereoisomers—a pair of enantiomers and a meso compound. Priorities are assigned in the usual way.

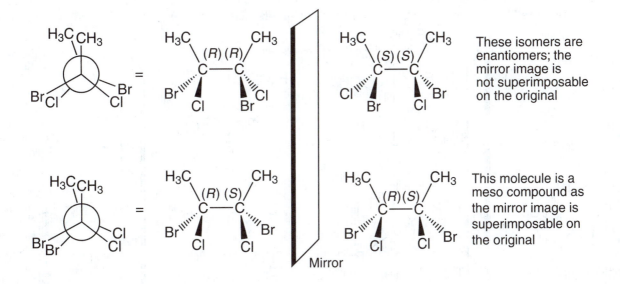

These isomers are enantiomers; the mirror image is not superimposable on the original

This molecule is a meso compound as the mirror image is superimposable on the original

Mirror

(d) This molecule contains stereogenic carbons and yields the full complement of $2^2 = 4$ stereoisomers. The molecules are shown in eclipsed forms for clarity and ease of analysis, but these are not the minimum energy arrangements, which will have staggered bonds.

(continued)

Problems 4.17 and 4.19 *(continued)*

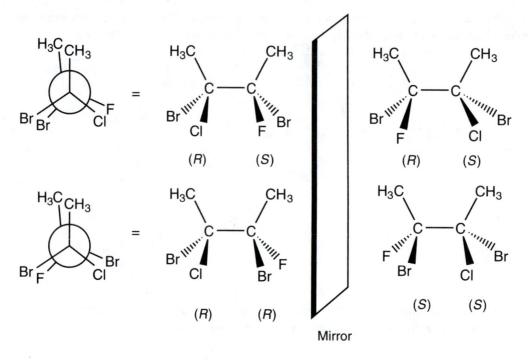

Mirror

Problem 4.18 If rotational barriers were high, then each staggered form would be isolable. The number of isomers increases dramatically, as each isolable staggered form will have an enantiomer. Now there are 12 total isomers, in six pairs of enantiomers. The answer is given in Newman projections.

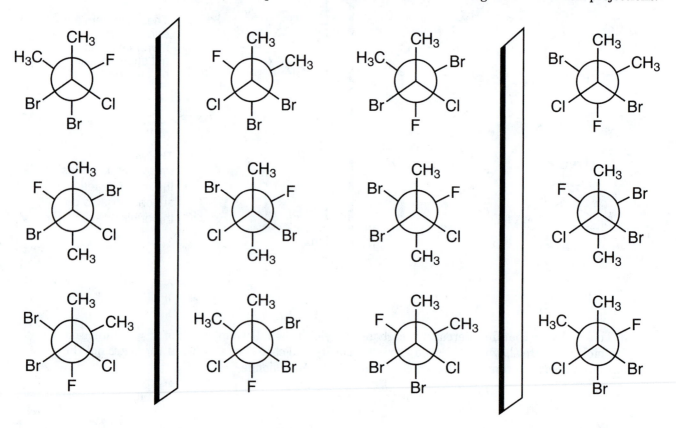

Problem 4.21 The (*R*) and (*S*) designations are assigned from the following priorities:

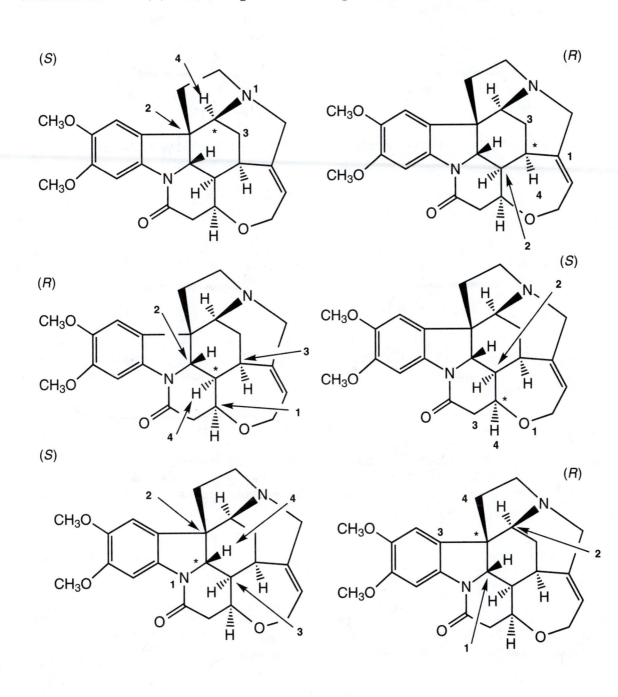

Problem 4.22

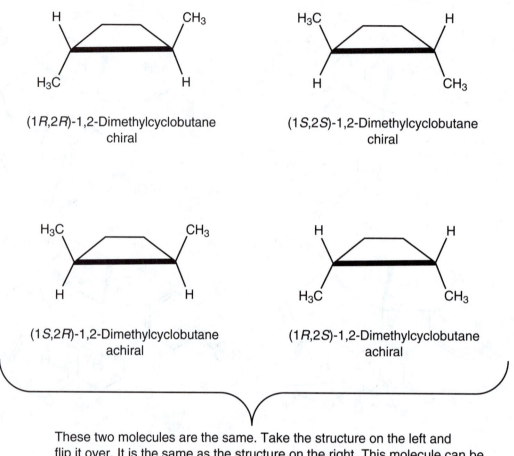

(1R,2R)-1,2-Dimethylcyclobutane
chiral

(1S,2S)-1,2-Dimethylcyclobutane
chiral

(1S,2R)-1,2-Dimethylcyclobutane
achiral

(1R,2S)-1,2-Dimethylcyclobutane
achiral

These two molecules are the same. Take the structure on the left and flip it over. It is the same as the structure on the right. This molecule can be adequately described as *cis*-1,2-dimethylcyclobutane. It is also meso.

Problem 4.23

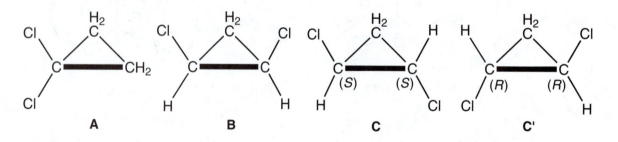

A   B   C   C'

Compound **A** is a structural isomer of **B**, **C**, and **C′**. Compounds **C** and **C′** are enantiomers; **B** and **C** and **B** and **C′** are pairs of diastereomers.

Problem 4.24

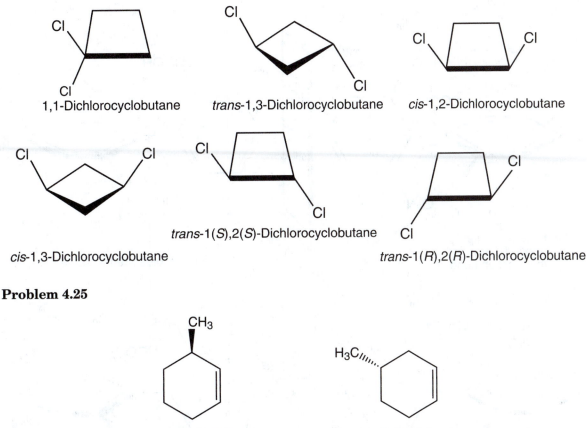

1,1-Dichlorocyclobutane   *trans*-1,3-Dichlorocyclobutane   *cis*-1,2-Dichlorocyclobutane

cis-1,3-Dichlorocyclobutane

trans-1(*S*),2(*S*)-Dichlorocyclobutane

trans-1(*R*),2(*R*)-Dichlorocyclobutane

Problem 4.25

(*R*)-3-Methylcyclohexene   (*S*)-4-Methylcyclohexene

(*R*)-3-Methylcyclohexene and (*S*)-4-methylcyclohexene have the same molecular formula (C_7H_{12}). But they do not have all connections to the same atoms: one is a 3-methyl and the other is a 4-methyl. These molecules are constitutional isomers.

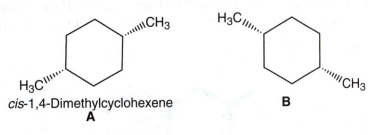

cis-1,4-Dimethylcyclohexene
A

B

Question 1: **A** and **B** have the same molecular formula.
Question 2: All the connections are to the same atoms.
Question 3: Compounds **A** and **B** are superimposable. These molecules are identical. There is no
stereogenic atom, so the molecule is not meso.

Problem 4.30

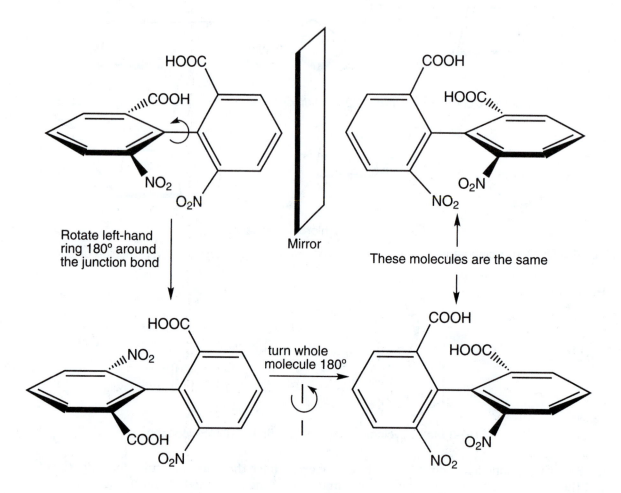

Problem 4.31   Were hexahelicene to be planar, the atoms of the "end" rings would have to occupy the same space. Accordingly, one ring (dark lines) slips over the other (dotted lines). A coil or helix is formed.

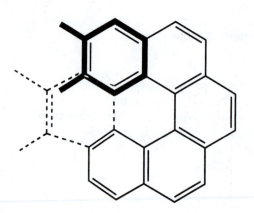

Helices are chiral and can spiral in a right-handed or left-handed way. The right-handed helix and the left-handed helix are nonsuperimposable mirror images:

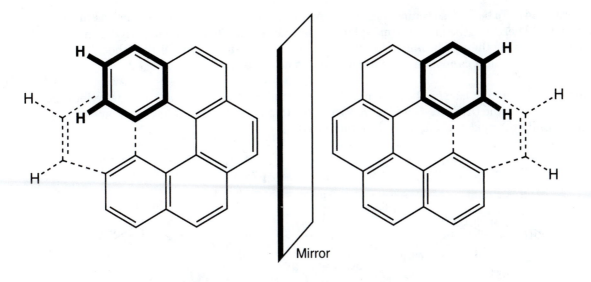

Mirror

Additional Problem Answers

Problem 4.32   This one is not bad at all; there are only two.

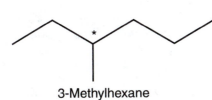

3-Methylhexane

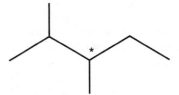

2,3-Dimethylpentane

Problem 4.33

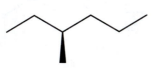

(*S*)-3-Methylhexane

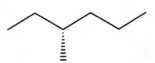

(*R*)-3-Methylhexane

(*R*)-2,3-Dimethylpentane

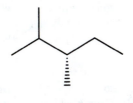

(*S*)-2,3-Dimethylpentane

Problems 4.34   To be a meso compound, there must be at least two stereogenic carbons in the octane isomer. Therefore, the molecule must be a dimethyl isomer. There are no diethyl isomers possible (try drawing them). The only trimethyl isomer with a plane of symmetry is the 2,3,4-trimethylpentane. But it cannot be meso because it has no stereogenic carbons.

The dimethyl-containing isomer of octane that has potential for being meso is 3,4-dimethylhexane. The meso isomer is the (*R*,*S*) version. The (*R*,*R*) and the (*S*,*S*) are chiral and have no plane of symmetry. Build the model, if you need to be convinced.

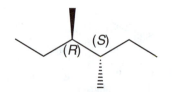   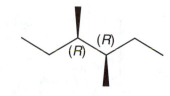   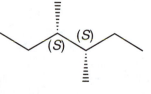

meso-3,4-Dimethylhexane   (*R,R*)-3,4-Dimethylhexane   (*S,S*)-3,4-Dimethylhexane
achiral   chiral   chiral

The (*R*,*S*)-dimethylhexane is meso because it is a molecule with stereogenic carbons and it has a plane of symmetry. It is achiral. To see the plane of symmetry, you need to rotate the C(3)—C(4) bond by 180°, as shown.

Plane of symmetry

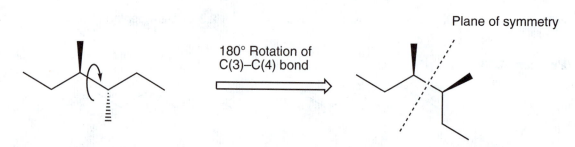

180° Rotation of
C(3)—C(4) bond

Problem 4.35   There's no big trick to this problem. You have three bromines to locate on a 4-carbon chain. Think of all the possible attachments in an orderly fashion. We could have all the bromines on the same carbon to give 1,1,1 (it's achiral). All three bromines on the other end of the butane gives 4,4,4, which is actually 1,1,1-tribromobutane again.

Now move one bromine to another carbon to give 1,1,2 (chiral), then 1,1,3 (chiral), then 1,1,4 (achiral). We could have two bromines on carbon number 2 to give 1,2,2 (achiral), and two bromines on carbon number 3 is 1,3,3, which is actually 2,2,1 using lowest numbers (achiral), and 1,4,4, which is actually 1,1,4 (achiral). The only other isomer that has two bromines on one carbon is 2,2,3 (chiral).

We can now consider the isomers that have bromines on three different carbons. That gives us 1,2,3 (chiral) and 1,2,4 (chiral). That gives us the five isomers that are chiral.

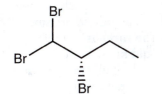

(*S*)-1,1,2-Tribromobutane

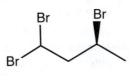

(*S*)-1,1,3-Tribromobutane

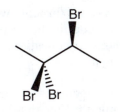

(*S*)-2,2,3-Tribromobutane

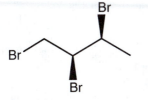

(2*S*,3*S*)-1,2,3-Tribromobutane
Note that there are two stereogenic carbons in this isomer.

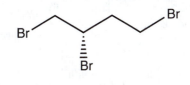

(*S*)-1,2,4-Tribromobutane

Problem 4.36 Drawing one meso molecule shouldn't be too hard. To be meso, the molecule must have stereogenic carbons and a plane of symmetry. That means one of the bromines must be on the central carbon. We can't have a plane of symmetry with two bromines on one side and one on the other. The only symmetrical options are 1,3,5-tribromopentane and 2,3,4-tribromopentane. As you can see, 1,3,5-tribromopentane has no stereogenic carbons. The 2,3,4-tribromopentane can be meso. Where is the second meso compound?

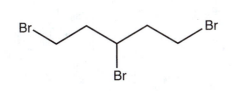

1,3,5-Tribromopentane
achiral

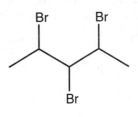

2,3,4-Tribromopentane
meso

Finding the two meso molecules is a tougher challenge. To be meso, the bromines on carbons number 2 and number 4 must be both out (or both back). But the bromine on the central carbon can be either direction. It can be coming out or going back. And that gives us the two meso isomers!

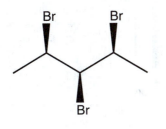

meso-2,3,4-Tribromopentane

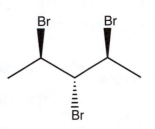

meso-2,3,4-Tribromopentane

(continued)

Problem 4.36 *(continued)*

We aren't through. We are asked to determine the number of stereogenic carbons for each isomer. It's tempting to say two for each, but that isn't quite correct. This is really an advanced organic chemistry question. So feeling challenged at this point is totally allowed. The carbons number 2 and number 4 in both isomers are stereogenic carbons. That much you should know. Carbon number 3 in both isomers is also a stereogenic carbon. Look at the molecule with all three bromines coming out of the plane of the paper. The group to the left of carbon number 3 has the (R) configuration and the group on the right of carbon number 3 has the (S) configuration. They are different groups. To determine the (R) and (S) configuration of carbon number 3, we give the (R) stereochemistry priority over the (S). Therefore the molecule on the left is (2R,3R,4S)-2,3,4-tribromopentane. The molecule on the right is (2R,3S,4S)-2,3,4-tribromopentane. Both molecules have three stereogenic carbons.

You might have guessed that the following isomer is chiral (not meso) and it has only two stereogenic carbons!

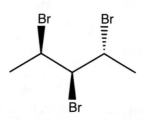

(2R,4R)-2,3,4-Tribromopentane

Problem 4.37   3,4-Dimethylheptane has two stereogenic carbons, so the maximum number of stereoisomers is $2^2 = 4$. These appear as two pairs of enantiomers. Me = methyl, Et = ethyl, and Pr = propyl.

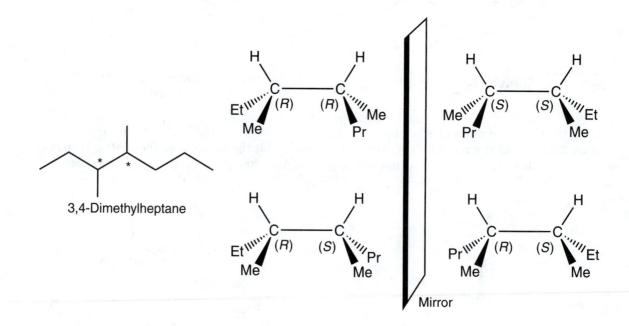

3,5-Dimethylheptane also has two stereogenic carbons, but there is a plane of symmetry in this molecule, and there are only three stereoisomers: one pair of enantiomers and a meso compound.

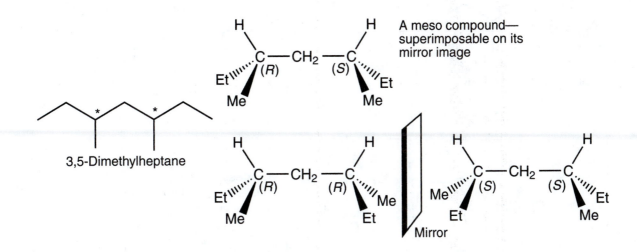

Problem 4.38 Each of the diastereomers will have a set of nine different carbons. So, barring accidental overlapping of signals, we should see a total of 18 different signals. The enantiomers will not have separable signals in an achiral solvent such as CDCl₃.

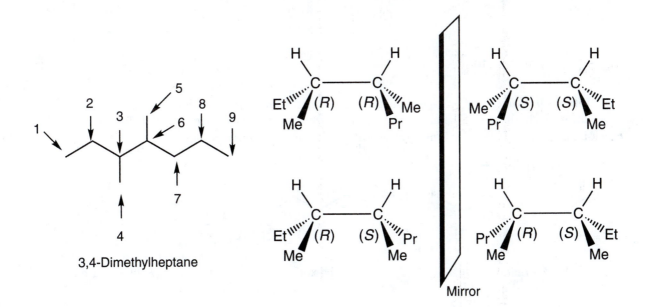

Problem 4.39   Here we also have two diastereomers, but one is a meso compound. There will be 5 signals from the meso compound and 5 more from the pair of enantiomers for a total of 10.

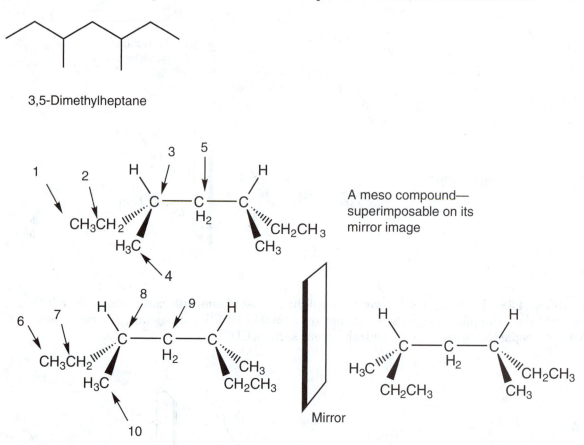

3,5-Dimethylheptane

A meso compound—
superimposable on its
mirror image

Mirror

Problem 4.40   We can draw 1-pentene and consider the six different hydrogens. Replacing each of those different hydrogens (H_a–H_f) with a methyl group (one at a time) gives six different isomers. A quick inspection tells us that 3-methyl-1-pentene is the only chiral isomer.

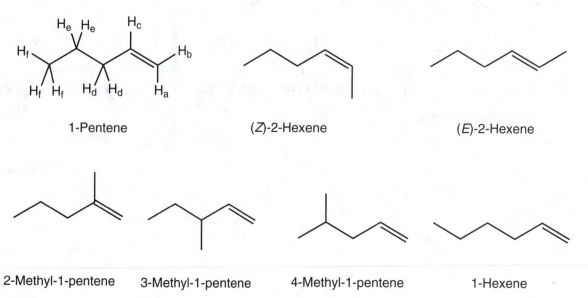

1-Pentene   (*Z*)-2-Hexene   (*E*)-2-Hexene

2-Methyl-1-pentene   3-Methyl-1-pentene   4-Methyl-1-pentene   1-Hexene

The two enantiomers of 3-methyl-1-pentene are the (S) and (R) structures drawn here.

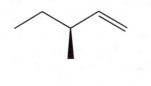 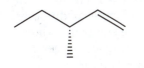

(S)-3-Methyl-1-pentene   (R)-3-Methyl-1-pentene

Problem 4.41

(a) The nitrogen is the only stereogenic atom in this molecule. We know that the amine nitrogen can undergo inversion at room temperature, but as drawn, the configuration is (S). The lone pair gets the lowest assignment because it has an atomic number of zero. The three atoms attached to the stereogenic nitrogen are all carbons. So we have to compare the next level of attachment. The ring carbon coming out at us has attachments of (C, C, C). The ring carbon going back has (C, H, H), and the carbon of the ethyl group to the right has (C, H, H). That means the carbon coming out toward us is number 1 in our (R/S) analysis. The other two carbons both have (C, H, H) attachment. So we have to go to the next level. The next carbon on the ring has (C, C, N). The next carbon on the ethyl group has (H, H, H). So we have the order as shown. Without moving anything, we can draw an arrow from 1 to 2 to 3 and see that the circle is clockwise as looking from above. But that is looking from above with number 4 (the lone pair) in front. So reverse your answer. The counterclockwise direction tells us this is the (S) enantiomer.

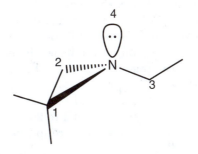

(b) The phosphorus atom is the only stereogenic atom in this molecule. The attachments are C, N, and two O's. The carbon is the lowest atomic number, so it is number 4. Nitrogen is next lowest, so it is number 3. The oxygens are the same, so we must go out to the next attachment, which is carbon for both. The carbon of the isopropyl group is attached to (C, C, H). The group to the right has its carbon attached to (C, H, H). So the isopropyl group is number 1. Without moving anything, we can draw an arrow from 1 to 2 to 3 and see that the circle is clockwise as looking from the left. But looking from the left has number 4 (the propyl group) in front. So reverse your answer. The counterclockwise direction tells us this enantiomer is (S).

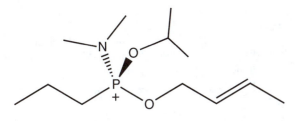 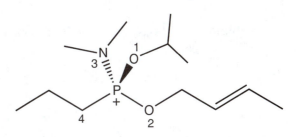

(continued)

Problem 4.41 *(continued)*

(c) The sulfur is the stereogenic atom in this molecule. It has a lone pair, an oxygen, and two carbons directly attached. The lone pair is number 4 and the oxygen is number 1. The two carbons need to be analyzed further to differentiate them. As shown in the middle structure, the aromatic ring carbon has three bonds to carbons (C, C, C). The propyl group carbon has attachments of (C, H, H). So the aromatic ring is number 2. Without moving anything, we can draw an arrow from 1 to 2 to 3 and see that the circle is clockwise as looking from the left with the number 4 group (the lone pair) in back. The clockwise direction tells us this enantiomer is (R).

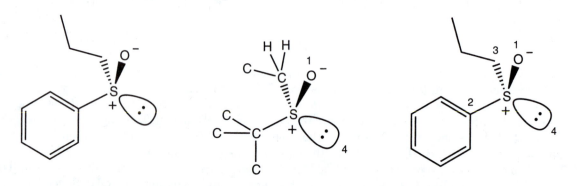

(d) The nitrogen is the stereogenic atom in this molecule. It is attached to an oxygen, an isopropyl group, an ethyl group, and an isobutyl group. The oxygen has the larger atomic number, so it is number 1. The other attachments are to carbon and must be compared by looking at their attachments as shown in the middle structure. The carbon coming out at us has (C, H, H). The carbon to the right of the isopropyl group has (C, C, H). The carbon going back has (C, H, H). So the carbon of the isopropyl group to the right is assigned number 2. Now we compare the next attachment for the two remaining groups. The carbon on the group going back (the ethyl group) is attached to (H, H, H). The carbon on the isobutyl group coming out has (C, C, H). So the isobutyl group is number 3 and the ethyl group is number 4.

Without moving anything, we can draw an arrow from 1 to 2 to 3 and see that the circle is clockwise as looking from the front with the number 4 group (the ethyl) in back. The clockwise direction tells us this enantiomer is (R).

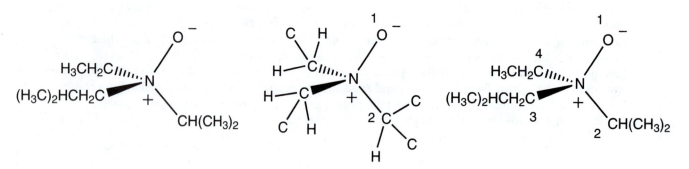

Problem 4.42   (a) No stereogenic carbons (*), two stereoisomers:

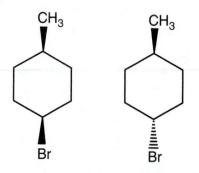

(b) Two stereogenic carbons (*), four stereoisomers:

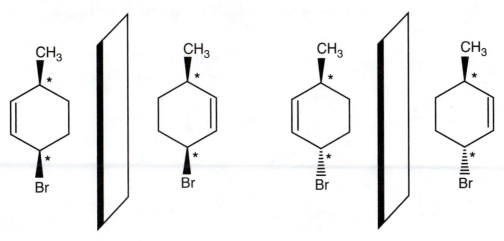

(c) Two stereogenic carbons (*), four stereoisomers:

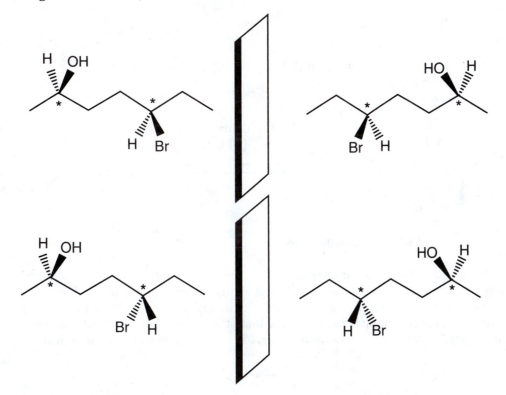

(d) One stereogenic carbon (*), two stereoisomers:

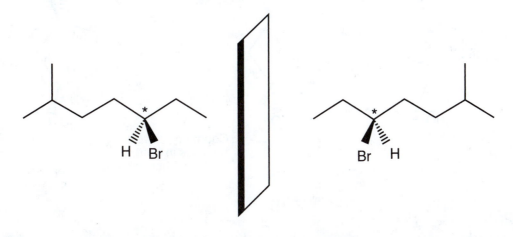

Problem 4.43

(a) Achiral. The plane of symmetry bisects the bromine, the ring, and the OCH₃ group.

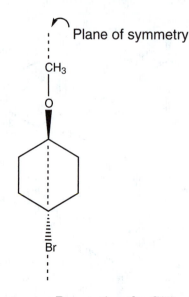

(b) Chiral. There is no plane of symmetry. By rotating the C(3)—C(4) bond 180°, you can see that the molecule lacks symmetry.

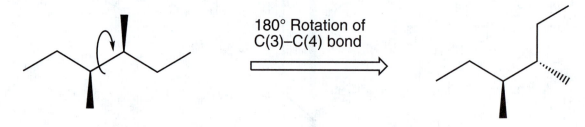

This molecule is (3S,4S)-3,4-dimethylhexane.

(c) Chiral. There is no plane of symmetry in this molecule. Let's determine the configuration of the two stereogenic carbons independently. On the left structure we see that the carbon with the OH group coming out at us has the (R) configuration. The middle structure shows the analysis of the carbon with the OH group going back. It has the lowest atomic number in front. The arrow from 1 to 2 to 3 is counterclockwise, but because number 4 is in front, this enantiomer is (R).

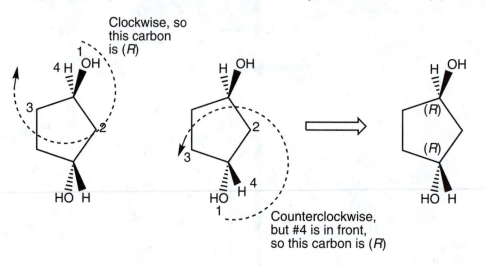

(d) Chiral. There is no plane of symmetry in this molecule. The two alkenes are different. The alkene on the left is (*E*) and the alkene on the right is (*Z*). Let's determine the configuration for the two stereogenic carbons separately. The carbon that has the attached fluorine going behind the plane of the paper has attachments of H, F, and two C's. The hydrogen is number 4. The fluorine is number 1. We have to compare the next attachments for the two carbons. This is shown on the second structure. The carbon on the right has (F, C, H). The carbon on the left has (C, C, C). We don't add the atomic numbers, we compare them. Because F is bigger than any carbon of (C, C, C), the number 2 designation goes to the group on the right. We can draw an arrow from 1 to 2 to 3 and see that it is counterclockwise, but the lowest atomic number is in front, so this an (*R*) configuration.

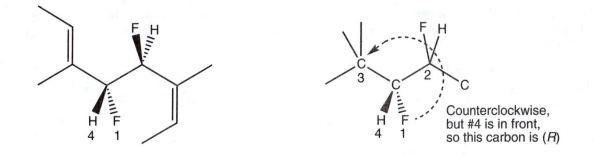

Now we can analyze the stereogenic carbon with the fluorine group coming out at us. The attachments are H, F, and two C's. The hydrogen is number 4. The fluorine is number 1. We can compare the next attachments for the two carbons as shown. The carbon on the left has (F, C, H). The carbon on the right has (C, C, C). Because F is bigger than any of the carbons of (C, C, C), the number 2 designation goes to the group on the left.

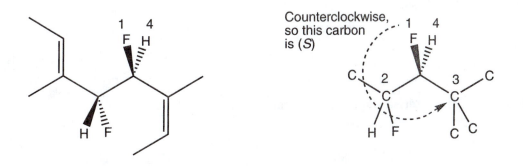

By the way, we know the IUPAC name for this molecule. It is (2*E*,4*R*,5*S*,6*Z*)-4,5-difluoro-3,6-dimethyl-2,6-octadiene.

(e) Achiral. There is a plane of symmetry through the middle of the cyclohexene. Even though the molecule is achiral, there are two stereogenic carbons. Let's analyze them separately. The carbon that is at the bottom of the cyclohexane has H, Cl, and two C's attached. The hydrogen is number 4 and the Cl is number 1. The two carbons need to be compared further. The carbon to the left has (C, H, H). The carbon to the right has (Cl, C, H). Because the Cl is the largest atomic number, the group to the right is number 2 and the group to the left is number 3. Drawing an arrow from 1 to 2 to 3 gives a counterclockwise direction, so this configuration is (*S*).

(continued)

Problem 4.43 *(continued)*

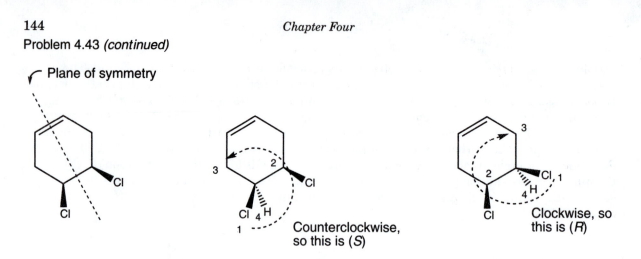

Counterclockwise, so this is (*S*)

Clockwise, so this is (*R*)

Analysis of the other stereogenic carbon tells us that it is an (*R*) configuration. This molecule is (4*R*,5*S*)-4,5-dichlorocyclohexene. It is a meso compound. It is the same as (4*S*,5*R*)-4,5-dichlorocyclohexene.

Problem 4.44

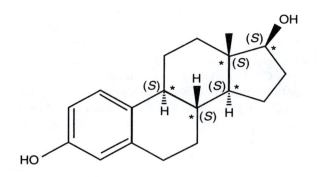

Problem 4.45   The asterisk shows the stereogenic carbons.

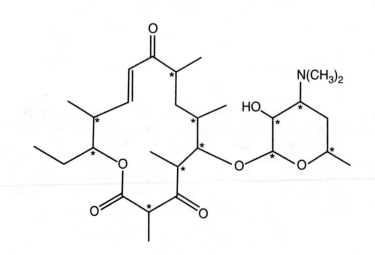

Problem 4.46 They are all chiral, except for (a), which is a meso compound.

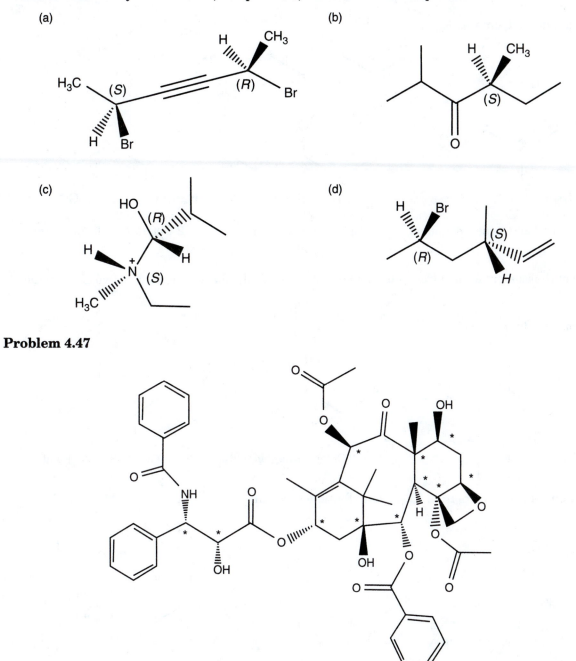

(a)

(b)

(c)

(d)

Problem 4.47

Problem 4.48 Two of the isomers formed from hexane are chiral.

From hexane,

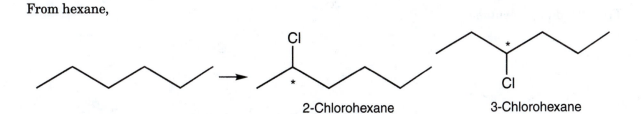

2-Chlorohexane   3-Chlorohexane

(continued)

Problem 4.48 *(continued)*

Three of the molecules formed from 2-methylpentane are chiral.

From 2-methylpentane,

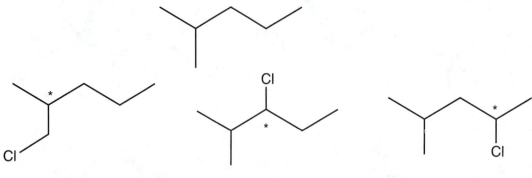

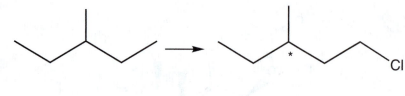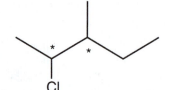

1-Chloro-2-methylpentane   3-Chloro-2-methylpentane   2-Chloro-4-methylpentane

Two of the isomers formed from 3-methylpentane are chiral, and one isomer has two stereogenic carbons.

From 3-methylpentane,

1-Chloro-3-methylpentane   2-Chloro-3-methylpentane

Only one of the isomers from 2,2-dimethylbutane is chiral.

From 2,2-dimethylbutane,

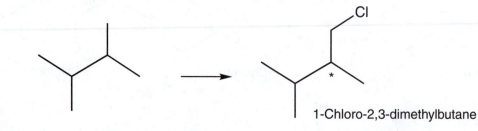

3-Chloro-2,2-dimethylbutane

Similarly, only one of the isomers from 2,3-dimethylbutane is chiral.

From 2,3-dimethylbutane,

1-Chloro-2,3-dimethylbutane

Problem 4.49 There is only one isomer to consider, 2-chloro-3-methylpentane. There will be four stereoisomers: two pairs of enantiomers.

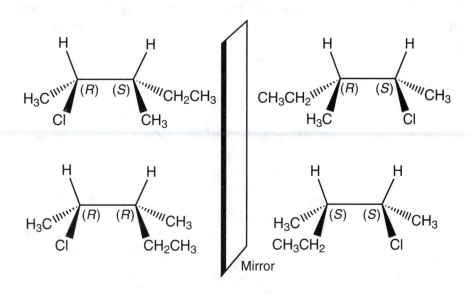

Problem 4.50 Only compound (b) can exist in (*E*) and (*Z*) forms. Compounds (a) and (d) have stereogenic carbon atoms:

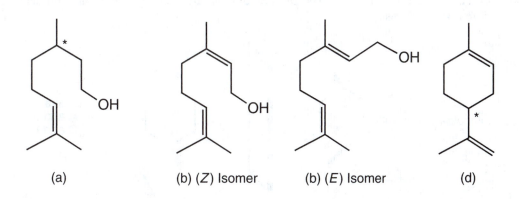

(a) (b) (*Z*) Isomer (b) (*E*) Isomer (d)

Technically, (d) could exist in (*E*) and (*Z*) forms. However, a trans double bond in a six-membered ring produces too much strain, and the isomer shown is the only one practically possible.

Problem 4.51 Base the tetrahedra on the stereogenic carbons. Determining priorities is certainly no problem in the acyclic molecules, but it may be more difficult in the cyclic species. *Remember*: The ring makes no special difference in this problem. For example, the priorities would be exactly the same in molecule **A** formed by eliminating a bond as shown. Why should anything be altered by closing the ring? It isn't.

(a)

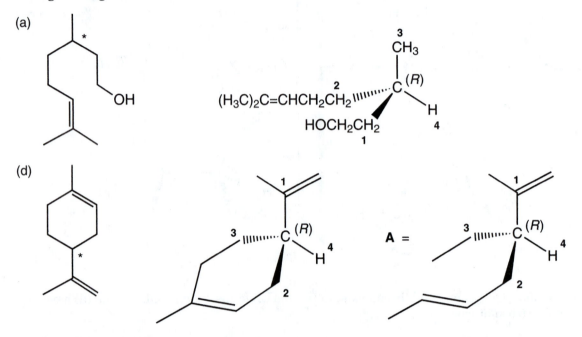

(d)

Problem 4.52 This molecule comes in cis and trans versions. In each case, there is a pair of enantiomers. There is a total of four stereoisomers, and the figure shows the enantiomeric and diastereomeric relationships.

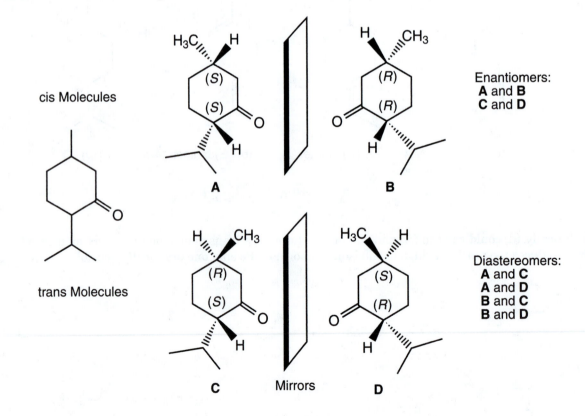

Problem 4.53   There are 2 degrees of unsaturation in C$_4$H$_5$Cl, which means we can have one of the following: two double bonds, a triple bond, a double bond and a ring, or two rings.

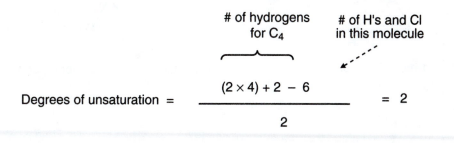

$$\text{Degrees of unsaturation} = \frac{(2 \times 4) + 2 - 6}{2} = 2$$

Here are five chiral molecules with this formula, with the stereogenic carbons indicated by an asterisk.

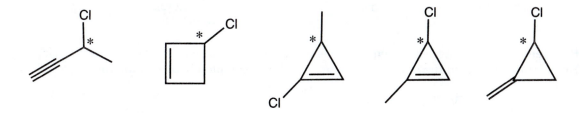

This is another chiral C$_4$H$_5$Cl molecule. It does not have a stereogenic carbon.

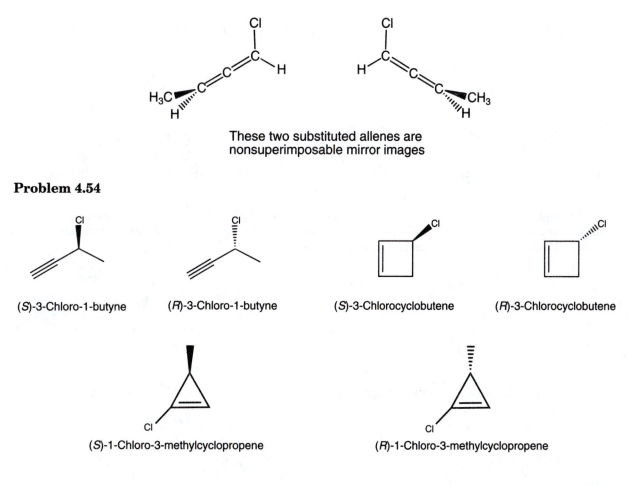

These two substituted allenes are nonsuperimposable mirror images

Problem 4.54

(S)-3-Chloro-1-butyne   (R)-3-Chloro-1-butyne   (S)-3-Chlorocyclobutene   (R)-3-Chlorocyclobutene

(S)-1-Chloro-3-methylcyclopropene   (R)-1-Chloro-3-methylcyclopropene

(continued)

Problem 4.54 *(continued)*

(*S*)-3-Chloro-1-methylcyclopropene    (*R*)-3-Chloro-1-methylcyclopropene

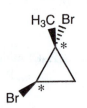

(*S*)-1-Chloro-2-methylenecyclopropane 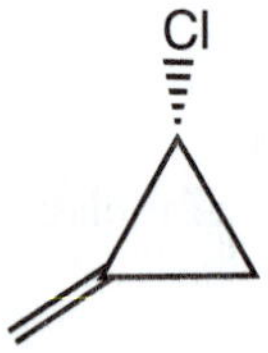   (*R*)-1-Chloro-2-methylenecyclopropane

Problem 4.55   We have shown one of the enantiomers for each chiral dibromomethylcyclopropane. You might have drawn the other enantiomer, which is a completely correct answer.

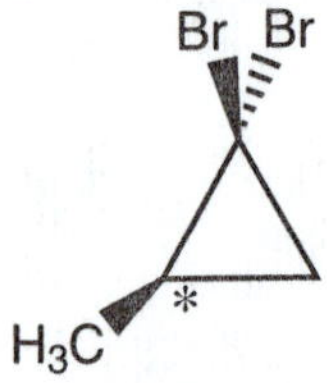

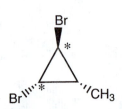

(*R*)-1,1-Dibromo-2-methylcyclopropane   (1*R*,2*R*)-1,2-Dibromo-1-methylcyclopropane

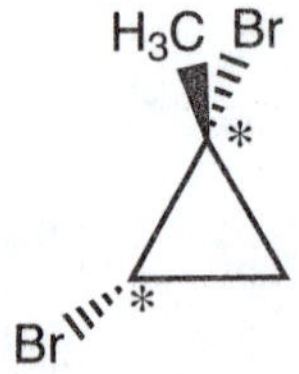

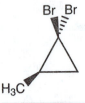

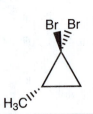

(1*R*,2*S*)-1,2-Dibromo-1-methylcyclopropane   (1*S*,2*S*)-1,2-Dibromo-3-methylcyclopropane

Problem 4.56   The (*R*) configuration of 1,1-dibromo-2-methylcyclopropane was drawn in the answer key for Problem 4.55. Both enantiomers are shown here. There is only one stereogenic carbon. There are no diastereomers of this molecule.

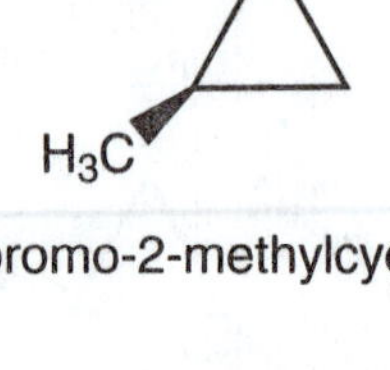

(*R*)-1,1-Dibromo-2-methylcyclopropane   (*S*)-1,1-Dibromo-2-methylcyclopropane

Problem 4.57

(a) (2*R*,4*S*)-2,4-dibromo-4-chlorohexane

(b) 4,4-dimethyl-1-pentene

(c) (*E*,5*R*,8*S*)-5-ethyl-8-iodo-3-methyl-2-nonene

(d) (2*R*,6*R*)-2-chloro-6-fluoroheptane

(e) (*Z*)-3-bromo-4-methyl-3-hexene

(f) (*R*)-2,2-dichloro-3-ethylhexane

(g) (*Z*,4*R*)-4-bromo-2-pentene

Problem 4.58   Had the poison been gathered from nature, it would have been optically active. The sleuths got out their trusty polarimeter and determined that the poison was racemic—it could not have come from a mushroom. Lathom was hanged. He should have studied his orgo more.

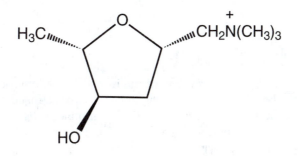

Problem 4.59   The stereochemical outcome in each reaction is inversion. Because the carbon goes from one tetrahedral configuration to another, it must invert along the way.

Problem 4.60   The planar carbon is sp^2 hybridized. It has an unoccupied p orbital. If a stereogenic carbon underwent a reaction by losing a bromide, it would no longer be stereogenic. It would be flat, and the molecule would no longer be chiral (if the reacting carbon was the only stereogenic carbon in the molecule).

Problem 4.61   The hydrogen is anti to the bromide leaving group. This relationship is critically important. It is only the electrons of the anti C—H bond that can move into the available σ* orbital at the backside of the C—Br bond.

Problem 4.62   Yes, the starting material is chiral. The configuration is (*S*). The product is achiral. This animation shows a chiral molecule undergoing a reaction that leads to an achiral product.

## 5

Rings

Problem 5.1 First look at planar cyclopentane in which the five top carbon–hydrogen bonds (as well as the bottom carbon–hydrogen bonds) are eclipsed. Once again, offset your eye just a bit so you can see the eclipsed bonds and atoms in the back. As in cyclobutane (Fig. 5.9), puckering of the five-membered ring relieves some of the eclipsing of the carbon–hydrogen bonds, and thus relieves torsional strain.

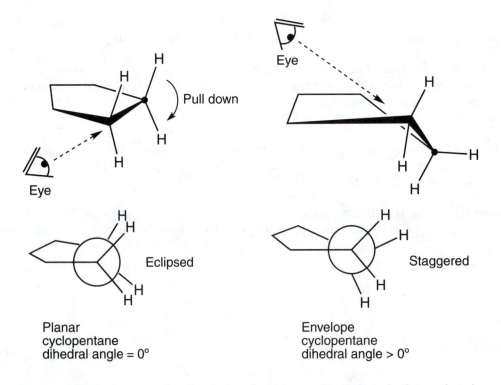

Planar
cyclopentane
dihedral angle = 0°

Envelope
cyclopentane
dihedral angle > 0°

Note how puckering opens up the dihedral angle between the carbon–hydrogen bonds

Problem 5.3 There is only one kind of carbon in cyclohexane, and thus only one signal in the ^{13}C NMR spectrum. Get out your models and be sure you see that the six carbons are all equivalent.

Problem 5.4   The answer would seem at first to be "two signals": one for the set of six equivalent axial hydrogens and another for the set of six equivalent equatorial hydrogens.

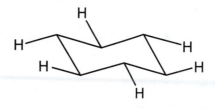

The set of six equivalent axial hydrogens   The set of six equivalent equatorial hydrogens

But if there is enough energy, these two sets interconvert, and there will be only one average signal in the ^{1}H NMR spectrum. In practice, at low temperature, two separate signals are observed, and at higher temperature (about room temperature), one averaged signal appears.

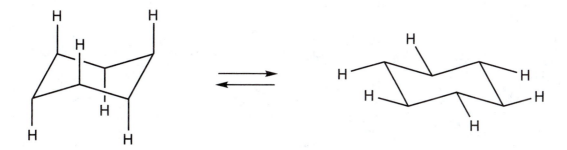

Problem 5.5   Presumably, taking the difference in heat of formation between two compounds that differ only by one methylene group will give the heat of formation of a single methylene. In practice, one takes the average of many such determinations. In this example, heptane ($\Delta H_f^\circ = -44.8$) – hexane ($\Delta H_f^\circ = -39.9$) gives –4.9; octane ($\Delta H_f^\circ = -49.8$) – heptane ($\Delta H_f^\circ = -44.8$) gives –5.0; nonane ($\Delta H_f^\circ = -54.5$) – octane ($\Delta H_f^\circ = -49.8$) gives –4.7; decane ($\Delta H_f^\circ = -59.6$) – nonane ($\Delta H_f^\circ = -54.5$) gives –5.1. The average of these determinations is –4.9 kcal/mol.

Problem 5.6   You might want to use your model set to help analyze this question. You will notice in column 4 of Table 5.1 that there is considerable strain associated with the small rings. Cyclopropane and cyclobutane have almost the same amount of strain (about 27 kcal/mol). Cyclopentane and cycloheptane have about 6.0 kcal/mol strain, but as you make a model, you will realize that this strain is not the same kind as in cyclopropane and cyclobutane. The 5- and 7-membered rings have very similar eclipsing interactions. The 8-membered to 12-membered rings (cyclooctane to cycloundecane), however, are higher in strain energy for a third reason. They have van der Waals strain. The 8-membered ring has 9.5 kcal/mol strain energy (see column 4 of Table 5.1). There is no ring strain for this ring size. The ring is sufficiently flexible to avoid eclipsing strain. Let's assume that most of the 9.5 kcal/mol is due to van der Waals strain. You should make a model, then you

(continued)

Problem 5.6 *(continued)*

will see that there are many conformations for cyclooctane. It is very floppy. The best bond angles (fewest gauche interactions) are obtained with an extended chair-like structure, which has three pairs of CH_2 groups interacting across the ring from each other. If that is the only cause of strain, then each CH_2—CH_2 across the ring interaction produces about 3.2 kcal/mol of strain.

Now consider cyclodecane, which has four pairs of CH_2 groups interacting with each other. We see from column 4 that cyclodecane has 12.1 kcal/mol of strain. We would have predicted about 12.8 kcal/mol strain (four pairs of CH_2 groups at 3.2 kcal/mol per pair). This tells us that we aren't too far off.

Problem 5.7   This figure resembles Figure 5.21. The three isomeric alkanes all react with oxygen to give 7 CO_2 + 8 H_2O. The product of combustion of all three hydrocarbons is exactly the same. Therefore, the differences in their heats of combustion give the differences in their energies.

$$C_7H_{16} + 11\ O_2 \longrightarrow 7\ CO_2 + 8\ H_2O$$

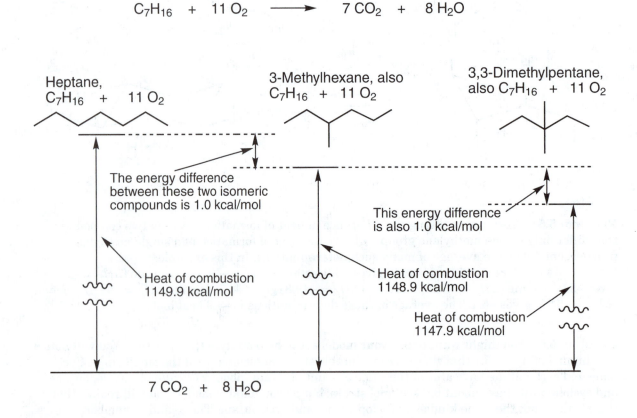

Problem 5.10   The gauche butane from Figure 5.28 has the ability to relieve some of the steric interactions by rotating the front carbon of the Newman projection slightly to counterclockwise, as shown below. The methyl–methyl interaction goes from having a 60° angle between the groups to something closer to a 65° angle.

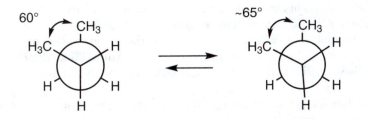

Rotating the front carbon much further leads to the higher energy eclipsing conformation. But going from a 60° angle to about 65° is energetically favorable.

Now look at the Newman projection for axial methylcyclohexane. As we rotate the front carbon counterclockwise, we reduce the methyl-ring gauche interaction. But at the same time we are increasing the ring–ring gauche interaction. So there is no reduction of overall torsional strain for this system.

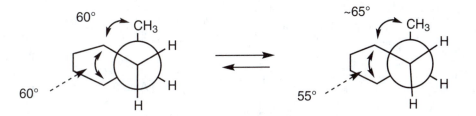

Problem 5.11 $\Delta G = -RT\ln K$, where R is about 2 cal/deg•mol, which is equal to about 0.002 kcal/deg•mol; $T = 25\ °C = 298\ K$; and $\Delta G = 2.8$ kcal/mol.

$2.8 = -(0.002)(298) \ln K$
$-4.70 = \ln K$
$9.1 \times 10^{-3} = K$

So, 2.8 kcal/mol translates into a mixture of products in which the ratio is 99.1 : 0.9.

Problem 5.13   First of all, we have to draw the molecule and be careful to keep the *tert*-butyl group equatorial. There will be a negligible amount present of the chair with the *tert*-butyl group axial. The *tert*-butyl group has two different kinds of carbon, and there are four different ring carbons. So we will see a total of six signals in the ^{13}C NMR spectrum.

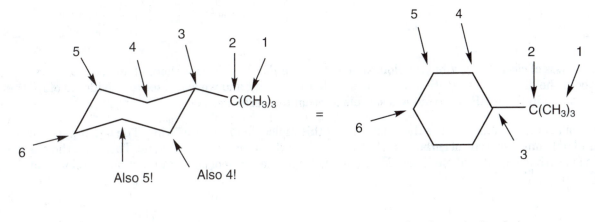

A three-dimensional view   A schematic, "top" view

Problem 5.14 The analysis of *cis*-1,3-dimethylcyclohexane requires that we think about the preferred conformation. Although there will be some diaxial conformer at room temperature, we expect that most of the molecules will be in the diequatorial conformation at room temperature. The two methyl groups in the axial position would produce more than 3.5 kcal/mol of steric strain (see Problem 5.21). So we will use the diequatorial isomer in our NMR analysis.

cis-1,3-Dimethylcyclohexane is a bit challenging to analyze in the chair conformation. It is much easier to see the symmetry for this molecule in the perspective or "top" view. There is a plane of symmetry through the center of the molecule. As a result, the methyl groups are equivalent (labeled number 1). The carbons of the ring that are attached to the methyl groups are equivalent (labeled number 2). The carbons labeled number 3 are also equivalent. The result is five different carbons in *cis*-1,3-dimethylcyclohexane.

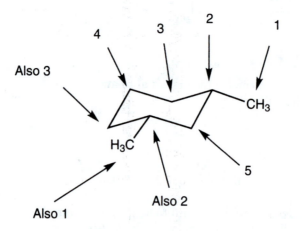

A chair conformation view

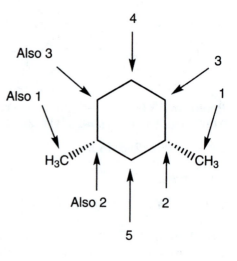

A "top" view with perspective

Analysis of *cis*-1,3-di-*tert*-butylcyclohexane is very similar. In this case there is no need to worry about the diaxial conformation. It is much higher in energy, and we won't see any evidence of its presence in the NMR spectrum of a sample at room temperature.

Once again, it is easier to see the symmetry of this molecule in the "top" view. The *tert*-butyl groups add only one new type of carbon to what we found in the dimethylcyclohexane. Therefore the answer is six carbon signals will be observed in the ^{13}C NMR spectrum of *cis*-1,3-di-*tert*-butylcyclohexane.

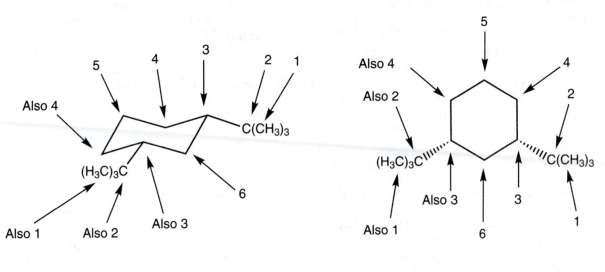

A chair conformation view   A "top" view with perspective

The all-*cis*-1,3,5-tri-*tert*-butylcyclohexane also has symmetry. The chair conformation view displays the symmetry fairly well. Perhaps you can see it better in the "top" view. You can see that there are only four different carbons in this molecule. So the ^{13}C NMR spectrum of all-*cis*-1,3,5-tri-*tert*-butylcyclohexane will have four signals.

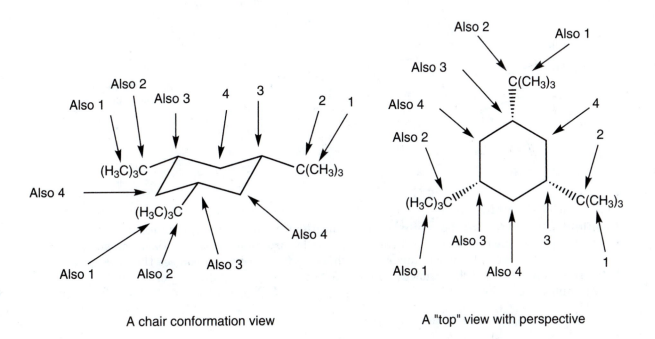

A chair conformation view   A "top" view with perspective

One point of this problem is to help you see the symmetry of the substituted cyclohexanes. You should also become comfortable twisting and turning the molecules so that the symmetry is easier to see.

Problem 5.15   In 1-isopropyl-1-methylcyclohexane, either the methyl or isopropyl group can be equatorial, but not both. Your intuition might tell you that the conformation with the larger isopropyl group equatorial will be preferred, but you will need to consult Table 5.3 to do the quantitative calculation. Table 5.3 tells us that a cyclohexane with an equatorial isopropyl group is more stable than a cyclohexane with an axial isopropyl group by 2.61 kcal/mol. Similarly, a cyclohexane with an equatorial methyl is 1.74 kcal/mol more stable than a cyclohexane with an axial methyl group. The ring flip of 1-isopropyl-1-methylcyclohexane involves the interconversion of equatorial and axial isopropyl and methyl groups.

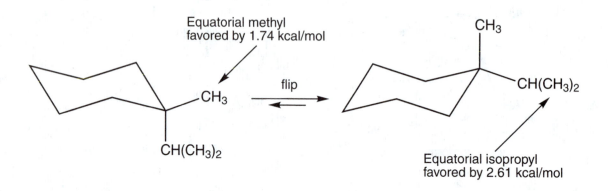

We calculate 2.61 – 1.74 = 0.87 kcal/mol in favor of the isomer with the equatorial isopropyl group.

Problem 5.16   As the Newman projection shows, this dihedral angle is 180°.

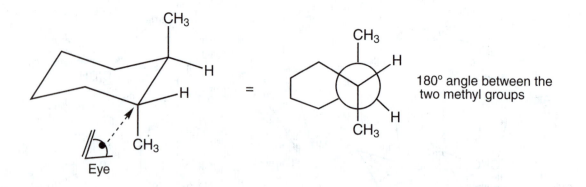

Problem 5.17   In *trans*-1,2-dimethylcyclohexane, the ring flip converts the diequatorial stereoisomer into the diaxial form. These two molecules are conformational diastereomers, not enantiomers. There can be no racemization in this ring flip. Be certain you see the difference between this process and that shown in Figure 5.39 for similar ring flipping of *cis*-1,2-dimethylcyclohexane.

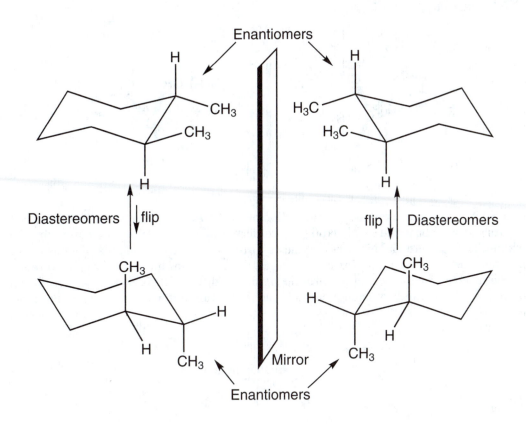

Problem 5.18 The two stereogenic carbons are shown in bold and lightface. The priorities used to determine (R) or (S) are also shown in bold and lightface, as is appropriate for each carbon.

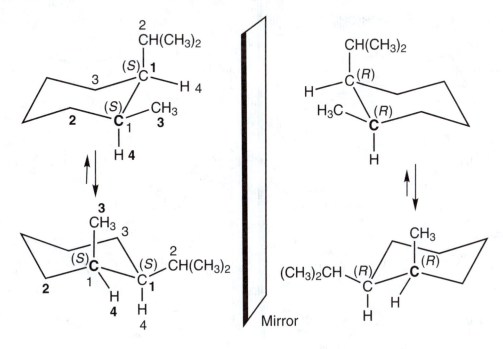

(continued)

Problem 5.18 *(continued)*

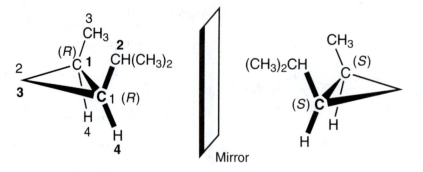

Problem 5.19 The axial, axial (aa) conformation of (1*S*,2*R*)-*trans*-1-isopropyl-2methylcyclohexane is chiral. Its enantiomer is the axial, axial (aa′) conformation of (1*R*,2*S*)-*trans*-1-isopropyl-2-methyl-cyclohexane. The diequatorial isomers (ee and ee′) are enantiomers. The aa and ee′ isomers are diastereomers, as are the aa′ and the ee molecules. Cyclohexanes aa and ee are diastereomers, but they can be interconverted by rotation around sigma bonds (ring flip). So they are conformational diastereomers. The aa′ and ee′ isomers are also conformational diastereomers for the same reason.

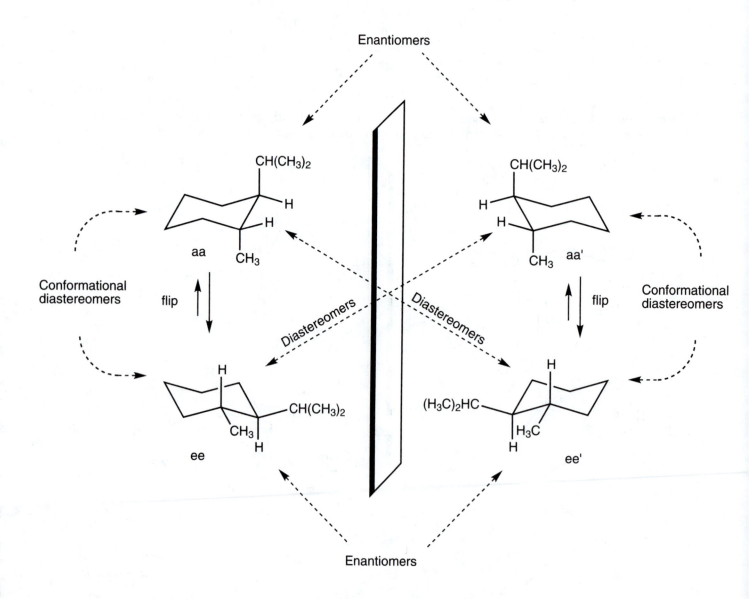

Problem 5.21 Table 5.3 shows that a cyclohexane with an equatorial methyl group is favored over a cyclohexane with an axial methyl group by 1.74 kcal/mol. In this ring flip, two equatorial methyl groups interconvert with two axial methyl groups. Accordingly, we expect the diequatorial isomer to be favored by at least 2 × 1.74 = 3.48 kcal/mol. There will be an additional serious destabilizing interaction between the two 1,3-diaxial methyl groups. The total energy difference between the two is about 5.5 kcal/mol.

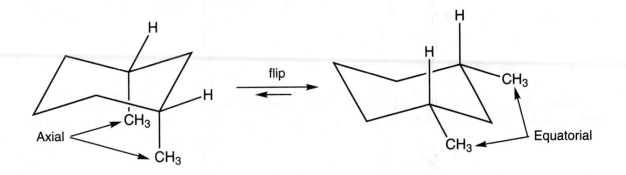

Problem 5.23 The *cis*-bicyclo[2.2.0]hexane can be represented without wedges and dashes or, perhaps more clearly, with the wedges and dashes.

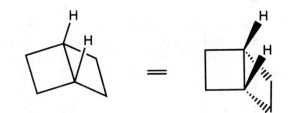

cis-Bicyclo[2.2.0]hexane

The trans isomer is more difficult to represent. It certainly is painful to the mind's eye. Perhaps you can see why it has never been isolated. The connection between the two atoms coming off the ring is too stretched.

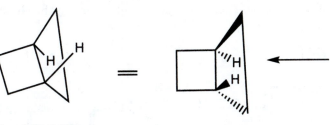

Trans connection requires a long bond between the two atoms coming off the ring

trans-Bicyclo[2.2.0]hexane

(continued)

Problem 5.23 *(continued)*

There are two fused bicyclooctanes. One is *trans*-bicyclo[4.2.0]octane. It is a stable molecule because there are two additional carbons connecting the atoms that connect to the four-membered ring. This significantly reduces the ring strain.

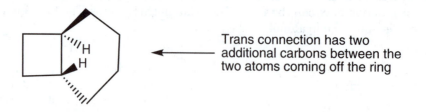

Trans connection has two additional carbons between the two atoms coming off the ring

trans-Bicyclo[4.2.0]octane

The other bicyclooctane isomer is a [3.3.0] structure. The drawing of *trans*-bicyclo[3.3.0]octane shows that the extra carbon allows for an easier connection between the atoms coming off the ring.

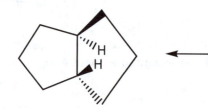

Trans connection has an extra carbon between the two atoms coming off the ring

trans-Bicyclo[3.3.0]octane

A fused four-membered ring and five-membered ring gives bicyclo[3.2.0]heptane. The trans fused ring has the ability to make a connection between the trans atoms coming off the ring. This molecule is 24 kcal/mol higher in energy than its cis isomer. There is ring strain owing to the presence of the four-membered ring. But it has been isolated.

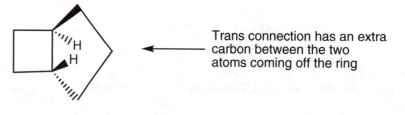

Trans connection has an extra carbon between the two atoms coming off the ring

trans-Bicyclo[3.2.0]heptane

Problem 5.24 Get out your models! In the trans compound, ring flip requires that the second ring be connected through a pair of axial bonds. There are not enough atoms in the chain to do this.

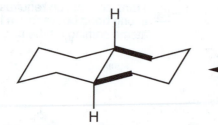

Too long a distance to span

For this molecule to "flip," the two darkened bonds must become axial; it is not possible to connect the second ring in this way

By contrast, in the cis compound, the junction is made through one axial and one equatorial bond. Ring flip interchanges these two and is quite easy.

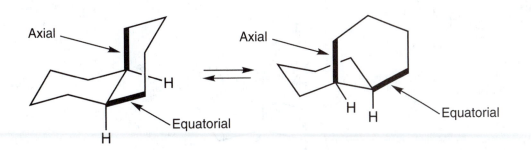

Problem 5.26 The directions for constructing bicyclic molecules are in the chapter. In (a), we connect the bridgeheads (always start with the bridgehead atoms) with two three-carbon bridges and one no-carbon bridge. In (b), all the bridges are the same, and in (c), there are a six-carbon bridge, a one-carbon bridge, and a no-carbon bridge. Be careful in this example to note that the bridge junction is trans. In (b), we place the fluorine at the 1-position, the bridgehead. In (c), we count around the longest bridge first, which makes the free cyclopropane position the 9-position.

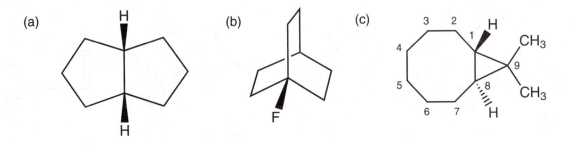

Problem 5.27

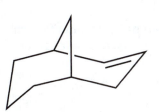

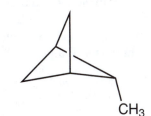

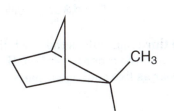

Bicyclo[3.3.1]non-2-ene   2-Methylbicyclo[1.1.1]pentane   5,5-Dimethylbicyclo[2.1.1]hexane

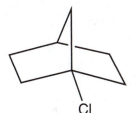

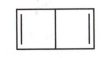

1-Chlorobicyclo[2.2.1]heptane   Bicyclo[2.2.0]hexa-2,5-diene

Problem 5.28   In the first molecule, all the carbons are different. There would be nine signals. If that double bond were not present, the situation would be very different. Then the molecule could be factored into two halves—there would be a plane of symmetry, and there would be only four signals.

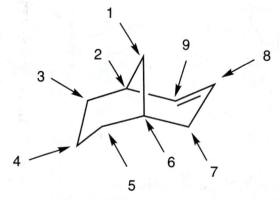

That is essentially what happens in three of the other molecules. In each, there is at least one plane of symmetry, and several carbons appear in pairs of equivalent positions:

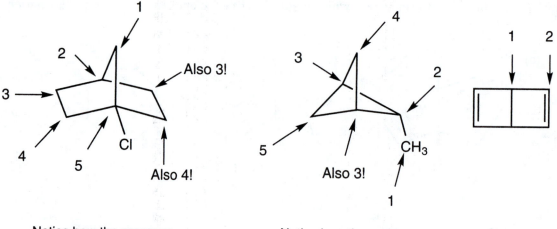

Notice how the presence
of the chlorine makes
2 and 5 and 3 and 4 different

Notice how the presence
of the methyl group
makes 4 and 5 different

The third molecule is the most difficult. The two methyl groups are different (one points toward a CH_2 and the other toward a CH_2—CH_2), but there is still a plane of symmetry, and four carbons appear as two equivalent pairs.

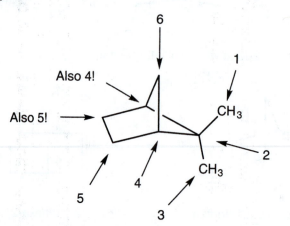

Problem 5.29 This molecule has all sorts of problems. Of course, it contains no fewer than three cyclopropane rings, and that alone will introduce severe strain. However, the factor that differentiates this molecule from the "merely" severely strained is that all four valences of the "bridgehead" carbons are aimed in the same direction! What kind of hybridization can be involved? That's no easy question, and the chemical world has spent quite some effort at puzzling out just why this molecule is as stable as it is. Although that's not a reasonable subject for an introductory book, two things about this problem are important. First, we can see why this molecule is so remarkable. Second, the isolation of this molecule shows how imperfect our understanding of chemistry still is! However, one must admit that it is a pleasure to see challenging curiosities continue to appear. There is much to learn, and many molecular marvels are still hiding from us.

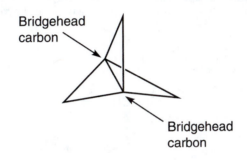

Problem 5.30 The empirical formula of diamond is C.

Problem 5.31 There are three possibilities for tetramantane. This problem is nearly impossible to do without models at this point in your career.

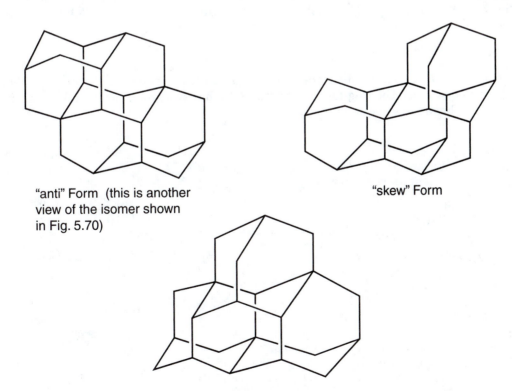

"anti" Form (this is another
view of the isomer shown
in Fig. 5.70)

"skew" Form

"iso" Form

Additional Problem Answers

This chapter deals almost exclusively with the structures of ring compounds, and six-membered rings (cyclohexanes) are examined especially closely. The structural consequences of the ring flip of cyclohexanes (really just a series of rotations around carbon–carbon bonds) are emphasized, and the following problems should give you lots of practice in drawing, flipping, and evaluating substituted cyclohexanes.

Problem 5.32 The three isomers of all-*cis*-tetramethylcyclohexane are found by trial and error. Start with the 1,2,3,4-tetramethylcyclohexane, then move one methyl group from C(4) to C(5). That gives 1,2,3,5-tetramethylcyclohexane. If we move the C(5) methyl to C(6), that will produce 1,2,3,4-tetramethylcyclohexane again. Next we can try moving two methyl groups. That gives 1,2,4,5-tetramethylcyclohexane.

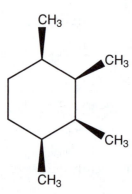

All-*cis*-1,2,3,4-
tetramethylcyclohexane

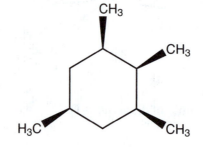

All-*cis*-1,2,3,5-
tetramethylcyclohexane

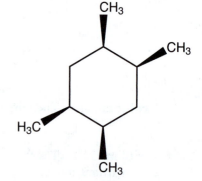

All-*cis*-1,2,4,5-
tetramethylcyclohexane

The chair structure for all-*cis*-1,2,3,4-tetramethylcyclohexane has two methyl groups axial and two methyl groups equatorial. The ring-flipped structure also has two groups axial and two groups equatorial. They have the same stability.

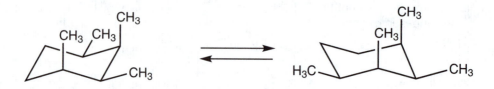

One of the chair structures of all-*cis*-1,2,3,5-tetramethylcyclohexane has three methyl groups equatorial and one methyl group axial. The ring-flipped structure would be much higher in energy because it has three methyls axial and one methyl equatorial.

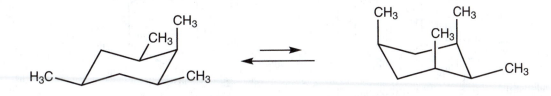

More stable conformation

The chair structures for the all-*cis*-1,2,4,5-tetramethylcyclohexane are also the same energy. Both chair conformations have two methyls axial and two methyls equatorial. They are identical structures.

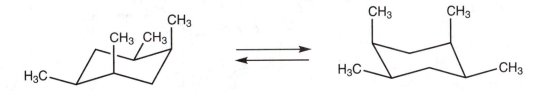

Problem 5.33 Both the 1,2,3,4-tetramethyl and the 1,2,4,5-tetramethyl isomers have two groups axial. The more stable isomer of the 1,2,3,5-tetramethylcyclohexane has only one methyl group axial. In addition, the methyl–methyl 1,3-diaxial interaction induces more van der Waals strain. Another consideration is the number of methyl–methyl gauche interactions. The 1,2,3,4-tetramethyl isomer has three methyl–methyl gauche interactions. Both the 1,2,3,5- and the 1,2,4,5-tetramethylcyclohexane isomers have two methyl–methyl gauche interactions. Therefore, the most stable isomer would be the all-*cis*-1,2,3,5-tetramethylcyclohexane.

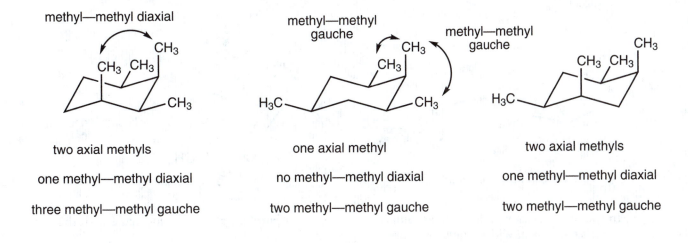

two axial methyls

one methyl—methyl diaxial

three methyl—methyl gauche

one axial methyl

no methyl—methyl diaxial

two methyl—methyl gauche

two axial methyls

one methyl—methyl diaxial

two methyl—methyl gauche

Problem 5.34

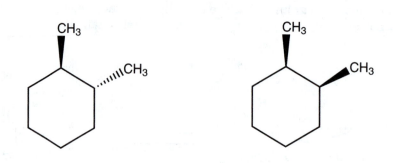

(*R,R*)-1,2-Dimethylcyclohexane   (*R,S*)-1,2-Dimethylcyclohexane

The bottle must contain (*R,S*)-1,2-dimethylcyclohexane because the (*R,R*) isomer would be optically active. The (*R,S*) isomer, however, is a meso molecule and therefore achiral. It would have no optical activity.

Problem 5.35

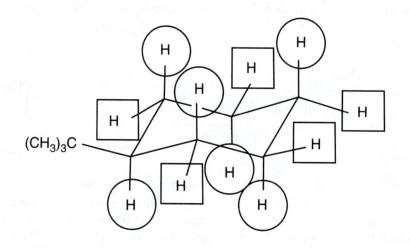

Problem 5.36   In the cis isomer, one methyl group must be equatorial and the other axial. Ring flip of one isomer produces an identical molecule. To see this, rotate the right-hand structure 180°, as shown.

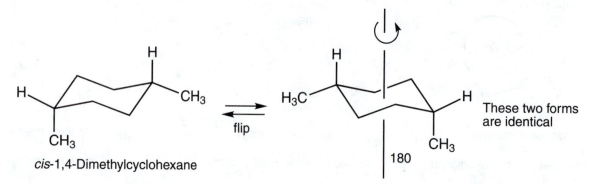

cis-1,4-Dimethylcyclohexane

In the trans isomer, both methyl groups are equatorial or both are axial. The molecule with the two groups equatorial will be far more stable.

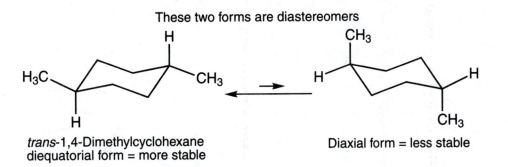

trans-1,4-Dimethylcyclohexane
diequatorial form = more stable

Diaxial form = less stable

Neither the cis nor trans compound is chiral. Both mirror images are superimposable on the original.

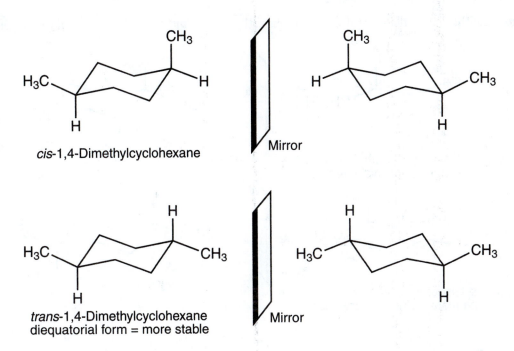

cis-1,4-Dimethylcyclohexane

Mirror

trans-1,4-Dimethylcyclohexane
diequatorial form = more stable

Mirror

Problem 5.37 In the cis form, it will be the diastereomer with the large isopropyl group equatorial that is favored.

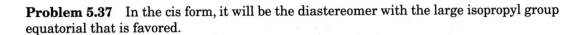

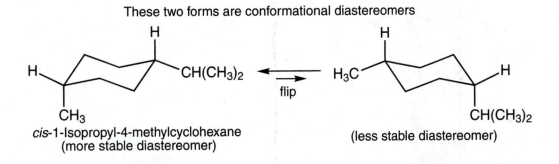

cis-1-Isopropyl-4-methylcyclohexane
(more stable diastereomer)

(less stable diastereomer)

(continued)

Problem 5.37 *(continued)*

Similarly, in the trans form, it is the diastereomer with both groups equatorial that is more stable.

These two forms are conformational diastereomers

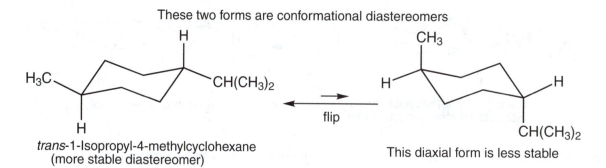

trans-1-Isopropyl-4-methylcyclohexane
(more stable diastereomer)

flip

This diaxial form is less stable

Once again, neither molecule is chiral, as in each case, the mirror image is superimposable on the original.

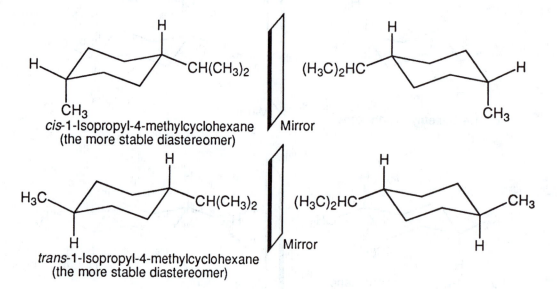

cis-1-Isopropyl-4-methylcyclohexane
(the more stable diastereomer)

Mirror

trans-1-Isopropyl-4-methylcyclohexane
(the more stable diastereomer)

Mirror

Problem 5.38

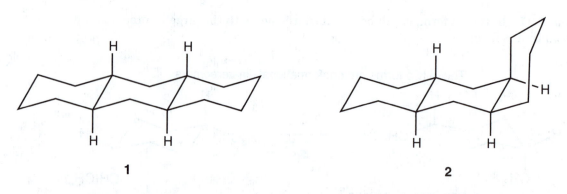

1

2

These molecules are diastereomers. There are four stereogenic carbons in each, and only one of the stereogenic carbons is switched (*S* to *R*) comparing compound **1** to compound **2**.

Isomer **1** is more stable than isomer **2** because isomer **1** has no axial groups. Isomer **2** has one carbon in an axial position.

Isomer **1** is achiral. It has a plane of symmetry. Isomer **2** is chiral. It has no plane or point of symmetry.

Problem 5.39 The more stable conformation of *cis*-1-isopropyl-3-methylcyclohexane is the one with the two alkyl groups equatorial. Ring flip generates the less stable conformation with the two alkyl groups axial. Each of these two diastereomers is chiral.

Neither mirror image is superimposable on the original—both diastereomers are chiral.

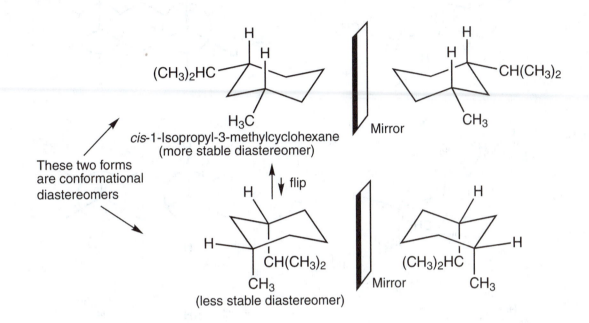

cis-1-Isopropyl-3-methylcyclohexane
(more stable diastereomer)

These two forms are conformational diastereomers

flip

(less stable diastereomer)

The situation is the same in the trans molecule. The more stable conformation has the large isopropyl group equatorial. Ring flip generates the less stable conformation with the large isopropyl group axial. Each of these two diastereomers is chiral.

Neither mirror image is superimposable on the original—both diastereomers are chiral.

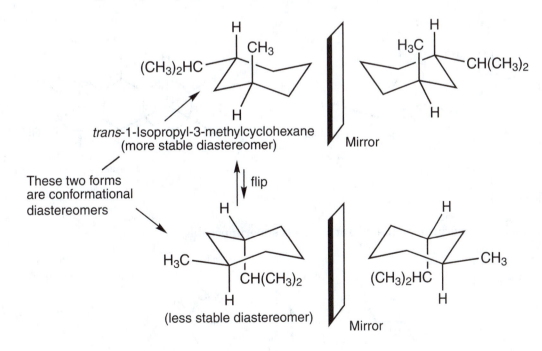

trans-1-Isopropyl-3-methylcyclohexane
(more stable diastereomer)

These two forms are conformational diastereomers

flip

(less stable diastereomer)

Problem 5.40

(a)

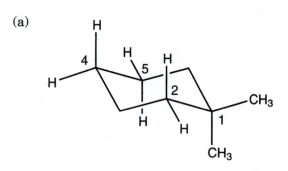
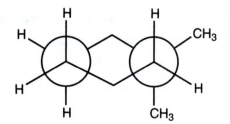

(b)

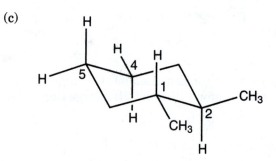
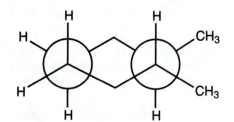

(c)

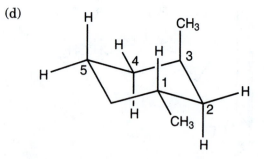
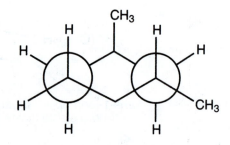

(d)

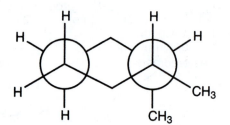

Problem 5.41

(a) This molecule is achiral as the mirror image is superimposable on the original.

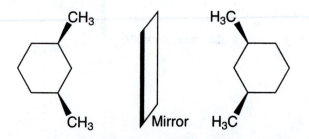

(b) This one is chiral—the mirror image is nonsuperimposable.

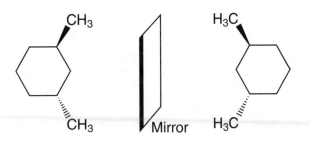

(c) This molecule is chiral. Its mirror image is nonsuperimposable. There is no plane of symmetry in the molecule.

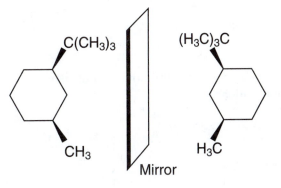

(d) This molecule is chiral. Its mirror image is nonsuperimposable. There is no plane of symmetry in the molecule.

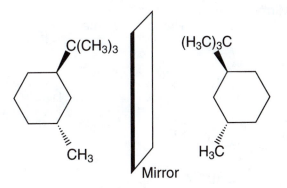

(continued)

(e) *cis*-1,4-Dibromocyclohexane is achiral. Its mirror image is identical to the original. There is a plane of symmetry in the molecule.

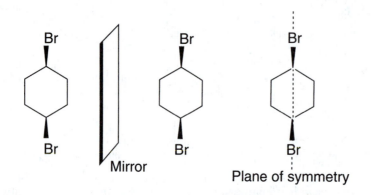

(f) *trans*-1,4-Dibromocyclohexane is achiral. Its mirror image is identical to the original. There is a plane of symmetry in the molecule.

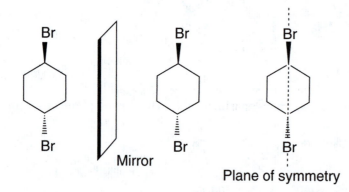

(g and h) *cis*-1-Chloro-4-methylcyclohexane and *trans*-1-chloro-4-methylcyclohexane are both achiral. They both have a plane of symmetry in the molecule.

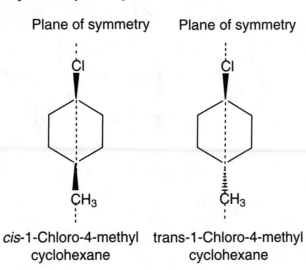

Plane of symmetry   Plane of symmetry

cis-1-Chloro-4-methyl   *trans*-1-Chloro-4-methyl
cyclohexane   cyclohexane

Problem 5.42

(a)

 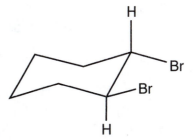

(b)

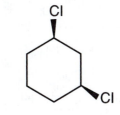 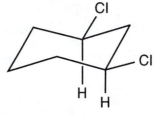

(c)

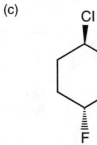 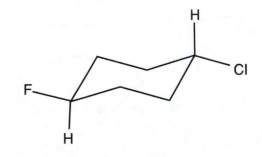

Problem 5.43

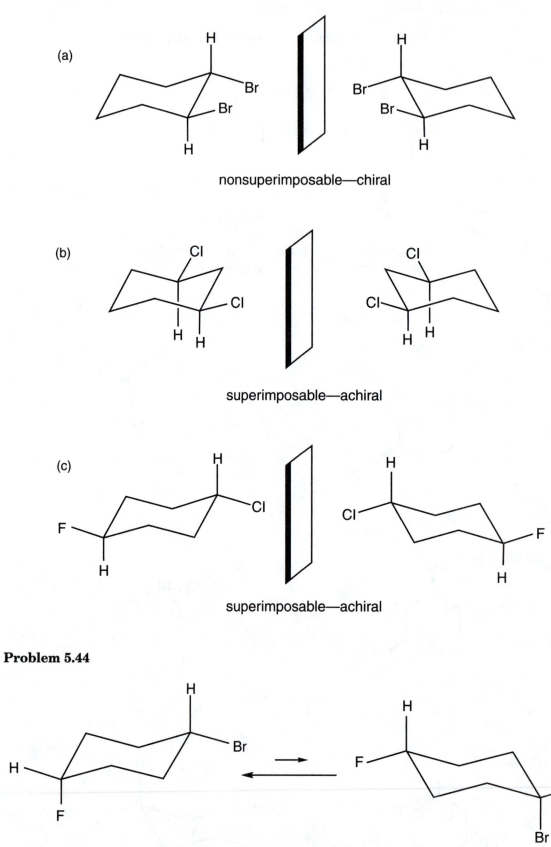

(a)

nonsuperimposable—chiral

(b)

superimposable—achiral

(c)

superimposable—achiral

Problem 5.44

On the left, the larger Br is equatorial. Thus, this form is favored at equilibrium.

Problem 5.45

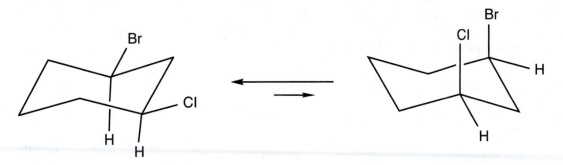

On the left, both large groups are equatorial, and this form is far more stable than the flipped form on the right. It is also chiral.

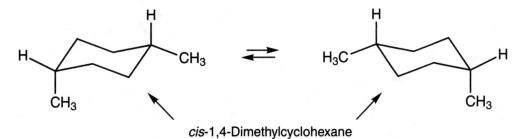

nonsuperimposable—chiral

Problem 5.46 *cis*-1,4-Dimethylcyclohexane "flips" into itself. Thus there can be no energy difference between the two isomers.

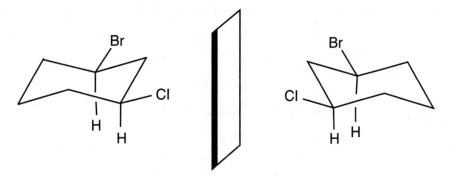

cis-1,4-Dimethylcyclohexane

In *trans*-1,4-dimethylcyclohexane, the molecule with two equatorial methyl groups will be more stable by twice the energy difference between equatorial and axial methylcyclohexane, or $2 \times 1.74 = 3.48$ kcal/mol (Table 5.3, p. 207).

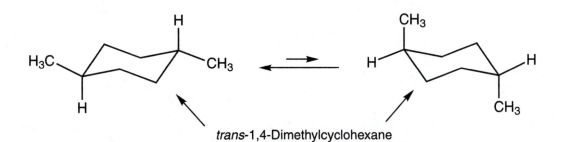

trans-1,4-Dimethylcyclohexane

Problem 5.47   Let's do the trans compound first. In this molecule, both substituents are equatorial or both are axial. Table 5.3 tells us that an axial methyl group is destabilizing by 1.74 kcal/mol and that an axial isopropyl group is destabilizing by 2.61 kcal/mol. Accordingly, the chair with both groups equatorial will be more stable than the chair with both groups axial by the sum of these numbers, 4.35 kcal/mol.

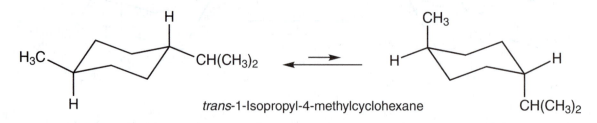

trans-1-Isopropyl-4-methylcyclohexane

In the cis isomer of 1-isopropyl-4-methylcyclohexane, one chair has the methyl group axial, and the other has the isopropyl group axial. The right-hand diastereomer will be more stable by 2.61 − 1.74 = 0.87 kcal/mol. It is energetically more favorable to have the larger isopropyl group equatorial.

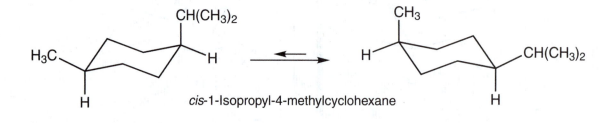

cis-1-Isopropyl-4-methylcyclohexane

Problem 5.48   Stereogenic atoms (nine, all carbons) are shown with an asterisk.

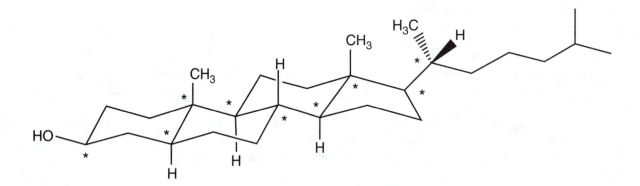

Problem 5.49 The question asks you to decide first what effect the increase in temperature can have. Cyclohexane chairs interconvert at room temperature and above, but the chair–chair flip can be stopped at low temperature. It would therefore seem that we must consider only one chair at low temperature and both chairs at high temperature.

(a) Low temperature. The chair form will be as follows, and there will be two different methyl groups: one axial, the other equatorial. There will be a total of six signals for the six different carbons.

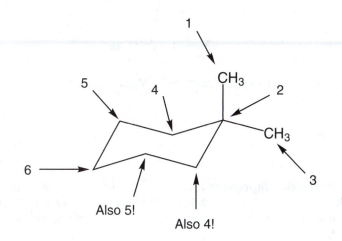

(b) High temperature. Now the two possible chairs will interconvert, and we must consider the averaged spectrum. There will now be only one methyl group, as the ring flip interconverts axial and equatorial positions.

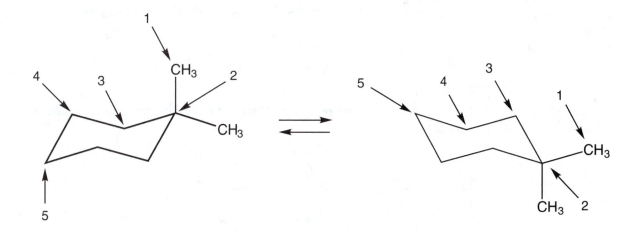

For our tally of different carbons, we can even use the averaged, flat structure:

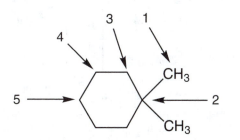

Problem 5.50

(a) *trans*-1,3-Dimethylcyclohexane will have eight different carbons at low temperature. Perhaps you can see that there are no two carbons that are equivalent when this molecule is "frozen" in one conformation. There is no symmetry.

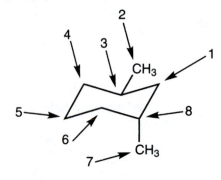

(b) At high temperature, the ring flipping will be fast, and that will make the methyl groups equivalent. If you look carefully at the two conformations **A** and **B** of *trans*-1,3-dimethylcyclohexane, you will see that they are identical. Rotate structure **B** 180° about the axis shown, and you will see that it is the same as **A**.

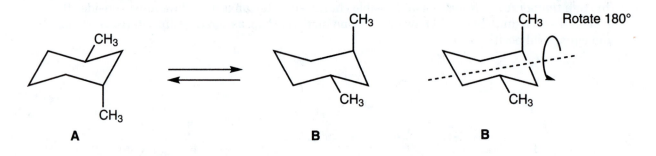

The drawing of the flat structure (top view with perspective) should make this point clearer. There will be five ^{13}C NMR signals at high temperature.

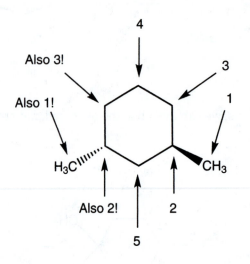

Problem 5.51 This one may also be hard to see. Adamantane has only two different carbons, the CH and the CH_2. Symmetry makes the four methine groups equivalent and the six methylene groups equivalent.

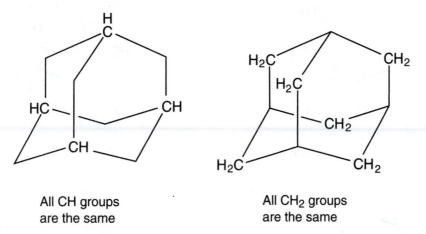

All CH groups
are the same

All CH_2 groups
are the same

Problem 5.52

(a) We start with a tough call. The conformation with two groups equatorial (Cl and CH_3) is the more stable one, even though the largest group, isopropyl, must be axial.

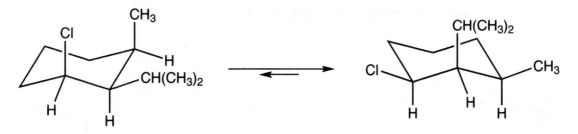

(b) This one is easy. The conformation with all three groups equatorial is the more stable one.

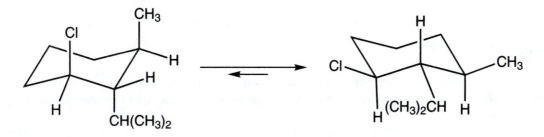

(c) Once again, the conformations with two groups equatorial are more stable than the conformations with two groups axial. Note that in (c), each conformational diastereomer has a gauche isopropyl–methyl interaction.

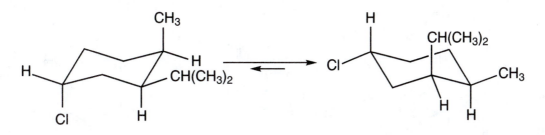

(continued)

Problem 5.52 *(continued)*

(d) Here the conformational diastereomer with isopropyl and methyl equatorial is favored, despite the gauche isopropyl–methyl interaction.

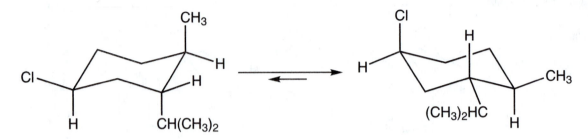

(e) The conformation with all three groups equatorial is the more stable one.

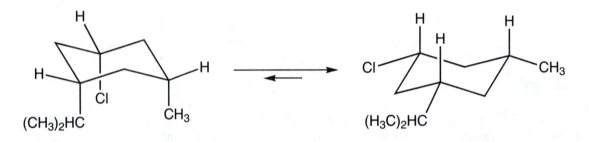

(f) The conformation with two groups equatorial is the more stable one.

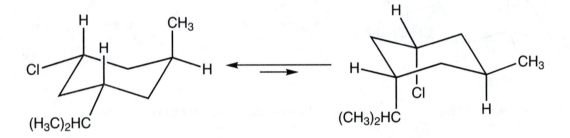

Problem 5.53

(a) All three bridges have two carbons, so the base name is bicyclo[2.2.2]octene. We start counting at the bridgehead and give the double bond the lowest possible number, so the final name is bicyclo[2.2.2]oct-2-ene.

(b) We count first around the larger ring, so this compound is 2,2-difluorobicyclo[2.1.0]pentane.

(c) This compound is a bicyclo[3.1.1]heptene. We count around the largest bridge first and give the OH the lowest number. The molecule is bicyclo[3.1.1]hept-3-en-2-ol.

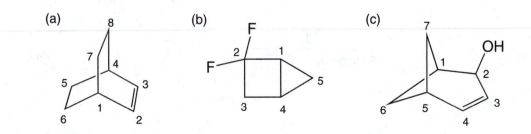

Problem 5.54 In (a), (b), and (c), remember to start counting at the bridgehead. In (c), no numbers are necessary for the methyl groups because this is the only possible hexamethyl compound.

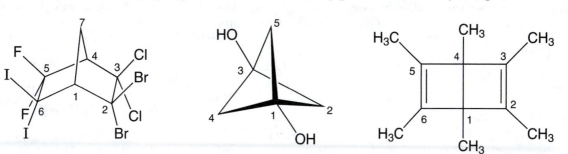

Problem 5.55 First, draw a chair with the oxygen taking the place of one of the ring carbons of cyclohexane. Then add the substituents, putting them all in the more favorable equatorial positions.

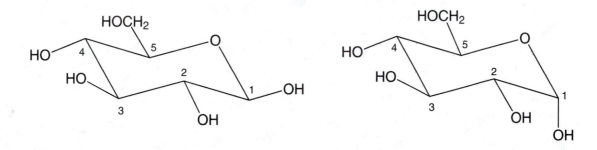

The OH at C(1) can be either "up," in the equatorial position, or "down," in the axial position.

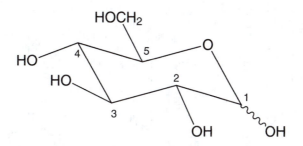

Problem 5.56 There are three different carbons in bicyclo[2.2.1]heptane, and a methyl group could be attached to each of them. At the 1- or 7-position, there is only one possible way of attaching a group, but at the 2-position, the methyl (or any group) could point toward the short bridge or the longer bridge. These positions are called "exo" and "endo," respectively. The endo and exo isomers are diastereomers.

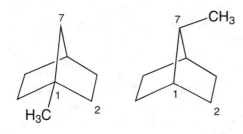

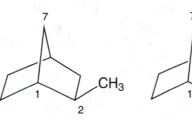

exo-2-Methylbicyclo-
[2.2.1]heptane

endo-2-Methylbicyclo-
[2.2.1]heptane

Problem 5.57

(a) No, there are no exo and endo forms here, as the two sides of the molecule are identical. There are only two structural isomers of methylbicyclo[2.2.2]octane.

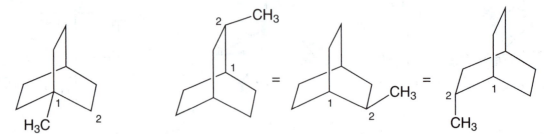

1-Methylbicyclo[2.2.2]octane   All three structures are 2-methylbicyclo[2.2.2]octane

(b) Yes, 2-methylbicyclo[2.2.2]octane is chiral. Its mirror image is not superimposable on the original.

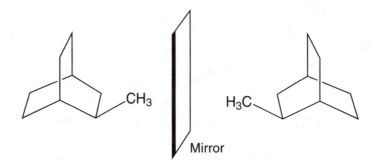

Mirror

Problem 5.58

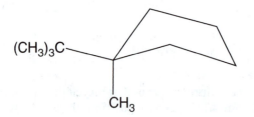

Problem 5.59 You will learn why this reaction is called S_N2 in Chapter 7. If you make a model of bromocyclopropane, you will find that the ring structure interferes less with the backside attack by a nucleophile compared to backside attack on 2-bromopropane. This phenomenon is referred to as a "tied-back" effect. The methyl groups don't block the backside as much if they are tied back, as they are in the ring compound. So, steric interactions are reduced. However, the ring carbon must go through an inversion in the S_N2 reaction. A carbon in a three-membered ring is already dealing with ring strain, and going through the sp^2-like transition state of the S_N2 reaction adds more strain. Thus, it is difficult to predict the correct answer. If you argue the S_N2 reaction for the bromocyclopropane is faster than for 2-bromopropane, then your reason is that the steric effects are greatly reduced as a result of the tied-back methyl groups. If you argue that the S_N2 reaction is going to be slower, then your reason is that ring strain in the transition state will be much higher for the bromocyclopropane.

Problem 5.60 The S$_N$2 reaction between axial bromocyclohexane and a nucleophile will undergo inversion to produce a product with the nucleophile in the equatorial position. In this reaction, the tied-back effect will reduce the steric interactions, and there is no added ring strain in the transition state. So it is likely that this displacement will be a faster reaction than the S$_N$2 reaction on 2-bromopropane.

Problem 5.61 The backside of the carbon–bromine bond for equatorial bromocyclohexane is not very accessible. This effect is much easier to see with a model set. It is difficult to get a nucleophile past the axial hydrogens that are on carbons 3 and 5. This reaction is slower than the S$_N$2 reaction on 2-bromopropane.

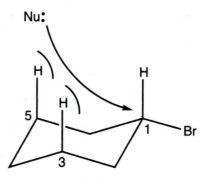

Problem 5.62 You will learn why this reaction is called an "E2" reaction in Chapter 8. You will notice that the hydrogen that is removed in the E2 reaction is on the adjacent carbon and in the anti position relative to the bromine. This geometry is a characteristic of all E2 reactions. Now look at the bromocyclohexane with the bromine in the equatorial position. Notice that there are no anti hydrogens on either of the adjacent carbons. So the E2 reaction cannot occur easily on this conformation of bromocyclohexane. The ring-flipped conformation has bromine in the axial position. Now there are anti hydrogens on carbon 2 and carbon 6. So this conformer undergoes the reaction easily.

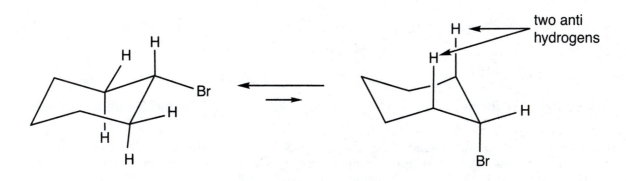

Substituted Alkanes: Alkyl Halides, Alcohols, Amines, Ethers, Thiols, and Thioethers

6

Problem 6.1 (a) This molecule is 3-chloropentane. Notice that there is no stereogenic carbon, even though the three-dimensional perspective is shown.

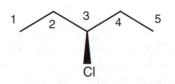 same as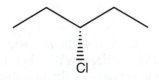

3-Chloropentane

(b) This molecule is 3-bromo-5-methylheptane. The longest carbon chain is seven carbons, so the root word is heptane. Numbering the longest chain from either end gives a 3,5-disubstituted system. So the alphabetical priority tells us the correct direction for numbering.

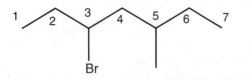 not this numbering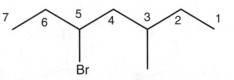

3-Bromo-5-methylheptane   not 5-bromo-3-methylheptane

(c) This molecule is 5-bromo-2-methylheptane. The longest carbon chain is seven carbons, so the root word is heptane. We number from the right in order to be 2,5-disubstituted rather than numbering from the left, which would give us 3,6-disubstituted.

5-Bromo-2-methylheptane   not 3-bromo-6-methylheptane

(d) This molecule is 1-fluorohexane. The longest carbon chain is six carbons, so we use the root word hexane. The only substituent is the flourine on carbon number 1.

(e) This molecule is 1,1,2-trichloroethene (or 1,1,2-trichloroethylene). There is no (*E*) or (*Z*) for this molecule because the carbon on one end of the alkene has two groups that are the same (two chlorines, in this case).

(f) This molecule is trichloromethane. Its common name is chloroform.

(g) This molecule is (*R*)-4-iodocyclohexene. The alkene is the priority group in this molecule. When we number a molecule containing a carbon–carbon double bond, the carbons of that double bond need to be numbered consecutively. That means this structure cannot be a 3-iodocyclohexene. That leaves us with two options. If we number counterclockwise, we obtain 4-iodocyclohexene. If we number the ring clockwise, we obtain 5-iodocyclohexene. Numbering counterclockwise results in a lower number for the only substituent.

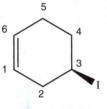

not 3-iodocyclohexene
because the alkene carbons
would be number 1 and number
6, which is not consecutive

4-Iodocyclohexene

not 5-iodocyclohexene

We don't need to indicate that this alkene is the (*Z*) isomer, although it would not be wrong to do so. The (*E*) cyclohexene is not a known molecule.

We also don't indicate that the priority group of the ring (the alkene in this case) is on carbon number 1. We would not call this molecule 3-iodo-1-cyclohexene. This is because the location of the priority group on a ring is defined as carbon number 1.

Finally, you noticed that there is a stereogenic carbon in the molecule. To determine the configuration, we can easily assign the highest and lowest atomic numbers. The iodo group on the stereogenic carbon is number 1 and the hydrogen on that carbon is number 4. The other two attachments are carbons. So we have to compare the next level of attachment. Both have (C, H, H). So we have to continue out to the next position on both lines. The line going to the left has (C, C, H) because of the double bond. That has a higher atomic number than the (C, H, H) of the other line. We can assign the number 2 and number 3 groups for this stereogenic carbon. We draw an arrow from group number 1 to number 2 to number 3 and see that the arrow is clockwise. This enantiomer is (*R*).

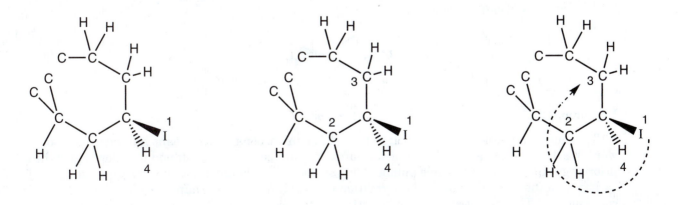

Problem 6.2

(a) (3*R*,4*S*)-Dichlorohexane

This molecule is meso. Its mirror image is identical. Either drawing is correct.

same as

Rotation of 180° around the C(3)—C(4) bond allows you to see the plane of symmetry in the molecule.

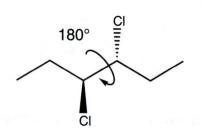

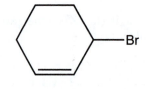

(b) 3-Bromocyclohexene

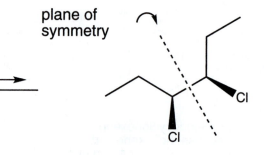

(c) (*Z*)-1-Bromo-1-butene

(d) Iodoethene

(e) 1,1-Dichlorocyclopropane

Problem 6.3

(a) This molecule is an alcohol. The longest chain is four carbons, which means butane is our root word. We drop the "-e" and add "-ol" for the suffix. We number from the left end in order to give the priority group the lowest possible number. This structure is 2-butanol. There is a stereogenic carbon with the three-dimensional information provided. We can determine the (*R*) or (*S*) configuration. The largest atomic number attached to the stereogenic carbon is oxygen and the lowest is hydrogen. The carbon to the right is attached to (C, H, H). The carbon to the left is attached to (H, H, H). Therefore the carbon to the right is number 2. The arrow from number 1 to number 2 to number 3 is counterclockwise. This molecule is (*S*)-2-butanol.

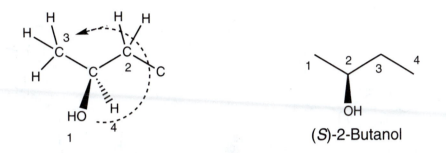

(*S*)-2-Butanol

(b) We recognize that the priority group in the molecule is the alcohol. We find the longest chain that includes the carbon attached to the priority group and use the appropriate root word (heptane, in this case). We drop the "-e" and add "-ol" in the suffix to give us heptanol. We number the chain from the end closest to the priority group so that the priority group gets the lowest possible number. This molecule is 5-bromo-3-heptanol.

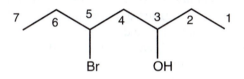

5-Bromo-3-heptanol

(c) The alcohol has a higher priority than an alkene. We find the longest chain that includes the carbon attached to the priority group, a heptane. Because there is a double bond in the chain, it is heptene. We have no prefix for an alkene in the IUPAC system. We drop the "-e" and add "-ol" to give heptenol. We number the chain from the end closest to the priority group so that the priority group gets the lowest possible number. (Don't forget to include the whole chain. It is a common error to number this incorrectly. The 1-methyl is part of the chain.) We must put the locating number for the alcohol next to the "-ol" in order to avoid confusion. This molecule is 6-hepten-2-ol.

6-Hepten-2-ol   not 1-methyl-5-hexen-1-ol

(d) The alcohol is the priority group. The longest chain that includes the carbon attached to the priority group is five carbons. This structure is a pentane. We drop the "-e" and add "-ol" to get pentanol. We number the chain from the end closest to the priority group so that the priority group gets the lowest possible number. There is an ethyl group on carbon number 2. You also notice that there is a stereogenic carbon with the three-dimensional perspective shown. That means we need to indicate whether this is the (*R*) or the (*S*) enantiomer. The stereogenic carbon is attached to three carbons and one hydrogen. The hydrogen gets number 4. The carbon attached to the alcohol has (O, H, H), and oxygen is a larger atomic number than (C, H, H) that the other two carbons have

(continued)

Problem 6.3 *(continued)*

for attachments. The carbon attached to oxygen is number 1. Comparing the carbon at the next level of attachment for the remaining two lines, we have (C, H, H) compared to (H, H, H). The propyl group is number 2 and the ethyl group is number 3. The arrow from number 1 to number 2 to number 3 is clockwise. This structure is (*R*)-2-ethyl-1-pentanol.

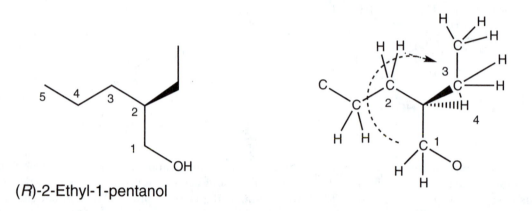

(*R*)-2-Ethyl-1-pentanol

(e) This molecule has an alcohol as the priority group. The longest chain is four carbons and contains an alkene. The root word is butene. We drop the "-e" and add "-ol", which gives us butenol. We number the chain so that the priority group gets the lowest possible number. This process gives us 3-chloro-2-buten-1-ol. We aren't finished. The double bond could be (*E*) or (*Z*). Because the atoms with the higher atomic numbers attached to each end of the alkene are on opposite sides of the alkene, this molecule is (*E*)-3-chloro-2-buten-1-ol.

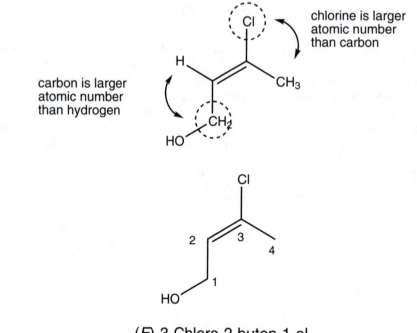

(*E*)-3-Chloro-2-buten-1-ol

(f) The priority group is the alcohol. There is also an alkene. This molecule is a cyclohexenol. As is always the case for a ring, carbon number 1 is the carbon bearing the priority group, which means that the only number needed in the name is the location of the alkene. This structure is 2-cyclohexenol.

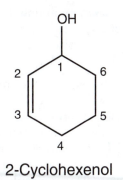

2-Cyclohexenol

(g) This molecule is 2,3-butanediol.

Problem 6.4

(a) The priority group is the amine. The longest chain including the carbon attached to the priority is a four-carbon chain. This molecule is a butane. Because it is an amine, we drop the "-e" and add "-amine" in the suffix. We number from the left end in order to give the priority group the lowest possible number, giving us 2-butanamine. There is a stereogenic carbon with the three-dimensional information provided. The largest atomic number attached to the stereogenic carbon is nitrogen and the lowest is hydrogen. The carbon to the right is attached to (C, H, H). The carbon to the left is attached to (H, H, H). Therefore the carbon to the right is number 2. The arrow from number 1 to number 2 to number 3 is counterclockwise. This molecule is (S)-2-butanamine.

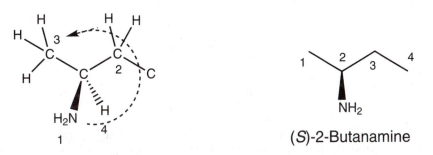

(S)-2-Butanamine

(b) There are no priority groups in this molecule. The longest chain is seven carbons, which means this structure is a heptane. Because there is no priority group, we don't change the suffix. We have a bromo group and a methoxy group on the chain. Numbering the chain from either end gives a 3,5-disub-stituted system. So the alphabetical rule tells us the correct direction for numbering. This molecule is 3-bromo-5-methoxyheptane.

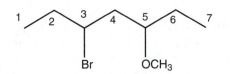

not this
numbering

3-Bromo-5-methoxyheptane

not 5-bromo-3-methoxyheptane

(c) The alkene is the priority group in this molecule. An ether is a subordinate group (see the inside front cover of the textbook). The longest chain containing the priority group is seven carbons. This structure is a heptene. We number the chain from the left end in order to give the priority group the lowest possible number. The only substituent on this chain is the ethoxy group on carbon number 6. This molecule is 6-ethoxy-1-heptene.

(continued)

Problem 6.4 *(continued)*

(d) The alcohol is a higher priority group than an amine. The longest carbon chain is four carbons long. This is a butanol. We number the chain so that the priority group gets the lowest possible number. This molecule is a 1-butanol. The substituents are a methoxy group on carbon number 3 and an amino group on carbon number 4. This structure is 4-amino-3-methoxy-1-butanol.

(e) The priority group in this molecule is the amine. The longest chain is four carbons and it includes a double bond. This structure is a butene, we drop the "-e" and add "-amine" to give butenamine. We number the chain so that the lowest possible number is given to the priority group, and we must indicate the location of the alkene. That means the location of the amine will need to be connected to the amine in order to avoid confusion. This structure is a 2-buten-1-amine. There is a methyl group on carbon number 3. Because the attachments to carbon number 3 (one end of the alkene) are the same, there is no (*E*) or (*Z*) for this molecule. This molecule is 3-methyl-2-buten-1-amine.

(f) The priority group in this molecule is the amine. It is attached to a five-carbon ring, which means this is a cyclopentanamine. The carbon attached to the amine must be carbon number 1 because it is the priority group. There is a methyl group on the nitrogen and a methyl group on carbon number 2 of the ring. This molecule is *N*,2-dimethylcyclopentanamine.

(g) The priority group in this molecule is the alcohol. The longest chain that includes the carbon with the priority group is four carbons. This is a butanol. We number from the right side to give the priority group the lowest possible number. This molecule has a methyl on carbon number 3 and an epoxy group, which is an ether, a subordinate group. The way to name an epoxide is to use epoxy as the prefix with the location indicating the two carbons that are attached to the group. This molecule is 2,3-epoxy-3-methyl-1-butanol.

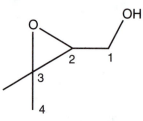

Problem 6.5 Angle and hybridization are intimately connected; if you know one, you know the other. The angle in methyl alcohol is almost exactly the tetrahedral angle, and oxygen must be almost exactly hybridized sp^3. We know the angle by comparison to water, for example.

Problem 6.6

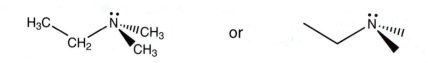

Problem 6.7 First of all, the carbon–nitrogen bond in aniline is an sp^2/sp^2 bond and will be shorter than the sp^3/sp^3 bond of methylamine. Second, in aniline there is overlap between the orbital containing the two nonbonding electrons on nitrogen and the orbitals of the ring. The resonance forms show this well. The result is double-bond character in the bond attaching the nitrogen to the ring. As double bonds are shorter than single bonds, double-bond character leads to a shortening of the carbon––nitrogen bond distance compared to that in alkylamines, in which there is no resonance effect.

"Double-bond character"

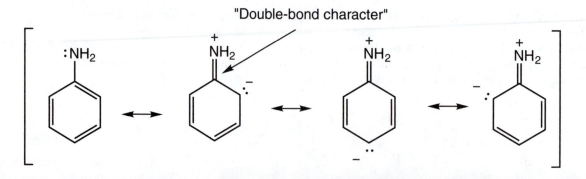

Problem 6.8 A stereogenic carbon has four bonds. There is no lone pair on such a carbon, so no inversion is possible. Bonding occurs because it is stabilizing. Remember the orbital interaction diagrams that show the lower energy bonding molecular orbitals. Doing an inversion on a carbon with four bonds would require breaking a bond and leaving atoms charged. There could be no mixing between the orbitals.

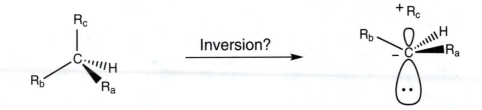

Problem 6.9 In aziridines, there is more angle strain in the transition state than there is for noncyclic amines. In an amine, the most favorable angle is approximately the tetrahedral angle, about 109°. In the sp^2-hybridized transition state, this angle is expanded from about 109° to 120°. One might say that there is approximately 11° worth of angle strain. In an aziridine, in which one angle is restricted to 60° through incorporation into the three-membered ring, the transition state is very highly strained. Now there is 60° of angle strain. Accordingly, it is much more difficult to reach the transition state for inversion in aziridines than it is in acyclic amines.

Transition state

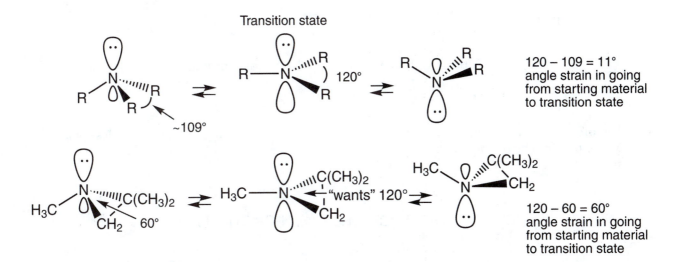

Problem 6.10 A subtle, tough question. In the starting aziridine, there is relatively poor overlap between the hybrid orbital on nitrogen containing the nonbonding electrons and the orbitals of the phenyl ring. There is some overlap, but the connection between the unsymmetrical hybrid and the $2p$ orbital on carbon is not ideal. In the transition state, the nitrogen is hybridized sp^2, and now there is excellent overlap, and hence good delocalization of the nonbonding electrons into the benzene ring. The transition state is stabilized, is lowered in energy, and becomes easier to reach from starting material.

Starting material   Transition state

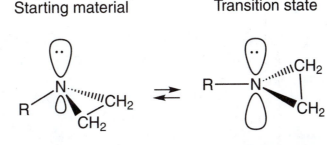

(continued)

Problem 6.10 *(continued)*

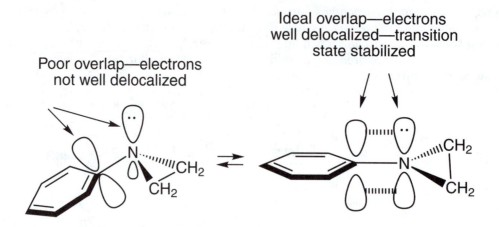

Poor overlap—electrons
not well delocalized

Ideal overlap—electrons
well delocalized—transition
state stabilized

Problem 6.11 Citric acid (lemon) acts as a Brønsted acid toward the Brønsted base, the amine (fish odor), in a proton-transfer reaction to form a salt. Unlike the amine, the salt is not volatile and cannot easily be detected by our sense of smell. The fish has been deodorized by the acid–base reaction.

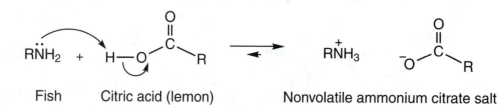

| Fish | Citric acid (lemon) | | Nonvolatile ammonium citrate salt |

Problem 6.12 Conjugate acids and conjugate bases are related through the gain or loss of a single proton. The conjugate acids of the molecules shown are always "the molecule plus one proton."

(a) $H_3\overset{+}{O}$ (b) HOH (c) $\overset{+}{N}H_4$ (d) $CH_3\overset{+}{O}H_2$ (e) $H_2C=\overset{+}{O}H$ (f) CH_4

Sometimes there are choices to be made as to where to put the proton in making the conjugate acid. In this set, only $H_2C=O$ really presents such a choice. Another conjugate acid is $H_3C—O^+$, although this molecule is much less stable than the one shown above. (Why?)

Problem 6.13 The conjugate bases of these molecules will always be "the molecule less one proton."

(a) HO^- (b) O^{2-} (c) $^-NH_2$ (d) CH_3O^- (e) $^-CH_3$ (f) $^-OSO_2OH$ (g) $^-OSO_2O^-$

Problem 6.16 The anilinium ion (pK_a = 4.6) is a much stronger acid than the ethyl ammonium ion (pK_a = 10.6) because the conjugate base of the anilinium ion, aniline, is stabilized by resonance, whereas the conjugate base of the ethyl ammonium ion, ethylamine, is not. Moreover, the electron-withdrawing benzene ring destabilizes adjacent electron deficiency.

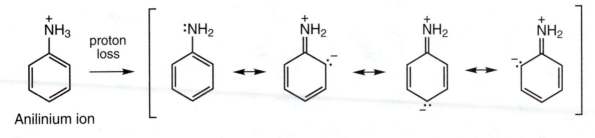

Anilinium ion

Resonance stabilization lowers the energy of aniline and makes deprotonation of the ammonium ion relatively easy

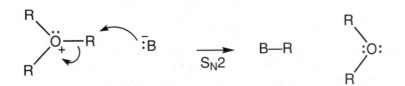

In this case, neither reactant nor product is stabilized by resonance

Problem 6.17 Oxonium ions are destroyed by nucleophiles through an S_N2 alkylation process.

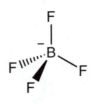

Most negatively charged counterions are nucleophiles, but $^-BF_4$ is an exception. There is no lone pair of electrons on boron and therefore no possible nucleophilicity.

Problem 6.18   Basically, the problem asks you to build ethane, or at least the carbon–carbon sigma bond. Here is the diagram. Because you know the bond strength of ethane (Ch. 2, p. 72), you know the stabilization of the two electrons in σ.

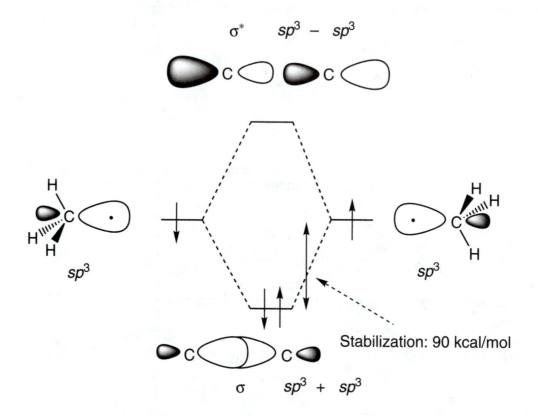

Stabilization: 90 kcal/mol

Problem 6.19   Let's work backward. Given the previous paragraph in the text, certainly the final step is the quenching of an organometallic reagent with D_2O. The organolithium or Grignard reagent is made from the bromide and either lithium or magnesium.

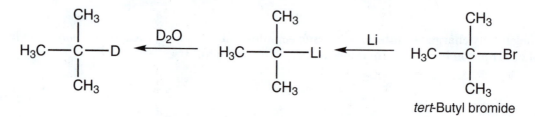

tert-Butyl bromide

Now, the only remaining question is how to convert the alkene into the bromide. For this you must remember material in Chapter 3. Simple addition of HBr to isobutene gives *tert*-butyl bromide (Chapter 3, p. 136).

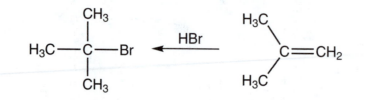

Basically, the second part of this problem involves the same chemistry as the first part. Add HBr to 1-butene to give 2-bromobutane, make the organometallic reagent, and quench with D_2O.

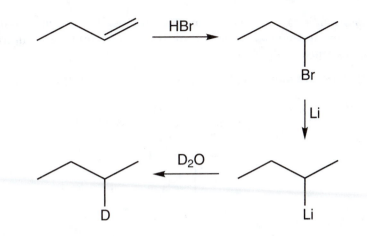

Problem 6.20

(a) This one is right out of the text. Treat the thiol with Raney nickel.

(b) Make the Grignard, then add water: 1. Mg, 2. H_2O.

(c) Just convert cyclohexene into bromocyclohexane (add HBr) and proceed as in part (b): 1. HBr, 2. Mg, 3. H_2O. There is a much simpler way, but you don't know it yet.

Additional Problem Answers

Problem 6.21

(a) 3-Methyl-1-butanol

(b) 3-Methyl-2-butanol

(c) 2-Methyl-2-butanol

(d) 2-Methyl-1-butanol

(e) 2-Fluoro-1-pentanol

(f) 4-Phenyl-1-pentanol

(g) *trans*-1,2-Cyclobutanediol
   or (1*R*, 2*S*)-1,2-Cyclobutanediol

(h) 4-Penten-1-ol

(i) (*E*)-3-Penten-1-ol

(j) *cis*-3-Ethoxycyclohexanol
   or (1*S*, 3*R*)-3-Ethoxycyclohexanol

(k) 1,3-Propanediol

(l) Propanethiol (propyl mercaptan)

(m) Dipropyl sulfide

Problem 6.22

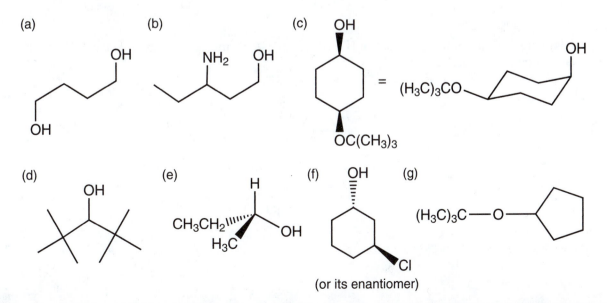

Problem 6.23   This problem includes some common names. It is worth memorizing these common names. There are systematic names for each one, of course, but you are likely to encounter the common names in the future.

(a) *tert*-Butyl alcohol is a common name; the IUPAC name is 2-methyl-2-propanol.

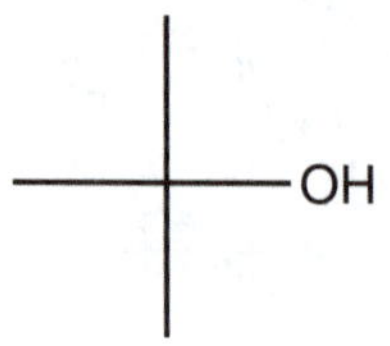

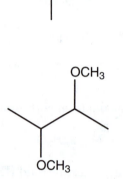

(b)

(c)

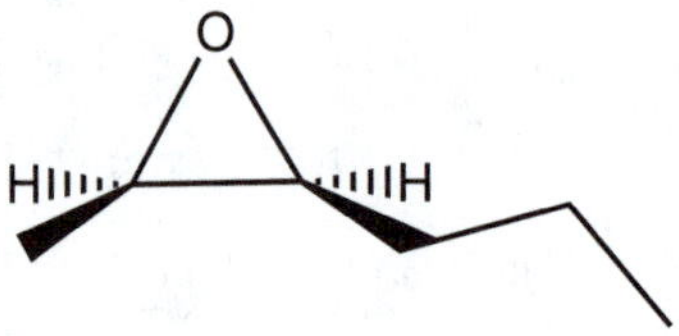

   (or its enantiomer)

(d) Tetrahydrofuran (THF) is a common name; the IUPAC name is oxolane. A more systematic IUPAC name is 1,4-epoxybutane.

(e) Ethylene oxide is a common name; the IUPAC name is oxirane. A more systematic IUPAC name is 1,2-epoxyethane.

Problem 6.24

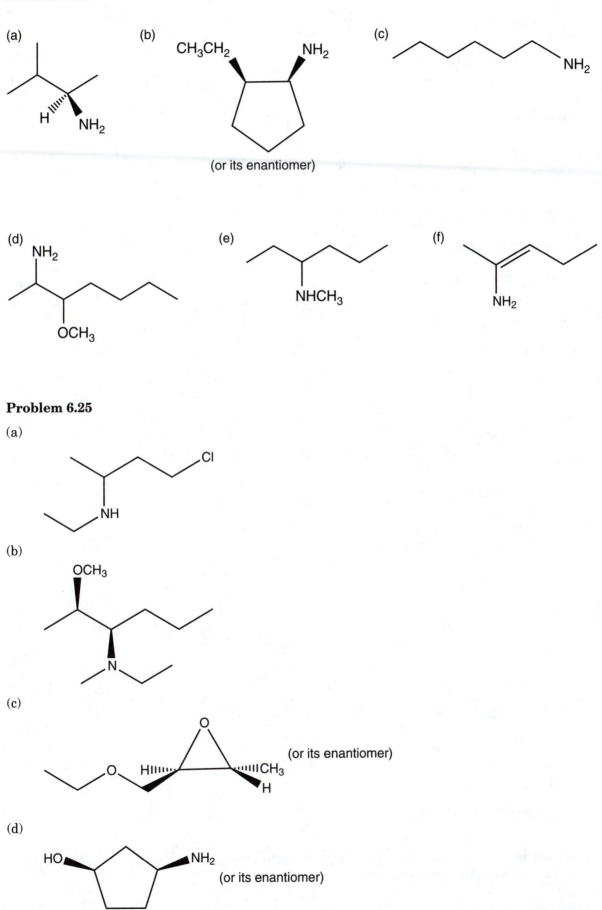

(a)

(b)

CH₃CH₂ ... NH₂

(or its enantiomer)

(c)

(d)

(e)

(f)

Problem 6.25

(a)

(b)

(c)

(or its enantiomer)

(d)

(or its enantiomer)

(continued)

Problem 6.25 *(continued)*

(e)

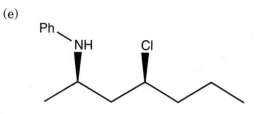

Problem 6.26

(a) Cyclohexanamine

(b) (*S*)-4-Methyl-2-pentanamine

(c) (3*S*, 5*S*)-Heptanediamine

(d) 2-Amino-4,5-dimethyl-1-hexanol

(e) 6-Methoxy-2-heptanamine

Problem 6.27

(a) All C, N, O sp^3.

(b) All ring C and N sp^2. CH$_2$ and O are sp^3.

(c)

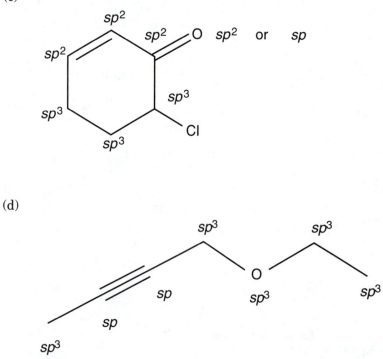

(d)

(e) Ring Cs and N are all sp^2.

Problem 6.28   All the molecules in Problem 6.27 can hydrogen bond to water. Molecules (a), (b), and (e) can hydrogen bond to themselves.

Problem 6.29

(a) In solution, at least, ethyl alcohol is a stronger acid than isopropyl alcohol (Table 6.6, p. 246). So we would expect that the primary alcohol would be the stronger acid here.

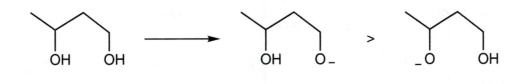

(b) In the gas phase, the acidity order for alcohols is the reverse of what obtains in solution. So in this case, it will be the tertiary alkoxide that will be preferentially formed.

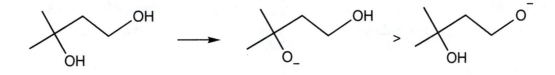

(c) Thiols are stronger acids than alcohols, so the SH will be preferentially deprotonated.

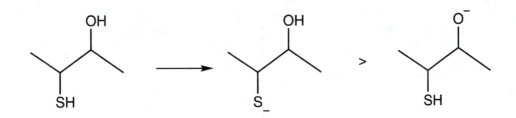

Problem 6.30 The carbon–chlorine bond is strongly polarized, with the negative end of the dipole on chlorine. There is a partial positive charge on carbon. That δ^+ will be stabilizing to a neighboring anion. The closer it is, the greater the stabilization. Accordingly, the stability of the related alkoxide, and thus the acidity order of the alcohols, is in the following order:

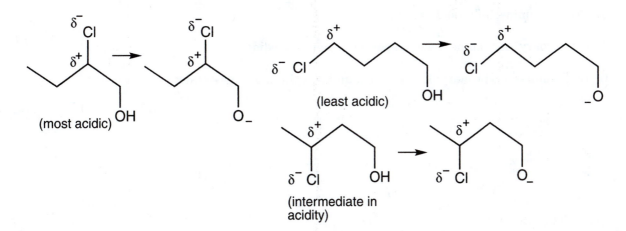

Problem 6.31 We think this is all of the possible isomeric alcohols of formula $C_6H_{14}O$.

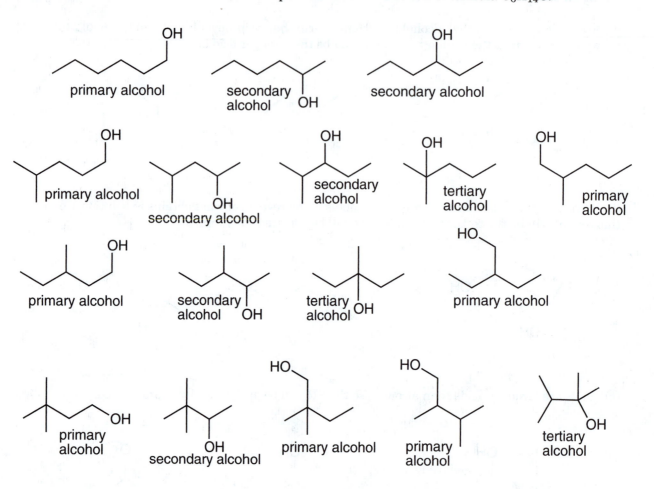

Problem 6.32

(a) Primary amine: 2-pentanamine or 2-aminopentane.

(b) Secondary amine: *N*-pentyl-1-pentanamine or dipentylamine.

(c) Tertiary amine: *N,N*-dipentyl-1-pentanamine or tripentylamine.

(d) Primary amine: 3-fluoro-2-pentanamine.

(e) Quaternary ammonium ion: tetrapentylammonium iodide.

(f) Both amines are primary: 1,3-pentanediamine or 1,3-diaminopentane.

Problem 6.33

(a) Primary amine: 3-buten-2-amine.

(b) Tertiary amine: ethylmethylphenylamine or *N*-ethyl-*N*-methylaniline or *N*-ethyl-*N*-methyl-benzenamine.

(c) Primary amine: 2-propyn-1-amine or propargylamine.

(d) Primary amine: 3-amino-2-butanol.

Problem 6.34   At low temperature, what you see is what you get. There are five different carbons in the molecule and, of course, five different signals in the ^{13}C NMR spectrum. At higher temperature, amine inversion sets in, and the two carbon-attached methyl groups and the two ring carbons become equivalent. There are now only three different carbons (shown as a, b, and c in the figure) and three signals in the ^{13}C NMR spectrum.

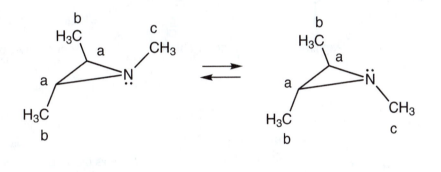

Problem 6.35   Lindane is 1,2,3,4,5,6-hexachlorocyclohexane. It is the stereoisomer shown. Lindane has been used as an insecticide, most often for fruit trees and seed crops. It has been banned from agricultural use in most countries since 2009, including the United States. There is a plane of symmetry in lindane. It is achiral.

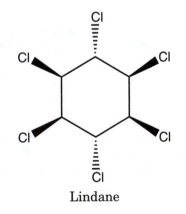

Lindane

(continued)

Problem 6.35 *(continued)*

Diastereomers of lindane are

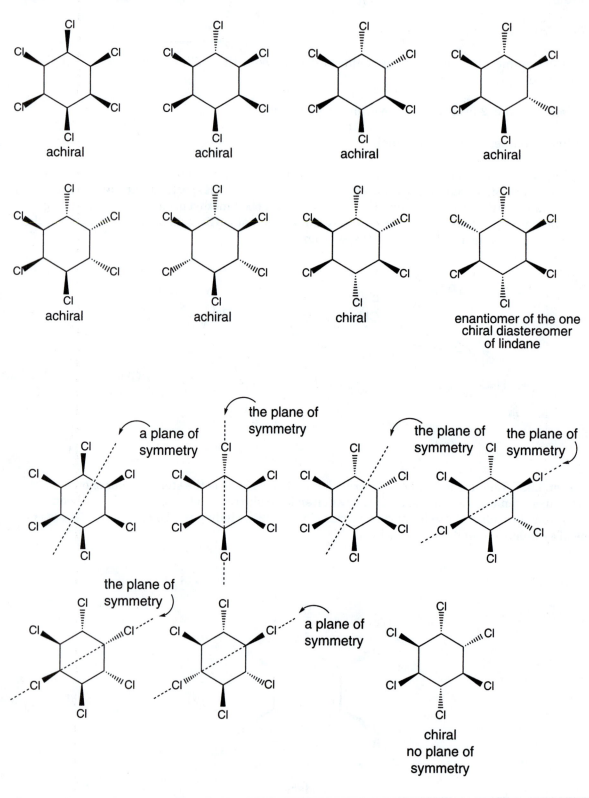

Problem 6.36

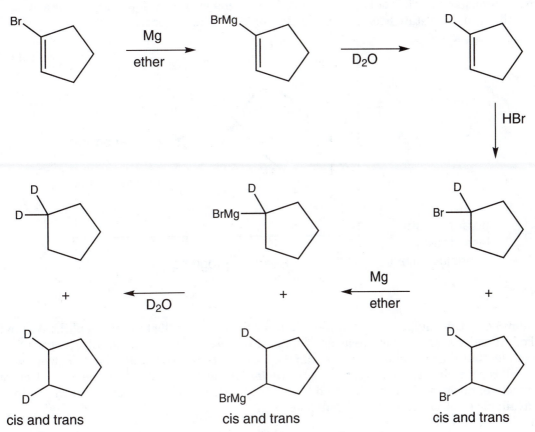

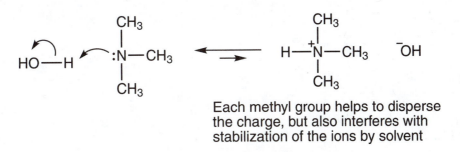

Problem 6.37 The lower the pK_b, the greater the base strength. So, methylamine is a stronger base than ammonia, and dimethylamine is a (slightly) stronger base still. Yet trimethylamine is a weaker base than the other methylated species and almost as weak a base as ammonia itself. Indeed, we might well have been disturbed at the small difference between methylamine and dimethylamine as the two are almost of equal base strength. The problem is one of solvation. Each methyl group increases base strength but decreases the possibilities for solvation of the charged intermediates (and the transition states leading to them). So, for trimethylamine, each methyl group stabilizes the intermediate by helping to disperse the charge, but, at the same time, it interferes with the stabilizing effects of solvation. It's a balancing act. See the discussion of these effects in terms of the pK_a values of the conjugate acids in this chapter (p. 250).

Each methyl group helps to disperse the charge, but also interferes with stabilization of the ions by solvent

Problem 6.38 None of the above. The Hofmann elimination reaction requires a quaternary ammonium ion.

Problem 6.39   A polar solvent would stabilize the very polar starting materials. The products are relatively nonpolar. So, one would predict that the reaction would be slower in a polar solvent, because the transition state leading to products would be a higher barrier.

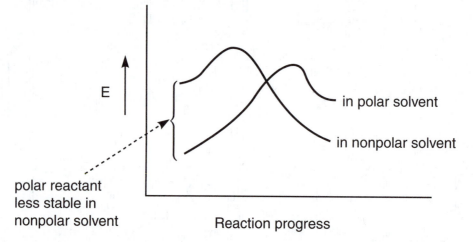

Problem 6.40   One might guess that the alkene or the amine is the best nucleophile. We have seen both functional groups behave as nucleophiles. We know that an alkene will react with electrophiles, but the alkene will not undergo the S_N2 reaction with the alkyl halide, whereas an amine will react well with the alkyl halide. So, we might conclude that the amine is a better (or more reactive) nucleophile. The calculated HOMO supports this idea. It shows that the amine has more available electrons than does the alkene.

Problem 6.41   According to the energy diagram, the most stable intermediate is the protonated alcohol that is formed first. The least stable intermediate is the secondary carbocation.

The process that is occurring in the first step of the reaction is called a protonation. The alcohol oxygen is being protonated by one of the acidic protons on the H_2SO_4. The first step occurs because the pK_a of sulfuric acid is about -3. The pK_a of the conjugate acid of the alcohol is about -2 (see Table 6.6). The equilibrium for this process lies to the right.

$$ROH \quad + \quad H_2SO_4 \quad \rightleftharpoons \quad R\overset{+}{O}H_2 \quad + \quad HSO_4^-$$

Substitution Reactions: The S$_N$2 and S$_N$1 Reactions

7

Problem 7.2 The HOMO is the filled 1s orbital on hydrogen and the LUMO is the empty 2p orbital of the methyl cation.

Problem 7.5 Here is a simple, plug-in-the-numbers problem. This problem is designed to help you to develop a feeling for the relationship between ΔG and K so that you can make quick estimates in either direction.

$\Delta G° = -2.3RT \log K$; at 25 °C, $2.3RT = 1.364$ kcal/mol
  if $K = 7$, $\Delta G° = -(1.364) \log 7 = -1.15$ kcal/mol
  if $K = 14$, $\Delta G° = -(1.364) \log 14 = -1.56$ kcal/mol

Problem 7.6

(a) Bonds broken:

		Bonds formed:	
H—Br	87.5 kcal/mol	C—H	101.1 kcal/mol
C=C	66 kcal/mol	C—Br	72.1 kcal/mol
	153.5 kcal/mol		173.2 kcal/mol

Difference in $H° = 173.2 - 153.5 = 19.7$ kcal/mol
Products more stable than starting materials

(b) Bonds broken:

		Bonds formed:	
H—OH	118.8 kcal/mol	C—H	101.1 kcal/mol
C=C	66 kcal/mol	C—OH	92.1 kcal/mol
	184.8 kcal/mol		193.2 kcal/mol

Difference in $H° = 193.2 - 184.8 = 8.4$ kcal/mol
Products more stable than starting materials

(c) Bonds broken:

		Bonds formed:	
H—F	136.3 kcal/mol	C—H	101.1 kcal/mol
C=C	66 kcal/mol	C—F	115 kcal/mol
	202.3 kcal/mol		216.1 kcal/mol

Difference in $H° = 216.1 - 202.3 = 13.8$ kcal/mol
Products more stable than starting materials

(continued)

Problem 7.6 *(continued)*

(d) Bonds broken: Bonds formed:

H—Cl	103.2 kcal/mol
C≡C	<u>66 kcal/mol</u>
	169.2 kcal/mol

C—H	101.1 kcal/mol
C—Cl	<u>83.7 kcal/mol</u>
	184.8 kcal/mol

Difference in $H°$ = 184.8 − 169.2 = 15.6 kcal/mol

Products more stable than starting materials

Problem 7.7   In every case, the reaction is R—H ⟶ R• + •H. There can be no difference in the energy of the hydrogen atom formed in every reaction, and the differences must lie in the R• partner, the radical. As it takes less energy to form the more substituted radical (R•), the more substituted radicals must be more stable than the less substituted radicals. One can even make a very rough estimate as to the amount of the increased stability.

$H—CH_3$ ⟶ H• + •CH_3 $\Delta H°$ = 105.0 kcal/mol

$H—CH_2CH_3$ ⟶ H• + •CH_2CH_3 $\Delta H°$ = 101.1 kcal/mol
(primary radical more stable than methyl radical by 3.9 kcal/mol)

$H—CH(CH_3)_2$ ⟶ H• + •$CH(CH_3)_2$ $\Delta H°$ = 98.6 kcal/mol
(secondary radical more stable than primary radical by 2.5 kcal/mol)

$H—C(CH_3)_3$ ⟶ H• + •$C(CH_3)_3$ $\Delta H°$ = 96.5 kcal/mol
(tertiary radical more stable than secondary radical by 2.1 kcal/mol)

Problem 7.9

$\Delta G^{\ddagger}$ reverse reaction) = $\Delta G^{\ddagger}$ (left to right reaction) + $\Delta G°$

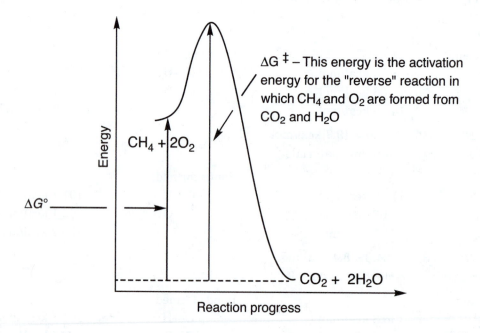

Problem 7.10   Inversion has occurred, but both starting material and product are (R) configuration. We can be convinced that inversion has occurred by comparing the starting material to the product. The iodide in the starting material has been replaced by a hydroxide. Notice that the hydroxide has attached to what was the backside of the C—I bond. The chloro and methyl groups have moved to the other side of the carbon. This reaction takes place with inversion. Because the

starting material is the (*R*) configuration, we expect the product of inversion to have the (*S*) configuration. But in this case, the product is also (*R*). There is no guarantee that inversion will result in a change in configuration in a substitution reaction. The (*R*) and (*S*) configurations are determined using the Cahn–Ingold–Prelog system (p. 156), which is based on the atomic numbers of the atoms attached to the stereogenic carbon. The nucleophile may not have the same position in the order of atomic numbers as the leaving group. In this example, the starting material has (I, Cl, C, H) and the product has (Cl, O, C, H).

Problem 7.11 An argument has just been made (p. 291) that steric factors are important in determining the rate of the S$_N$2 reaction. The more substituted the substrate, the slower the reaction. In tertiary substrates, the three R groups guard the rear of the C—L bond so efficiently that the S$_N$2 reaction completely fails. There should be no surprise that the size of the entering nucleophile is important as well. The larger the nucleophile [$(CH_3)_3C$—$O^- > CH_3$—O^-], the greater the steric interactions, and the slower the reaction.

Problem 7.12

(a) This is an S$_N$2 reaction. The starting material is a secondary iodide and the azide anion ($^-N_3$) is an excellent nucleophile. The product is 3-azidopentane.

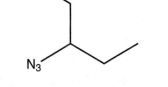

(b) This is an S$_N$2 reaction. The starting material is a primary bromide. A bromide is a good leaving group and the cyanide anion (^-CN) is an excellent nucleophile. The iodide is a better leaving group than a bromide, but we will only see S$_N$2 reactions on sp^3-hybridized carbons. See Problem 7.67. So there is no substitution of the vinyl iodide.

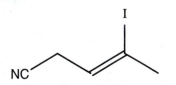

(c) This is an S$_N$2 reaction. The starting material is a secondary bromide and the thiolate (CH_3S^-) is an excellent nucleophile. The product is a thioether.

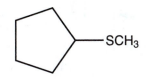

(d) This is an S$_N$2 reaction. There are two leaving groups in this molecule, but only the primary bromide will react with a nucleophile. The backside of the bromide on the aromatic ring is not accessible by any nucleophile. No inversion can occur at the sp^2-hybridized carbon in a ring. Methoxide (CH_3O^-) is a good nucleophile.

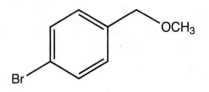

Problem 7.13 Strain in the starting material will raise its energy, and any change of this kind will lower the activation energy for a reaction if there is no change in the energy of the transition state, a most unlikely prospect.

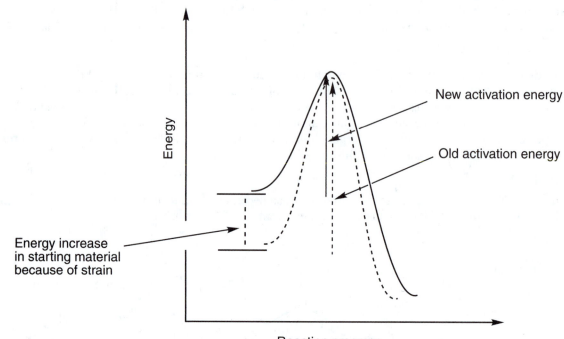

In fact, in this case the energy of the transition state is raised *more* than is the energy of the starting material. The hint tells us to consider the structures of both starting material and transition state. Let's focus on deviations from ideal angles. In the starting material, the cyclopropane "wants" something like 109°, but is restricted to 60°. There are some 49° of angle strain. In the transition state, the sp^2 carbon "wants" 120°, but is restricted to 60° (see Fig. 7.44). There are 60° of angle strain, and the transition state is more destabilized than the starting material by angle strain effects. The result is an increase in activation energy even though the starting material is destabilized by angle strain.

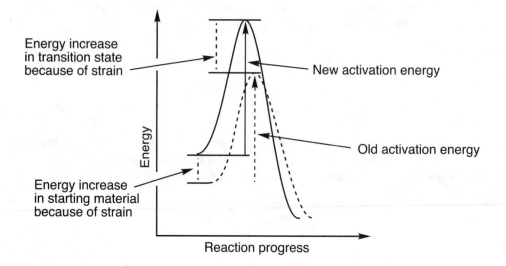

Problem 7.15 It is amazing how hard this question is. Perhaps our focus on displacement reactions makes it difficult to see the good old-fashioned Brønsted acid–base chemistry. The strong base removes the proton from the alcohol to give the alkoxide ion, RO⁻. You need to be able to consider more than one reaction at the same time.

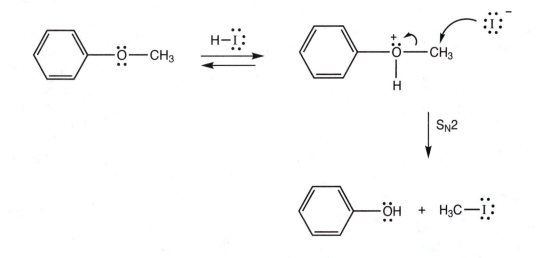

Problem 7.16 Protonation by hydrogen iodide yields an intermediate in which only one S_N2 reaction is possible. There can be no S_N2 reaction at the carbon of the benzene ring—the ring blocks access for the iodide ion.

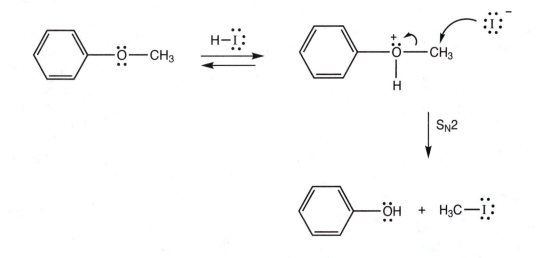

Problem 7.17 The mechanism is a standard ether cleavage reaction applied in a cyclic system. It is complicated by the transformation of the initial product, an iodo alcohol, into a diiodide.

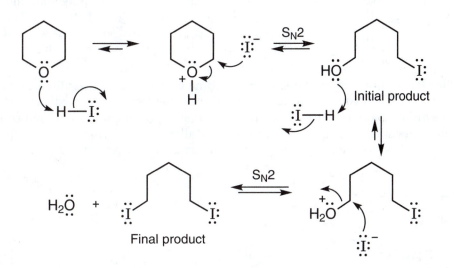

Problem 7.18   This problem is nothing more than a simple S_N2 displacement of a good leaving group by an alkoxide.

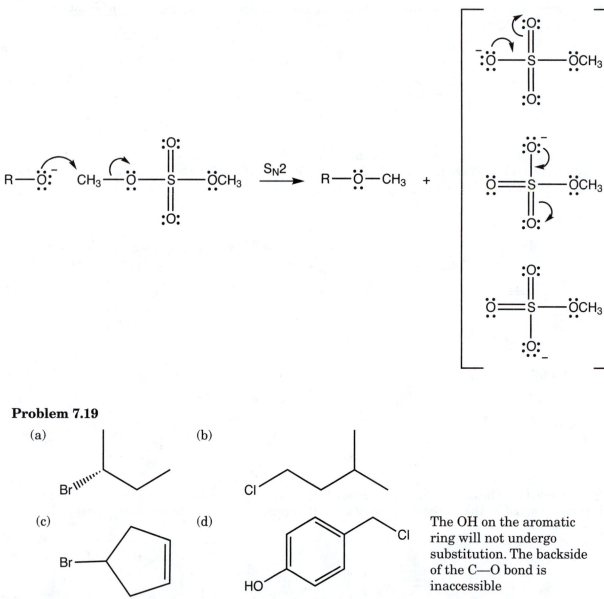

Problem 7.19

(a)

(b)

(c)

(d)

The OH on the aromatic ring will not undergo substitution. The backside of the C—O bond is inaccessible

Problem 7.20   In this problem, you are asked to outline an arrow formalism for a reaction you have not yet seen, although you will certainly see this reaction later in the next chapter. You need to reason backward from the product, visualize what must have happened, and then supply the arrows. The elements of HBr are eliminated to give the alkene 2-methylpropene. The formula C_4H_8 suggests this much. Here is the product along with an arrow formalism for the process.

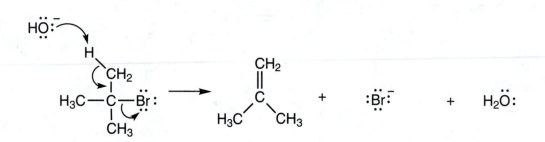

Problem 7.21

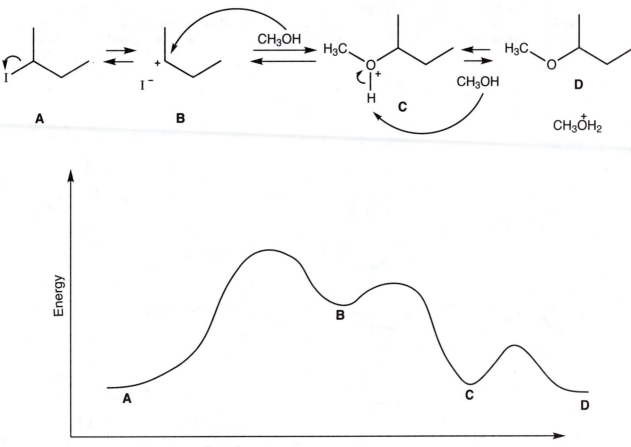

Reaction progress

Problem 7.22   The S_N1 reaction is usually more favorable for tertiary halides such as the one shown at the top of Figure 7.76. However, this particular secondary halide leads to an especially stable carbocation. Localized charge is bad, and delocalized charge is better. The phenyl (cyclohexatriene) ring provides a way to delocalize the charge generated on ionization. This delocalization makes this phenyl secondary carbocation relatively stable, and therefore relatively easy to form.

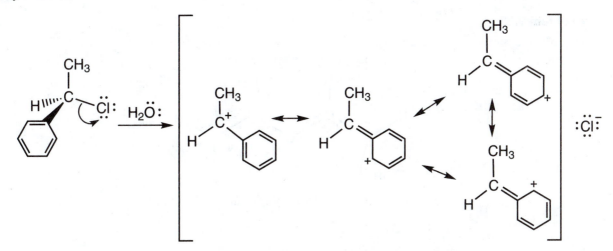

Several resonance forms make for a stable species

Problem 7.23   Polar solvents stabilize ions. The more stable an ion is, the less reactive it is, and the longer it lives. Long life allows for movement of the two ions away from their positions in the original ion pair. So, the more polar the solvent, the more likely it is that the symmetrical ion of Figure 7.75 will be formed from the unsymmetrical ion pair of Figure 7.77. A polar solvent favors racemization, the inevitable result of formation of the symmetrical carbocation of Figure 7.75.

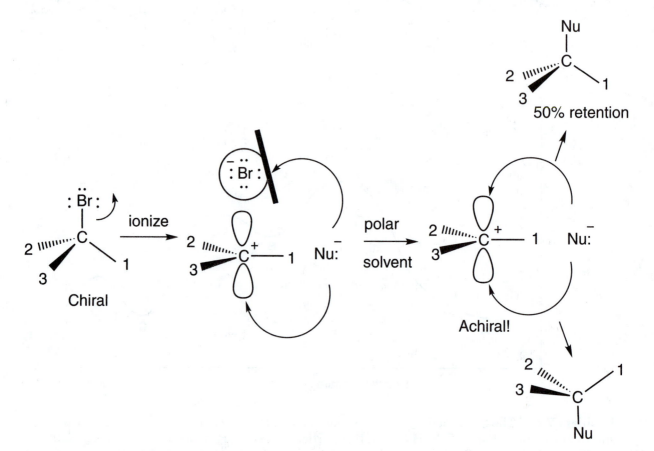

Problem 7.25

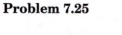

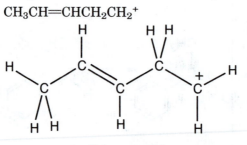

Primary cation

CH$_3$CH$_2$CH=CHCH$_2$$^+$

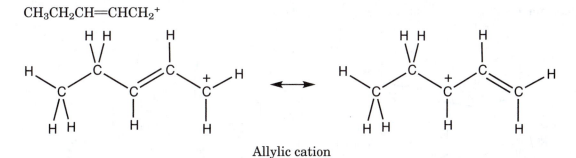

Allylic cation

The allylic cation is more stable because resonance allows for sharing of electrons. As a result of resonance, the carbocation is equally shared by two atoms. A primary allylic cation is more stable than a primary cation.

Problem 7.26

(a) This reaction is an S$_N$2 displacement. Iodide is an excellent nucleophile, and it will displace the secondary chloride. There are many solvents that will work. THF is a commonly used solvent. Acetone would also work well. Water could be used, but it would slow down the rate by solvating the iodide.

(b) This reaction is an S$_N$2 displacement. The phenoxide anion (PhO$^-$) is a good nucleophile. THF is the best solvent for this reaction.

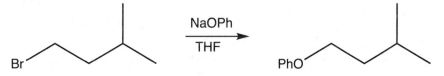

(c) This is an S$_N$1 solvolysis. We can tell that by noting the loss of stereochemistry. We will see in Chapter 8 that this reaction will also give elimination products. Methanol (CH$_3$OH) is a good solvent and nucleophile for this reaction. But we would not want to use water. Water would compete in the substitution reaction, giving 2-cyclohexenol rather than the desired ether.

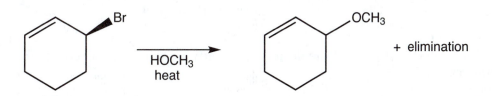

(d) This is an S$_N$2 reaction. Hydroxide is a good nucleophile. Hydroxide is also a base, but there are no elimination products possible for this primary iodide. THF is a fine choice for solvent in this reaction. Water would also work. We would not want to use methanol because the hydroxide could deprotonate the methanol to give methoxide, which would compete with the hydroxide for the substitution.

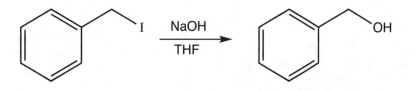

Problem 7.28

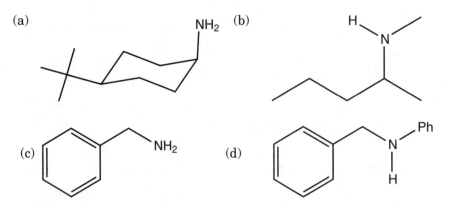

Problem 7.29   Epoxides react with nucleophiles to open the three-membered ring. Here ethylamine acts as nucleophile in the S_N2 reaction. Protonation gives the first product.

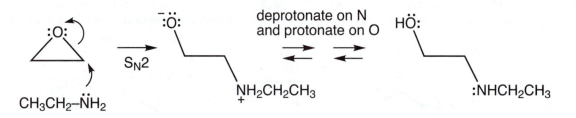

The first-formed product is still a strong enough nucleophile to react with the epoxide, and so a second step takes place.

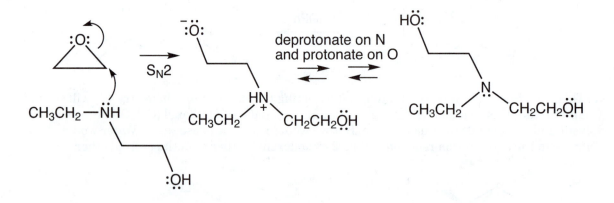

Although this product is still a nucleophile, it is apparently too sterically hindered to react further with another epoxide.

Problem 7.30

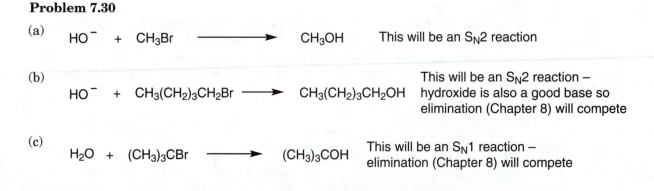

(a) HO^- + CH_3Br ⟶ CH_3OH This will be an S_N2 reaction

(b) HO^- + $CH_3(CH_2)_3CH_2Br$ ⟶ $CH_3(CH_2)_3CH_2OH$ This will be an S_N2 reaction – hydroxide is also a good base so elimination (Chapter 8) will compete

(c) H_2O + $(CH_3)_3CBr$ ⟶ $(CH_3)_3COH$ This will be an S_N1 reaction – elimination (Chapter 8) will compete

(d)

$$H_2O \quad + \quad CH_3(CH_2)_3CH_2Br \longrightarrow$$

No reaction in this case –
this primary bromide can only do S_N2 chemistry,
but water is not nucleophilic enough to be
useful for S_N2 reactions

Problem 7.31   There are many correct solutions to this problem. Here is one way to do it. We can use ethylene oxide as an electrophile, as we saw in Problem 7.29, where we opened the epoxide ring with an amine. We know how to make an acetylide (p. 316), and that can be used to attack the epoxide. The best way to solve a synthesis problem like this is to think backward. The 5-bromo-2-pentyne can be made from the corresponding alcohol (3-pentyn-1-ol). That alcohol can be made from the addition of an acetylide of propyne to ethylene oxide.

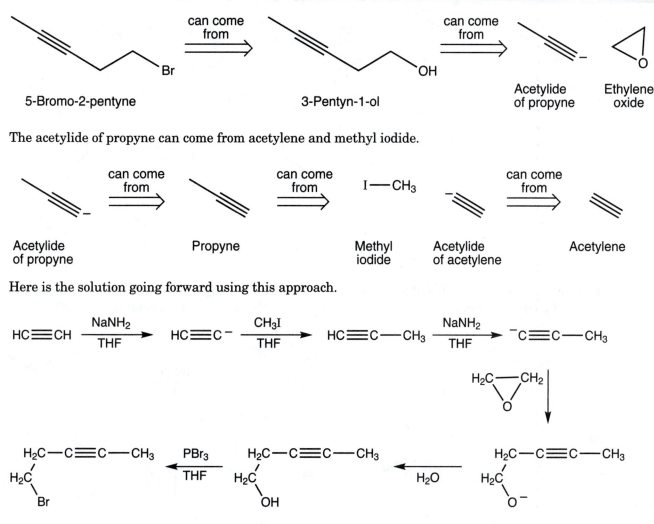

The acetylide of propyne can come from acetylene and methyl iodide.

Here is the solution going forward using this approach.

Problem 7.32

(a) 2-Bromobutane will be faster in the S_N1 reaction (secondary carbocation more stable than a primary carbocation) and 1-bromobutane faster in the S_N2 reaction (primary bromide less sterically hindered than the secondary bromide).

(b) Cyclopentyl chloride will be faster in the S_N1 reaction and 1-chloropentane faster in the S_N2 reaction for the same reasons as in (a).

(c) 1-Iodopropane will be faster in the S_N2 reaction because iodide is a better leaving group than is chloride. Neither molecule will undergo the S_N1 reaction because a primary carbocation must be formed.

(continued)

Problem 7.32 *(continued)*

(d) *tert*-Butyl iodide will be faster in the S$_N$1 reaction because the tertiary carbocation is more stable than a secondary carbocation and therefore formed more easily. Isopropyl iodide will be faster in the S$_N$2 reaction because it is less sterically hindered. Approach to the rear of the carbon–iodine bond is less hindered in the secondary system than in the tertiary system.

(e) Neither *tert*-butyl iodide nor *tert*-butyl chloride will react at all in the S$_N$2 reaction. *tert*-Butyl iodide will be more reactive in the S$_N$1 reaction because iodide is a better leaving group than is chloride.

Problem 7.33

(a)

(b)

(c)

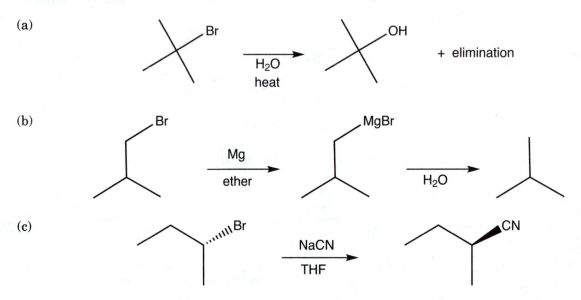

(d) A double inversion is needed here. We don't have a way of doing a substitution with retention. So we can't go directly from the alcohol to the cyano (CN) substituted product retaining the (S) stereochemistry.

(e)

(f)

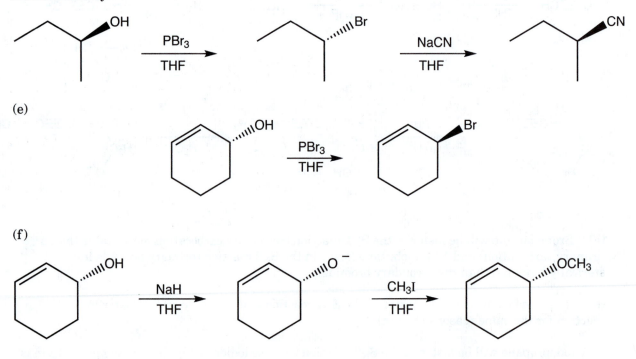

(g) Using excess amine in the S_N2 reaction gives the neutral cyclopentanamine product.

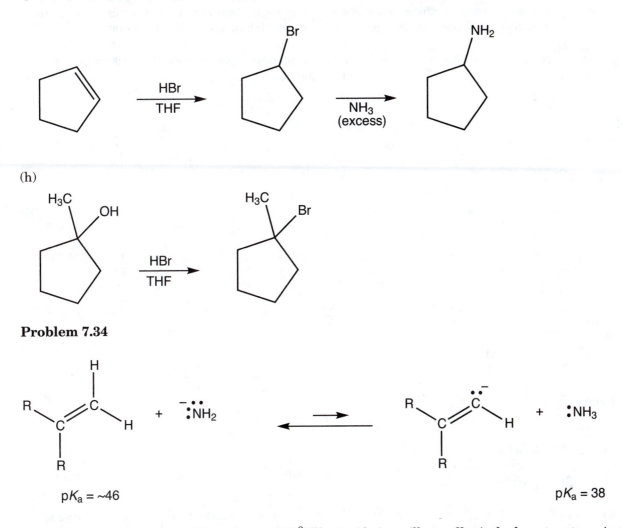

(h)

Problem 7.34

pK$_a$ = ~46

pK$_a$ = 38

Ammonia is the stronger acid by a factor of 10^8! The amide ion will not effectively deprotonate a vinyl hydrogen.

Additional Problem Answers

Problem 7.35

(a) The phosphine is the stronger nucleophile. Third-row atoms are more polarizable than second-row atoms, and hence better nucleophiles (see Figure 7.50, p. 295).

(b) The amide ion is a far better nucleophile than the neutral amine (see Figure 7.50, p. 295).

(c) Same answer as part (a). The thiolate is the better nucleophile.

Problem 7.36   Just ask yourself, what replaces the I in the starting material? The nucleophile must be that atom or group of atoms. You also have to be careful to adjust for the proper charge. In (a), the nucleophile must be neutral because the entering group winds up positive. In all the other cases, the nucleophile must be negative.

(a) :NH$_3$ (b) $^-$N$_3$ (c) $^-$:$\ddot{\text{S}}$H (d) $^-$:CN: (e) $^-$:$\ddot{\text{O}}$ (f) $^-$:$\ddot{\text{O}}$

Problem 7.37   Bromide is by no means a strong enough nucleophile to displace the poor leaving group hydroxide. Indeed, the only possible reaction is a very, very slight deprotonation of the OH, but even this reaction will be poor as HBr is a much stronger acid (pK_a = −9) than a simple alcohol (pK_a ~17).

However, when the alcohol is treated with HBr, the first step converts the OH into a far better leaving group, $^+OH_2$, and now the bromide can do the displacement.

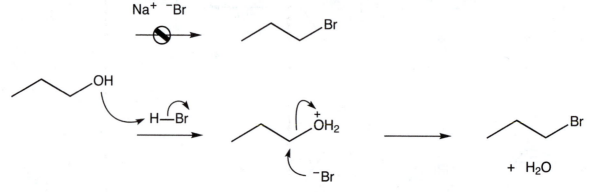

Problem 7.38

(a) HCl
(b) 1. H_3O^+/H_2O 2. NaH   3. CH_3I
(c) 1. HI   2. excess NH_3
(d) 1. HCl   2. Li   3. H_2O

Problem 7.39

(a) HCl
(b) 1. NaH   2. CH_3I
(c) 1. HI   2. excess NH_3
(d) 1. PBr_3 2. NaSH

Problem 7.40   The sulfonate anions are well stabilized by resonance, as the negative charge is shared by three oxygen atoms. The transition state for S_N1 ionization or S_N2 displacement will have a partial negative charge (δ^-) developed on the sulfonate oxygen, and the transition state will be stabilized by delocalization.

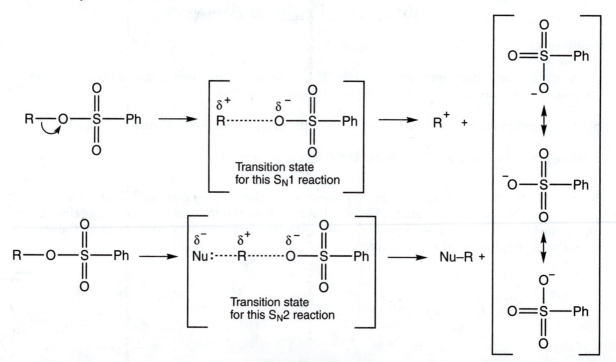

Problem 7.41

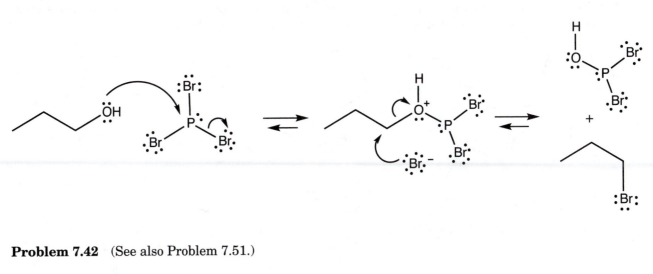

Problem 7.42 (See also Problem 7.51.)

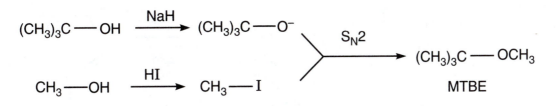

Problem 7.43 There is nothing difficult or profound here, despite the apparent complexity of the molecule. This substitution pattern allows an experimenter to determine the stereochemical course of the reaction. If this methyl transfer is an S_N2 reaction, inversion must take place. The rest of the molecule is really no more than a complicated R group.

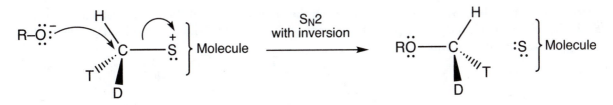

Problem 7.44

(a) Amine **1** can react with ethyl iodide in S_N2 fashion to give a pair of diastereomeric ammonium ions.

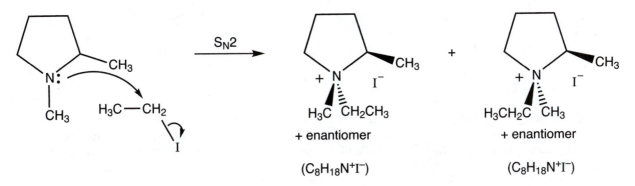

(continued)

Problem 7.44 *(continued)*

(b) When amine **2** undergoes the same kind of S_N2 reaction, a proton can be removed from the initially formed ammonium ion to give an amine. Amine inversion will interconvert the two possible isomers, and only one set of signals will be observed in the NMR spectrum.

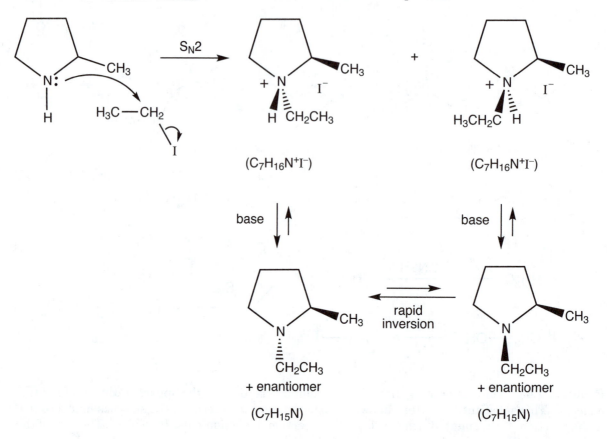

Problem 7.45   Here we have a strong nucleophile, or, really, two of them, in the same molecule, the dithiolate. Surely a double S_N2 displacement of the pair of bromides must lead to one product. The two displacements go one at a time.

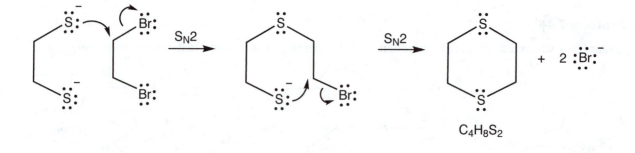

The other compound still contains two sulfurs and two bromines, so the double displacement shown above must have been shortcut somehow. Perhaps after one displacement, the second loss of bromide takes place with another molecule of dibromide—the presence of six carbons shows that such an idea is quite reasonable.

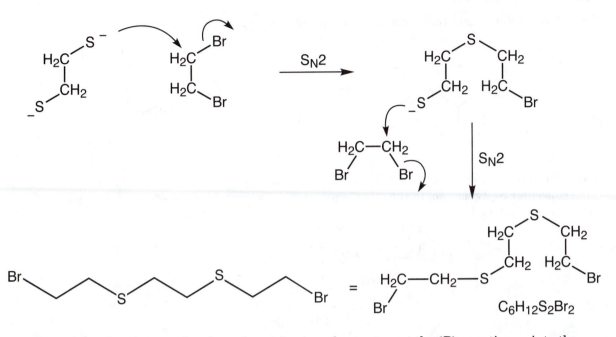

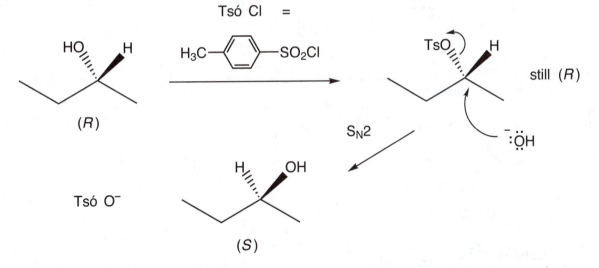

Problem 7.46   Here we need to do an inversion in order to convert the (R) enantiomer into the (S) enantiomer, but OH is surely not a good leaving group. Before we can displace it, we need to transform the leaving group. Page 299 gives one technique—formation of the tosylate from the alcohol. Now displacement by hydroxide is easy, and the S$_N$2 reaction provides the necessary inversion. Hydroxide will cause an elimination, as we will see in Chapter 8.

Problem 7.47   This problem should be fairly easy, except for (d) and (e), which ask you to devise intramolecular S$_N$2 reactions. Nothing is fundamentally changed from the early parts of this problem, but it always seems difficult in the beginning to think "intramolecularly." In parts (a) and (b), there is nothing more than a straightforward displacement of a good leaving group (here chosen as iodide, but many others would do as well) by the appropriate nucleophile.

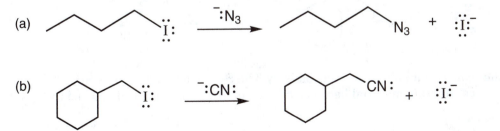

(continued)

Problem 7.47 *(continued)*

Part (c) is similar, except that a charged species, an ammonium ion, is formed.

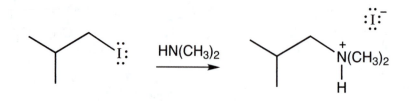

Part (d) can be answered reasonably in two ways. In one, we use a cyclic amine to do an intermolecular S_N2 reaction. In the other, we design an intramolecular version of the S_N2 reaction closely resembling (c).

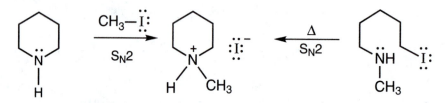

In part (e), we must design another intramolecular S_N2 reaction. The nucleophile required must be formed first from an alcohol by treatment with a base such as sodium hydride.

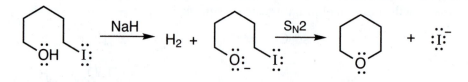

Problem 7.48   Parts (a–c) are straightforward displacements of a leaving group by a nucleophile. Part (a) leads to a pyridinium ion.

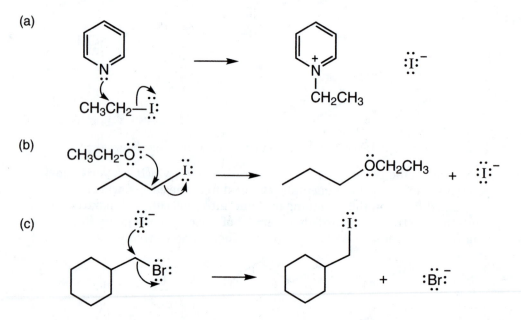

Part (d) requires you to be careful about stereochemistry. The S_N2 reaction always occurs with inversion, so the cis starting material becomes a trans product as iodide is displaced by cyanide.

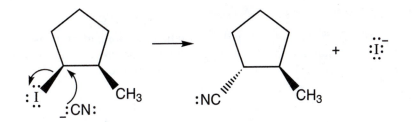

In part (e), there can be no reaction, as the substrate is tertiary and can undergo no S$_N$2 reaction.

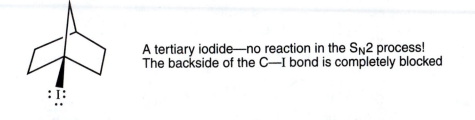

A tertiary iodide—no reaction in the S$_N$2 process!
The backside of the C—I bond is completely blocked

Problem 7.49   The ether cleavage reaction proceeds through protonation of the ether oxygen and displacement by halide ion. In the first example, this can happen in two ways. In the prescence of excess HI, the alcohol products will also be converted into alkyl iodides.

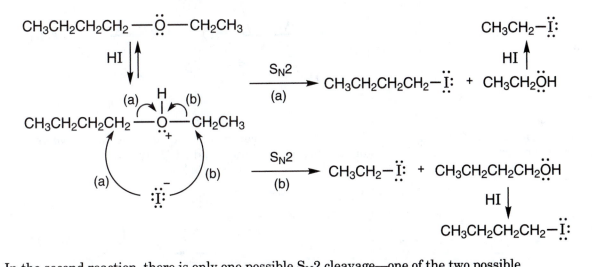

In the second reaction, there is only one possible S$_N$2 cleavage—one of the two possible displacements involves a tertiary carbon, and the S$_N$2 reaction never succeeds under such circumstances.

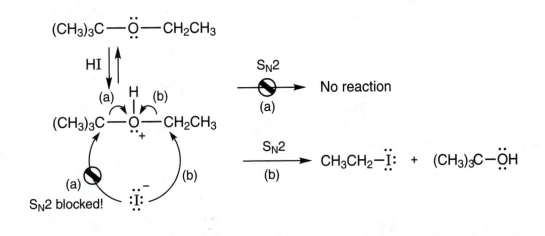

Problem 7.50

In part (a), a pair of diastereomeric alcohols will be produced as ionization produces a carbocation to which water can add at either the "top" or "bottom" lobe of the $2p$ orbital. The intermediate oxonium ions are deprotonated by water to give the alcohol products.

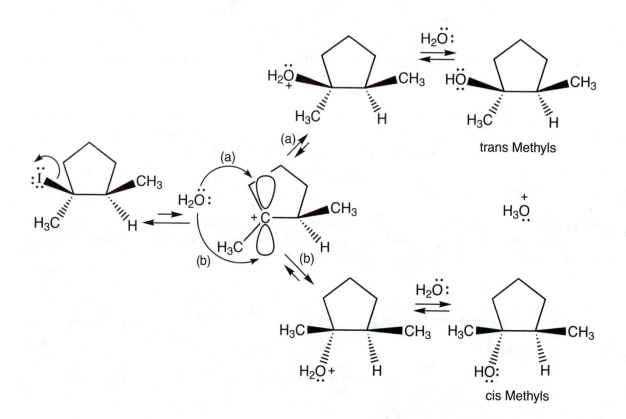

In part (b), ionization to a secondary carbocation leads to capture by the ethyl alcohol (solvolysis) to give an oxonium ion that will be deprotonated to give the ether product, 2-ethoxybutane.

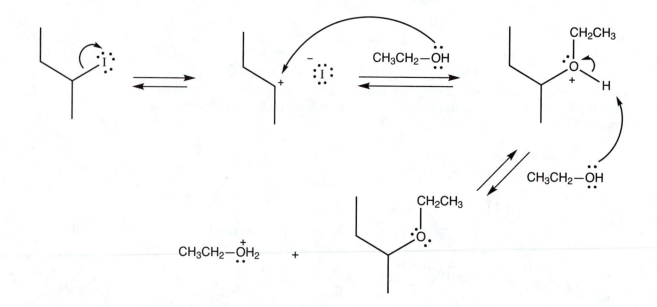

In part (c), a resonance-stabilized carbocation is formed on ionization. This cation can be captured at the two carbons sharing the positive charge to give two products after deprotonation.

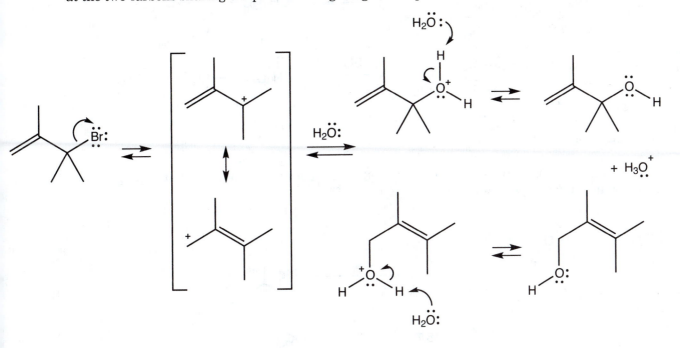

In part (d), ionization will give the *tert*-butyl cation. This intermediate can be captured by any available nucleophile. As there are three nucleophiles present, there will be three products.

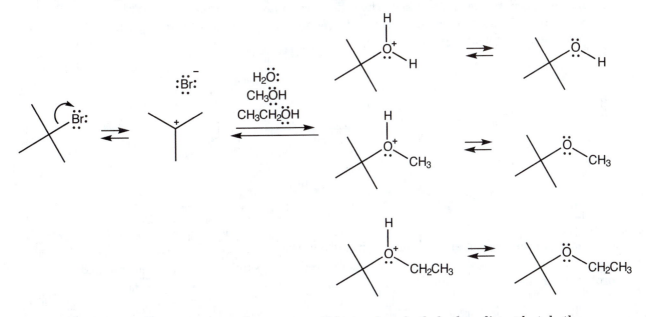

Problem 7.51   In principle, three ethers are possible; *tert*-butyl ethyl ether, di-*tert*-butyl ether, and diethyl ether.

$$CH_3CH_2—\ddot{O}—C(CH_3)_3 \quad (CH_3)_3C—\ddot{O}—C(CH_3)_3 \quad CH_3CH_2—\ddot{O}—CH_2CH_3$$

In practice, only two of these can be made from the starting materials given. *tert*-Butyl iodide cannot be used as a substrate in the S_N2 reaction because it is a tertiary halide and the S_N2 reaction will not proceed with tertiary halides (an elimination will occur instead, as we will see in Chapter 8). We are restricted to using ethyl iodide as the halide, and this leads to only two of the three possible ethers.

(continued)

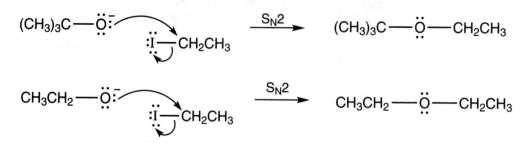

Problem 7.52

(a) The equilibrium for this reaction favors the left by an enormous factor of 10^{18}! It is *not* practical to protonate an alcohol with water.

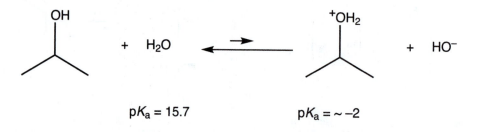

(b) This equilibrium is even less likely. The difference in pK_a is a factor of 10^{40}. This is so unlikely that we wouldn't draw the forward arrow in the equilibrium. We don't think of an alcohol as a base in comparison to ammonia (or an amine).

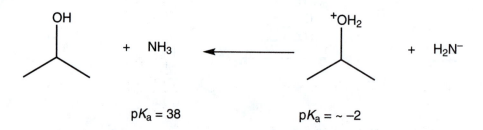

(c) This is another case of an equilibrium that is not very reversible. The estimated pK_a for this cation is –11, which means that the alkene is favored by a factor of 10^{27}. It is not practical to protonate an alkene with water.

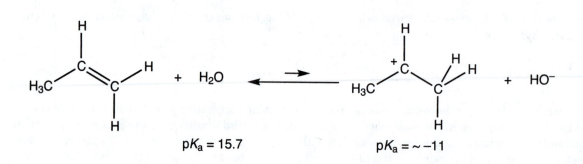

(d) The equilibrium favors the alkene and protonated water, but by a factor of about 10^9, which means that there will be some carbocation formed (one part in about a billion). Notice that this reaction becomes more favorable with a stronger acid, such as HCl (pK_a of –8) or HBr (pK_a of –9).

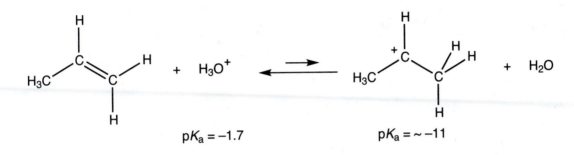

Problem 7.53 A sulfonate is a better leaving group than an acetate because it is a more stabilized conjugate base. There are three resonance structures for the oxygen anion of the sulfonate compared to the two resonance structures for the oxygen anion of the acetate leaving group.

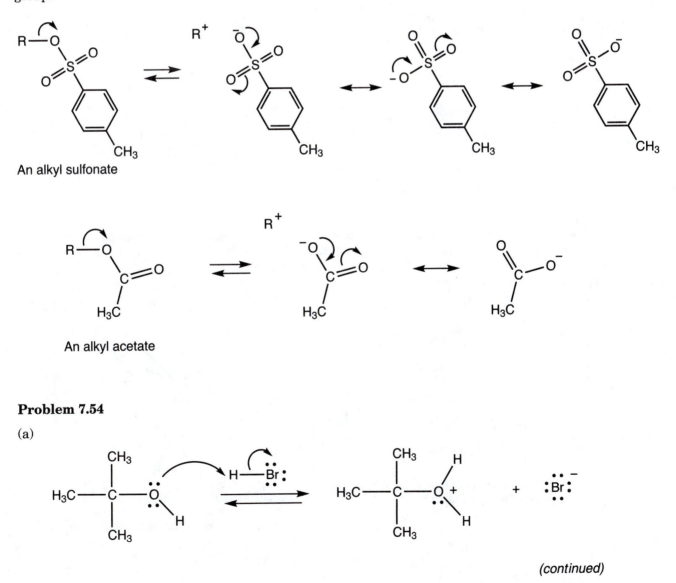

An alkyl sulfonate

An alkyl acetate

Problem 7.54

(a)

(continued)

Problem 7.54 *(continued)*

(b)

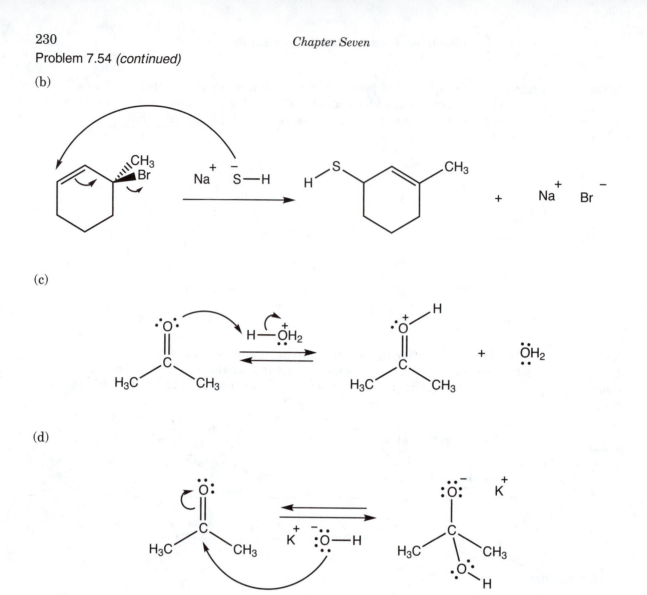

(c)

(d)

Problem 7.55   Recall that in the transition state for the S$_N$2 reaction, the carbon at which displacement occurs is approximately sp^2 hybridized. The transition state for S$_N$2 displacement in allyl systems benefits from delocalization through overlap of the alkene π orbitals with the $2p$ orbital at the "central" carbon.

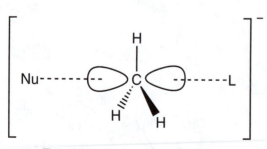

Transition state for typical S$_N$2
displacement reaction of CH$_3$—L

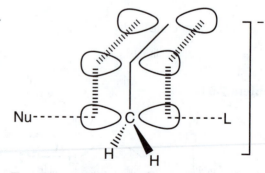

Transition state for S$_N$2 displacement
reaction in an allyl system; note the
overlap with the π orbitals of the double
bond

Problem 7.56

(a) The primary bromide of 1-bromobutane will react faster than the secondary bromide in 2-bromobutane in an S$_N$2 reaction because the backside of the C—Br bond in the secondary position is more sterically hindered. It is more difficult for the nucleophile to approach the empty C—Br antibonding orbital on the secondary carbon.

(b) A protonated alcohol is a very good leaving group. An alcohol is not a leaving group. Therefore, it is the protonated alcohol that will react more rapidly in an S$_N$2 reaction.

(c) A chloride is a better leaving group than a fluoride. Therefore, 1-chlorobutane will react faster than 1-fluorobutane in an S$_N$2 reaction.

(d) The 3-bromocyclopentene will react faster than bromocyclopentane in an S$_N$2 reaction because the 3-bromocyclopentene has the bromo group in an allylic position (p. 291). Leaving groups in the allylic position are especially easily displaced in S$_N$2 reactions because the transition state for displacement is delocalized (see Problem 7.55).

Problem 7.57

(a) The secondary bromide (2-bromobutane) will react much faster than the primary bromide (1-bromobutane) in an S$_N$1 reaction. Remember that it is very difficult to make a primary carbocation. That intermediate is too high in energy, too unstable, for a typical organic reaction.

(b) The secondary bromide (3-bromopentane) will react faster than the secondary chloride (3-chloropentane) in an S$_N$1 reaction because the bromide is a better leaving group. The C—Br bond is weaker than the C—Cl bond.

(c) A tosylate is a better leaving group than a chloride. Therefore, the structure on the right will react faster in an S$_N$1 reaction. The tosylate is a better leaving group because it can accommodate the negative charge so well. There are three resonance structures of the tosylate anion (see Problem 7.53).

(d) The allylic bromide of 3-bromocyclopentene is more reactive than the secondary bromide of bromocyclopentane in an S$_N$1 reaction because the carbocation intermediate that is allylic is more stable. Formation of the more stable carbocation has a lower energy transition state than formation of the higher energy, less stable carbocation.

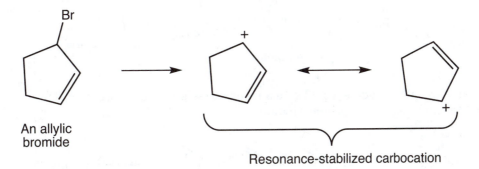

An allylic bromide

Resonance-stabilized carbocation

Problem 7.58

(a) The primary bromide will surely be more reactive than the tertiary bromide in the S_N2 reaction.

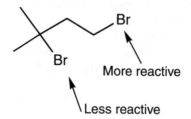

(b) It will be the allylic iodide that is the more reactive, as the transition state for S_N2 displacement will benefit from delocalization (Table 7.3, p. 291, Problem 7.55, p. 328).

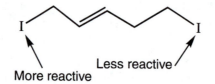

(c) Bromide is a better leaving group than chloride and will be more reactive in the S_N2 reaction.

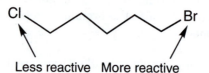

Problem 7.59

(a) The tertiary bromide will ionize to a relatively stable tertiary carbocation. It will be much more reactive in the S_N1 reaction than the primary bromide.

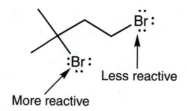

(b) Ionization of the allylic iodide would give a resonance-stabilized carbocation. This reaction will be preferred to ionization of the primary iodide.

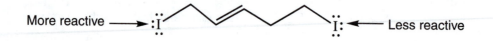

(c) Ionization of either bromide would lead to a tertiary carbocation. However, only one of the carbocations can become planar. As carbocations are most stable when planar, it will be that bromide that is lost more easily.

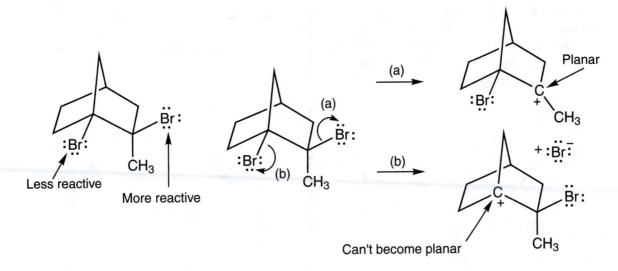

Problem 7.60 The first thing to do is to draw the two possible iodides. What does that squiggly bond really mean? The two possibilities are

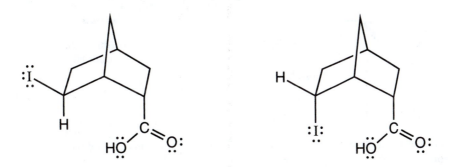

Notice that only one of the salts has a carboxylate in position to do a backside S_N2 displacement of iodide. This displacement leads to the compound $C_8H_{10}O_2$. The other alkoxide would have to do a frontside S_N2 reaction, and, as we know, this never happens. So, the molecule that undergoes the intramolecular S_N2 reaction has the iodine in the "exo" position, and the molecule that cannot do the S_N2 reaction must have the iodine "endo."

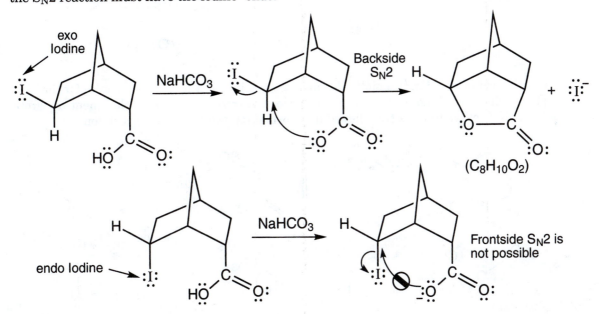

Problem 7.61   The same point is made in this problem as in the last one. The impossibility of a frontside S$_N$2 reaction renders one stereoisomer unreactive. The isomer that can undergo a backside S$_N$2 reaction does so to produce the new compound. The compound that reacts is the trans isomer and the one that does not is the cis isomer.

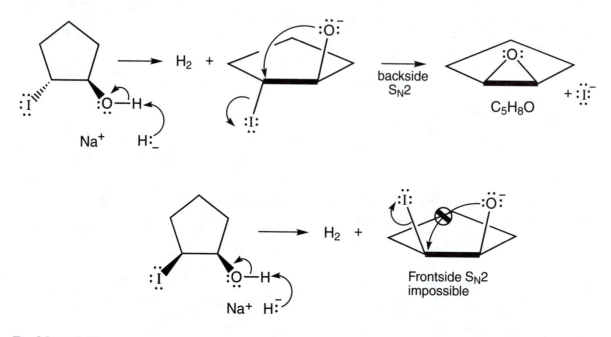

Problem 7.62

(a) The "obvious" mechanism is an intramolecular S$_N$2 reaction to transfer the methyl group from oxygen to carbon.

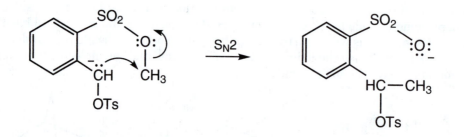

(b) If the reaction is bimolecular, two molecules must be involved in the transition state for the reaction. The mechanism cannot be the simple unimolecular *intramolecular* displacement we wrote in (a). Perhaps the methyl group is transferred in a normal *intermolecular* S$_N$2 reaction.

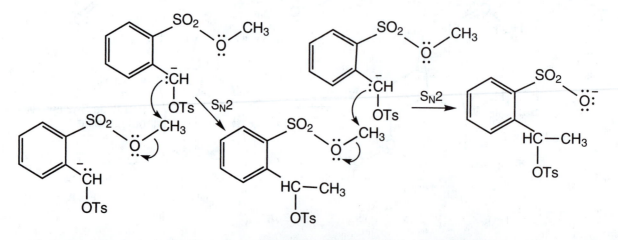

(c) Why is this intermolecular process more favorable energetically than the intramolecular reaction? There is a good nucleophile and an excellent leaving group. However, the transition state for the S$_N$2 reaction is ideally linear, and a linear arrangement of nucleophile, substrate carbon, and leaving group cannot be attained in the cyclic process. Apparently, this poor alignment is enough to make the bimolecular reaction more favorable.

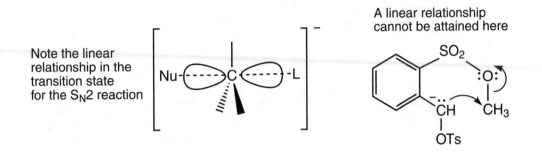

Problem 7.63 No! The methoxide ion is not strong enough to deprotonate propane. Methanol is a much, much stronger acid than propane.

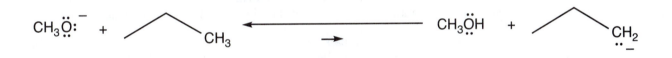

Problem 7.64 In the ionization, the angles around the central carbon expand from approximately 109° (*sp*³) to approximately 120° (*sp*²). The large *tert*-butyl groups are further apart in the planar, *sp*² carbocation than they are in the starting material. The larger the R group, the more strain relief there will be on ionization, and the faster the reaction.

(continued)

Problem 7.64 *(continued)*

Here is an energy versus reaction progress diagram for the two reactions:

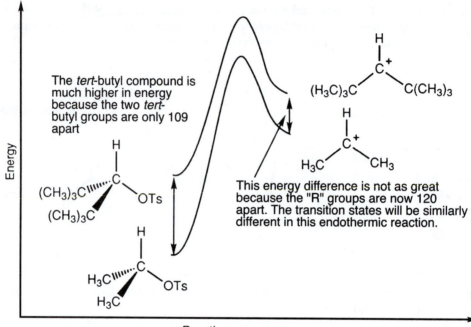

Problem 7.65 The stabilization by resonance of the cation formed from ionization of the starting halides depends on overlap between an exo-ring 2*p* orbital and the π orbitals of the ring.

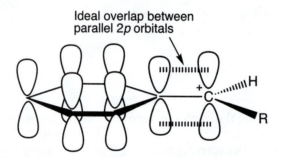

However, when "R" is large, it "bumps" with the nearby hydrogens. This destabilizing interaction can be relieved by rotation about a carbon–carbon bond, but this rotation decreases resonance stabilization in the carbocation and makes the intermediate more difficult to form.

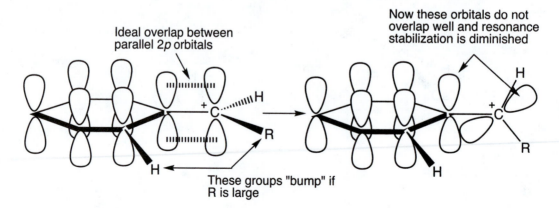

Accordingly, the bigger the "R" group, the more difficult it is to form the carbocation in this example.

Problem 7.66 We want to make (*S*)-2-iodobutane from (*R*)-2-butanol. This looks rather easy, if we can use phosphorus triiodide. But you may have noticed that we haven't used this reagent in the text. That's because it is violently reactive with water. It's fine to use it as your answer to this problem, but in the lab, you might want to consider an alternative route. The reaction we hoped you would consider is the conversion of the alcohol to the tosylate, which occurs with no change of stereochemistry, and then an S_N2 reaction using the iodide anion as the nucleophile.

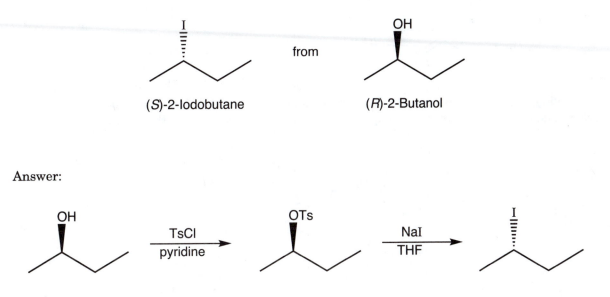

(*S*)-2-Iodobutane   from   (*R*)-2-Butanol

Answer:

OH   TsCl / pyridine → OTs   NaI / THF → I

Making the (*R*)-2-iodobutane from the (*R*)-2-butanol is more difficult. One way to perform a substitution with retention is to do two substitutions that are both with inversion. That is the way to go in this case.

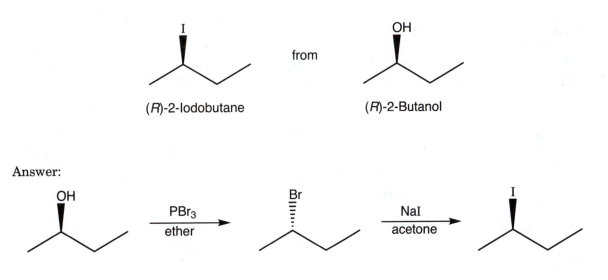

(*R*)-2-Iodobutane   from   (*R*)-2-Butanol

Answer:

OH   PBr_3 / ether → Br   NaI / acetone → I

Problem 7.67   Vinyl halides don't undergo S$_N$2 reactions. Here are a few logical explanations.

(1) The carbon–halide bond of an *sp*² carbon is stronger than the *sp*³ carbon–halogen bond we have been breaking in a typical S$_N$2 reaction. The bond is stronger partly because it is shorter.

(2) Resonance! Perhaps the S$_N$2 reaction doesn't occur because there is resonance between the halide and the alkene. It's not a carbon–halogen single bond; it has appreciable double bond character.

(3) The vinyl carbon–halogen antibonding orbital is more hindered than a typical *sp*³- hybridized carbon. The methyl group in this example blocks access to the backside of the C—Br.

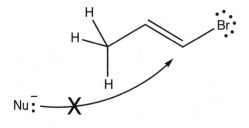

(4) The antibond of a stronger bond is higher in energy and more unlikely to mix with the typical nucleophile. Remember that the orbital mixing works best if the HOMO and the LUMO are similar in energy (p. 36).

Problem 7.68   The amine group in each molecule is first converted into a diazonium ion. In order to see what happens next, we need better three-dimensional drawings of these diazonium ions. So, first draw out the chair conformations of these molecules, and be careful to keep the large *tert*-butyl group in the lower energy equatorial position. In each case, nitrogen will be displaced by a group or bond in the proper position (180°, antiperiplanar) to assist departure by displacement from the rear.

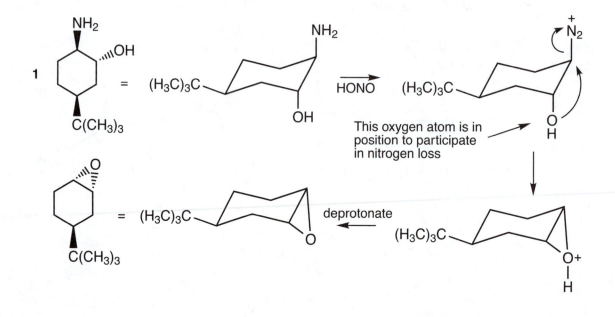

This oxygen atom is in position to participate in nitrogen loss

deprotonate

In the second case, the hydroxyl oxygen is no longer in a position to displace nitrogen. Instead, the axial carbon–hydrogen bond is in position. A hydride shift, which we will learn more about in Chapter 8, takes place to give the resonance-stabilized cation **A**. Deprotonation gives the ketone.

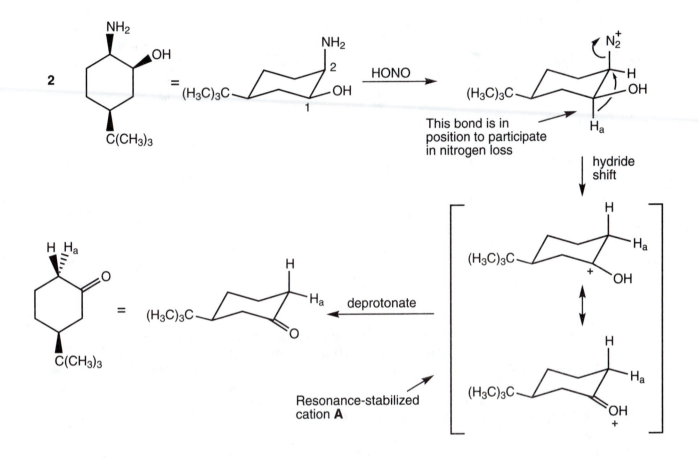

It is H_a that is able to shift from C(1) to C(2). Because H_a shifts with its electrons, we call it a hydride shift. It is this hydrogen that shifts because the electrons in the C(1)—H bond are aligned perfectly to feed into the antibond of the C(2)—N bond. Remember that N_2^+ is the world's best leaving group. When H_a moves to C(2), there is inversion at C(2). The hydrogen that was originally on C(2) moves from the equatorial position to the axial position as the carbon goes through inversion.

Problem 7.69   This procedure takes advantage of iodide being both an excellent nucleophile and an excellent leaving group (Tables 7.5 and 7.6). Initially, the small amount of iodide displaces chloride (S_N2) to give an intermediate iodide that is more reactive in the S_N2 reaction than the original chloride. The nucleophile now displaces iodide (S_N2 again) to generate a molecule of product and *regenerate* iodide ion. The iodide now can recycle, displacing another chloride, etc.

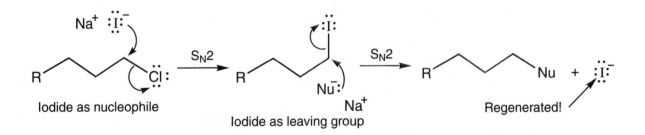

Problem 7.70   Here is the reaction:

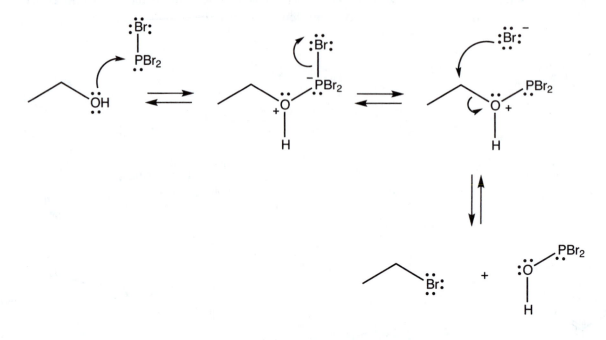

The animation shows two intermediates. The first intermediate has a phosphorus atom with an expanded sphere (10 electrons in the outer shell). It is entirely possible that the alcohol does an S_N2 reaction on the phosphorus, thus bypassing the first intermediate. In that case, there would be only one intermediate.

The fastest step is probably the last one—the bromide attacking the carbon and displacing the $HOPBr_2$.

Problem 7.71   Other intermediates that are likely to be present in this reaction are $ROPBr_2$, $(RO)_2PBr$, $(RO)_3P$, $ROP(OH)_2$, $(RO)_2POH$, and the protonated versions of each of these.

Problem 7.72   The animation suggests that the last step is an S_N2 reaction. The bromide attacks from the backside of the carbon–oxygen bond. This pathway is not possible for a tertiary alcohol. The tertiary alcohol does react with PBr_3 to give the tertiary bromide, but it is undoubtedly through an S_N1 pathway.

Elimination Reactions: The E1 and E2 Reactions

8

Problem 8.1

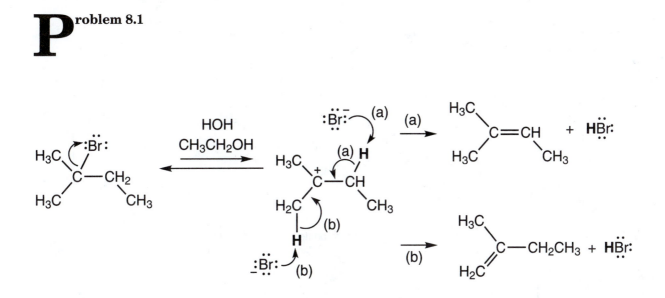

A key point here is the assistance in loss of H⁺ by a base, here shown as bromide ion.

Problem 8.2 The S$_N$1 reaction involves ionization followed by product formation through addition of a nucleophile. The carbocation can be captured by any of the nucleophiles present. If bromide ion captures the carbocation, starting material is regenerated. The other nucleophiles present are ethyl alcohol (CH$_3$CH$_2$OH) and water. Reaction of these molecules with the carbocation leads ultimately to an alcohol and an ethyl ether:

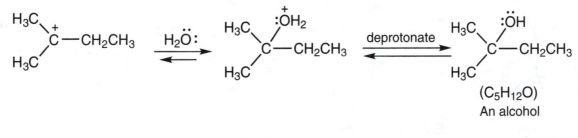

(continued)

Problem 8.2 *(continued)*

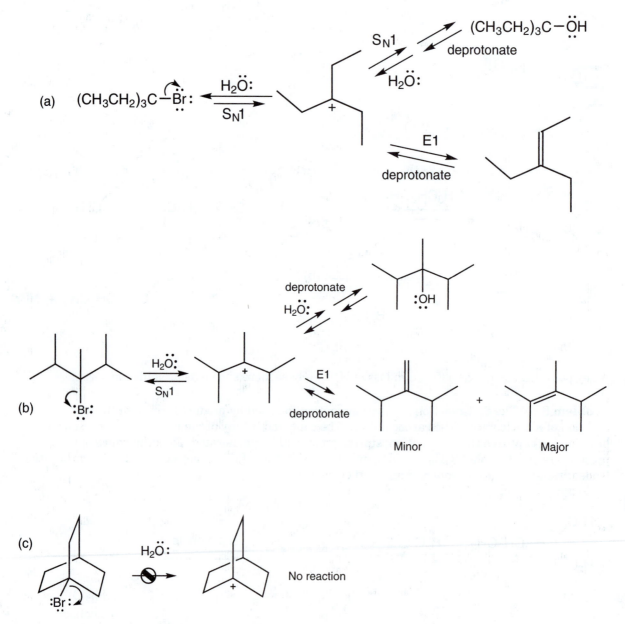

Problem 8.3 The interesting case here is the last, bicyclic molecule. Ionization to the bridgehead cation cannot take place, even though the ion is tertiary. The bicyclic system ensures that the intermediate cation cannot become planar, and that destabilization is enough to stop the reaction. The lesson here is that you can never suspend thought. You have to look critically at each reaction intermediate in every reaction.

Problem 8.4 These are E1 reactions. The major product in each case will be the thermodynamically most stable product (Saytzeff product). We will expect minor amounts of the less stable alkenes. There are also minor products in (a), (b), and (c) that we can only predict after we have learned about carbocation rearrangements (Section 8.5).

(a) The (*E*)-3-methyl-2-pentene is slightly more stable than the (*Z*)-3-methyl-2-pentene.

(b) For this reaction we expect one major product and two minor products.

(c) You should be able to predict the major product for this reaction. We also expect 3-methylcyclopentene.

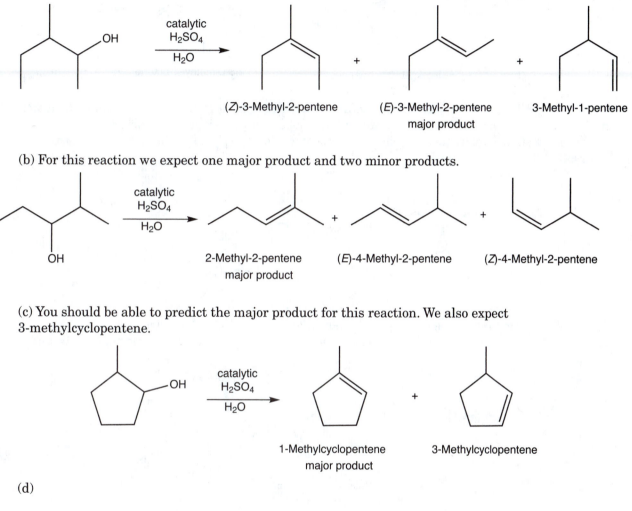

(d)

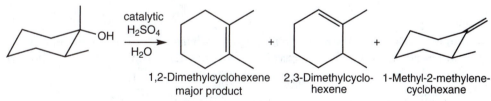

Problem 8.5 There is a second E2 elimination possible in which one of the six equivalent methyl hydrogens is lost.

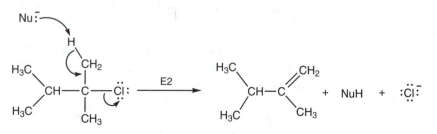

Problem 8.7   These are E2 reactions.

(a) This molecule is a primary bromide, but the strong base is bulky. We only expect an E2 reaction. S_N2 will not compete.

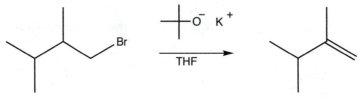

2,3-Dimethyl-1-butene

(b) This molecule is a tertiary bromide reacting with a strong base. S_N2 will not compete. The major product will be the Saytzeff product. There will likely be some minor products that result from attack at the other β hydrogens. The minor products would be (*E*)- and (*Z*)-3,4-dimethyl-2-pentene and 2-ethyl-3-methyl-1-butene.

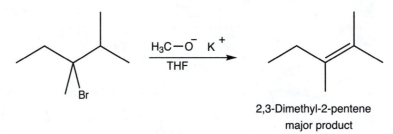

2,3-Dimethyl-2-pentene
major product

(c) This E2 reaction involves a secondary bromide and a strong, bulky base. There is only one product expected in this elimination. That is because there is only one β hydrogen that can be anti to the leaving group.

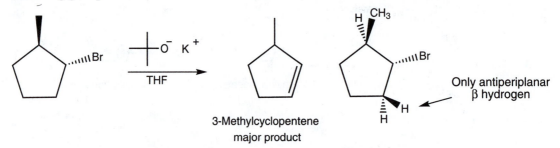

3-Methylcyclopentene
major product

Only antiperiplanar
β hydrogen

(d) This reaction involves a tertiary bromide and a strong base, therefore it will be an E2 reaction. There are two antiperiplanar β hydrogens available for this dehydrohalogenation. The major product is pre-dicted to be the more substituted alkene (the Saytzeff product).

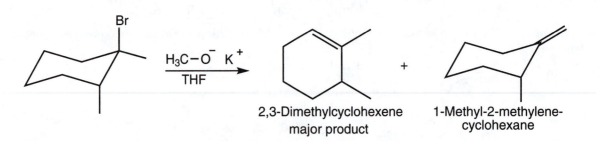

2,3-Dimethylcyclohexene
major product

1-Methyl-2-methylene-
cyclohexane

Problem 8.8

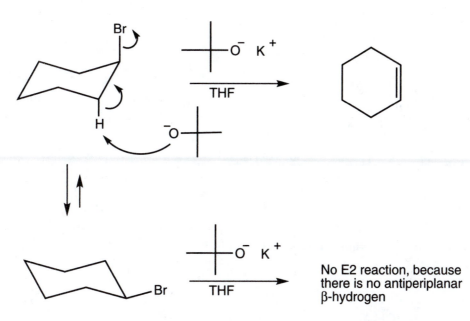

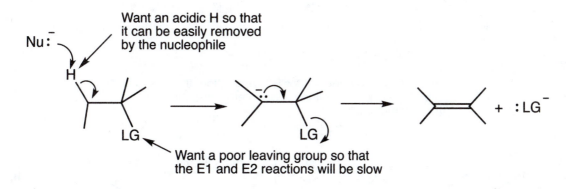

No E2 reaction, because there is no antiperiplanar β-hydrogen

Problem 8.9   For a good E1cB reaction, a poor leaving group and a highly acidic hydrogen are needed.

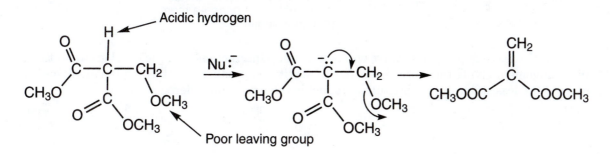

The following molecule fills the bill. Methoxide is a poor leaving group, and the two $COOCH_3$ (methyl ester) groups will stabilize the negative charge on an adjacent carbon by resonance, thus making the loss of a proton easier. There are many other possible answers, of course.

Problem 8.10   The possible products are

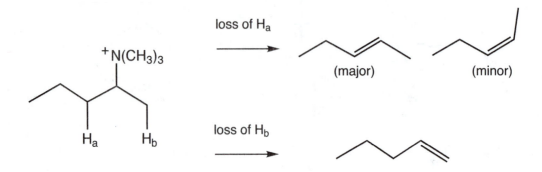

loss of H$_a$

(major)   (minor)

loss of H$_b$

Here are the Newman projections for loss of H$_a$ and H$_b$:

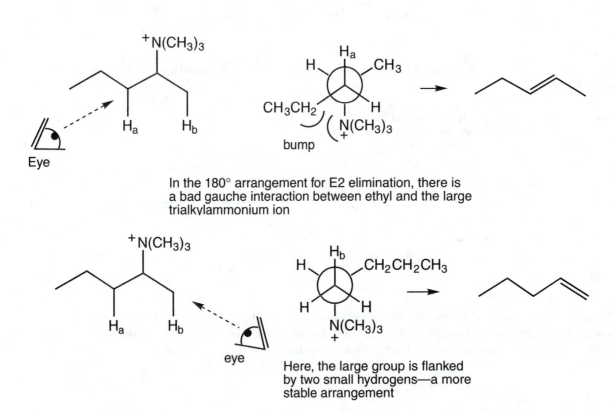

Eye

In the 180° arrangement for E2 elimination, there is
a bad gauche interaction between ethyl and the large
trialkylammonium ion

eye

Here, the large group is flanked
by two small hydrogens—a more
stable arrangement

The larger the leaving group, the worse will be the steric interaction with the ethyl group in the
180° arrangement giving the more stable, more substituted product. By contrast, the arrangement
leading to the less stable, less substituted product is relatively unhindered.

Problem 8.12   The activation energy for product Y going to starting material is higher than the activation energy for product X going to starting material. That is because product Y is *more* stable than product X *and* the transition states differ in energy *less* than do the products. If the transition state from product X were higher in energy than the transition state from product Y by an amount larger than the $\Delta G°$, then the activation energy for X going back to starting material would be higher than the activation energy for product Y going back to starting material.

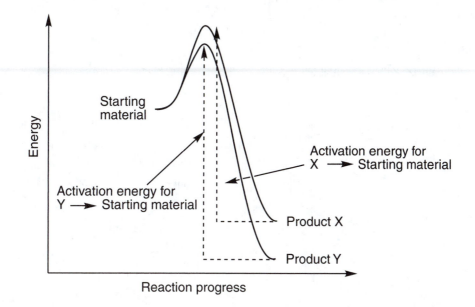

Problem 8.14   On page 286 of the text, we saw the essentially degenerate S_N2 reaction between methyl iodide and radioactive iodide (*I⁻). Here's a completely degenerate S_N2 reaction. One would not be able to monitor this reaction, as the product is exactly the same as the starting material.

Problem 8.15

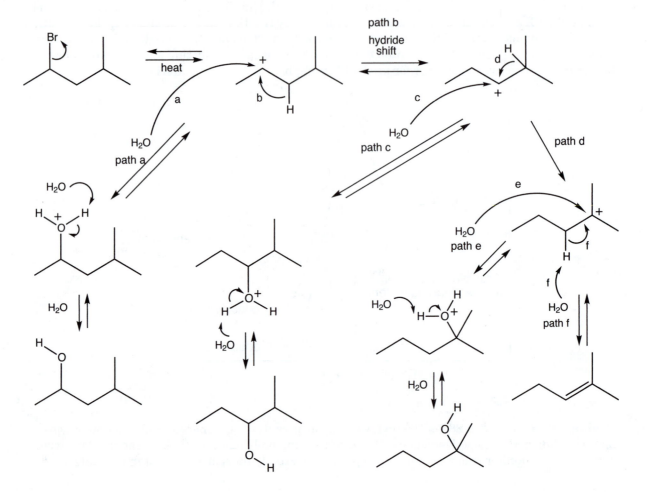

The alkene product can be formed from the tertiary carbocation in path f, as shown in the preceding scheme. It is also possible to form the same alkene from the second carbocation, as shown in the following scheme. Each reaction pathway regenerates H_3O^+.

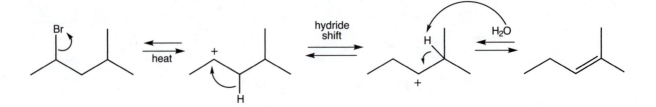

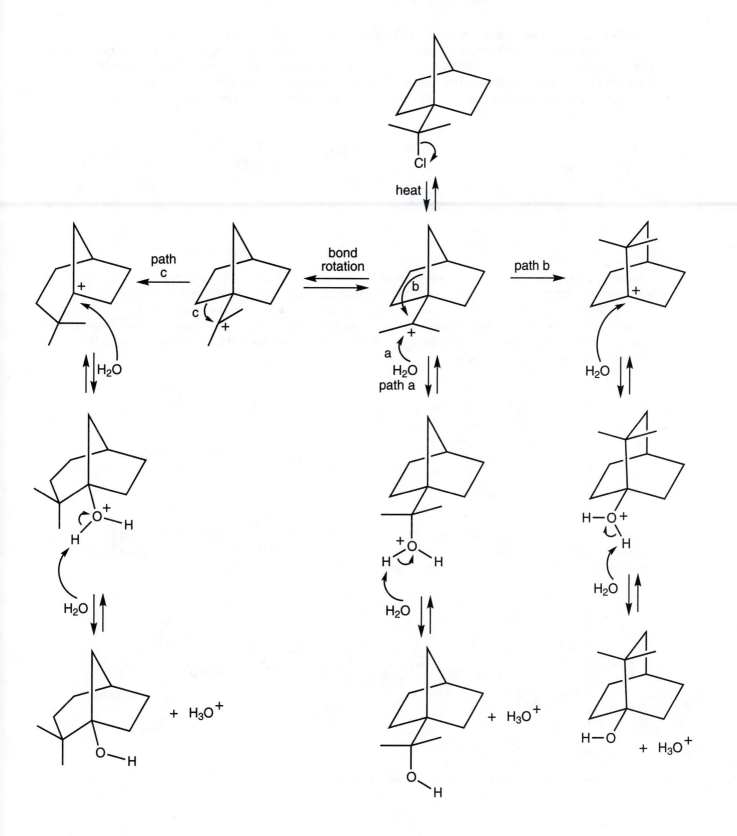

Problem 8.16   There are two hydride shifts that occur in the first part of Problem 8.15. These are shown in more detail here. In the first hydride shift, one of the hydrogens on carbon 3 moves to the carbocation carbon (path b). The hydrogen moves with its pair of electrons from the C—H bond (the nucleophile) to the empty orbital of the carbocation (the electrophile). In this case, two secondary carbocations interconvert. Therefore, the rearrangement is reversible.

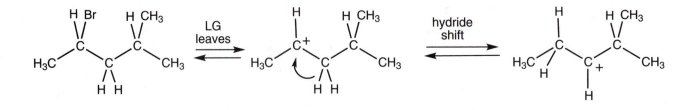

The second hydride shift for this reaction is shown next. In this case, a secondary carbocation becomes a tertiary carbocation as a result of the hydride shift. The tertiary carbocation is much more stable, so we don't consider this a reversible reaction. Once again the migrating hydrogen and its electrons from the C—H bond move to the empty *p* orbital of the adjacent carbocation. We see that the shifting hydride (with its pair of electrons) is the nucleophile. The carbocation (with the empty orbital) is the electrophile.

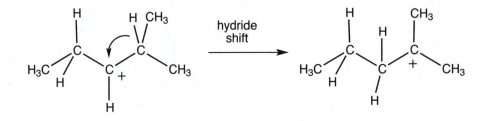

Problem 8.17   There is nothing tricky here. The only problem is to make the connection between carbon disulfide (CS_2), which you have not encountered before, and carbon dioxide (CO_2), which you have. Addition to carbon disulfide is followed by an S_N2 reaction to make the xanthate.

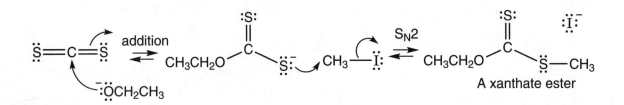

Now a reaction very much like that of Figure 8.43 occurs.

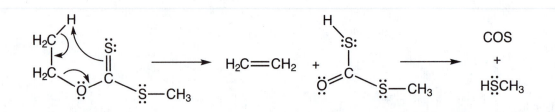

Problem 8.18

(a) These are both examples of thermal elimination reactions. In part (a), it is the negative oxygen atom that acts as base and removes a proton.

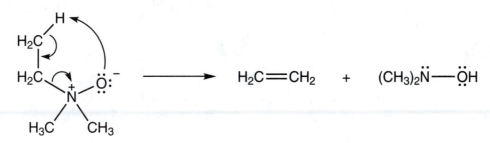

Reaction (b) is just a double ester elimination.

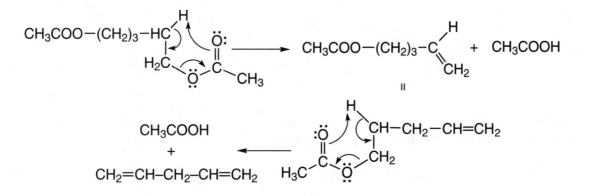

Problem 8.19   The mixing of the three orbitals involved in the migration of a hydrogen from one carbon to an adjacent carbon gives the molecular orbitals shown in Figure 8.50. The three orbitals for the cation contain two electrons, and both of those electrons are in the bonding orbital. The radical would have three electrons involved in the migration. The electron in the antibonding orbital destabilizes the transition state sufficiently to make this rearrangement relatively uncommon in radical intermediates.

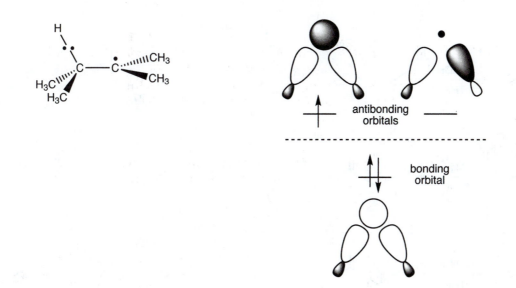

(continued)

Problem 8.19 *(continued)*

The anion would have four electrons involved in the migration. This scenario is even worse than the radical case. With two electrons in the antibonding orbitals, there is no net favorable mixing of the orbitals for this transition state. We will learn that there is an even more unfavorable interaction of this molecular orbital picture when we get to the aromatic ring chapter (Chapter 14).

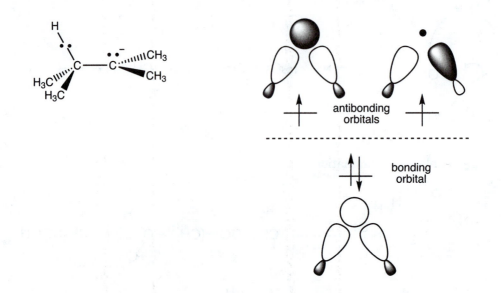

Migration of an alkyl group involves a similar set of orbital interactions. However, it is an sp^3 orbital rather than an s orbital that is involved in mixing.

Additional Problem Answers

Problem 8.20

(a)

(b)

(c)

(d)

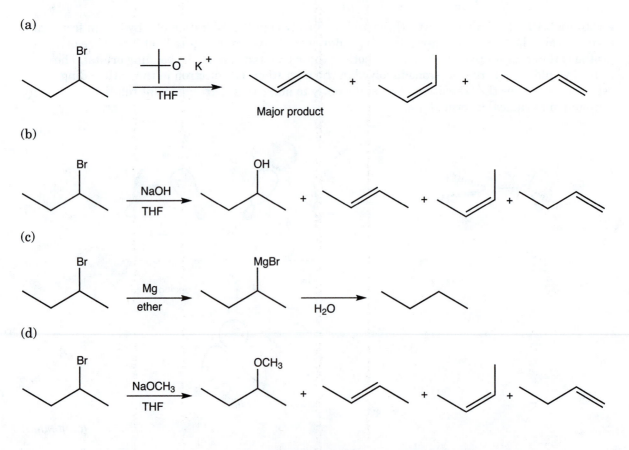

Problem 8.21

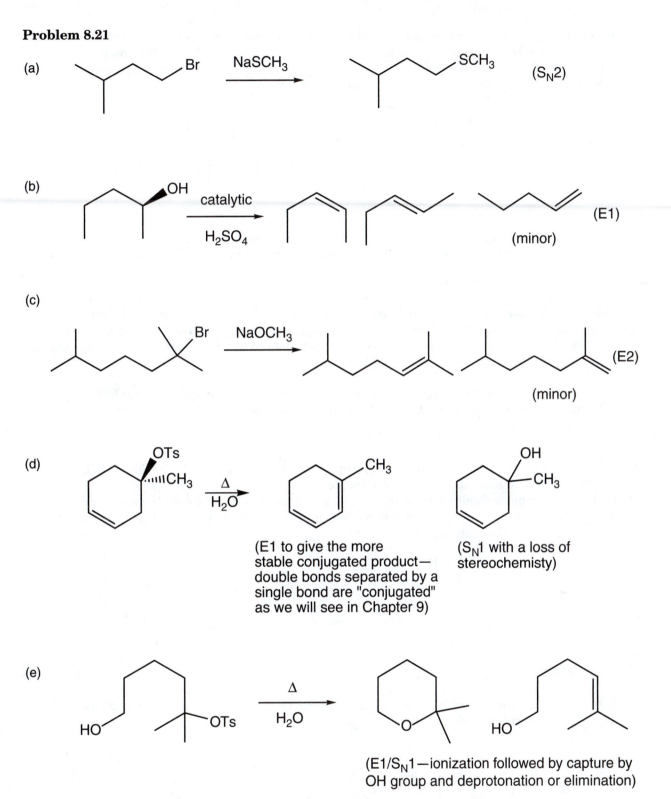

Problem 8.22   These reactions all seem to be elimination reactions, and the presence of the strong base *tert*-butoxide must lead us to think of the E2 process. So, the two possible alkenes are methylenecyclohexane and 1-methylcyclohexene. Which we get will depend on the position and structure (Saytzeff or Hofmann) of the leaving group.

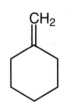
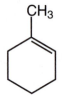

Methylenecyclohexane   1-Methylcyclohexene has seven
has only five ^{13}C NMR signals   different carbons and seven ^{13}C NMR signals

Compound (a) has a Hofmann leaving group, F, and must give mostly the less-substituted alkene, methylenecyclohexane. Compound (b) can only give methylenecyclohexane because of the position of the leaving group—no other product is possible. So (a) and (b) are the two compounds that give the alkene with only five ^{13}C NMR signals.

Compounds (c) and (d) have Saytzeff leaving groups and must give mostly the most highly substituted alkene possible, in this case, 1-methylcyclohexene, which shows seven ^{13}C signals in its NMR spectrum.

Problem 8.23   An E2 reaction requires a strong base. We typically use hydroxide (LiOH, NaOH, or KOH) or alkoxide (NaOR). A protonated alcohol is a strong acid. The pK_a of a protonated alcohol is about −2. If we try to use a strong base in the presence of a strong acid, we will only have proton transfer. An example is shown here. The base deprotonates the acidic hydrogen rather than initiating the E2 reaction.

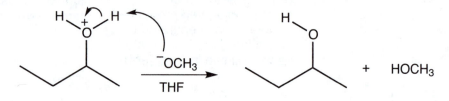

Problem 8.24

Part (a) is a straightforward E2 elimination with a Saytzeff leaving group. Any strong base will do, and isopropoxide will work well.

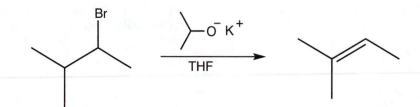

Part (b) is trickier, as we need to do a Hofmann elimination and the molecule contains a Saytzeff leaving group. Clearly, we need to switch leaving groups somehow. Why not displace the bromide with trimethylamine, giving us the ammonium ion, a Hofmann leaving group? Now we can do our E2 reaction and the major product will be the desired less-substituted alkene.

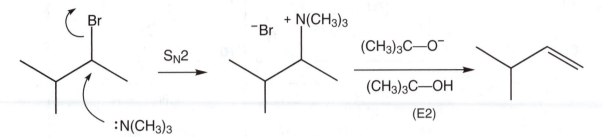

Part (c) is hard, as again you need to do two reactions in sequence. The simplest way to produce the final product is to add DBr to isobutene, so this part of the problem quickly resolves into a search for a way to make isobutene from *tert*-butyl bromide.

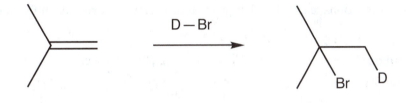

An E2 elimination will produce the required alkene.

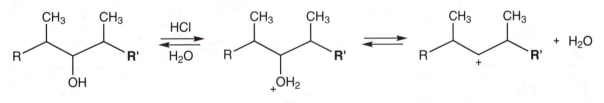

Problem 8.25   There is quite a number of products possible! First of all, what's likely? Protonation of the alcohol, followed by loss of water, to give the carbocation starts us off on an $S_N1/E1$ process.

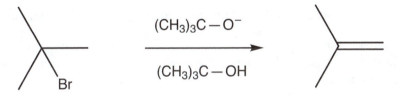

There are two nucleophiles present, chloride ion and water. So, two possible products are the starting alcohol and the corresponding chloride, **A**. This is the S_N1 part of this reaction.

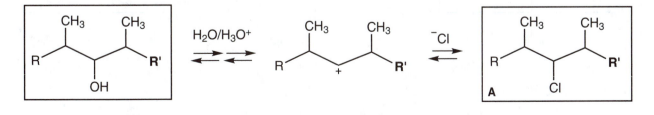

(continued)

Problem 8.25 *(continued)*

The E1 part involves deprotonation of the carbocation to give an alkene. Deprotonation can occur in two directions, and there are two stereoisomers of each alkene. Now we have products **B–E**.

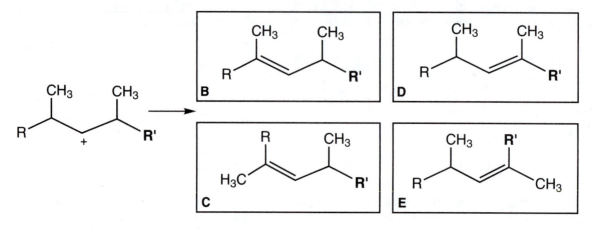

But remember—if a hydride shift to give a more stable carbocation can occur, it will. In this case, there are two such reactions, each of which generates a tertiary carbocation from a secondary carbocation.

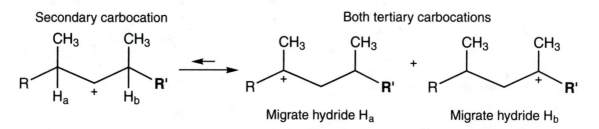

Each new carbocation can react with the two nucleophiles present, water and chloride ion. Four new products, **F–I**, are formed.

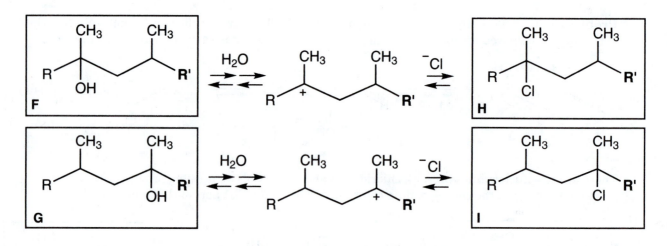

Finally, each of the tertiary carbocations can lose a proton in an E1 reaction to give two more new products, **J** and **K**, as well as alkenes **B, C, D**, and **E**.

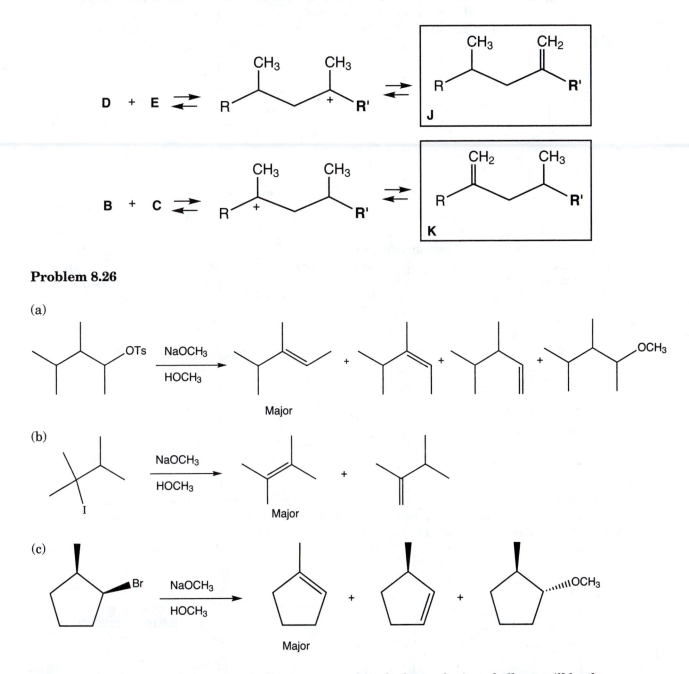

Problem 8.26

(a)

(b)

(c)

(d) Because the *tert*-butoxide base is bulky, we expect that the less-substituted alkene will be the major product. It is a sterically congested hydrogen that would have to be removed in order to give the more-substituted alkene.

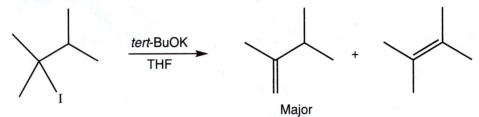

Problem 8.27

(a) This reaction should lead to three products. Two are straightforward. Ionization will give initially the secondary carbocation, and there are two possible losses of hydrogen to give an alkene. It will be the more stable alkene (the trisubstituted one in this case) that dominates (Saytzeff elimination).

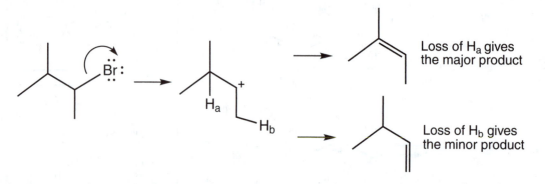

Loss of H_a gives the major product

Loss of H_b gives the minor product

However, we can also expect rearrangement through hydride shift to give a more stable tertiary carbocation, and there are also two different hydrogens (H_c and H_d) that can be lost to give an alkene. One of the product alkenes is the same as the major one formed from the secondary carbocation, but the other one is new. All three will be produced.

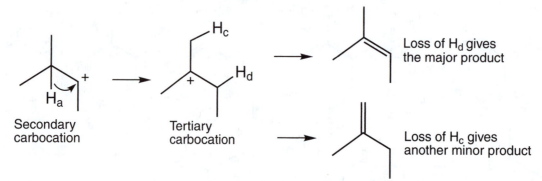

Secondary carbocation

Tertiary carbocation

Loss of H_d gives the major product

Loss of H_c gives another minor product

(b) This reaction is a straightforward Saytzeff elimination. The tertiary carbocation will be formed on ionization, and two protons can be removed to give different alkenes. The more substituted, more stable alkene will be the major product.

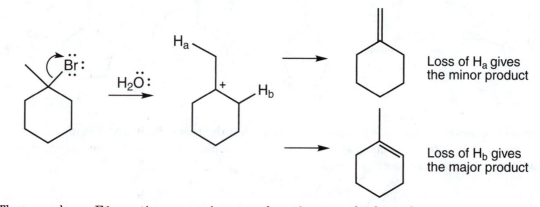

Loss of H_a gives the minor product

Loss of H_b gives the major product

(c) There can be no E1 reaction, as a primary carbocation must be formed.

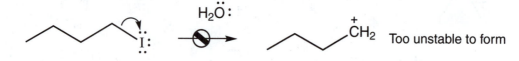

Too unstable to form

(d) Even though the leaving group is a "Hofmann" leaving group, this E1 reaction will give the more substituted alkene (Saytzeff elimination) as the major product. The intermediate is the same as in (b), and there can be no substantial change in product distribution.

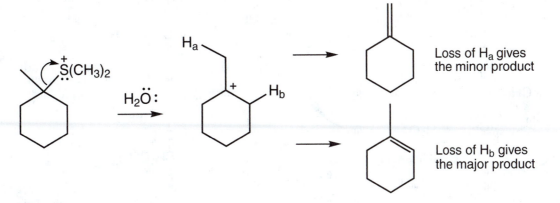

Loss of H$_a$ gives the minor product

Loss of H$_b$ gives the major product

Problem 8.28

(a) Start with a good three-dimensional drawing of menthyl chloride. In the more stable chair form, there is no hydrogen oriented at the optimal 180° angle for E2 elimination. In the less stable chair form, there is only one hydrogen in the proper 180° orientation to the chloride for an E2 elimination. Thus, there is only one product.

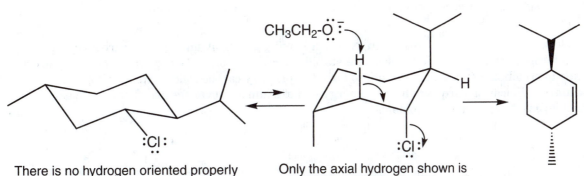

There is no hydrogen oriented properly for the E2 reaction in this chair form

Only the axial hydrogen shown is oriented properly for the E2 reaction

The more stable chair conformation for neomenthyl chloride has the chlorine axial. In this molecule, there are *two* axial hydrogens that can be lost in a "perfect" 180° E2 reaction. As usual, the more stable product is the major one formed (Saytzeff elimination).

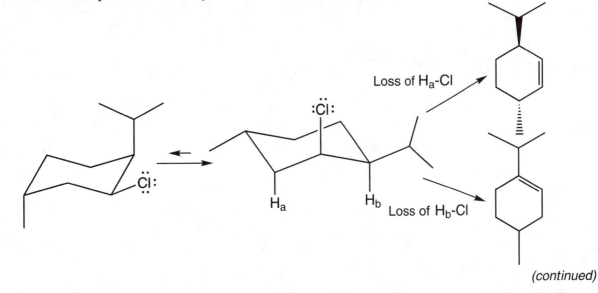

Loss of H$_a$-Cl

Loss of H$_b$-Cl

(continued)

Problem 8.28 *(continued)*

(b) This reaction is E1, and ionization will generate a planar carbocation that can lose either of two protons to give the two products shown in the problem. The more stable, trisubstituted alkene will be favored (Saytzeff elimination).

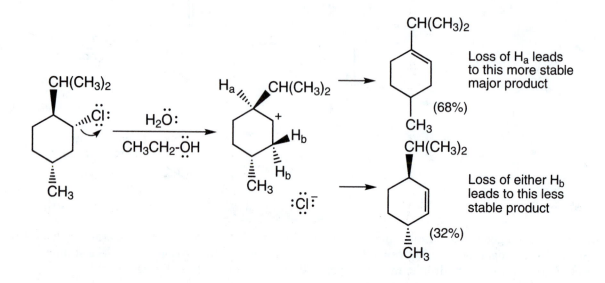

Loss of H_a leads to this more stable major product

(68%)

Loss of either H_b leads to this less stable product

(32%)

Problem 8.29 These reactions are examples of the Tiffeneau–Demyanov ring expansion. The first step in each case is the conversion of the amino group into a diazonium ion.

(a) Here, displacement of nitrogen, an excellent leaving group, by an adjacent carbon–carbon single bond forms the ring-expanded, resonance-stabilized carbocation **A**. Deprotonation of **A** by any base leads to cyclohexanone. Be careful not to form the primary carbocation **B**. *Remember*: Simple primary carbocations are very high in energy and are almost never formed. They serve as mechanistic "stop signs" in problems.

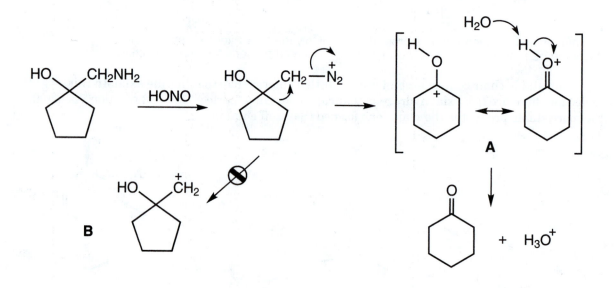

(b) In the second example, two different modes of ring expansion (a and b) are possible, as two different carbon–carbon bonds are in position to help displace the leaving group.

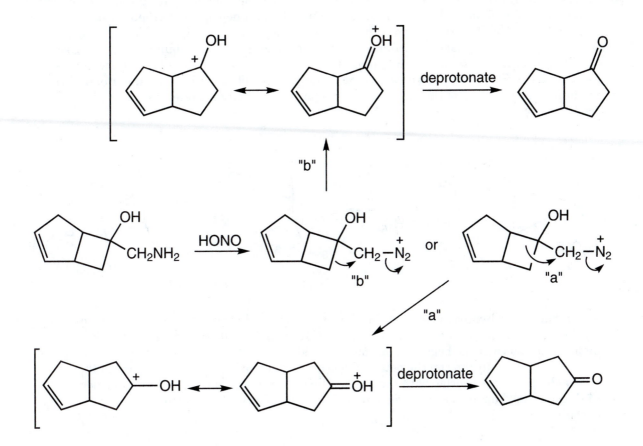

Problem 8.30 This is an E2 reaction. The rate of an E2 reaction of a secondary iodide is

Rate = $\upsilon = k[\text{R—I}][\text{base}]$

If you double the concentration of base, the rate of the reaction will double. The rate of an E1 reaction of a secondary iodide is

Rate = $\upsilon = k[\text{R—I}]$

Therefore, if you double the concentration of base in an E1 reaction, there will be no change in the rate of the reaction.

Problem 8.31   Task one is to see how the unwanted compound **B** is formed. It is pretty clearly the result of an S_N2 displacement with inversion. The S_N2 and E2 reactions are both sensitive to steric effects. However, the steric requirements for deprotonation (E2) are far less severe than those for the S_N2 reaction, which requires the approach of a nucleophile to the rear of the departing carbon-leaving group bond. In this case, our hopes of making alkene are being thwarted by the S_N2 displacement of bromide to give the dialkoxy compound. If we increase the size of the reactants, we should be able to prejudice the reaction in favor of the sterically less demanding E2 reaction. We can't do anything about the size of the substrate, the bromide, but we could increase the size of the base. *tert*-Butoxide, a branched alkoxide base, would be a good choice.

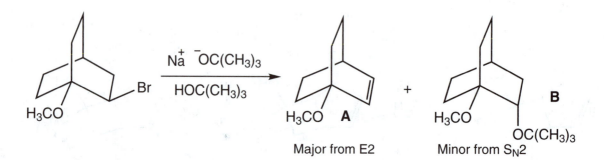

Problem 8.32   The hydrogens that are on the beta carbons (there are two beta carbons in this molecule) are in the anti position and have LUMO character. If you look at the 2-bromopropane shown at the beginning of the animation, the carbon–bromine bond is pointing down and the anti hydrogens on the beta carbons are both pointing up. The electrons in these carbon–hydrogen bonds overlap with the antibonding σ* orbital of the carbon–bromine bond when they are lined up in this arrangement. This overlap of orbitals is called hyperconjugation.

Problem 8.33   There are three hydrogens that have the most LUMO character and are therefore most acidic. It is the hydrogens that are lined up with the *p* orbital of the carbocation that are acidic. This is because the electrons in these sigma bonds are best aligned with the empty *p* orbital. The electrons of the aligned sigma bonds feed into the empty *p* orbital, and that makes the hydrogens on those aligned bonds more acidic.

Problem 8.34   The steps involved in this E1 rearrangement reaction are shown on the next page. The methyl group that shifts in the third step of the reaction is the methyl group that is aligned with the empty *p* orbital of the carbocation. It is those electrons that can mix with the empty *p* orbital. The nucleophile is technically the electrons in the sigma bond and the electrophile is the *p* orbital of the carbocation. We might loosely say that the migrating methyl group is the nucleophile and the carbocation is the electrophile.

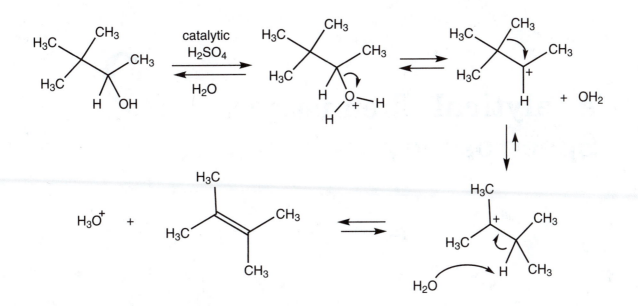

Problem 8.35 Note that there are several weak bases that could be involved in the deprotonation step. Water has been shown as the weak base in the scheme shown for the previous reaction. Another molecule of alcohol could be the weak base. Even the HSO_4^- is basic enough to deprotonate the hydrogen on the β carbon.

It is the β hydrogen that will give the most substituted alkene that is most acidic. The orbital that lines up best will be attached to the hydrogen that is most acidic. The transition state will be lowest for deprotonation of the aligned hydrogen because it is the hydrogen that is giving its electrons most effectively.

Analytical Chemistry: Spectroscopy

9

Solving spectral problems involves having a reasonable familiarity with the spectral techniques themselves, and then lots and lots of practice. You need to know what kinds of things each spectral technique can tell you: Mass spectrometry is different from nuclear magnetic resonance spectroscopy, for example. These early, in-chapter problems try to highlight what the techniques can do. The Additional Problems (Section 9.13) will go on to start you off on the actual solving of problems involving interpretation of spectra.

Problem 9.1 In Chapter 4, we discussed "resolution," which involves converting a pair of enantiomers (same physical properties) into a pair of diastereomers (different physical properties) by reaction with a single enantiomer of another molecule. The diastereomers can be separated and then the pair of enantiomers regenerated.

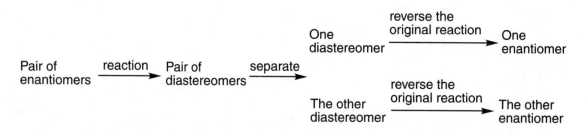

In Chapter 4, the technique of column chromatography was also introduced (p. 176). The chromatographic technique for separating enantiomers is a variation on the general theme. Diastereomeric *compounds* are not formed, but diastereomeric *complexes* are. The pair of enantiomers to be separated is passed though a column in which the stationary phase is constructed from a single enantiomer. As the pair to be separated passes through the column, the components will be adsorbed differently on the stationary phase because the complexes formed are diastereomeric. In this technique, the strong, covalent bonding of resolution is replaced with weaker partial bonding, or complexing, as the enantiomers are adsorbed on the material of the column.

Problem 9.3 You know that the atomic weight of oxygen is 16. So you can start with the molecule C_7H_{16}, which has a molecular weight of 100, and replace one C and four H's with an oxygen. That will keep the MW at 100 because one C contributes 12 to the MW and four H's will contribute 4 to the MW (12 + 4 = 16). Therefore, $C_6H_{12}O$ has a molecular weight of 100 (so will $C_5H_8O_2$). The formula $C_6H_{12}O$ has 1 degree of unsaturation. Here are some isomers:

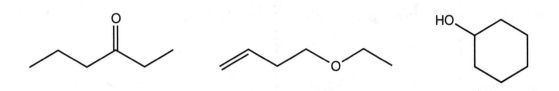

A molecule with a molecular weight of 100 and containing N will require a little more thinking. First, let's try doing the same kind of substitution we did with the oxygen. In this case, nitrogen has an atomic weight of 14. If we can start with C_7H_{16}, we will need to substitute one carbon and two hydrogens (12 + 2 = 14) with one nitrogen. That gives us a formula of $C_6H_{14}N$. But the degrees of unsaturation for this formula are $[(2 \times 6) + 2 - 14 + 1]/2$, which is equal to 1/2. There is no neutral molecule with $C_6H_{14}N$. Try drawing such a molecule and you will find that nothing works. It turns out that a neutral molecule with one nitrogen will have an odd MW. In fact, an organic molecule with an odd number of nitrogens will have an odd MW. An organic molecule with an even number of nitrogens will have an even MW. So let's try putting in two nitrogens. That will be the equivalent of a molecular weight of 28. We can get 28 by switching two carbons (24) and four hydrogens (4). So $C_5H_{12}N_2$ has a molecular weight of 100. The degrees of unsaturation for such a molecule are $[(2 \times 5) + 2 - 12 + 2]/2 = 1$. Here are some isomers:

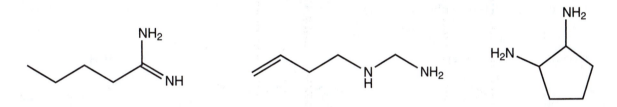

Problem 9.4

(a) The benzylic hydrogen (H_a) on the methyl group is most likely to be lost.

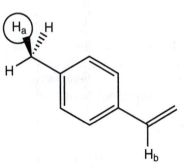

The hydrogen labeled H_b is not a likely candidate because the resulting empty sp^2 orbital does not "see" into the aromatic ring. There are no resonance structures for that potential carbocation. Also, vinyl cations are even less stable than primary cations.

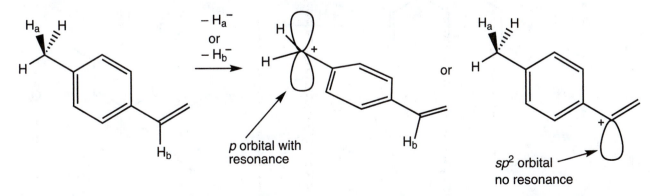

(continued)

Problem 9.4 *(continued)*

Loss of the H_a benzylic hydrogen gives a carbocation that has many stabilizing resonance structures.

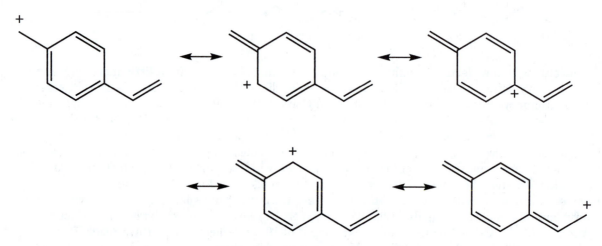

(b) The secondary benzylic cation would be more stable than a primary benzylic cation. We predict loss of one of the CH_2 benzylic hydrogens.

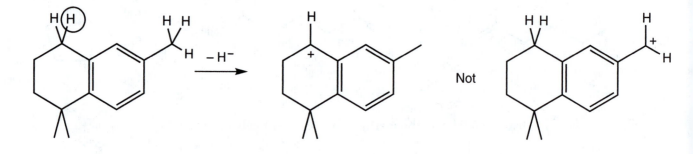

(c) The carbocation on the carbon adjacent to a carbonyl is destabilized because the carbonyl carbon is significantly electron deficient. We can predict that the more stable carbocation would be the benzylic CH_2 on the left. However, the carbon bonded to bromine is also somewhat electron deficient. In this case, it is more important to notice all the issues than it is to get the right answer.

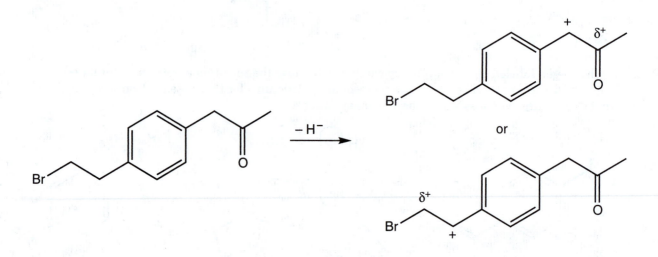

Problem 9.5 In any alkene, the bonding π molecular orbital is filled with two electrons and the antibonding π* orbital is empty. When a π electron is ejected in the mass spectrometer, it will come from the highest occupied MO, which is the filled π orbital.

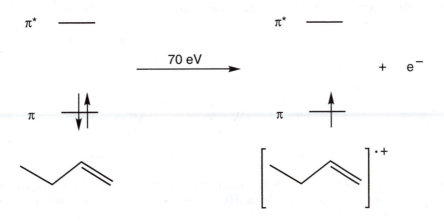

Problem 9.7

E = Nhc/λ and the quantity Nhc = 28.6×10^3 (p. 382).

So, E = $(28.6 \times 10^3)/200$ = 143 kcal/mol at 200 nm and E = $(28.6 \times 10^3)/800$ = 35.8 kcal/mol at 800 nm.

Problem 9.8

(b)

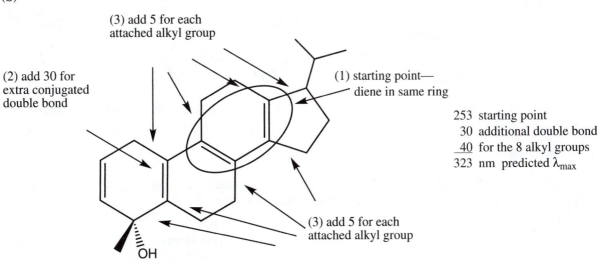

(3) add 5 for each attached alkyl group

(2) add 30 for extra conjugated double bond

(1) starting point— diene in same ring

253 starting point
 30 additional double bond
 40 for the 8 alkyl groups
323 nm predicted λ_{max}

(3) add 5 for each attached alkyl group

Problem 9.9 In Problem 9.9(b), the giveaway is the broad band above 3000 cm^{-1}. This band can only result from an OH. There is a strong band at 1720 cm^{-1}; a carbonyl group is present. The combination of these two bands means that you should think of a carboxylic acid, RCOOH. The appropriate strong C—O stretching vibrations are present between 1400 and 1200 cm^{-1}. As there are only two carbons in this molecule, the structure must be acetic acid, CH$_3$COOH.

In Problem 9.9(c), the band at 2220 cm^{-1} is in a region of few absorptions in the IR. As the compound contains nitrogen, it is very likely that the compound is a nitrile. There are C—H stretching bands above 3000 cm^{-1}. These bands reveal the presence of either "olefinic" or "aromatic"

(continued)

Problem 9.9 *(continued)*

hydrogens. There is a strong band at 1710 cm^{-1}, indicating that a carbonyl group is present. The band at about ~2740 cm^{-1} tells us that the carbonyl group is an aldehyde. The sequence of medium-to-weak bands at about 1600 cm^{-1} indicates that the molecule is aromatic. A reasonable guess would be an aromatic ring with an aldehyde and a nitrile. This one turns out to be a 1,4-disubstituted aromatic ring as indicated by the strong band at about 830 cm^{-1}.

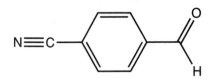

Problem 9.13   We can have a chiral environment in an NMR tube by using a chiral solvent. There aren't many chiral solvents available. A deuterated solvent would be best, so that the NMR spectrum of the solvent doesn't overwhelm the desired sample spectrum.

Another method is adding a small amount of a chiral reagent into the NMR tube. This practice is fairly common. The material added is called a chiral shift reagent. Here are examples of a possible solvent and a useful shift reagent.

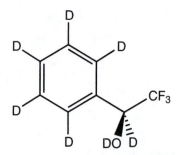

A possible chiral NMR solvent

A commercially available chiral shift reagent

Problem 9.14

(a) The underlined hydrogens in 1-butanamine are enantiotopic. Replace one underlined H by an X and we get the enantiomer of replacing the other underlined H by an X.

Enantiomers

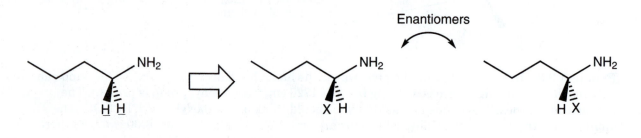

(b) The underlined hydrogens are diastereotopic. Replacing one underlined H by an X gives a diastereomer of the molecule obtained by replacing the other underlined H by an X.

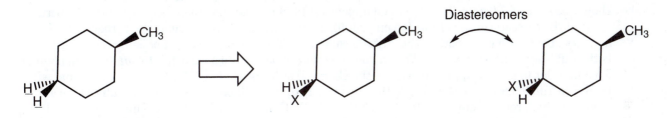

(c) The underlined hydrogens are diastereotopic. Replacing one underlined H with an X gives a diastereomer of the molecule obtained by replacing the other underlined H with an X.

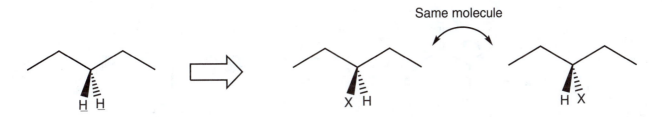

(d) The underlined hydrogens in pentane are homotopic. Replacing one underlined H with an X gives the same molecule as obtained by replacing the other underlined H with an X.

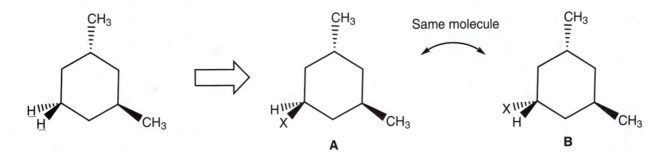

(e) This question is more difficult. It turns out that the underlined hydrogens in *trans*-1,3-dimethylcyclohexane are homotopic. Replacing one underlined H with an X gives the same molecule as obtained by replacing the other underlined H with an X.

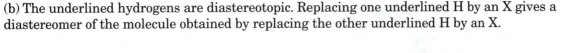

A   B

(continued)

Problem 9.14 *(continued)*

The first structure (**A**) can be rotated by 180° to give the second structure (**B**).

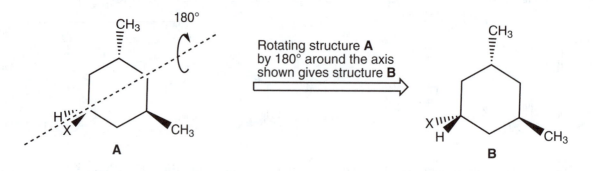

Problem 9.15   The vinyl hydrogens should be about δ 5.5 ppm. The H$_B$ signal will be even further downfield because of the contribution of the resonance structure shown here. The position of H$_B$ should be about δ 7.0 ppm. The H$_C$ signal appears shifted further downfield because of the inductive effect of the carbonyl group (Table 9.5 on p. 400), about δ 2.7 ppm, and without the carbonyl, a CH appears about δ 1.5 ppm (both of these shifts come from Table 9.5). The effect of being α to a carbonyl is the difference (2.7 – 1.5 = 1.2). We can add the effect of being α to a carbonyl to our starting point of δ 5.5 in order to get our approximation of the chemical shift for H$_C$ of δ 6.7 ppm. The next signal will be H$_D$. In Table 9.5 we see that a CH$_3$ that is α to a carbonyl will be at δ 2.0 ppm. Finally, H$_A$ is an allylic methyl group. From Table 9.5 we see that an allylic methyl will have a chemical shift of δ 1.7 ppm. A good estimate for this molecule is H$_B$ at δ 7.0, H$_C$ at δ 6.7, H$_D$ at δ 2.0, and H$_A$ at δ 1.7. The actual values obtained from the literature are H$_B$ at δ 6.9, H$_C$ at δ 6.1, H$_D$ at δ 2.2, and H$_A$ at δ 1.9. We were pretty close, but most important, we have the right order of the signals.

Problem 9.17

(a)   (b)

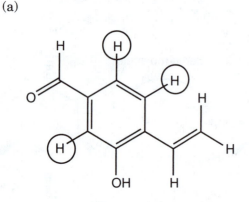   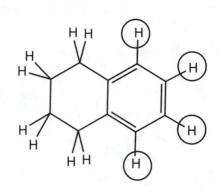

(c)

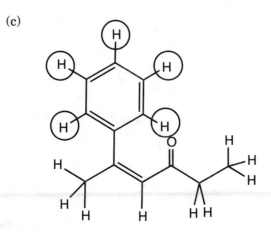

Problem 9.18 The only coupling that we will typically consider is 3-bond coupling. That means that we only need to consider the coupling (1) between H_a and the vinyl methyl, which will make H_a a quartet, and (2) between H_a and H_b, which will make H_a a doublet. Because H_a will be coupling to both H_b and the vinyl methyl group at the same time, the signal will be a doublet of quartets. You will see on page 411 of the text that the coupling between H_a and H_b (about 16 Hz) is larger than the value for the coupling between H_a and the methyl group (about 7 Hz).

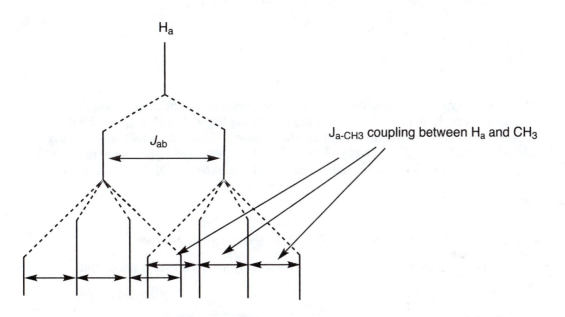

Problem 9.19 In this case, there will be overlapping, and a pattern of five lines in the ratio 1:2:2:2:1 will appear. The single H_b will produce two lines, each of which will be split into a 1:2:1 triplet by the pair of H_c's. With the coupling constants set as they are in the problem, the lines of the triplets reinforce each other.

(continued)

Problem 9.19 *(continued)*

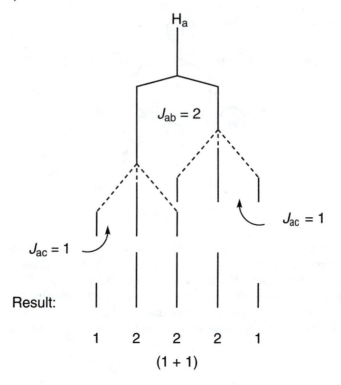

Problem 9.20 2-Methoxypropane should present no problems. The underlined hydrogen is flanked by six equivalent hydrogens and so must appear as a septet (n +1 rule). 1,3-Dichloro-2-methoxypropane is much harder. It might seem that there should be five lines as there are four equivalent hydrogens adjacent to the underlined hydrogen. But those hydrogens come in diastereomeric pairs, so the correct answer is that the signal would be a triplet of triplets.

Problem 9.21

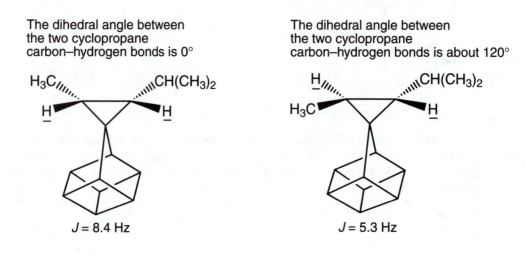

The dihedral angle between the two cyclopropane carbon–hydrogen bonds is 0°

J = 8.4 Hz

The dihedral angle between the two cyclopropane carbon–hydrogen bonds is about 120°

J = 5.3 Hz

A look at the Karplus curve (Figure 9.53) shows that the cis (0°) compound should have a larger coupling constant than the trans (120°) compound. This is exactly how the stereochemical assignment was made in the research paper describing these two compounds.

Problem 9.22 Here is an "arrow-pushing" problem deep in the middle of a spectroscopy chapter. It is a reminder that you never outgrow your need to push arrows.

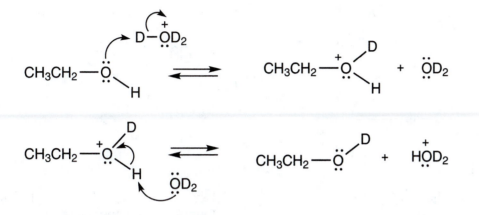

The arrow pushing for NaOD-catalyzed exchange of diethylamine is shown here. We don't really expect these conditions to lead to rapid exchange. The hydroxide base is not strong enough to deprotonate the amine (pKa ~ 36) appreciably. Acidic conditions will be more effective for the exchange of the N—H.

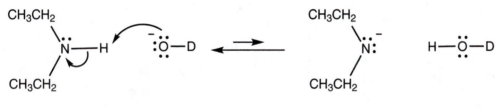

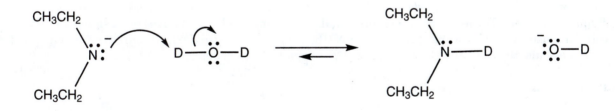

Problem 9.23 If we let the coupling constants be equal, the triplet of quartets (12 lines) we "expect" simplifies to six lines because of overlaps. The line intensities are not drawn to scale for clarity.

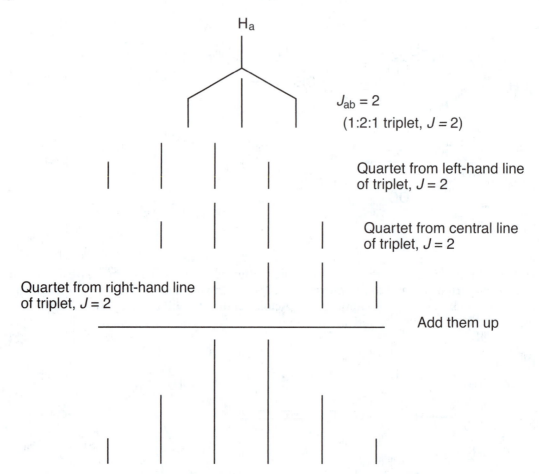

$J_{ab} = 2$
(1:2:1 triplet, $J = 2$)

Quartet from left-hand line of triplet, $J = 2$

Quartet from central line of triplet, $J = 2$

Quartet from right-hand line of triplet, $J = 2$

Add them up

Problem 9.24 Use the correlation chart. The electron-withdrawing carbonyl group (recall the δ+ on the carbonyl carbon) reduces the shielding around the hydrogens on the adjacent carbon, which come into resonance at relatively low field.

Problem 9.26 Yes, we could use ^{13}C NMR for Problem 9.25. Compound **A** will have three carbon signals, compound **B** will have five carbon signals, and compound **C** will have seven signals. We could not use the aromatic signals of the ^{1}H NMR spectrum to distinguish the three compounds. In each molecule, the aromatic C—H signal is a 2H singlet due to symmetry.

Problem 9.27

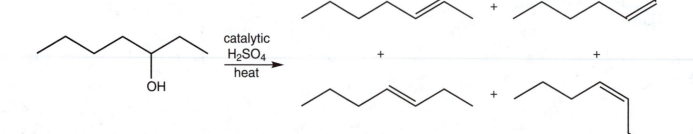

Perhaps a ^{13}C NMR expert could differentiate these molecules. But it isn't going to be an easy task, for sure. All four isomers will show seven signals in their ^{13}C NMR spectra. Just separating the molecules in order to get a clean spectrum for each would be a significant challenge.

Problem 9.31   In Problem 9.13 you considered two ways to differentiate enantiomers in an NMR sample. We can use a chiral solvent or add a chiral shift reagent. In a chiral environment, the enantiotopic hydrogens of ethanol will be different. Using a chiral solvent is very expensive and does not always give significant differences in signals of enantiotopic groups. Adding a chiral reagent to the NMR sample is also expensive, although much more reliable. A third method is similar to resolution (p. 174 of the text). We can make the hydrogens diastereotopic by using a chiral derivatizing agent as shown below.

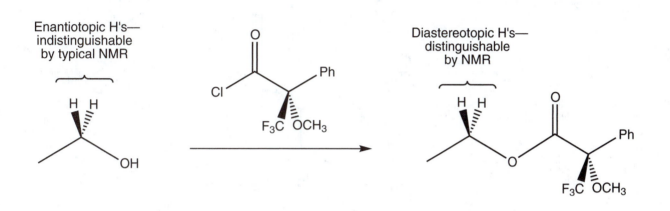

Problem 9.33   The first compound is easy. The *tert*-butyl and methyl groups will be largely in equatorial positions. There is only a single hydrogen on the carbon adjacent to the methyl group, and by the $n + 1$ rule the methyl hydrogens will apear as a doublet.

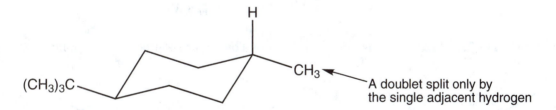

The second compound is more interesting. The methyl groups are diastereotopic and must give rise to different signals. Each methyl group will appear as a doublet in the NMR. Use the technique of replacing each methyl group with X to see this.

(continued)

Problem 9.33 *(continued)*

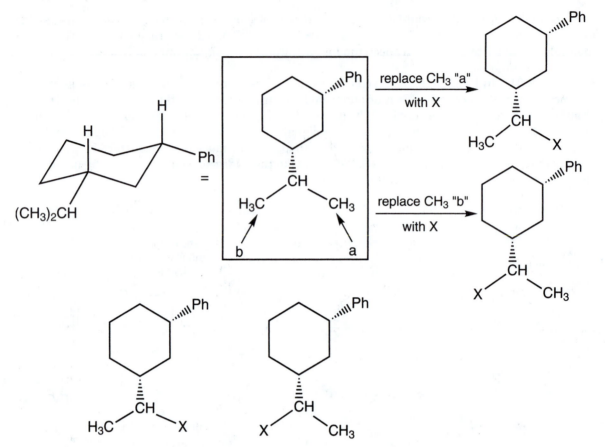

These are not the same! They are diastereomers and the methyl groups are diastereotopic

Additional Problem Answers

Problem 9.34 The S_N2 product is 2-methoxybutane. The major E2 product is (*E*)-2-butene.

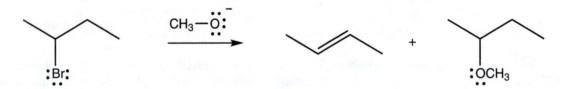

The easiest IR data to use would be the sp^2 C—H stretch for the alkene at approximately 3010 cm^{-1}. The ether would not have such a band. The C=C stretch will not be visible in (*E*)-2-butene because there is no net change in the dipole in this symmetrical molecule. So we can't use the C=C stretch to distinguish the two products. There is an alkene C—H bend at 975–965 cm^{-1} for a trans alkene that might be useful for distinguishing these two products.

Problem 9.35

(a) This cyclic ester has one double bond and a ring. Therefore there are 2 degrees of unsaturation.

(b) This aromatic ketone has four double bonds and one ring. There are 5 degrees of unsaturation.

(c) This molecule has an alkene. There is only 1 degree of unsaturation.

(d) The triple bond contributes 2 degrees of unsaturation. Therefore, the alkyne and the alkene in the same molecule give 3 degrees of unsaturation.

(e) There are two alkenes and one ring in this molecule. It has 3 degrees of unsaturation.

(f) There are no degrees of unsaturation in 3-methoxyhexane.

(g) Pyridine has 4 degrees of unsaturation (three double bonds and one ring).

Problem 9.36   All the structures are consistent with the O—H or N—H stretching band centered roughly at 3235 cm^{-1}, so no distinction is possible through use of this band. (Actually, the IR maven might object that primary amines and amides should show two bands in this region, but this seems a fine point.) Structures **A** and **C** can be eliminated because of the absence of a C=O stretch in the 1700 cm^{-1} region. So, the choice is between **B** and **D**. The band at 2240 cm^{-1} is consistent with a cyanide group (C≡N triple bond stretch), and so the structure must be **B**.

Problem 9.37

Compound **1**   The IR spectrum of compound **1** exhibits an intense carbonyl stretch at 1750 cm^{-1}. By itself, this does not help much; however, the presence of a C—O stretch at 1200 cm^{-1} suggests the possibility of an **ester**. Anhydrides and carboxylic acids also show C—O stretches, but these can be ruled out by the absence of a second carbonyl stretch or the absence of an O—H stretch, respectively. Furthermore, the frequency of the carbonyl stretch for compound **1** suggests that it is probably not a conjugated ester (i.e., not an α,β-unsaturated or aryl ester).

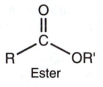

Ester

Compound **2**   The IR spectrum of compound **2** displays a strong carbonyl stretch at 1710 cm^{-1}. The important additional bands in this spectrum are the weak C—H stretching absorptions at 2810 and 2715 cm^{-1}, which are characteristic of **aldehydes**. (The higher-frequency band is often obscured by overlapping aliphatic C—H stretches.) In addition, the frequency of the carbonyl stretch for compound **2** suggests that this compound is probably not a conjugated aldehyde.

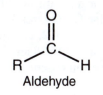

Aldehyde

Compound **3**   The IR spectrum of compound **3** shows two carbonyl stretches at 1855 and 1785 cm^{-1}. This pair of bands suggests that the compound is an **anhydride**. Anhydrides should also exhibit a C—O stretching band. Does this one?

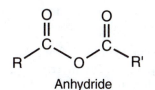

Anhydride

(continued)

<u>Compound 4</u>   The IR spectrum of compound **4** exhibits a strong carbonyl stretch at 1720 cm^{-1}. The diagnostic absorption in this spectrum is the very broad, intense absorption at 3600–2500 cm^{-1}, characteristic of the O—H stretch of **carboxylic acid** dimers.

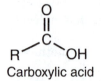

Carboxylic acid

<u>Compound 5</u>   The IR spectrum of compound **5** displays a strong carbonyl stretch at 1710 cm^{-1}. This spectrum must belong to the **ketone**. Note the absence of a second carbonyl stretch, the aldehyde C—H stretching doublet, an O—H stretch, or a C—O stretch. The frequency of the carbonyl stretch suggests that compound **5** is probably not a conjugated ketone.

Ketone

Problem 9.38

<u>Spectrum 1</u>   The choices for spectrum **1** are alcohol, carboxylic acid, or phenol.
All three classes would be expected to show an O—H stretch. For the alcohol and phenol, this absorption occurs in the 3400–3200 cm^{-1} region of the IR spectrum, whereas for the carboxylic acid, a broad, intense O—H stretch occurs in the 3300–2400 cm^{-1} region. The IR spectrum of compound **1** is most consistent with the latter values. Even stronger evidence for the carboxylic acid is the strong carbonyl stretch at 1720 cm $^{-1}$. This band would not be expected for the alcohol or phenol. Therefore, the IR spectrum of compound **1** is certainly that of a **carboxylic acid**.

Carboxylic acid

<u>Spectrum 2</u>   The carbonyl stretch at 1715 cm^{-1} could be diagnostic for an aldehyde, an ester, or a ketone. So, we will have to look elsewhere for a means of differentiating between these classes. An ester would be expected to display an intense C—O stretch in the 1300–1000 cm^{-1} region of the IR spectrum. Such a band is absent in the spectrum of **2**. The important additional bands in the spectrum of **2** are the weak C—H stretching absorptions at 2810 cm^{-1} and 2700 cm^{-1}, character-istic of **aldehydes**. (The higher-frequency band is often obscured by overlapping aliphatic C—H stretches.)

Aldehyde

<u>Spectrum 3</u>   The weak band at 2220 cm^{-1} could be from the C≡C stretch of a 1-alkyne or the C≡N stretch of a nitrile (cyanide). However, symmetrically disubstituted alkynes do not exhibit a C≡C stretch because this vibration does not result in a dipole moment change. 1-Alkynes also

display a characteristic ≡C—H stretch at about 3300 cm⁻¹, absent in the IR spectrum of compound **3**. Accordingly, this leaves the **nitrile** as the best fit for spectrum **3**.

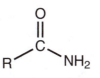

Nitrile

<u>Spectrum 4</u> The intense bands at 3350 cm⁻¹ and 3180 cm⁻¹ are consistent with the N—H stretches of either a primary amine or primary amide, but not with a nitro compound. To differentiate between the amine and the amide, look for the presence or absence of a carbonyl stretch. The IR spectrum **4** shows an absorption centered at 1650 cm⁻¹, the result of a combination of a carbonyl stretch and an N—H bend. Therefore, the IR spectrum **4** is probably that of a **primary amide**.

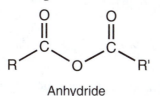

Primary amide

<u>Spectrum 5</u> The carboxylic acid can be eliminated immediately because of the absence of the diagnostic broad, intense O—H stretch in the 3300–2400 cm⁻¹ region of the IR spectrum. Both anhydrides and esters display a strong C—O stretch in the 1300–1000 cm⁻¹ region (present in this case at 1040 cm⁻¹). However, esters normally exhibit only a single carbonyl stretch, whereas anhydrides display two carbonyl stretches. The presence of two carbonyl absorptions at 1810 cm⁻¹ and 1750 cm⁻¹ in the IR spectrum **5** is strong evidence for an **anhydride**.

Anhydride

Problem 9.39 Alkene isomers are traditionally distinguished through the use of the ═C—H out-of-plane bending vibrations. The third spectrum has a band at 697 cm⁻¹, the position typical of cis alkenes. This molecule is *cis*-2-octene.

Spectrum **2** has a strong band at 966 cm⁻¹, a typical position for a trans alkene. This molecule is *trans*-2-octene. Notice that in the IR spectrum for this relatively symmetrical molecule, there is no visible band for the C═C stretch. By contrast, in the IR spectrum of the less symmetrical cis isomer **3**, this band appears at ~1650 cm⁻¹.

Spectrum **1** must belong to 1-octene by a process of elimination, but let's look at the spectrum anyway to make certain. The two bands for a vinyl group (HC═CH₂) appear at 991 cm⁻¹ and 909 cm⁻¹, and there is a peak for the C═C stretch at about 1640 cm⁻¹.

Problem 9.40 The keys to this problem are the bands in spectrum **2** at 3333 cm⁻¹ and 2128 cm⁻¹ for the stretching of the alkyne terminal carbon–hydrogen bond and the triple bond, respectively. In the first spectrum, there is no visible band for either the alkyne carbon–hydrogen bond or the carbon–carbon triple bond. The first molecule is 6-phenyl-2-hexyne, a disubstituted acetylene too symmetrical to show a C≡C stretch. The second is the 1-alkyne, 5-phenyl-1-pentyne.

Problem 9.41 Ethanol has a broad O—H band at ~3400 cm⁻¹, and this band will disappear during the reaction. The aldehyde product has a strong C═O stretch at about 1730 cm⁻¹, and this band will appear during the oxidation of ethanol. One could monitor this biochemical process by following either IR band. The O—H band might be complicated by the presence of water, so the carbonyl band at ~1730 cm⁻¹ is the more logical choice.

Problem 9.42   As Table 9.1 shows, natural bromine has two common isotopes, ^{79}Br and ^{81}Br, in nearly equal amounts. The two ions come from the two molecules, $C_8H_7{}^{79}BrO_2$ (m/z = 214) and $C_8H_7{}^{81}BrO_2$ (m/z = 216).

Aldehydes generally show very intense (M − 1) peaks as a hydrogen atom is easily lost to give the resonance-stabilized acylium ion.

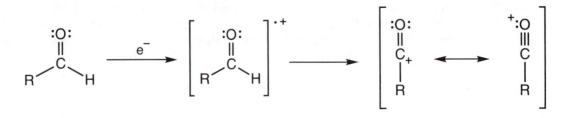

Of course, in this case the acylium ion still contains bromine, and two peaks will be observed, $C_8H_6{}^{79}BrO_2$ (m/z = 213) and $C_8H_6{}^{81}BrO_2$ (m/z = 215).

Problem 9.43   Our first task is to derive the empirical formula from the elemental analysis data. If you are unsure of these calculations, be sure to review the problems in Chapter 3 (p. 148). We can find the percentage of oxygen easily, 100 − (80.00 + 6.70) = 13.30.

C $\dfrac{80.00}{12.01}$ = 6.66   H $\dfrac{6.70}{1.01}$ = 6.63   O $\dfrac{13.30}{16.00}$ = 0.831

Now we divide by the smallest value, 0.831:

C $\dfrac{6.66}{0.831}$ = 8.01   H $\dfrac{6.63}{0.831}$ = 7.98   O $\dfrac{0.831}{0.831}$ = 1.00

So, the empirical formula is C_8H_8O and formula weight is 120 g/mol.

A quick look at the mass spectrum shows that the highest peak, likely to be the molecular ion, is at m/z = 120. Accordingly, the empirical formula and the molecular formula are the same, C_8H_8O.

Next, it is useful to work out the degrees of unsaturation (Ω) in this molecule. Remember, Ω = [(2n + 2) − (No. of hydrogens)]/2.

Here, Ω = [18 − 8]/2 = 5. There is a total of five rings and/or π bonds in this molecule. For this small a molecule, this many degrees of unsaturation almost certainly means there will be a benzene ring (four unsaturations) in the molecule.

Now it is finally time to analyze the IR spectrum. The presence of a strong band at 1690 cm^{-1} suggests a carbonyl group. Taken together with the formula and the notion that a benzene ring is present, this limits the choices to the following five possibilities.

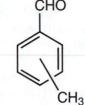

Three possible isomers
of tolualdehyde

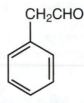

Phenylacetaldehyde

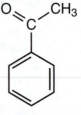

Acetophenone

The strong bands at 758 cm^{-1} and 690 cm^{-1} indicate a monosubstituted benzene, eliminating the three isomers of tolualdehyde. All aldehydes are eliminated by the lack of the two characteristic bands for the aldehyde C—H stretch at 2900–2700 cm^{-1}. Only acetophenone, the last possibility, remains.

Problem 9.44 First, we need to know the amount of oxygen in this molecule. A little addition and subtraction does the job.

C 70.60
H 5.90   100 − 76.50 = 23.50 = % oxygen in the compound
  76.50

The calculation of the empirical formula is now as follows:

C $\dfrac{70.60}{12.00} = 5.88$ H $\dfrac{5.90}{1.01} = 5.84$ O $\dfrac{23.50}{16.00} = 1.47$

Now we divide by the smallest value, 1.47:

C $\dfrac{5.88}{1.47} = 4.00$ H $\dfrac{5.84}{1.47} = 3.97$ O $\dfrac{1.47}{1.47} = 1.00$

So, the empirical formula is C_4H_4O and formula weight is 68 g/mol.

The mass spectrum shows that the molecular formula is twice the empirical formula as the molecular ion is 136, (2 × 68). So the molecular formula is $C_8H_8O_2$.

The degrees of unsaturation can now be calculated: $\Omega = (18 − 8)/2 = 5$. A benzene ring is suggested and would account for 4 degrees of unsaturation. The IR spectrum reveals a strong band at 1685 cm^{-1}. This absorption shows that a carbonyl group is present as the fifth unsaturation. The pair of peaks at 2820 cm^{-1} and 2720 cm^{-1} is diagnostic for aldehydes. The low-frequency position of the carbonyl band suggests that the carbonyl group is conjugated and, in combination with the presence of a benzene ring, indicates a benzaldehyde. An atom inventory shows that we are now short CH_3O, probably a methoxy group.

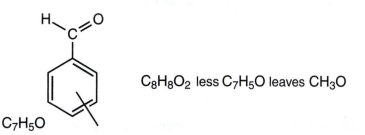

$C_8H_8O_2$ less C_7H_5O leaves CH_3O

It is tempting at this point to imagine that compound **A** is a methoxybenzaldehyde, and the only real question left is the position of substitution on the ring. The single band at 833 cm^{-1} is diagnostic for para substitution, and, indeed, this compound is *p*-methoxybenzaldehyde.

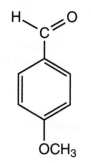

Problem 9.45   It is our first task to determine the molecular formulas for isomers **A, B,** and **C**. Is there any oxygen? No, the percentages of C, H, and N add up to 100.

C 71.17
H 5.12
N 23.71
 100.00

There is no oxygen present!

To determine the molecular formula, apply the usual calculation:

C $\frac{71.17}{12.01}$ = 5.92   H $\frac{5.12}{1.01}$ = 5.07   N $\frac{23.71}{14.01}$ = 1.69

C $\frac{5.92}{1.69}$ = 3.50   H $\frac{5.07}{1.69}$ = 3.00   N $\frac{1.69}{1.69}$ = 1.00

Therefore, the empirical formula = $2(C_{3.50}H_{3.00}N_{1.00})$ = $C_7H_6N_2$, and the formula weight is $(C_7H_6N_2)$ = 118 g/mol.

So, the empirical formula = the molecular formula, as the molecular ion in the mass spectrum is 118.

Now let's determine the degrees of unsaturation, Ω.

$$\Omega = \frac{2(7 + 2) + 2 - 2 - 6}{2} = 6$$

Six degrees of unsaturation suggests the possibility of a benzene ring.

Now it is time to see where we are by doing an atom inventory:

$C_7H_6N_2$ Formula
$- C_6$ Benzene ring
CH_6N_2 Remaining

The IR spectra of isomers **A, B,** and **C** all show an absorption at about 2220 cm^{-1}. This band could be the result of a C≡C stretch of an alkyne or the C≡N stretch of a nitrile. Assuming that our conjecture that these isomers contain a benzene ring is correct, the alkyne C≡C stretch can be ruled out because the atom inventory shows only one residual carbon atom. Also note that the nitrile C≡N accounts for the remaining two degrees of unsaturation.

—C≡N   Two degrees of unsaturation

The IR spectra of isomers **A, B,** and **C** also exhibit doublets at about 3490 cm^{-1} and 3390 cm^{-1}. These are N—H stretches indicative of a primary amine, which shows two N—H stretches.

R—NH$_2$ primary amine

All the available information suggests that isomers **A, B,** and **C** are the three possible aminobenzonitriles; that is,

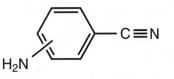

Isomer **A**. The IR spectrum of isomer **A** exhibits three peaks at 860 cm^{-1}, 780 cm^{-1}, and 680 cm^{-1}, which indicates 1,3- or meta-disubstitution. Thus isomer **A** is 3-aminobenzonitrile.

Isomer **B**. The IR spectrum of isomer **B** displays a band at 835 cm^{-1}, which suggests 1,4- or para-disubstitution. Isomer **B** is 4-aminobenzonitrile.

Isomer **C**. Finally, the IR spectrum of isomer **C** shows only an intense peak at 750 cm^{-1}, indicative of 1,2- or ortho-disubstitution. Isomer **C** is 2-aminobenzonitrile (anthranilonitrile).

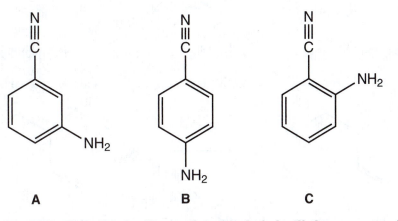

Problem 9.46 The IR band for the O—H stretch in an alcohol will sharpen upon dilution because the amount of hydrogen bonding diminishes. There are fewer neighboring molecules with which hydrogen bonding can occur. The typical broad band of the O—H stretch between 3200 and 3600 cm^{-1} is due to the many different types of hydrogen-bonded species (dimers, trimers, tetramers, etc.). The O—H stretching frequency of the isolated monomer appears at approximately 3600 cm^{-1}.

The O—H stretching frequency for *tert*-butyl alcohol is often sharp because the *tert*-butyl makes the area around the O—H group congested. There is less hydrogen bonding possible. The O—H stretching frequency appears at approximately 3600 cm^{-1}.

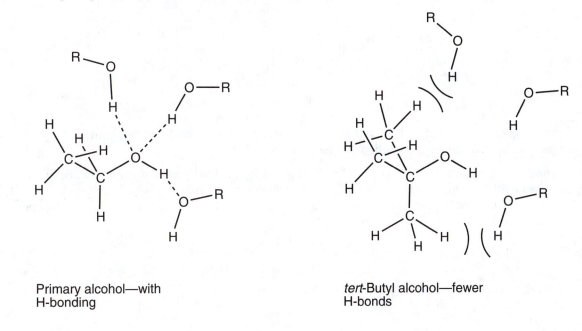

Primary alcohol—with
H-bonding

tert-Butyl alcohol—fewer
H-bonds

Problem 9.47   The O—H stretch for a carboxylic acid is 3200–2800 cm^{-1}. The O—H stretch for an alcohol is 3550–3300 cm^{-1}. One reason for the difference between the two O—H stetches is that the OH hydrogen of the carboxylic acid is more acidic than the OH of an alcohol. Therefore, it is more strongly involved in hydrogen bonding.

Another reason for the difference is that the oxygen of the O—H in the carboxylic acid has a different hybridization. The carboxylic acid is stabilized by resonance. Therefore, the hybridization of the O—H oxygen is close to sp^2. There is no resonance for the alcohol oxygen, so its oxygen is close to sp^3.

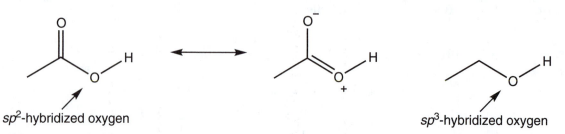

sp^2-hybridized oxygen sp^3-hybridized oxygen

Problem 9.48   Benzyl bromide will react with NaOH in an S$_N$2 reaction to give benzyl alcohol.

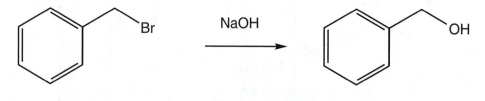

Both the starting material and product should show IR bands for a monosubstituted aromatic ring (Table 9.4, p. 391). The major difference will be that the starting bromide will not show an OH stretching band at 3500–3300 cm^{-1} and a C—O stretching absorption at 1260–1000 cm^{-1}, whereas the product alcohol will.

Problem 9.49   IR spectroscopy is a great tool that can be used to distinguish between an alkyl halide (such as 3-chloropentane) and an alcohol (such as 3-pentanol). The broad band at 3400 cm^{-1} that is a characteristic of the O—H stretch of an alcohol will be observed for the 3-pentanol but not for 3-chloropentane.

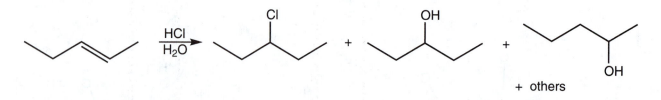

+ others

IR spectroscopy is not a good tool for distinguishing between 3-pentanol and 2-pentanol. NMR would be a better tool for such a challenge (see Problem 9.64).

Problem 9.50   The formula is C$_2$H$_4$Br$_2$. The C$_2$H$_4$ part is constant and has a mass of 28. Each bromine has a 50:50 chance to be 79 or 81. Thus, both bromines can be 79, both bromines can be 81, and there are two ways that one bromine can be 79 and one 81. So we should see three molecular ions in the ratio 1:2:1.

28	28	28	28
79	81	79	81
79	79	81	81
186	188	188	190

Problem 9.51 We will use the technique outlined in the chapter (p. 401). Simply replace the hydrogens in question one by one with X and determine the relationship between the two "new compounds." If they are identical, the two hydrogens are homotopic; if the "new compounds" are enantiomers, the two hydrogens are enantiotopic; and if they are different compounds entirely, the two hydrogens are diastereotopic. The first three cases are quite easy:

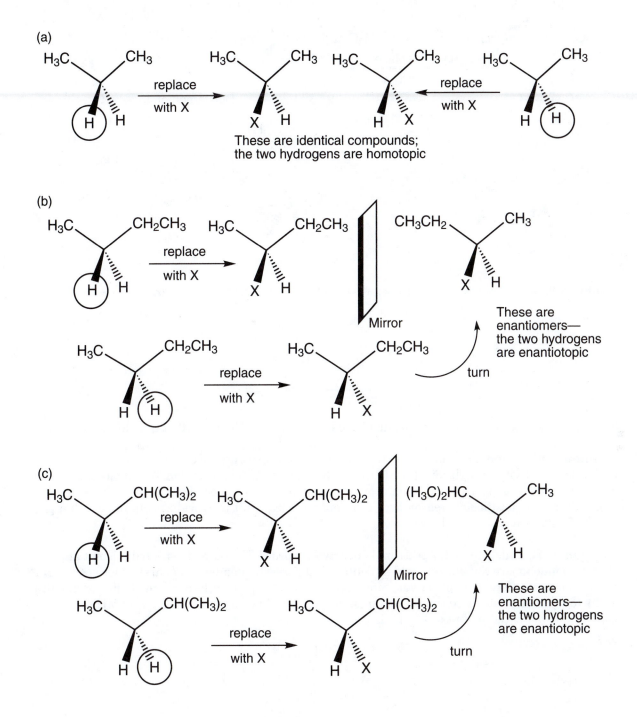

(a)

These are identical compounds; the two hydrogens are homotopic

(b)

Mirror

These are enantiomers—the two hydrogens are enantiotopic

turn

(c)

Mirror

These are enantiomers—the two hydrogens are enantiotopic

turn

(continued)

Problem 9.51 *(continued)*

Part (d) is the most interesting case. The "replacement technique" leads to diastereomers. The two hydrogens in question are diastereotopic.

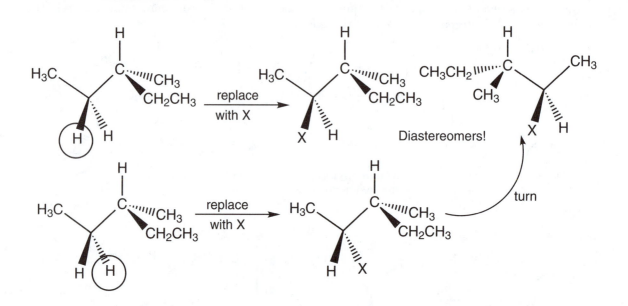

Problem 9.52 Silicon is less electronegative than carbon; thus the dipole arrow runs as follows:

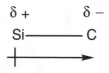

The "extra" electrons around carbon shield the hydrogens, which are shifted upfield as a result.

Problem 9.53 There are four factors that can have an impact on the chemical shift: (1) electronegativity, (2) hybridization, (3) aromaticity, and (4) delocalization. A cyclopropane has no electronegative groups. The ring is not aromatic, and there is no delocalization possible in the molecule, so the most logical reason for a different chemical shift is hybridization. In fact, that is the explanation found in the literature.

Problem 9.54 The 300-MHz spectrum is nearly first order. The 60-MHz spectrum is not and is therefore much more complicated and difficult to interpret. Remember, for a first-order spectrum to appear for two different coupled hydrogens such as H_a and H_b, the difference in Hz in the chemical shifts (δ_{ab}) must be about 10 times greater than their coupling constant (J_{ab}). At 60 MHz this condition is not satisfied, but at 300 MHz it is.

$$H_a \qquad H_b$$

$$CH_3CH_2CH_2CH_2Br$$

Problem 9.55 There are only two kinds of hydrogen in 2-chloropropane. H_a will appear as a septet (six adjacent equivalent hydrogens, $n + 1$ rule) at about $\delta = 3.8$ ppm, the position appropriate for a hydrogen on a methine carbon bearing a chlorine atom. The six methyl hydrogens (H_b) will be a doublet (only one adjacent hydrogen) at about $\delta = 1.5$ ppm, the position for a methyl hydrogen β to a chlorine.

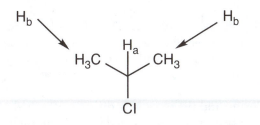

Problem 9.56 The first step for predicting a ^{1}H NMR spectrum is to identify the different signals. For 1,3-dichloropropane, there will be only two signals. The signals will be from the four equivalent H_a's and the two H_b's. The chemical shift for H_a will be δ 3.5 ppm because these hydrogens are on a C that is directly attached to a chlorine. The H_a CH_2 on the left is too far from the chlorine on the right to be further deshielded. The limit for deshielding by electronegativity is three bonds. So the hydrogens on the α carbon (CH_2 directly attached to the electronegative group) will be strongly deshielded, and the hydrogens on the β carbon (CH_2 once removed from the electronegative group) will be weakly deshielded. The hydrogens on a γ carbon (CH_2 twice removed) are not substantially shifted by the electronegative group. They are too far away. The H_b hydrogens are once removed from both chlorines. The chemical shift for a CH_2 once removed from a chlorine is about δ 1.8 ppm. In order to determine the chemical shift of H_b, we have to add the **effect** of the second chlorine. We cannot simply add the chemical shifts. To determine the **effect** of being a CH_2 once removed from a chlorine, we simply compare such a CH_2 with and without the chlorine. A typical CH_2 appears at a chemical shift of about δ 1.3 ppm. The CH_2 once removed from a chlorine is about δ 1.8 ppm. Therefore, the effect of the chlorine is the difference, which is 1.8 ppm – 1.3 ppm = 0.5 ppm. That is the value we can add to take into account the second chlorine. We predict a chemical shift of δ 1.8 ppm for the chlorine on one side, and we add 0.5 ppm for the chlorine on the other side to get δ 2.3 ppm for our predicted chemical shift of H_b.

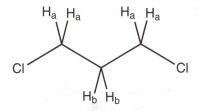

For the coupling of the two signals, we predict that H_a will be a triplet because they will "see" two H_b hydrogens. The H_b signal will be a quintet (or pentet) because there are four equivalent vicinal H_a hydrogens.

Problem 9.57

(a) The chemical shift for the H_a signal will be the furthest downfield, which means the shift of δ 5.72 ppm. H_a will be a triplet because there are two hydrogens (H_b) on the adjacent carbon.

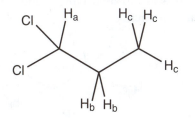

(continued)

Problem 9.57 *(continued)*

The chemical shift for H$_b$ is the next most deshielded signal, which means it must be at δ 2.22 ppm. We would predict a shift of about δ 2.3 ppm (see Problem 9.56). The coupling for H$_b$ will be a quintet (or pentet) because the rapid rotation around the single bonds will result in nearly the same coupling constant with the four hydrogens (one H$_a$ and three H$_c$). Predicting a doublet of quartets is also an acceptable answer.

The chemical shift for the CH$_3$ will be δ 1.10 ppm. The CH$_3$ is twice removed from the chlorines, which is too far away to have much deshielding. A typical CH$_3$ appears at a chemical shift of about δ 0.9 ppm (see Table 9.5). The methyl signal will be coupled to the H$_b$ hydrogens on the adjacent carbon. Therefore it will be a triplet.

(b) The CH$_2$ between the chlorine and the oxygen is directly attached to the two electronegative groups. It will be the signal furthest downfield, which is δ 5.5 ppm. The H$_a$ signal will be a singlet because there are no hydrogens on the adjacent atoms.

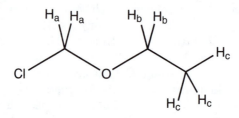

We predict the H$_b$ hydrogens to be at δ 3.5 ppm because that CH$_2$ is directly attached to an oxygen, which produces the same chemical shift as a CH$_2$ directly attached to a chlorine. The fact that the H$_b$ hydrogens are at δ 3.75 ppm in this molecule suggests that this particular oxygen is a little bit more deshielding. This is undoubtedly due to the resonance contribution shown here. A positively charged oxygen is more electronegative. The coupling for H$_b$ would be a quartet because there are three hydrogens on the adjacent carbon.

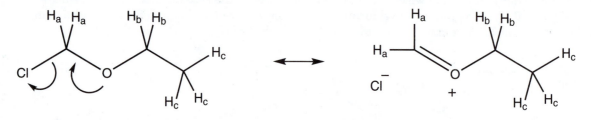

We can assign the H$_c$ signal to the δ 1.28 ppm. The methyl group is once removed from the electronegative oxygen. The H$_c$ signal will be a triplet because there are two H$_b$ hydrogens on the adjacent carbon.

(c) The chemical shift for H$_a$ would be furthest downfield. Because the methylene group is next to an iodine, we predict that shift to be about δ 3.2 ppm (see Table 9.5). That fits nicely with the observed δ 3.16 ppm signal. The CH that is once removed from the iodine should be at δ 1.5 ppm because it is a methine and a little further downfield because the iodine is electronegative and deshielding on the once-removed position. That fits well with the δ 1.72 ppm signal. The six H$_c$ hydrogens are equivalent. They are more removed from the iodine and should appear at δ 0.9 ppm, if the iodine has no effect on them. It isn't too surprising to find that there is a small downfield shift on H$_c$, so δ 1.03 ppm is a reasonable chemical shift.

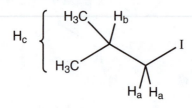

The signal for H_a would be a doublet because of the single H_b that is on the only adjacent carbon. The H_c methyls would also be a doublet because of the single H_b. The H_b signal could appear as a nonet (nine lines) if the rotation around the C—C bonds makes the coupling constant between H_a and H_b (J_{ab}) the same as the coupling constant between H_b and Hc (J_{bc}). In that case, H_b would be split by the eight vicinal hydrogens. It is possible that the H_b splitting will appear as a triplet of septets, although that would be difficult to ascertain.

(d) The first thing to notice about 1,3-dibromopropane is that the molecule is symmetrical. There are four equivalent H_a hydrogens. Based on Table 9.5, the chemical shift for H_a should be δ 3.4 ppm, the chemical shift for a methylene attached to a bromine. The δ 3.59 ppm signal indicates that each of the methylenes is shifted by the attached bromine in addition to the bromine that is two carbons away. The H_b hydrogens should be at a chemical shift of about δ 2.3 ppm because the methylene, which has a chemical shift of δ 1.3 ppm, is once removed from two bromines. Each bromine has about a δ 0.5 ppm effect when once removed. Thus, δ 1.3 + 0.5 + 0.5 = 2.3 ppm. The observed δ of 2.36 ppm fits nicely with where we would predict the H_b methylene signals to appear.

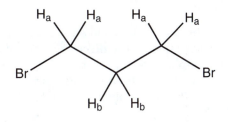

The coupling for the H_a signal would be a triplet due to the two H_b hydrogens that are vicinal. The H_b signal would be a quintet (five lines) because of the four equivalent vicinal H_a hydrogens.

Problem 9.58 For each of these problems, it is good to keep the following chemical shifts in mind. A CH_3 directly attached to an oxygen has a typical chemical shift of δ 3.2 ppm. A CH_2 directly attached to an oxygen has a chemical shift of about δ 3.5 ppm. A CH directly attached to an oxygen has a chemical shift of approximately δ 3.8 ppm. The chemical shift for being β (once removed) from an oxygen is about δ 1.9 ppm for a CH, δ 1.6 ppm for a CH_2, and δ 1.3 ppm for a CH_3. We will ignore the effect of an electronegative group for hydrogens on the γ carbon (twice removed).

(a)

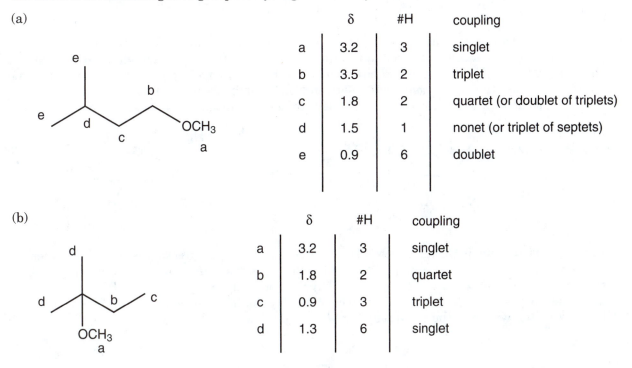

	δ	#H	coupling
a	3.2	3	singlet
b	3.5	2	triplet
c	1.8	2	quartet (or doublet of triplets)
d	1.5	1	nonet (or triplet of septets)
e	0.9	6	doublet

(b)

	δ	#H	coupling
a	3.2	3	singlet
b	1.8	2	quartet
c	0.9	3	triplet
d	1.3	6	singlet

(continued)

Problem 9.58 *(continued)*

(c) The methyl groups e and f are not equivalent because of the stereogenic carbon in the molecule. They are diastereotopic. Each will have a slightly different shift. The result is that the signal for H_d will be very messy.

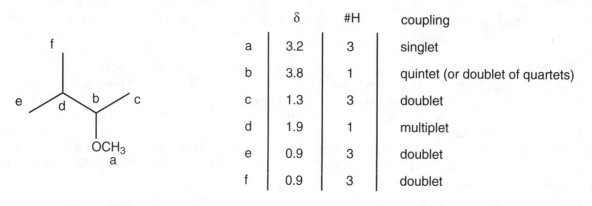

	δ	#H	coupling
a	3.2	3	singlet
b	3.8	1	quintet (or doublet of quartets)
c	1.3	3	doublet
d	1.9	1	multiplet
e	0.9	3	doublet
f	0.9	3	doublet

Problem 9.59   A vinyl hydrogen will have a chemical shift of about δ 5.5 ppm. In methoxyethene, the H_a vinyl hydrogen will also feel the effect of the directly attached oxygen. Therefore, the δ 6.4 ppm signal must belong to H_a. The coupling for H_a will be a doublet because of H_b and a doublet because of H_c. Because the coupling to each of those hydrogens will be different (trans coupling is larger than cis coupling), the signal will be a doublet of doublets.

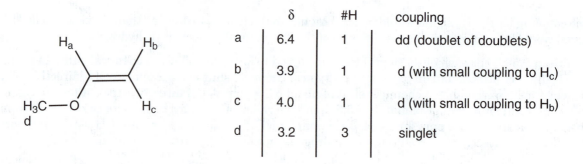

	δ	#H	coupling
a	6.4	1	dd (doublet of doublets)
b	3.9	1	d (with small coupling to H_c)
c	4.0	1	d (with small coupling to H_b)
d	3.2	3	singlet

A methyl group attached to an ether oxygen appears at δ 3.2 ppm. Therefore H_d will be at δ 3.2 ppm, and it will be a singlet because there are no hydrogens on the adjacent atom. The chemical shift for the H_b and H_c hydrogens will be upfield from the typical δ 5.5 ppm chemical shift of a vinyl hydrogen because the oxygen contributes electron density to the β carbon of the alkene (see the resonance structure). Therefore we expect these hydrogens to appear around δ 4.0 ppm.

Deciding which signal is H_b and which is H_c is not critical. The hydrogen that is cis to the oxygen might have a little more deshielding, but the difference is very small. We don't usually try to assign the chemical shifts of such hydrogens. This is why the question asks you to assign these signals "as accurately as possible." Both H_c and H_b will couple to H_a. So they will be doublets. A well-tuned high-field NMR will show the small coupling (J = ~ 1 Hz) between H_c and H_b.

Problem 9.60

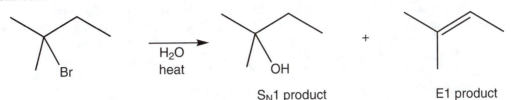

SN1 product   E1 product

The ¹H NMR spectrum of the alkene product will have a vinyl hydrogen near δ 5.5 ppm. It will appear as a quartet. The alcohol product will have a CH_2 quartet near δ 1.8 ppm. There are other differences that can be used to distinguish these two products.

Problem 9.61   The four compounds, **A**, **B**, **C**, and **D**, divide into pairs. Compounds **B** and **D** have only one kind of hydrogen attached to the four-membered ring and thus must share structures 2 and 3. Compounds **A** and **C** have two kinds of ring hydrogen and thus share structures 1 and 4. Compound 2 is more polar than compound 3 and thus must have the cis OH groups (dipoles reinforce) and is **B**. Compound 3 must be **D**. Compound 4 is more polar than compound 1 and thus must be **C** with cis OH groups. By elimination, compound 1 must be **A**.

Problem 9.62   The figure shows the spectrum of ultrapure ethyl alcohol. The methylene group appears as a doublet of quartets (or, equivalently, a quartet of doublets).

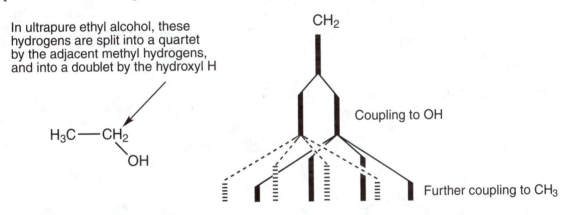

In ultrapure ethyl alcohol, these hydrogens are split into a quartet by the adjacent methyl hydrogens, and into a doublet by the hydroxyl H

CH_2

Coupling to OH

Further coupling to CH_3

H_3C—CH_2
    OH

Problem 9.63   The clue in this problem is the words "scrupulously dry." Alcohols generally show no coupling of the OH hydrogens because of the rapid exchange catalyzed by small amounts of water. If essentially all the water is removed, one sees the first-order spectrum (p. 414). In **A**, we see a doublet, and there is only one hydrogen in either molecule that can appear as a doublet, the OH of 1-phenylethanol. The OH of the other molecule, 2-phenylethanol, appears as a triplet as it is split by the two equivalent adjacent methylene hydrogens, as in **B**.

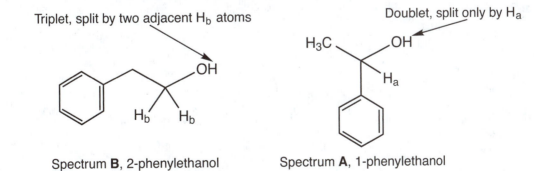

Triplet, split by two adjacent H_b atoms

Doublet, split only by H_a

Spectrum **B**, 2-phenylethanol   Spectrum **A**, 1-phenylethanol

When water is added, the acid-catalyzed exchange reaction effectively decouples the OH hydrogen from all adjacent hydrogens. The OH appears as an average signal, a broad singlet.

Problem 9.64 To distinguish between 3-pentanol and 3-chloropentane, one would need to look for the alcohol signal. There will not be any major difference in the chemical shifts of the other hydrogens. Sometimes the OH hydrogen is difficult to see in a ^{1}H NMR spectrum. To determine if the unknown material is an alcohol, you can (1) obtain the spectrum of your compound and then (2) add a drop of D_2O to your sample and obtain a second NMR spectrum of your compound. Now compare the spectra. If there are no significant shifts between the two, it is not an alcohol. If one of the signals shifts significantly (> 0.2 ppm), it is an alcohol.

To distinguish between 3-pentanol and 2-pentanol, look at the methyl signals. The 3-pentanol methyls will be 6H, δ ~0.9 ppm, triplet. The 2-pentanol methyls will be 3H, δ 0.9 ppm, triplet and 3H, δ ~1.2 ppm, doublet.

Problem 9.65 We use Equation (9.7) (p. 396), $\upsilon = \gamma B_o/2\pi$

for ^{1}H:

$\nu = (2.7 \times 10^8 \text{ rad T}^{-1}\text{s}^{-1})(4.7 \text{ T})/2(3.14 \text{ rad})$
$\nu = 2.02 \times 10^8 \text{ s}^{-1} = 200 \text{ MHz}$

for ^{2}H:

$\nu = (0.41 \times 10^8 \text{ rad T}^{-1}\text{s}^{-1})(4.7 \text{ T})/2(3.14 \text{ rad})$
$\nu = 3.07 \times 10^7 \text{ s}^{-1} = 31 \text{ MHz}$

for ^{13}C:

$\nu = (0.67 \times 10^8 \text{ rad T}^{-1}\text{s}^{-1})(4.7 \text{ T})/2(3.14 \text{ rad})$
$\nu = 5.01 \times 10^7 \text{ s}^{-1} = 50 \text{ MHz}$

Problem 9.66 This problem should be especially easy.

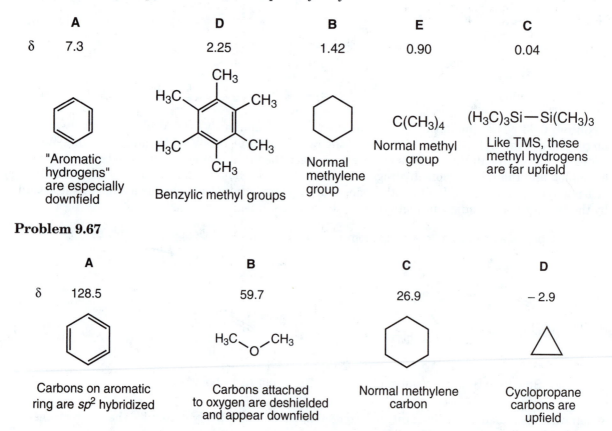

	A	D	B	E	C
δ	7.3	2.25	1.42	0.90	0.04

A
"Aromatic hydrogens" are especially downfield

D
Benzylic methyl groups

B
Normal methylene group

E
$C(CH_3)_4$
Normal methyl group

C
$(H_3C)_3Si—Si(CH_3)_3$
Like TMS, these methyl hydrogens are far upfield

Problem 9.67

	A	B	C	D
δ	128.5	59.7	26.9	− 2.9

A
Carbons on aromatic ring are sp^2 hybridized

B
Carbons attached to oxygen are deshielded and appear downfield

C
Normal methylene carbon

D
Cyclopropane carbons are upfield

Problem 9.68   Oxygen is more electronegative than nitrogen and will withdraw electrons more effectively, making the adjacent carbon more electron deficient (greater δ^+). All other effects being more or less equal, the adjacent hydrogens are more shielded in the amines than they are in the alcohols and will appear at higher field.

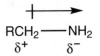

$$RCH_2 \overset{\delta^+}{\underline{}} \overset{\delta^-}{NH_2}$$

$$RCH_2 \overset{\delta^+}{\underline{}} \overset{\delta^-}{OH}$$

The smaller dipole means that the hydrogens are more shielded than in the alcohols. They will appear upfield of the CH_2–O.

The greater dipole means that the hydrogens are deshielded more than the amines. They will appear downfield of the CH_2–N.

Problem 9.69   The two hydrogens are diastereotopic, and thus there should be two signals. Use the technique first outlined on page 401. To see how many signals should appear, replace each in turn with an "X" and see what the relationship is between those two hypothetical compounds.

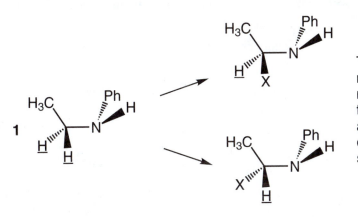

These are stereoisomers, but not mirror images; these two molecules are diastereomers. The two hydrogens replaced with "X" are diastereotopic, and must give different signals in the NMR spectrum.

If we obtain the NMR spectrum of compound **1** at room temperature, which is the normal temperature for an NMR experiment, we find that the underlined hydrogens are equivalent. The reason for this observation is that the amine undergoes rapid inversion. The underlined hydrogens are no longer diastereotopic when there is rapid inversion of the nitrogen.

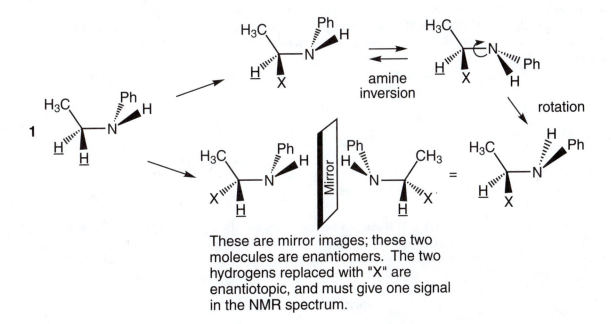

These are mirror images; these two molecules are enantiomers. The two hydrogens replaced with "X" are enantiotopic, and must give one signal in the NMR spectrum.

Problem 9.70   All of the hydrogens of cyclooctane appear as a single peak. This must mean that the hydrogens are equivalent. It is only possible if there is rapid conformational interconversion of the ring structure.

Problem 9.71   Not at all. Both kinds of ring carbon in pyrrole are partially negative. Those additional electrons will serve to shield nearby hydrogens and shift them upfield.

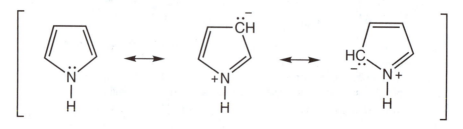

Problem 9.72   Vicinal (1,2) diols react in acid to give carbonyl compounds in what is called the pinacol rearrangement. If the 1,2-diol is cyclic, this reaction becomes a ring contraction process. The problem gives clues that an aldehyde is the product. Note the IR stretching frequency at 1729 cm^{-1} and the signal in the ^{1}H NMR spectrum at δ 9.5 ppm, a position diagnostic for aldehydes.

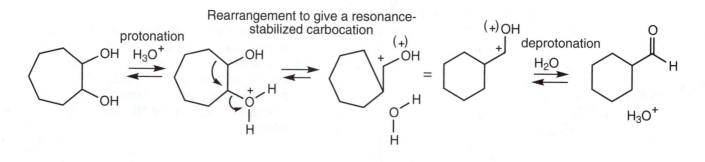

Problem 9.73   The assigned structure can't be correct. There is no hydrogen in the proposed structure that should absorb upfield of about δ ~ 7 ppm. This spectrum shows no fewer than three hydrogens upfield of this value. Indeed, one could also have pointed out that in the assigned structure there is no possible absorption for a single hydrogen. It's wrong.

Problem 9.74   Perhaps hydrogen chloride adds to the double bond of the proposed structure to give a chloride. If so, there should be a low-field dd exactly as appears in the spectrum. The lone hydrogen adjacent to the chlorine is split into doublets by the two different (diastereotopic) adjacent hydrogens of the methylene group. The new methylene group would be a complex multiplet, and the signal centered roughly at δ 3.4 ppm certainly qualifies for that description.

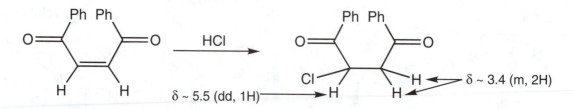

Problem 9.75   Compounds **A** and **B** should be easy because there are only single peaks. One, **B**, is in the range for hydrogens attached to a double bond, and the other, **A**, is in the range for saturated methylene groups. Compound **C** shows only two signals, one for the hydrogens attached to the double bonds, and the other for the allylic methylene groups. The last, most complicated spectrum must be **D**. The assignments follow.

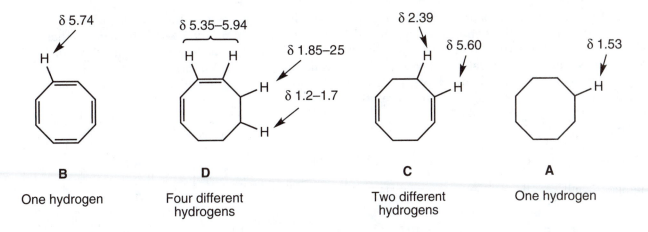

B

One hydrogen

D

Four different hydrogens

C

Two different hydrogens

A

One hydrogen

Problem 9.76 Hydration gives the more substituted alcohol, in this case *tert*-butyl alcohol. The hydroboration–oxidation sequence gives the less substituted alcohol, in this case isobutyl alcohol.

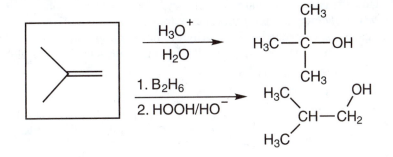

Spectrum **1** must be the more symmetrical *tert*-butyl alcohol. The hydroxyl hydrogen appears at δ 1.61 ppm and the nine methyl hydrogens at δ 1.28 ppm. The more complicated spectrum is that of isobutyl alcohol. The assignments are as follows:

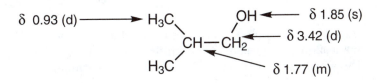

Addition of D_2O will replace the OH hydrogens with D and make HOD. The peak for the ROH hydrogen will become a peak for the HOD hydrogen, which will appear at a different chemical shift.

Problem 9.77 These methyl groups are diastereotopic. Use the "replacement technique" to see this. The two methyl groups are different, and each will give rise to a separate doublet signal.

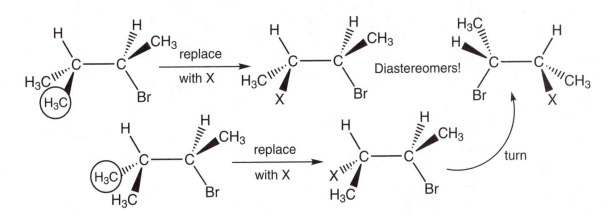

Problem 9.78   This problem should also be easy. There is no need to think about complete assignments of the signals. Instead one takes advantage of symmetry. 1,4-Difluorobenzene has only two different carbons and must have spectrum **C**. Similarly, 1,2-dichlorobenzene has three different carbon atoms and must have spectrum **B**. The remaining isomer, 1,3-dichlorobenzene, has four different carbons and must have spectrum **A**. The complete assignments are shown, but symmetry alone is enough to make the structural assignments.

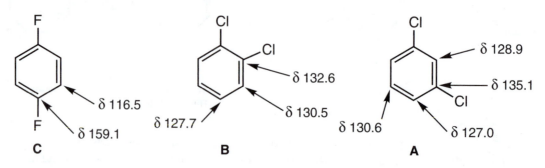

Problem 9.79   How far can we get using symmetry? The answer is "somewhere, but not all the way." We can tell which isomer is 4-methylcyclohexanone, but that's all. 4-Methylcyclohexanone has only five different carbons, whereas both 2-methylcyclohexanone and 2,3-dimethylcyclopentanone have seven different carbons. We can tell which isomer is 4-methylcyclohexanone, but we can go no further than this.

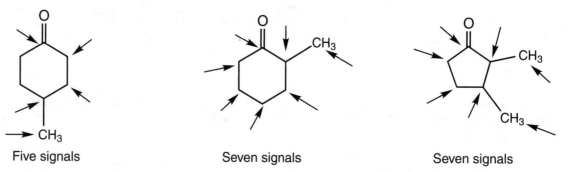

Five signals   Seven signals   Seven signals

An examination of the hydrogen-coupled ^{13}C NMR spectra should allow us to distinguish the last two compounds. For example, 2-methylcyclohexanone will show one ^{13}C signal as a quartet, the methyl carbon attached to three hydrogens, whereas 2,3-dimethylcyclopentanone will show two such quartets.

Problem 9.80   Hydration gives the more substituted alcohol, in this case 2-methyl-2-butanol. The hydroboration–oxidation sequence gives the less substituted alcohol, in this case 3-methyl-2-butanol.

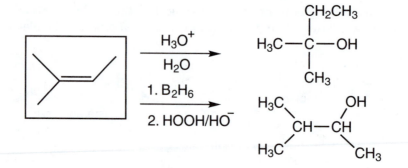

The more symmetrical compound, 2-methyl-2-butanol, gives the simpler spectrum **1**. A critical point to notice is the six-hydrogen <u>singlet</u> for the two equivalent methyl groups. Note especially in spectrum **2** the downfield multiplet for the hydrogen adjacent to the oxygen atom.

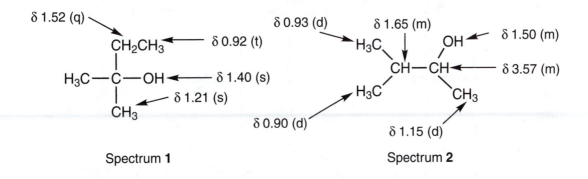

Spectrum **1**   Spectrum **2**

Problem 9.81   The three isomers of pentane are pentane, isopentane, and neopentane. Neopentane is easy to assign, as it is the only isomer with only two different carbons. Isopentane has four different carbons and pentane three different carbons. So pentane is **A**, and isopentane is **B**. Note that it is not necessary in this case to be able to assign all the signals, just to count the numbers of different carbons. This practice is common for ^{13}C NMR spectra.

Pentane **(A)**   Isopentane **(B)**   Neopentane **(C)**

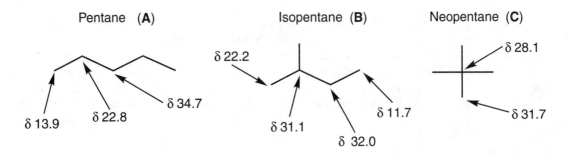

Problem 9.82   Isomer (b) shows a singlet for the two equivalent "aromatic hydrogens" and must be spectrum **1**. Isomer (a) shows two singlets for the two nonequivalent "aromatic hydrogens" and must be spectrum **2**. The other two isomers each show a typical AB quartet, but the spectrum for (c) is much more narrowly split than that for (d). Four-bond coupling is typically much smaller than adjacent, three-bond coupling. Compound (c) must have spectrum **3**, and compound (d) has spectrum **4**.

Problem 9.83

<u>Compound **A**</u> In this problem, we are given no information with which to determine the molecular formula. Although we have no elemental analysis data, we do have a mass spectrum, and this tells us the molecular weight of the compound, as the molecular ion is 148. We can also gain information from looking at the fragment ions, 105 and 77. These result from sequential losses of masses 43 (148 – 105) and 28 (105 – 77). At this point we can't be certain, but masses of 43 and 28 suggest C_3H_7 and CO, respectively.

The ^1H NMR spectrum shows five aromatic hydrogens (δ 7–8 ppm), and this in turn will surely make us think of a monosubstituted benzene. The septet integrating for a single proton looks very much like an "isopropyl" hydrogen [C<u>H</u>(CH$_3$)$_2$], and its chemical shift implies that it is adjacent to an electronegative atom.

(continued)

Problem 9.83 *(continued)*

The IR spectrum shows a strong band at 1675 cm^{-1}, possibly a conjugated ketone, and is consistent with the loss of CO shown by the mass spectrum. Collecting the pieces leads to the following parts:

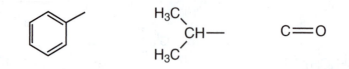

Do these parts add up to a molecule with a molecular weight of 148? Yes, just zip them together to give isobutyrophenone:

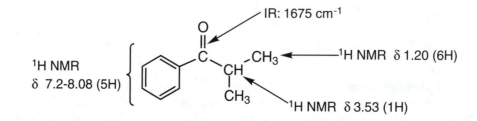

Note also that the mass spectrum becomes quite logical. If the isopropyl group is lost to give an acylium ion, we see loss of C_3H_7. Subsequent loss of CO leads to the fragment of molecular weight 77.

Compound **B** We know this molecule is isomeric with compound **A**, isobutyrophenone, so the formula is $C_{10}H_{12}O$. The IR spectrum shows a strong band at 1705 cm^{-1}, at substantially higher frequency than the carbonyl stretch of compound **A** and indicative of an unconjugated carbon–oxygen double bond. The ^{1}H NMR spectrum shows five "aromatic" hydrogens, and so a mono-substituted benzene ring is apparently present, as in compound **A**. The combination of an upfield triplet (3H) and a somewhat downfield quartet (2H) is always strongly suggestive of an ethyl group (CH_2CH_3). There is also an uncoupled 2H signal.

Now we can put the pieces together:

The combination is benzyl ethyl ketone:

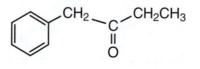

Problem 9.84 We have the formula $C_{11}H_{12}O_4$, so the first task is to work out the number of degrees of unsaturation (Ω) for **A** and **B**.
In this case,

$$\Omega = [2(11) + 2 - 12]/2 = 6$$

Six degrees of unsaturation strongly suggests the possibility of a benzene ring.

This surmise is supported by the ^{1}H NMR spectra. The ^{1}H NMR spectra for both compounds **A** and **B** are similar in several respects. Both spectra exhibit three aromatic hydrogens and three uncoupled methyl signals. Two of the methyl groups appear at about δ 3.9 ppm, indicating that they are attached to an electronegative substituent (for example, oxygen). The other methyl group appears at about δ 2.5 ppm, suggesting attachment to a carbonyl group or the benzene ring. At the very least, we might surmise that compounds **A** and **B** are trisubstituted benzene derivatives; that is,

The IR spectra of compounds **A** and **B** reveal the presence of an ester group as indicated by a strong carbonyl stretch at 1720 cm^{-1} and an intense C—O stretch at 1240–1245 cm^{-1}. The relatively low-frequency position of the carbonyl absorptions suggests the possibility of conjugation. Now let's do an atom inventory to see where we are:

$C_{11}H_{12}O_4$		Formula
$^-$ C_6 H_3		Trisubstituted benzene ring
$^-$ C_3 H_9		Three methyl groups
$^-$ C	O_2	One ester
C	O_2	Remaining

The presence of a residual CO_2 suggests the presence of a second ester group, accounting for the last degree of unsaturation. This result is also consistent with the presence of two low-field methyl groups in the ^{1}H NMR spectra of **A** and **B**. Putting all the structural fragments together gives trisubstituted benzenes, in which the three substituents are two methyl esters and a methyl group. The ^{1}H NMR aromatic resonances of compounds **A** and **B** can be used to deduce the exact substitution patterns.

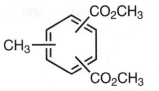

Compound A Two of the aromatic hydrogens are identical and have a chemical shift of δ 8.05 ppm, whereas the other aromatic hydrogen appears at δ 8.49 ppm. The small 2-Hz coupling constant is indicative of meta coupling. The only possible arrangement of substituents with only meta coupling is dimethyl 5-methylisophthalate. Note that H_4 and H_6 are identical and are each adjacent to one ester group. The H_2 is adjacent to two ester groups and accordingly appears at the lower field.

^{1}H NMR (CDCl$_3$): δ 2.46 (s, 3H)

3.94 (s, 6H)

8.05 (d, *J* = 2 Hz, 2H)

8.49 (t, *J* = 2 Hz, 1H)

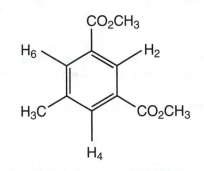

Dimethyl 5-methylisophthalate

A

(continued)

Problem 9.84 *(continued)*

<u>Compound **B**</u> The 8 Hz coupling constant is indicative of 1,2 coupling, whereas the 2 Hz coupling constant is appropriate for 1,3 coupling. This coupling pattern suggests a 1,2,4-trisubstituted benzene ring; that is,

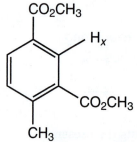

^{1}H NMR (CDCl$_3$): δ 7.28 (d, *J* = 8 Hz, 1H)   H$_a$

8.00 (dd, *J* = 8 Hz, *J* = 2 Hz, 1H)   H$_m$

8.52 (d, *J* = 2 Hz, 1H)   H$_x$

The aromatic hydrogen H$_x$, which exhibits only 1,3 coupling, is furthest downfield and should be flanked by both ester groups. Dimethyl 4-methylisophthalate is the only reasonable stucture for compound **B**.

Dimethyl 4-methylisophthalate

B

Problem 9.85   Our first task is to determine the molecular formulas of **A** and **B**. We can get the percentage of oxygen in the compounds by subtraction, and then work out the formula in the usual way.

C 63.15   100.00
<u>H 5.30</u>   <u>− 68.45</u>
  68.45   31.55% O

C   63.15/12.01 = 5.26   C   5.26/1.97 = 2.67
H   5.30/1.01 = 5.25   H   5.25/1.97 = 2.66
O   31.55/16.00 = 1.97   O   1.97/1.97 = 1.00

$(C_{2.67}H_{2.66}O_{1.00}) \times 3 = C_{8.01}H_{7.98}O_{3.00}$

Empirical formula = $C_8H_8O_3$

As we know the molecular weight from freezing point depression to be about 150 g/mol, we know that the empirical formula = molecular formula.

Second, let's work out the number of degrees of unsaturation, Ω. In this case,

$\Omega = [2(8) + 2 - 8]/2 = 5$

Although the presence of 5 degrees of unsaturation might suggest the presence of a benzene ring, the IR and NMR spectra of compounds **A** and **B** do not support this hypothesis, particularly in the case of compound **B**.

The presence of three oxygen atoms and two high-frequency carbonyl bands in the IR spectra of compounds **A** and **B** suggests the possibility of an anhydride; that is,

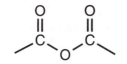

Two degrees of unsaturation

(A true IR maven would also recognize that the anhydrides are cyclic because the lower-frequency carbonyl band is more intense than the higher-frequency band.)

Now let's do an atom inventory:

$C_8H_8O_3$	Formula
$- C_2 \quad O_3$	Anhydride
$\overline{C_6H_8}$	Remaining

There must be at least one carbon–carbon double bond (from the reaction that converts an alkene into an alkane), leaving 2 degrees of unsaturation unaccounted for, and suggesting the possibility of one or two rings. (Alternatively, a disubstituted alkyne and one ring could be present.)

The ^{1}H NMR spectra of compounds **A** and **B** should help with the structural assignments.

Compound **A** The ^{1}H NMR spectrum of compound **A** suggests the presence of two vinylic hydrogens (δ 5.65–6.25 ppm), two deshielded hydrogens (δ 3.15–3.70 ppm), possibly adjacent to the carbonyl groups, and four slightly deshielded hydrogens (δ 2.03–2.90 ppm), probably allylic. One possible way to join these fragments is shown. Note that the relative numbers of hydrogens given by the integral need to be multiplied by two to bring the hydrogen count to eight.

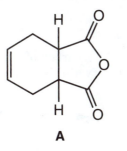

A

The 5 degrees of unsaturation are two carbonyl groups, one carbon–carbon double bond, and two rings. The type of ring juncture (cis or trans) is not obvious at this time.

Compound **B** The ^{1}H NMR spectrum of compound **B** shows the absence of vinylic hydrogens and the presence of four slightly deshielded hydrogens (δ 2.15–2.70 ppm)—possibly allylic—and four hydrogens with a "normal" aliphatic chemical shift (δ 1.50–2.10 ppm). It should also be noted that the carbonyl absorptions in the IR spectrum of compound **B** are about 15–20 cm^{-1} lower in frequency than those for compound **A**, indicating the possibility of conjugation. Putting all this information together suggests the following structure for compound **B**.

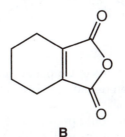

B

(continued)

Problem 9.85 *(continued)*

<u>Compound C</u> The reaction that converts an alkene into an alkane does so by adding hydrogens to the same face of the alkene. The experiments in this case imply that compound **C** is *cis*-1,2-cyclohexanedicarboxylic anhydride. This structure is consistent with the ^{1}H NMR spectrum, which exhibits eight hydrogens with a "normal" aliphatic chemical shift (δ 1.10–2.30 ppm) and two deshielded hydrogens (δ 3.05–3.55 ppm) adjacent to the carbonyl groups. Notice also that as **C** must have a cis ring juncture (from the reaction of **B**), then compound **A**, which gives **C**, must also have a cis ring junction.

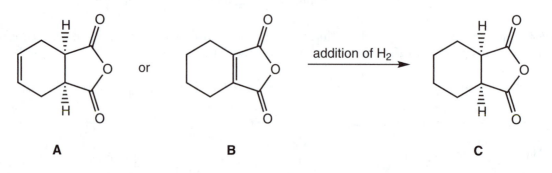

A   **B**   **C**

Problem 9.86

(a) The ^{13}C NMR chemical shift of the β carbon of α,β-unsaturated ketones such as **A** can be rationalized on the basis of the lower electron density at this position. The resonance formulation shows it well.

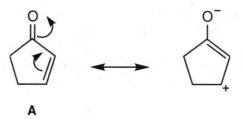

A

Similarly, the ^{13}C NMR chemical shift of the β carbon of vinyl ethers such as **B** occurs further upfield because of the greater electron density at this position.

B

(b) The ^{1}H NMR chemical shifts of the ring hydrogens of aniline (**C**) can also be correlated by considering the appropriate resonance contributors.

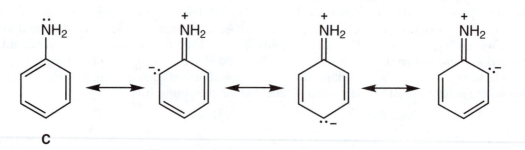

C

Therefore, the C(2), C(4), and C(6) hydrogens of **C** appear at a higher field than the C(3) and C(5) hydrogens because of the greater electron density at these positions.

A similar treatment of acetophenone (**D**) would predict that the C(2), C(4) and C(6) hydrogens would be deshielded because of the lower electron density at these positions.

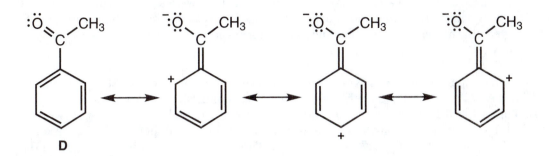

D

However, **D** shows only two deshielded aromatic hydrogens, not three, as predicted by this resonance argument. Clearly, there must be another factor (or factors) operating here. We already know that the hydrogens on a benzene ring are deshielded because of the induced field of the π ring current. The carbon–oxygen double bond in **D** also has an induced field of circulating π electrons. Because the carbonyl bond and the benzene ring of **D** are coplanar, the ortho protons lie in the deshielding zone of the adjacent carbonyl group. As a result, the two ortho protons are shifted further downfield than the other three ring protons. The ortho protons of **D** may also be inductively deshielded by the adjacent carbonyl group.

For chlorobenzene (**E**), the ^{1}H NMR chemical shifts of the aromatic protons can be rationalized by a consideration of inductive and resonance effects. In this case, these two effects, which oppose each other, are apparently roughly in balance. By contrast, in alkyl chlorides, the chlorine substituent is deshielding because only the inductive effect is operative, as there can be no resonance effect.

The ^{13}C NMR chemical shifts of **C** and **E** are, in general, consistent with the proton chemical shifts for these compounds. For example, note the upfield chemical shifts of C(2) and C(4) in aniline (**C**). Finally, note that the ^{13}C NMR chemical shifts for C(1) of compounds **C** and **E** are good indications of the inductive effect of the attached substituent.

Problem 9.87 First, let's determine the molecular formula for compound **1**.

C 81.76
H 10.98
92.74

100.00
−92.74
7.26 %O

C 81.76/12.011 = 6.807
H 10.98/1.008 = 10.89
O 7.26/15.999 = 0.454

C 6.807/0.454 = 14.99
H 10.89/0.454 = 23.99
O 0.454/0.454 = 1.00

Empirical formula = $C_{15}H_{24}O$

Formula weight ($C_{15}H_{24}O$) = 220 g/mol

Because the mass spectrum of compound **1** displays a molecular ion with m/z = 220, the empirical formula equals the molecular formula.

Now, how many degrees of unsaturation are present?

$\Omega = [2(15) + 2 - 24]/2 = 4$

Four degrees of unsaturation suggests the possibility of a benzene ring, which accounts for all of the degrees of unsaturation.

(continued)

Problem 9.87 *(continued)*

Next, examine the spectral data. The IR spectrum of compound **1** shows a sharp, medium-intensity band at 3660 cm^{-1}. The high frequency and sharpness of this absorption suggest the possibility of a "free" O—H stretch, that is, no hydrogen bonding. This information will prove to be particularly useful later on.

All 24 hydrogens are apparent in the ^{1}H NMR spectrum of compound **1**. However, there are only seven signals in the ^{13}C NMR spectrum, which indicates a reasonably high degree of symmetry. The 18H singlet at δ 1.43 ppm in the ^{1}H NMR spectrum of **1** is highly suggestive of two identical *tert*-butyl groups. This supposition is further supported by the singlet at δ 34.2 ppm and the quartet at δ 30.4 ppm in the ^{13}C NMR spectrum. The 3H singlet at δ 2.27 ppm in the ^{1}H NMR spectrum is indicative of a methyl group attached to the benzene ring. This methyl carbon dutifully appears as a quartet at δ 21.2 ppm in the ^{13}C NMR spectrum. The 1H singlet at δ 5.00 ppm in the ^{1}H NMR spectrum could be due to a phenolic hydrogen. Finally, the 2H singlet at δ 6.98 ppm in the ^{1}H NMR spectrum is consistent with two identical phenyl hydrogens. The doublet at δ 125.5 ppm in the ^{13}C NMR spectrum supports this assignment. It is also worth noting that there are only three other signals for the phenyl carbons, all of which are singlets. Also note the deshielded signal at δ 151.5 ppm, which corresponds to the phenyl carbon directly attached to the oxygen. Thus, compound **1** must be a tetrasubstituted benzene derivative with a high degree of symmetry.

Given the deduced structural fragments and the high degree of symmetry, there are really only two structures, **A** and **B**, that we need to consider for compound **1**.

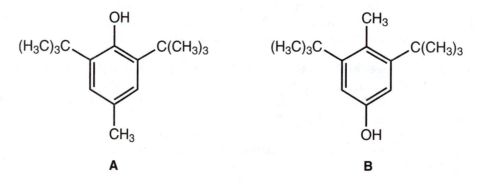

Both these structures would exhibit only seven signals in the ^{13}C NMR spectrum. Structure **A** appears to be the better candidate for at least two reasons. First of all, the phenyl hydrogens ortho to the OH group in structure **B** would be appreciably shielded in the ^{1}H NMR spectrum, as would the corresponding carbons in the ^{13}C NMR spectrum. (Why?) This is clearly not the case. Second, we return to the IR spectrum of compound **1**. Recall that the O—H stretch of **1** is quite sharp and appears at a very high frequency, suggesting the absence of hydrogen bonding. It is obvious that intermolecular hydrogen bonding involving the OH group would be sterically inhibited by the large *tert*-butyl groups in structure **A** but not in structure **B**. Thus, compound **1** is 2,6-di-*tert*-butyl-4-methylphenol, **A**, commonly called butylated hydroxytoluene (BHT), which is used as a stabilizer (antioxidant) at concentrations of 0.025% in solvents such as THF.

Problem 9.88   The loss of CO from compound **A** is similar to the elimination reaction we saw in Figure 8.46 (p. 359). The MS data confirm that there is loss of CO in going from **A** to **B**. Now we need to check to see if the spectral data are consistent with this proposed structure of compound **B**.

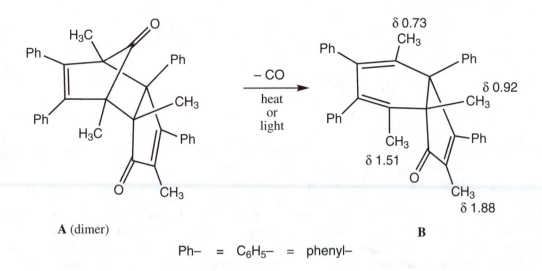

A (dimer)   B

Ph– = C₆H₅– = phenyl–

IR and NMR Spectra of Compound B The presence of a band at 1704 cm^{-1} in the IR spectrum of compound **B** is consistent with the C=O stretch of a conjugated ketone. The ^{1}H NMR spectrum is also consistent with the proposed structure. For example, there are 20 aromatic hydrogens, which we expect from the four phenyl groups. There are four different methyl singlets as expected, but assigning the methyl signals is not going to be easy. Three of the methyl groups are allylic and should be about δ 1.7 ppm (see Table 9.5). The methyl group that is α to the ketone might be expected to be the furthest downfield of the four. So let's assign it as the δ 1.88 ppm. The upfield methyl ought to be the bridgehead methyl that isn't allylic. We can guess that it is at δ 0.73 or 0.92 ppm. The two remaining methyls look very similar, but one is at δ 1.51 ppm and the other at δ 0.92 or 0.73 ppm. The factor that we haven't included in our thinking is the way aromatic rings can influence signals of hydrogens that are forced to be near the center of the aromatic ring. Such hydrogens will be shifted upfield (p. 406) due to the ring current. The methyl group, pointing to the back of the structure as drawn above, is surrounded by three phenyl rings. Perhaps one of those rings is forced by steric constraints to be oriented so that the methyl group is in the center of the ring and shifted upfield. Using molecular modeling confirms that the allylic methyl is likely to be most shielded by the aromatic rings.

Problem 9.89 First, let's determine the molecular formula of compound **A**. From the mass spectral data, we know that compound **A** has a molecular weight of 152 g/mol. From the ^{13}C NMR spectrum, there must be a minimum of eight carbons. In addition, we know from the ^{1}H NMR spectrum that there are at least eight hydrogens. Any molecule of the formula C₈H₈ has a molecular weight of 104 g/mol, which leaves 48 g/mol unaccounted. The simplest fragments that could account for this missing mass are C₄ or O₃. As the IR spectrum of compound **A** suggests the presence of at least two oxygen atoms (the peak at 3205 cm^{-1} could be an O—H stretch and the peak at 1675 cm^{-1} could be a C=O stretch), let's assume that the missing fragment is O₃ and see where this leads. Therefore, the tentative molecular formula of compound **A** is C₈H₈O₃. Now let's work out the number of degrees of unsaturation:

$$\Omega = [2(8) + 2 - 8]/2 = 5$$

Five degrees of unsaturation suggests the possibility of a benzene ring. Compound **A** is soluble in 5% aqueous NaOH solution, but not in 5% aqueous NaHCO₃ solution. Carboxylic acids and phenols are strong enough acids to react with NaOH to yield water-soluble salts. However, unlike carboxylic acids, most phenols are too weakly acidic to undergo reaction with NaHCO₃. Thus, these solubility data suggest that compound **A** is a substituted phenol.

(continued)

Problem 9.89 *(continued)*

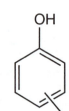

The presence of the broad peak at 3205 cm^{-1} in the IR spectrum of compound **A** is consistent with the O—H stretch of the proposed phenol. In addition, the presence of the strong band at 1675 cm^{-1} suggests the possibility of a C=O stretch, accounting for the last degree of unsaturation. The frequency of the carbonyl stretch further indicates that the carbonyl group is probably conjugated. What type of carbonyl group? An anhydride can be ruled out because of the absence of a second carbonyl stretch in the infrared; an aldehyde can be eliminated because of the absence of the C—H stretching doublet; and a carboxylic acid can be ruled out from the solubility data. These eliminations leave the possibility of either an ester or a ketone. There are several bands in the 1300–1100 cm^{-1} region of the IR spectrum that could be an ester C—O stretch. However, the ketone cannot be ruled out on this basis. (The ^{13}C NMR spectrum of compound **A** is very helpful in this regard, as we will see shortly.)

Now is a good time for an atom inventory.

$C_8H_8O_3$	Formula
− C_6	Benzene ring
− H O	OH of phenol
− C O	Carbonyl group
C H$_7$O	Remaining

Now let's look at the proton and carbon NMR spectra of compound **A**. First of all, let's deal with the ketone versus ester problem. The carbonyl carbon of ketones (and aldehydes) has a ^{13}C NMR chemical shift in the δ 190–220 ppm range, whereas the carbonyl carbon of esters appears in the δ 150–180 ppm range. The ^{13}C NMR spectrum of compound **A** fails to show a signal for a ketone but does exhibit a singlet in the ester range (δ 170.7 or 162.0 ppm).

It is obvious from both the ^{1}H and ^{13}C NMR spectra of compound **A** that the remaining unassigned carbon (from the atom inventory) is a deshielded CH$_3$ group (the singlet at δ 3.92 ppm in the proton spectrum and the quartet at δ 52.1 ppm in the carbon spectrum). If we put all the structural fragments together, we obtain a methyl hydroxybenzoate; that is,

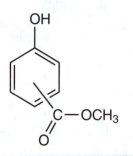

Note that this preliminary structure also accounts for the four aromatic hydrogens in the δ 6.85–7.83 ppm region of the ^{1}H NMR spectrum of compound **A**. Is compound **A** the 1,2 (ortho); 1,3 (meta); or 1,4 (para) isomer? There are several ways to approach this question; happily, all of

them give the same answer. First, the IR spectrum of compound **A** shows a single aromatic C—H bending absorption at 757 cm^{-1}, which is indicative of an ortho-disubstituted benzene. Second, note that all four aromatic hydrogens in the δ 6.85–7.83 ppm region of the ^{1}H NMR spectrum show ortho coupling (J = 8 Hz). The only disubstituted isomer in which four different aromatic hydrogens all have at least one adjacent ortho proton is the ortho isomer. Finally, and more esoterically, the chemical shift of the phenolic O—H hydrogen in the ^{1}H NMR spectrum of compound **A** is very informative. The O—H hydrogen of phenols normally has a chemical shift of δ 4.0–7.5 ppm. However, an ortho carbonyl group shifts the phenolic hydrogen downfield to δ 10.0–12.0 ppm because of intramolecular hydrogen bonding. The phenolic hydrogen of compound **A** appears at δ 10.8 ppm. Thus compound **A** is methyl salicylate (methyl 2-hydroxybenzoate).

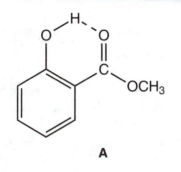

A

The ^{1}H NMR chemical shifts and coupling constants for compound **A** are summarized.

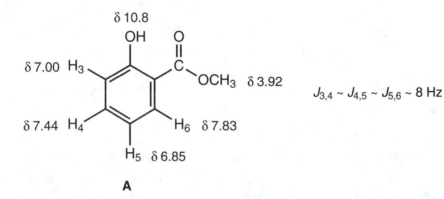

A

The H(3) is shielded by the adjacent OH group and is split into a doublet by H(4). (Remember that J_{meta} and J_{para} were not observed at 300 MHz.) H(4) has a "normal" aromatic chemical shift and is split into a "triplet" by H(3) and H(5). H(5) is shielded by the para OH group and is split into a "triplet" by H(4) and H(6). Finally, H(6) is deshielded by the adjacent ester carbonyl bond and is split into a doublet by H(5).

A tentative assignment of the ^{13}C NMR resonances of compound **A** is as follows:

δ 52.1(q) CH$_3$
112.7(s) C-1
117.7(d) C-3 (or C-5)
119.2(d) C-5 (or C-3)
130.1(d) C-6 (or C-4)
135.7(d) C-4 (or C-6)
162.0(s) C-2 (or C=O)
170.7(s) C=O (or C-2)

A

Problem 9.90

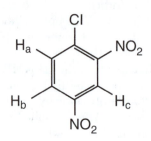

In the ^{1}H NMR spectrum for 1-chloro-2,4-dinitrobenzene (ignoring long-range 1,3 and 1,4 coupling),

H_a should appear around δ 7.5 ppm; it will not be strongly deshielded by the nitro groups that are 1,3 to H_a; H_a will be a doublet because there is only one vicinal hydrogen (H_b).

H_b should appear around δ 8.0 ppm because it will be deshielded by both nitro groups (resonance) and by the 1,2 nitro group (induction); H_b will be a doublet because it has only one vicinal hydrogen (H_a).

H_c should appear around δ 8.5 ppm because it will be strongly deshielded by both nitro groups as a result of resonance and induction and will be a singlet because there are no vicinal hydrogens.

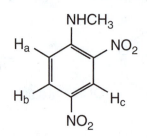

In the ^{1}H NMR spectrum for *N*-methyl-2,4-dinitroaniline (ignoring long-range coupling),

H_a should appear around δ 7.0 ppm; it will be strongly shielded by the amine group that is 1,2. The nitro groups that are 1,3 to H_a will have little effect. H_a will be a doublet because it will "see" one vicinal hydrogen (H_b).

H_b should appear around δ 8.0 ppm because it will be deshielded by both nitro groups by resonance and by the ortho nitro group by induction. The 1,3 amino group will not have any effect; H_b will be a doublet because it "sees" one vicinal hydrogen (H_a).

H_c should appear around δ 8.5 ppm because it will be strongly deshielded by both nitro groups as a result of resonance and induction; the 1,3 amino group will have no impact; H_c will be a singlet because there are no vicinal hydrogens.

The chemical shift for NH is difficult to predict, but adding D_2O to the NMR sample will cause the signal to be replaced by a DOH signal, which will have a different chemical shift; the NH will likely appear as a broad singlet.

NCH_3, the methyl group, will be about δ 2.5 ppm as a result of being attached to a nitrogen, and the aromatic ring will probably add to the downfield shift.

The biggest difference in the IR spectra of the starting material and the product will be the N—H stretch that will appear in the product around 3300 cm^{-1}.

Problem 9.91 The second carbocation in the reaction scheme is more stable because it is a tertiary carbocation. The reaction of that carbocation with water is shown here.

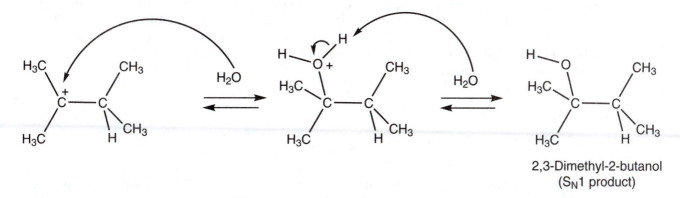

2,3-Dimethyl-2-butanol
(S$_N$1 product)

The major E1 product is 2,3-dimethyl-2-butene, and its ^{1}H NMR spectrum will be a singlet around δ 1.7 ppm. The ^{1}H NMR spectrum for 2,3-dimethyl-2-butanol will clearly be more complex. One could also use IR spectral data to differentiate between an alcohol and an alkene by looking for the broad band at 3400 cm^{-1} for the alcohol.

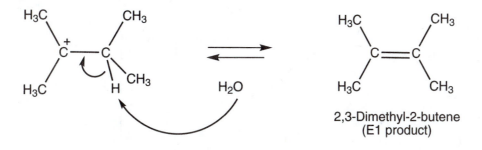

2,3-Dimethyl-2-butene
(E1 product)

Problem 9.92 Our first approximation for the ^{1}H NMR spectrum of 2-pentyne would be as shown in the table that follows. We can start by assuming that an alkyne will be much like an alkene. Therefore, the methyl hydrogens on C(1) would be about the same chemical shift as an allylic methyl (δ 1.7 ppm). The CH$_2$ group ought to be the same as an allylic methylene (δ 2.0 ppm), and the C(5) methyl might be about δ 1.0 ppm. If we limit our thinking to vicinal (three-bond) coupling, then we expect the coupling listed in our chart.

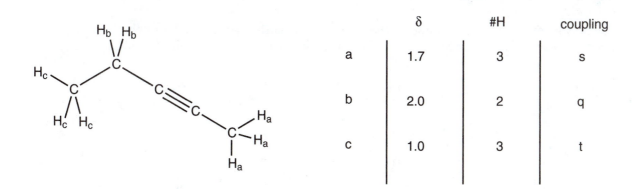

	δ	#H	coupling
a	1.7	3	s
b	2.0	2	q
c	1.0	3	t

(continued)

Problem 9.92 *(continued)*

Here are the actual ^{1}H NMR data for 2-pentyne: Ha (δ 1.77 ppm, 3H, t, J = 2.5 Hz), H$_b$ (δ 2.13 ppm, 2H, quartet of quartets, J = 2.5, 7.3 Hz), H$_c$ (δ 1.11 ppm, 3H, t, J = 7.3 Hz). We were close on the chemical shifts, but we missed the long-range coupling that is observed in alkynes. The five-bond coupling through an alkyne is an impressive long-range interaction. The J_{ab} varies from 2 to 3 Hz! So our H$_a$ signal will actually be a triplet (J = 2.5 Hz) because H$_a$ couples to the two H$_b$ hydrogens. The H$_b$ couples to the three H$_a$ hydrogens (J = 2.5 Hz) and to the three H$_c$ hydrogens (J = 7.3 Hz) so that it is a quartet of quartets.

There is much more guessing in predicting chemical shifts for the ^{13}C NMR for 2-pentyne because Table 9.6 has larger ranges and less specificity. We can guess that the methyl on the alkyne would be further downfield in the range of methyls (δ 0–30 ppm). The two alkyne carbons will be very similar, and we might as well pick the middle of the range (δ 70–90 ppm) for them. The methylene might be near the downfield limit of the methylene range (δ 15–55 ppm). The C(5) methyl will be near the starting point for a typical methyl, which is about 10 ppm. The coupling data will simply be a reflection of the number of hydrogens attached to each carbon and using the n + 1 rule.

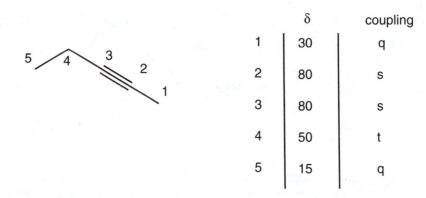

	δ	coupling
1	30	q
2	80	s
3	80	s
4	50	t
5	15	q

Here are the actual ^{13}C NMR data for 2-pentyne: C(1) δ 3.4 ppm, q; C(2) δ 74.8 ppm, s; C(3) δ 80.8 ppm, s; C(4) δ 12.6 ppm, t; C(5) δ 14.4 ppm, q. We were close enough on the chemical shifts, except for the C(1) and C(4). These signals must be shifted upfield because of the alkyne ring current (Fig. 9.42). The coupling was correct!

The only significant IR band would be the weak carbon–carbon triple bond stretch at ~2200 cm^{-1}. The mass spectrum would show a molecular ion at 68 because the molecular formula is C_5H_8.

Electrophilic Additions to Alkenes

10

Problem 10.1 As always, our estimate of the exothermicity or endothermicity of the reaction is made by comparing the bond energies of the bonds broken and made in the reaction. In this case, the bonds broken are the π bond of the alkene and the σ bond of hydrogen chloride. The bonds made are the carbon–hydrogen and carbon–chlorine bonds in the product chloride.

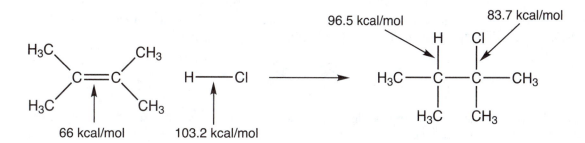

The bonds to be broken are worth 169.2 kcal/mol, but the bonds made are worth 180.2 kcal/mol. Accordingly, the reaction is exothermic by about 11 kcal/mol (180.2 − 169.2). $\Delta H \cong -11$ kcal/mol.

Problem 10.3 The orbital in Figure 10.12 that can accept electrons from a nucleophile is the lowest unoccupied molecular orbital (LUMO). It is the orbital shown below.

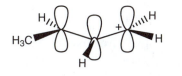

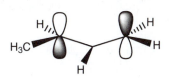

| The orbitals involved in the mixing of this allylic cation | The LUMO for this molecule with all the atoms still shown | The LUMO of just the allyl system |

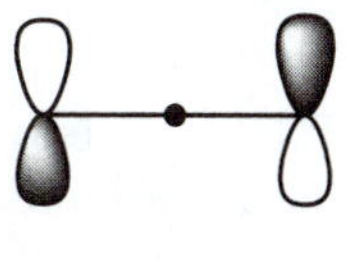

The reason the nucleophile reacts with the LUMO is that it is the most available empty orbital. The nucleophile can add to either end of the allylic system. If the nucleophile is chloride, for example, the product can be either 1-chloro-2-butene (chloride attacking the carbon on the right side of the LUMO shown above) or 3-chloro-1-butene (chloride attacking the carbon on the left side of the LUMO shown above).

(continued)

Problem 10.3 *(continued)*

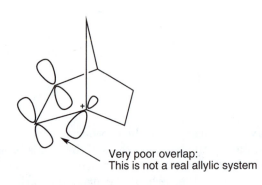

1-Chloro-2-butene   3-Chloro-1-butene

Problem 10.4   Here, *on paper*, the resonance double arrow is as correct as shown in the problem. However, geometry raises its ugly head, and there is really little or no overlap between the *p* orbitals making up this deceptive, "imitation" allylic system. Because there is no overlap, there is no real resonance.

Very poor overlap:
This is not a real allylic system

Problem 10.5

(a) ^{13}C NMR: Heptane has four unique carbons. The symmetry of heptane was covered in Chapter 2 (p. 92). There will be four signals in the ^{13}C NMR spectrum of heptane.

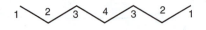

^{1}H NMR: Heptane has four hydrogen signals. The same symmetry applies to this molecule. The hydrogens on the carbons labeled 1 are equivalent. The hydrogens on the carbons labeled 2 are equivalent and different from the hydrogens on the carbons labeled 3. The methylene on carbon 4 is unique.

(b) ^{13}C NMR: (*E*)-2-Heptene has seven unique carbons. There will be seven signals in its ^{13}C NMR spectrum.

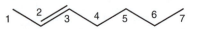

^{1}H NMR: There are seven different hydrogen signals in (*E*)-2-heptene. The hydrogens on each carbon are unique.

(c) ^{13}C NMR: (2*E*,4*E*)-2,4-Heptadiene has seven unique carbons. There will be seven signals in its ^{13}C NMR spectrum.

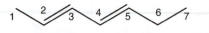

^{1}H NMR: There are seven different hydrogen signals in this molecule as well.

(d) ^{13}C NMR: This heptadienyl cation is a combination of three resonance structures. We can write a dotted line structure that represents all three resonance structures and allows us to see the symmetry in the molecule.

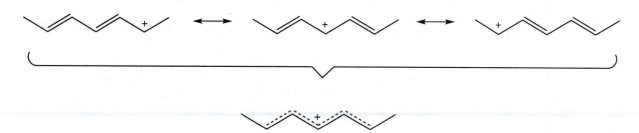

There are four unique carbons in this heptadienyl cation, therefore there will be four signals in its ^{13}C NMR spectrum.

^{1}H NMR: There will be four hydrogen signals in the ^{1}H NMR spectrum of this cation. There is a plane of symmetry through the central carbon. So the hydrogens on the left methyl group are equivalent to the hydrogens on the right methyl group. The hydrogen on the carbons numbered 2 are equivalent, as are the hydrogens on the carbons numbered 3. The hydrogen on the middle carbon is unique.

(e) ^{13}C NMR: The heptadienyl anion is also a combination of three resonance structures. We can write a dotted line structure that represents all three resonance structures and allows us to see the symmetry in the molecule. Again there are four unique carbons in this heptadienyl ion, therefore there will be four signals in its ^{13}C NMR spectrum.

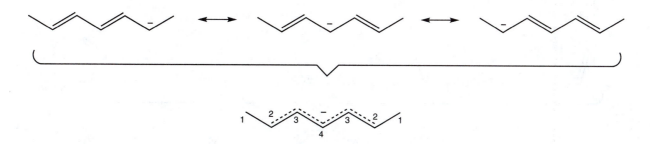

^{1}H NMR: This structure also has four different hydrogens because of the symmetry in the molecule. The hydrogens on the left methyl group are equivalent to the hydrogens on the right methyl group. The hydrogen on the carbons numbered 2 are equivalent, as are the hydrogens on the carbons numbered 3. The hydrogen on the middle carbon is unique. There will be four signals in the ^{1}H NMR spectrum of this anion.

(continued)

Problem 10.5 *(continued)*

(f) ^{13}C NMR: The five carbons in cyclopentane are equivalent. There will only be one signal in its ^{13}C NMR spectrum.

^{1}H NMR: There will only be one signal in the ^{1}H NMR spectrum of cyclopentane.

(g) ^{13}C NMR: Cyclopentene has three unique carbons. There will be three signals in its ^{13}C NMR spectrum.

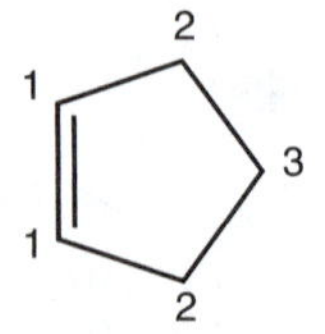

^{1}H NMR: There is a plane of symmetry in this molecule that allows us to see that the hydrogens on the double bond of cyclopentene are equivalent. The hydrogens on the carbons numbered 2 are equivalent to each other. The hydrogens on the carbon numbered 3 are unique. There are three signals in the ^{1}H NMR spectrum of cyclopentene.

(h) ^{13}C NMR: Cyclopentadiene also has three unique carbons. There will be three signals in its ^{13}C NMR spectrum.

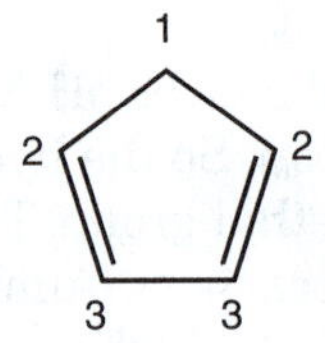

^{1}H NMR: There are also three signals in the ^{1}H NMR spectrum of cyclopentadiene.

(i) ^{13}C NMR: The cyclopentadienyl cation is very unstable, as we will see in Chapter 14. There are five resonance structures we can draw for this molecule. The composite structure shows that every carbon is equivalent in this cation. There would only be one signal in the ^{13}C NMR spectrum of the cyclopentadienyl cation.

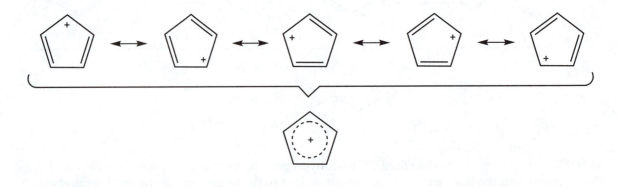

^{1}H NMR: There would only be one signal in the ^{1}H NMR spectrum of this cation.

(j) ^{13}C NMR: The cyclopentadiene anion is a very stable anion, as we will see in Chapter 14. There are five resonance structures we can draw for this molecule. The composite structure shows that all five carbons are equivalent. There is only one signal in the ^{13}C NMR spectrum of the cyclopentadiene anion.

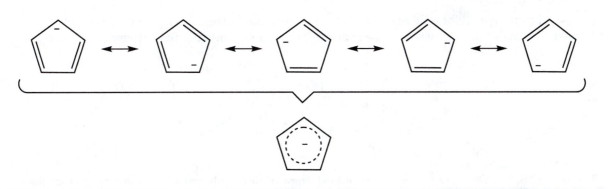

^{1}H NMR: There is only one signal in the ^{1}H NMR spectrum of the cyclopentadiene anion.

Problem 10.6 The answers for HCl are shown. The reactions will be the same with HBr, with the Br instead of the Cl in the product.

(a) The electrophile adds to the C(1) carbon, which forms the secondary carbocation. The chloride anion reacts with the cation to form 2-chlorobutane.

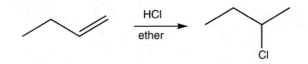

(b) The electrophile will add to either carbon of the alkene because the intermediate carbocation in both cases is a secondary carbocation. There will be no predictable preference. The chloride will capture whichever carbocation is formed and the ratio of the products will be close to 50:50.

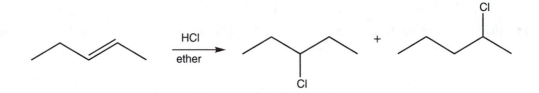

(c) The electrophile adds to C(3) of 2-methyl-2-pentene. That forms the intermediate tertiary carbocation, which reacts with the chloride anion.

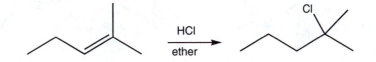

(d) Treatment of 4,4-dimethylcyclohexene with HCl could give several products. There are minor products that can arise from hydride shifts and even methyl shifts. 3-Chloro-1,1-dimethylcyclohexane and 4-chloro-1,1-dimethylcyclohexane would be the major products.

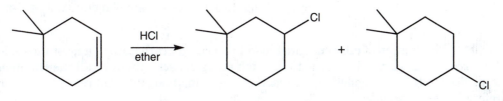

(continued)

Problem 10.6 *(continued)*

(e) 1,4,4-Trimethylcyclohexene reacts with the electrophile so that the tertiary carbocation is formed. The major product would come from chloride anion capturing the cation intermediate.

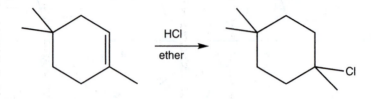

Problem 10.7 "Yes and no" might be the best answer. Certainly one can draw a resonance form for each carbon–hydrogen bond of the three methyl groups, making a total of nine. However, orbitals must overlap for there to be real delocalization. All nine carbon–hydrogen bonds cannot be properly lined up at the same time. A look at one methyl group makes the point.

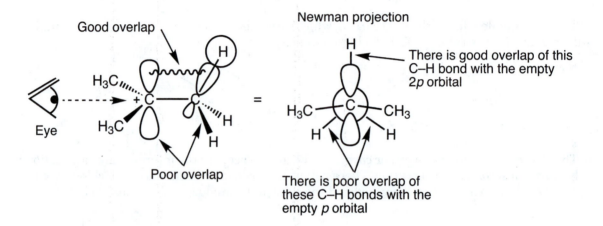

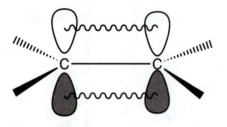

Problem 10.8 It is weaker. Overlap is less good, as the orbitals involved are not optimally lined up.

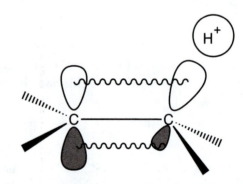

The π bond in ethylene is made up of 2*p*/2*p* overlap. The two 2*p* orbitals are parallel and overlap well

In the "hyperconjugative" resonance form for the ethyl cation, the "double" part of the double bond is not made up of 2*p*/2*p* overlap, but by 2*p*/*sp*^{3} overlap. These orbitals do not overlap as well as the two 2*p* orbitals, and the bond will be weaker

Problem 10.9 Protonation of ethylene must lead to a primary carbocation, a most unstable species. This reaction is very difficult due to the high energy of activation (recall the Hammond postulate, p. 351). In this very endothermic reaction, the transition state will resemble the product and therefore be very high in energy.

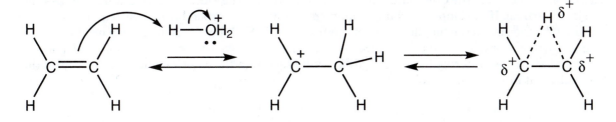

The structure that has the proton shared between the two carbons of ethylene is more stable because the positive charge is shared between three atoms.

Problem 10.10 This reaction involves a carbocation intermediate. We have seen the same intermediate in Problem 8.15 (p. 355). The initial tertiary carbocation that is formed can be captured by the chloride ion to give the first product listed. The hydrogens on the alkene (vinylic hydrogens) are shown in the first step for clarity.

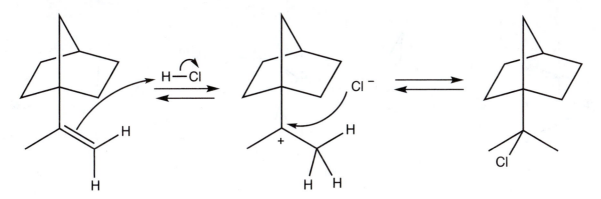

As we saw in Problem 8.15, the cation can also undergo rearrangement. Notice that there is symmetry in the first intermediate. Migration of either of the two different carbons attached to the carbon adjacent to the carbocation will result in two different tertiary carbocation intermediates. Reaction between either of those new carbocations and chloride ion will lead to the other two products listed.

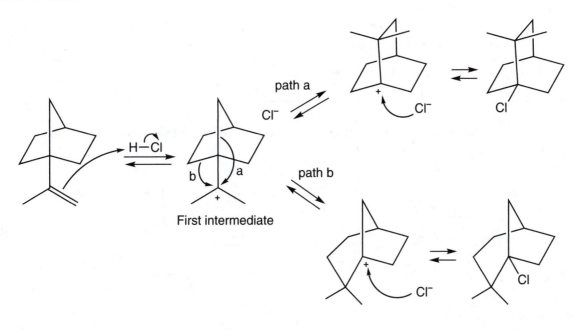

First intermediate

Problem 10.11   It is the orientation of the empty *p* orbital (the electrophile) that determines the product outcome. If the empty *p* orbital is aligned with the one carbon bridge (path a), the shift that occurs will give rise to the bicyclo[2.2.2]octane framework. Path b will occur if the *p* orbital is aligned with either of the two-carbon bridges. That path will give rise to the bicyclo[3.2.1]octane intermediate. These intermediate carbocations will react with the chloride nucleophile to form the products that are shown.

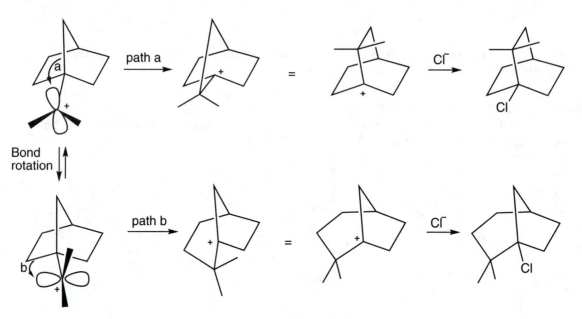

Problem 10.12   Protonation of 2-ethyl-1-butene takes place to give the more stable, tertiary carbocation rather than the less stable primary carbocation.

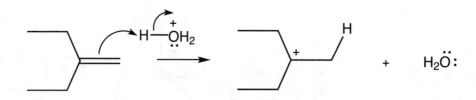

The tertiary carbocation can add to another alkene, again producing the more stable tertiary carbocation.

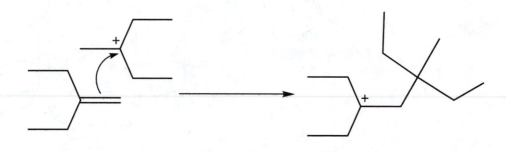

There are three different products that can be formed by loss of a proton from this new cation:

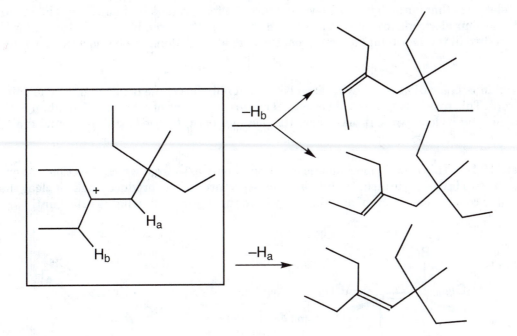

Problem 10.13

Polystyrene

Polyisobutylene

Polybutylene

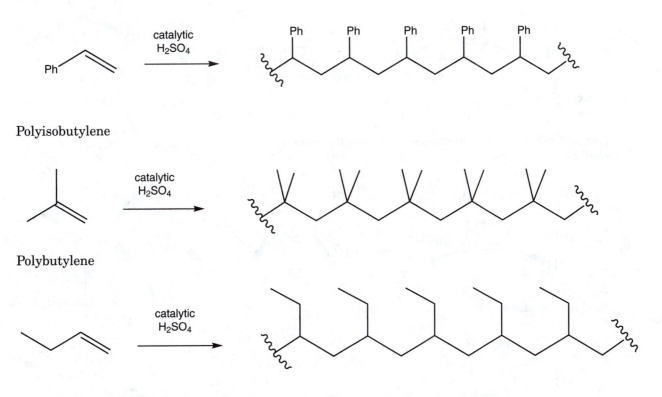

Problem 10.15 Boron has five positive charges in its nucleus. The boron in BH_3 has two electrons in its 1s orbital and three electrons involved in the bonds to the three separate hydrogens. Therefore, the five electrons that boron has as its own, balance the five protons in its nucleus. The boron in $R—BH_3$ has four electrons involved in bonds, which means the boron has six electrons that it senses as its own (two in the 1s orbital and four from the bonds). Therefore, the boron will have a formal charge of -1.

The boron in BH_3 (or BF_3) is sp^2 hybridized because it only has three bonds and no lone pairs of electrons. This hybridization is the lowest energy arrangement of orbitals for an atom with three bonds and no lone pairs. It maximizes the distance between the electrons around the central atom.

Problem 10.18 This is not a mechanistic question; it only asks from which direction (top or bottom) the two bromines attach. In this case, the structures of the products make it clear that one bromine comes from the "top" and the other from the "bottom," in what is called anti addition.

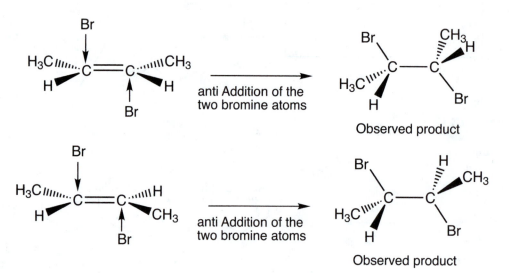

Had addition of the two bromines been from the same side, in syn fashion, just the opposite results would have been seen.

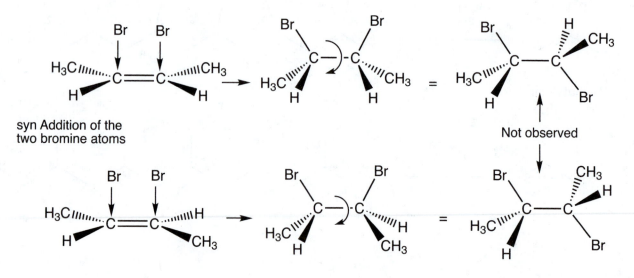

Problem 10.19 There are many possibilities, although most will require some sort of isotopic label. The question states that any appropriate starting material is available, so that's not a problem. Here's one in which a deuterated cycloalkene is used:

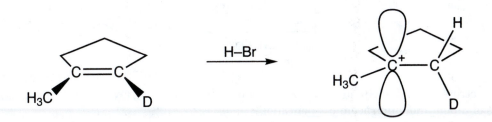

Addition can occur to either lobe of the 2p orbital to give either syn or anti addition. Both products will be formed. A careful determination of the structure of the product will allow a determination of the direction of addition.

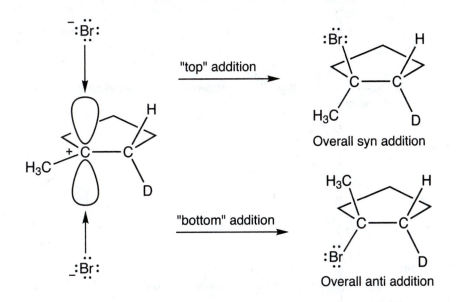

Overall syn addition

Overall anti addition

Problem 10.21 The boron atom in a boronic ester, B(OR)$_3$, remains a Lewis acid, as it still has an empty 2p orbital. Hydroxide is a strong nucleophile and adds to the Lewis acidic boron atom. Loss of ⁻OR completes the first stage. Two repetitions lead to boric acid.

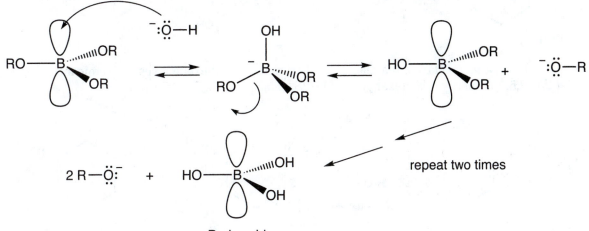

Boric acid

Problem 10.22

(a) 1-Butanol is the only product in this reaction.

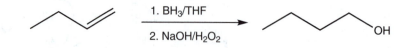

(b) The electrophile (BH$_3$) will add to either carbon of the 2-pentene. Oxidation leads to similar amounts of 3-pentanol and racemic 2-pentanol.

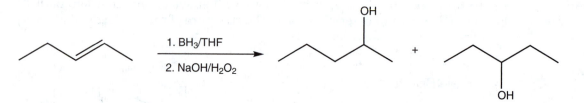

(c) The reaction of BH$_3$ with 2-methyl-2-pentene, followed by oxidation of the boron, gives a racemic mixture of 2-methyl-3-pentanol.

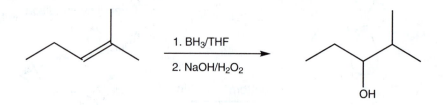

(d) Electrophilic addition of BH$_3$ to 4,4-dimethylcyclohexene, followed by oxidation with hydrogen peroxide, gives 4,4-dimethylcyclohexanol and a racemic mixture of 3,3-dimethylcyclohexanol.

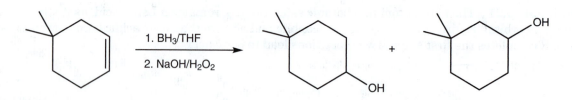

(e) The hydroboration/oxidation reaction of 1,4,4-trimethylcyclohexene gives a racemic mixture of (1R,2R)-2,5,5-trimethylcyclohexanol and (1S,2S)-2,5,5-trimethylcyclohexanol.

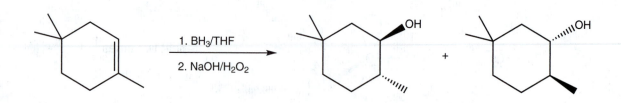

Problem 10.23 There are several straightforward parts to this problem, which only require that you remember one synthetic procedure.

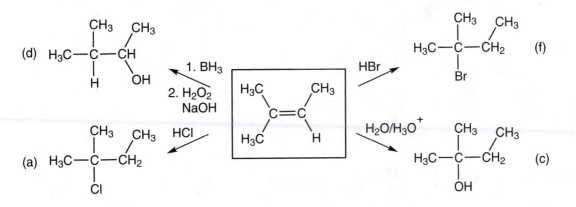

The remaining three molecules demand one more level of sophistication as they cannot be made directly but require a second transformation involving a molecule already constructed.

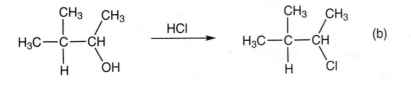

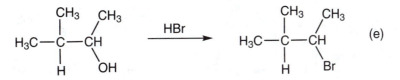

Additional Problem Answers

Problem 10.24 The π molecular orbital system for allyl consists of three orbitals. These were discussed in the chapter, and by now it should be possible to write them out without the construction process. It is especially useful to recall that the number of new nodes will increase 0 ... 1 ... 2. In the π system of the cation, there are only two electrons. In the neutral radical, there must be one more, and in the negatively charged anion, two more.

(continued)

Problem 10.24 *(continued)*

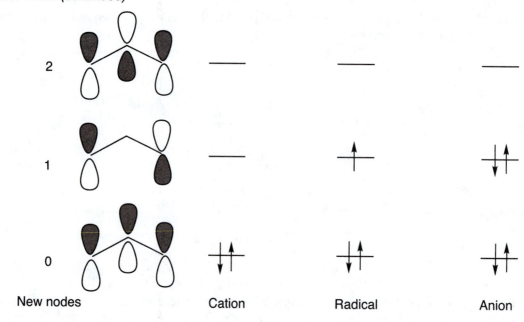

New nodes   Cation   Radical   Anion

Problem 10.25

(a) Initial addition of the electrophile gives a secondary carbocation. That cation can be captured by chloride to give the first product. Hydride shift will occur to give the more stable tertiary carbocation, which results in formation of the 1-chloro-1-ethylcyclohexane as a second product.

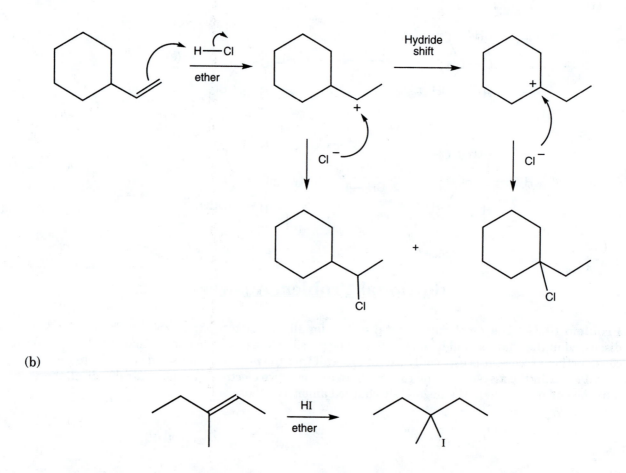

(b)

(c) The products for this reaction are shown.

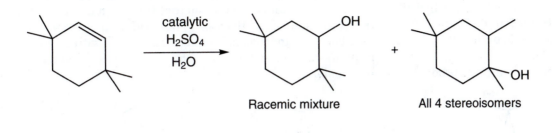

Racemic mixture   +   All 4 stereoisomers

3,3,6,6-Tetramethylcyclohexene is symmetrical. Protonation of the alkene can occur on either carbon. Water will capture the resulting carbocation **A** to give a protonated alcohol (path a) that produces 2,2,5,5-tetramethylcyclohexanol after deprotonation. A methyl shift (path b) in the initial secondary carbocation (**A**) will give a tertiary carbocation (**B**). Capture of the tertiary carbocation gives a protonated alcohol, which gives the 1,2,4,4-tetramethylcyclohexanol isomers upon deprotonation. There are two stereogenic carbons in the second product. We expect the (*R*,*R*), (*R*,*S*), (*S*,*R*), and (*S*,*S*) stereoisomers to be formed.

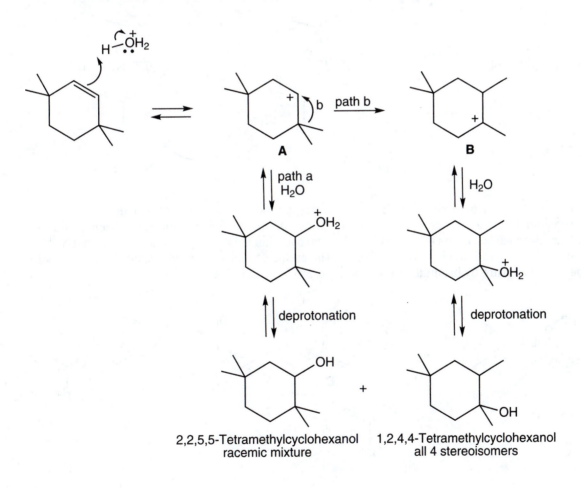

2,2,5,5-Tetramethylcyclohexanol   1,2,4,4-Tetramethylcyclohexanol
racemic mixture   all 4 stereoisomers

(continued)

Problem 10.25 *(continued)*

(d) The hydroboration of this alkene will occur from the less hindered side of the double bond. The carbon with two methyl groups on the top face of the alkene is very bulky. The addition of the electrophile (boron) and the hydride is a syn addition and can occur so that the boron is attached to either of two original alkene carbons. The second step of the reaction is the oxidation of the boron, which then leads to the alkyl shift from the boron to an oxygen. This reaction occurs with retention of configuration. Therefore, the oxygen remains on the less hindered face of what was the alkene. The two possible alcohol products are enantiomers.

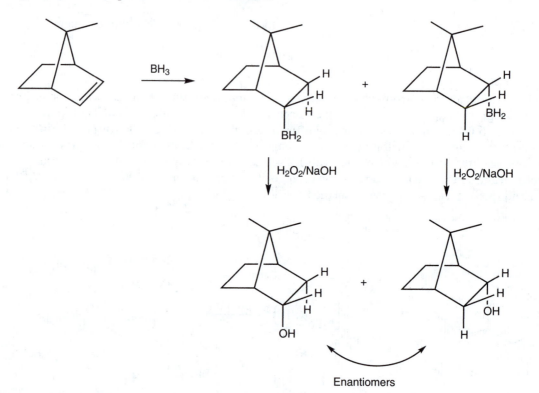

(e) Addition of HBr to 1-phenylpropene will give a single product. To come to this conclusion, we need to consider the two carbocations (**C** and **D**) that could arise from the initial addition of the electrophile. Both carbocations are secondary, but carbocation **C** is benzylic. The benzylic cation is resonance stabilized and much more stable than the cation **D**. The only product of this reaction will be formed from capture by bromide of the intermediate carbocation **C**.

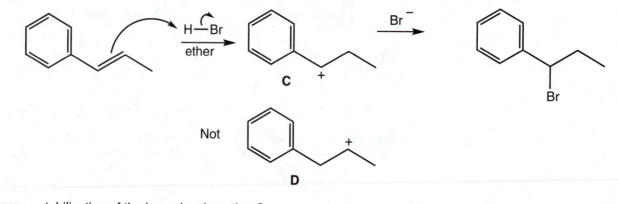

Resonance stabilization of the benzyl carbocation **C**

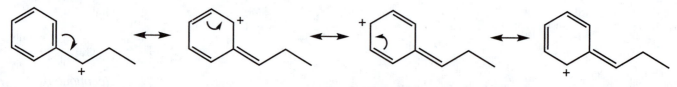

Problem 10.26

(a)

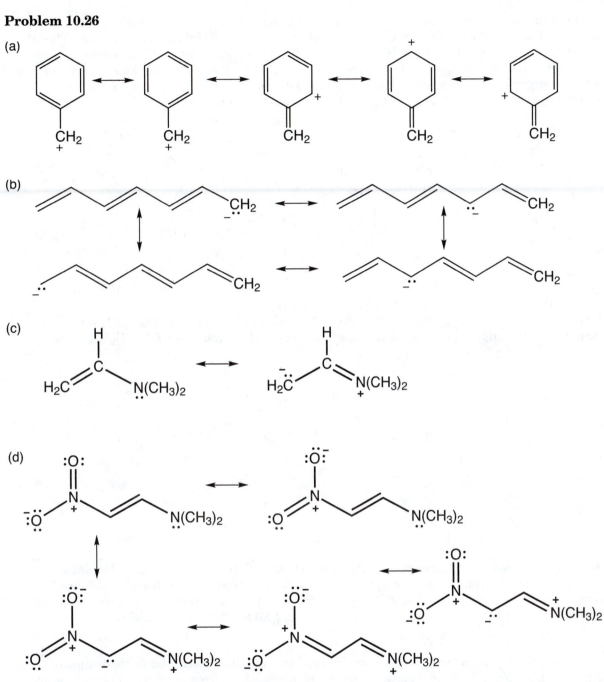

(b)

(c)

(d)

Problem 10.27 The C_5H_{10} molecule will have 1 degree of unsaturation. It will be either an alkene or a ring. We see from the ^1H NMR data that there are three hydrogens in the vinyl hydrogen region (around δ 5.5 ppm). Because there is only one double bond possible (from the degrees of unsaturation) and we have three vinyl hydrogens, we can conclude that the double bond is monosubstituted. That conclusion gives us this partial structure:

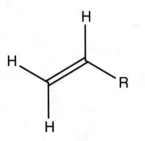

(continued)

Problem 10.27 *(continued)*

Because the double bond is the only functional group in the molecule, we expect that the allylic position should be the group most deshielded (other than the vinyl hydrogens, of course). The ^{1}H NMR data tell us that there is a CH$_2$ at δ 2.0 ppm. That chemical shift is exactly where we expect an allylic CH$_2$ group. It appears as a quartet, so it "sees" three hydrogens. We know it will "see" the vinyl hydrogen on the adjacent carbon in one direction, so there has to be a CH$_2$ in the other direction in order to "see" three hydrogens. That analysis gives us this piece:

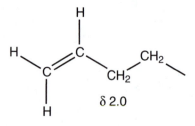

Now we have the molecule almost determined. The signal at δ 1.4 ppm that is a CH$_2$ sextet must "see" five hydrogens on the adjacent carbons. That means there must be a CH$_2$CH$_2$CH$_3$ in the molecule. That finishes the structure.

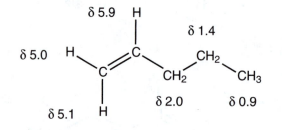

There are six unique hydrogens in the molecule, so there will be six signals in the ^{1}H NMR spectrum. The spectral data obtained on a 500 MHz NMR spectrometer will show the 2-bond coupling between the nonequivalent H's on C(1) and the long-range coupling between the C(1) H's and the C(3) H's, in which case the vinyl hydrogens at δ 5.0 and 5.1 ppm are multiplets.

Problem 10.28 You know that there are two products one can expect from the hydration of 3-methyl-1-pentene. They are 3-methyl-2-pentanol, which comes from normal addition of water across the double bond, and 3-methyl-3-pentanol, which involves a hydride shift in the initially formed carbocation to produce a more stable carbocation followed by addition of water.

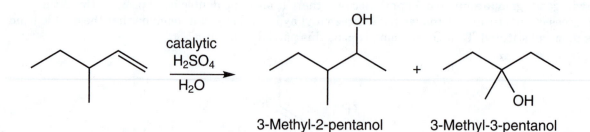

3-Methyl-2-pentanol   3-Methyl-3-pentanol

The ^{1}H NMR spectrum has a 6H triplet and a 3H singlet. Only 3-methyl-3-pentanol would give such signals.

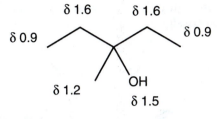

The ^{1}H NMR spectrum of 3-methyl-2-pentanol would be far more complex.

Problem 10.29 First, draw out the resonance forms for the two compounds, so we can see the overall symmetry.

In compound (a), the charge will be delocalized to three (two different) carbons in the ring. There will be five different carbons and therefore five signals in the ^{13}C NMR spectrum.

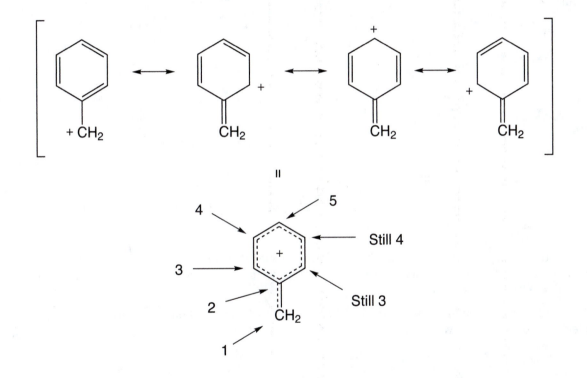

In compound (b), there will also be delocalization to make some of the carbons equivalent. There will be four signals.

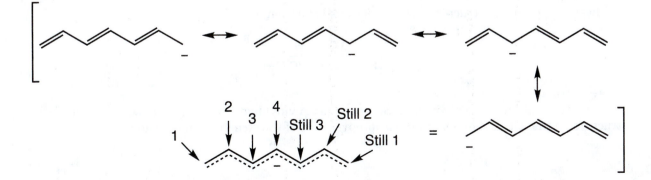

Problem 10.30   In (a) once again, resonance makes the system more symmetric than it looks at first glance. There are only four different carbons, and so only four signals appear in the ^{13}C NMR spectrum. In (b), there is no resonance stabilization (be sure you see this!) and therefore all the carbons are different. There will be six signals.

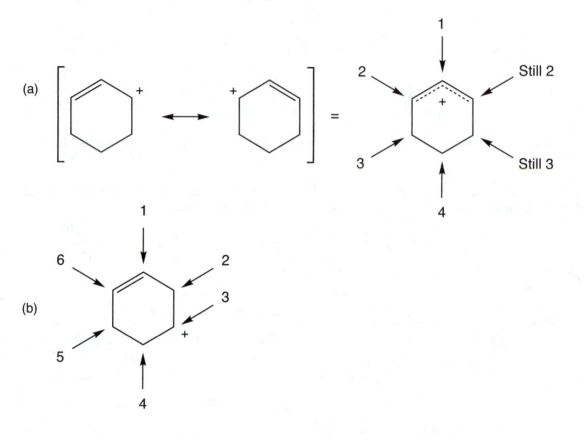

Problem 10.31

(a) These are not resonance structures. The structure on the left is an allylic carbocation. The structure on the right is not. Atoms must be moved to go from one structure to the other. Therefore they aren't resonance structures.

(b) These are resonance structures. Only electrons are moved to go from one structure to the other. The electron pushing is shown.

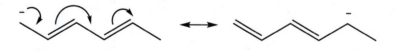

(c) These are not resonance structures. They are identical molecules. One structure need only be flipped over in order to draw the second.

(d) These are resonance structures. Only electrons have been moved.

(e) These two molecules are not resonance structures. A vinyl hydrogen has been moved in the structure on the left. It shows up in the isopropyl group of the structure on the right.

Problem 10.32   First, draw good Lewis structures for this pair.

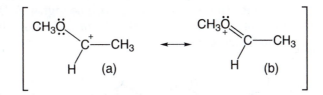

This problem is tougher than it first appears to be. At first glance, one might assume that resonance form (a) would be the more important one because (b) has a positive charge on oxygen, a more electronegative atom than carbon. However, form (b) has one more bond than (a), and in (a) the carbon atom does not have a filled octet and is electron deficient. In (b) both the oxygen and carbon have complete octets of electrons, making this structure the more important resonance form.

Problem 10.33

(a) Addition of the electrophile to the dimethyl-substituted carbon gives a carbocation that is resonance stabilized, as shown. This intermediate is relatively stable. It will be captured by the chloride to give 1-chloro-1,1-dimethoxy-2-methylpropane.

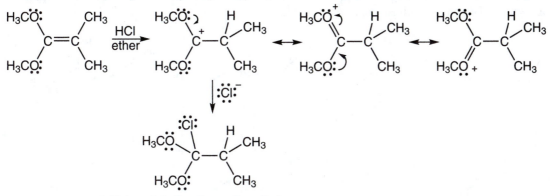

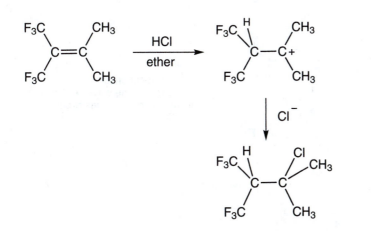

1-Chloro-1,1-dimethoxy-2-methylpropane

(b) The regioselectivity of this reaction is controlled by induction. The alkene will react with the electrophile to protonate one carbon or the other. Protonation will occur on the carbon attached to two trifluoromethyl groups producing the tertiary carbocation shown. Protonation on the carbon with two methyl groups would lead to a much higher energy carbocation over a prohibitively high transition state.

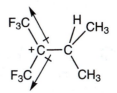

This carbocation is not formed because of induction

(continued)

Problem 10.33 *(continued)*

(c) The outcome of this reaction is also controlled by induction. The two possible carbocations for the first step of the reaction are shown. The bromine is electronegative and it will destabilize the carbocation that is closer, which means the transition state will be higher. As a result, it is the carbocation that is more distant from the bromo group that is formed as the first intermediate. Capture by chloride gives 1-bromo-3-chlorobutane.

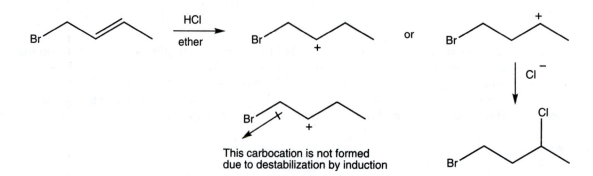

This carbocation is not formed
due to destabilization by induction

(d) In this case it is resonance that controls the reaction. We first draw the two possible carbocations and then we decide if one is more stable than the other. In this case, one of the carbocations is resonance stabilized. The preferred path will include the more stable cation, and the major product will be 1-bromo-1-chloropropane.

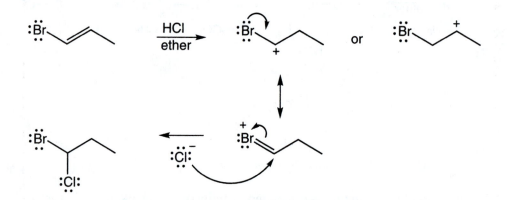

(e) Protonation of 1-bromo-4-fluoro-2-butene will form one of two carbocations. The carbocation that is closer to the fluorine will be more destabilized as a result of induction, since fluorine is more electronegative than bromine. Therefore, the transition state leading to the carbocation that is closer to the bromine will be lower and lead to the major product, 1-bromo-2-chloro-4-fluorobutane.

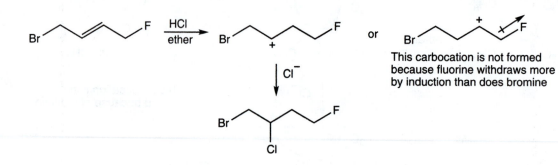

This carbocation is not formed
because fluorine withdraws more
by induction than does bromine

Problem 10.34

(a) Cyclohexene can make polycyclohexene.

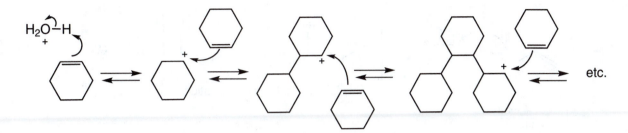

(b) Polyisopentylene comes from polymerization of 2-methyl-1-butene.

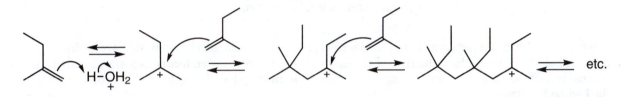

(c) Vinyl alcohol polymerizes to give polyvinyl alcohol.

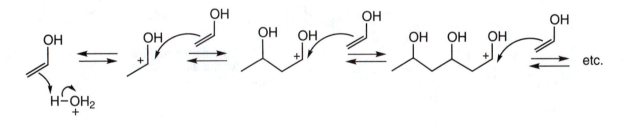

(d) One could make this polydichloroethylene from (*E*)- or (*Z*)-1,2-dichloroethylene.

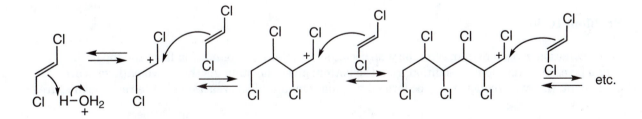

Problem 10.35 The first problem is to decide in which direction butadiene will protonate. There really is no choice; one protonation leads to a resonance-stabilized carbocation, the other to a simple localized primary carbocation.

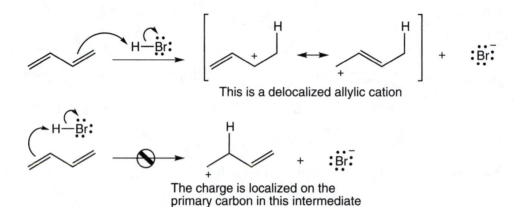

Two carbons share the positive charge in the resonance-stabilized, allyl cation. Be careful in the drawing below not to fall into the trap of thinking of the two resonance forms as separate entities. Two carbons share the charge in a *single* structure, summarized at the right of the figure with the "dashed bond" structure.

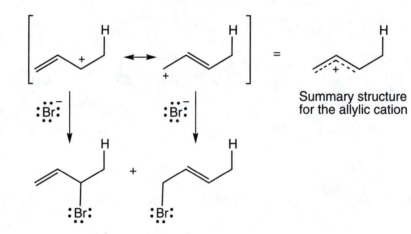

Problem 10.36

(a) Protonation gives a planar, secondary carbocation. Addition of bromide ion from the top of the cation gives one enantiomer; addition from the bottom gives the other. As these two pathways are equivalent, the two products must be formed in equal amounts. This reaction gives a racemic mixture.

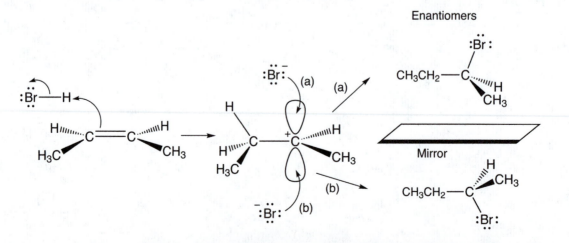

(b) There is something else that the carbocation can do besides react with the bromide ion. Like all other cations, it can lose a proton to give an alkene, if there is a proton attached to an adjacent carbon atom. In this case there are two such protons, H_a and H_b. H_a can be lost from two rotational isomers of the carbocation to give the starting alkene, *cis*-2-butene and its diastereomer, *trans*-2-butene. Alternatively, H_b can be lost to give 1-butene. Any base in the reaction can assist in proton removal. The figure shows bromide acting as the base.

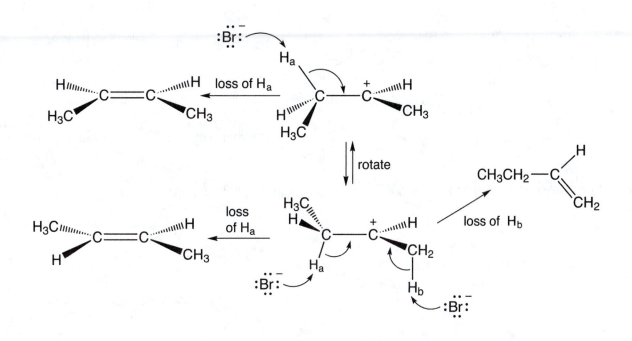

Problem 10.37 Compound **A** is formed by a straightforward hydration in Markovnikov fashion:

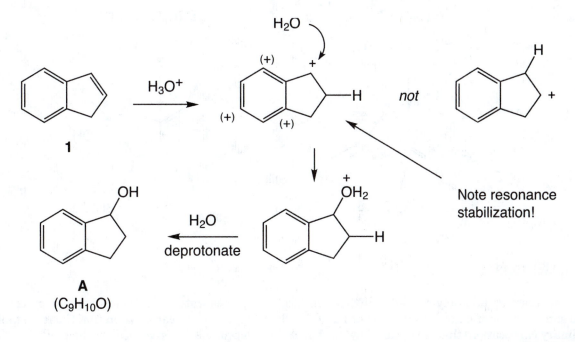

(continued)

Problem 10.37 *(continued)*

All nine carbons in **A** are different, so the ^{13}C NMR spectrum will show nine signals, and this must be the first route.

The second route uses hydroboration/oxidation conditions and therefore must give anti-Markovnikov addition. Here is how it works.

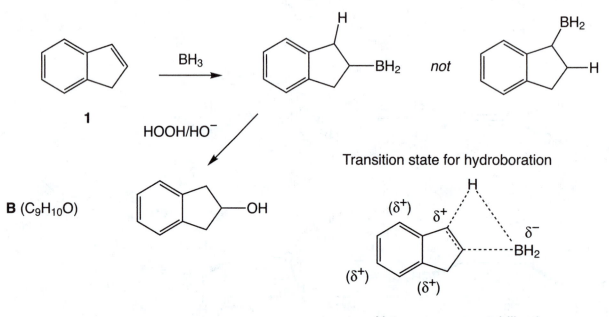

B ($C_9H_{10}O$)

Transition state for hydroboration

Note resonance stabilization

Compound **B** has a plane of symmery and thus only five different carbons. Its ^{13}C NMR spectrum will contain only five signals, in contrast to the nine of compound **A**.

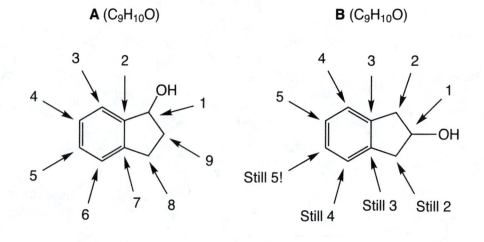

A ($C_9H_{10}O$)   **B** ($C_9H_{10}O$)

Problem 10.38

(a) This reaction is closely related to hydration. Alcohol replaces water, but the mechanistic process is the same. Protonation by either H_3O^+ or $CH_3O^+H_2$ leads to a tertiary carbocation (not the less stable primary carbocation) that is captured by alcohol. A final deprotonation gives the product.

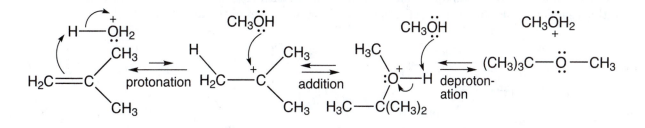

(b) This problem is just a common example of (a), the acid-catalyzed addition of an alcohol to an alkene. In this case, protonation occurs to give the resonance-stabilized carbocation, not the localized secondary carbocation.

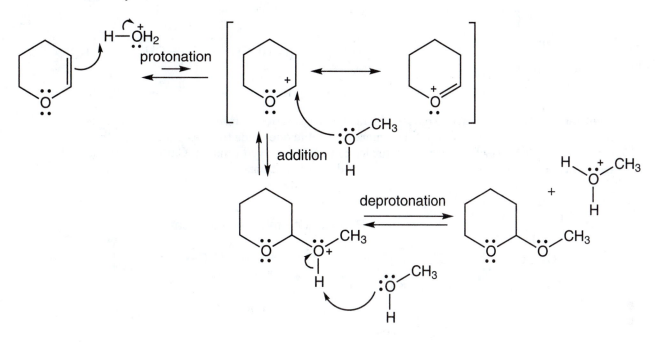

Problem 10.39 The first part of this reaction is exactly like hydration except that hydrogen sulfide (H_2S) plays the part of water. The product is a thiol (RSH). Notice the pattern of three steps: protonation, addition, and deprotonation, which is present in many addition reactions.

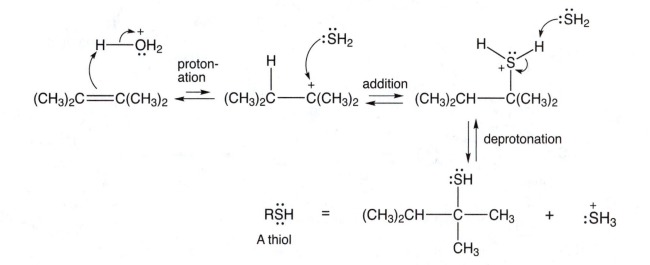

(continued)

Problem 10.39 *(continued)*

As this product, RSH, builds up, it can begin to compete with H_2S for the carbocation. The product of this reaction is the thioether (RSR), the other product shown in the problem.

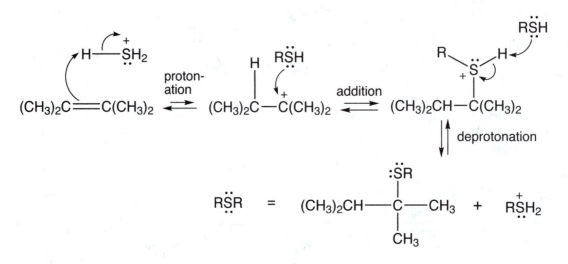

Problem 10.40   There are two possibilities for protonation of the carbon–carbon double bond in this molecule, called a "ketene." The more favorable one leads to a resonance-stabilized carbocation. Capture of this ion by water leads to the observed product. Capture of the less stable carbocation by water would give the compound that is not observed in the reaction.

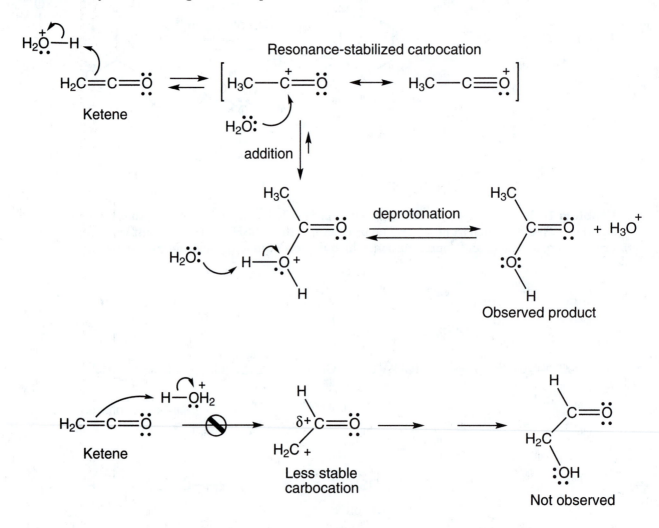

Problem 10.41 Both of these reactions involve multiple hydroborations, one or more of which are intramolecular.

(a) Let R = (CH₃)₂CHC(CH₃)₂, and draw the molecule in a suggestive way. In doing this, we are only taking advantage of the free rotations about carbon–carbon single bonds.

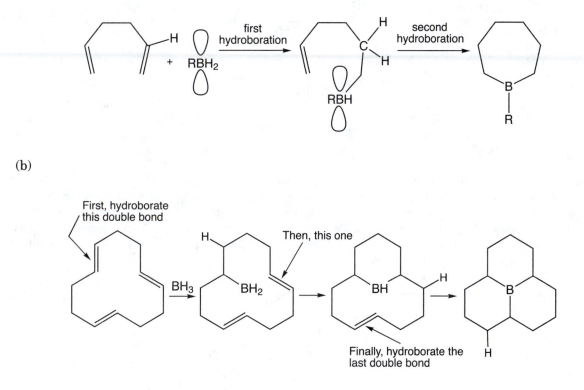

(b)

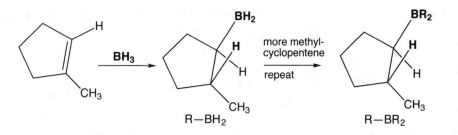

Problem 10.42 We know that the original hydroboration occurs in a syn fashion.

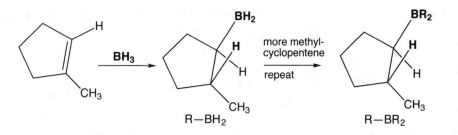

As the product alcohol has the same stereochemistry as the original borane, the overall replacement of BR₂ with OH must occur with retention, whatever the mechanism.

(continued)

Problem 10.42 *(continued)*

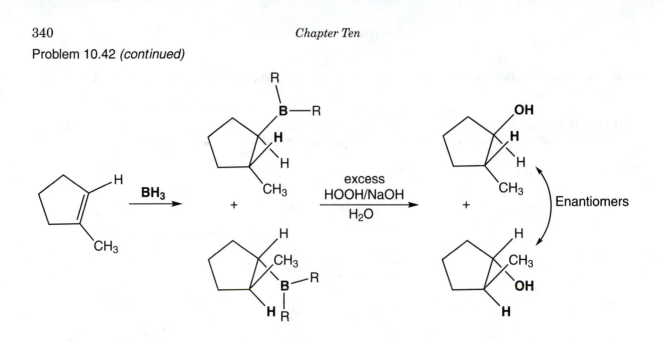

Problem 10.43 Remember: Always do synthesis problems backward. So, here the question to ask is, How would I make the target molecule from anything in a single step; that is, what might be the *immediate* precursor(s) of the target molecule? In this case, displacement of a good leaving group by the excellent nucleophile ⁻SH seems a good idea. This process yields another target. Apply the same process repeatedly and you will eventually get to the starting material. Here is an outline of one possible route:

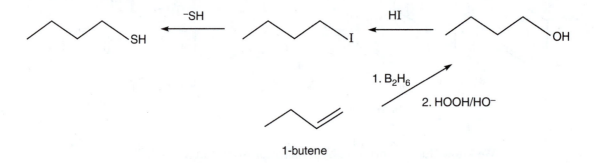

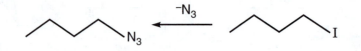

1-butene

Problem 10.44 See the admonition to work backward in Problem 10.43. Here, we can use the same route to the iodide and then displace it with the excellent nucleophile azide.

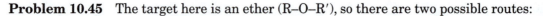

Problem 10.45 The target here is an ether (R–O–R'), so there are two possible routes:

RO⁻ + R'–LG, or R'O⁻ + R–LG. In this case the first of the two routes shown in the drawing is far better than the other. Why? The second will lead to lots of E2 elimination from the secondary iodide. We could also use different good leaving groups such as tosylates, but as we started with iodides above, we'll stick with them.

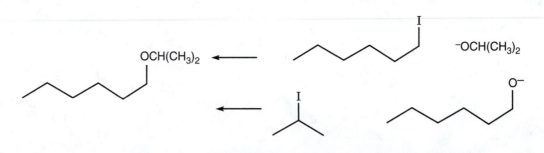

Now we need to make our reagents, isopropoxide and 1-iodohexane:

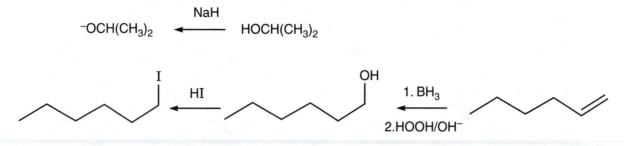

Problem 10.46   3-Methyl-2-butanol is a secondary alcohol, and it could react with hydrogen bromide and hydrogen chloride by either an S_N1 or S_N2 process (or both). In any case, the first step is protonation of the alcohol.

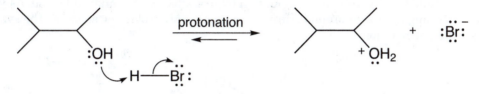

If the leaving group water is displaced by halide ion (shown here as bromide) in an S_N2 process, there should be no problem. (We might well expect this to be a relatively slow S_N2 reaction—Why?)

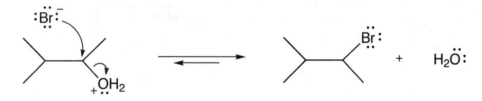

It is also possible that water could leave the protonated alcohol in an S_N1 reaction to give a secondary carbocation. If this cation is captured by halide ion, there is no problem. However, the secondary carbocation could also rearrange to the more stable tertiary carbocation through a hydride shift. If this happens, capture by halide will give 2-halo-2-methylbutanes.

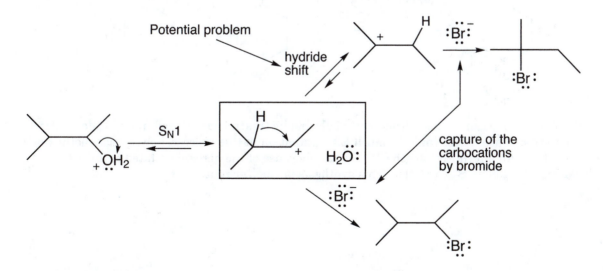

So, depending on the extent of the S_N1 component of these reactions, the desired 2-halo-3-methylbutanes could be contaminated with the isomeric 2-halo-2-methylbutanes.

Problem 10.47 Two parts of this problem are extremely easy and require only that you remember a pair of simple Markovnikov addition reactions.

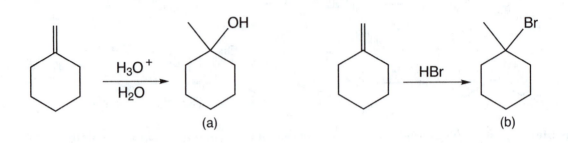

(a)   (b)

Another part, (c), involves hydroboration/oxidation to give anti-Markovnikov addition. As yet, you have no way of adding hydrogen bromide directly in an anti-Markovnikov sense, so the synthesis of (d) requires a further reaction of (c):

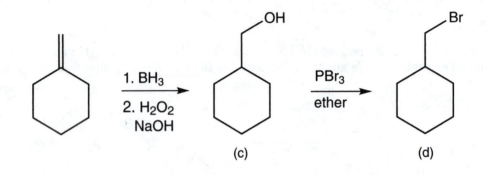

(c)   (d)

Similarly, (d) can be used to make (e) and (f) through S$_N$2 reactions.

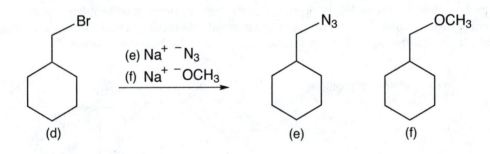

(d)   (e)   (f)

Sodium methoxide can be made from methyl alcohol by treatment with sodium hydride (NaH) or another strong base. There could be much E2 elimination in this process, however, as methoxide is a strong base. An alternative synthesis of (f) involves use of the alkoxide related to (c) as the displacing agent in an S$_N$2 reaction. This synthesis is a better route to (f).

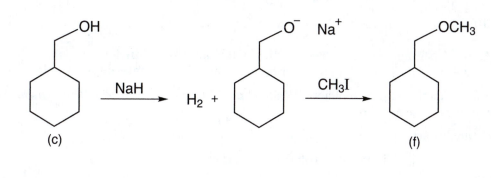

(c)   (f)

Problem 10.48

(a)

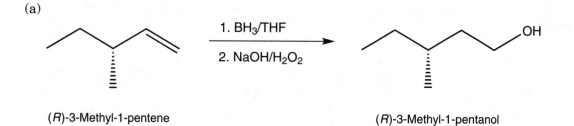

(*R*)-3-Methyl-1-pentene   (*R*)-3-Methyl-1-pentanol

(b) Because the borohydride reaction occurs by syn addition, the reaction gives only (2*S*,3*S*)-3-methyl-2-pentanol and its enantiomer, (2*R*,3*R*)-3-methyl-2-pentanol.

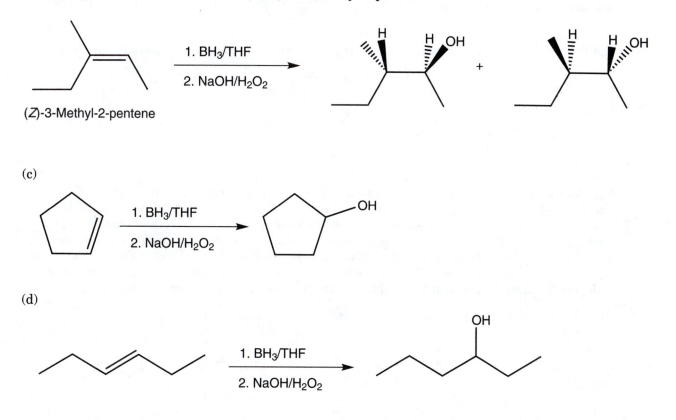

(*Z*)-3-Methyl-2-pentene

Problem 10.49

(a) (*R*)-3-Methyl-1-pentanol is chiral. The starting material, (*R*)-3-methyl-1-pentene, is also chiral. The product is formed as a single enantiomer because the starting material is a single enantiomer.

(b) The (2*S*,3*S*)-3-methyl-2-pentanol and (2*R*,3*R*)-3-methyl-2-pentanol products are chiral. The reaction gives a racemic mixture.

(c) Cyclopentanol is achiral.

(d) The product, 3-hexanol, is chiral. It is formed as a racemic mixture of both the (*R*) and the (*S*) enantiomers.

Problem 10.50 The boron–oxygen bond in B(OR)$_3$ has double bond character because of the resonance that is available between the empty *p* orbital of the boron and one lone pair from each of the oxygens.

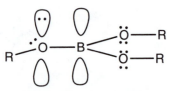

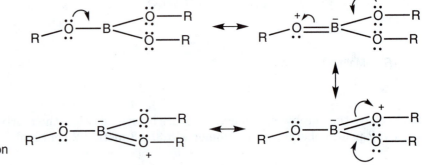

Each oxygen has a lone pair
in a *p* orbital that can overlap
with the empty *p* orbital of boron

The B(OR)$_4$$^-$ is not able to delocalize its electrons because there is no empty *p* orbital on boron.

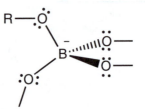

Problem 10.51   Formation of (*E*)-2-pentene could be done from 2-bromopentane using an E2 reaction, which would probably give the best yield of the desired product. An E1 reaction would certainly give a mixture of alkenes and S$_N$1 products. We can obtain the 2-bromopentane from 1-pentene. Although we could dehydrate 1-pentanol to obtain 1-pentene, the dehydration reactions are often accompanied by rearrangement. It would be better to make the 1-pentene from a controlled reaction such as an E2 pathway. That means we want to make 1-pentene from 1-bromopentane. We can make the 1-bromopentane from 1-pentanol using S$_N$2 chemistry to avoid carbocations.

Here is our analysis using arrows that tell us where each compound comes from.

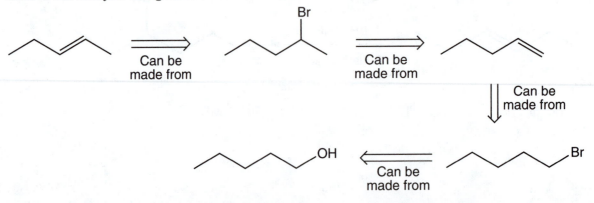

Putting it all together in a forward sense, we have:

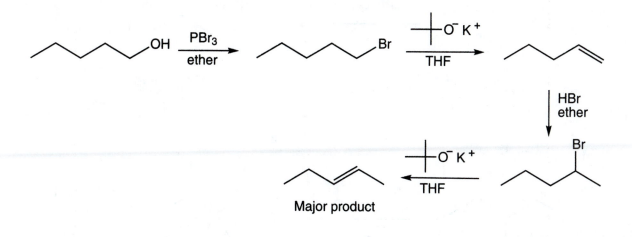

Major product

Problem 10.52 The selectivity for each of these reactions is poor because the carbocation intermediates would have a similar likelihood of formation.

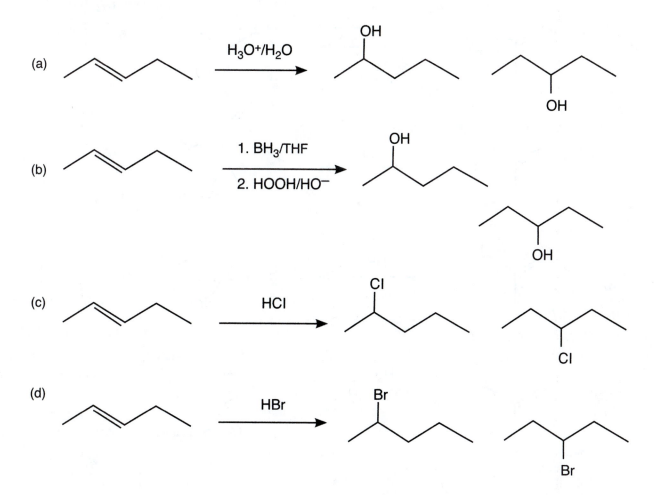

(continued)

Problem 10.52 *(continued)*

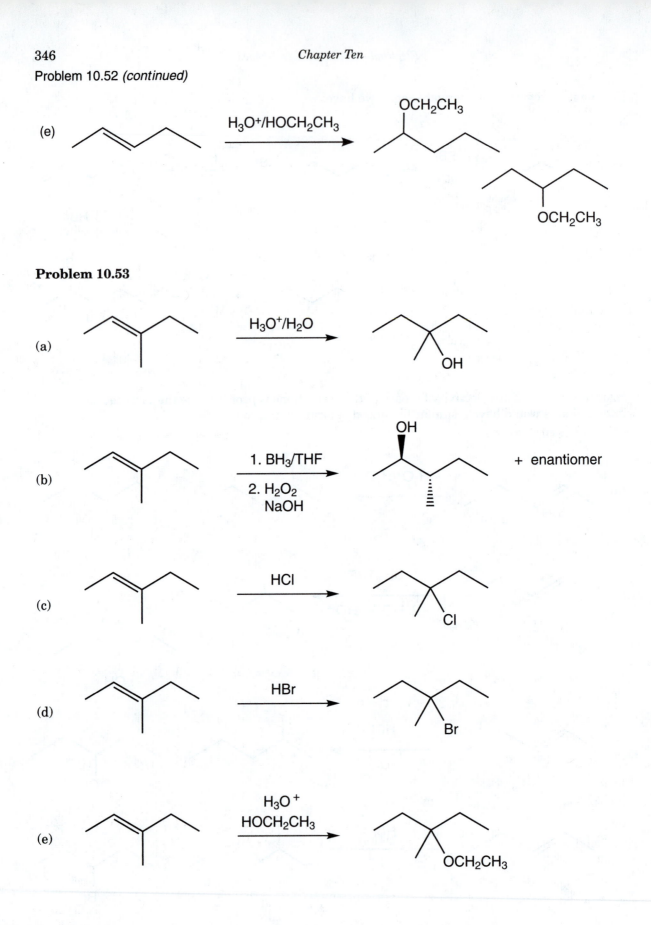

Problem 10.53

Problem 10.54 The best first step is protonation of the right-hand double bond so as to give the tertiary carbocation, the most stable carbocation possible in this system.

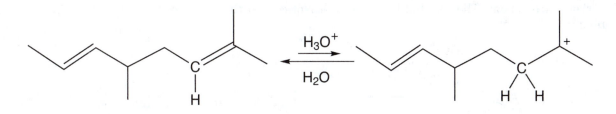

The product is a cyclic molecule, so clearly a ring must be closed. In this case, the carbocation (Lewis acid) adds to the internal double bond (Lewis base) to form a new carbon–carbon bond and close a six-membered ring. Don't be offended at the simplicity of this analysis; "the starting material is acyclic, but the product is cyclic; therefore a ring must be closed." It is extraordinarily important to think this way when doing problems.

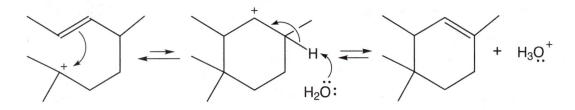

Here an important and vexing point arises. In the last step, there is another possible proton loss to give **A**, a molecule not formed in the reaction.

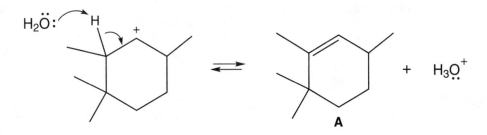

How is one to know which one to choose? The answer to this kind of question is frustrating, because sometimes the only answer is, Because of the structure of the product. There is no obvious (to us, anyway) reason that the final deprotonation goes the way it does. Can one just say, Because the product formed is more stable than the one that isn't formed? Well, you can say that, but it doesn't add anything, at least unless you can explain *why* the product shown is more stable. We are left to reason backward. Another way of putting this is to point out that it is not a fair question to ask which alkene would be formed from the cyclized carbocation. It is fair to show the product and ask how it is formed.

Problem 10.55 Once again, start by protonating to give the most stable possible carbocation.

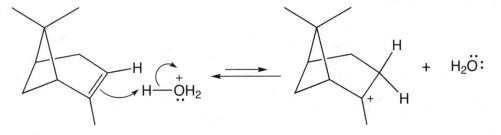

(continued)

Problem 10.55 *(continued)*

Now look at the products. Clearly a ring has been opened. And it is not difficult to see why. The starting material contains a four-membered ring, and the strain of that ring is eliminated (no pun) if the ring opens. Just as a carbon–hydrogen bond adjacent to a carbocation can break, so can a carbon–carbon bond. That is what happens here. Notice that the carbocation produced is still tertiary.

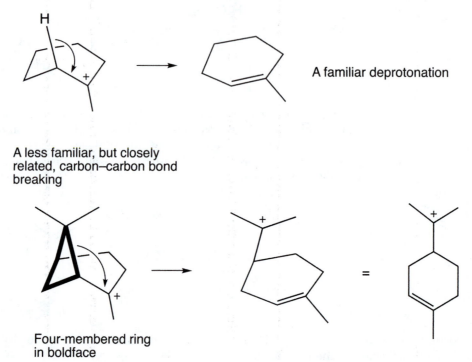

A familiar deprotonation

A less familiar, but closely related, carbon–carbon bond breaking

Four-membered ring in boldface

Now what? There are two possible proton losses (H_a and H_b), and they lead to the two observed products.

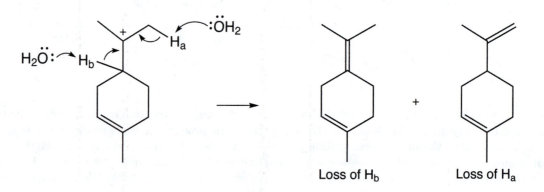

Loss of H_b Loss of H_a

Problem 10.56 For the third time, the first step is protonation to give the tertiary carbocation. The "easy" product is formed by capture of this ion by chloride ion.

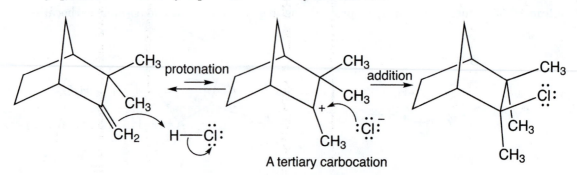

protonation

addition

A tertiary carbocation

Finding the mechanism for the "hard" product is really difficult at this point. Here's how. First of all, recognize that something really strange has gone on. Apparently methyl groups have been wandering all over the place. Let's try to find another mechanism because these wholesale migrations are surely unlikely. Remember that carbons can migrate in carbocationic reactions (Wagner–Meerwein rearrangement). In the first-formed carbocation, there is a carbon atom beautifully poised for migration. The figure shows the lineup between the empty orbital of the carbocation and the carbon–carbon bond.

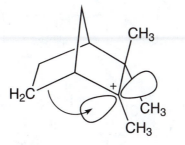

Migration of carbon does the trick, but it takes some spatial reorganizing to see it. Some labels are left in the drawing to show the relationship between the two pictures of the new cation.

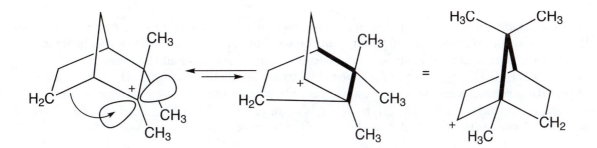

Capture by chloride gives the product.

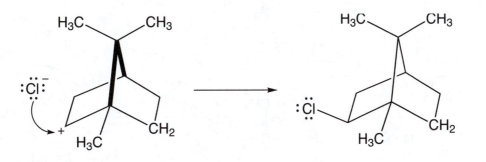

Why give a problem this hard? First, it "looks forward" to Chapter 24, where we shall see that the situation is even more complicated than it seems here. Second, it is not *that* hard, and it provides useful practice in spatial manipulation. Finally, and probably most important, problems like this are fun and mimic quite well what "real" chemists have to think about. When we find a strange compound in a reaction, we must seek a reasonable route for its formation, and sometimes that involves new chemistry that's hard for anyone to see at first. There is no reason to deny you this pleasure.

Problem 10.57   This problem begins with protonation of the alcohol. There is only one possible elimination, and this process generates one of the two products. The only difficulty in this problem is the second structure. Migration of a carbon–carbon bond gives a tertiary carbocation **A**, and now a proton can be removed to give the ring-contracted product.

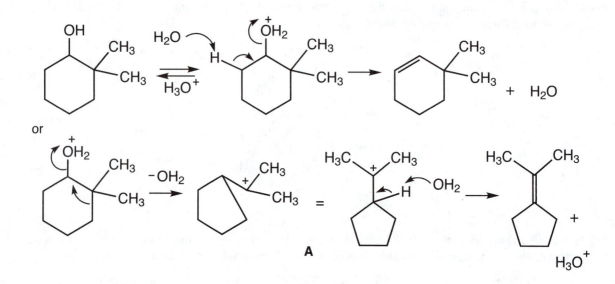

Problem 10.58   Surely the first step in the treatment of a diol in acid must be protonation of one OH. Loss of the good leaving group water (just the beginning of an $S_N1/E1$ sequence) leads to a tertiary carbocation. Loss of water seems to be a good step. Why? Look at the formula of the products, $C_6H_{12}O$. One oxygen atom has been lost from the starting material. The question now is, What further reactions are possible? If water re-adds, completing the S_N1 sequence, we simply regenerate starting material. The E1, however, is more promising as it does lead to a compound of the proper formula. The alkene **A** must be one of the "products" we are supposed to find.

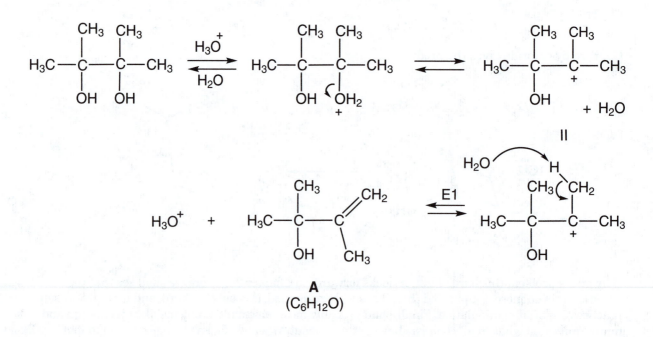

Another possibility is capture of the carbocation, not by the external nucleophile, water, but by the internal nucleophile, the internal OH. This process leads to a protonated epoxide, and subsequent deprotonation gives the epoxide itself, **B**, another $C_6H_{12}O$.

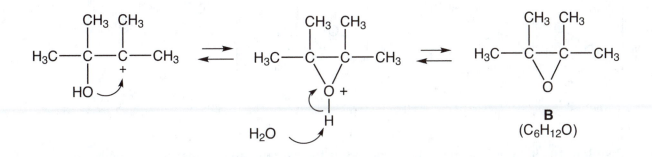

The mechanism of the pinacolone formation must have protonation and dehydration as the first two steps—there just isn't much else possible. Whenever you see a carbocation, and especially when the unexpected has happened, you are well advised to consider the possibility of rearrangement. We have seen this kind of reaction many times in carbocation chemistry. In fact, it appears whenever it is possible to migrate a group to give a more stable cation. In this molecule, there is only one possible migrating group, one of the adjacent methyl groups. The thing to do now is to write the product of this rearranged ion **C** and try to see if it looks more or less stable than the starting carbocation. We might suspect that we are on the right track because we have generated the appropriate carbon skeleton of pinacolone.

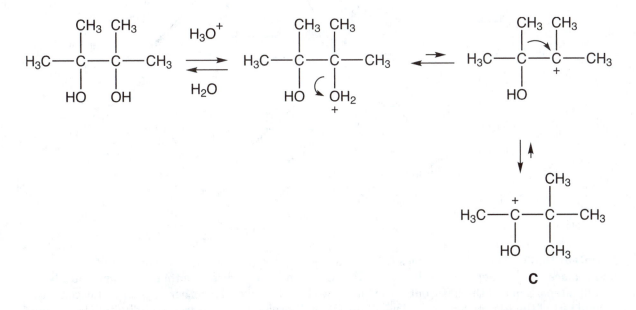

Cations flanked by heteroatoms bearing pairs of electrons are resonance stabilized, and it is this factor that makes the rearranged ion more stable than the starting cation. Deprotonation of the rearranged ion **C** by water gives the final product. This reaction, called the pinacol rearrangement, takes its name from the trivial name of the diol used in this example, pinacol, and is common whenever a 1,2-diol is treated with acid.

(continued)

Problem 10.58 *(continued)*

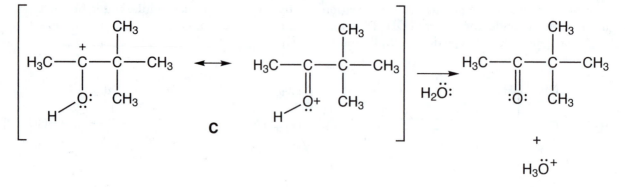

Problem 10.59 The first step of this sequence is the hydroboration/oxidation. Remember that this is a syn addition with the alcohol ending up on the less substituted carbon (anti-Markovnikov). The tosyl chloride reagent converts the alcohol into the tosylate, which is an excellent leaving group. In the last step, the E2 elimination will require an anti-periplanar β-hydrogen. This results in selective formation of product **C**. The only available anti-periplanar β-hydrogen is on the carbon to the right of the carbon bearing the tosylate, as drawn below.

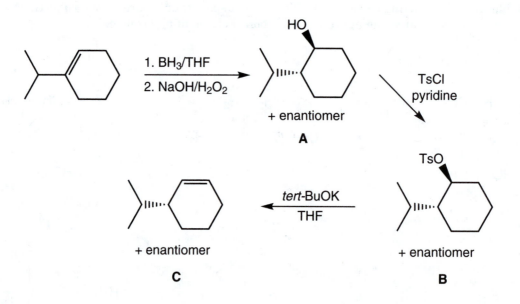

Problem 10.60 This answer comes to you courtesy of Professor Grant Krow of Temple University and replaces our earlier, less good version. This problem is tough because there is a large number of "small" steps. Some of the basic outlines should be clear, however. The ether oxygen will become the second OH of the top benzene ring. The allylic OH will probably be lost in an elimination reaction that helps build the lower left ring in apomorphine. Somehow the bridge in morphine must become the new azacyclohexane ring. The real trick in this problem is to see how to do that transformation. What can happen in acid? Protonation of the alcohol, that's what. Loss of water leads to an allylic cation, nicely resonance stabilized.

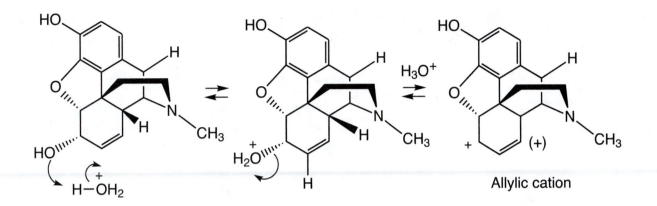

Deprotonation completes what is nothing more than an E1 elimination. Next, we protonate the ether oxygen and break a bond to make another resonance-stabilized cation.

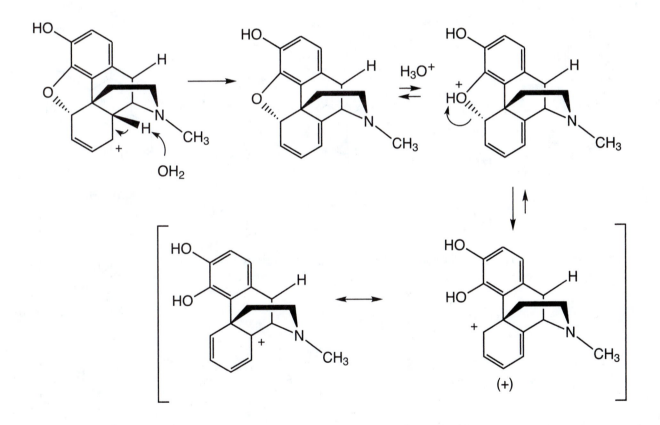

The bridge is now set to do its first migration to yield yet another beautifully resonance-stabilized cation. Look at all those atoms sharing the positive charge!

(continued)

Problem 10.60 *(continued)*

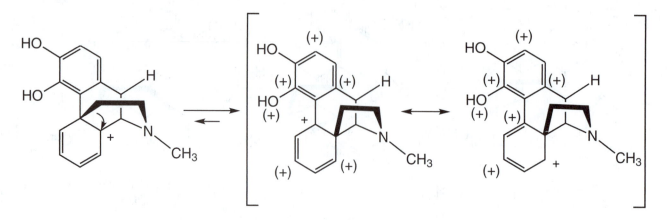

A second migration of the bridge establishes the final skeleton of the product; indeed in this rather economical mechanism (at least compared to our first effort!), we are now only a deprotonation away from the end:

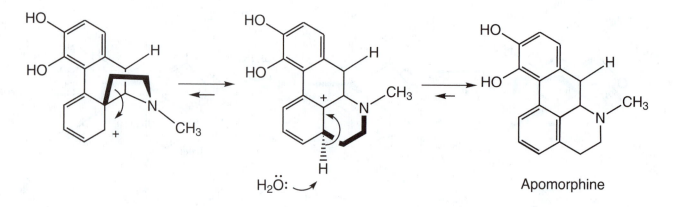

Apomorphine

Problem 10.61 Here are the top views of allyl and pentadienyl:

	+ − + − +
+ − +	
	+ − o + −
+ o −	+ o − o +
	+ + o − −
+ + +	+ + + + +
Allyl	Pentadienyl

What generalizations can you make? Some things you know already. For example, you know that the number of molecular orbitals must match the number of atomic orbitals being combined. You also know that the number of nodes must increase monotonically (one more node each time) starting from the lowest molecular orbital with no nodes other than the node present in any orbital made from 2*p* orbitals.

Here are some generalizations:

(1) As long as you start at the left of the orbital with (+), the right-hand ends of the orbitals alternate sign, (+), (–), (+), (–), and so on. Presumably, if nature is a kind, symmetrical creature, this observation will always be true.

(2) The highest molecular orbital is always alternating (+) and (–). You already know the lowest energy molecular orbital is "all +." So, you can specify the lowest and highest molecular orbitals with no trouble.

(3) The middle molecular orbital is always (+), (0), (–), (0), (+), and so on.

(4) The pattern of zeros (atoms at which the sign of the wave function is zero) is symmetrical, and the number of zeros increases by one to the middle molecular orbital, then decreases by one.

Perhaps you can find other "rules." These observations, together with some faith that nature is symmetrical at this level, allow us to write molecular orbitals for acyclic, "linear" fully conjugated systems quite easily. But don't extend this business too far! These rules won't work for systems containing even numbers of carbons or cyclic molecules. For such systems, you will need other observations.

Problem 10.62   Let's apply the observations from Problem 10.61. First, there will be seven molecular orbitals and the nodes will increase monotonically. Next, the bottom (lowest energy) and top (highest energy) molecular orbitals must be as in drawing (a). We can next add the middle molecular orbital as in (b). Next, add the ends of all the molecular orbitals (c), and fill in the orbital second lowest in energy. It has only a single node and is always easy to draw (d). Now the work begins. The next-to-highest molecular orbital will have a zero in the same place as the next-to-lowest, and the two other molecular orbitals will each have two "zeros." Given that information, it is not so hard to fill the rest in (e). At the end, check to see that the proper nodal pattern (0 . . . 1 . . . 2 . . . 3 . . .) appears.

(a)	(b)	(c)	(d)	(e)	Nodes
+ – + – + – +	+ – + – + – +	+ – + – + – +	+ – + – + – +	+ – + – + – +	6
		+ – + –	+ –	+ – + 0 – + –	5
		+ +	+ +	+ – 0 + 0 – +	4
	+ 0 – 0 + 0 –	+ 0 – 0 + 0 –	+ 0 – 0 + 0 –	+ 0 – 0 + 0 –	3
		+ +	+ +	+ + 0 – 0 + +	2
		+ –	+ + + 0 – – –	+ + + 0 – – –	1
+ + + + + + +	+ + + + + + +	+ + + + + + +	+ + + + + + +	+ + + + + + +	0

Problem 10.63   The central carbon is sp^2 hybridized in the intermediate. The intermediate is planar and the central carbon is attached to three other carbons.

The calculated lowest unoccupied molecular orbital shows that there is considerable sharing of the positive charge throughout the molecule, particularly with the carbon–hydrogen bonds that are parallel to the empty p orbital.

An approaching nucleophile can attack the empty p orbital from above or below (S_N1 reaction). It might also deprotonate one of the acidic hydrogens, those that have so much of the LUMO density. Remember that the LUMO is where the positive charge resides, and the hydrogens that have positive charge are by definition acidic.

Problem 10.64   There needs to be a high concentration of alkene. An alkene needs to collide with the carbocation before any other nucleophile (such as water) does.

The number of monomers needed to make a polymer is not a fixed value. Most useful polymers are of the order several thousand monomers long. This value (the number of monomers in the polymer) is referred to as the *degree of polymerization*.

Problem 10.65   Carbocations are unstable. It isn't possible for a carbocation to just hang around. Therefore, a polymer is most unlikely to be charged.

One of the properties of milk jugs is that they don't react with the milk. Milk is mostly water. If the polymer had carbocations, then these charged carbons would react with the water in the milk.

A final step that you likely thought of is elimination. In the absence of more alkene to make a longer polymer, the carbocation probably eliminates. There might also be traces of water present in the polymer-forming process, and, as noted above, the carbocation would react with those water molecules.

Problem 10.66   The main difficulty is not knowing what the reacting borane species actually is. We know the borane is highly solvated by the tetrahydrofuran (THF), but at what point does the alkene displace the THF? We don't know how strong the borane complex with THF is in the presence of the alkene.

The alkyl group that shifts is the alkyl group that is anti to the oxygen–oxygen bond. The σ bond of the migrating alkyl group overlaps with the σ* antibond of the weak oxygen–oxygen bond (hyperconjugation again). It is only the group in the anticonformation that can overlap efficiently with the antibond. The Newman projection looking down the oxygen–boron bond (boron in the back) might help you see this relationship.

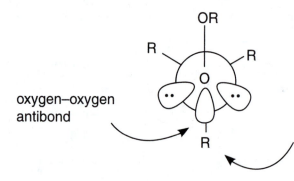

oxygen–oxygen antibond

only this R group will migrate because it can "see" into the oxygen–oxygen antibond

## 11

More Additions
to π Bonds

Problem 11.2 Another easy one, possibly even too easy for you to trust your answer. The bulky adamantyl groups simply shield the rear of the carbon–bromine bonds in the bromonium ion, making S_N2 attack difficult.

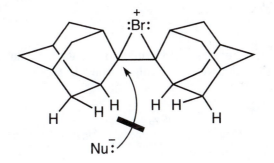

Problem 11.3 In this case, formation of the open carbocation competes favorably with formation of the bromonium ion. The reason is that the open cation is exceptionally well stabilized by resonance. The open cation can lead to both cis and trans products, whereas the bromonium ion is destined to give only trans dibromide. Note again the convention we use in which a charge in parentheses is used to show the various atoms sharing the charge. If it is not obvious that the charge is delocalized to these positions, draw out the resonance forms.

(continued)

Problem 11.3 *(continued)*

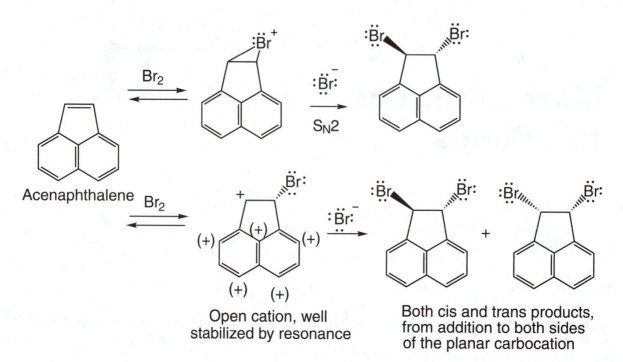

Acenaphthalene

Open cation, well
stabilized by resonance

Both cis and trans products,
from addition to both sides
of the planar carbocation

Problem 11.4

(a) The bromination of an alkene results in addition of Br to each of the alkene carbons.
Presumably the addition is anti, but we are unable to know in this case.

$$CH_3\text{—}CH_2\text{—}CH_2\text{—}CH_2\text{—}CH\text{=}CH_2 \xrightarrow[\text{THF}]{Br_2} CH_3\text{—}CH_2\text{—}CH_2\text{—}CH_2\text{—}CHBr\text{—}CH_2Br$$

(b) We do not know if the initial 3-hexene in this problem is the (Z) or (E) isomer. Therefore, we
cannot specify the stereochemistry of the dibromide product. But we predict that the bromination
occurred via anti addition.

$$CH_3\text{—}CH_2\text{—}CH\text{=}CH\text{—}CH_2\text{—}CH_3 \xrightarrow[\text{THF}]{Br_2} CH_3\text{—}CH_2\text{—}CHBr\text{—}CHBr\text{—}CH_2\text{—}CH_3$$

(c) Chlorination results in addition of Cl to each of the alkene carbons. We assume the addition is
anti, but we have no way to know when halogenating a terminal alkene.

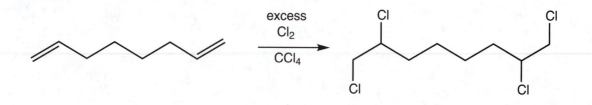

(d) An alkyne adds Cl_2 across the two π bonds to give the tetrachloride. Although we are unable to verify the stereochemistry of addition, we expect each Cl_2 adds anti.

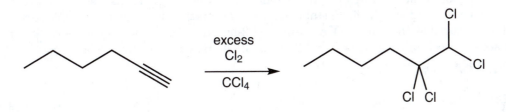

(e) The π bonds of an aromatic ring are not as reactive with electrophiles as an alkene. So the bromination occurs only on the propene carbons. Even in excess Br_2, the reaction will give only 1,2-dibromo-2-phenylpropane.

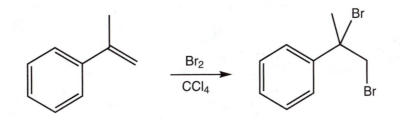

(f) The Br_2 electrophile adds to the alkene to form a bromonium ion. Water will react with the bromonium ion by adding from the backside of the weaker bond of the bromonium ion intermediate. The result is addition of water at the more substituted carbon. After deprotonation, the halohydrin is formed. Notice that the alkyl chloride is unaffected by the reaction.

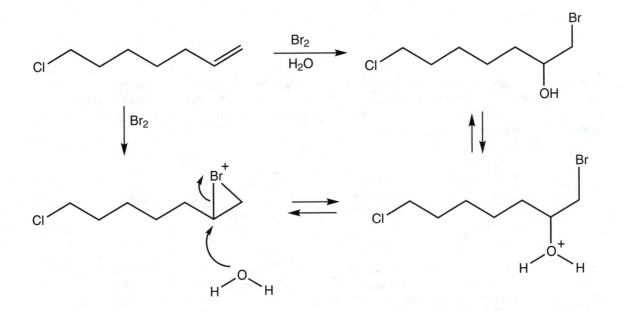

Problem 11.6

(a) The first step of the acid-catalyzed hydration of 3-methyl-1-butene forms a secondary carbocation, which will rearrange to form the more stable tertiary carbocation. It is the more stable carbocation that will be captured by the water to form 2-methyl-2-butanol. We also know that water can react with the secondary cation to give 3-methyl-2-butanol. The E1 product can also be formed.

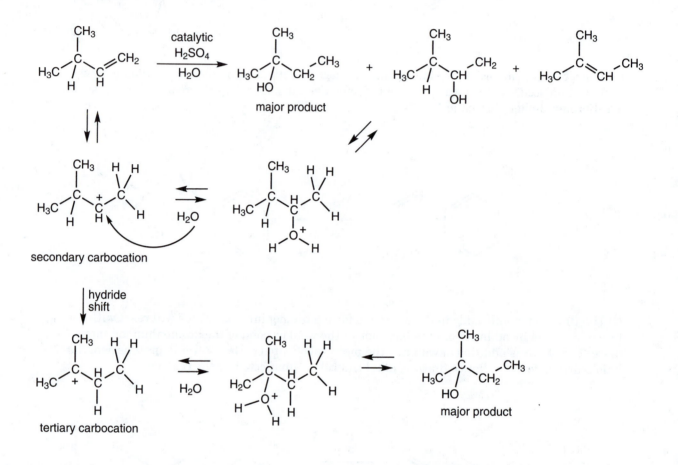

(b) The hydroboration/oxidation reaction does not involve a free carbocation. So there will not be any rearrangement. The product is a result of addition of boron and hydride across the alkene, presumably via syn addition. The boron is the electrophile. The oxidation step results in replacement of the boron by the OH group.

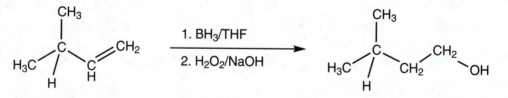

(c) The oxymercuration reaction does not involve a free carbocation. The initial mercurinium ion is opened by water attacking the more substituted carbon. The product is a result of anti addition of an OH group and mercury to the carbons of the alkene. The demercuration step replaces the mercury with hydrogen.

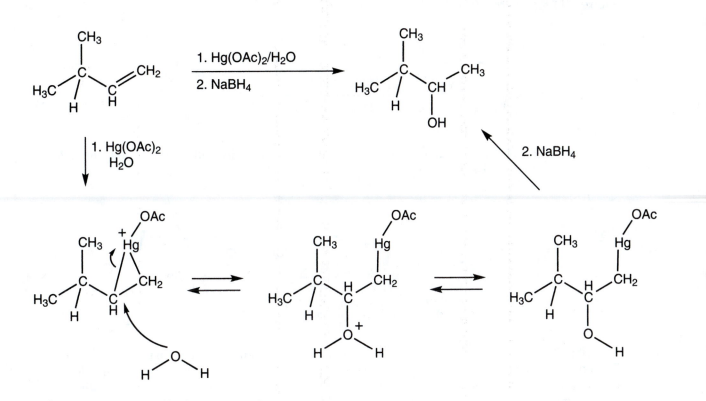

(d) No rearrangement is predicted in this reaction. The intermediate tertiary carbocation is relatively stable.

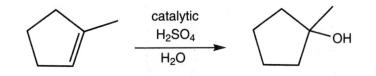

(e) The syn addition results in a racemic mixture of *trans*-2-methylcyclopentanol. Syn addition means that the groups that are added are adding to the same face (top or bottom) of the alkene. In this case, the groups that add to the alkene are boron and hydrogen. The boron is replaced by an OH in the second step of the reaction. The H and the OH have been added to the same face of the alkene in both products, giving the two enantiomers shown.

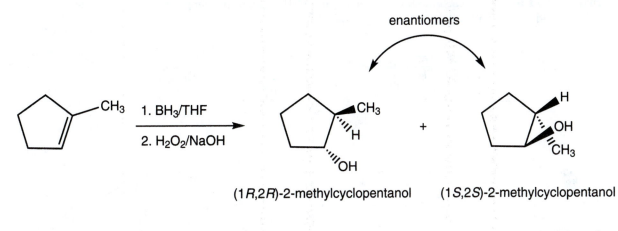

(1*R*,2*R*)-2-methylcyclopentanol   (1*S*,2*S*)-2-methylcyclopentanol

(continued)

Problem 11.6 *(continued)*

(f)

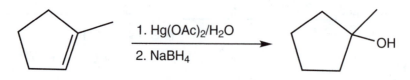

Problem 11.7 The new oxygen atom will be delivered to the other side of the double bond. So, the product, formed in 85% yield, is:

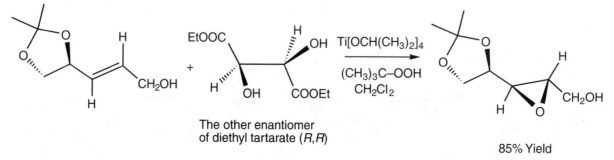

The other enantiomer
of diethyl tartarate (*R,R*)

85% Yield

Problem 11.8 The difference appears because of the strain energy of the three-membered ring. The energy of the starting material is raised, and therefore the activation energy for reaction declines. Cyclopropane is strained by some 27 kcal/mol, and an oxirane cannot be very different. Both of these species will be more reactive than their unstrained counterparts, and ring openings of epoxides by nucleophiles are common.

A very rare reaction;
alkoxide is a poor
leaving group

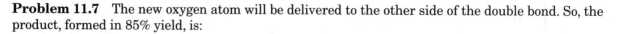

The strain in the three-membered ring
raises the energy of the starting material
and makes this ring opening possible

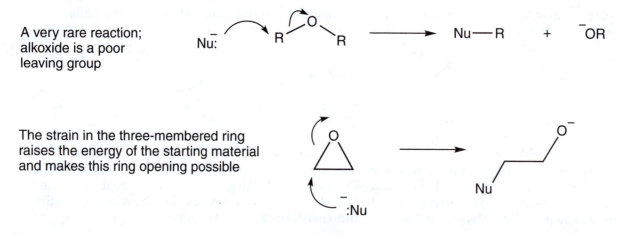

Problem 11.9 In the second step of this sequence, the Grignard reagent adds to the less substituted carbon of the epoxide.

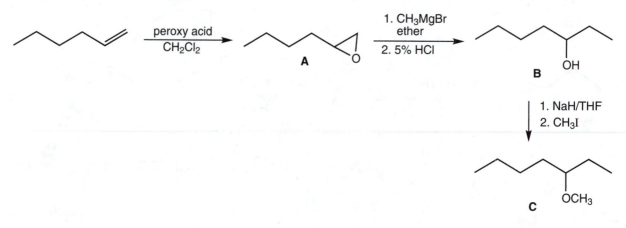

Problem 11.10   There are many resonance structures that can be drawn for diazocyclopenta-diene. The first task is to draw the full structure of the diazo compound itself. The five resonance structures that we can draw by moving the anion around the five-membered ring are structures that contribute most for reasons we will see in Chapter 14.

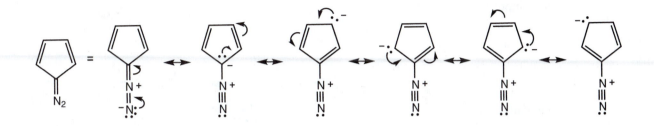

Problem 11.11   Although chloroform is not a strong acid, it can be deprotonated in strong bases such as potassium *tert*-butoxide (or sodium hydroxide).

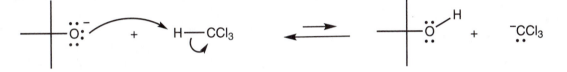

The trichloromethyl anion can lose a chloride ion to give the carbene.

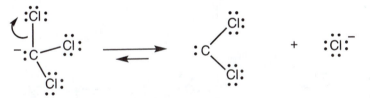

Problem 11.12   In the singlet state, the empty 2*p* orbital on carbon overlaps with a filled 3*p* orbital on chlorine. This resonance between the empty orbital and a filled orbital is not as stabilizing in the triplet because it is between an orbital with one electron and a filled orbital. The triplet will have three electrons in the π system, which means two electrons will be in the bonding orbital and one electron will be in the antibonding orbital. It is that one electron in the antibonding orbital that makes the triplet less stable than the singlet.

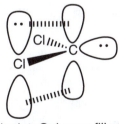

Singlet: Only one filled orbital on one of the Cl atoms is shown; the overlap with the empty 2*p* orbital on carbon is shown

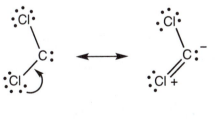

Resonance: We can represent the electron delocalization for the singlet carbene by drawing its resonance structure

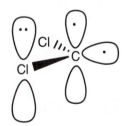

Triplet: The 2*p* orbital on carbon is half filled; overlap with the filled 3*p* orbital on either adjacent Cl atom is not as stabilizing as in the singlet

Problem 11.13   The resonance forms shown in Figure 11.45 emphasize the 1,3-dipolar nature of these reagents. Each of the resonance structures drawn for the 1,3-dipoles have all atoms with a filled octet. Other resonance structures can be drawn.

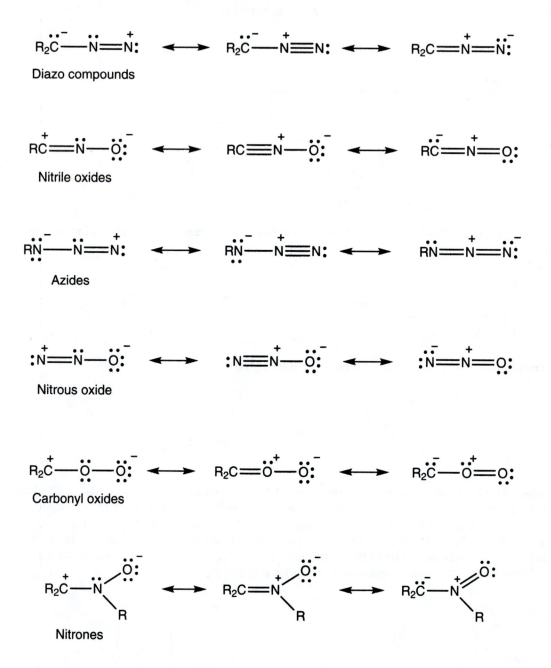

Problem 11.14

The reaction of an azide with (*E*)-2-butene:

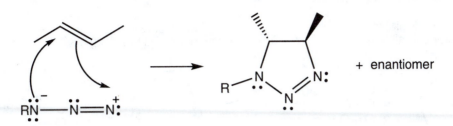

The reaction of a nitrile oxide with (*E*)-2-butene:

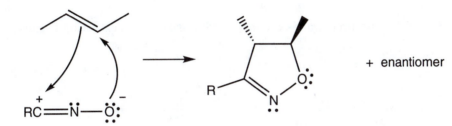

The reaction of a carbonyl oxide with (*E*)-2-butene:

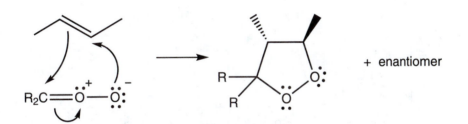

Because the 1,3-dipole reagents are unsymmetrical, reaction of one of them with (*E*)-2-pentene will result in two products in each case. For example, here is a reaction with an azide with (*E*)-2-pentene:

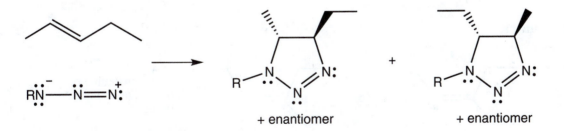

Because of symmetry, reaction of one equivalent of dipolar reagent with 2-butyne will give only one product. For example, here is the reaction of a nitrile oxide with 2-butyne:

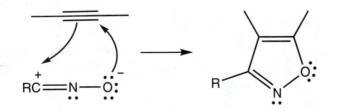

Problem 11.15 Some approximations must be made here. In this answer, we use the bond energy of ethane for the carbon–carbon bonds, the bond energy of a simple alcohol for the carbon–oxygen bonds, and the bond energy for di-*tert*-butyl peroxide for the oxygen–oxygen bonds. The final ozonide is much more stable than the primary ozonide.

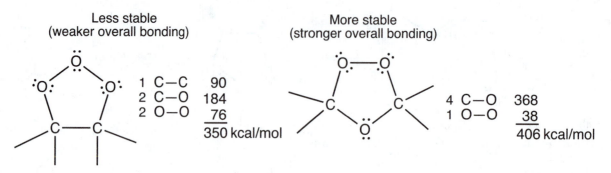

Problem 11.16 A full diagram includes the intermediates as well as starting material and products.

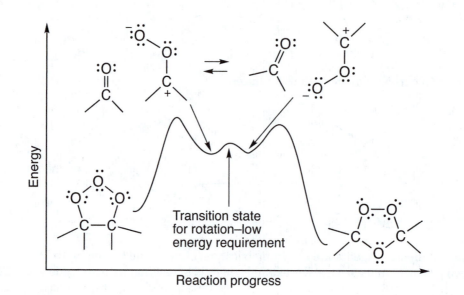

Problem 11.17

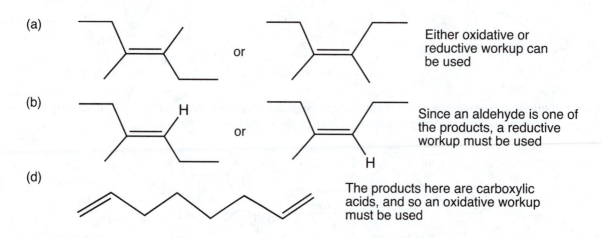

(a)   or   Either oxidative or reductive workup can be used

(b)   or   Since an aldehyde is one of the products, a reductive workup must be used

(d)   The products here are carboxylic acids, and so an oxidative workup must be used

Problem 11.18 The stereochemical relationship of the methyl groups does not change. The cis alkene gives a cis-substituted five-membered ring. There can be no steps in the reaction that allow rotation. If there were, a different diol would also be isolated. The initial reaction of OsO₄ with *cis*-2-butene must be one-step (concerted):

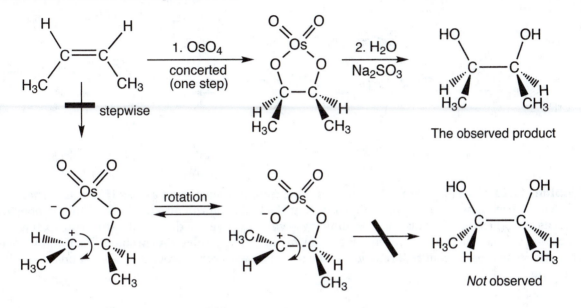

Problem 11.19 Racemic 2,3-butanediol must be the mixture of (R, R)- and (S, S)-2,3-butanediol, because the (R, S)- and (S, R)-2,3-butanediol are equivalent and not chiral. The (R, R) and (S, S) enantiomers come from a syn-dihydroxylation of (E)-2-butene.

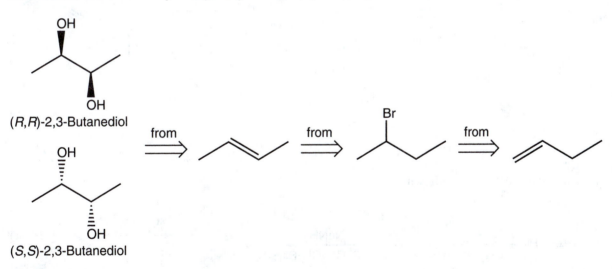

Therefore, the question we need to answer is, How can we make (E)-2-butene? We know that an E2 reaction will work well, which requires 2-bromobutane. We can make the 2-bromobutane from 1-butene.

The reactions starting with 1-butene are

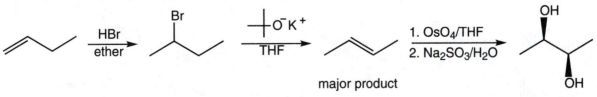

Problem 11.20   The first step of the reaction will form the unstable vinyl carbocation. This ion might be a cyclic species, as shown in Figure 11.60. The second addition of HCl will proceed via the more stable carbocation because it is stabilized by resonance.

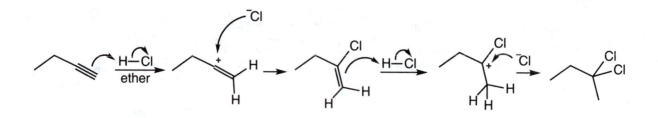

Problem 11.21   The 3,3-dibromohexane can come from the reaction of excess HBr with either 3-hexyne or from 2-hexyne. But you will notice that the 2-hexyne will also produce 2,2-dibromohexane, whereas the 3-hexyne—because of its symmetry—will only give the desired 3,3-dibromohexane. Therefore, we want to think about making 3-hexyne in order to solve this problem. Fortunately, the 3-hexyne comes easily from S_N2 reaction of the acetylide with ethyl bromide on both ends of the acetylene.

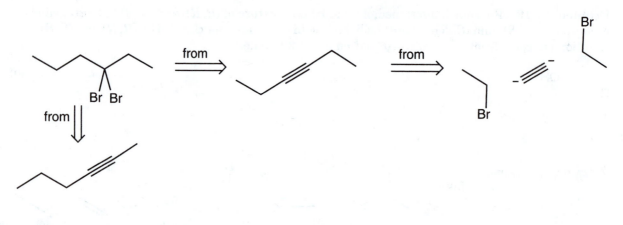

The forward reactions are shown.

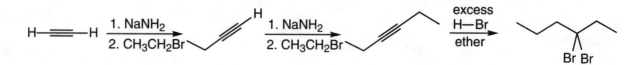

Problem 11.22 Protonation gives an intermediate, probably a cyclic cation of some kind, which can be captured by water to give the enol after deprotonation. Continued reaction of the enol will eventually produce the ketone.

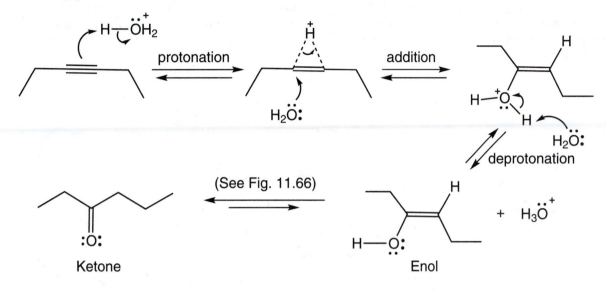

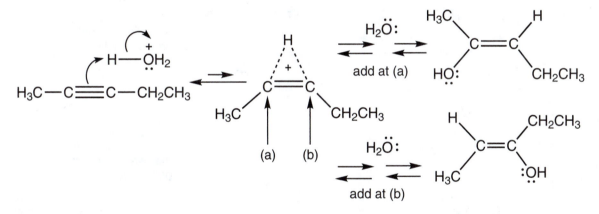

Problem 11.24 In the cyclic intermediate for this reaction, both carbons will share the positive charge by approximately the same amount because the two carbons are equally substituted. Accordingly, addition of the nucleophile will occur at both carbons to give, ultimately, two enols in roughly equal amounts.

Conversion of the enols into ketones will give *two* ketones (for a mechanism, see Fig. 11.66). This reaction is *not* a useful synthetic process.

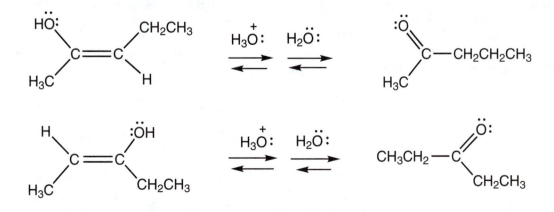

Problem 11.25   Follow through the two mechanisms, look for the point at which the pathways diverge, and try to see why the pathway to the ketone is favored. In this case, it is the initial protonation of the alkyne that determines the final product. Protonation gives the more substituted, more stable vinyl cation (or a cyclic intermediate in which the more substituted position bears most of the positive charge). Addition of water to give the enol is followed by conversion of the enol into the final product, the ketone (see Fig. 11.66 for mechanistic details of the formation of the ketone from the enol).

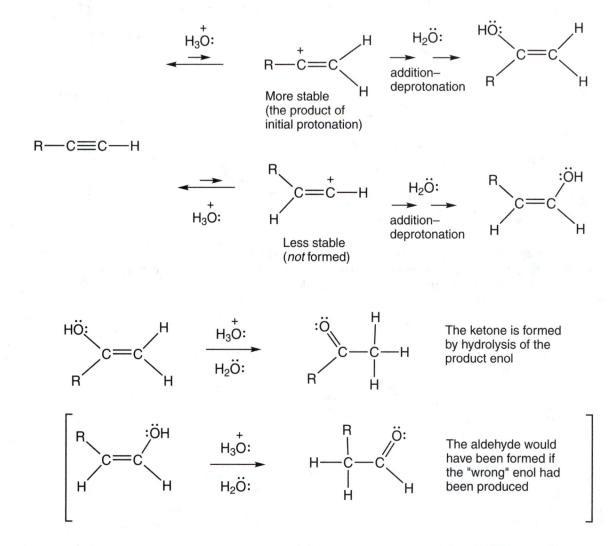

Problem 11.26   The best way to learn mechanisms is to write them backward. This problem presents an opportunity to do that for the mechanism of ketone formation from enols by asking you to do the reverse reaction. Remember, if you have written the mechanism in one direction, you have automatically written it in the other. All you need to do is to read the first one backward. So you could take the mechanism of Figure 11.66 as your answer.

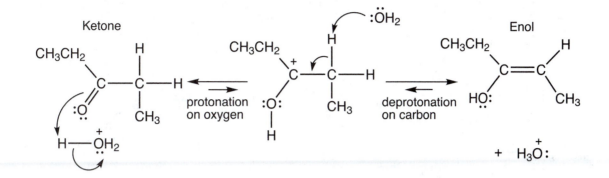

Problem 11.27 Why do we press this point so relentlessly? You should get into the habit of asking this question about any new reaction. The analysis is always the same: Compare the bond energies of the bonds broken and made in the reaction. The bonds broken are the π bond of ethylene and the σ bond of hydrogen. Two new carbon–hydrogen σ bonds are made. The reaction is exothermic by about 30 kcal/mol, and ΔH is negative for this reaction.

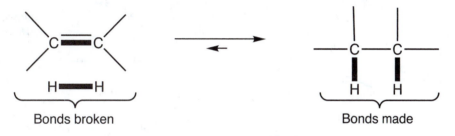

Bonds broken   Bonds made

π bond = 66 kcal/mol   σ bond = 104 kcal/mol   Two C—H σ bonds = about 200 kcal/mol

200 – 170 = 30 kcal/mol exothermic, ΔH = – 30 kcal/mol

Problem 11.28

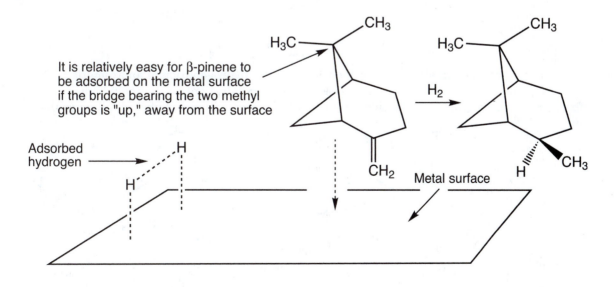

It is relatively easy for β-pinene to be adsorbed on the metal surface if the bridge bearing the two methyl groups is "up," away from the surface

Adsorbed hydrogen

Metal surface

(continued)

Problem 11.28 *(continued)*

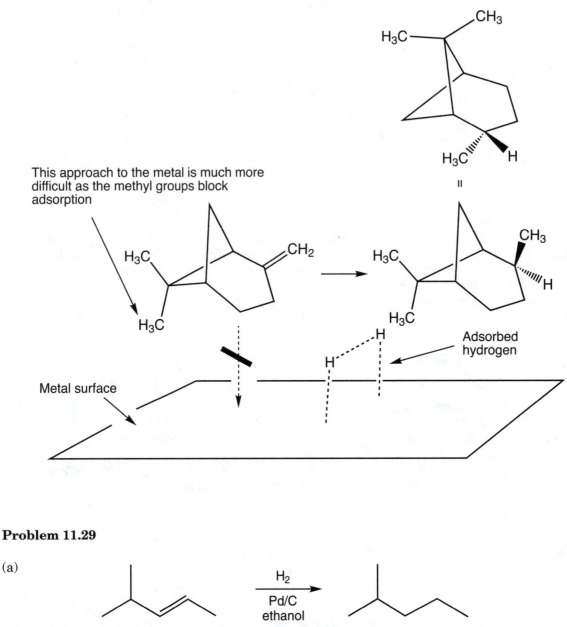

This approach to the metal is much more difficult as the methyl groups block adsorption

Metal surface

Adsorbed hydrogen

Problem 11.29

(a)

$$\xrightarrow[\text{ethanol}]{\text{H}_2 \atop \text{Pd/C}}$$

(b) The hydrogenation of an alkene is much easier than the hydrogenation of an aromatic ring. The π bonds of the aromatic ring will hydrogenate with the Pd catalyst only if we apply higher temperatures and pressure. Even with excess H₂, this reaction will give 2-phenylbutane as the product.

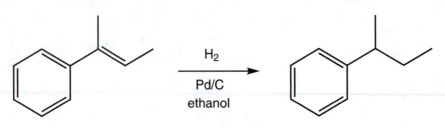

$$\xrightarrow[\text{ethanol}]{\text{H}_2 \atop \text{Pd/C}}$$

(c) An alcohol does not participate (is not changed) in the hydrogenation reaction.

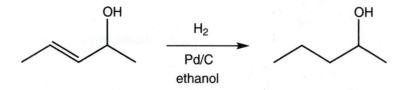

(d) We can confirm that hydrogenation is a syn addition using 1,2-dimethylcyclopentene. The product is achiral. Adding hydrogen to the top face of the alkene will give the same product, which is *cis*-1,2-dimethylcyclopentane.

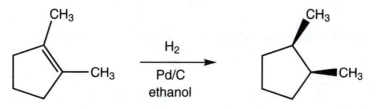

(e) The hydrogenation of this methylene-substituted cyclohexane will occur from the less hindered side of the alkene. In this case, that side is the top of the molecule. The methyl groups that are axial on either side of the alkene will block that side of the alkene. Adding hydrogen to the top face will push the new methyl group into the equatorial position.

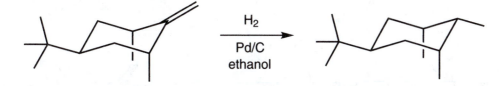

(f) Both alkenes will be hydrogenated in (*E*)-2-methyl-1,4-hexadiene. It would be difficult to control monohydrogenation of this reagent. It is usually steric factors that regulate hydrogenation, and both alkenes in this molecule are both disubstituted.

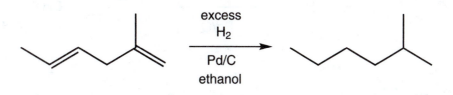

Problem 11.30 The reaction does not occur spontaneously because there is an activation energy barrier (Chapter 7, p. 281). It is not enough that the starting material, here alkene plus hydrogen, be higher in energy than the product, here alkane. There must be energy enough to surmount the barrier. The role of the catalyst is to provide a lower energy mechanistic pathway leading to product; a trail on a lower slope of the activation mountain (Fig. 10.39, p. 461).

(continued)

Problem 11.30 *(continued)*

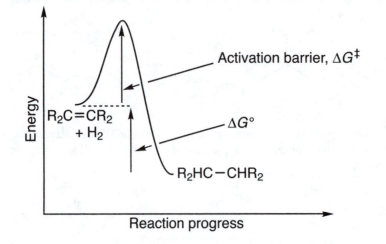

Problem 11.31   This problem is just an exercise in remembering reagents.

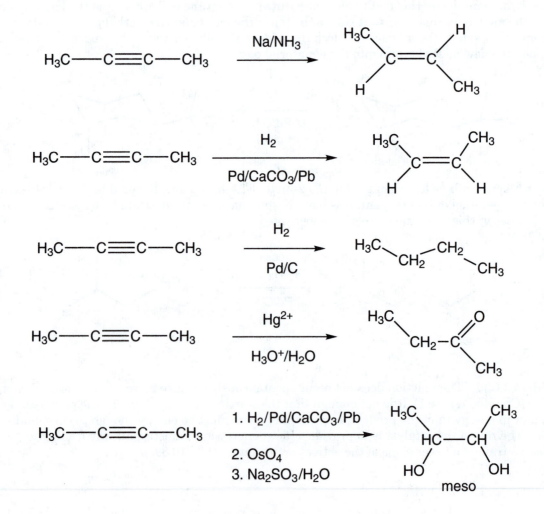

Additional Problem Answers

Problem 11.32 We should be able to make the vinyl bromide 2-bromo-1-butene from the corresponding 1-butyne using one equivalent of HBr. The 1-butyne can be obtained from S$_N$2 reaction between acetylide anion and ethyl bromide. The ethyl bromide can be obtained from ethylene, and that in turn can come from acetylene.

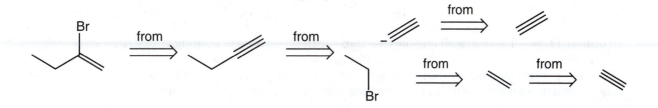

The reactions are shown in the following scheme:

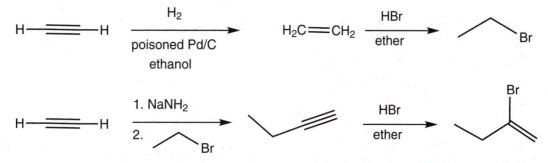

Problem 11.33 Displacement of a good leaving group such as bromide or iodide by azide ion would work. The halide can be made from 1-butene through the addition of HBr or HI to 1-butene.

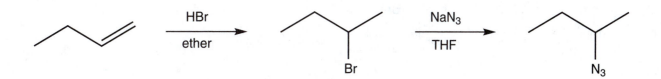

Problem 11.34 Opening of cyclohexene epoxide must give the trans compound because the S$_N$2 reaction always goes with inversion.

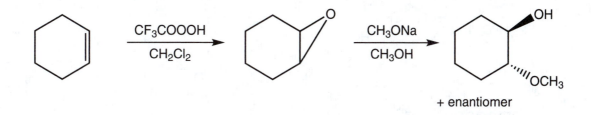

+ enantiomer

Problem 11.35   The use of poisoned palladium to catalyze hydrogenation of an alkyne results in syn addition to give the cis alkene. The cis alkene does not react further. Therefore, we predict that the poisoned palladium will not hydrogenate the trans alkene either.

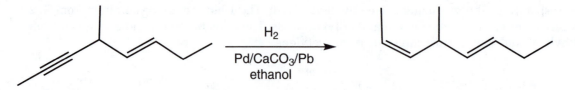

Problem 11.36   These are all one- or two-step syntheses. There is nothing tricky here, only remembering the reactions. If you forget, fall back on a mechanistic analysis, and this should lead you to the product in most cases. Remember, we start with achiral materials, and in *all* cases racemic mixtures must result. In the answers, only one enantiomer is shown.

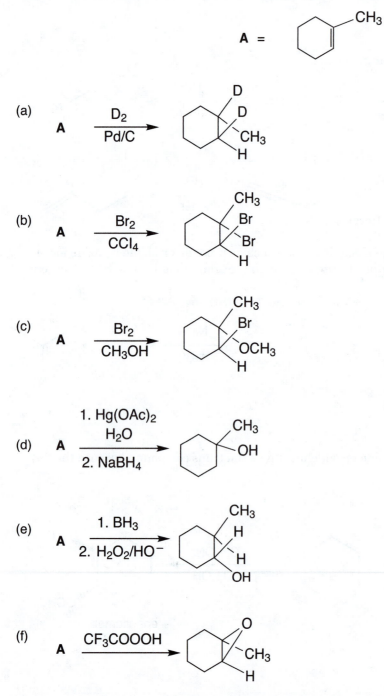

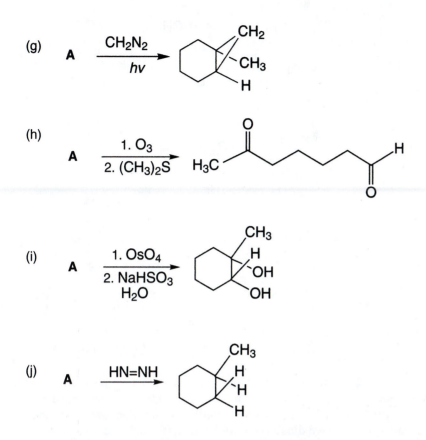

(g) A $\xrightarrow[hv]{CH_2N_2}$

(h) A $\xrightarrow[\text{2. (CH}_3)_2S]{\text{1. O}_3}$

(i) A $\xrightarrow[\substack{\text{2. NaHSO}_3 \\ \text{H}_2\text{O}}]{\text{1. OsO}_4}$

(j) A $\xrightarrow{\text{HN=NH}}$

Problem 11.37 Deuterium can be delivered from either side of the planar double bond. The two enantiomers of the methylcyclohexane product result.

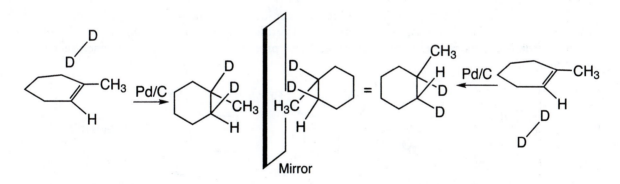

Problem 11.38 In each case, an intermediate bromonium ion is opened by bromide addition at the more substituted carbon. In (b), (c), and (d) ring-opening will occur at both possible positions.

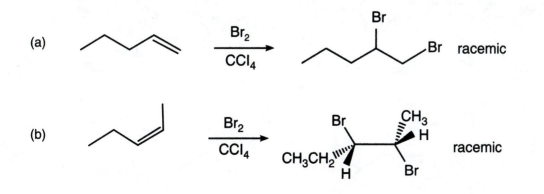

(a) $\xrightarrow[\text{CCl}_4]{\text{Br}_2}$ racemic

(b) $\xrightarrow[\text{CCl}_4]{\text{Br}_2}$ racemic

(continued)

Problem 11.38 *(continued)*

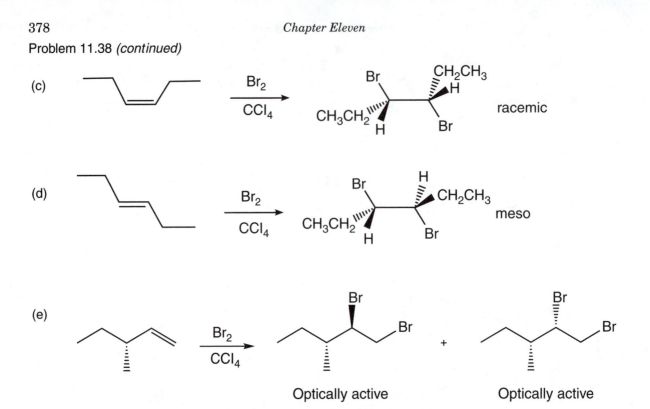

(c)

(d)

(e)

Optically active   Optically active

Problem 11.39 The products in Problem 11.38(a), (b), and (c) are chiral and are formed as racemic mixtures in each case. The single product in 11.38 d is a meso compound, therefore it is achiral. The starting material in 11.38 e is an optically active molecule with (*R*) configuration. Because the reaction generates a new stereogenic center, there will be two diastereomers formed, (2*R*,3*R*)-1,2-dibromo-3-methylpentane and (2*S*,3*R*)-1,2-dibromo-3-methylpentane. Each will be formed as a single enantiomer.

Problem 11.40 In each case, an intermediate bromonium ion is opened either by bromide ion or water to give the products shown. As all starting materials are achiral, there can be no net optical activity in the products. Since water is the solvent, we predict the major product in each case will be the halohydrin.

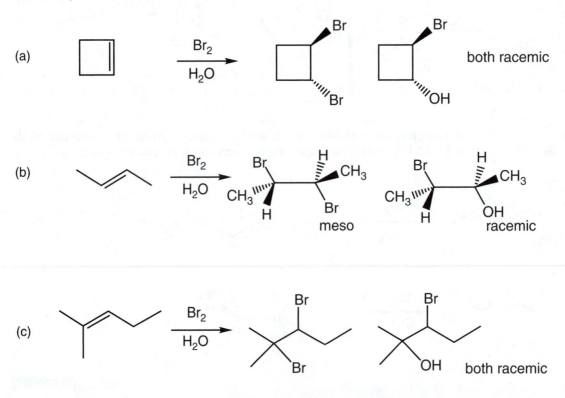

(a)   both racemic

(b)   meso   racemic

(c)   both racemic

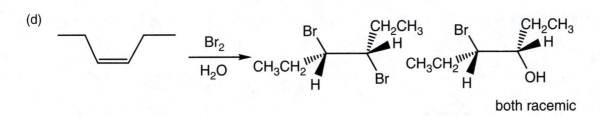

(d)

both racemic

Problem 11.41

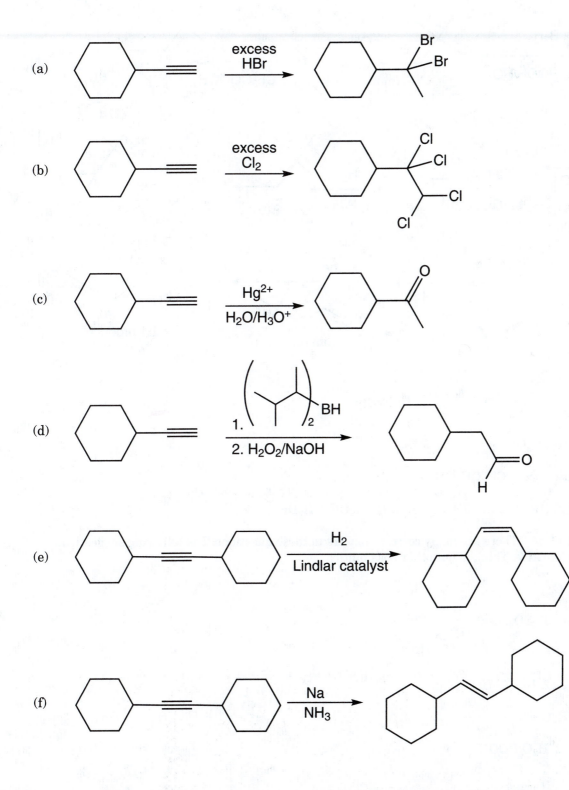

(a) excess HBr

(b) excess Cl₂

(c) Hg^{2+} / H_2O/H_3O^+

(d) 1. $\left(\right)_2 BH$ 2. $H_2O_2/NaOH$

(e) H_2 / Lindlar catalyst

(f) Na / NH_3

Problem 11.42

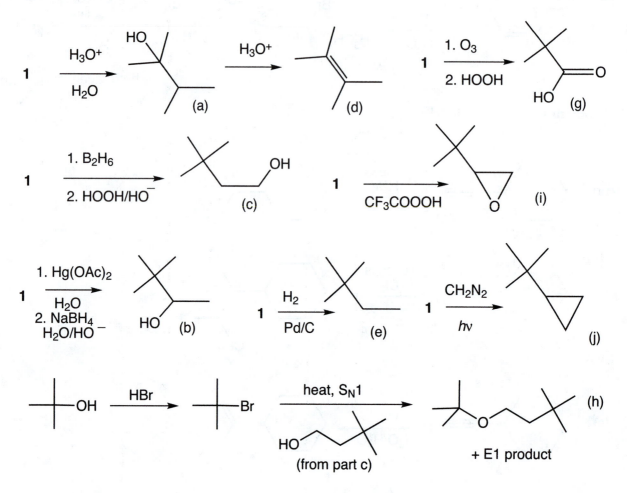

Problem 11.43

(a) HBr/ether

(b) Br$_2$/CH$_3$CH$_2$OH

(c) catalytic H$_2$SO$_4$/CH$_3$CH$_2$OH

(d) 1. BH$_3$/THF   2. NaOH/H$_2$O$_2$ 3. NaH   4. CH$_3$CH$_2$Br

Problem 11.44   There are many correct answers to these questions. The following suggestions are not exhaustive. Think about other possibilities.

(a) H$_2$/Pd/C

(b) catalytic H$_2$SO$_4$, heat

(c) Start with (b): 1. BH$_3$ 2. HOOH/HO$^-$

(d) 1. BH$_3$ 2. HOOH/HO$^-$

(e) H$_3$O$^+$/H$_2$O

(f) 1. O$_3$ 2. H$_2$O/HOOH

(g) Start with (d): PBr$_3$

(h) 1. CF$_3$COOOH   2. $^-$OH/H$_2$O

(i) Br$_2$/CH$_3$OH

(j) Start with (b): 1. O$_3$ 2. HOOH

Problem 11.45   Once again, there are many possible correct answers. Here are some. Mrs. Tao's eels, though incredibly delicious, are ineffective.

(a) 1. BH$_3$ 2. HOOH/HO$^-$

(b) H$_3$O$^+$/H$_2$O

(c) HBr

(d) Compound (a) and PBr$_3$

(e) 1. OsO$_4$ 2. Na$_2$SO$_3$/H$_2$O

(f) 1. H$_2$/CaCO$_3$/Pb/Pd   2. CH$_2$N$_2$, light or heat

(g) 1. NH$_3$/Na   2. CF$_3$COOOH

(h) H$_3$O$^+$/H$_2$O

(i) 1. H$_2$/CaCO$_3$/Pb/Pd   2. O$_3$ 3. H$_2$/Pd

(j) 1. NH$_3$/Na   2. O$_3$ 3. Zn

Problem 11.46   Formation of a bromonium ion could, in principle, occur in two ways: syn to the methyl group or anti to the methyl group. Opening, again in principle, could occur through an S$_N$2 reaction in two ways for each bromonium ion. Here are the four possibilities:

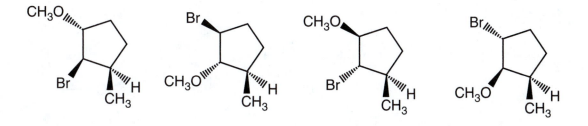

Steric factors will favor the anti bromonium ion. Opening by methanol, followed by deprotonation, will occur at the less sterically hindered position with inversion to give the product shown.

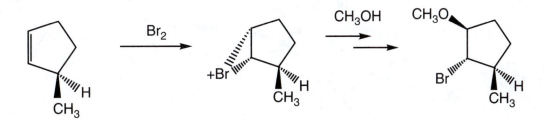

Problem 11.47   Each of the potential products shown in Problem 11.46 would be formed as a single enantiomer. The stereogenic carbon in the starting material is the (*R*) configuration, and it does not change throughout the reaction. Forming a racemic mixture would mean that the isomer with methyl in the (*S*) configuration would also have to be present.

Problem 11.48   In the addition of Br_2 to *cis*-2-butene, an open carbocation must give two diastereomeric products.

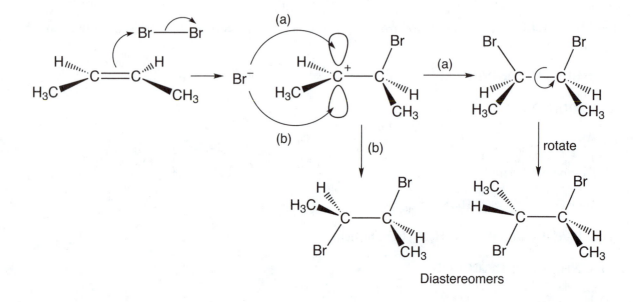

Diastereomers

As this result is not observed, this mechanism must be wrong. Formation of a bromonium ion that opens up in an S_N2-like fashion solves the problem. The bromonium ion must open by addition of bromide from the rear, and this reaction leads to racemic dibromide when *cis*-2-butene is the starting material.

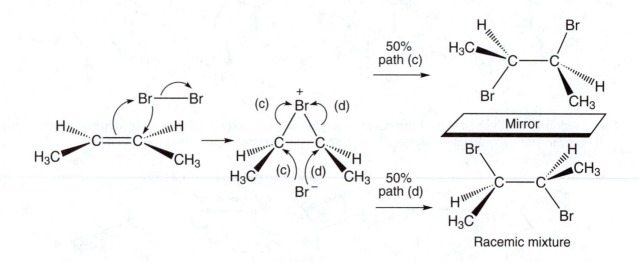

Racemic mixture

Similarly, in the addition to *trans*-2-butene, an open carbocation must lead to two products, a result not observed experimentally:

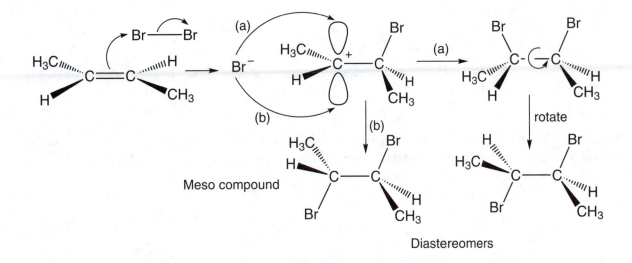

Meso compound

Diasteromers

A bromonium ion shows how the experimentally observed single product, a meso compound, is formed.

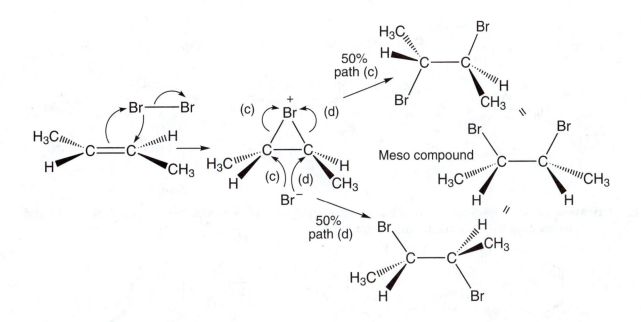

Meso compound

Problem 11.49   In the acid-catalyzed reaction, the first step is the protonation of the epoxide oxygen to give a resonance-stabilized cation. The two bonds from carbon to oxygen are different as the tertiary carbon will have a greater share of the positive charge than will the primary carbon.

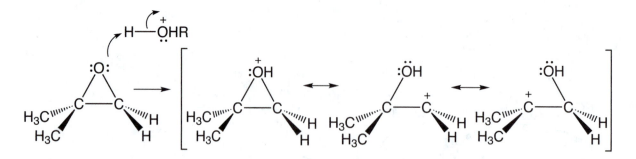

The alcohol will add predominately to break the longer and weaker, more substituted carbon–oxygen bond. Deprotonation of the intermediate oxonium ion leads to the product and regenerates the catalyst, the protonated alcohol.

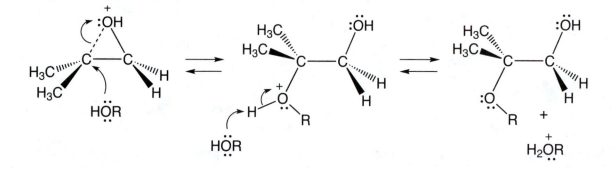

The base-catalyzed reaction is a straightforward S_N2 displacement of a leaving group by a negatively charged nucleophile. The epoxide opens from the less hindered side. Proton transfer completes the reaction and regenerates a molecule of alkoxide.

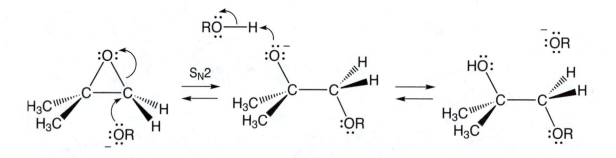

Problem 11.50   The cyclic osmium intermediate will be preferentially formed anti to the *tert*-butyl group, so opening will give the product shown:

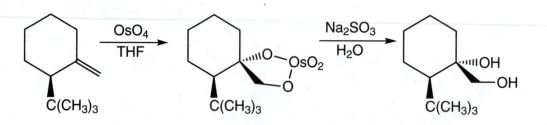

Addition of BH_3 will be from the side away from the the *tert*-butyl group. Accordingly, the final product will have the cis stereochemistry shown:

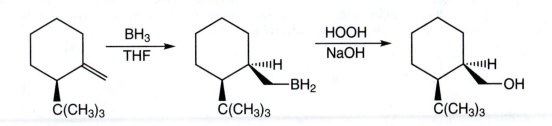

The mercurinium ion will form anti to the *tert*-butyl group. Opening by water at the more hindered position followed by borohydride reduction will lead to the compound with the OH cis to the *tert*-butyl group.

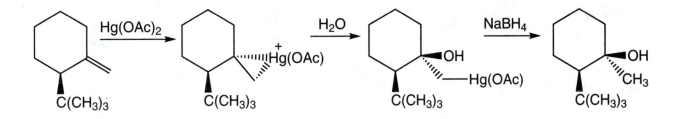

Hydrogenation will occur preferentially from the side away from the large *tert*-butyl group.

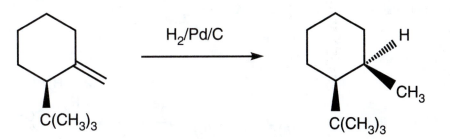

The major—not exclusive—product will come from addition of water to the intermediate carbocation from the side away from the large *tert*-butyl group. There could be a hydride shift—what product would that give?

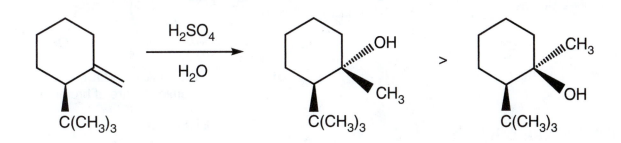

Problem 11.51   First of all, what are the basic mechanisms of bromohydrin formation and epoxide formation from the bromohydrin? The first part is a straightforward addition reaction covered many times already. In the second and more difficult part of the problem, alkoxide formation is followed by intramolecular displacement of bromide in S_N2 fashion. Thus, this process involves *two* inversions of configuration. Two inversions result in net retention of configuration as cis alkene becomes the cis epoxide. A mechanistic analysis should make this clear.

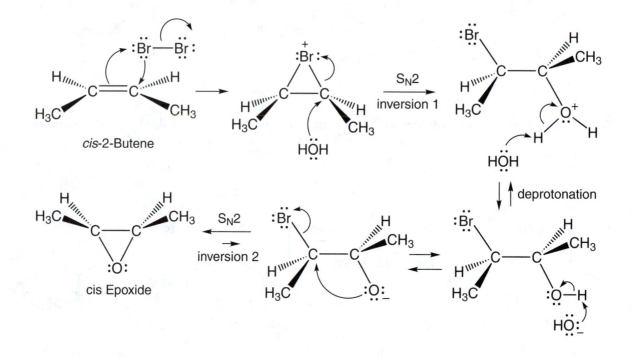

Problem 11.52   Formation of the bromonium ion and opening by water leads to a racemic mixture of bromohydrins (3-bromo-2-butanol, **A**).

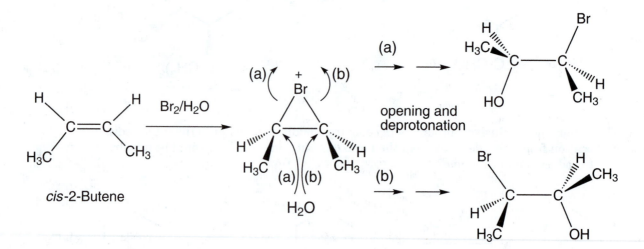

Product **A**, C_4H_9OBr, is a racemic mixture of bromides

Treatment of **A** with NaH forms the alkoxides, and intramolecular S_N2 displacement gives the same meso epoxide, **B**, in each case.

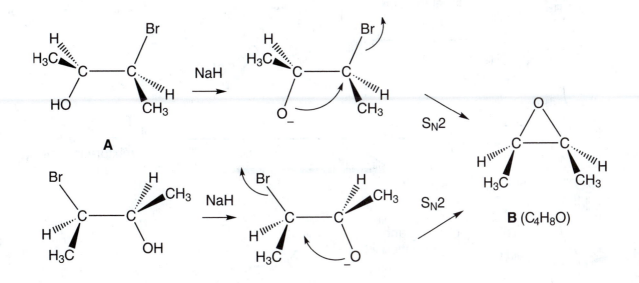

The keys to seeing this step of the problem are remembering that NaH is a useful base for forming alkoxides, and the recognition that an HBr is lost in going from C_4H_9OBr to C_4H_8O.

The *trans*-2-butene undergoes exactly the same reaction sequence, but the stereochemical relationships are different. The bromonium ion can be formed by addition of the bromine from either the top or bottom of the alkene to give a pair of enantiomeric ions.

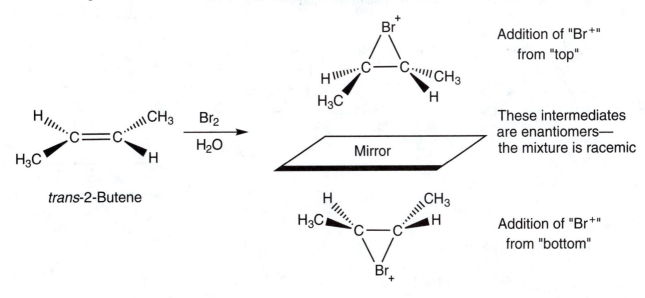

Addition of "Br⁺"
from "top"

These intermediates
are enantiomers—
the mixture is racemic

Addition of "Br⁺"
from "bottom"

(continued)

Problem 11.52 *(continued)*

Opening of these bromonium ions with water leads to compound **C,** which is a pair of enantiomeric bromohydrins.

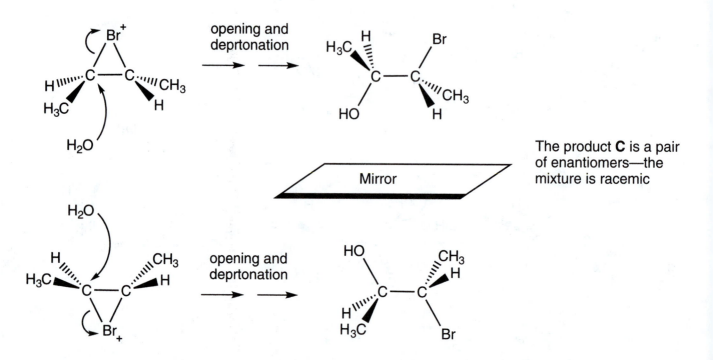

The product **C** is a pair of enantiomers—the mixture is racemic

Treatment of **C** with NaH results in intramolecular S_N2 displacement of bromide to give the epoxide **D** as a racemic mixture. The epoxide has two equivalent methyl groups that will be coupled to the hydrogen on the epoxide ring giving rise to a doublet at δ 1.30 ppm. The hydrogens in compound **D** on the ring are equivalent. They will not couple with each other. The chemical shift for these hydrogens will be upfield from the otherwise predicted value (δ ~3.8 ppm) because they are on a three-membered ring. They give rise to the quartet at δ 2.70 ppm.

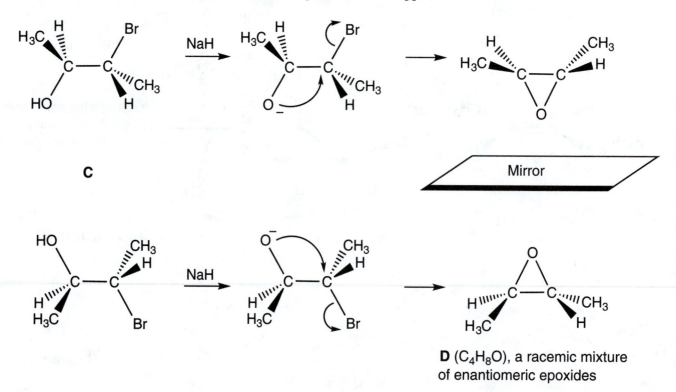

D (C_4H_8O), a racemic mixture of enantiomeric epoxides

Problem 11.53 The more substituted an alkene, the better a nucleophile it is. So, it is the tetrasubstituted double bond that is the more reactive site in this diene.

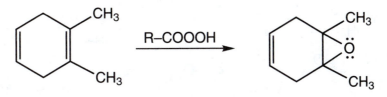

But wait! There is another effect we haven't considered. Is not the tetrasubstituted double bond more hindered sterically? Yes it is, and the steric difficulties should slow the reaction. So, there are two factors working in opposite directions. It is the electronic effect that dominates in this case. A good answer to this question mentions both effects.

Problem 11.54 The alcohol is 2-butanol, and here are the transformations:

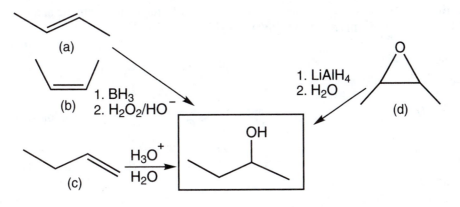

Problem 11.55

(a) Epoxidation is followed by S$_N$2 opening of the three-membered ring from the rear with hydroxide acting as nucleophile. Notice that in this reaction, the epoxide is opened from the sterically less-encumbered side to give a trans diol.

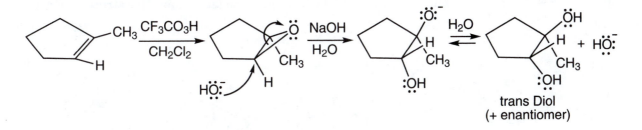

(b) Osmium tetroxide (usually in THF) reacts with an alkene via syn addition to make a five-membered ring intermediate that gives a cis diol after hydrolysis.

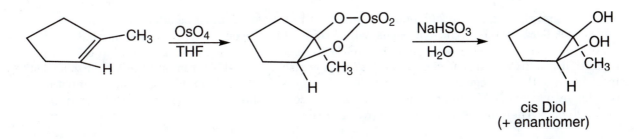

Problem 11.56 In this intramolecular version of the periodate cleavage reaction, both new carbonyl groups appear in the same molecule.

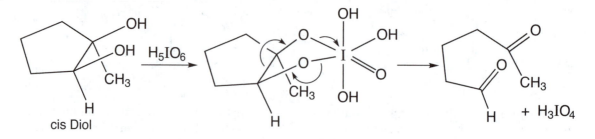

Ozonolysis of 1-methylcyclopentene with a reductive workup [(CH₃)₂S or H₂/Pd] would lead to the same product.

In the trans diol, the two hydroxy groups are too far apart to form the cyclic intermediate easily. Reaction of trans diols with periodate is slow at best.

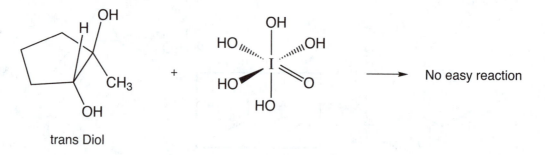

Problem 11.57 Compound **A** (C₇H₁₄) has 1 degree of unsaturation. We know that it will react with catalytic H_2SO_4 and H_2O, so it must be an alkene rather than a ring, which would not react with acid and water. We can see from the NMR spectral data for **A** that there are three vinyl hydrogens. The are no other groups in the molecule that can shift hydrogens downfield to the δ ~5 ppm region. So we know that compound **A** is a terminal alkene. There are two hydrogens at δ 2.05 ppm, which is the chemical shift expected for allylic hydrogens. So far we have this structure for **A**:

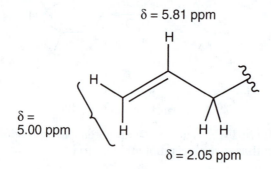

We know that the signal at δ 2.05 ppm is a quartet. That means it must be adjacent to three hydrogens. Because the allylic hydrogens will be coupled by the one vinyl hydrogen at δ 5.81 ppm, there must be a CH₂ on the other side. That CH₂ on the other side must be the signal at δ 1.27 ppm because it is a 2H signal. The signal at δ 1.27 ppm is also a quartet, so it must be adjacent to a CH in addition to the hydrogens at δ 2.05 ppm. The CH signal is at δ 1.56 and it is a nonet. It must be coupled by a total of eight hydrogens. It must be adjacent to the two methyl groups (6 H) that are at δ 0.90 ppm. Therefore, **A** must be 5-methyl-1-hexene.

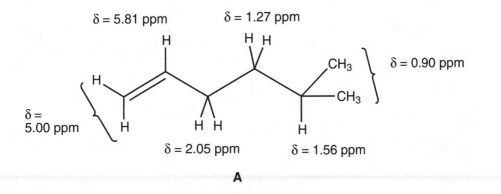

δ = 5.81 ppm

δ = 1.27 ppm

δ = 0.90 ppm

δ = 5.00 ppm

δ = 2.05 ppm

δ = 1.56 ppm

A

Now we can determine the structure of **B**, because we know that an alkene gives an alcohol when it reacts with catalytic H_2SO_4 and H_2O. Compound **B** must be 5-methyl-2-hexanol. We can see if the NMR spectral data match this structure. The signal for the OH hydrogen should shift when D_2O is added to the sample. Therefore it is the signal at δ 2.29 ppm. The C(2) hydrogen should be strongly deshielded by the oxygen, and we predict its signal to be at δ 3.8 ppm. It must be the 1H signal at δ 3.75 ppm. It should be a sextet because it is adjacent to the C(1) methyl and the C(3) methylene—it "sees" five hydrogens. We would not expect this signal to be a clean sextet, because the five hydrogens are not equivalent. The C(3) hydrogens are diastereotopic, which should add to the complexity of this spectrum.

The signal at δ 1.49 ppm must be a CH_2 once removed from the electronegative oxygen, and the signal at δ 1.18 ppm must be a CH_3 also slightly deshielded by the oxygen. The coupling for the signal at δ 1.49 ppm must be complex due to the hydrogens being diastereotopic (nonequivalent). That makes it difficult to predict the rest of the molecule, but we know there is an isopropyl group because of the 6H doublet at δ 0.89 ppm. That allows us to assign the remaining signals as shown.

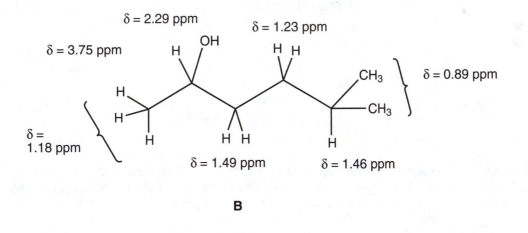

δ = 2.29 ppm

δ = 3.75 ppm

δ = 1.23 ppm

δ = 0.89 ppm

δ = 1.18 ppm

δ = 1.49 ppm

δ = 1.46 ppm

B

Problem 11.58 The epoxide of disparlure must be obtained from epoxidation of the cis-substituted alkene. We know that we can obtain the cis alkene from the corresponding alkyne. Alkynes can be alkylated by S_N2 reactions between the acetylide and primary halides. The structures are shown here:

(continued)

Problem 11.58 *(continued)*

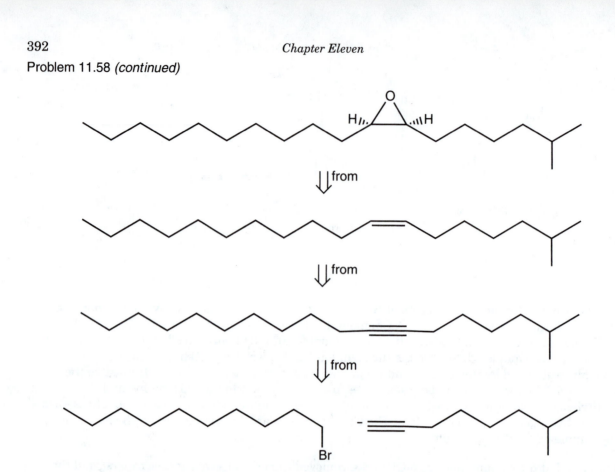

Now we know the path forward. We can deprotonate the 7-methyl-1-octyne, alkylate it with 1-bromodecane, hydrogenate with poisoned Pd, and then epoxidize with a peroxy acid.

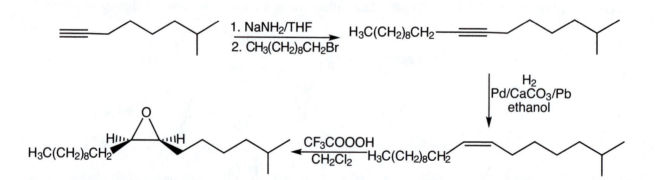

Problem 11.59 The ozonolysis data let us piece together the structures of **3** and **4**. Ozonolysis results in the conversion of a carbon–carbon double bond into a pair of carbon–oxygen double bonds. Accordingly, the structures of the products tell us the structures of **3** and **4**.

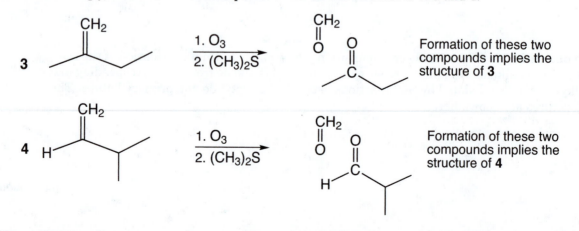

Given that **3** and **4** are formed by partial hydrogenation of **1**, we now know **1**. Full hydrogenation of **1** gives **2**.

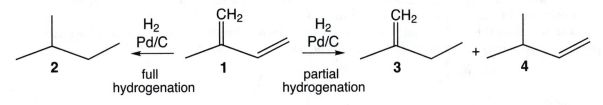

Problem 11.60 For α-terpinene (**1**), there is only one way to put the pieces formed on ozonolysis back together. (*Remember:* With an oxidative workup, carboxylic acids, not aldehydes, are formed.)

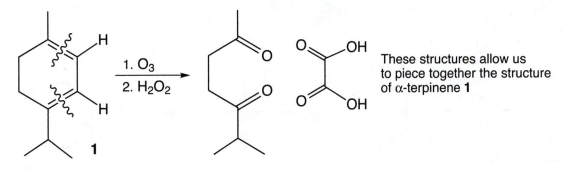

These structures allow us to piece together the structure of α-terpinene **1**

For γ-terpinene (**2**), there are two ways to put the pieces back together. However, only one of the possible cyclohexadienes would yield 1-isopropyl-4-methylcyclohexane on hydrogenation.

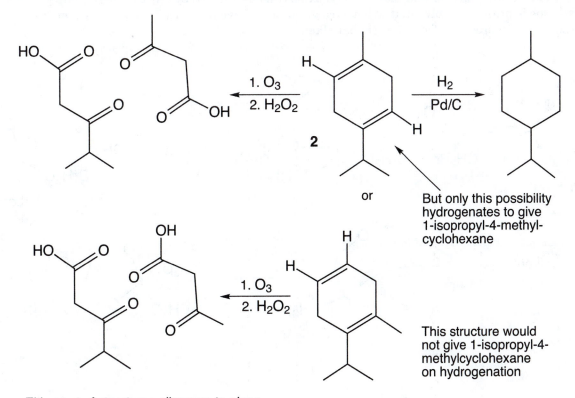

But only this possibility hydrogenates to give 1-isopropyl-4-methyl-cyclohexane

This structure would not give 1-isopropyl-4-methylcyclohexane on hydrogenation

Either set of structures allows us to piece together a possible structure for γ-terpinene

Problem 11.61   The third step in this reaction sequence uses an acid-catalyzed dehydration to go from the alcohol to the alkene. This reaction gives the most substituted alkene as the product. It is the desired alkene. You might have been tempted to use an E2 reaction to make the alkene. But you would need to convert the OH group into a leaving group first. If you make it a tosylate (OTs), for example, then the E2 reaction would not give the desired product. The only available anti-periplanar β hydrogen results in formation of the less substituted alkene.

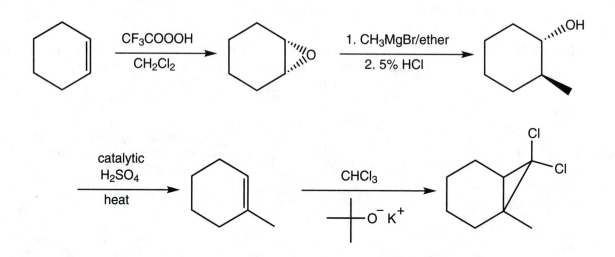

Problem 11.62   Once formed, the carbonyl oxide can be captured by *any* carbonyl group present. If acetone is present as solvent, it will have a great advantage in the capture process over the carbonyl group produced from the original alkene.

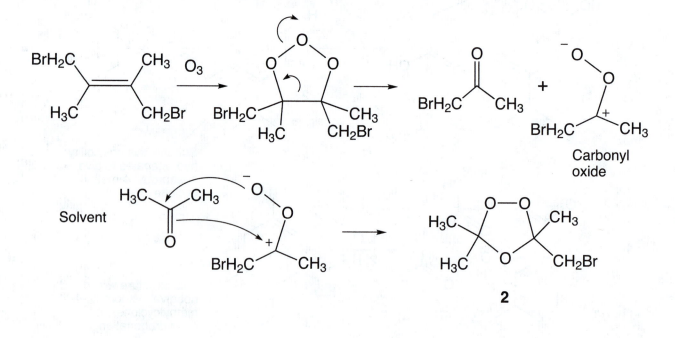

2

Problem 11.63 First, form the primary ozonide and allow it to break down to give the carbonyl oxide and, in this case, acetone.

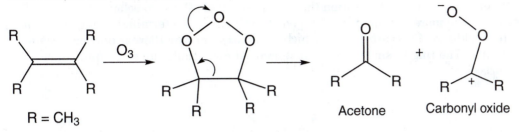

R = CH₃

If the carbonyl oxide diffuses away from the acetone, it can collapse to the dioxirane.

Alternatively, it might be captured by another carbonyl oxide molecule, which eventually leads to the six- and nine-membered rings.

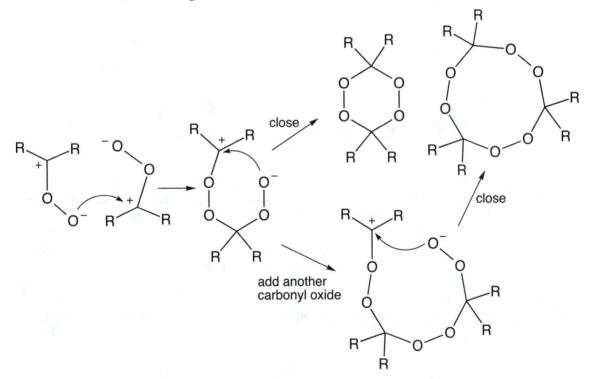

Problem 11.64 The arrow formalism is easy. In principle, but not actually, the compound with the Ph and CH₃ cis might have been formed.

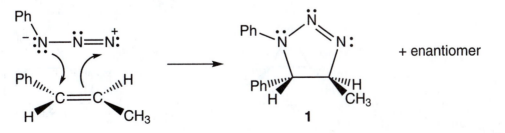

+ enantiomer

(continued)

Problem 11.64 *(continued)*

The formation of the single stereoisomer **1** with retention of stereochemistry (the trans alkene gives a trans five-membered ring) means that the mechanism of formation of **1** must involve no free carbocation (**2**, for example) capable of rotation that would scramble the stereochemistry originally present in the alkene. The more electrophilic nitrogen of the azide is the terminal nitrogen, and it probably starts to bond to the C(2) carbon first, which would explain why there is only one regioisomer in this reaction. The ring closing process must occur at essentially the same time, since there is no free carbocation.

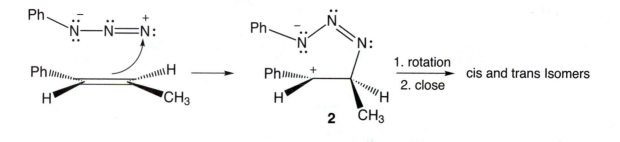

Problem 11.65   In the concerted mechanism, nitrogen is lost in a single step to give aziridine **2**. The trans stereochemistry present in **1** will be preserved in **2**.

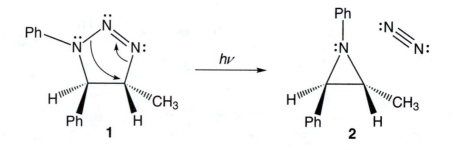

There are many possible two-step (nonconcerted) mechanisms. All must go through an intermediate in which the stereochemical relationships originally present in the alkene and triazoline **1** can be lost. Both **2** and the stereoisomeric **3** will be formed. Here is one possibility.

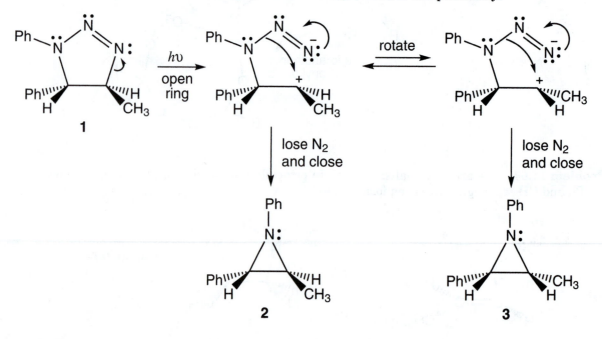

So the formation of **3** becomes diagnostic for the nature of the mechanism of aziridine formation. A one-step mechanism cannot form **3**; a two-step mechanism must form **3**.

Problem 11.66   Azides behave much like diazo compounds. Just as diazo compounds can lose nitrogen to give a carbene (divalent carbon), so azides can lose nitrogen to give "nitrenes" (monovalent nitrogen).

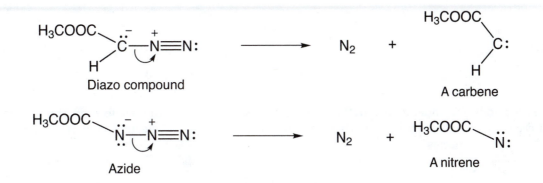

Carbenes add to alkenes to give cyclopropanes. Nitrenes behave in similar fashion, forming the three-membered rings called aziridines.

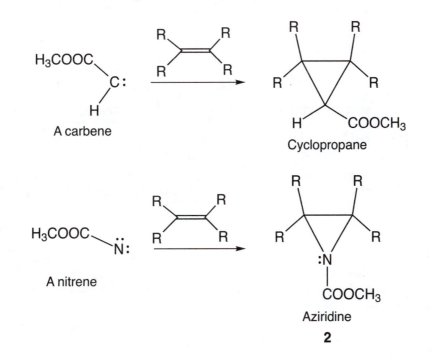

Problem 11.67   Retention of stereochemistry means that aziridine formation must take place in a single step.

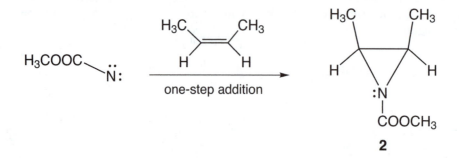

A two-step addition would give both **2** and the stereoisomeric **3**. As only **2** is formed, the reaction must occur in a single step. In turn, that implies a singlet, all-paired electron structure for the nitrene reacting species.

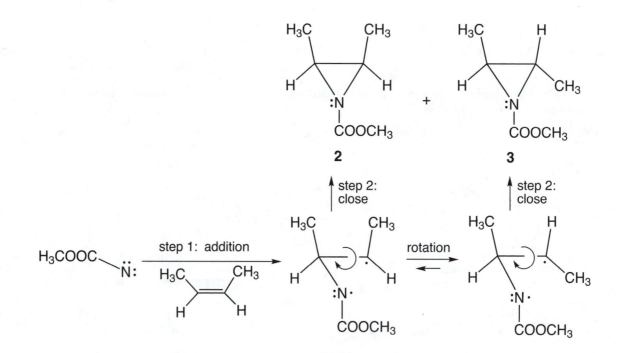

Problem 11.68

(a) In a one-step reaction, there can be no intermediate; the product must be formed directly. There really isn't much to draw in this case.

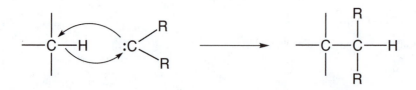

A two-step mechanism must involve hydrogen abstraction followed by recombination:

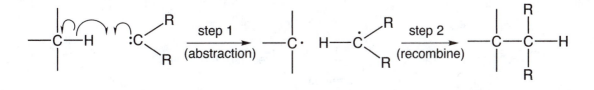

(b) The Doering and Prinzbach experiment shows that the insertion reaction must occur in one step. A two-step process would generate a resonance-stabilized intermediate allyl radical, and the position of the label would be scrambled. As this does not happen, the reaction must be one step.

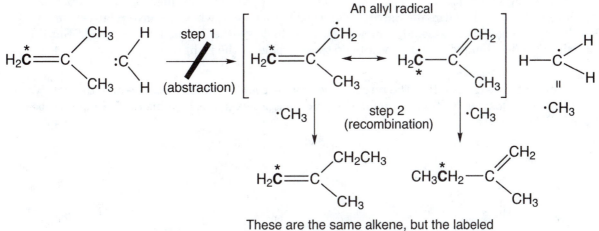

An allyl radical

These are the same alkene, but the labeled carbon is in a different position

Problem 11.69 Both hydroboration/oxidation and mercury-catalyzed hydration of 2-pentyne must give a mixture of two ketones, 2-pentanone and 3-pentanone. Neither process is a useful preparative method. By contrast, either procedure applied to the symmetrical alkyne 3-hexyne would give the same product, 3-hexanone. Either would be a reasonable preparative method.

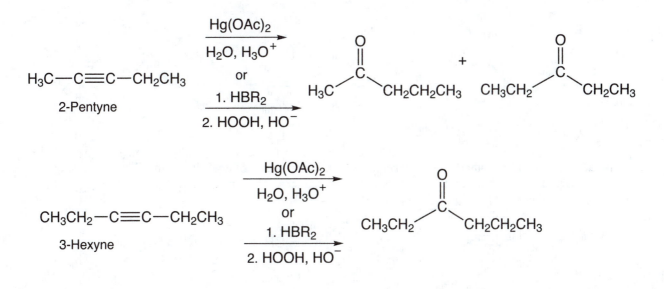

Problem 11.70   In all synthesis problems, it is imperative that you work backward. This is called "retrosynthetic analysis" by those who do it for a living, and *every* successful synthetic chemist analyzes problems this way. There is even a special, "retrosynthetic" arrow that points to the immediate precursors to the target. Thus,

Target $\Longrightarrow$ Precursor molecules

(retrosynthetic
arrow)

Who are we to disagree? We won't. In practice, the rather fancy term "retrosynthetic analysis" simply means, "search for the *immediate* precursor for the target molecule." *Do not* attempt to see all the way back to the ultimate starting material. At this point, when we have relatively few synthetic methods in our arsenal, it may be possible to do synthetic problems "forward." But this practice is bad technique, and it is best to do these problems the right way.

(a) In this case, we can see that the target, 2-hexanone, could be made from the mercury-catalyzed hydration of 1-hexyne. (2-Hexyne would also give some 2-hexanone, but this route would not be practical. Why?)

However, 1-hexyne has two more carbons than our allowable starting materials, so we need a synthesis of this molecule. The problem has been reduced to finding a synthesis for 1-hexyne. Here is a suggestion: An S_N2 reaction between an acetylide anion and bromobutane would do the trick.

The acetylide can be formed from the reaction of a large excess of acetylene, an allowed starting material, and sodium amide (the excess acetylene reduces the formation of the diacetylide).

$HC\equiv C:^-$ $\Longrightarrow$ $HC\equiv CH$ + $NaNH_2/NH_3$

We still need to make bromobutane, however, because this material isn't an allowed starting material. We might reduce 1-butyne to 1-butene, hydroborate, and then form the bromide.

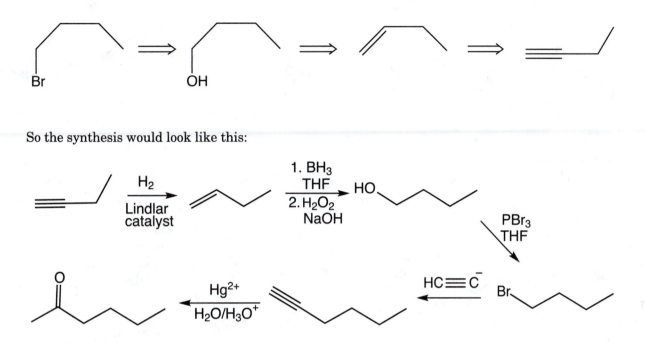

So the synthesis would look like this:

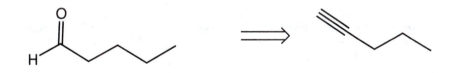

(b) The target molecule, pentanal, should be available from 1-pentyne through a hydroboration-oxidation sequence with diisoamylborane, $HB[CHCH_3CH(CH_3)_2]_2$. 1-Pentyne can be made through a sequence similar to that used for 1-hexyne in (a).

Here is the synthesis:

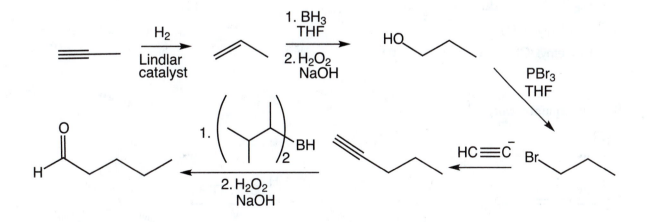

(continued)

Problem 11.70 *(continued)*

There are other possibilities. For example, we could use the 1-hexyne we made in part (a). The retrosynthetic analysis looks like this:

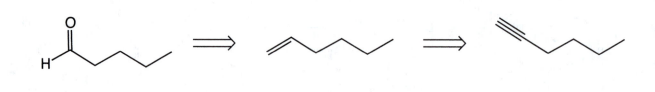

In other words,

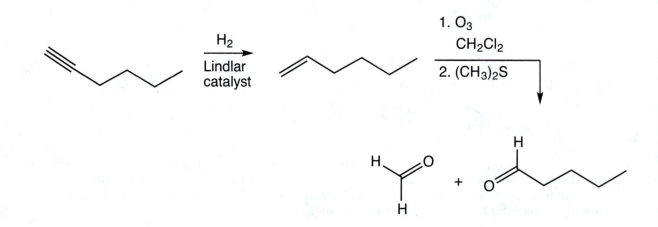

(c) Retrosynthetic analysis suggests that our target epoxide should be available from 3-octyne.

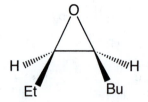

Bu = butyl, $CH_2CH_2CH_2CH_3$
Et = ethyl, CH_2CH_3

In conventional terms,

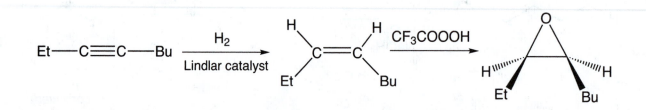

3-Octyne can be prepared by alkylation of the acetylide of 1-butyne with bromobutane, a molecule we made in (a).

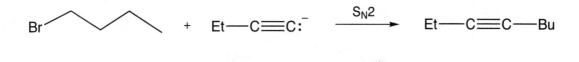

The acetylide comes from the corresponding acetylene:

Et—C≡CH $\xrightarrow{\text{NaNH}_2/\text{NH}_3}$ Et—C≡C:⁻   Na⁺

(d) The trans epoxide must come from epoxidation of a trans alkene. In turn, the trans alkene comes from a stereospecific reduction of an alkyne, in this case, 4-octyne.

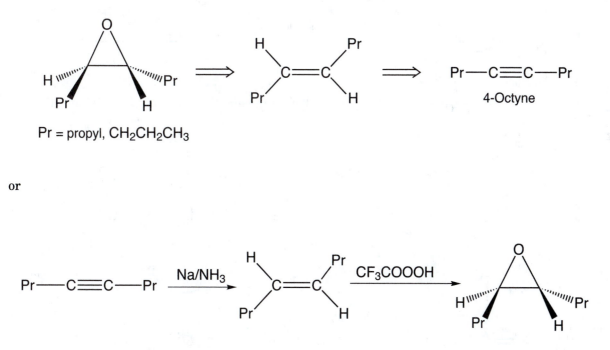

Pr = propyl, CH₂CH₂CH₃

or

4-Octyne can be made from the acetylide of 1-pentyne and propyl bromide. Both reagents were made in (b).

Problem 11.71   In addition reactions of bromine, there is always a competition between formation of a cyclic bromonium ion and an open carbocation. Anything that will favor one over the other may tip the balance in favor of that mechanism. In this case, formation of an open ion is favored by resonance stabilization.

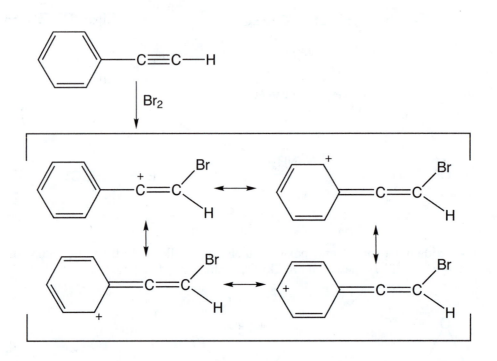

Now, addition of the bromide ion can be to either lobe of the empty $2p$ orbital, thus giving both stereoisomers.

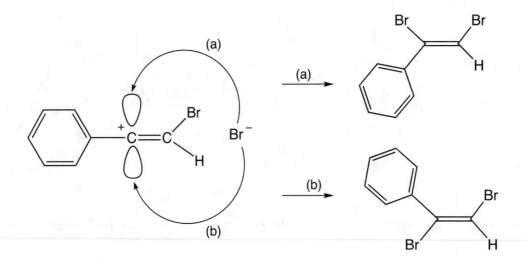

If a cyclic bromonium ion had been the intermediate, only the trans dibromide could have formed. Because both cis and trans diastereomers are formed, the open cation must be involved.

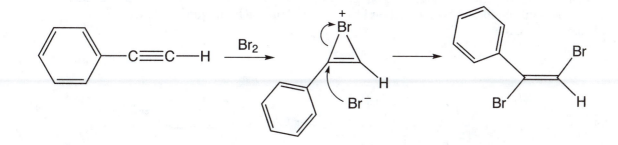

Problem 11.72   The intermediate carbocation is relatively stable. Notice that the energy of the intermediate is relatively close to that of the starting material.

Resonance structures (**A–F**) are shown here. Based on the LUMO representation, it appears that structures **C** and **D** contribute most. Perhaps structure **F** contributes least. It is clear that the structures **B**, **E**, and **F** contribute less than do **A**, **C**, and **D**. Notice that **A**, **C**, and **D** maintain three double bonds in one six-membered ring (aromaticity). Structures **B**, **E**, and **F** do not have any aromatic rings. This lack is a likely reason for the higher weighting factor of **A**, **C**, and **D**.

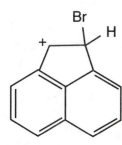

A

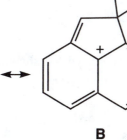

B

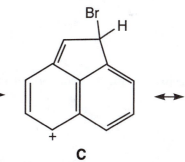

C

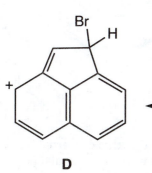

D

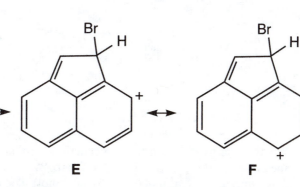

E

F

Problem 11.73 The LUMO density for the bromonium ion in this reaction is located on the bromine of the bromonium ion (labeled 1 on structure in the figure), on the left-side carbon of the three-membered ring (labeled 2), on the right-side carbon of the three-membered ring (labeled 3), and on the hydrogen that is on the adjacent carbon to the three-membered ring (labeled 4). The reaction of bromide with the LUMO at position 1 is just the reverse reaction of the formation of the bromonium ion. Reaction at the LUMO on either of the carbons (labeled 2 or 3) gives the enantiomeric trans dibromides shown. Reaction at the LUMO on the hydrogen labeled 4 gives 3-bromocyclohexene.

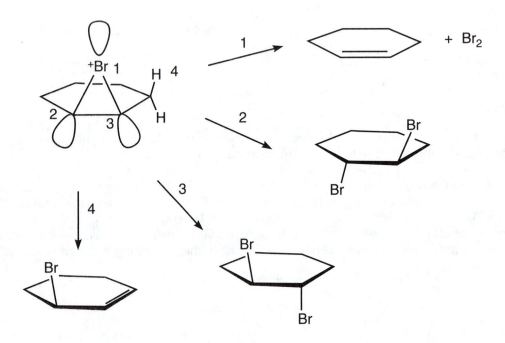

The mechanism for the reaction with the hydrogen labeled 4 is shown.

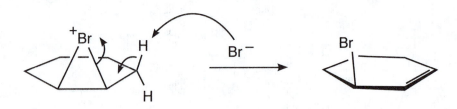

Problem 11.74 Yes, the energy diagram for this reaction is consistent with a concerted reaction. There are no intermediates in the diagram. In Table 7.2 (p. 277), we can see that the alkene π bond and the O—O bond are worth 117 kcal/mol (66 + 51 kcal/mol). The epoxide has two C—O bonds at 92 kcal/mol each. Even with the strain of the three-membered ring (27 kcal/mol), the reaction with hydrogen peroxide should be favored. In practice, hydrogen peroxide works better with electron-poor alkenes such as the enones we will see in Chapter 19.

Radical Reactions $\quad$ **12**

R adical reactions provide new opportunities and new difficulties. Radical reactions allow you to reverse the regiochemistry of addition of hydrogen bromide to alkenes, for example, or to introduce functionality at the position adjacent to a double bond. The following problems allow you to practice such things. Many more radical reactions are known than are discussed in this chapter, and these problems allow you to find a few. Radical chemistry, though different from the polar chemistry we have emphasized so far, nonetheless can be understood through an analysis of structure and orbitals. Keep in mind that chain processes abound in radical chemistry. In such reactions, a chain-carrying radical is produced in one of the propagation steps. In a sense, this is analogous to the regeneration of a catalyst at the end of a polar reaction. There are clues to the presence of a radical reaction: The presence of peroxides or *N*-bromosuccinimide (NBS) is one clue, for example.

Problem 12.1

(a) The nucleophile is the oxygen of the alcohol, specifically one of the lone pairs of electrons. The electrophile is the acidic hydrogen of HBr, specifically the antibond of the H—Br sigma bond.

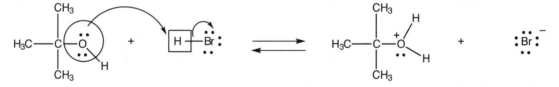

(b) The nucleophile is the π bond of the alkene. The electrophile is the acidic hydrogen of HBr, specifically the antibond of the H—Br sigma bond.

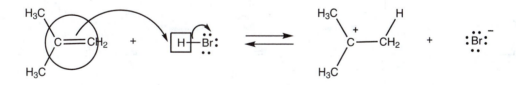

(c) The nucleophile is the iodide ion, specifically one of the lone pairs of electrons of the iodide. The electrophile is the carbon bonded to the Br, specifically the antibonding orbital of the C—Br bond.

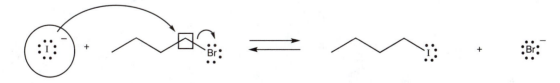

Problem 12.2

(a)

(b) This radical is called a benzyl radical. It is a particularly stable radical.

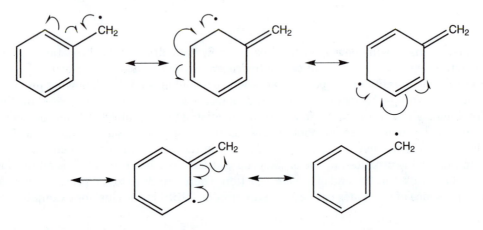

(c) The carbon radical will be in a *p* orbital in order to delocalize with one of the lone pairs of the oxygen. The resonance structure we draw for that mixing has nine electrons around the oxygen. It is a π bond with three electrons (two electrons in the π bonding orbital and one electron in the π antibonding orbital).

$$H_3C—CH_2—CH—O—CH_3 \longleftrightarrow H_3C—CH_2—CH=O—CH_3$$

(d) Notice that all five carbons share the radical.

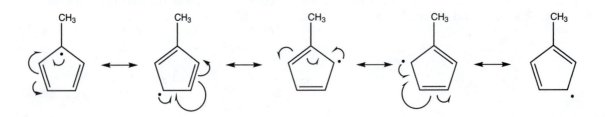

Problem 12.4   The carbon–carbon bond is weaker (90.1 kcal/mol) than the carbon–hydrogen bond (101.1 kcal/mol) and therefore is easier to break.

Problem 12.5 In the transition state for abstraction of hydrogen from butane by a methyl radical, the carbon–hydrogen bond of butane is partially broken and the methyl–hydrogen bond is partially made.

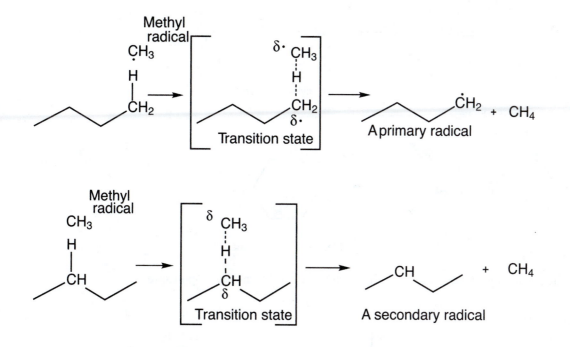

Each of these two reactions is somewhat exothermic, as the carbon–hydrogen bond in methane is stronger than either carbon–hydrogen bond in butane. Accordingly, in each transition state the hydrogen atom is slightly less than halfway transferred (the transition state is slightly "starting material-like"). Recall the Hammond postulate. Do you see why in the transition state for breaking of a secondary carbon–hydrogen bond the hydrogen is slightly less transferred than in the transition state for breaking a primary carbon–hydrogen bond?

Problem 12.6

(a) Disproportionation of this radical will give 1-hexene and the (*E*) and (*Z*) isomers of 2-hexene.

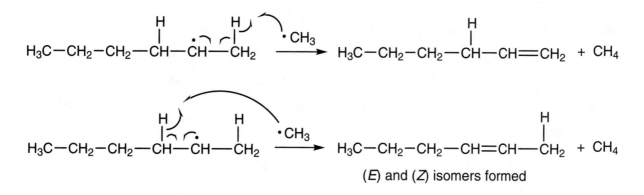

(*E*) and (*Z*) isomers formed

(continued)

Problem 12.6 *(continued)*

(b) Disproportionation of this radical will give 1-methylcyclohexene and methylenecyclohexane.

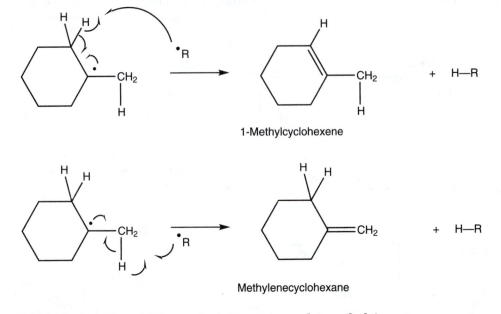

1-Methylcyclohexene

Methylenecyclohexane

(c) This radical will give (*E*) and (*Z*) 4-methyl-2-pentene and 2-methyl-2-pentene.

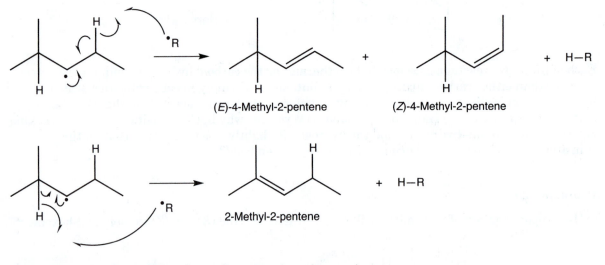

(*E*)-4-Methyl-2-pentene   (*Z*)-4-Methyl-2-pentene

2-Methyl-2-pentene

(d) This benzyl radical will undergo disproportionation to give styrene.

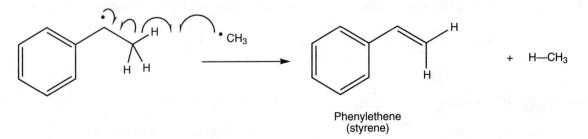

Phenylethene
(styrene)

Problem 12.7 In a β-cleavage reaction, one molecule cleaves into two, causing an increase in entropy. Recall that $\Delta G° = \Delta H° - T\Delta S°$, which means that the entropy factor will become more important as the temperature increases. We see from Table 7.2 that the BDE associated with formation of a secondary radical is 98.6 kcal/mol, and the BDE for formation of the methyl radical is 105 kcal/mol, which means that the products are about 6.4 kcal/mol more stable based on just the stability of the radical intermediates. We do break a σ bond and form a π bond in this β cleavage. In Table 7.2, we see that breaking the C—C bond costs 89 kcal/mol and forming the π bond gains 66 kcal/mol, which means the starting material is 23 kcal/mol more stable based on the bonds broken and formed.

Bond strengths:	favored by 23 kcal/mol	
Radical stability:		favored by 6.4 kcal/mol
Entropy:		favored at higher temp

It must be the entropy factor that favors the β-cleavage reaction.

Problem 12.8 The metal catalyst provides a surface from which the carbon radicals can abstract hydrogen atoms. Carbon radicals gain hydrogen atoms to become alkanes.

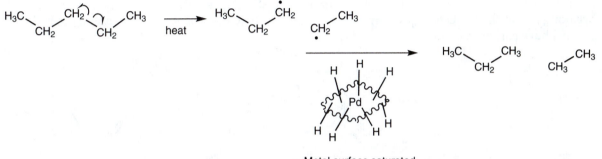

Metal surface saturated
with hydrogen

Problem 12.10 There are, of course, numerous possibilities, and the text will go on to discuss a number of them. The stability of radicals increases with substitution, so one might imagine that the bond dissociation energy of carbon–carbon bonds would decrease as substitution increases. One answer to this problem is as simple as "propane!"

$$H_3C-\{-CH_3 \longrightarrow 2 \quad \cdot CH_3$$

$$H_3C-\{-CH_2CH_3 \longrightarrow \cdot CH_3 + \cdot CH_2CH_3$$

BDE (kcal/mol)
90.1
89.0

Problem 12.11 Lowering of the product energy is accompanied by a lowering of the energy of the transition state leading to product and, therefore, of the activation energy for the reaction.

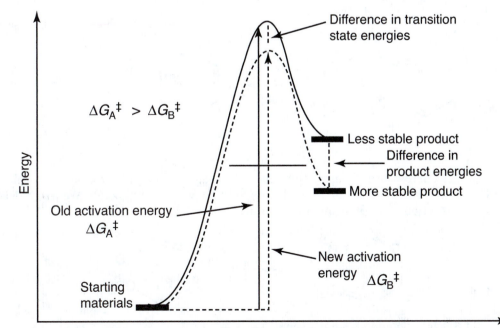

Problem 12.12

(a) The most likely C—C bond cleavage would be the one between C(4) and C(5) of 2-pentene. Cleavage of that bond would give the resonance-stabilized allylic radical and a methyl radical.

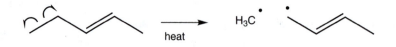

(b) Cleavage between C(3) and C(4) of 4,4-dimethylcyclohexene would form the allyl radical and a tertiary carbon radical. These are the most stable radical intermediates that can be formed by breaking one of the C—C bonds. Cleavage at the other allylic position would give an allylic radical and primary carbon radical.

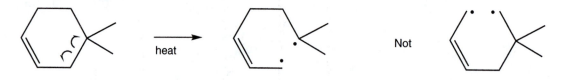

(c) Thermolysis of 6-methyl-3-heptene would give the most stable radicals by cleavage of the C(5)—C(6) bond. That generates an allylic radical and a secondary radical. Cleavage at the other allylic position would give an allylic radical and the less stable methyl radical.

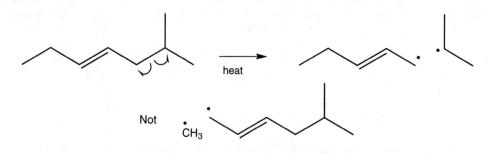

(d) Cleavage of the C(1)—C(2) bond in 1-phenylpropane gives the very stable benzyl radical and the ethyl radical.

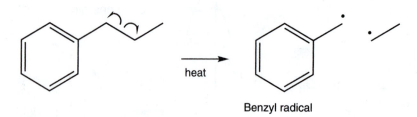

Benzyl radical

Product 12.13   The arrow formalism for this reaction is easy, although little else about this reaction is. The arrows run from the breaking bonds to the new (forming) bonds. We will meet this molecule and this reaction again in exquisite detail in Section 23.6.

1,1,6,6-Tetradeuterio-1,5-hexadiene   3,3,4,4-Tetradeuterio-1,5-hexadiene

Problem 12.14 Azobisisobutyronitrile (AIBN) decomposes to give two resonance-stabilized radicals, whereas simple azo compounds do not. This stabilization of the products by resonance makes it easier to break the bonds in the starting materials.

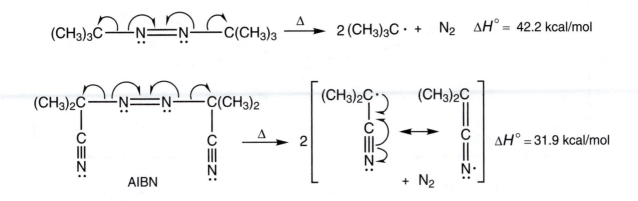

Problem 12.15 The interconversion between the sp^3-hybridized radicals must go though the sp^2 hybrid. The energy diagram shown here is correct if sp^3 is the more stable hybridization for the radical. For allyl and benzyl radicals that can be involved in resonance, the sp^2 form is more stable than the pyramidal sp^3 hybridized atom, as shown on the next page.

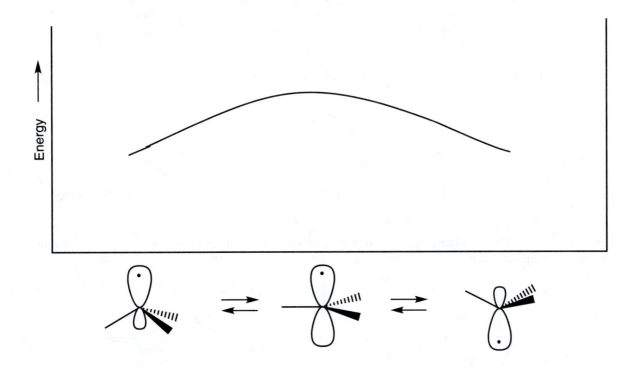

(continued)

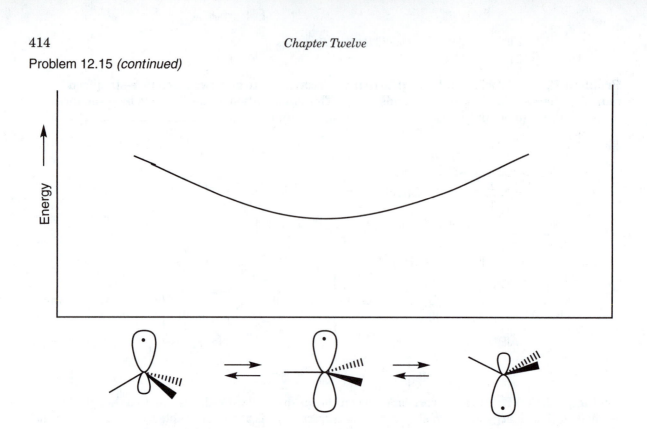

Problem 12.16   Table 12.1 shows the BDE for breaking the C—H bond of propane to give the secondary radical. That value is 98.6 kcal/mol. The closest entry in Table 12.1 for making the primary radical of propane is the value of 101.1 kcal/mol for the ethyl radical. If breaking the primary C—H bond requires about 2.5 kcal/mol more energy than breaking the secondary C—H bond, then we can estimate that the secondary C—H bond is about 2.5 kcal/mol weaker.

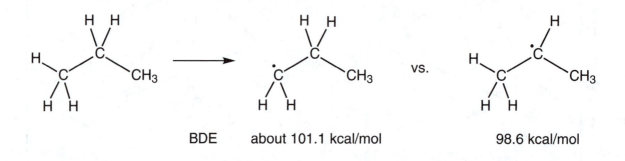

Problem 12.17   The "obvious," "just push 'em together" dimer suffers from severe steric problems. The large groups attached to the central carbons bump into each other.

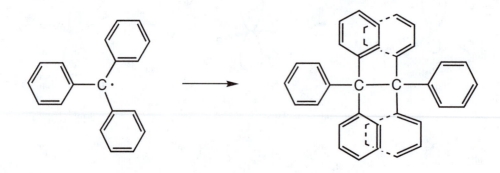

Approach to an "outside" position is not so hindered, and the arrow formalism shows the formation of the real dimer.

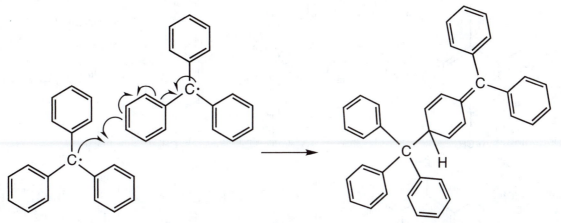

Problem 12.18   The bond formed by abstraction of an H atom is an O—H bond, with an approximate bond energy of 105 kcal/mol (Table 7.2, p. 277). Abstraction of Br would lead to an O—Br bond, which is much weaker, about 56 kcal/mol. Formation of the O—H bond is exothermic, whereas formation of the O—Br bond is endothermic.

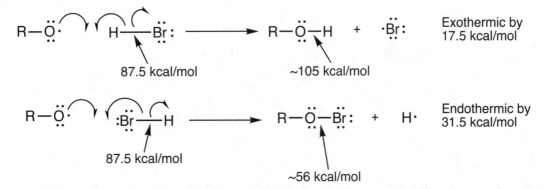

Problem 12.19   One possible reaction between BHT and a radical is hydrogen abstraction from the benzylic position to form a benzylic radical. An alternative hydrogen abstraction pathway would involve abstraction of the O—H hydrogen, resulting in formation of the phenoxy radical.

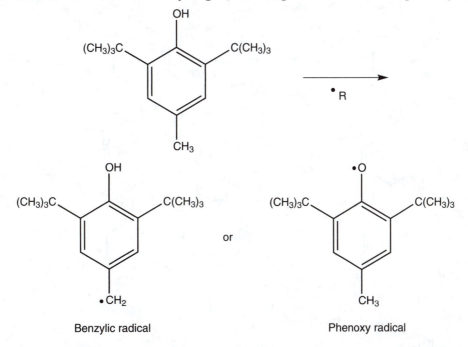

Benzylic radical   or   Phenoxy radical

(continued)

Problem 12.19 *(continued)*

The benzylic radical has several resonance structures that would help stabilize such an intermediate.

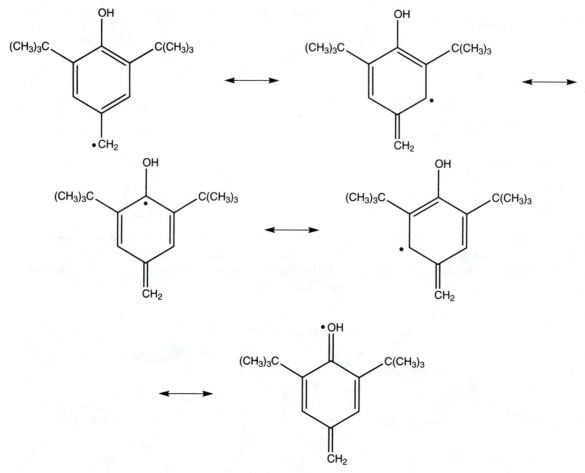

However, the phenoxy radical allows for more substituted radicals in the resonance structures. In addition, the phenoxy radical avoids the steric interaction between the O—H hydrogen and the bulky *tert*-butyl groups. It turns out that the phenoxy radical is more stable.

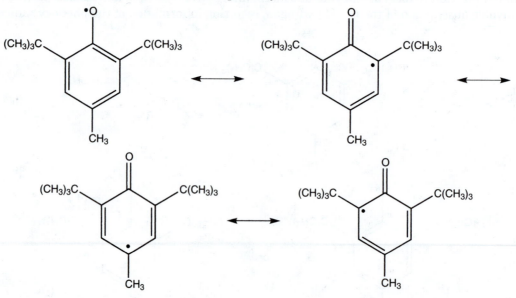

Problem 12.20

(a) Use a retrosynthesis to analyze this question. What do you need to make the desired ether? It could come from *tert*-butyl alcohol (using NaH and CH₃I). Or it could come from 2-methylpropene (using catalytic acid and CH₃OH). The alkene can come from the 1-bromo-2-methylpropane starting material.

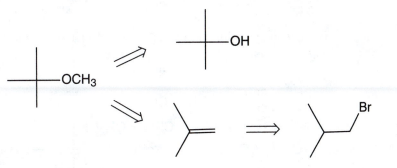

The forward solution is the following:

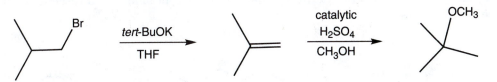

(b) The 2-methyl-1-propanethiol could come from 1-bromo-2-methylpropane via an S$_N$2 reaction with NaSH. We can make the primary bromide from 2-methylpropene, and the alkene can come from the *tert*-butyl bromide.

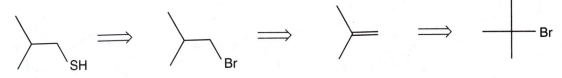

The forward solution is:

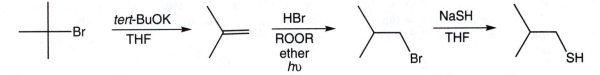

(c) The methoxy ether can be formed by an S$_N$2 reaction on a primary bromide, or it could come from an alkoxide reacting with methyl iodide.

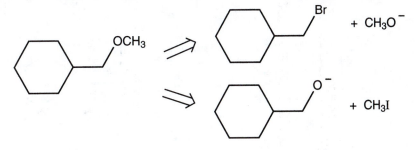

One way to make the ether:

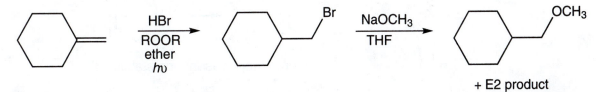

(continued)

Problem 12.20 *(continued)*

(d) Another way to make the ether with no E2 complications:

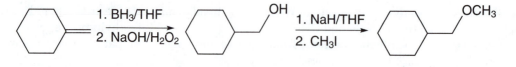

Problem 12.21   Many exist. The destruction of any chain-carrying radical will terminate the chain reaction. Here are three examples.

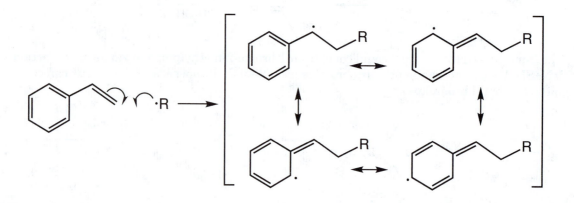

Problem 12.22   Addition of a radical to styrene gives an exceptionally well resonance stabilized and hence relatively stable radical.

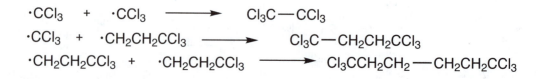

Problem 12.23   An initiator radical (IN·) can add to styrene to give two possible intermediate radicals. Resonance stabilization makes the radical formed from addition at the β position much more stable than that produced from addition at the α position.

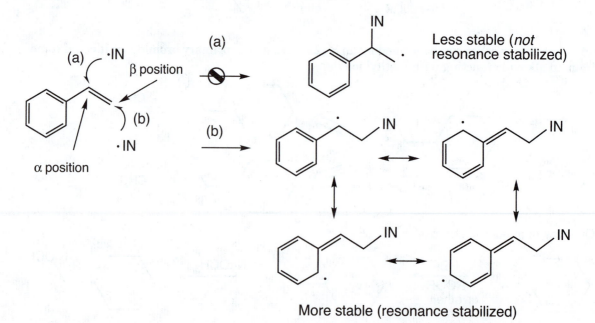

More stable (resonance stabilized)

Now, this new radical can add to styrene as the initiator radical did. The same two choices exist: addition along (a) or (b). Again, and for the same reasons, (b) is favored because it gives the resonance-stabilized intermediate.

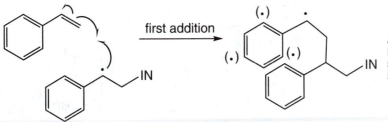

first addition

A resonance-stabilized radical with the positions sharing the single electron shown by (•)

Repeated additions in the (b) sense lead to the polymer

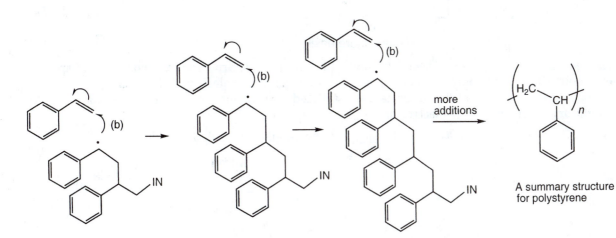

more additions

A summary structure for polystyrene

Problem 10.13 shows the polymerization of styrene under acid-catalyzed conditions. The pathway is the same, except the intermediates are cations rather than radicals. Cationic polymerization is more sensitive to traces of water and other impurities than radical polymerization. Radical polymerization is typically the process that industry uses to make polystyrene.

Problem 12.24   Addition of a radical to the vinyl halide can lead to either of two new radicals. In the more stable radical, the free electron is adjacent to a halogen and is resonance stabilized. The other possible radical is not resonance stabilized and is less stable.

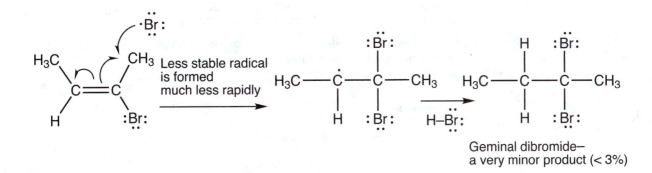

Less stable radical is formed much less rapidly

Geminal dibromide— a very minor product (< 3%)

(continued)

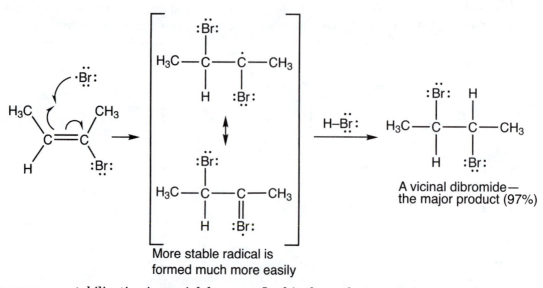

More stable radical is
formed much more easily

This resonance stabilization is special, however. In this three-electron system, two electrons are well stabilized in a low-energy bonding molecular orbital, but one electron must occupy an antibonding molecular orbital. This system is stabilized overall, but the situation is not as straightforward as many two-electron cases.

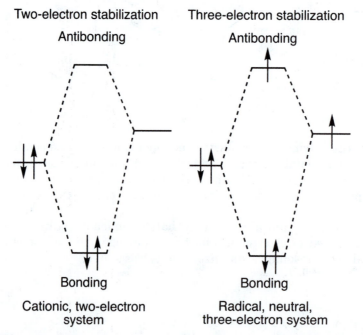

Two-electron stabilization

Three-electron stabilization

Cationic, two-electron
system

Radical, neutral,
three-electron system

Problem 12.25 Abstraction of the methyl hydrogen leads to an unstabilized primary radical, whereas abstraction from the methylene position leads to a resonance-stabilized radical. This resonance stabilization is a three-electron π system.

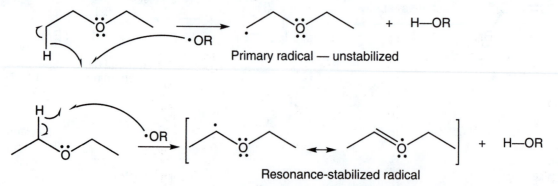

Problem 12.26 When the carbon–hydrogen bond in methane breaks, a methyl radical and a hydrogen atom are produced. Methyl chloride gives a hydrogen atom and a chloromethyl radical, which is stabilized by resonance and, therefore, more stable than the unstabilized methyl radical. Accordingly, the bond is easier to break. Are you getting the idea that one of the generic answers to questions in organic chemistry is, It's resonance stabilized and therefore more stable? Well, you're right.

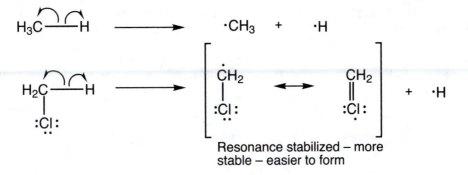

Resonance stabilized – more
stable – easier to form

Problem 12.27

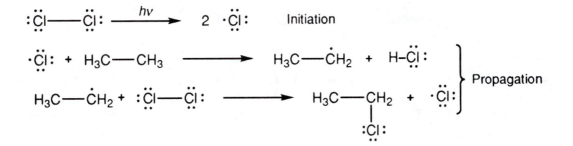

There are many possible termination steps.

Problem 12.28 The first step is exothermic by 2.1 kcal/mol (103.2 – 101.1 = 2.1). The hydrogen–chlorine bond is stronger than the carbon–hydrogen bond in ethane.

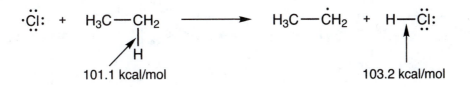

The second propagation step is also exothermic, this time by 25.8 kcal/mol (84.8 − 59 = 25.8).

Problem 12.29

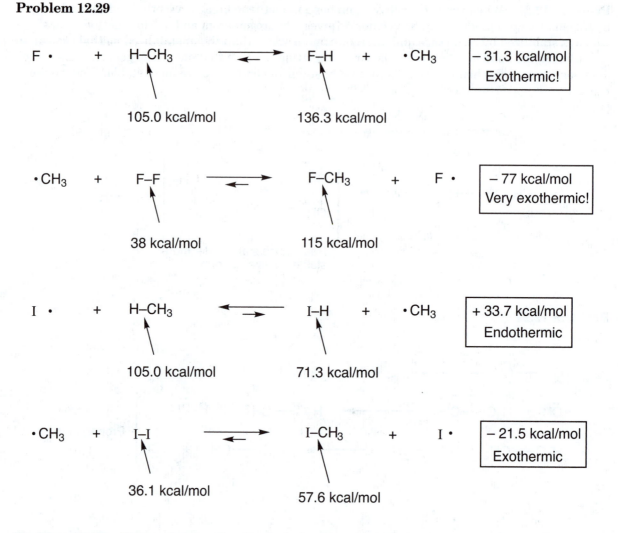

Problem 12.30   The mechanism for photohalogenation of labeled propene involves an unsymmetrical allyl radical. Reaction of the allyl radical with Br_2 gives two products. Both products are 3-bromopropene, but they differ in the location of the ^{13}C label. One has the ^{13}C on carbon 3 and the other has it on carbon 1.

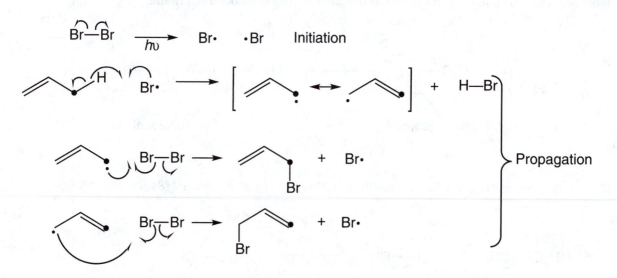

The molecular orbital (MO) analysis gives us the same answer, of course. It is just another (and probably more accurate) way to analyze the reaction. Allyl is a three *p*-orbital system. The molecular orbitals obtained from mixing the three *p* orbitals can be represented as shown in the following diagram. See p. 447 in the text for a discussion of the allyl orbitals. There are three π electrons in the allyl radical. The orbital that will be used by the allyl radical to make a bond with bromine is the SOMO (singly occupied molecular orbital). It should be clear that the orbital density for the SOMO will be localized on carbons number 1 and number 3. This is where the radical will begin to form a new bond to the Br_2 in the propagation step.

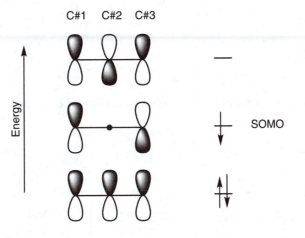

Problem 12.31   Aha! This is a great question. If you have been wondering about this issue, then you are on top of the subject. First, let's review the HBr/ROOR/hν reaction. The reaction of 2-methyl-propene, for example, gives 1-bromo-2-methylpropane.

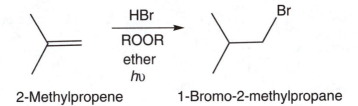

2-Methylpropene   1-Bromo-2-methylpropane

We know that the bromine radical adds to the alkene in the first propagation step of this reaction so that the tertiary radical intermediate is formed. Rather than add to the alkene, why not abstract an allylic hydrogen to give the very stable allyl radical? The answer is that, in fact, allylic hydrogen abstraction is almost always favored and it does happen.

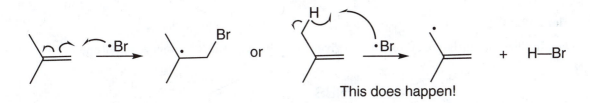

This does happen!

The difference is that in the HBr reaction, the allylic radical can only abstract an H from H—Br, which just reforms the starting material. Only in Br_2 can the allylic position react to give the allylic bromide product.

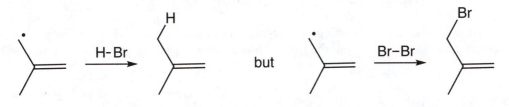

Problem 12.32 The carbonyl oxygen is a base and can be protonated by hydrogen bromide.

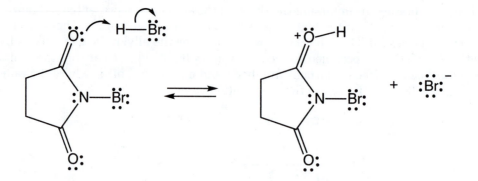

Now bromide can form Br$_2$ by attacking the protonated NBS at bromine:

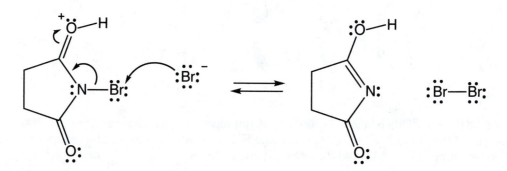

Succinimide itself is produced by a series of proton transfers. Can you write a mechanism for this reaction? (See Section 11.8, p. 523.)

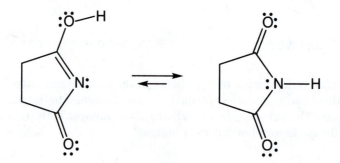

Problem 12.33

(a) The allylic bromination of cyclopentene gives only the racemic 3-bromocyclopentene.

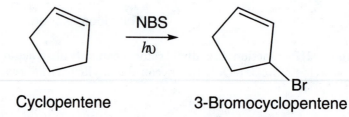

Cyclopentene   3-Bromocyclopentene

(b) Allylic bromination of 1-butene will give racemic 3-bromo-1-butene and the isomer (*E*)-1-bromo-2-butene because the intermediate allylic radical has a resonance structure that can also add bromine.

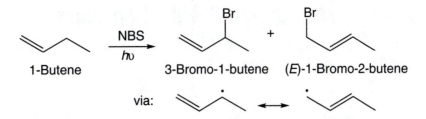

(c) Allylic bromination of (*E*)-3-heptene gives three products. The initial radical can be formed on C(2) or C(5) of the 3-heptene. Both of these initial radicals are resonance stabilized and that means in principle that there can be four products. The reason that there are only three is that the resonance structure for the radical on C(5) is symmetrical. Capture at either end of the allylic radical gives (*E*)-5-bromo-3-heptene.

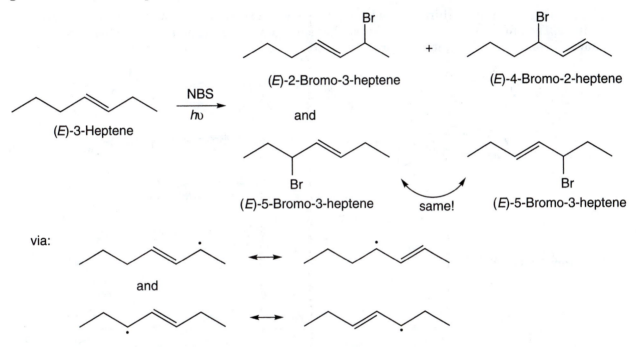

(d) This reaction is an example of bromination at the benzylic position. It is a very clean reaction. Only one product is formed. The radical intermediate certainly has resonance stabilization, but the only reactive sight is the benzylic position because it is the only product that maintains the aromatic ring. We will discuss the amazing stability of the aromatic ring in Chapter 14.

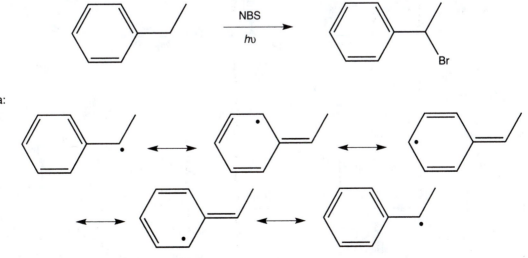

Additional Problem Answers

Problem 12.36 The allylic bromination of cyclopentene gives 3-bromocyclopentene. This molecule is chiral. Carbon 3 is a stereogenic carbon.

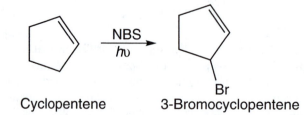

The starting material is achiral, so we know that we will form a racemic mixture of 3-bromocyclopentene. The bromine will add to the flat allylic radical intermediate on either the top face, as shown, forming the (*S*) enantiomer, or on the bottom face, forming the (*R*) enantiomer.

Problem 12.37

Problem 12.38

Problem 12.39

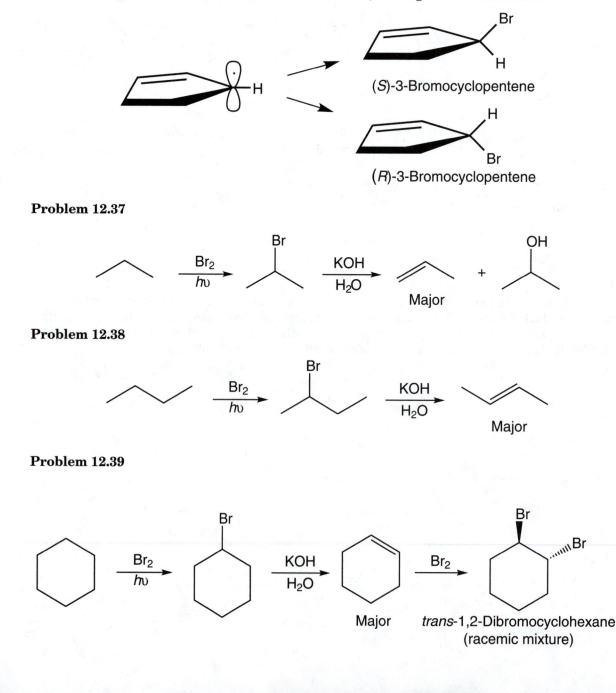

Problem 12.40

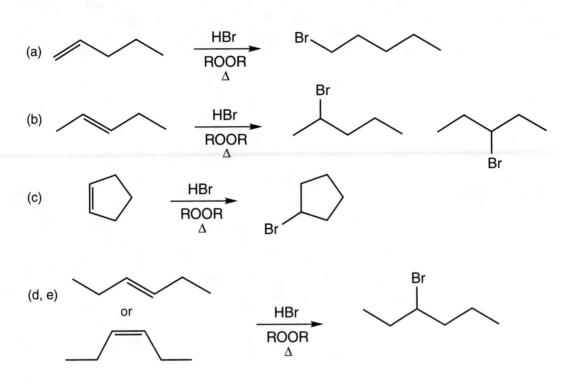

(a)

(b)

(c)

(d, e)   or

Problem 12.41

(a) There are three possible monochlorides that can be formed in this reaction.

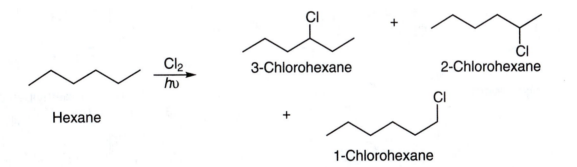

Hexane

3-Chlorohexane   +   2-Chlorohexane

+

1-Chlorohexane

(b) Cyclopentane will give only one monochloride in this reaction. All carbons of cyclopentane are identical.

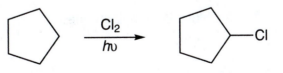

Cyclopentane   Chlorocyclopentane

(continued)

Problem 12.41 *(continued)*

(c) There are only four different carbons in methylcyclopentane. Chlorination at these different carbons gives (chloromethyl)cyclopentane, 1-chloro-1-methylcyclopentane, 1-chloro-2-methyl-cyclopentane, and 1-chloro-3-methylcyclopentane. But the disubstituted ring compounds (1-chloro-2-methylcyclopentane and 1-chloro-3-methylcyclopentane) can be formed as cis or trans isomers. The 1,2- and 1,3-disubstituted cyclopentanes will be formed as racemic mixtures.

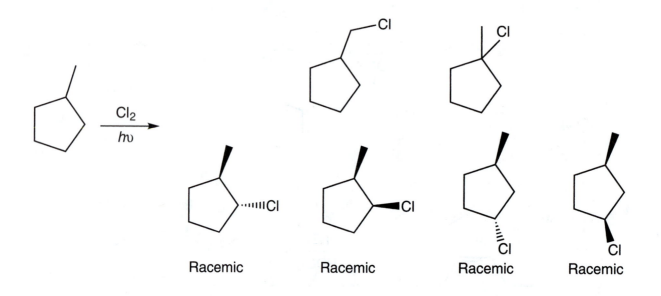

Racemic   Racemic   Racemic   Racemic

(d) There are five monochlorides that can be formed in this reaction. The three that are chiral would be formed as racemic mixtures.

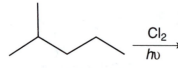

2-Methylpentane

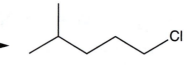

1-Chloro-4-methylpentane

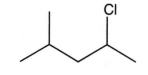

2-Chloro-4-methylpentane
(racemic)

+

3-Chloro-2-methylpentane
(racemic)

+

2-Chloro-2-methylpentane

+

1-Chloro-2-methylpentane
(racemic)

Problem 12.42

Problem 12.43

Only this one is chiral and it will
be formed as a racemic mixture

Problem 12.44

(a)

HBr / ROOR / ether / hʋ

Radical reaction; a racemic
mixture of both cis and
trans isomers will be formed

(b)

HBr / ether

Ionic addition of HBr
(Markovnikov addition)

(c)

Br₂ / CCl₄

Anti addition of Br₂ via bromonium
ion; racemic mixture formed

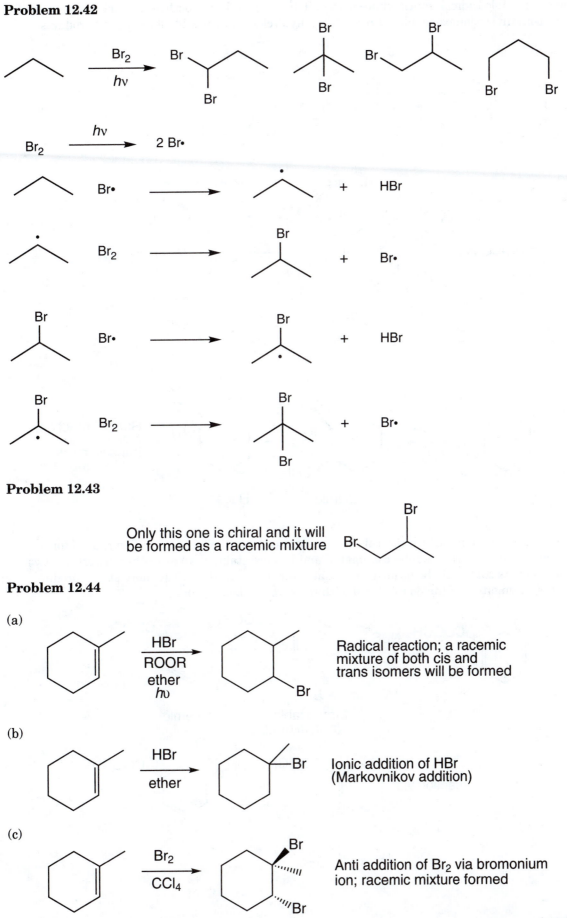

(continued)

Problem 12.44 *(continued)*

(d) The most stable radical intermediate is the allylic radical that is both secondary and tertiary. So 3-bromo-1-methylcyclohexene and 3-bromo-3-methylcyclohexene would be the major products.

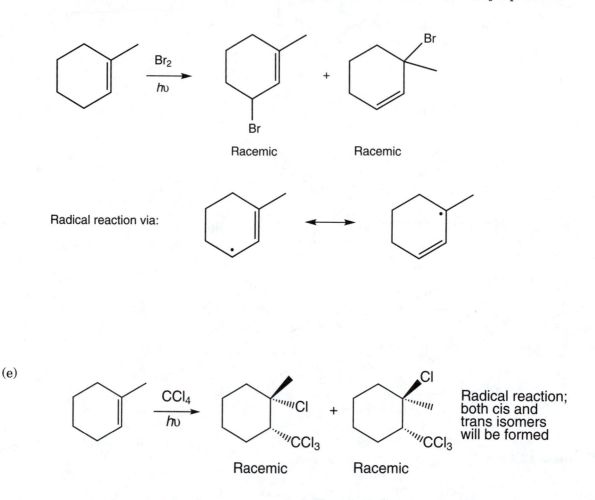

Racemic   Racemic

Radical reaction via:

(e)

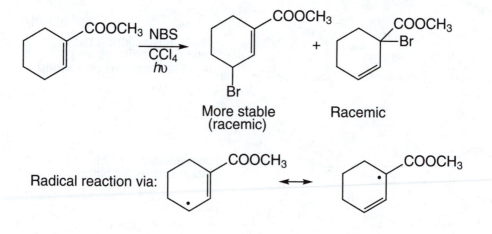

Racemic   Racemic

Radical reaction; both cis and trans isomers will be formed

(f) First we need to decide which radical intermediate is most likely involved. The radical formed at C(3) of the cyclohexene ring is fully conjugated and is more stable than any other. There are two bromide products that could be formed from that radical intermediate. The more stable product retains the conjugation of the double bond with the ester carbonyl group.

More stable (racemic)   Racemic

Radical reaction via:

(g) The alkyne reacts with HBr forming 2-bromo-1-butene. A second addition occurs via the bromine-stabilized carbocation. The product is the achiral geminal dibromide.

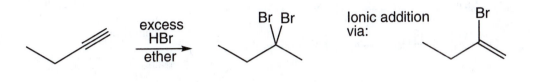

(h) This reaction is a radical process that gives the vicinal dibromide. The bromine radical formed from the initiation steps adds to the alkyne to give the most stable vinyl radical intermediate. The vinyl radical will abstract hydrogen from H—Br. The newly formed alkene (cis and trans) will react with another bromine radical to give a radical on the carbon bonded to bromine, which allows resonance stabilization. The last step will be the carbon radical abstracting a hydrogen from H—Br. The final product would be a racemic mixture.

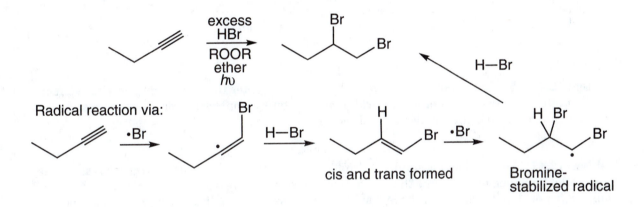

Problem 12.45 The two products formed in this reaction will be 4-bromocyclohexene and 3-bromocyclohexene.

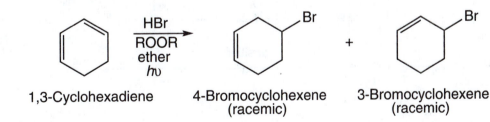

Problem 12.45 *(continued)*

The mechanism for the reaction is shown here:

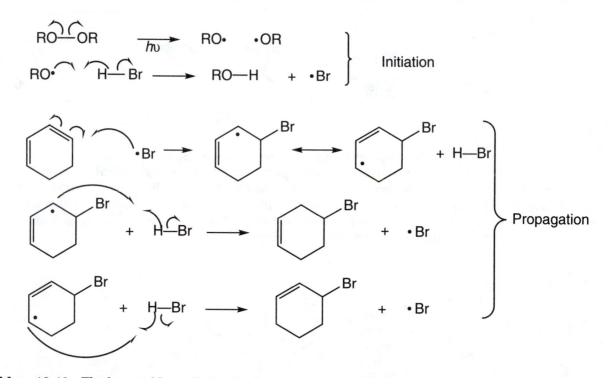

Problem 12.46 The less stable methyl radical recombines faster than the more substituted, more stable, and thus longer lived isopropyl radical. The more bulky isopropyl radical has more difficulty in recombining than does the smaller methyl radical.

Problem 12.47 This problem is similar to Problem 12.46, although the steric argument is a little more subtle. First, one might argue that formation of the more substituted, more stable radicals should be faster than formation of the less substituted, less stable radicals. That idea would be right, as the transition state for radical formation will have partially formed radicals and will benefit energetically from substitution.

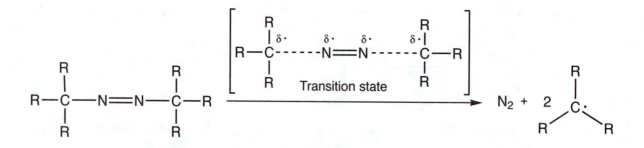

But this analysis ignores an even more subtle steric argument. There will be an energy incentive for the larger R groups to go from sp^3 to sp^2 hybridization as the azo compound is transformed into nitrogen and a pair of radicals. The R–C–R angle will change from about 109° to about 120°. The more relatively large methyl groups, the stronger this effect, and the faster the rate of radical formation.

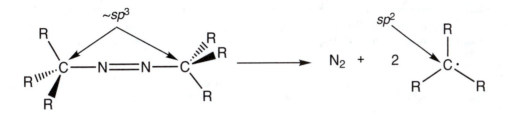

Problem 12.48 The major product for the bromination of (*R*)-3-methylhexane is 3-bromo-3-methylhexane. This product is chiral. It will be formed as a racemic mixture because the radical intermediate will lose its stereochemical purity. The radical easily goes from sp^3 to sp^2. Once it becomes planar it has no preference for returning to the (*R*) configuration or becoming the (*S*) enantiomer. Reaction of the radical with Br_2 will occur from either face of the tertiary radical to give the racemic mixture.

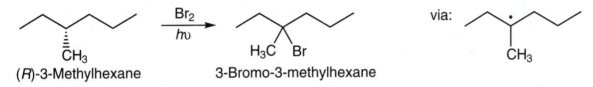

Problem 12.49 There are only two possibilities for hydrogen atom abstraction from ethyl chloride. A simple primary radical would be produced from abstraction of a methyl hydrogen, whereas a resonance-stabilized radical appears when a hydrogen atom is removed from the carbon already attached to one chlorine.

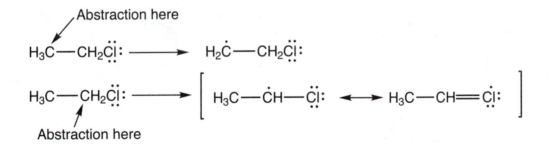

The more stable radical will be formed preferentially and will abstract a chlorine atom to yield a geminal dichloroethane, 1,1-dichloroethane, not the vicinal dichloride, 1,2-dichloroethane.

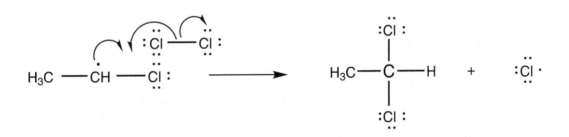

Problem 12.50 Decomposition of NBS generates a bromine atom. The bromine atom then abstracts a hydrogen atom from the methylene group adjacent to both the ring and the carbonyl group. This radical is resonance stabilized. The positions sharing the free electron are indicated by (·).

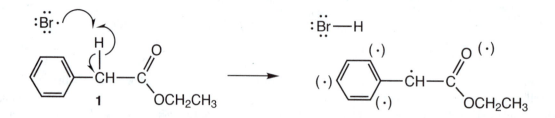

The hydrogen bromide then reacts with NBS to form the low steady-state concentration of bromine necessary to carry the reaction as shown on p. 577. Bromine reacts with the newly formed radical to give the brominated product and regenerate a chain-carrying bromine atom.

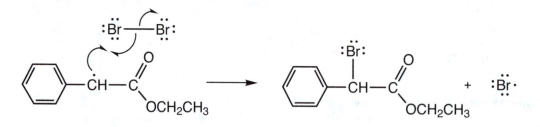

Problem 12.51

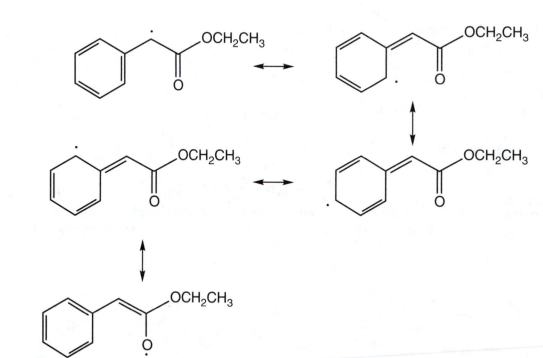

Problem 12.52

(a) Only one allylic bromination product possible.

(b) Two allylic bromination products possible.

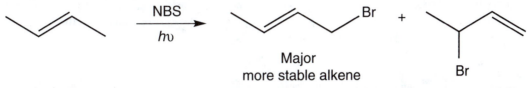

Major
more stable alkene

(c) The allylic hydrogen that would be abstracted is the one that leads to the bromine-stabilized allylic radical. That species has two resonance structures, and each carbon sharing the radical would react with bromine to give the dibromides shown. The 1,3-dibromide is stabilized by resonance and is more stable than the 1,1-dibromide product. The transition state leading to the 1,3-dibromide will be lower and, as a result, the 1,3-dibromide will be the major product.

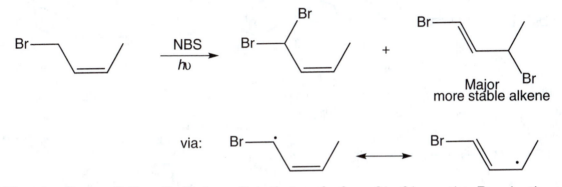

(d) There is only one allylic radical intermediate that can be formed in this reaction. Bromination can occur at either carbon of the allylic radical. The more substituted alkene is the more stable.

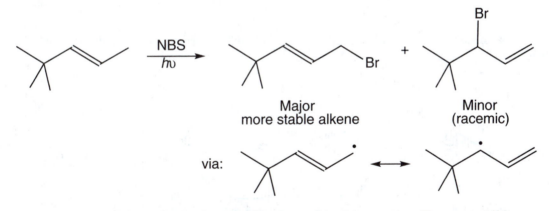

(e) There are several allylic radicals that could be formed in this reaction. The most stable intermediate is the allylic radical that is secondary and tertiary. Two products can arise from that intermediate. The more stable product is the more substituted alkene.

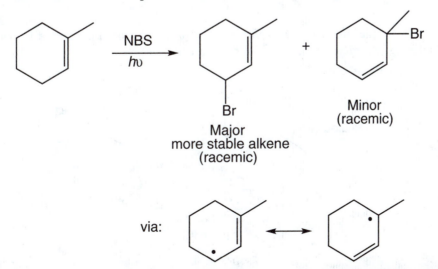

Problem 12.53 The key to this problem is the recognition that the strained three-membered ring will react with a chlorine atom much as will a double bond. An addition reaction gives the radical **B** that then abstracts a chlorine atom from Cl_2. This reaction makes a product molecule and generates a new chlorine atom to carry the chain.

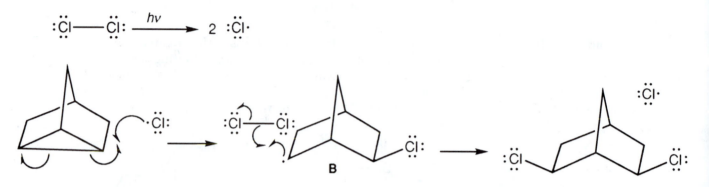

Problem 12.54 Chlorospiropentane is formed by a straightforward photochlorination chain reaction.

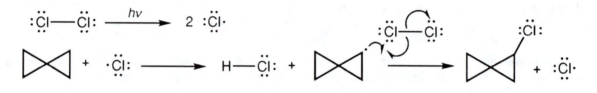

The strained three-membered ring can also be opened by a chlorine atom (see Problem 12.53). Abstraction of a chlorine atom from chlorine leads to one product and a new chlorine atom.

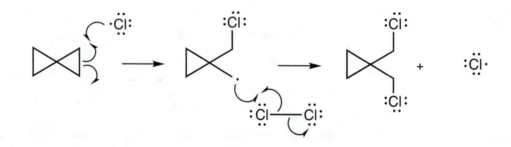

The final product is by far the hardest to rationalize. It comes from opening of the second three-membered ring (a β cleavage), followed by chlorine abstraction.

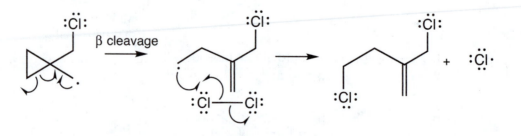

Problem 12.55 The initial addition of an iodine atom to *cis*-2-butene is endothermic and reversible. Addition leads to a radical in which rotation about a carbon–carbon single bond is fast. Reversal of the addition leads to both *cis*- and *trans*-2-butene.

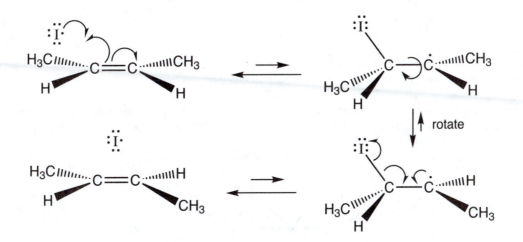

Problem 12.56

(a) Somehow, the radical $F_3CCl_2C^{\cdot}$ must be formed, as the only reasonable mechanism for formation of **2** involves addition of this radical to the alkene, followed by chlorine abstraction.

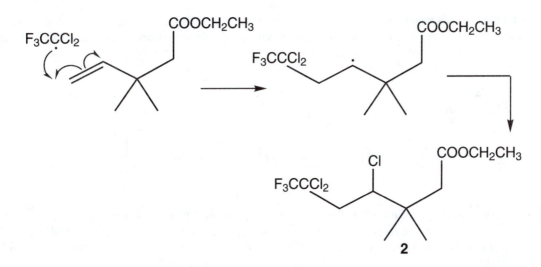

The problems now are to find a suitable source of $F_3CCl_2C^{\cdot}$ and the chlorine atom. An examination of Table 7.2 (p. 277) reveals that carbon–chlorine bonds are weaker than carbon–fluorine bonds. Accordingly, one mechanism that has been proposed is a Cu-catalyzed redox-transfer chain mechanism. This initiation step involves abstraction of a chlorine atom from the polyhalide by CuCl.

$$Cu^{I}Cl \ + \ F_3CCCl_3 \ \longrightarrow \ Cu^{II}Cl_2 \ + \ F_3C\dot{C}Cl_2$$

$Cu^{II}Cl_2$ is presumably a more reactive chlorine donor than the polyhalide. Abstraction of chlorine from $Cu^{II}Cl_2$ gives **2** and $Cu^{I}Cl$. The $Cu^{I}Cl$ is then recycled to begin the process anew.

(continued)

Problem 12.56 *(continued)*

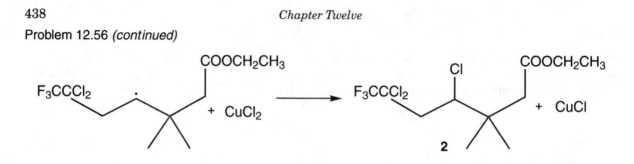

(b) Three steps need to occur in the conversion of **2** into **1**: (1) cyclopropane formation, (2) loss of hydrogen chloride, and (3) hydrolysis of the ester to the carboxylic acid.

(c) There are four possible stereoisomers of carboxylic acid **1**. There is a pair of enantiomeric cis isomers and a pair of enantiomeric trans isomers. The cis (1*R*,3*S*) isomer is shown below:

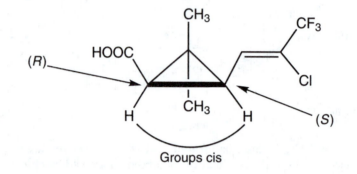

Groups cis

Problem 12.57 This reaction is a standard radical chain process. In the initiation steps, the weak oxygen–oxygen bond breaks to give a pair of alkoxy radicals. Abstraction of a chlorine atom from PCl_3 gives a dichlorophosphorous radical, $\cdot PCl_2$.

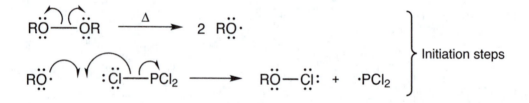

The propagation steps include addition of $\cdot PCl_2$ to 1-octene and abstraction of a chlorine atom from PCl_3. These reactions form a molecule of product and regenerate the chain-carrying radical, $\cdot PCl_2$. Note the regiochemistry of the reaction. The $\cdot PCl_2$ radical adds so as to generate the more stable secondary radical.

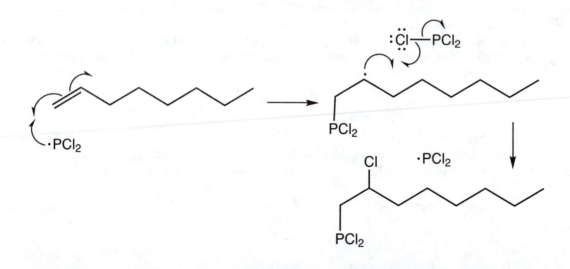

Problem 12.58 Decomposition of the peroxide occurs by breaking the weak oxygen–oxygen bond to give a pair of carboxy radicals:

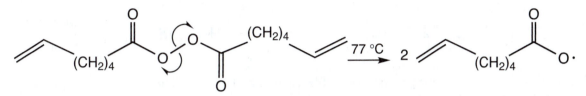

One product comes from simple abstraction of hydrogen from toluene ($PhCH_3$) by this radical.

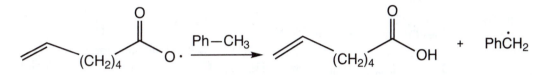

The carboxy radical can also lose carbon dioxide to give radical **A**, the source of most of the other products.

Disproportionation of **A** produces 1-hexene and 1,5-hexadiene, and dimerization of **A** gives the C_{12} product.

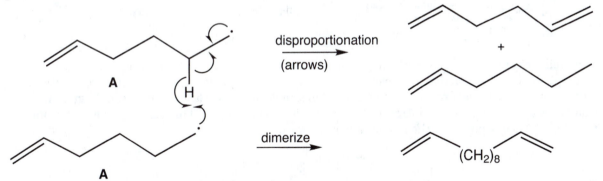

The radical **A** can also undergo an intramolecular cyclization. In fact, this can occur in two ways to give either a five- or six-membered ring. Abstraction of hydrogen yields the cycloalkanes.

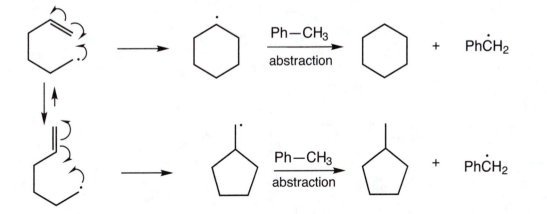

(continued)

Problem 12.58 *(continued)*

In several of these reactions, the benzyl radical, PhĊH$_2$, appears. If it dimerizes, the final product, PhCH$_2$CH$_2$Ph, is formed.

$$2 \text{ PhĊH}_2 \longrightarrow \text{PhCH}_2\text{CH}_2\text{Ph}$$

Problem 12.59 The acyclic product comes from a straightforward radical chain addition reaction. Note that in the last step, the chain-carrying trichloromethyl radical is regenerated.

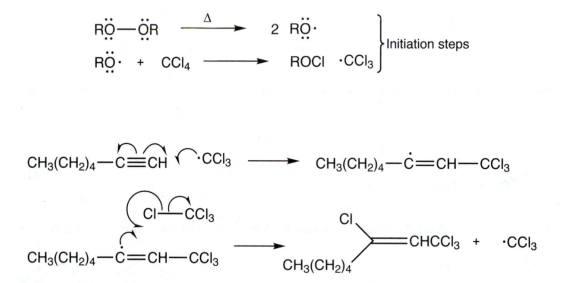

The cyclic product is more difficult. What do we have to do? The two major things that must be accomplished in this reaction are formation of a five-membered ring and loss of a chlorine atom. Formation of the cyclic product starts with an intramolecular hydrogen abstraction to give **A**. Radical **A** then undergoes an intramolecular addition to produce **B**. This reaction completes one major goal: the formation of the five-membered ring.

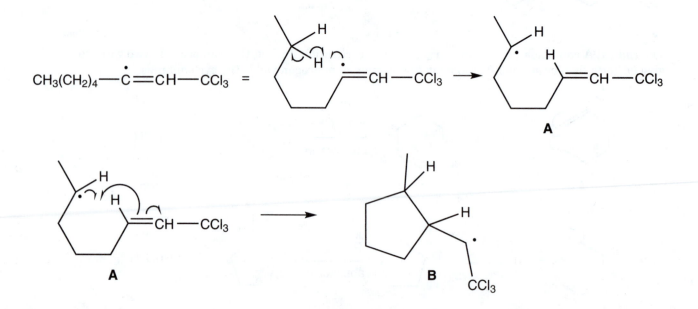

Loss of a chlorine atom finishes the mechanism.

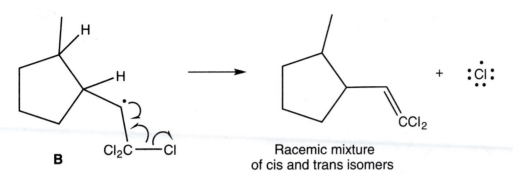

B — Racemic mixture of cis and trans isomers

Problem 12.60 This is another free radical chain reaction. An alkoxy radical formed in the first initiation step abstracts a hydrogen atom from isobutene to generate a resonance-stabilized methallyl radical.

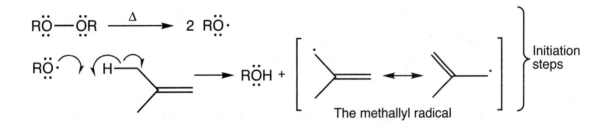

The methallyl radical adds to trichloroethylene to give either radical **A** or **B**. Formation of **A** will be preferred both because it is more stable than **B** (resonance stabilization by two chlorines instead of one) and on the grounds that approach of the methallyl radical to the end of trichloroethylene bearing only one large chlorine will be preferred as a result of steric congestion.

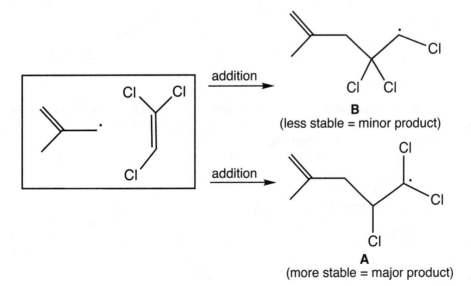

B (less stable = minor product)

A (more stable = major product)

Loss of a chlorine atom from **A**, the predominant radical intermediate, leads to **1**, the major product of the reaction. The minor products come from chlorine loss from the less favored radical intermediate **B**.

(continued)

Problem 12.60 *(continued)*

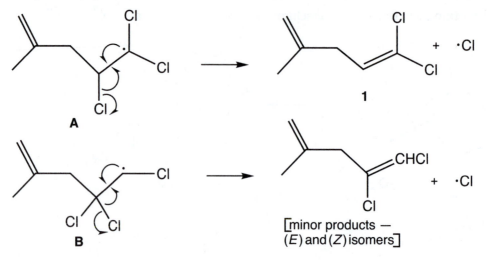

A   **1**

B   [minor products —
                                          (*E*) and (*Z*) isomers]

Problem 12.61   The initiation steps are easy.

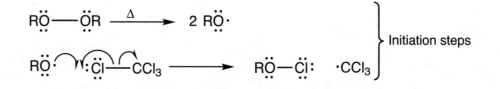

Initiation steps

Addition of the trichloromethyl radical to the exocyclic double bond accomplishes one goal, the formation of a CH_2CCl_3 group.

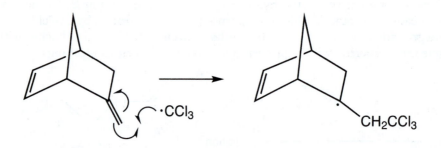

A second goal, the construction of the three-membered ring, is attained through an intramolecular addition reaction. The reaction is completed by abstraction of a chlorine atom from CCl_4 to give **1** and a chain-carrying trichloromethyl radical.

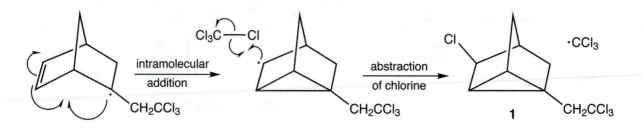

intramolecular
addition

abstraction
of chlorine

Problem 12.62 We'll start by borrowing the molecular orbitals (MOs) for HHH from Problem 1.64 (p. 50).

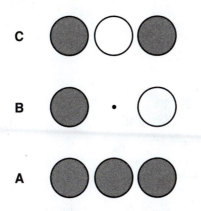

Here are the molecular orbitals for cyclic H_3. We'll build them just as we made the molecular orbitals on p. 50.

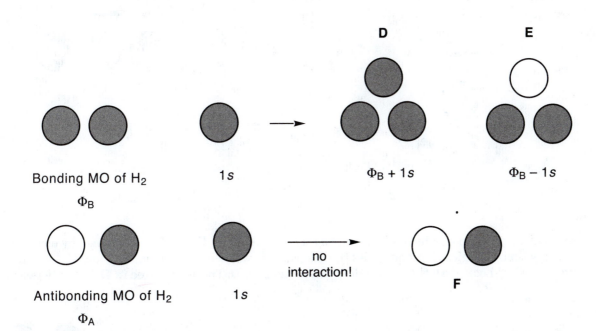

(continued)

Problem 12.62 *(continued)*

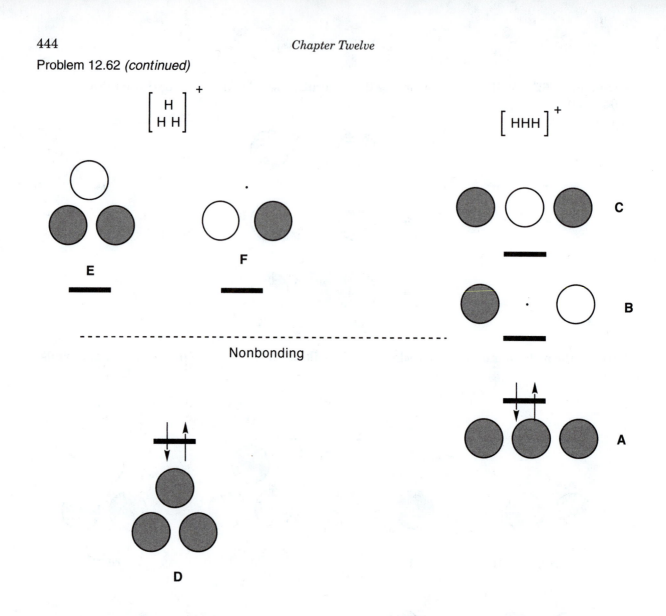

So, the critical question is, which is lower in energy, **D** or **A**? Only **D** and **A** are occupied by electrons, so it is only their energies that matter. As the diagram shows, **D** is lower in energy, and H_3^+ will be triangular. Why? Look at all the 1,2-bonding interactions in **D**. There are three in **D**, but only two in **A**.

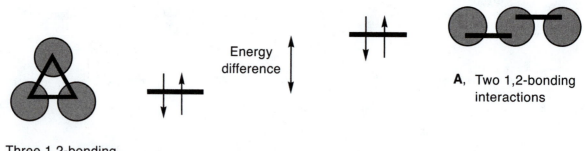

D, Three 1,2-bonding interactions

A, Two 1,2-bonding interactions

Problem 12.63

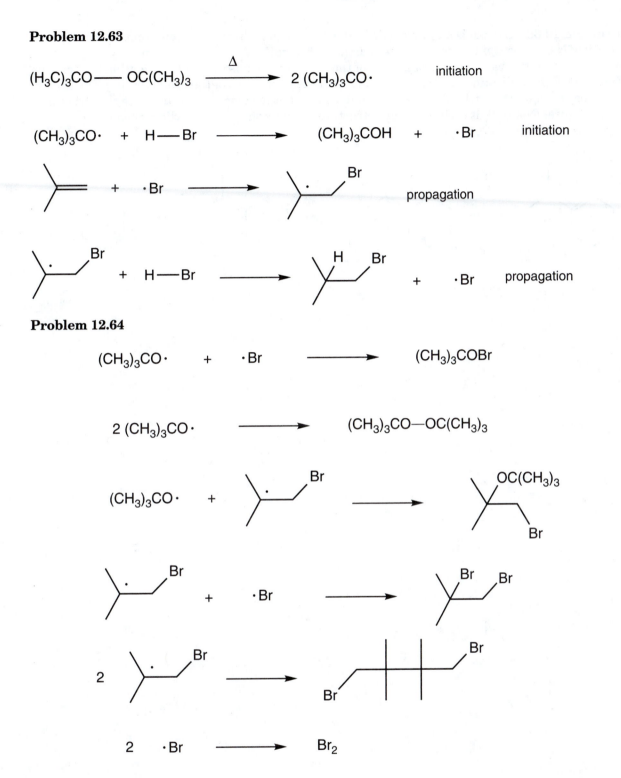

Problem 12.65 The central carbon is sp^2 hybridized in the animation. The sp^2 hybridization maximizes overlap with the sigma bonds on the adjacent carbons, in other words, hyperconjugation. The SOMO track shows the orbital mixing with the sigma bonds on the methyl groups. We have learned that the radical carbon is probably a rapidly inverting pyramid, somewhere between sp^2 and sp^3 hybridization.

Problem 12.66 Oxygen is a diradical. It is able to abstract a hydrogen atom from the benzylic position. No, the benzylic hydrogen is not particularly acidic (pK_a of about 41) and oxygen is not a strong base. In this case, the hybridization of the radical carbon is sp^2, and it is not likely to change because sp^2 hybridization maximizes overlap with the π electrons of the aromatic ring. The carbon must be sp^2 in order to delocalize the radical density into the adjacent π system (see the structures in the figure). Notice that each resonance structure requires that the benzylic carbon be sp^2 hybridized.

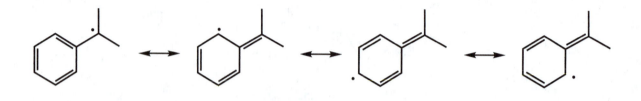

Dienes and the Allyl System: 2p Orbitals in Conjugation

13

This chapter opens a three-chapter sequence in which we will explore the consequences of overlap of more than two 2p orbitals, "conjugation." There are both structural and chemical consequences, and the following problems deal with both areas. Two exceptionally important reactions emerge: the formation of allyl cations through protonation of 1,3-dienes and the Diels–Alder reaction. The former opens up synthetic possibilities and leads to a discussion of thermodynamic and kinetic control of reactions. The latter is arguably the most important synthetic reaction in the chemist's arsenal and leads in a single step to all manner of complex compounds containing six-membered rings. The following problems give lots of practice in both areas.

Problem 13.2

(a) 1,3-Dibromopropadiene is chiral. The mirror image is non-superimposable on the original structure. Notice that there is no plane of symmetry in this molecule.

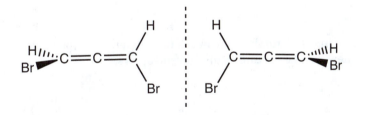

(b) 2-Methyl-2,3-pentadiene is achiral. The mirror image is superimposable on the original structure. The plane of the paper is a plane of symmetry for this molecule.

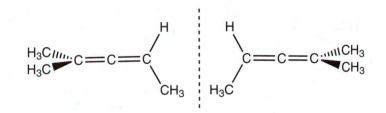

(continued)

Problem 13.2 *(continued)*

(c) 1,3-Dichloro-1,2-butadiene is chiral. The mirror image is non-superimposable on the original structure. There is no plane of symmetry.

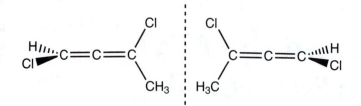

(d) 1,1-Dichloro-1,2-pentadiene is achiral. The mirror image is superimposable on the original structure. There is a plane of symmetry, which is the plane of the paper.

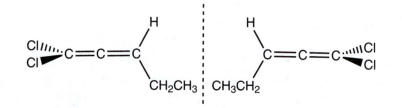

Problem 13.3   All these molecules are related to allenes.

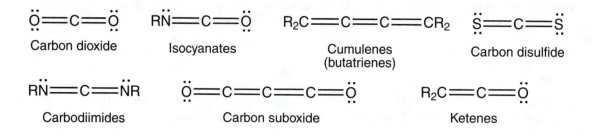

Carbon dioxide   Isocyanates   Cumulenes   Carbon disulfide
                                            (butatrienes)

Carbodiimides   Carbon suboxide   Ketenes

Problem 13.4   Just as in ethylene (but not in allene), the end groups of this molecule are coplanar. Therefore, there can be cis and trans stereoisomers.

In this molecule, the R–C–R groups are coplanar

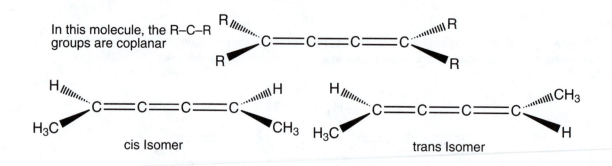

cis Isomer   trans Isomer

These flat molecules are all achiral, just like ethylene. Their mirror images are always superimposable on the originals.

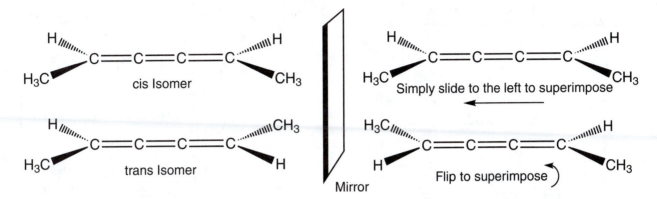

Problem 13.5 The 3,3-dibromohexane can come from 3-hexyne. We know that an alkyne reacts with excess HBr to give the geminal dibromide (p. 520). 3-Hexyne can come from alkylation of 1-butyne with ethyl iodide. The 1-butyne can come from 2-butyne by the process shown in Figure 13.7.

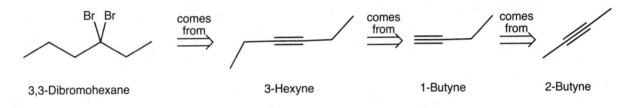

The forward reactions are shown. Because the 3-hexyne is symmetrical, treatment with excess HBr gives only one product.

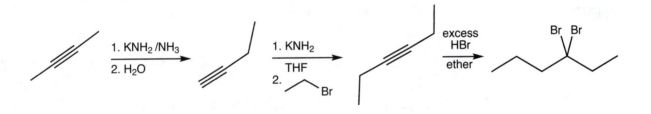

Problem 13.6 This is just a two-stage version of the mechanism outlined in Figures 13.7–13.13. Strong base removes a proton from the carbon adjacent to the triple bond to give resonance-stabilized anion **A**. Reprotonation can occur in two places, one of which gives an intermediate allene. The allene is also acidic, and strong base can remove a proton to produce a second resonance-stabilized anion, **B**. Reprotonation, followed by a third removal of a proton, gives **C**, which protonates to give a second allene. One more cycle of deprotonation to anion **D**, followed by reprotonation, generates 1-hexyne.

(continued)

Problem 13.6 *(continued)*

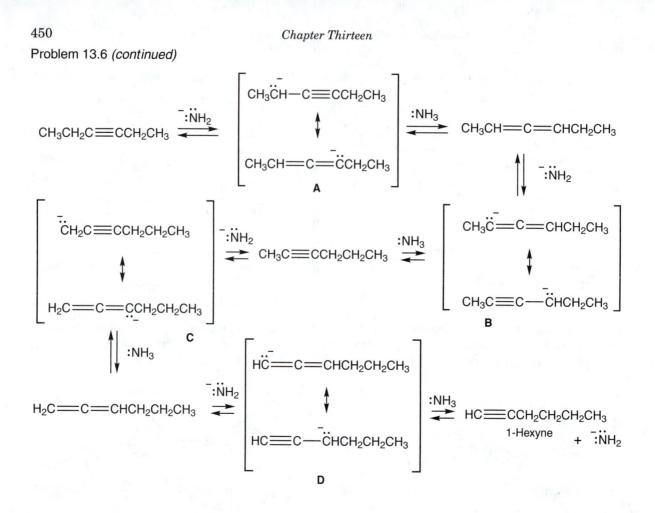

Now the terminal acetylene can lose its acetylenic hydrogen to generate an acetylide, and the reaction stops at this critical point. The 1-alkyne is a *much* stronger acid than any of the other species in equilibrium. Removal of the acetylenic proton is greatly favored thermodynamically. When the solution is quenched through addition of water, the acetylide protonates to produce 1-hexyne under conditions (neutral) to which it is stable.

$$HC\equiv CCH_2CH_2CH_2CH_3 \underset{:NH_3}{\overset{^-:\ddot{N}H_2}{\rightleftharpoons}} {}^-C\equiv CCH_2CH_2CH_2CH_3$$

$$:\ddot{C}\equiv CCH_2CH_2CH_2CH_3 \underset{}{\overset{H_2O}{\rightleftharpoons}} HC\equiv CCH_2CH_2CH_2CH_3$$

Problem 13.8 Each carbon in 1,3-butadiene is sp^2 hybridized. Therefore, each C—C bond is made by mixing an sp^2 orbital from one carbon with the sp^2 orbital of the adjacent carbon.

1,3-Butadiene

Showing the *p* orbitals for the π bonds

sp^2/sp^2 σ Bonds

Problem 13.9

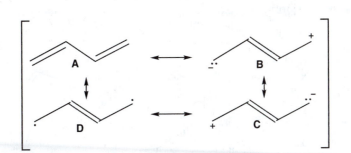

Forms **B**, **C**, and **D** all have one fewer bond than form **A**. In addition, forms **B** and **C** have separated charges. Only **A** has a filled octet for each carbon. Form **A** is by far the major contributing structure.

Problem 13.10

(a) This reaction will be complicated by the reversible nature of alcohol dehydration. The most stable product is the result of two hydride shifts. It is the most stable because it has the most substituted conjugated diene.

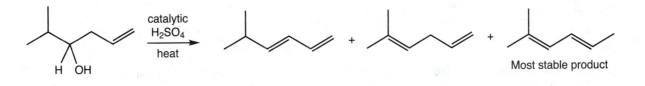

The mechanism for the formation of the most stable product is shown here.

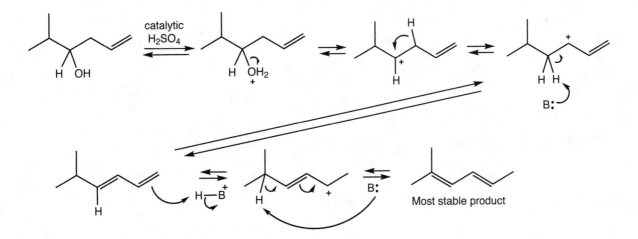

(continued)

Problem 13.10 *(continued)*

(b) An E1 reaction is reversible. With time the most stable diene, which is the conjugated diene with the most substitution in this example, will be formed. The mechanism for its formation is shown.

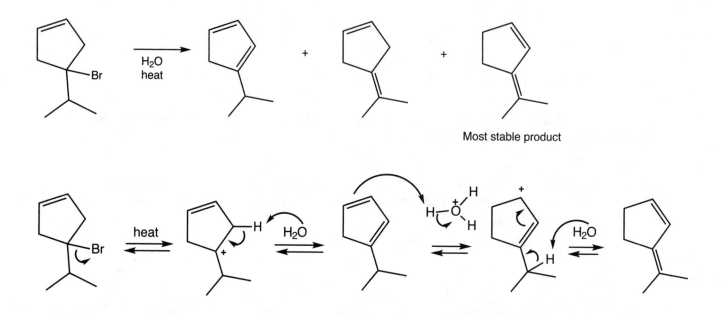

Most stable product

(c) This reaction is an E2 process that requires a β anti-periplanar hydrogen. There are two such hydrogens available, but the allylic hydrogen will be more acidic and more accessible (it is a secondary hydrogen and the other one is tertiary). In addition, the conjugated diene is thermodynamically more stable than the unconjugated diene.

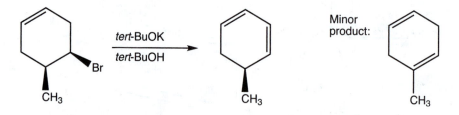

(d) This E1 reaction will favor the conjugated system. The product is called an enone.

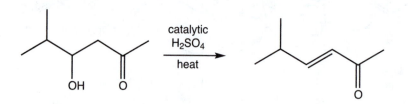

Problem 13.12   The equation to use is $\Delta G = -2.3RT \log K$.

So, $-2.5 = -1.364 \log K$ [at 25 °C, $2.3RT = 1.364$ kcal/mol (Footnote 2, Chapter 7, p. 276)]

$\log K = 1.83$

$K = 68$, and there is only 1.4% of the s-cis compound present at equilibrium.

Problem 13.13 The transition state for rotation about one of the "double" bonds in a 1,3-butadiene contains a localized radical and a delocalized allylic radical. It is this delocalization of the allylic radical in the transition state that makes this rotation more favorable than the related motion in a simple ethylene. In an alkene, two localized radicals appear in the transition state for rotation about a carbon–carbon bond.

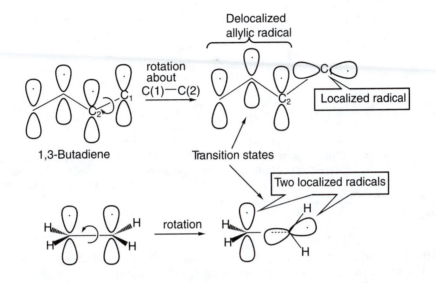

Problem 13.14 Based on the information from Problem 13.13, the rotation around the C(2)—C(3) bond of the (Z,Z)-2,4-hexadiene conjugated system should be about 52 kcal/mol (217 kJ/mol). We might expect the energy to be somewhat less than 52 kcal/mol as there is more destabilizing steric interaction with the C(1) methyl than is present in the 1,3-butadiene, the data for which are given in Problem 13.13. It is easier to isomerize when the starting (Z) isomer is higher in energy (less stable).

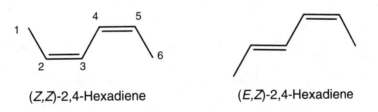

If we want the K value to be 5:95 (which is 0.053), and ΔG is 52 kcal/mol, we can use the R value of 1.98×10^{-3} kcal/mol·kelvin and the formula $\Delta G = -RT \ln K$.

The math goes like this:

52 kcal/mol ÷ 1.98×10^{-3} kcal/mol·kelvin = $-T \ln 0.053$
26,300 kelvin = $-T(-2.94)$
8,940 kelvin = T
8,670 °C = T

Therefore, a very high temperature would be required to have a ratio of 95:5 for the $E{:}Z$ isomers. However, at 150 °C, the equilibrium for the E/Z isomerization is about $10^{-27}{:}1$!

Problem 13.15   The same resonance-stabilized carbocation is formed on S$_N$1 solvolysis of both chlorides. The positive charge is shared by two carbons. The nucleophile, water, can add to each of these carbons to give, after proton loss, two alcohols. As the same carbocation intermediate is formed from each chloride, the two alcohols must be produced in the same ratio.

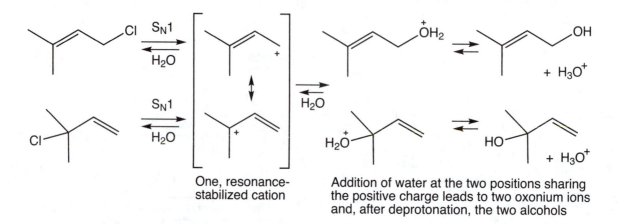

One, resonance-stabilized cation

Addition of water at the two positions sharing the positive charge leads to two oxonium ions and, after deprotonation, the two alcohols

Problem 13.18   The initial protonation gives the resonance-stabilized cation shown. The kinetic product will be the one that results from 1,2-addition for the reasons summarized in Figure 13.38, p. 611. Given the broad hint in the problem, you surely won't pick the product of 1,4-addition as the thermodynamic product. In this case, the kinetic and thermodynamic products are the same. The question is why. Compare the products formed in 1,2- and 1,4-additions. The product of 1,2-addition contains a trisubstituted double bond, whereas the product of 1,4-addition has only a less stable disubstituted double bond.

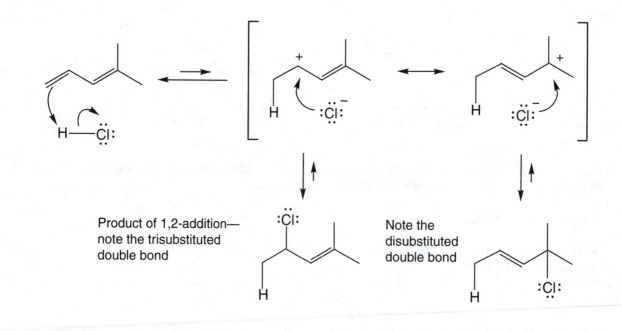

Product of 1,2-addition—note the trisubstituted double bond

Note the disubstituted double bond

Problem 13.19 There are two π electrons in the allyl cation, three π electrons in the allyl radical, and four π electrons in the allyl anion. The resonance picture shows the individual electronic descriptions contributing to the real, delocalized structures.

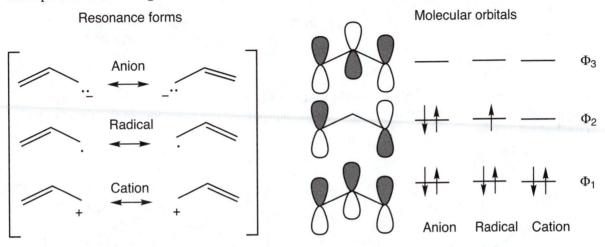

Problem 13.20 The two compounds ionize to the same carbocation intermediate. Although the two transition states are different, the ion will be partially developed in the transition state, and the rates of chloride loss will be quite similar.

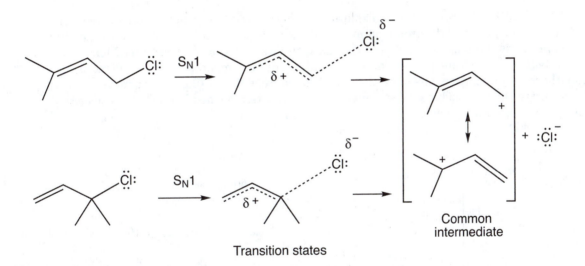

Both compounds ionize faster than the others in Table 13.2 because of the tertiary δ⁺ in the transition state. A tertiary partial positive charge is more stable than a secondary or primary partial positive charge. For example,

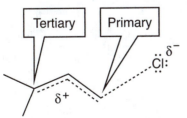

In this transition state, the positive charge is shared between a tertiary and primary position; this is the more stable transition state

(continued)

Problem 13.20 *(continued)*

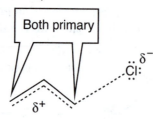

In this transition state, the positive charge is shared between two primary positions; this is the less stable transition state

There are, of course, many possible molecules that would ionize faster. Perhaps the simplest modification would be to add two methyl groups as shown. Now the resonance-stabilized cation is tertiary at both ends.

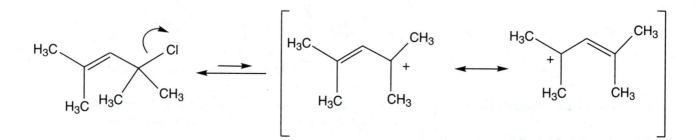

Problem 13.21   The problem is to explain why there should be a connection between the stability of the products (thermodynamics) and the energies of the transition states leading to the products (kinetics). Why should thermodynamics and kinetics be related? Ionization is surely an endo-thermic process, and, as noted by the Hammond postulate, the more endothermic the reaction, the more the transition state will look like the product. Accordingly, the thermodynamic stability of the allyl cation formed is related to the ease of formation, a kinetic property.

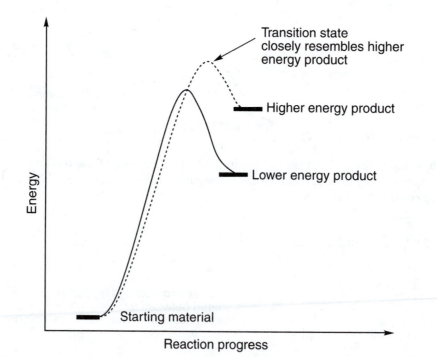

There is another way to put it. The transition states for ionizations to give carbocations will contain partially developed positive charges. A partially developed tertiary positive charge will be more stable than a partially developed secondary positive charge, and this difference will be reflected in the relative energies of the transition states.

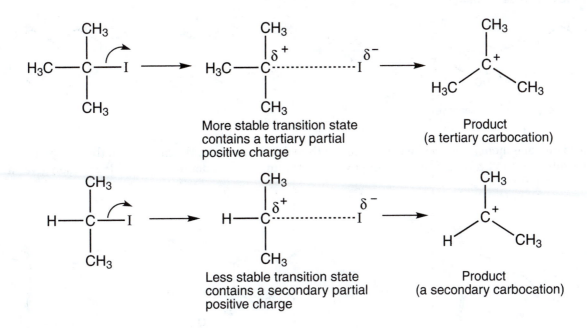

Problem 13.22

(a) Because the stereochemistry of the esters on the ring is not specified, we could use either the (E) or the (Z) starting dienophile. The answer that is shown uses the (Z) dienophile, which gives the cis product (both enantiomers). If we start with the (E) dienophile, we would obtain the trans product (both enantiomers).

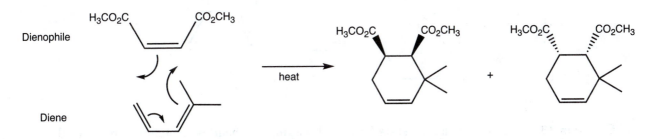

(b) This reaction will give a mixture of stereoisomers.

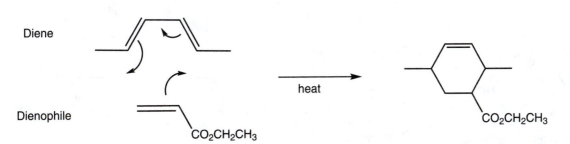

(c) The best Diels–Alder reaction for making this product uses dicyanoethylene, which is a great dienophile. The stereochemistry of the product is not specified. The (E) dieneophile is used this time to show the trans substitution (both enantiomers) in the product. If we used the (Z) dieno-phile, then we would obtain the cis isomer (which is achiral).

(continued)

Problem 13.22 *(continued)*

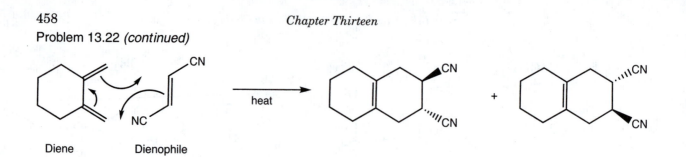

Diene   Dienophile

(d) Only one dienophile can be used to make this bicyclic product. We cannot use the (*E*) dienophile because we can't have a five-membered ring containing a trans double bond. This reaction gives a single product.

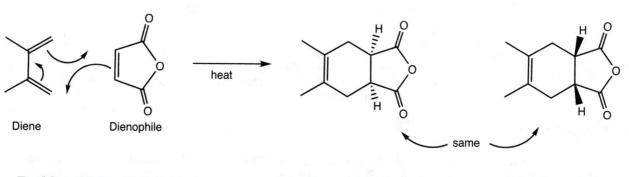

Diene   Dienophile

same

Problem 13.23   Nothing to it.

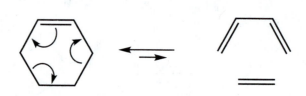

Problem 13.24   In the transition state for bond breaking, the bond is stretching, and radical centers are developing on the bonded carbons (δ·). This process is illustrated for carbon–carbon bond breaking in cyclohexane, a reasonable model for ethane.

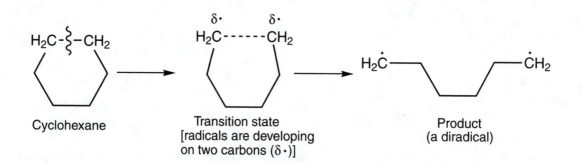

Cyclohexane   Transition state   Product
                     [radicals are developing   (a diradical)
                     on two carbons (δ·)]

In cyclohexene, one of the developing radicals is a resonance-stabilized allyl radical, and it is more stable than the simple localized radicals formed in carbon–carbon bond breaking in cyclohexane. Accordingly, the product diradical, and the transition state leading to it, are more stable than those in the cyclohexane reaction. The energy required to break the σ bond in cyclohexene is lower, and this is reflected in the bond dissociation energy quoted in the problem (80 kcal/mol).

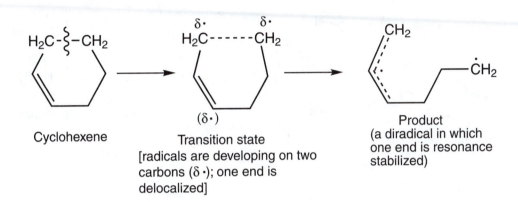

Cyclohexene

Transition state
[radicals are developing on two
carbons (δ ·); one end is
delocalized]

Product
(a diradical in which
one end is resonance
stabilized)

Problem 13.26 Vinylcyclobutane lies higher in energy than cyclohexene, largely because of the strain energy in the molecule. In the transition state for formation of vinylcyclobutane (see dashed lines), some of this strain will be present. Accordingly, it, too, lies relatively high in energy.

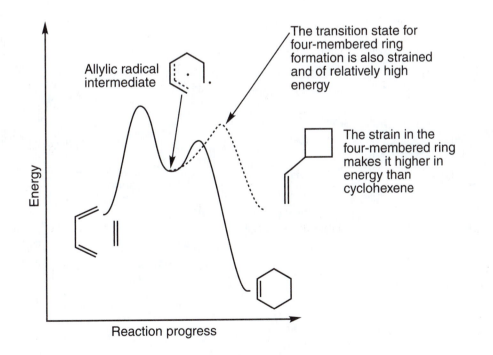

Problem 13.27   A one-step addition to the trans,trans diene must give a single product, the one shown in the problem figure in which the R groups are cis to each other.

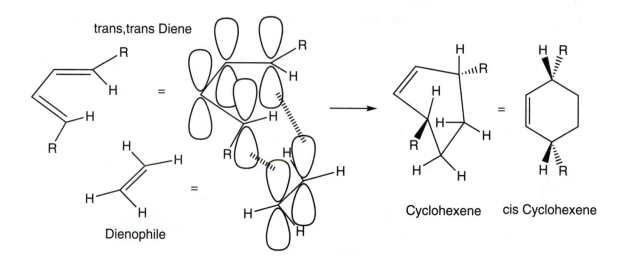

trans,trans Diene

Dienophile

Cyclohexene   cis Cyclohexene

This product is exactly what is seen experimentally, so the one-step, concerted mechanism is in accord with experiment. Let's now see if the two-step process works out. If one bond is made before the other, there must be an intermediate diradical in which the second bond can be formed from the same or opposite side as the first. Two different stereoisomers will result. That there should be two isomers depending on the direction of the second bond formation is relatively easy to understand; what's hard to see is just how the groups on the diene appear in the cyclohexene product. Follow the diagram, but do look at models as well.

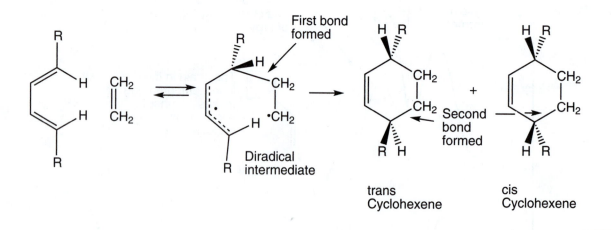

First bond formed

Diradical intermediate

Second bond formed

trans Cyclohexene   cis Cyclohexene

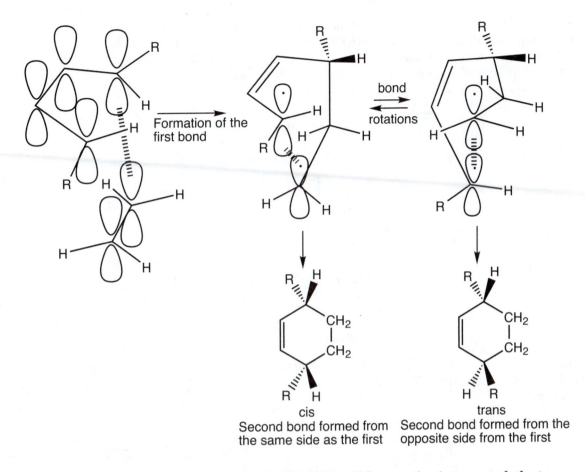

Formation of the first bond

bond rotations

cis
Second bond formed from the same side as the first

trans
Second bond formed from the opposite side from the first

Once again, the experimental results show that the Diels–Alder reaction is concerted; the two new bonds are formed at the same time, and there is no intermediate (Fig. 13.53, p. 620).

Problem 13.30 An initiating radical (In·) is formed in the first step. It adds to a butadiene to form a resonance-stabilized allylic radical in the second step. This allylic radical then adds to another butadiene to form a new allylic radical. This process is repeated many times to form the polymer.

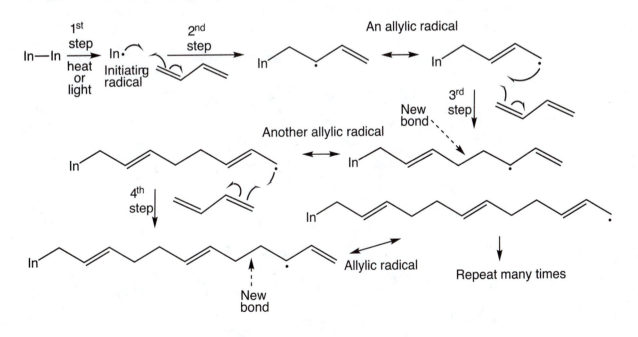

Problem 13.31 In the rubber molecule shown on page 625, all the double bonds are (*Z*). There is another form with all the double bonds (*E*), which is called gutta percha. Of course, there could also be all sorts of polymers with some double bonds (*E*) and others (*Z*). These molecules may be reasonable answers to the problem as set, but they do not occur in nature.

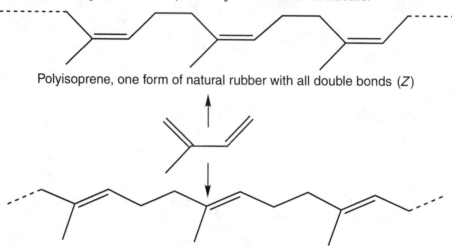

Polyisoprene, one form of natural rubber with all double bonds (*Z*)

Polyisoprene with all double bonds (*E*), gutta percha

Problem 13.32 Loss of the diphosphate ion, shown here as an S_N1 reaction, leads to a resonance-stabilized diphosphate anion. The resonance stabilization of the anion makes diphosphate a good leaving group.

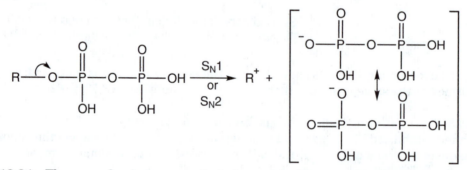

Problem 13.34 There are five isoprene units in laserpitin. The oxygens aren't part of the carbon framework. Each circled group has five carbons that are ultimately derived from isoprene.

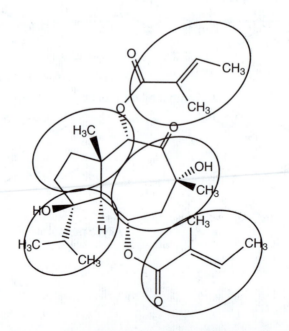

Problem 13.35   Each rearrangement step is tertiary carbocation to tertiary carbocation.

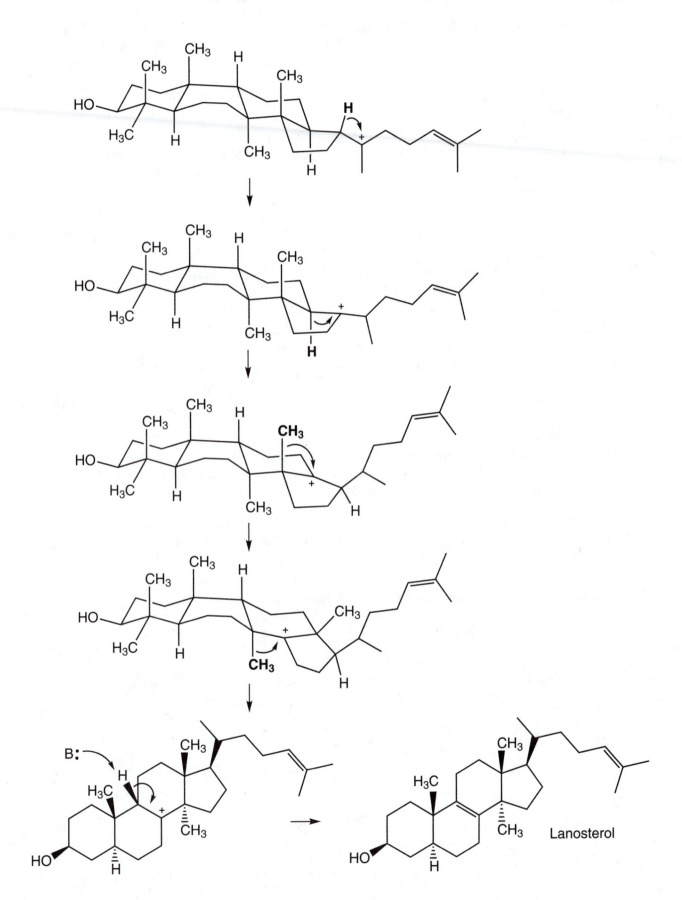

Lanosterol

Additional Problem Answers

Problem 13.36   In the first ion, there are only three atoms that share the negative charge. The resonance structures are shown. The carbonyl group is not involved in resonance stabilization of the anion.

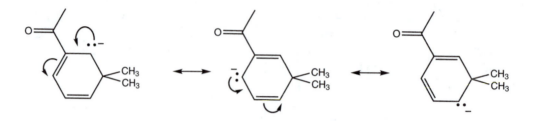

In the cyano-substituted cyclohexadienyl anion, the negative charge is located on an atom that does allow for delocalization out of the ring onto the more electronegative nitrogen.

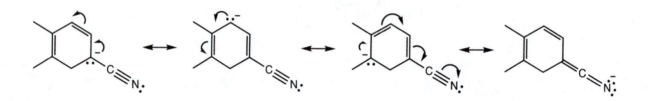

In the straight-chain ion, there are only three resonance structures if the nitrogen is sp^2 hybridized. They are shown here. The lone pair of the nitrogen is not able to delocalize into the p orbital of the π system because it is perpendicular to the dienyl system.

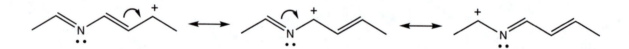

The last ion is the cyclopentenyl cation. It has four resonance structures.

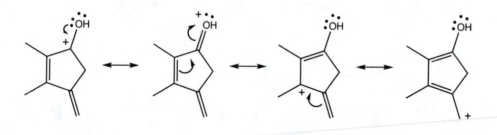

Problem 13.37   In each of the pairs, it is the second, right-hand compound that is chiral. In the following figure, to see that the two compounds of the upper pair are identical (superimposable), and the lower pair is non-superimposable, rotate 90° along a longitudinal axis. By all means make models! This one is not easy to see.

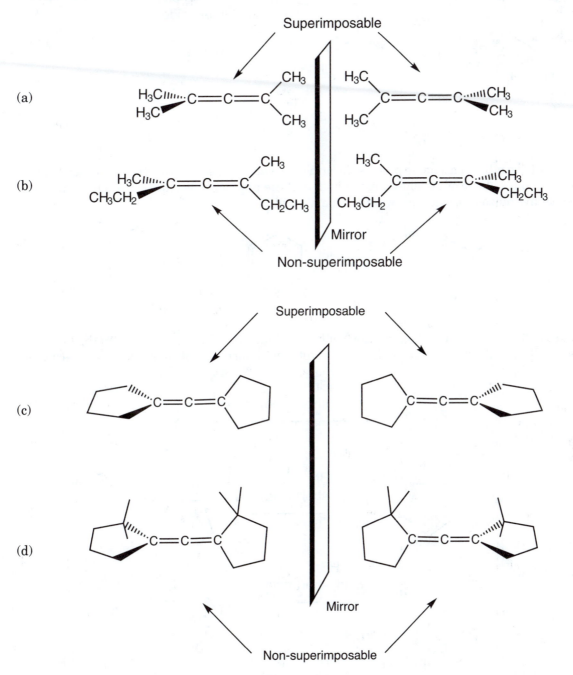

Problem 13.38   Azide is a great nucleophile, so a reasonable plan is to displace a good leaving group with it. The trick is to be sure that addition to 1,3-butadiene puts that leaving group in the right position. We need to do a 1,2-addition, not a 1,4-addition, and therefore we must use gentle conditions that ensure kinetic control.

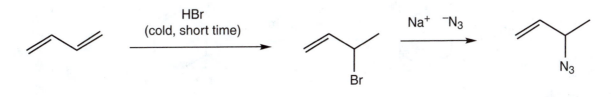

Problem 13.39 There is an easy "double E2" elimination to give 1,3-cyclohexadiene, which is certain to be the major product.

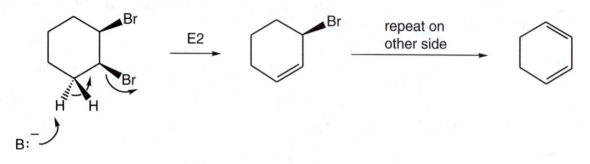

There really is no other reasonable process. In principle, you could eliminate HBr two times to give either 1,2-cyclohexadiene or cyclohexyne, but both of these products have hideous angle strain (note the *sp*-hybridized carbon that "wants" to be linear).

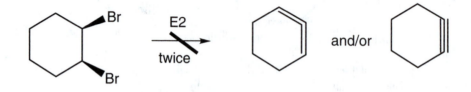

Problem 13.40 In each case, the conditions of kinetic control will make 1,2-addition the favored process. In all the following reactions, the initial protonation forms the most stable possible cation.

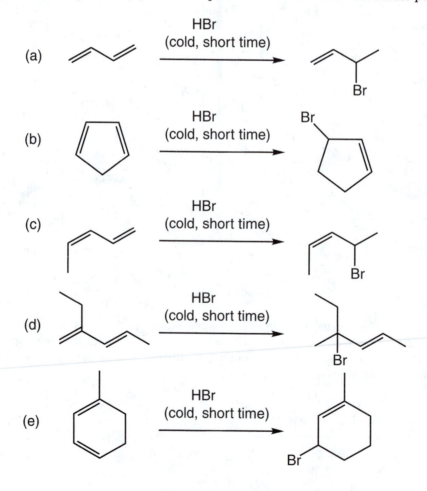

Problem 13.41 In part (a), the 1,4-addition product is more stable than the 1,2-addition product. In part (c), the (Z) double bond from 1,2-addition is less stable than the (E) double bond from 1,4-addition. The reactions of (b) and (d) will have no product selectivity because the 1,2- and 1,4-addition products have the same level of substitution. In (e), the trisubstituted double bond makes 1,2-addition dominate.

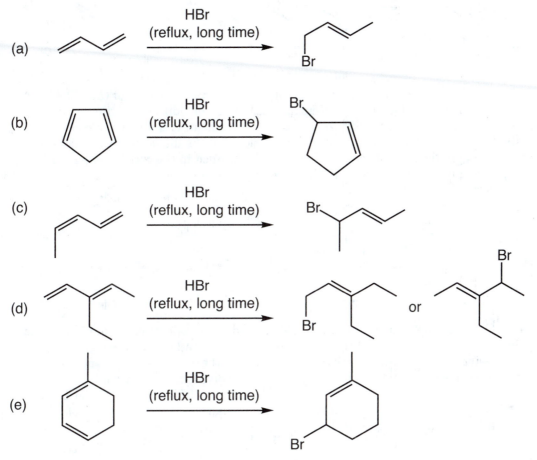

Problem 13.42

(a) A diene reacts with the electrophilic H—Cl to give a resonance-stabilized allylic cation. With the cold conditions, the kinetic product will predominate. The kinetic product results from 1,2-addition.

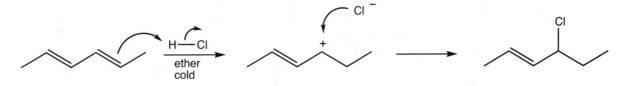

(b) Hydrogenation (with one equivalent of H_2) of a conjugated diene can result in hydrogen adding across one of the double bonds or adding to the end atoms of the conjugated system.

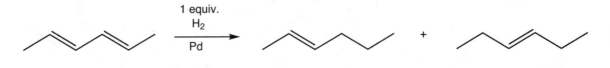

(continued)

Problem 13.42 *(continued)*

(c) Reaction of a diene with one equivalent of the electrophilic Cl_2 (or Br_2) gives both 1,2- and 1,4-addition products. Adding more Cl_2 (or Br_2) gives the tetrachloride.

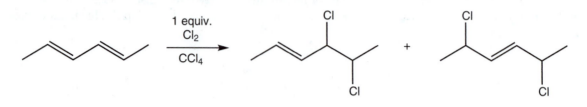

(d) The Diels–Alder reaction requires the diene to be in the s-cis conformation. For (*E,E*)-2,4-hexa-diene, this conformation is available in low concentrations. The cycloaddition with the dienophile (the alkyne in this example) in a concerted process that requires heat. It is a reversible reaction, but the more stable product is the Diels–Alder adduct. Notice that the diene stereochemistry (methyl groups pointing in the same direction—both out) is retained in the cyclohexene product (methyl groups pointing in the same direction—both up or both down).

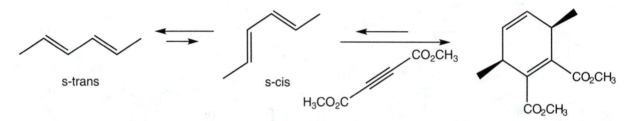

Problem 13.43 Compounds **2** and **3** are the 1,2- and 1,4-addition products of 1,3-butadiene and bromine. Compounds **1** and **4** involve addition of one bromine atom and the methoxy group of methyl alcohol. In principle, two mechanisms can be written for these additions, depending on how the intermediate cation is envisioned. If a bromonium ion is involved, direct S_N2 additions of the bromide ion or methyl alcohol to the bromonium ion give the 1,2-addition products **2** and **1**. The 1,4-addition products **3** and **4** could be formed by what is called an S_N2' reaction with bromide ion or methyl alcohol acting as nucleophile. (See Problems 13.62–13.64.)

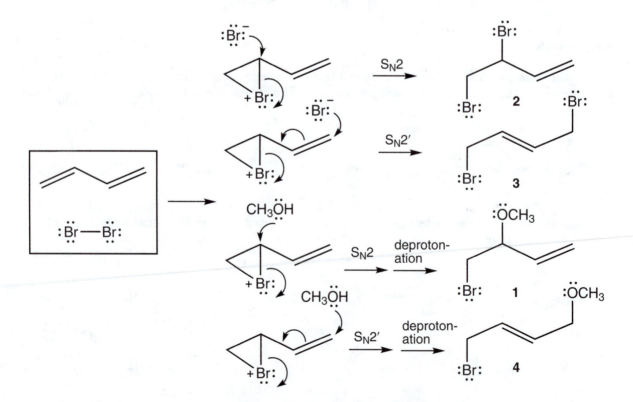

Alternatively, the intermediate could be seen as a resonance-stabilized allylic carbocation. Addition of the two nucleophiles, bromide and methyl alcohol, leads to the observed four products.

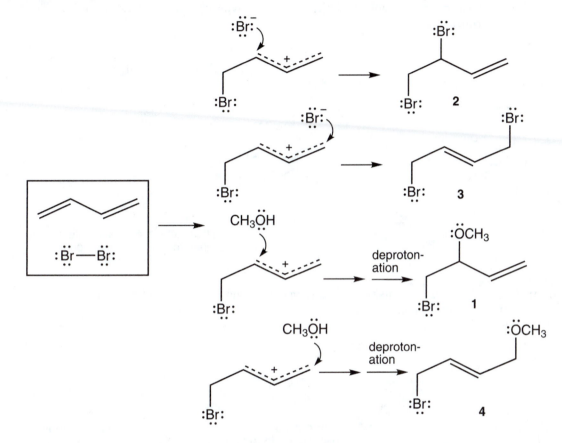

It is not possible to distinguish between these two possibilities without further experiments. It is worth noting that the ratios of the two types of 1,2- and 1,4-addition products are appreciably different; structure **2**/structure **3** = 2.2, structure **1**/structure **4** = 15. This result must be ascribed to the greater nucleophilicity of bromide ion relative to that of methyl alcohol.

Problem 13.44 As with 1,3-butadiene, the kinetic (more easily) formed product will be that from 1,2-addition. Under thermodynamic conditions, the more stable 1,4-adduct prevails. The problem is that there are two possible 1,2-addition products. The issue can be decided by examining the two possible allyl cations formed by protonation at the different ends of the 1,3-diene.

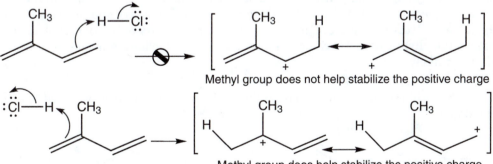

Problem 13.44 *(continued)*

Addition of chloride at the relatively nearby position sharing the positive charge leads to the product of 1,2-addition (kinetic control). Addition at the relatively remote position sharing the positive charge gives the more stable, but kinetically disfavored, product of 1,4-addition (thermodynamic control).

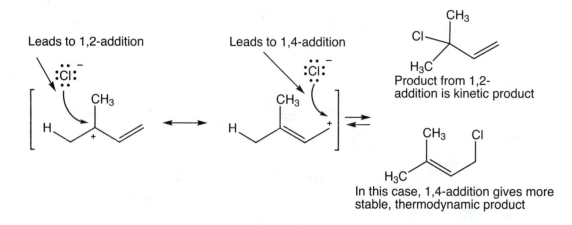

Problem 13.45   This question asks if the allylic cation shown below is likely to undergo an alkyl shift to give the tertiary cation.

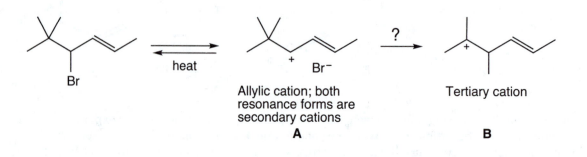

To answer this question, we need to compare **A** (an allylic cation that is secondary in both resonance forms) to **B** (a tertiary cation). Whichever is more stable will be the intermediate involved in S_N1 or E1 chemistry of this allylic bromide. We can get a good idea about the stability of these cations by comparing the rate of formation of similar cations listed in Table 13.2. However, the table does not list an allylic cation that is a combination of two secondary cations. Fortunately, we have the allylic cation that is tertiary and primary, which would be a close comparison. Formation of the tertiary cation has a relative rate of 3×10^4, and an allylic cation that is tertiary and primary has a relative rate of about 5×10^6 (taking the average of the two tertiary/primary entries). It is about 100 times faster to make the allylic cation. We can conclude that the intermediate **A** is more stable and that an alkyl shift is not likely.

Problem 13.46   "All" cyclohexenes come from the Diels–Alder reaction, so we need to do the following reaction:

So, all we need to do is to convert *cis*-2-butene into 1,3-butadiene. Here's how to do it.

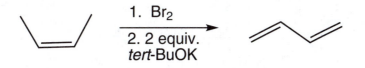

Problem 13.47 This synthesis problem should be relatively easy. The only real difficulty might be that it is a cascade problem; you have to be able to make one of the products in order to get the others. In this case, the critical compound is 3-hexyne. It comes from formation of the acetylide from 1-butyne followed by an S_N2 reaction on ethyl iodide.

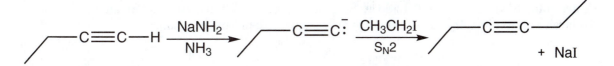

Once 3-hexyne is made, the others all follow in straightforward fashion.

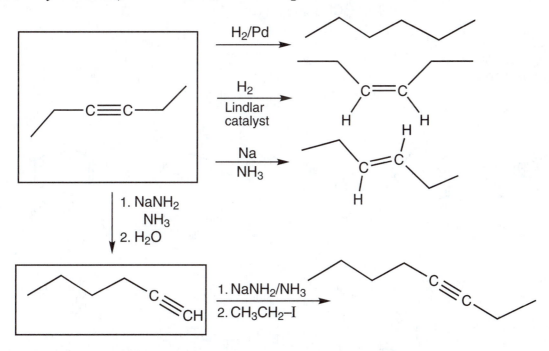

Problem 13.48 Converting 2-butyne into 1-butyne involves a "zipper" reaction.

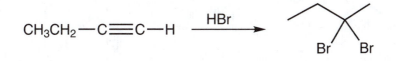

As the problem suggests, going the other way poses more problems. Here is one suggestion. First, add two molecules of HBr in polar, Markovnikov fashion to give 2,2-dibromobutane.

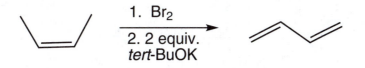

(continued)

Problem 13.48 *(continued)*

Now a pair of E2 reactions will give mostly 2-butyne. There are other possible products, 1-butyne and 1,2-butadiene, but they should be less favored.

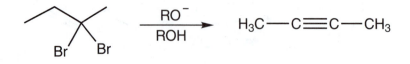

Problem 13.49   Here is the target:

$$(CH_3)_3C-C\equiv C-CH_2CH_2CH_2CH_3$$

We are going to use an acetylide to do an S_N2 displacement. Because we can't do an S_N2 displacement at a tertiary carbon, there is only one way to do this reaction.

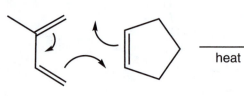
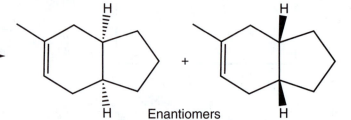

Problem 13.50

(a)

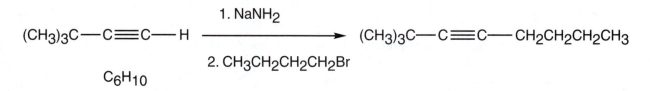

(b)

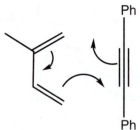
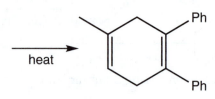

(c)

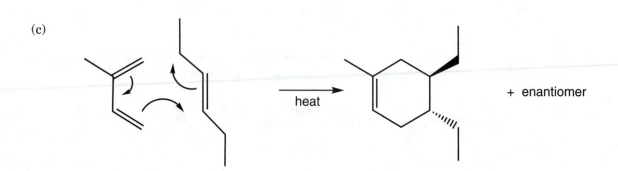

(d)

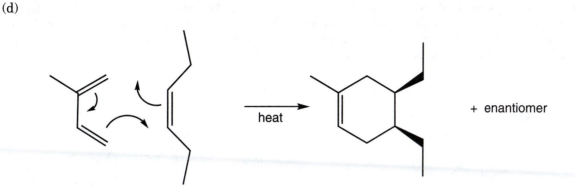

heat

+ enantiomer

Problem 13.51 This problem was designed to illustrate the scope and variety of compounds available from the Diels–Alder reaction.

(a) In this reaction, a 76:24 mixture of endo and exo products is obtained. Be sure you see how these two products arise from a relatively more stable endo transition state and a relatively less stable exo transition state. In doubt? See Figure 13.57 (p. 622).

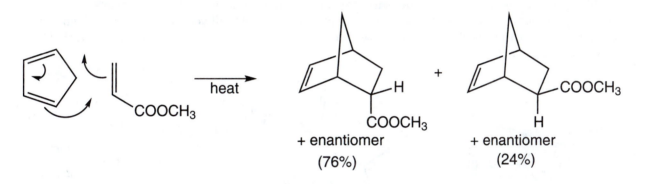

(b) The only unusual thing about this variation of the Diels–Alder reaction is the use of an alkyne as a dienophile rather than the more usual alkene. The product is a 1,4-cyclohexadiene rather than a cyclohexene. Note that only one π bond of the alkyne is involved in this Diels–Alder reaction.

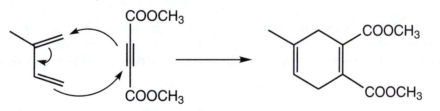

(c) In this variation, both the diene and the dienophile are cyclic compounds. The first complication is to decide whether the reaction will pass over an endo or exo transition state. In fact, the major cycloadduct has the endo stereochemistry, and this reaction is normal in that respect.

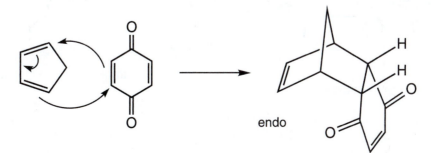

endo

(continued)

Problem 13.51 *(continued)*

You may have noticed that the product of this Diels–Alder reaction still has a double bond with two attached carbonyl groups. This carbon–carbon double bond is also a quite reactive dienophile. When this reaction is run in the presence of two equivalents of 1,3-cyclopentadiene, a 2:1 adduct is readily formed. The second addition occurs from the more available face of the enone.

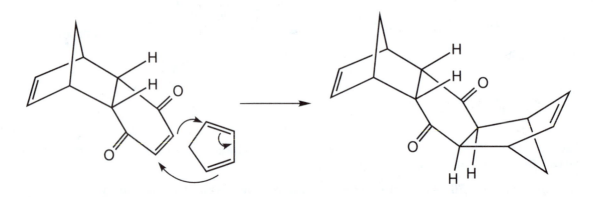

(d) This reaction illustrates another useful variation of the Diels–Alder reaction in which the primary adduct loses a small molecule such as carbon monoxide, carbon dioxide, or nitrogen. In this problem, the loss of carbon monoxide is indicated by the molecular formula. In problems such as these, watch for losses such as this one. The formula will usually tell you, but if it is not given, you must be alert for the possibility, as it is often encountered both in the real world and the world of examination problems.

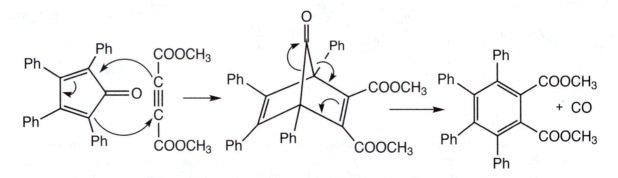

(e and f) The Diels–Alder reaction has been used to deduce the position of double bonds in steroids. In example part (e), the diene is locked into an s-cis arrangement, and the Diels–Alder reaction occurs. In part (f), the diene is s-trans, and the Diels–Alder reaction is not possible. Be sure you see why.

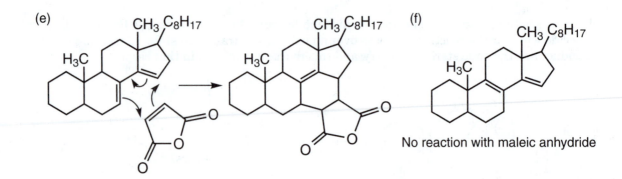

No reaction with maleic anhydride

(g) In this example, we first have to realize that a "1,3-diene" is present. True, it contains two nitrogen atoms, but it remains a 1,3-diene nonetheless. Otherwise the first reaction is quite simple.

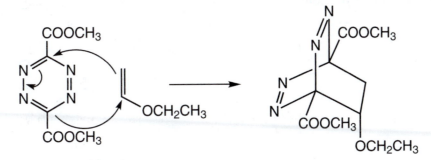

Now, as in part (d), a small molecule is lost, this time N_2. Elimination of another molecule, ethyl alcohol, also occurs to give the final product.

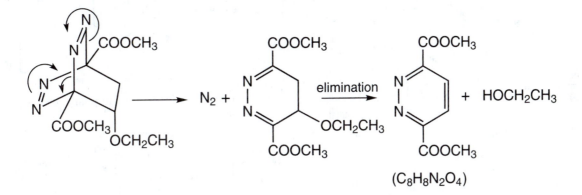

$$(C_8H_8N_2O_4)$$

Problem 13.52 This reaction involves a Diels–Alder cycloaddition with the nitrile acting as dienophile. In this case, two initial cycloadducts are possible because both the nitrile and the diene are unsymmetrical. Loss of carbon monoxide from the primary adducts gives the two observed pyridines.

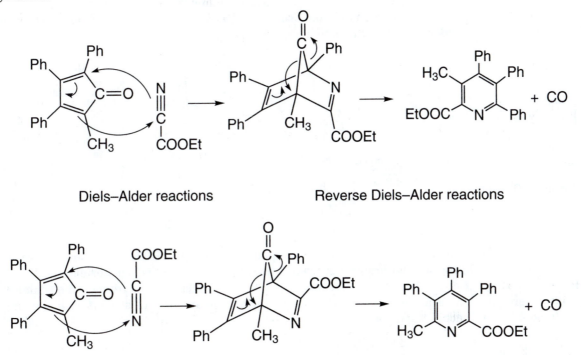

Diels–Alder reactions   Reverse Diels–Alder reactions

Problem 13.53   A polar, stepwise mechanism would surely put the minus charge near the CN (so it will be stabilized by resonance), leaving the plus charge on carbon, where it, too, is resonance stabilized. There are two ways to do this:

A

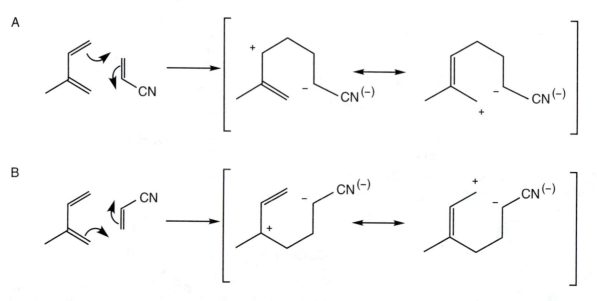

B

Path **B** will be favored because the carbocation has a resonance form in which the positive charge is borne on a tertiary carbon. Closure would give the product shown.

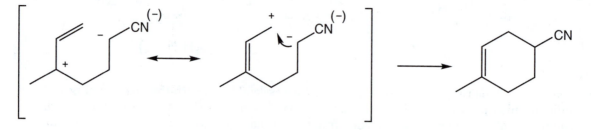

Problem 13.54   This problem is a classic case of thermodynamic versus kinetic control, quite similar to the situation encountered in Section 13.10 (Fig. 13.35, p. 609) for the addition of chlorine to 1,3-butadiene. Consider the general reaction shown here.

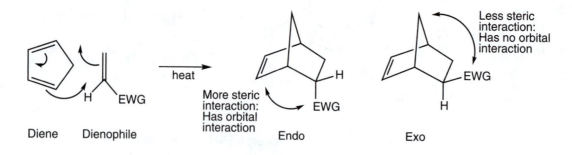

The exo compound is more stable than the endo compound because the EWG (electron-withdrawing group) points toward the smaller bridge, not toward the larger bridge. This is a steric issue. However, the transition state leading to the less stable endo product benefits from (1) molecular orbital interaction between the EWG and the developing alkene and/or (2) a solvent exclusion as the dienophile approaches the diene (p. 622).

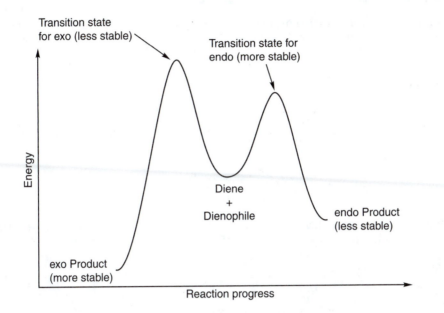

At high temperature, all barriers are passable, and the major product will be the more stable exo adduct. At low temperature, the reaction is less reversible, and the major product will be determined by the relative energies of the transition states, which typically favor the endo product in the Diels–Alder reaction.

Problem 13.55   First, we will determine the molecular formulas of compounds **B** and **C**. Compound **B** is a dehydration product of α-terpineol (**A**) (i.e., $C_{10}H_{18}O - H_2O = C_{10}H_{16}$). If compound **B** reacts with maleic anhydride to give a 1:1 adduct, compound **C** has a molecular formula of ($C_{10}H_{16} + C_4H_2O_3$) = $C_{14}H_{18}O_3$. The molecular weight for this formula is 234 g/mol. Thus, our surmise that compound **C** is a 1:1 adduct of **B** and maleic anhydride is consistent with the mass spectrum of **C**. This proposal is also supported by the presence of 18 hydrogens in the ^{1}H NMR spectrum of **C**.

Now we need to work out the number of degrees of unsaturation in **C**.

$$\Omega = [2(14) + 2 - 18]/2 = 6$$

So far in these problems, 6 degrees of unsaturation has generally suggested the possibility of a benzene ring, thus accounting for 4 degrees of unsaturation. Could this problem be an exception to this expectation? There seems no easy way to get an aromatic compound in this reaction.

As was already noted, compound **C** is a 1:1 adduct of **B** and maleic anhydride. Futhermore, **B** is a dehydration product of α-terpineol (**A**). It is worth speculating at this point about the structure of compound **B**. Simple dehydration of compound **A** could afford dienes **D** and **E**; perhaps one of these is compound **B**. This possibility seems unlikely, as neither **D** nor **E** is a conjugated diene, and therefore, neither can react in Diels–Alder fashion with maleic anhydride.

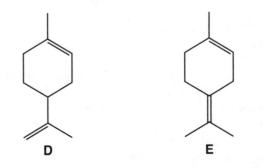

D   **E**

(continued)

Problem 13.55 *(continued)*

However, a more complicated dehydration of **A** could afford the conjugated diene, α-terpinene. You should be able to write a mechanism for this dehydration.

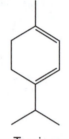

α-Terpinene

Clearly, α-terpinene and maleic anhydride could undergo a Diels–Alder reaction to give a 1:1 cycloadduct, and this could be compound **C**. Let's see if the rest of the spectral data are consistent with this proposal.

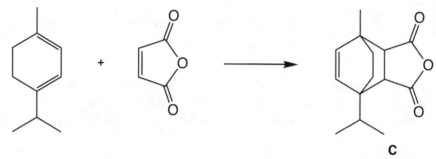

The two carbonyl stretches at 1840 cm^{-1} and 1780 cm^{-1} in the IR spectrum of **C** are consistent with the presence of a cyclic anhydride.

The NMR spectral data are also consonant with the proposed structure for **C**. The chemical shift assignments are summarized. Note that the isopropyl methyl groups are diastereotopic, as required by the intrinsic asymmetry of the proposed structure. Also note that the 6 degrees of unsaturation for compound **C** are three rings, 1 C=C and 2 C=O bonds.

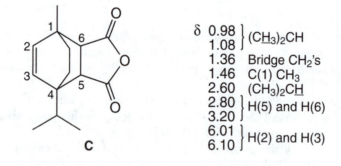

δ		
0.98	}	(CH$_3$)$_2$CH
1.08		
1.36	Bridge CH$_2$'s	
1.46	C(1) CH$_3$	
2.60	(CH$_3$)$_2$CH	
2.80	} H(5) and H(6)	
3.20		
6.01	} H(2) and H(3)	
6.10		

Two questions still remain concerning the stereochemistry of cycloadduct **C**.

(1) Are H(5) and H(6) cis or trans to each other?
(2) Is the cycloadduct endo or exo?

One of these questions can be answered with the available spectral data, whereas the other cannot. As you might expect from what you already know about the Diels–Alder reaction, the cis stereochemistry present in maleic anhydride is preserved in the cycloadduct; that is, H(5) and H(6) have a cis relationship. This assignment is supported by the magnitude of their coupling constant ($J = 9$ Hz). From the Karplus curve, it is known that vicinal coupling constants are dependent on the dihedral angle (ϕ) between the vicinal protons—with large coupling constants for dihedral angles of 0° and 180° and smaller coupling constants for dihedral angles approaching 90°.

Typical values for cis ($J_{exo,exo}$ and $J_{endo,endo}$; $\phi \sim 0°$) and trans ($J_{exo,endo}$; $\phi \sim 120°$) vicinal coupling constants for bicyclo[2.2.2]octenes are shown.

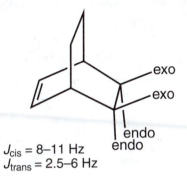

J_{cis} = 8–11 Hz
J_{trans} = 2.5–6 Hz

Although it is not possible to deduce the exo versus endo stereochemistry of cycloadduct **C** from the available spectral data, **C** has endo stereochemistry. The actual stereochemistry has been determined by using more complex spectral techniques, by chemical transformations, and by independent synthesis of the exo isomer.

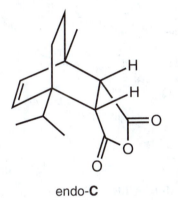

endo-**C**

Problem 13.56 In part (a), the Diels–Alder reaction will lead to cyclohexene, not the vinyl-cyclobutane (p. 618). ^{13}C NMR spectroscopy could easily distinguish the two possibilities, as cyclohexene has only three different carbons and vinylcyclobutane has five.

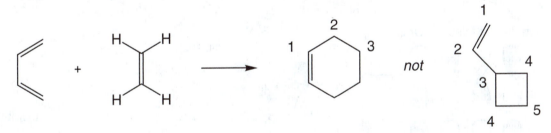

In part (b), the cis product will be formed, *not* the trans, as the Diels–Alder reaction occurs in a single step. However, ^{13}C NMR spectroscopy will be ineffective in distinguishing the two possibilities, as they each have four different carbons.

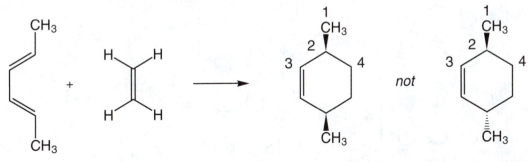

(continued)

Problem 13.56 *(continued)*

There is a similar problem in part (c). First draw the two products, the endo and exo isomers. The endo adduct will be favored kinetically, but ^{13}C NMR spectroscopy will not be able to distinguish the favored endo product from the unfavored exo, as each has eight different carbon atoms.

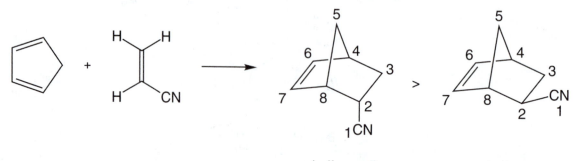

endo (favored)   exo (unfavored)

Problem 13.57   First of all, draw an arrow formalism for the reaction of an allyl species with 1,3-butadiene. *Remember*: The two participants in the reaction approach each other in parallel planes. There is no attempt in this schematic to show the orbital phases; this is a mapping exercise only.

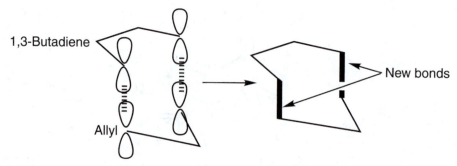

In each reaction there are two HOMO–LUMO interactions possible, HOMO (diene)–LUMO (dienophile) and HOMO (dienophile)–LUMO (diene). Look at both interactions in each case. For the participants in the reactions, 1,3-butadiene, the allyl cation, and the allyl anion, the molecular orbitals are

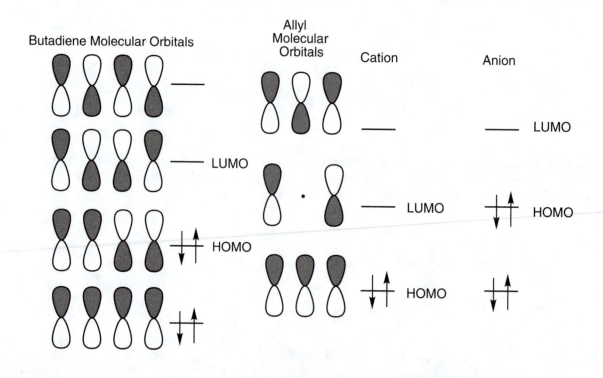

For the reaction to be successful, there must be bonding interactions at the points of formation of both new bonds. Here are the HOMO–LUMO interactions for the reaction of the allyl cation with 1,3-butadiene, showing only the points at which new bonds are made.

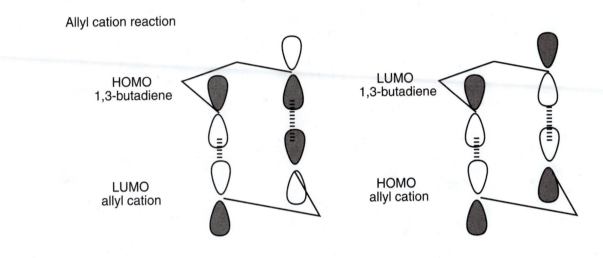

In each case, both new interactions are bonding. The new bonds can be made.

The situation turns out quite otherwise in cycloadditions of the allyl anion. The two HOMO–LUMO interactions are shown below.

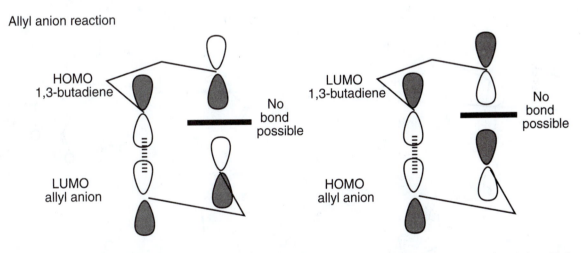

It is not possible to make two bonds in either HOMO–LUMO interaction. The reaction of the allyl anion fails.

Problem 13.58   Work through what the products must be and you will see how it comes out. In each case, there are two Diels–Alder products possible: one endo and one exo.

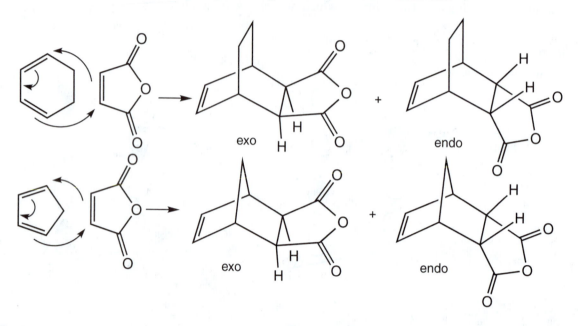

However, when the products are hydrogenated, each of the adducts from 1,3-cyclohexadiene gives the same product. If the drawing does not make this obvious to you, be sure to make models and convince yourself that the "two" hydrogenated compounds are really the same thing. The hydrogenated adducts from reaction of 1,3-cyclopentadiene are still different.

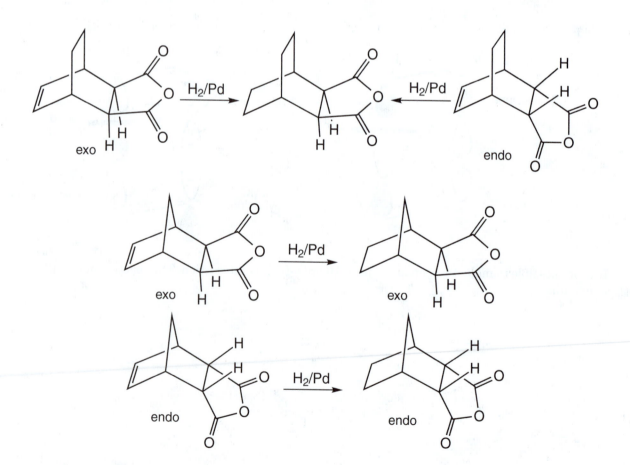

Problem 13.59 All Diels–Alder reactions require an s-cis diene. The first thing to do in solving this problem is to draw the s-cis forms of the diene in each case.

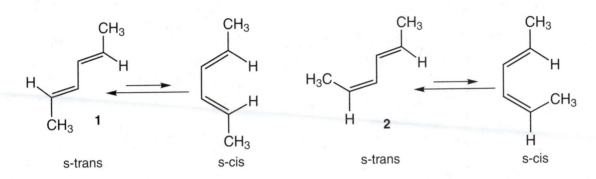

In **2**, a relatively large methyl group must be "inside," and this is surely destabilizing. So, at 35 °C there are no significant amounts of the required s-cis diene for **2**. Reaction of maleic anhydride takes place with the s-cis form of **1** to give product. Even this reaction is more complicated than it might seem at first. The stereochemical relationships of the substituents on the diene must be preserved in this one-step reaction, so the methyl groups must be cis in the product. However, addition can take place through either an endo or exo transition state to give endo or exo product. As usual (Chapter 13, p. 622), the endo transition state is lower in energy than the exo transition state, and the endo product is formed as the major product.

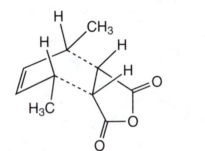

endo Transition state

35 °C

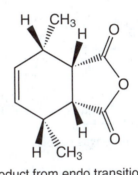

Product from endo transition state is the major product

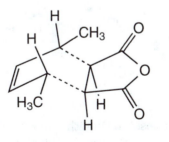

exo Transition state

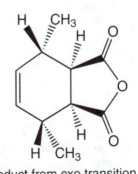

Product from exo transition state is a minor product

For **2**, higher temperature is required to form significant amounts of the s-cis form. When reaction occurs, once again the stereochemical relationship of the methyl groups in the s-cis conformation of the diene will be preserved in the adducts. In this case, they must be trans. As the methyl groups are trans, with one "up" and the other "down," there is only one product possible. The exo and endo transition states lead to the same product in this case.

(continued)

Problem 13.59 *(continued)*

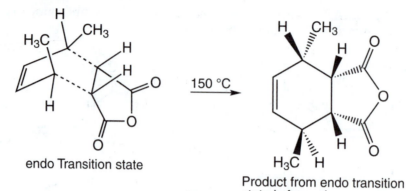

endo Transition state

150 °C →

Product from endo transition
state is formed

In **3**, both methyl groups must be "inside." We predict that this arrangement is too destabilizing
and that no significant amount of Diels–Alder product would be formed at 150 °C. Though this
diene was not examined by Alder, there is ample evidence from other examples that this notion is
correct.

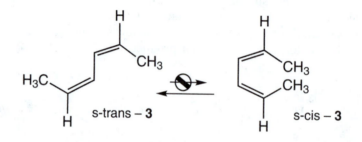

s-trans – **3**   s-cis – **3**

Problem 13.60   If you proposed a Diels–Alder reaction of furan with dimethylmaleic anhydride,
followed by catalytic hydrogenation of the carbon–carbon double bond, you'd be in very good
company. This route was proposed as early as 1928.

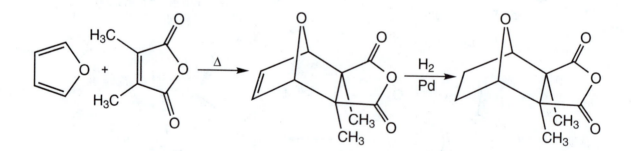

Unfortunately, the required Diels–Alder reaction was not successful. Dimethylmaleic anhydride is
a poor dienophile, presumably because of steric destabilization introduced by the two methyl
groups. In addition, cycloadducts of furan are prone to undergo retro Diels–Alder reactions, and it
may be that product is formed but is not thermodynamically stable relative to starting material.

Reactions such as the Diels–Alder reaction, which proceed with a net decrease in volume (two molecules are made into one), can be accelerated under high pressure. This Diels–Alder reaction fails even at pressures up to 40 kbar. However, this synthetic problem was finally solved by William Dauben and his co-workers at Berkeley. Dauben's group used a dienophile with fewer steric problems, along with high pressure. This is *not* a route you were expected to think up! If you came up with the reasonable ideas shown earlier, you did just fine.

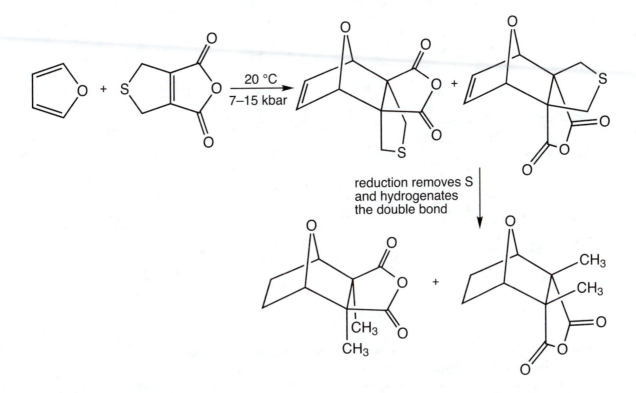

The desired exo compound could be crystallized from the mixture and isolated in 51%–63% yield.

Problem 13.61 (a) This is a simple reverse Diels–Alder reaction brilliantly camouflaged by the molecular architecture. The lesson here is that every time you see a cyclohexene, THINK REVERSE DIELS–ALDER! *All* cyclohexenes are conceptually related to a 1,3-diene and a dienophile. The thermodynamic driving force for this reaction is the strain relief in opening one three-membered ring. In the answer, the product is first drawn without moving any atoms, then "relaxed" to the real structure.

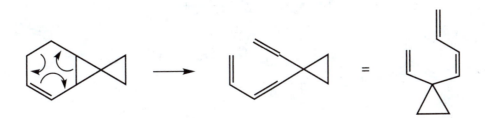

(continued)

Problem 13.61 *(continued)*

(b) This part also involves a reverse Diels–Alder reaction. In this case, the anthracene formed by the reverse reaction is captured in a "forward" Diels–Alder by maleic anhydride. A good clue here is the structure of the second product. Ask yourself how it can arise from starting material, and the reverse Diels–Alder reaction should appear. Alternatively, work backward. Ask what compounds can react to give the final adduct. The answer is maleic anhydride and anthracene. Again the need to make anthracene from starting material should suggest a reverse Diels–Alder reaction.

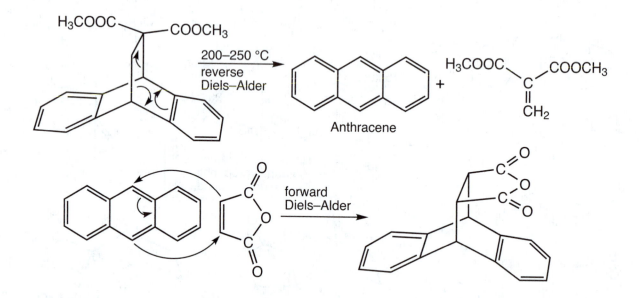

Anthracene

(c) Here we have an intramolecular Diels–Alder reaction. These reactions are notoriously hard to see, especially at the beginning. The reaction is certainly easier to see if the starting material is drawn in a "suggestive" way (as it almost never is in problems!). *Aside:* Notice that the one-step nature of the Diels–Alder reactions allows us to fix the stereochemistry at four different atoms (the termini where the new σ bonds are made) in this reaction. That's quite a synthetic advantage.

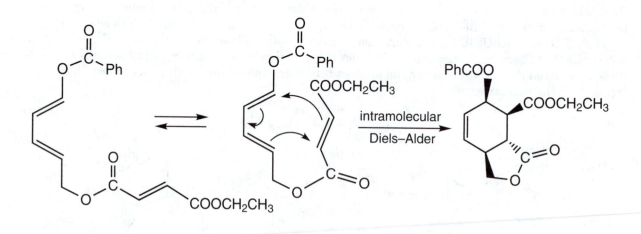

(d) This final transformation is also an example of the intramolecular Diels–Alder reaction, with the furan ring acting as the diene. Once again, it is necessary to draw the molecule in an arrangement that shows the proximity of the diene and dienophile. The product of the Diels–Alder reaction is first drawn without moving any atoms, then relaxed to a more realistic picture. This technique is very useful.

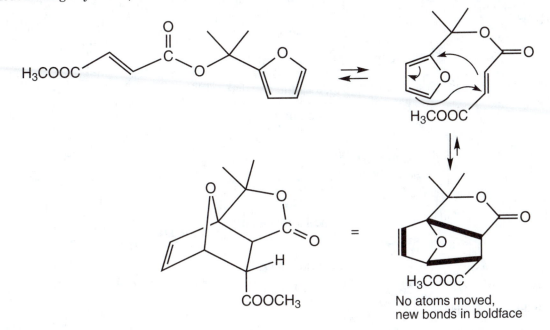

Problem 13.62 We need a label, and almost any will do. A methyl group would work. For example,

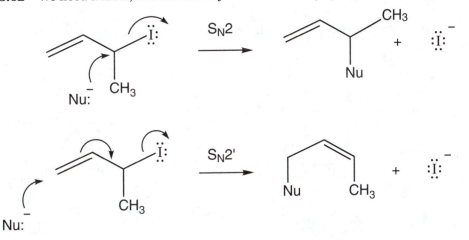

Now the two products are not the same, and we can tell which reaction is operative.

Problem 13.63 An easy one for a change. Any of a large number of possible labeling experiments will work. Here is one suggestion.

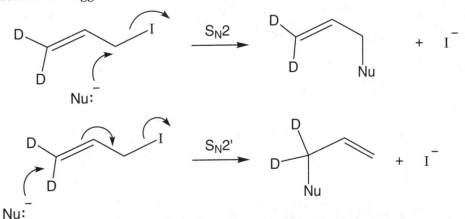

Problem 13.64 The data in the problem show that addition of the nucleophile is from the same side as the chlorine. The two products result from the presence of two rotational isomers of the starting material. In each case, diethylamine adds from the same side as chlorine.

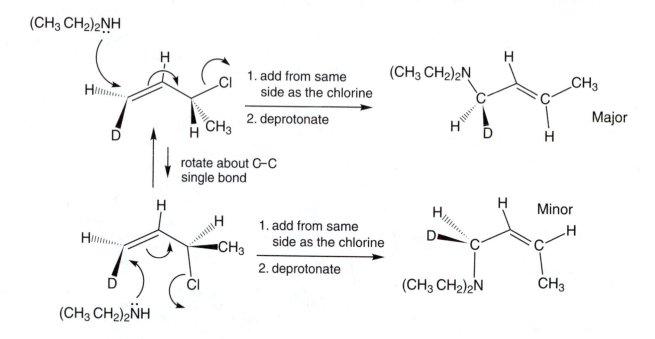

If addition had been from the opposite side, different products would have been observed.

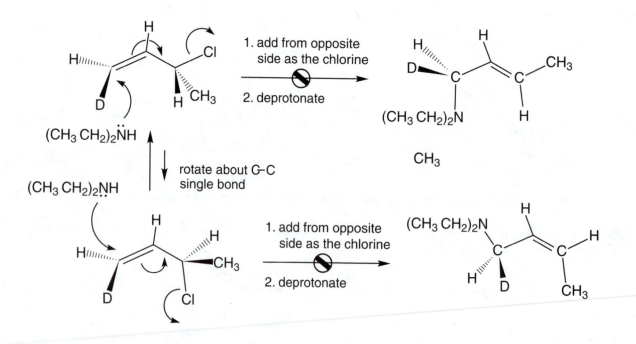

Problem 13.65 It is only necessary to interact the orbitals closest in energy: $\pi \pm \pi$ and $\pi^* \pm \pi^*$ in this case. The same four molecular orbitals result as in the procedure outlined in Figure 13.16, but they are not formed in the same order. The following figure shows the orbitals along with the nodes used to order them in energy.

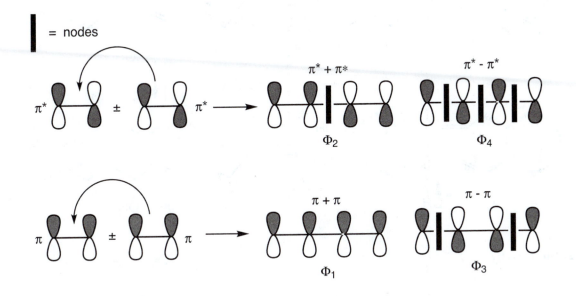

Problem 13.66 The six-membered ring is shown in bold.

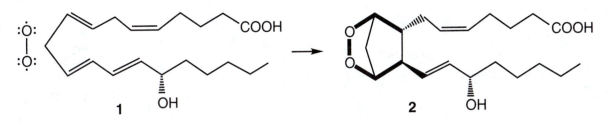

Oxygen must be reacting with compound **1** to give **2**.

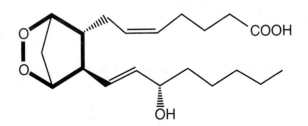

However, there are two reasons the reaction is not likely to be a simple Diels–Alder-related cyclo-addition. First, oxygen is a triplet molecule, with two unpaired electrons. It cannot form **2** directly. Recall our discussion of singlet and triplet carbenes in Chapter 11 (p. 506ff). Second, a concerted cycloaddition would not give the observed trans stereochemistry of **2**. The mechanism probably involves a stepwise process as shown in the figure. Note that there must be an inversion of spin somewhere in the mechanism, and that steric factors will lead to trans stereochemistry in the ring closure step.

(continued)

Problem 13.66 *(continued)*

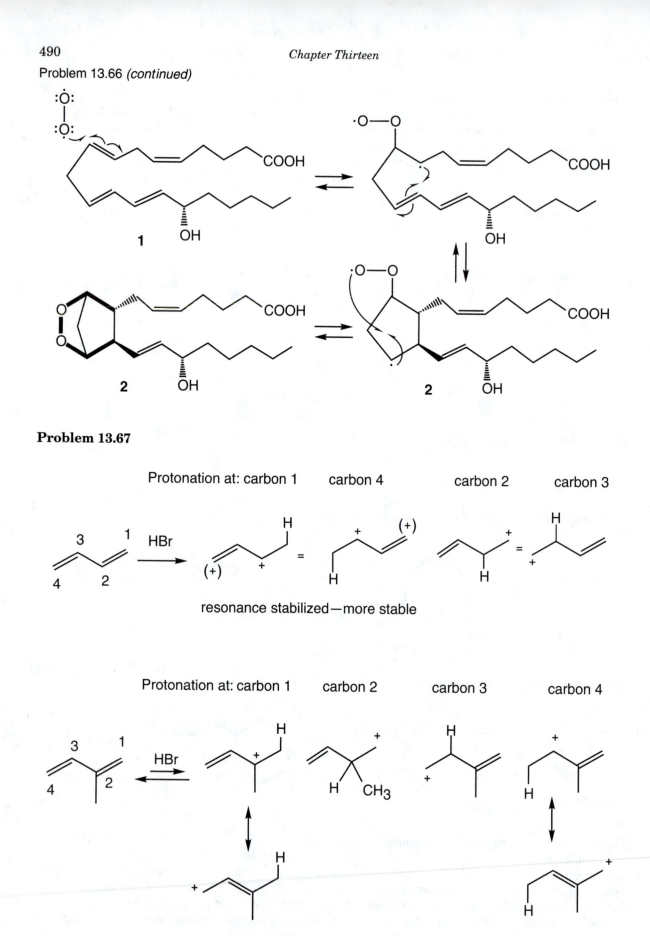

Problem 13.67

Protonation at: carbon 1   carbon 4   carbon 2   carbon 3

resonance stabilized—more stable

Protonation at: carbon 1   carbon 2   carbon 3   carbon 4

This one is the most stable.
Note the tertiary carbocation

Problem 13.68   No, the LUMO density is about the same size on the two carbons that share the positive charge.

Problem 13.69   The product of the 1,4-addition is lower in energy because it has a more substituted double bond than the 1,2-addition product. The 1,4-addition product is not always the most stable. When we start with *trans*-1,3-pentadiene, the 1,4-addition and 1,2-addition products are both *trans*-4-bromo-2-pentene.

Problem 13.70   The nucleophile is the HOMO and it is located on the diene. The electrophile is the LUMO and it is located on the alkene (the dienophile). The diene is the nucleophile because the π electrons are more available from the HOMO of the conjugated system, particularly with the electron-donating group on the diene. The dienophile is the electrophile because the π orbitals that are empty are lower in energy. The carbonyl on the alkene is an electron-withdrawing group, which lowers the energy level of the empty orbital.

Aromaticity

14

This chapter consists almost entirely of an exploration of the structural implications of the stabilizing effects of aromaticity. Reactivity does creep in, but only in a small way, as an introduction to the elaborate substitution reactions of Chapter 15. The following problems allow you to explore resonance and molecular orbital theory in order to find examples of molecules that possess the stabilizing quality called aromaticity. You will also get plenty of practice in drawing strange molecules of quite remarkable and varied structures. In addition, you will have a chance to work a bit on reactions. You will see the classic aromatic substitution process, as well as reactions that destroy the usually robust aromatic ring and reactions at the position adjacent to an aromatic ring, the "benzyl" position.

Problem 14.1 Cyclohexatriene and Ladenburg benzene each shows only a single ^{13}C NMR signal for its six equivalent carbons.

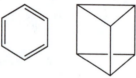

All carbons are
equivalent in each
of these molecules

Dewar benzene and bicyclopropenyl each have two different carbons, and benzvalene has three.

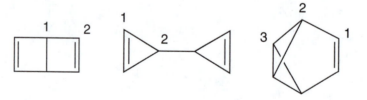

Problem 14.2 A Kekulé representation predicts two isomers of any 1,2-disubstituted benzene. In one, the two substituents, here methyl groups, are on the same double bond; in the other, they are on different double bonds.

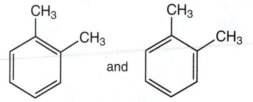

and

As there is only one real molecule, this poses severe problems for a 1,3,5-cyclohexatriene picture of benzene.

Problem 14.3 There are three contributing resonance structures of benzaldehyde (**A**) that introduce charge into the structure. There is one resonance structure (**A′**) that simply moves the aromatic ring double bonds. The important point is to recognize that the carbonyl group of benzaldehyde can withdraw electrons from the aromatic ring by resonance. The 1,3,5-cyclohexatriene (bond length differences exaggerated) would not have the ability to have further resonance through the ring.

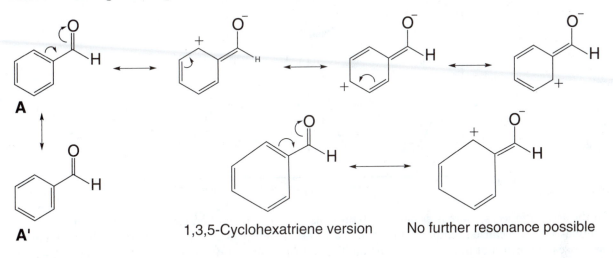

A

A′   1,3,5-Cyclohexatriene version   No further resonance possible

Problem 14.5 The ortho and para carbons of anisole will bear a partial negative charge because of the resonance shown here. The hydrogens that are on the carbons with a partial negative charge will be shielded. That means that they will be upfield from the typical δ 7.3 ppm for a hydrogen on an aromatic ring. Notice that the hydrogens on the meta carbons are not affected by the resonance, and so the meta hydrogens appear at the standard chemical shift.

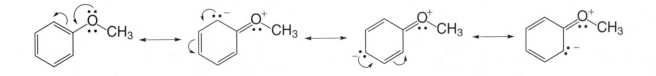

Problem 14.6 In this case, in the real Dewar benzene, the central bond is a σ bond. The two "hinge" carbons are hybridized approximately sp^3, and the molecule is shaped like an open book.

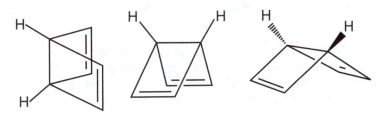

Three views of Dewar benzene

Problem 14.7 In the "tub-flip" process, the first step is conversion of one tub into the other by passing over a planar, bond-localized transition state. This transition state is not a regular octagon as the relatively long single bonds and relatively short double bonds are maintained. This process has an activation energy of 14.6 kcal/mol. The double bonds do not change position (see figure).

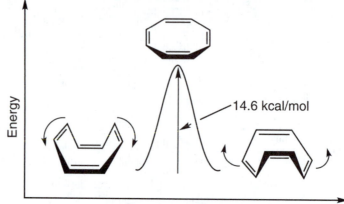

In the second process, there is a "bond switching" as well. The transition state in this reaction is the symmetrical, delocalized regular octagon. The activation energy is a bit higher, 17.0 kcal/mol. The two processes differ in that in the upper reaction, the eight-membered ring contains alternating single and double bonds. In the lower reaction, the transition state is the symmetrical, regular octagon.

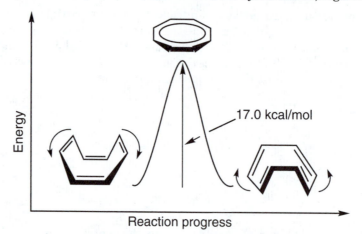

Problem 14.8

(a) Cyclononatetraene does not have a p orbital at every atom in the ring. This planar structure would be nonaromatic.

(b) The 14-membered ring that is completely conjugated will have a p orbital at every atom. It is cyclic, and there are 14 π electrons (each p orbital will have one electron). This fits the $4n + 2$ pattern (if $n = 3$, then $4n + 2 = 14$). So this molecule will be aromatic as long as the steric interactions of the hydrogens pointing into the ring do not force the molecule out of planarity.

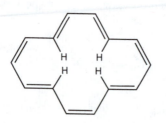

(c) This molecule is bicyclo[3.3.0]octatetraene. There is a *p* orbital at every atom. The molecule is flat and cyclic. There are eight π electrons, so the molecule is antiaromatic (eight electrons is $4n$, where $n = 2$ in this case). In fact, this molecule is so unstable it has never been isolated.

(d) This molecule is bicyclo[4.2.0]octatetraene. There is a *p* orbital at every atom. The molecule is flat and cyclic. There are eight π electrons through the two rings, so the overall molecule is antiaromatic. However, the left ring is a benzene with six π electrons, and it will be aromatic. The result is that the two π electrons in the four-membered ring on the right don't delocalize. The molecule has an aromatic six-membered ring.

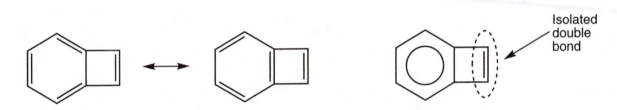

Isolated double bond

(e) This molecule is cyclononatriene. The nine-membered ring does not have a *p* orbital at every atom. The three π bonds potentially could "see" through space to overlap and give some delocalization and aromaticity, but the hydrogens on C(1) and C(6) will keep the triene from being planar. So the ring is nonaromatic.

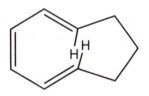

(f) This molecule is methylenecyclopentadiene. It has a *p* orbital at each atom in the ring. However, there are only five π electrons in this ring. Each ring carbon provides one π electron. This molecule does not have $4n$ (4, 8, 12, ...) nor does it have $4n + 2$ (2, 6, 10, 14, ...) π electrons. Therefore it is nonaromatic. There is an aromatic resonance structure that would contribute to this molecule. But the instability of a primary cation and the introduction of charge would outweigh the gain of aromaticity.

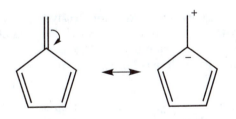

(g) This molecule is naphthalene (bicyclo[4.4.0]decapentaene). It has a *p* orbital at every atom around the outside ring (10 atoms). It is able to be planar, and there are 10 π electrons, which fits $4n + 2$ ($n = 2$) in this case. It is aromatic.

(h) This molecule is cyclodecapentaene. It is very similar to naphthalene, but the hydrogens on the trans double bonds are pointing into the ring and force the ring to be nonplanar. So on paper, this ring is aromatic (*p* orbital at every atom, cyclic, and 10 π electrons), but in reality the molecule is not planar and therefore is nonaromatic.

Problem 14.9   These are the 10 resonance structures that have the cation on different carbons. If you want to show resonance structures that just move the double bonds around the rings, then there are at least 30 more representations!

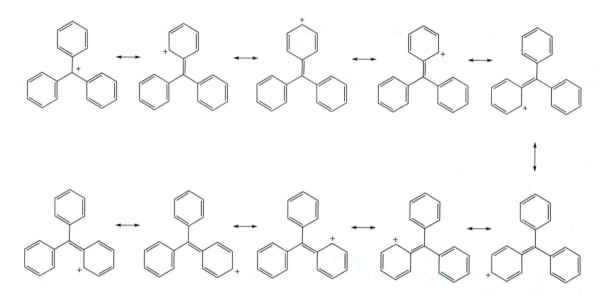

Problem 14.10

Induction: Chloroacetic acid is more acidic than acetic acid because the conjugate base of chloroacetic acid is stabilized by induction provided by the chloro group. The conjugate base of acetic acid is not as stable.

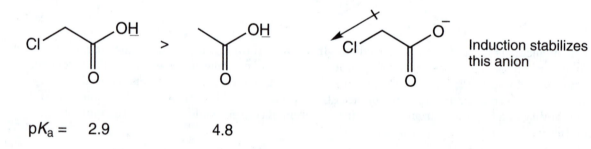

pK$_a$ =   2.9   4.8

Size: The SH hydrogen of a thiol is more acidic than an OH hydrogen of an alcohol because of the size of the atom to which the hydrogen is bonded. The S—H bond is not as strong as the O—H bond. Perhaps this is because of poor orbital overlap between the $3sp^3$ electrons of sulfur and the $1s$ electrons of hydrogen. In addition, the sulfur is better able to stabilize the conjugate base.

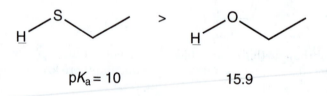

pK$_a$ = 10   15.9

Hybridization: The hydrogen on an *sp*-hybridized carbon is more acidic than a hydrogen on an sp^2-hybridized carbon, which is more acidic than a hydrogen on an sp^3-hybridized carbon. The conjugate base of the *sp* carbanion is more stable than the sp^2 carbanion, which is more stable than an sp^3 carbanion.

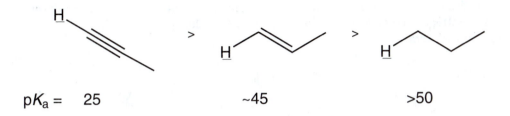

$pK_a =$ 25   ~45   >50

Aromaticity: The diallylic hydrogen on cyclopentadiene is more acidic than the diallylic hydrogen on 1,4-pentadiene or the allylic hydrogen of cyclopentene because the conjugate base of cyclopentadiene is aromatic.

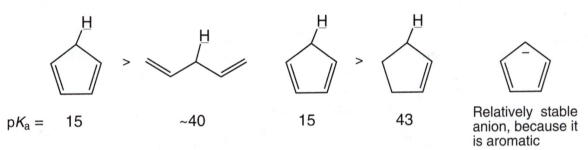

$pK_a =$ 15   ~40   15   43   Relatively stable anion, because it is aromatic

Resonance: The OH hydrogen of phenol is more acidic than a typical alcohol OH hydrogen as a result of resonance. The conjugate base of the phenoxide is stabilized by resonance. Aromaticity does not contribute to the acidity of phenol.

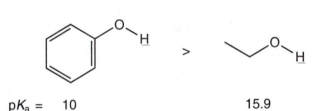
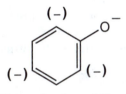

$pK_a =$ 10   15.9   This anion is stabilized by resonance

Electronegativity: If we compare hydrogens attached to atoms in the same row (i.e., C, N, O, F), we find that hydrogens on more electronegative atoms are more acidic than hydrogens on less electronegative atoms. The more electronegative is A in HA, the more acidic is HA.

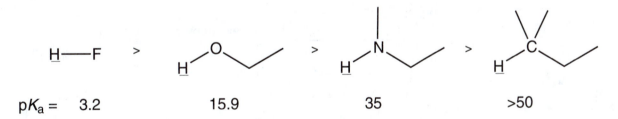

$pK_a =$ 3.2   15.9   35   >50

Problem 14.11   The ^{1}H NMR spectrum for cyclopentadiene anion shows a single signal. All of the hydrogens in this molecule are equivalent. Equivalent hydrogens do not couple, therefore the signal will be a singlet. Because there is considerable shielding for any anion, and each carbon of cyclopentadienyl anion shares the negative charge, the ^{1}H NMR chemical shift for the cyclopentadiene anion singlet is δ 5.6 ppm rather than the δ 7.3 ppm that we would expect for the aromatic ring hydrogens. The tropylium cation also has only one signal, and it is a singlet because the hydrogens in the tropylium cation are equivalent. The ^{1}H NMR chemical shift for the cation is δ 9.2 ppm, which reflects the downfield shift because of aromaticity and a downfield shift that results from a cationic charge on each carbon, which deshields each hydrogen.

Problem 14.12 The cyclopropenyl cation is an aromatic system (planar, cyclic, fully conjugated), with $4n + 2$ ($n = 0$) π electrons. Accordingly, despite the obvious strain and the presence of a positive charge, it is easy to form. For a cation, it is a very stable species.

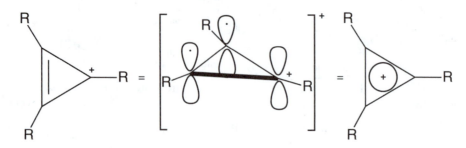

A Frost circle shows the molecular orbital system. For the cyclopropenyl cation, there are only two π electrons. The bonding molecular orbital is full, and there are no electrons in the antibonding orbitals.

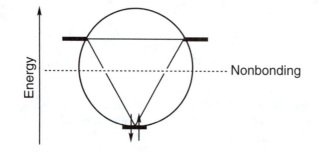

The cyclopropenyl anion presents a very different situation. Now there are four electrons, not a $4n + 2$ number, and antibonding molecular orbitals must be occupied. This system would not be expected to be stable (and it most definitely is not).

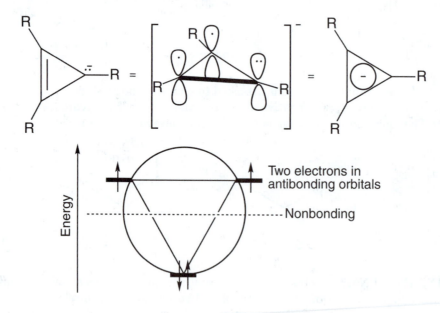

Problem 14.13 A trans double bond in a six-membered ring is too strained to be stable. Not only is one hydrogen "inside" and bumping into the other atoms, but the trans double bond is hideously twisted and thus most unstable. Try to make a model.

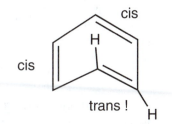

Problem 14.14 The first molecule, the all-cis isomer, has only one kind of carbon and therefore only one ^{13}C NMR signal. The second has an "inside" carbon and five different "outside" carbons. The di-trans molecule has two "inside" carbons and only two different "outside" carbons.

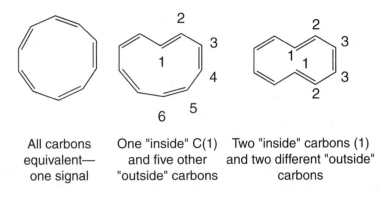

All carbons equivalent— one signal

One "inside" C(1) and five other "outside" carbons

Two "inside" carbons (1) and two different "outside" carbons

Problem 14.15 Let's assume that the double bonds in each cyclodecapentaene are in full resonance and the molecules are flat. As you read on in the text, you will see that this is not the case. But we can predict the spectra for the idealized molecules. In the all-(Z) isomer, the hydrogens are all equivalent. There would be no coupling between the equivalent hydrogens. The singlet would be between δ 5.5 ppm (the chemical shift of a vinyl hydrogen) and δ 7.3 ppm (the chemical shift of the standard aromatic ring).

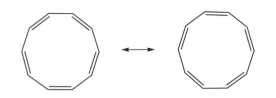

The (E,Z,Z,Z,Z) isomer has a single hydrogen that resides inside the ring (H_a). It would appear significantly upfield. We can predict its chemical shift to be around δ −3 ppm. It would be a triplet because it is vicinal to two H_b hydrogens. We will not try to predict the long-range coupling. The chemical shift for the other hydrogens (H_b to H_f) should be between δ 5.5 ppm and δ 7.3 ppm. The H_b hydrogens will appear as a doublet of doublets because H_b will couple differently with H_a (180° coupling) and H_c (0° coupling). The H_c, H_d, and H_e hydrogens would technically be doublets of doublets, but each would likely appear as a triplet because the angles and distances would be the same. The H_f hydrogen would be a triplet.

(continued)

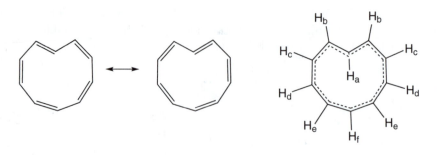

The flat, fully conjugated (*E,Z,E,Z,Z*) isomer would have two equivalent hydrogens (H$_a$) inside the ring. The chemical shift for H$_a$ hydrogens would be about δ −3 ppm, and the signal would be a triplet because H$_a$ is vicinal to two H$_b$ hydrogens. The chemical shift for the H$_b$ and H$_c$ hydrogens would be between δ 5.5 ppm (vinyl hydrogen) and δ 7.3 ppm (typical aromatic hydrogen). The coupling for H$_b$ hydrogens would be a doublet of doublets because H$_b$ would couple differently with H$_a$ (180° coupling) and H$_c$ (0° coupling).

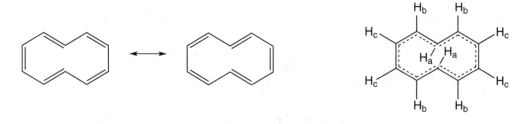

Problem 14.16   Here is the reaction shown for only one cyclodecapentaene. This picture is only a crude approximation as the cyclodecapentaenes are not really planar. Although it looks as if the "reach" across the ring to make the new bond is a long one, in reality the nonplanar structures bring the orbitals quite close together and make the bond-making process easy.

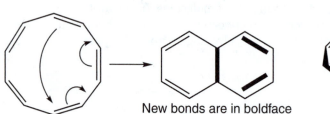

New bonds are in boldface

Side view showing two orbitals bending in to make the cross-ring bond

Problem 14.17   The first two parts are easy, and it is relatively simple to find that there are more than three possible disubstituted versions. For Ladenburg benzene (prismane), there are only three positional possibilities, but one is chiral. So none of these molecules can be the real benzene.

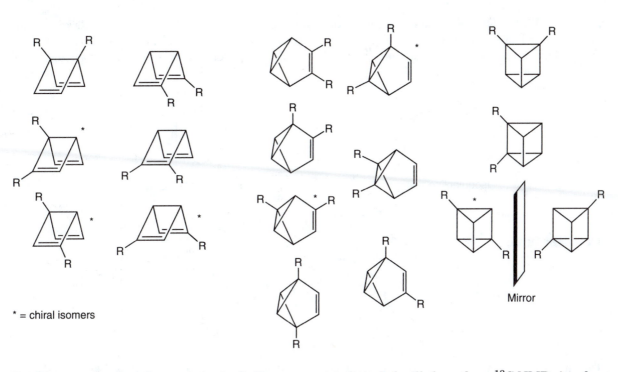

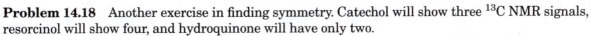

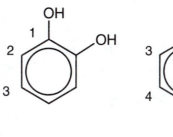

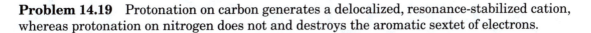

* = chiral isomers

Mirror

Problem 14.18 Another exercise in finding symmetry. Catechol will show three ^{13}C NMR signals, resorcinol will show four, and hydroquinone will have only two.

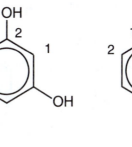

Catechol   Resorcinol   Hydroquinone

Problem 14.19 Protonation on carbon generates a delocalized, resonance-stabilized cation, whereas protonation on nitrogen does not and destroys the aromatic sextet of electrons.

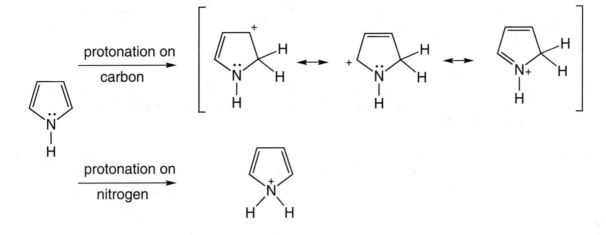

Problem 14.20   Quinoline has a nitrogen replacing the naphthalene C(1) carbon and its hydrogen. Isoquinoline has the nitrogen replacing the C(2) carbon and its hydrogen. Putting the nitrogen at the bridgehead is possible, but as there is no hydrogen there to replace, it is not an isomer, and it will be a cation.

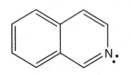

Isoquinoline

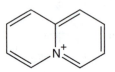

Not an isomer of quinoline

Problem 14.21

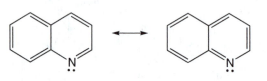

Problem 14.23   A tricky one. If you just draw circles, you are in big trouble. First of all, this is an isomer problem much like the ones in the early chapters, and you must evolve a system to be sure you have all the possible ring systems. We like to start with a linear array, and then increasingly complicate things.

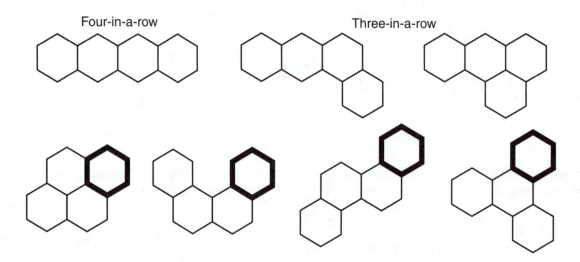

Two-in-a-row (but watch out for repeats of three-in-a-row isomers); in this scheme, one new ring is added at the upper right corner (boldface) and the fourth ring is moved around the frame

Now add double bonds. If you draw them as circles, all of the above frames will work, but this result isn't correct! As in Problem 14.22, there is one frame for which it is impossible to fit the proper number of double bonds into the ring system. Therefore, it cannot be aromatic.

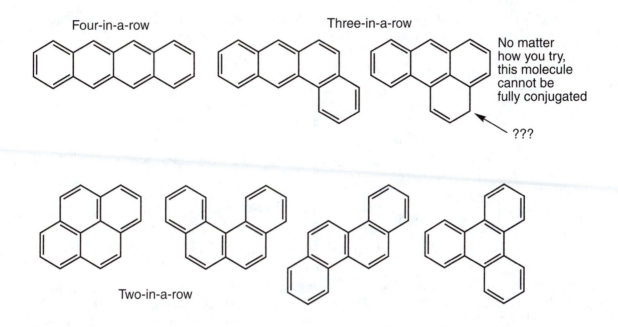

Four-in-a-row   Three-in-a-row

No matter how you try, this molecule cannot be fully conjugated

???

Two-in-a-row

Problem 14.24 Hexahelicene is drawn by adding a sixth ring to the five-ring aromatic at the indicated points.

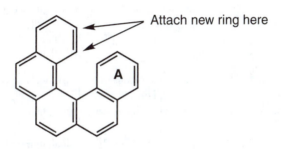

Attach new ring here

A

The problem is that the new ring overlaps with ring **A**. Hexahelicene must twist so that the new ring and ring **A** lie on top of each other. The molecule must be curled in either a right- or left-handed screw form.

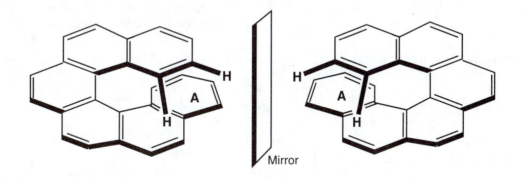

Mirror

(continued)

Problem 14.24 *(continued)*

"Twistoflex" is also a molecule that adopts a screw shape, but this time it is along a longitudinal axis, not the vertical direction as with hexaihelicene.

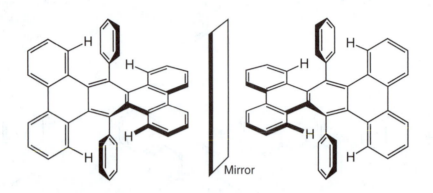

Problem 14.25   In both molecules, we can imagine an ionization to give a carbocation. For an adjacent benzene ring to stabilize a positive charge, the orbitals of the ring must overlap with the empty orbital on the benzylic carbon adjacent to the ring. For the trityl system, this is quite easy.

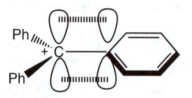

In the trityl cation, overlap between the ring orbitals and the 2*p* orbital on the benzyl position is easy; in this figure only one benzene ring is shown in detail, and only one of the six 2*p* orbitals making up the benzene π orbitals is drawn in

The situation is quite different for triptycenyl. Although the positively charged carbon is flanked by three benzene rings, just as is the case in trityl, the orbitals do not overlap well. They are essentially perpendicular, and the benzene rings offer no stabilization to the adjacent empty orbital.

There is essentially no orbital overlap here; this cation is not resonance stabilized!

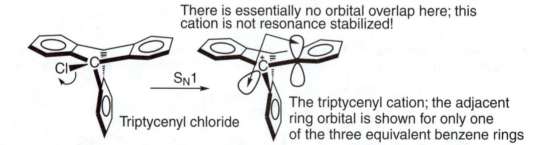

Triptycenyl chloride

The triptycenyl cation; the adjacent ring orbital is shown for only one of the three equivalent benzene rings

Problem 14.26   The mechanism is exactly parallel to that for allylic bromination using NBS (p. 577). The first step is formation of HBr and bromine atoms.

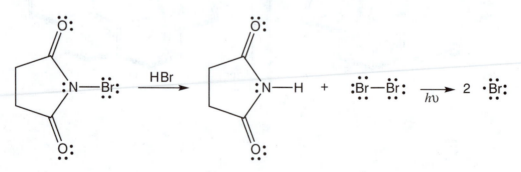

Next, the bromine atom abstracts a benzylic hydrogen from the methyl group of toluene to produce hydrogen bromide and a resonance-stabilized benzyl radical.

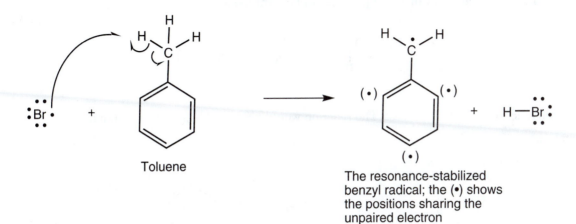

Toluene

The resonance-stabilized benzyl radical; the (•) shows the positions sharing the unpaired electron

The benzyl radical now abstracts a bromine atom from bromine to give benzyl bromide and a bromine atom that goes on to propagate the chain reaction by abstracting a hydrogen from toluene.

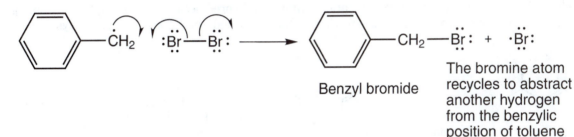

Benzyl bromide

The bromine atom recycles to abstract another hydrogen from the benzylic position of toluene

Problem 14.27 The NBS reaction involves a radical hydrogen atom abstraction. The hydrogen atom abstraction from C(1), the benzylic position, produces a more stable intermediate than abstraction from the C(2) position. The benzylic radical is stabilized by resonance. The tertiary radical is only stabilized by hyperconjugation.

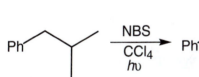

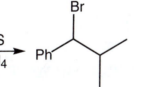

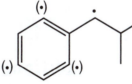

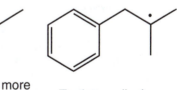

2-Methyl-1-phenyl-propane

1-Bromo-2-methyl-1-phenylpropane

Benzylic radical

more stable than

Tertiary radical

Problem 14.28

(a) Potassium permanganate oxidizes the alkyl side chains of an aromatic ring as long as there is at least one benzylic hydrogen.

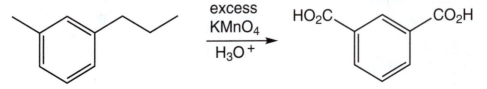

(continued)

Problem 14.28 *(continued)*

(b) This reaction is a bimolecular elimination (*E*2). The major product is (*E*)-1-phenylpropene because it is the more stable product and therefore has a lower transition state leading to it. There might be a minor amount of the (*Z*) isomer and perhaps 3-phenylpropene.

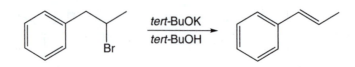

(c) This reaction is an S_N2 process. The reaction would presumably go with inversion.

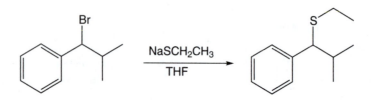

(d) This reaction is an S_N1 solvolysis. The relatively stable benzylic cation will be formed when the chloro group leaves. The water can be added in an S_N1 reaction and then be deprotonated to produce the major product. Rearrangement of the benzylic cation will not be favored because the methyl shift would give the less stable tertiary cation. Once the tertiary cation is formed, the E1 and S_N1 pathways will compete.

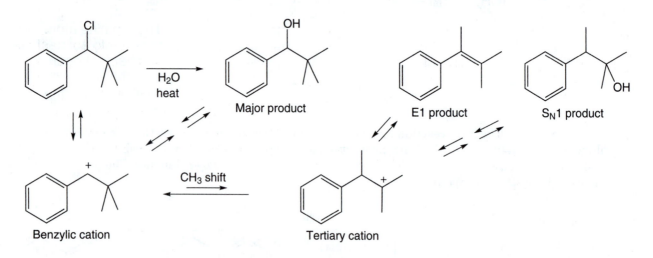

Additional Problem Answers

Problem 14.31

(a) *m*-bromofluorobenzene or 1-bromo-3-fluorobenzene

(b) *o*-iodotoluene or 2-iodotoluene

(c) *o*-nitrophenol or 2-nitrophenol

(d) *p*-dibromobenzene or 1,4-dibromobenzene

(e) 1-bromo-2-chloro-3-iodobenzene

(f) 2,5-dichloroaniline

Problem 14.32 Nothing tricky here—just two repeats as (b) and (c) are the same, as are (e) and (f).

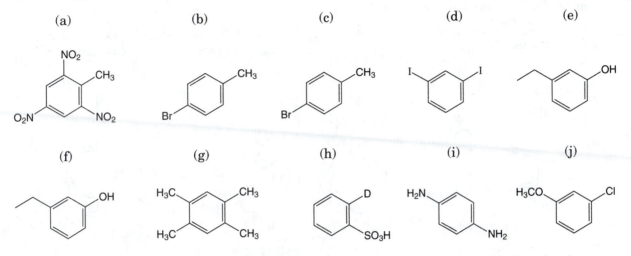

(a) (b) (c) (d) (e)

(f) (g) (h) (i) (j)

Problem 14.33 A *potential* problem with the bromonium ion that would be formed in bromination of the alkyne is that it would be antiaromatic. The bromine in the three-membered ring might be *sp²* hybridized and therefore would have a 4*p* orbital that would be able to delocalize with the remaining π bond. There would be four π electrons in the planar ring system, which would make the intermediate antiaromatic and very unstable.

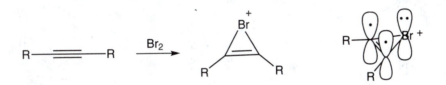

Problem 14.34 Here are the Lewis structures.

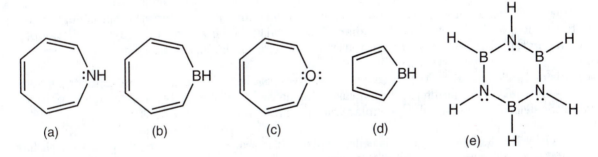

(a) (b) (c) (d) (e)

Compound (a), azepin, has eight π electrons and is not aromatic.

Compound (b), borepin, is aromatic. Remember that there is an empty 2*p* orbital on boron. This molecule has a *p* orbital at every atom, and its six π electrons will fit nicely into the set of three bonding molecular orbitals.

Compound (c), oxepin, has only eight π electrons despite the apparent count of 10. Two of the nonbonding electrons on oxygen are orthogonal to the π system. It is not aromatic.

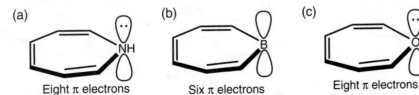

(a) (b) (c)

Eight π electrons   Six π electrons   Eight π electrons

(continued)

Problem 14.34 *(continued)*

Compound (d) has only four π electrons and cannot be counted as aromatic. The set of three bonding molecular orbitals will not be optimally filled.

Compound (e) is probably the most interesting of them all. Like the six-membered ring benzene, there is a set of six parallel 2p orbitals and six electrons to fill them. This molecule, borazine, is aromatic.

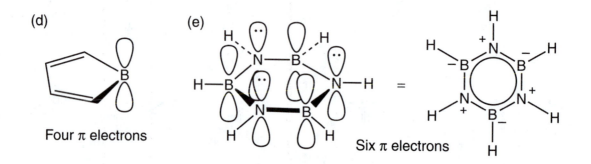

Four π electrons   Six π electrons

Problem 14.35   There are three, and they are all aromatic. All the lone-pair electrons on the nitrogens are orthogonal to the π system.

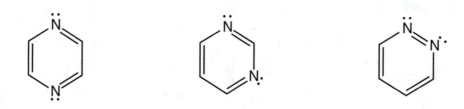

Problem 14.36

(a) This tricyclic molecule has 12 π electrons around the outer ring. Perhaps there is enough flexibility for the molecule to avoid being planar. The central carbon is sp^3 hybridized, so it will not be involved in the π system. To the extent that the outer ring is planar, this molecule is antiaromatic.

(b) This molecule is a 16 π electron ring system. It will not be stable as a planar conjugated system because it would be antiaromatic in that case. The molecule will certainly have enough flexibility to avoid the delocalization. It will be nonplanar and nonaromatic.

(c) This bicyclo[5.3.0]decapentaene is called azulene. It is aromatic. There are 10 π electrons, it has a p orbital at every atom in the ring, and it is planar.

(d) This molecule has an interesting structure. Drawn as shown on the left, both rings are antiaromatic. One has eight π electrons and the other has four π electrons. But look at the other two resonance structures. Counting the π electrons around the outer ring gives us 10 π electrons, which is aromatic ($4n + 2$, $n = 2$). There is a p orbital at every atom. You should predict that this molecule is aromatic or at least has some degree of aromaticity.

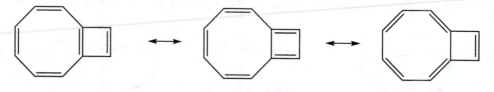

(e) Although there is a *p* orbital at every atom in the ring, the *p* orbitals don't all overlap with each other. The allene portion of the ring can have one of its double bonds aligned with the π system of the ring, but the other double bond of the allene will have its π electrons perpendicular to the rest of the ring system. So we can't count both double bonds of the allene, and there isn't a usable *p* orbital at each atom of the ring. This ring is nonaromatic.

(f) The methylene-substituted cycloheptatriene has a *p* orbital at each atom, but there are only seven π electrons in the ring system. This molecule is nonaromatic. An aromatic resonance structure might contribute slightly. It would require the formation of an unstabilized primary carbanion in order to have six π electrons in the ring system.

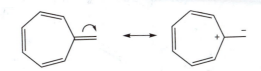

(g) This cyclodecatriene does not have a *p* orbital at every atom. It is nonaromatic.

(h) This bicyclic molecule has one ring that is aromatic. The ring on the left has six π electrons and a *p* orbital at every atom. The ring on the right does not have a *p* orbital at every atom and is therefore nonaromatic. We would describe this molecule as aromatic. It has at least one ring that is aromatic.

Problem 14.37

(a) The cyclobutenyl cation does not have a *p* orbital at every atom. It is nonaromatic.

(b) The cyclobutenyl dication has a *p* orbital at every atom in the ring and has two π electrons. It is very stable for a dication. It is an aromatic ring.

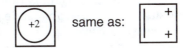

(c) The cyclopentadienyl radical has a *p* orbital at every atom, but there are only five π electrons in the ring. It is nonaromatic.

(d) Cyclononatetraene anion has a *p* orbital at every atom, and there are 10 π electrons in the ring. The ring size might make it difficult to maintain planarity, but the inability to be planar is difficult to predict in this particular case. We should predict that this intermediate is aromatic.

(e) The cyclohexadienyl radical does not have a *p* orbital at every atom. It is nonaromatic.

(f) The cyclohexadienyl diradical has a *p* orbital at every atom and 6 π electrons in the ring. It is aromatic.

(g) This molecule is called benzyne. We will learn more about it in Chapter 15. The six π electrons in the *p* orbitals of the ring are able to delocalize through the ring. Both the empty orbital and the orbital containing the lone pair are perpendicular to the ring, as shown below. The charges do not significantly impact the aromaticity of the ring.

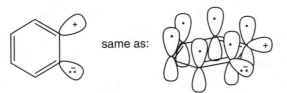

(h) The cyclopropenyl anion has a *p* orbital at every atom. It is a planar ring, and there are 4*n* π electrons. It is antiaromatic.

Problem 14.38   There are, of course, many possible answers to each part of this problem. The point is to construct molecules with $4n + 2$ π electrons. In the answer to (a), boron contributes no π electrons but does maintain orbital connectivity, the cycle of conjugation, because of its empty $2p_z$ orbital. In (b), one of the oxygen lone pairs is in the π system, but the other is in the perpendicular σ system. The same is true of (c), and the nitrogen lone pair is also orthogonal to the π system. In (d), only the lone pair on the NH is in the π system. In (e), each nitrogen contributes an electron pair to the π system, and each boron maintains conjugation through its empty $2p_z$ orbital. You saw this molecule earlier in Problem 14.34e.

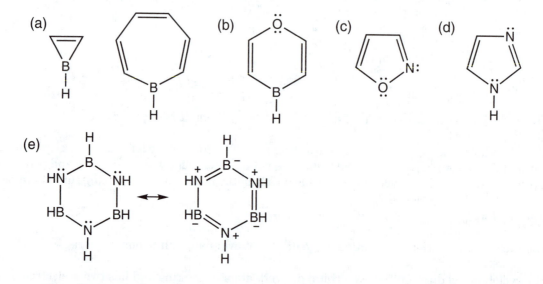

Problem 14.39   Well, it's at least very hard to do what this problem asks. The difficulty is that either there are too many π electrons, as in (a), or conjugation isn't maintained, as in (b). The question doesn't say "only" oxygen and nitrogen, so a possible answer is to introduce a boron to complete the conjugation, as in (c). Now there are six π electrons in a fully conjugated system. The two electrons on nitrogen and one electron pair on oxygen are in the σ system.

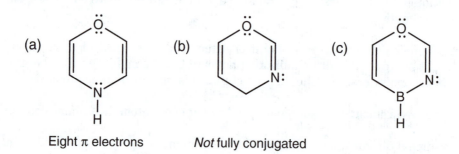

Problem 14.40   The 5-bromo compound is all too prone to elimination to give benzene. The great stability of benzene makes the transition state for elimination lie relatively low in energy. The 1-bromo compound cannot do an elimination—the bromine is in the wrong position.

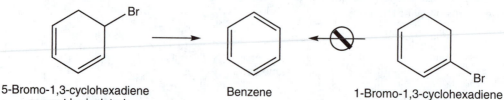

Problem 14.41 Both azirine isomers would surely be highly strained, as the three-membered ring bond angles will be far from the ideal sp^3 or sp^2 angles. In addition, isomer **1** will be destabilized by the presence of two electrons in degenerate antibonding orbitals, making it an antiaromatic diradical. By contrast, isomer **2** is not even a conjugated system, as the sp^3 carbon does not maintain the required orbital connectivity. Moreover, the nitrogen lone-pair electrons are in an approximately sp^2 orbital perpendicular to the π system. Thus, the two π electrons of **2** are nicely accommodated in the π bonding molecular orbital.

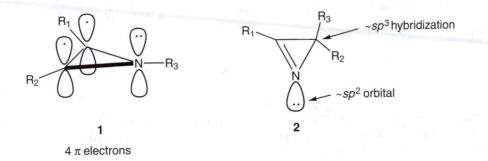

Problem 14.42 The dipole moment of THF reflects the relative electronegativity of oxygen. Of course, the oxygen in furan is also more electronegative than carbon, but this effect is attenuated by resonance forms such as **A**, in which the dipole is in the other direction. The net result is a reduced dipole.

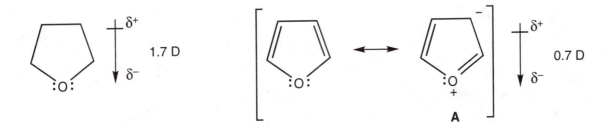

Problem 14.43 If we estimate the heat of hydrogenation for cyclooctatetraene as four times that of cyclooctene, then we predict 4 × 23 = 92 kcal/mol, not much below the experimental number of 101 kcal/mol. No case can be made that cyclooctatetraene is especially stable. By contrast, this molecule appears to be less stable than anticipated.

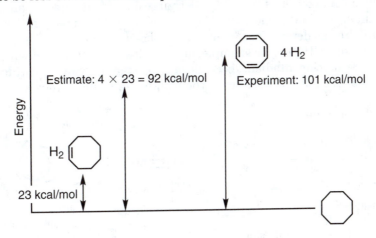

Of course, experiment bears this out: Cyclooctatetraene is a tub-shaped, decidedly nonplanar molecule and reacts as a simple polyene.

Problem 14.44   There are several clues as to what is going on. The presence of hydrogen bromide might evoke thoughts of an elimination reaction once bromine has added, for example.

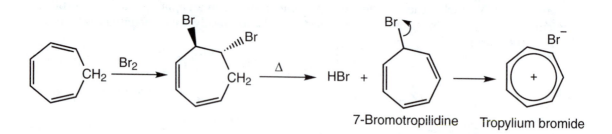

7-Bromotropilidene is the result, and this molecule easily ionizes to give the very stable tropylium bromide, the yellow solid. Reaction with water gives tropyl alcohol, a molecule that can react again with tropylium bromide to give ditropyl ether.

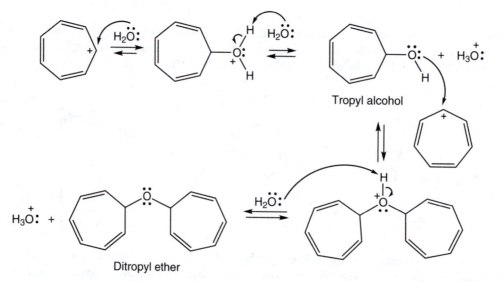

Problem 14.45   Look at the resonance description of diazocyclopentadiene. This molecule is sharply stabilized by delocalization of the negative charge on carbon. There is an aromatic cyclopentadienide anion within this molecule. This compound is still a *potential* explosive, however, and there are several famous detonations involving this beautiful material. Very unstable diazo compounds, diazomethane (CH_2N_2), for example, are handled in solution and with great caution. Compounds such as diazocyclopentadiene, because of their stability, can be distilled and handled in bulk. When the occasional explosion does occur, the results can be disastrous.

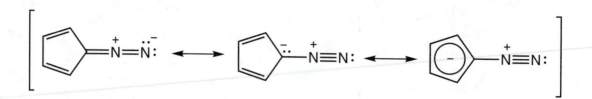

Problem 14.46 Dihydroxylation with OsO_4 will result in reaction at the alkene, but not on the aromatic ring. Keep in mind that the aromatic ring π system is not as reactive with electrophiles as is the alkene.

Problem 14.47

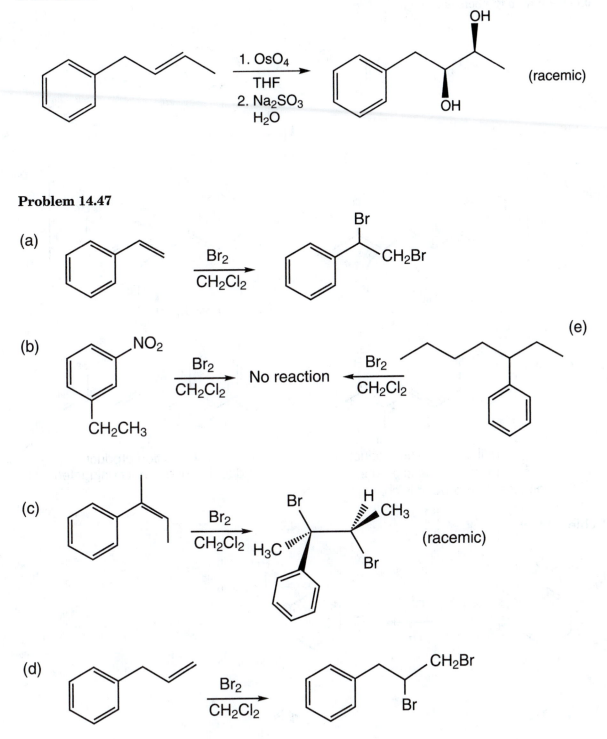

Problem 14.48 Kinetic control will lead to 1,2-addition as shown. Note that the structure of the product tells you which double bond in the diene undergoes addition. Thermodynamic control will also lead to 1,2-addition in this case because 1,4-addition would lose the conjugation of the remaining double bond with the benzene ring.

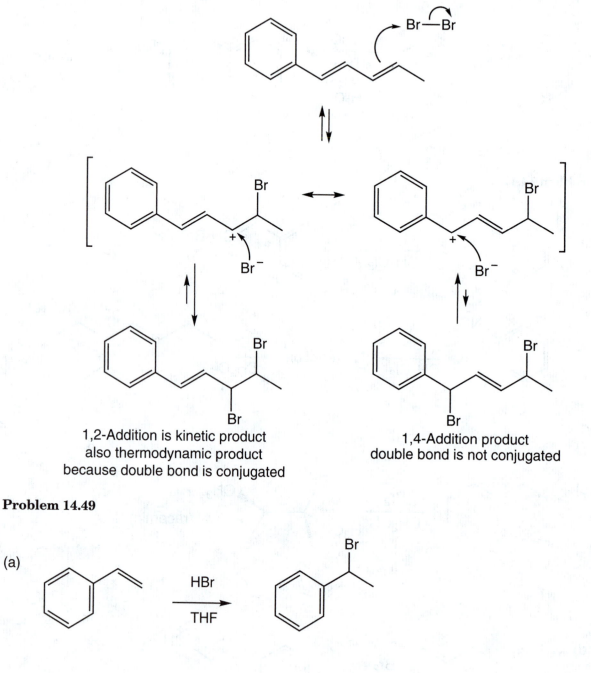

1,2-Addition is kinetic product
also thermodynamic product
because double bond is conjugated

1,4-Addition product
double bond is not conjugated

Problem 14.49

(a)

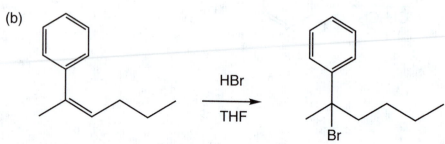

(b)

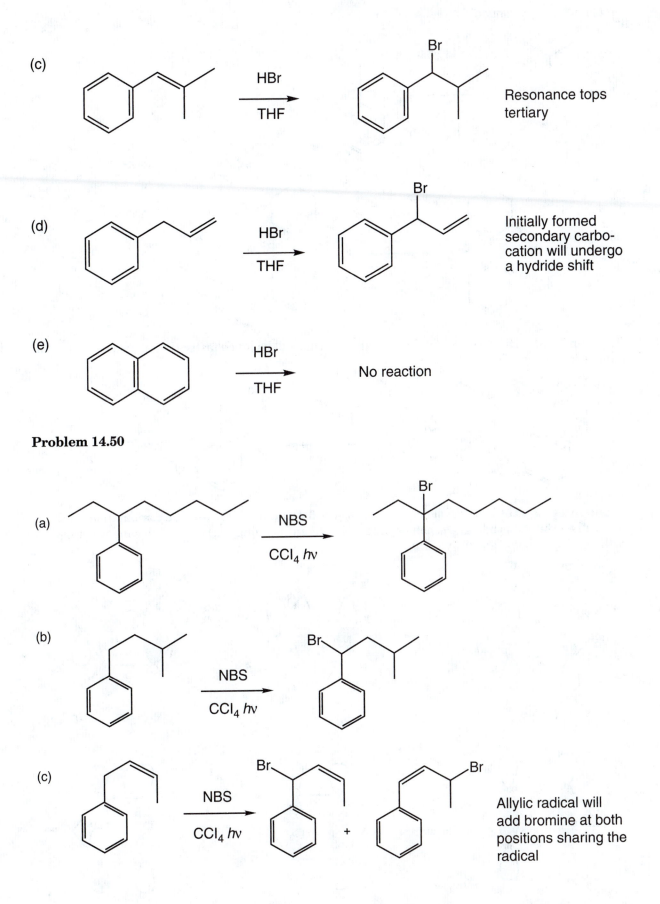

(c)

HBr
THF

Resonance tops
tertiary

(d)

HBr
THF

Initially formed
secondary carbo-
cation will undergo
a hydride shift

(e)

HBr
THF

No reaction

Problem 14.50

(a)

NBS
CCl$_4$ *hv*

(b)

NBS
CCl$_4$ *hv*

(c)

NBS
CCl$_4$ *hv*

+

Allylic radical will
add bromine at both
positions sharing the
radical

(continued)

Problem 14.50 *(continued)*

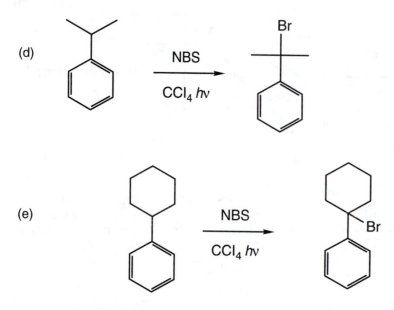

(d)

NBS

CCl$_4$ *hv*

(e)

NBS

CCl$_4$ *hv*

Problem 14.51 Displacement of the benzylic chlorine will be much faster than displacement of the remote secondary chlorine (p. 680).

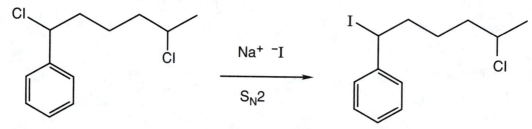

Na$^+$ $^-$I

S$_N$2

Problem 14.52 The benzene molecular orbitals, **1–6**, fall right out if we take combinations of the three allyl orbitals (bonding, Φ_B, nonbonding, Φ_N, and antibonding, Φ_A) in the following ways: Φ_B with Φ_B, Φ_N with Φ_N, and Φ_A with Φ_A.

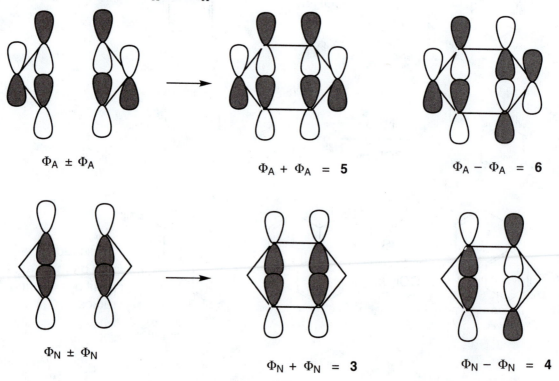

$\Phi_A \pm \Phi_A$　　　　　　$\Phi_A + \Phi_A = 5$　　　　　$\Phi_A - \Phi_A = 6$

$\Phi_N \pm \Phi_N$　　　　　　$\Phi_N + \Phi_N = 3$　　　　　$\Phi_N - \Phi_N = 4$

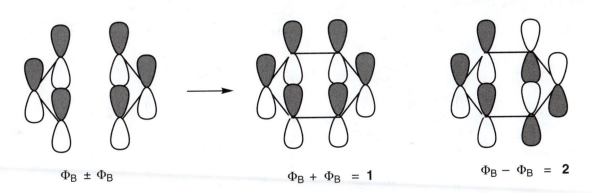

$$\Phi_B \pm \Phi_B \qquad\qquad \Phi_B + \Phi_B = \mathbf{1} \qquad\qquad \Phi_B - \Phi_B = \mathbf{2}$$

They can be ordered in energy by counting the nodes.

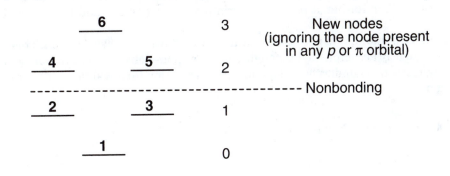

Problem 14.53

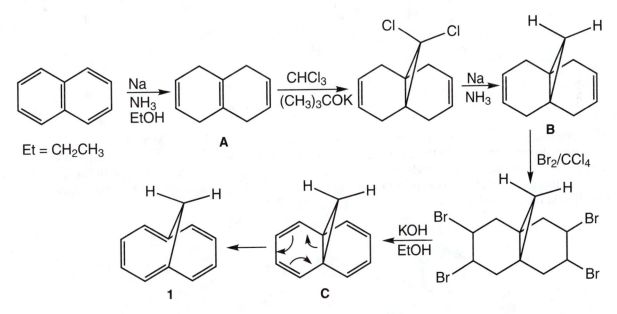

Et = CH_2CH_3

In this problem (or similar problems), it is not necessary to supply the exact reagents as long as you propose the same type of reagents. Here the first step is a double Birch reduction (p. 685) of naphthalene to give 1,4,5,8-tetrahydronaphthalene (**A**). The next two steps result in cyclopropanation of the more substituted carbon–carbon double bond. In the first of these two steps, dichlorocarbene, generated from chloroform and potassium *tert*-butoxide (p. 506), adds to the double bond. In the second step, the chlorines are removed in a reduction step to give **B**. You might wonder why this roundabout route was used. Why did Vogel and Roth not cyclopropanate the double bond using diazomethane?

(continued)

Problem 14.53 *(continued)*

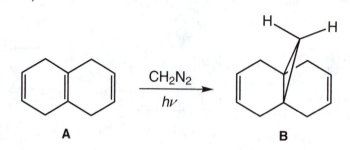

It is a question of selectivity. Dichlorocarbene adds selectively to the more substituted double bond, but methylene, generated by photolysis of diazomethane, is far more reactive and does not select one double bond over the other. Rather, it reacts indiscriminately to give two addition products.

With cyclopropane **B** in hand, it is necessary to introduce additional unsaturation. This transformation is conveniently accomplished through a bromination–dehydrobromination sequence to give **C**. Finally, tetraene **C** undergoes a smooth transformation (arrows) to the desired bridged cyclodecapentaene **1**.

Problem 14.54   Here are the Lewis structures.

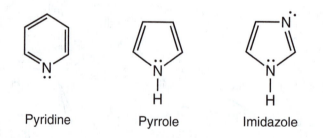

All three of these molecules are heteroaromatic compounds. To find a sensible answer, it is important to examine closely the lone-pair electrons on the nitrogen atoms in these compounds. In pyridine, the nitrogen lone-pair electrons are not part of the $(4n + 2)$ π system but lie in the plane of the ring in an sp^2 orbital that is part of the σ system. Accordingly, aromaticity is not disrupted when pyridine is protonated.

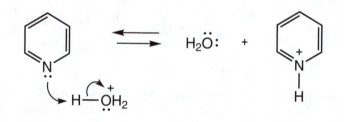

By contrast, as we saw in Problem 14.19, the nitrogen lone-pair electrons are part of the aromatic sextet in pyrrole. Protonation on nitrogen would result in a loss of aromaticity and is likely to be a very endothermic process.

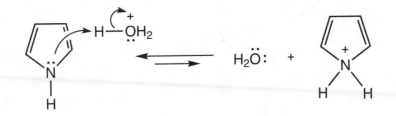

When pyrrole does protonate, it does so on carbon. Although aromaticity is still lost, at least a resonance-stabilized cation is formed.

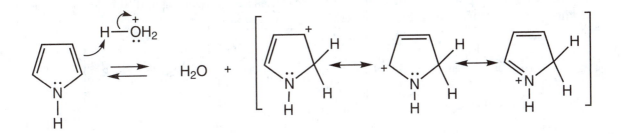

The two nitrogen lone pairs of imidazole are quite different. One nitrogen lone pair is part of the six π electron system and the other is not, as it lies in the σ system. Accordingly, protonation occurs on the latter lone pair as aromaticity is not lost.

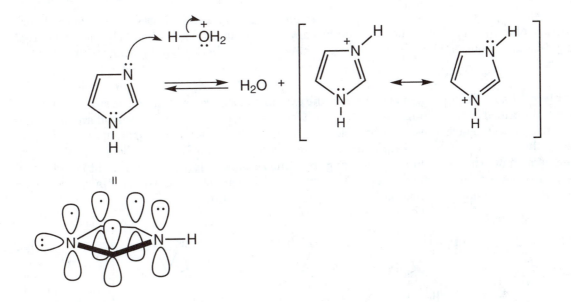

Problem 14.55   There are only two kinds of hydrogen in cyclopropene. Neither position is very acidic, but it will be easier to deprotonate at the vinyl position (1) than to make the very unstable (see Problem 14.12) cyclopropenyl anion by deprotonating at (2).

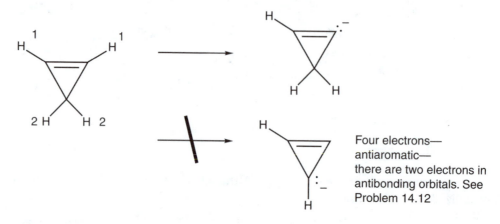

Four electrons—antiaromatic—there are two electrons in antibonding orbitals. See Problem 14.12

Problem 14.56   This problem is another case where circles are misleading and Kekulé forms are more appropriate. Anthracene can be represented by four resonance structures.

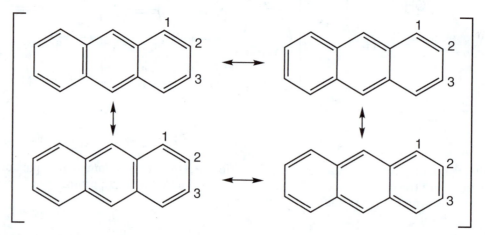

If we make the useful assumption that each resonance structure contributes equally to the hybrid structure of anthracene, then the 1,2-bond has approximately 3/4 double-bond character, whereas the 2,3-bond has only 1/4 double-bond character. Accordingly, the 1,2-bond is shorter than the 2,3-bond.

Problem 14.57   This problem is just a specific example of the general aromatic substitution process described in the chapter (p. 682). The nitronium ion, ($^+NO_2$), is a Lewis acid and first adds to the benzene ring to give a resonance-stabilized cyclohexadienyl cation. Removal of a proton by any base in the system completes the picture and regenerates the aromatic ring. Even though the arrow formalism shows this proton loss as deriving from one resonance form, do not fall into the trap of thinking of these forms as having separate existence. Write arrow formalisms for proton loss using the other two resonance forms.

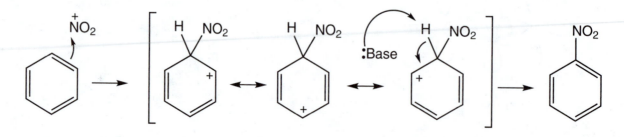

Problem 14.58 This molecule is just a fancy benzene. It has a 14 π electron ($4n + 2$, $n = 3$) perimeter and is aromatic. Aromatic substitution with $^+NO_2$ (nitration) proceeds as in Problem 14.57. In this case, the resonance structures of the intermediate are indicated in shorthand with (+) showing the positions sharing the positive charge. It is worth taking the time to write out the resonance forms. Removal of a proton regenerates the aromatic system.

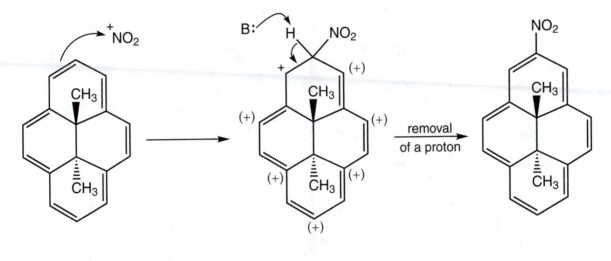

Problem 14.59 The person who reasoned that there would be 16 π electrons probably counted the alkyne as contributing four π electrons. If you count the π bonds, there are eight in the molecule. But you can't include both π bonds of an alkyne. Only one of the π bonds will be conjugated with the π system. The other π bond of the alkyne will be perpendicular, as shown below. So there are only 14 π electrons in this aromatic ring.

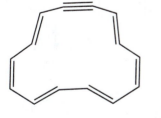

Looking down on the
p orbitals in the ring

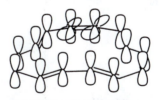

14 Electrons are in this
π system, therefore aromatic
($4n + 2$, $n = 3$)

Not part of the
π system!

Problem 14.60   This problem is an example of the Birch reduction with a slight, added twist. As implied in Figure 14.81, transfer of an electron from sodium to a benzene with an electron-withdrawing substituent produces a resonance-stabilized radical anion **A**. Apparently, activation by the carbonyl substituent permits reduction even in the absence of a proton donor alcohol. Further electron transfer affords anion **B**, which, in the absence of an alcohol, can be alkylated by methyl iodide in an S_N2 reaction.

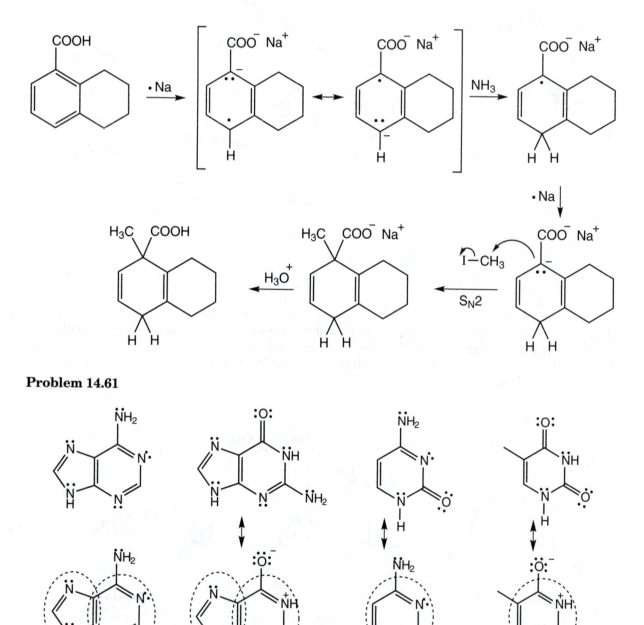

Problem 14.61

Adenine—both rings
aromatic

Guanine—both rings
aromatic

Cytosine—
aromatic ring

Thymine—
aromatic ring

Problem 14.62

(a) Anthracene adds across the 9,10-positions to form a single adduct. If you drew the product of addition across the 1,4-positions, that's not a bad answer, but it is the symmetrical compound that is really formed. See (b) for an analysis of why this occurs. Naphthalene adds across the 1,4-positions to give exo and endo adducts.

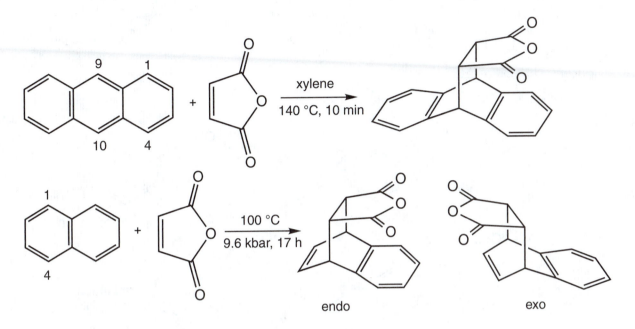

(b) The resonance energies of naphthalene and anthracene are 61 kcal/mol and 84 kcal/mol, respectively, as calculated from heats of combustion data. When anthracene undergoes the Diels–Alder reaction across the 9,10-positions, two intact benzene rings remain in the product. In the process, anthracene loses only about 25% of its resonance energy as one anthracene (84 kcal/mol resonance energy) becomes two benzene rings ($2 \times 32 = 64$ kcal/mol resonance energy). Addition across the 1,4-positions would leave behind a naphthalene, with a resonance energy (61 kcal/mol) less than that of two benzenes (about 64 kcal/mol). In contrast, when naphthalene undergoes the Diels–Alder reaction, about 45% of its resonance energy is lost as naphthalene (61 kcal/mol resonance energy) becomes one benzene ring (32 kcal/mol resonance energy). Although this is simply a statement about the thermodynamics of product energies and thus not directly relevant to questions of rate (kinetics), loss of aromatic resonance may already be considerable in the transition states leading to the products. Thus, a higher activation energy would be expected for the reaction with naphthalene.

Problem 14.63 Structure (a) would give *p-tert*-butylbenzoic acid upon oxidation with $KMnO_4$. The *tert*-butyl group is attached to the ring through a quaternary carbon and is not oxidized. Both side chains in (b) would be oxidized to acid groups. The product would be benzene-1,4-dicarboxylic acid (terephthalic acid). Finally, compound (c) would be untouched by permanganate, as it contains only a quaternary carbon adjacent to the ring.

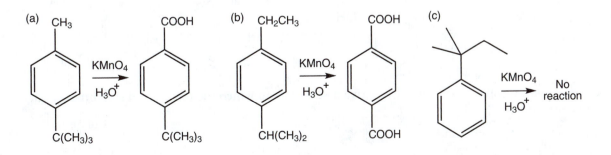

Problem 14.64

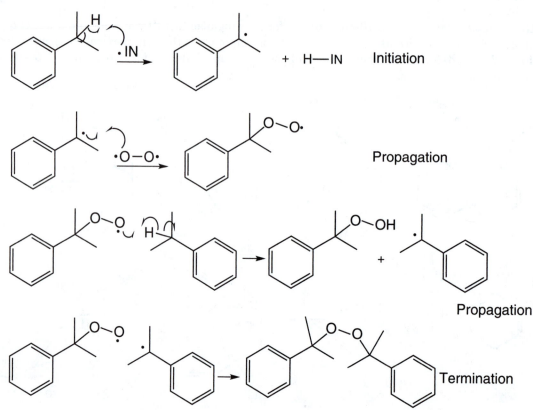

Problem 14.65   SOMO stands for singly occupied molecular orbital. The positions that are ortho and para to the alkyl group share the SOMO character. These locations contribute to the resonance hybrid structure of the benzyl radical. Perhaps this animation helps you understand that the benzyl radical is a mixure of the contributing structures (a hybrid species) rather than a set of equilibrating different structures.

Problem 14.66

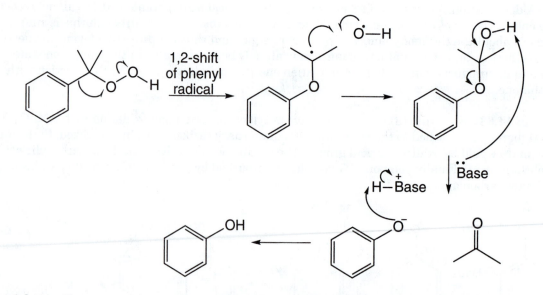

You might see if you can write an alternative mechanism in which the rearranged step produces a cation–anion pair.

Problem 14.67 The bromination probably occurs through syn addition because the bromide anion is born on the same side as the initial addition to the alkene. There is less distance for the bromide to travel to give the syn-addition product than to give the anti-addition product. The naphthylmethyl cation should be more stable. There are more resonance structures, and more electron delocalization almost always results in more stability.

Substitution Reactions of Aromatic Compounds

15

In this chapter, we see many reactions in which the aromatic ring is preserved and a few in which it is destroyed. The emphasis in the following problems is upon electrophilic aromatic substitution and the interplay between substituents on the ring. Synthesis appears in an important way as well, and retrosynthetic analysis emerges as an effective way to solve problems involving synthesis.

Mechanism problems are becoming increasingly important. In the answers to these problems, we emphasize the importance of analyzing problems before setting out to write structures and arrows.

Problem 15.1 Hydrogenation of benzene leads first to cyclohexadiene, which is hydrogenated further to cyclohexene. Cyclohexene, in turn, finally gives cyclohexane. However, the rates of hydrogenation of the nonaromatic cycloalkenes are much faster than that of the much more stable, and therefore much less reactive, benzene. In practice, as soon as a molecule of cyclohexadiene is formed, it is converted into cyclohexene, and then into cyclohexane much faster than more cyclohexadiene is produced. The nonaromatic intermediates are present, but their concentrations can never build up.

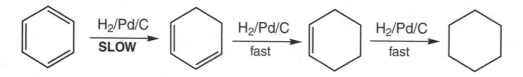

Problem 15.2 This molecule has a long carbon chain that would be lipophilic and a polar sulfate group that would be hydrophilic. That means it will be a good molecule for forming micelles. We will learn about micelles in Chapter 17. But for now, you know the long carbon chain tails will be attracted to each other and away from a polar solvent such as water. The lipophilic tails can gather around grease spots, and the polar sulfate groups will allow for the grease spot to be carried off with water.

The degradation of this molecule might occur in many ways. One method that we have learned about is the oxidation of side chains on the aromatic ring (p. 681). This process would make water-soluble carboxylic acids. The same organisms that decompose animal fat would probably chew up the detergent.

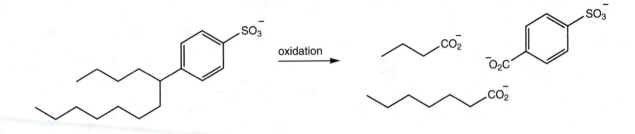

Problem 15.4 The mechanism exactly parallels that of Figure 15.27. You need only substitute Br for Cl.

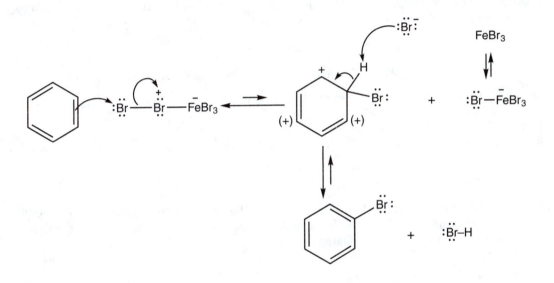

Problem 15.5 The S_N2 reaction lives! Old material doesn't go away; this fundamental reaction is as important now as it was earlier. Nucleophilic attack of benzene on the *tert*-butyl bromide–AlBr₃ complex is an S_N2 reaction, and these do not occur at tertiary carbons.

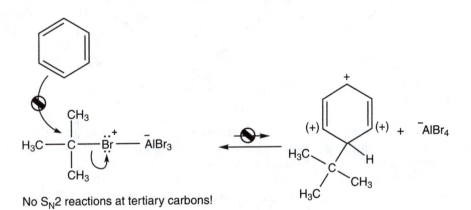

No S_N2 reactions at tertiary carbons!

(continued)

Problem 15.5 *(continued)*

This reaction must go through formation of the free *tert*-butyl cation.

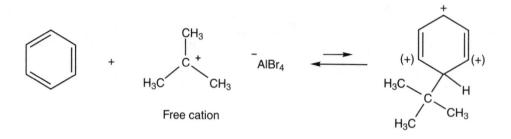

Free cation

Problem 15.6   The cyclohexadienyl cation intermediate in substitution of benzene has three resonance forms in which the carbocation is secondary.

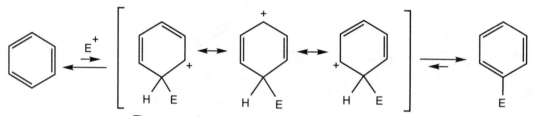

Three secondary carbocationic resonance forms

The intermediate from toluene has two forms in which the carbocation is secondary and one form in which the carbocation is tertiary. The intermediate from toluene is more stable and will be formed faster. So, toluene substitutes faster than benzene.

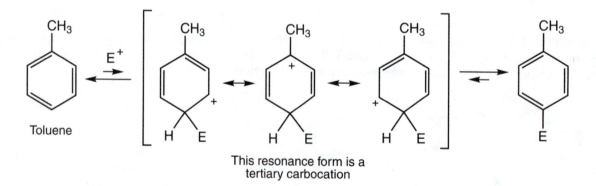

This resonance form is a
tertiary carbocation

Problem 15.7   The solution might be to use a vast excess of benzene. That way the advantage of toluene would literally be diluted. The electrophile, E^+, would be far more likely to encounter, and react with, a benzene molecule than a toluene molecule.

Problem 15.8   *tert*-Butylbenzene will be the product whenever the *tert*-butyl cation is formed in the presence of benzene. Aluminum chloride forms a complex with the alkyl bromide that has partial positive charge on a primary carbon. A hydride shift in this complex produces the much more stable tertiary carbocation. Standard electrophilic substitution of benzene leads to *tert*-butylbenzene.

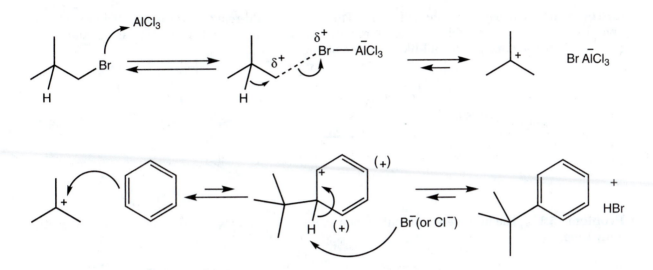

Problem 15.9 These are Friedel–Crafts alkylations. There is potential for polysubstitution in each case. If benzene is used as the solvent, the monosubstituted aromatic ring will be the major product.

(a) This reaction gives 2-phenylbutane.

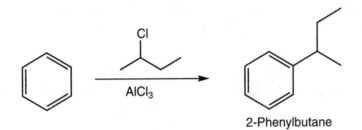

2-Phenylbutane

(b) This time there will be rearrangement. We don't know which product will be major, but it is likely to be 2-phenylbutane.

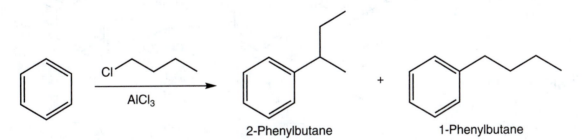

2-Phenylbutane   1-Phenylbutane

(c) Alkyl bromides can also be used in Friedel–Crafts alkylations. The major product would be 3-phenylpropene. If benzene is not the solvent, polypropylene might be the major product.

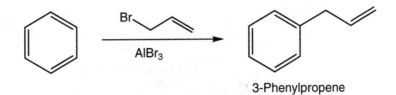

3-Phenylpropene

(continued)

Problem 15.9 *(continued)*

(d) This reaction also gives 2-phenylbutane. The alkene would react with the acid (Chapter 3) to give the *sec*-butyl cation, which would react with the benzene solvent. If benzene is not the solvent, then polybutylene would likely be the major product.

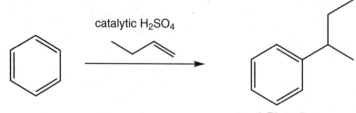

2-Phenylbutane

Problem 15.11   The first step is formation of a complex as the Lewis acid $AlBr_3$ reacts with acetic anhydride.

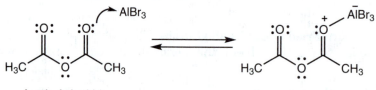

Acetic Anhydride

The complex can break down to give an acylium ion that goes on to react with benzene in a typical aromatic substitution reaction.

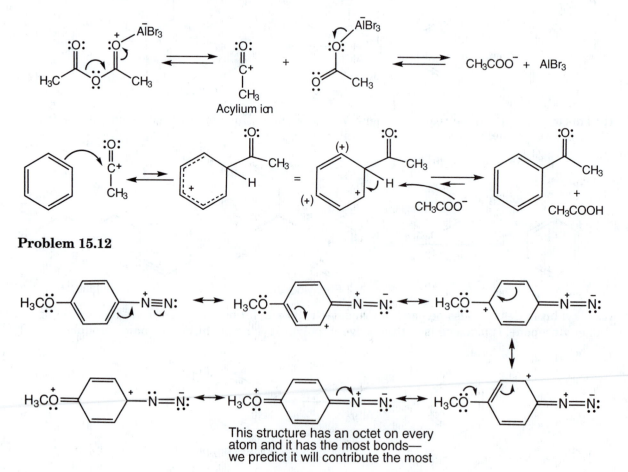

Problem 15.12

This structure has an octet on every atom and it has the most bonds— we predict it will contribute the most

Problem 15.13   In order to displace the leaving group N_2 in an S_N2 reaction, the nucleophile must approach from the rear of the breaking bond. The ring makes this process impossible. These reactions cannot be S_N2 displacements.

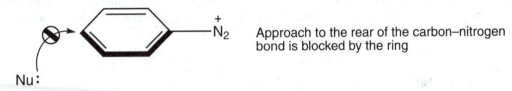

Approach to the rear of the carbon–nitrogen bond is blocked by the ring

Problem 15.14   The orbital containing the odd electron in Ph· does not overlap with the π system. There are no resonance forms in which this electron is delocalized.

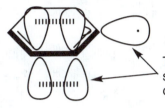

There is no overlap between the σ system and the π system; only two orbitals of the π system are shown

Problem 15.15   A resonance formulation of the benzenediazonium ion shows that the terminal nitrogen is a Lewis acid. It can act as an E^+ reagent and add to the highly activated ring of aniline. This reaction gives a cyclohexadienyl cation intermediate that can be deprotonated by any base in the system to give the observed side product, an azo compound.

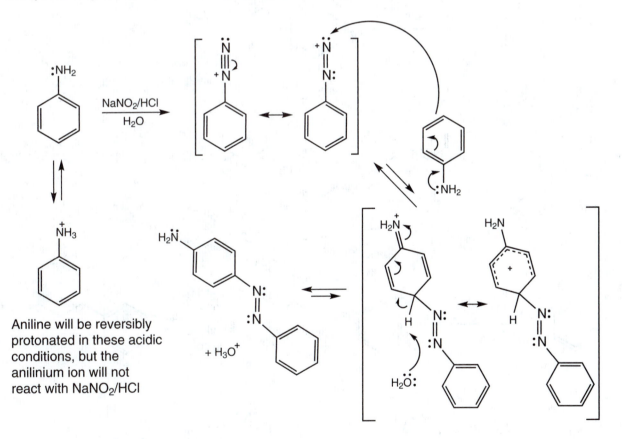

Aniline will be reversibly protonated in these acidic conditions, but the anilinium ion will not react with $NaNO_2/HCl$

Problem 15.16   This problem is really just an exercise in remembering reagents.

(a) Benzene and HNO_3/H_2SO_4 gives nitrobenzene, which can be reduced with Sn/HCl to aniline.

(b, c, and d) Aniline is first treated with $NaNO_2$/HCl to give the diazonium ion. Treatment of the diazonium ion with $^-BF_4$ gives (b), treatment with KI gives (c), and treatment with acidic water gives (d).

(e) Start with the aniline made in part (a).

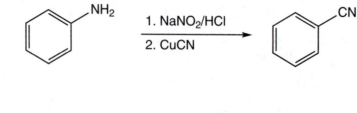

(f)

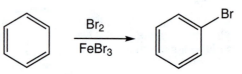

(g)

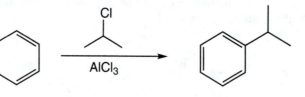

(h)

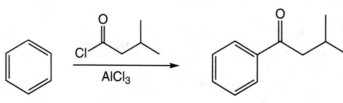

(i)

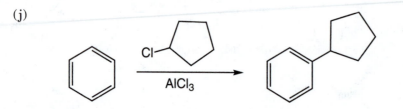

(j)

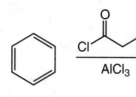 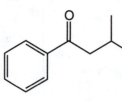 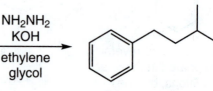

Problem 15.17 Phenol ionizes to give the resonance-stabilized phenoxide ion. By contrast, cyclohexanol gives a localized, much higher energy alkoxide.

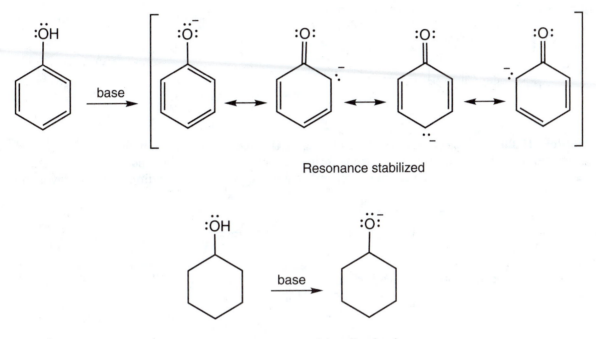

Resonance stabilized

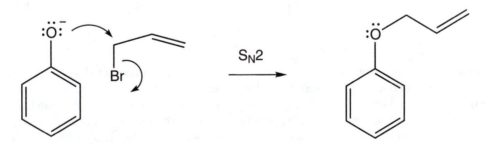

A localized anion

Problem 15.18 The arrow formalism shows a straightforward S_N2 reaction with deprotonated phenol, phenoxide (see Problem 15.17), acting as the nucleophile.

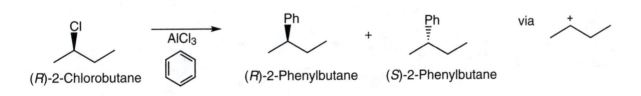

Problem 15.19 If a full carbocation is formed in the reaction, the product would be a racemic mixture of 2-phenylbutane. The carbocation is planar and the nucleophile, benzene in this case, could add to the carbocation from either face.

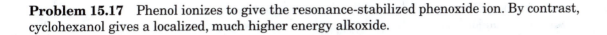

(R)-2-Chlorobutane   (R)-2-Phenylbutane   (S)-2-Phenylbutane

(continued)

Problem 15.19 *(continued)*

If the benzene attacks before the full carbocation is formed, then we expect that the (*S*)-2-phenyl-butane would be the major product, although the mixture may be something like 60:40.

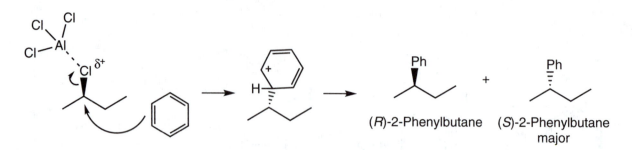

Problem 15.20   Substitution at the 2-position goes through an intermediate carbocation that is better stabilized by resonance than the one formed by substitution at the 3-position. This greater resonance stabilization of the intermediate, and of the transition state leading to it, will favor the former pathway.

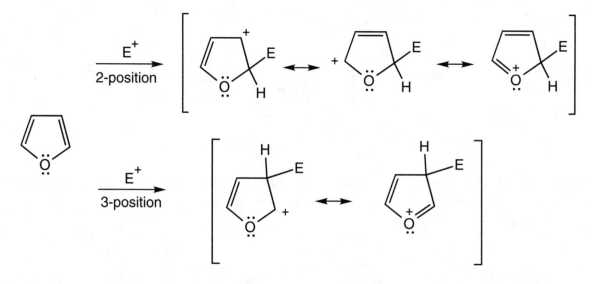

Problem 15.21   Reaction of an alkene with bromine leads to formation of the bromonium ion. If furan reacts like an alkene, the bromonium ion intermediate that is shown would lead to the trans dibromide intermediate. The typical E2 elimination is not possible for this dibromide because there are no antiperiplanar hydrogens. There is an electron pair on the oxygen that is antiperiplanar to the C(2) bromine. The oxonium ion intermediate would result from the trans dibromide. This process would lead to 3-bromofuran. This pathway cannot be involved because it leads to the wrong product.

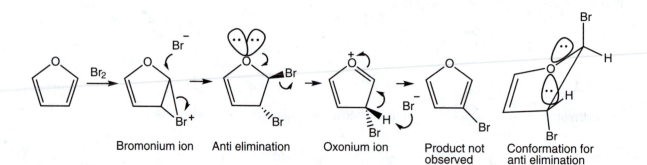

Problem 15.22   The Friedel–Crafts reaction using a vinyl chloride is not effective. There are at least two possible reasons. One is that the chlorine bond is too strong to break with the electrophile (AlCl₃). The Cl—C=C system has an especially strong bond because of the mixing of a lone pair on chlorine with the π system of the alkene. A second possible reason is that the vinyl and phenyl carbocations are too high in energy, which means we can't use the vinyl or aryl chloride as an electrophile.

The Friedel–Crafts reaction doesn't work in the presence of an amine because the amine complexes too tightly with the aluminum trichloride. This complexation makes it impossible for the aluminum trichloride to do its job of pulling off the chloride. The amine is a better nucleophile than the chloride, and it outcompetes the chloride for the electrophile.

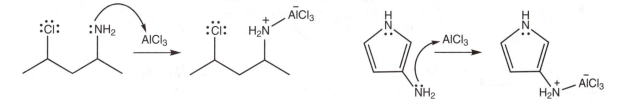

Problem 15.25   The obvious answer is that substitution in the ortho position encounters steric problems that are absent in the para position.

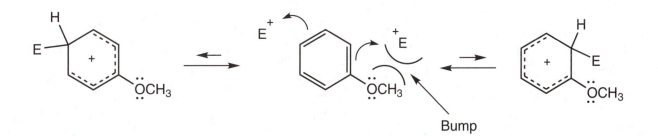

In answering Problem 15.25, you almost certainly hit upon this sensible reason for the dominance of para substitution over ortho, even in the face of a 2:1 statistical factor favoring ortho. Steric considerations will surely favor the para position. The problem is that sometimes the ortho position is *favored* over the para, and the statistical advantage doesn't seem adequate to explain this preference. The answer almost certainly lies in a molecular orbital treatment of the reaction. We can write resonance forms for the intermediate cations in para and ortho substitution, as in Figure 15.61 and the answer to Problem 15.24. When we do this, no distinction between the intermediates for ortho and para substitution is obvious, as both intermediates have similar resonance structures. In reality, this treatment is too simple. The orbital lobes at the ortho and para positions are not all of equal size. There are weighting factors (coefficients) that give the relative magnitude of the wave function at the positions sharing the charge, and it is sometimes

(continued)

Problem 15.25 *(continued)*

the case that the sum of the coefficients at the ortho position (remember the twofold statistical advantage) is greater than that at the para position, thus favoring ortho substitution. In some cases, ortho substitution is favored by chelation of the electrophile with the lone pairs of substituted groups. The ortho:para ratio is a very complicated area, and there is more research to be done.

Problem 15.26   The potential problem here is in translation from the thermodynamic notion that the intermediate formed from anisole is lower in energy than the one formed from benzene to the kinetic statement that the lower-energy intermediate (from anisole) will be formed faster than the higher-energy intermediate (from benzene). We dealt with this situation in some detail in Chapter 8 (p. 351). This connection between thermodynamics and kinetics is usually justified because those factors that stabilize the intermediate are also present to some extent in the transition state leading to the intermediate. Accordingly, the lower-energy intermediate is likely to have the lower-energy transition state leading to it. The following diagram summarizes the situation.

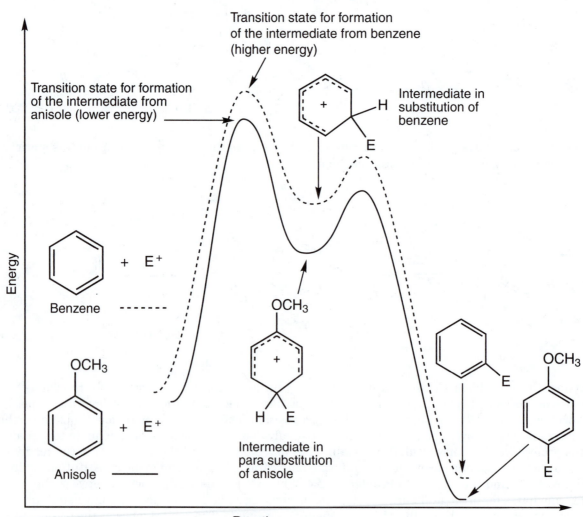

Problem 15.27　The situation is completely analogous to that for para versus meta substitution. In the cyclohexadienyl cation intermediate formed through ortho substitution, the methyl group operates to stabilize the positive charge on the ring. In the intermediate from meta substitution, the methyl group does not help.

Meta substitution of toluene

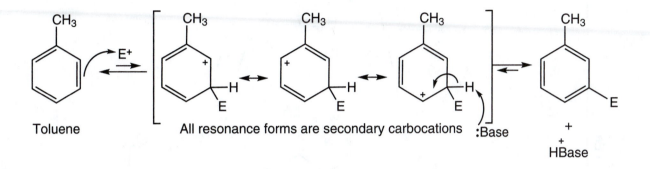

Toluene　　　　　　All resonance forms are secondary carbocations　:Base

+

+
HBase

Ortho substitution of toluene

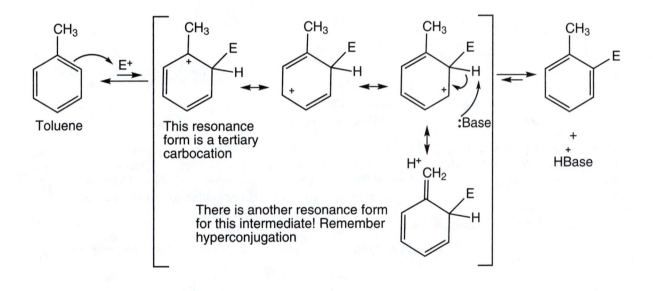

Toluene

This resonance form is a tertiary carbocation

There is another resonance form for this intermediate! Remember hyperconjugation

:Base

+

+
HBase

See the following page for an energy versus reaction progress diagram.

(continued)

Problem 15.27 *(continued)*

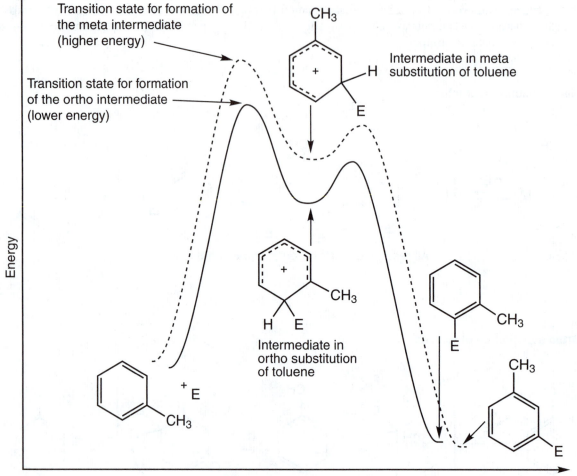

Reaction progress

Problem 15.28   In each case, the heteroatom with a lone pair of electrons can share the positive charge in the carbocation intermediate only if substitution is in the ortho or para position. These intermediates will be of lower energy and therefore formed faster (careful! see the answer to Problem 15.26). In the figure, only ortho substitution is shown for bromobenzene and only para substitution for aniline.

Ortho substitution of bromobenzene

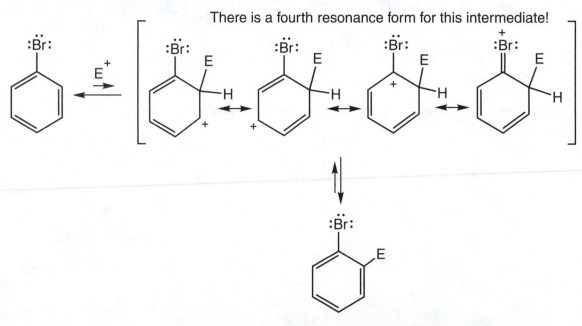

Para substitution of aniline

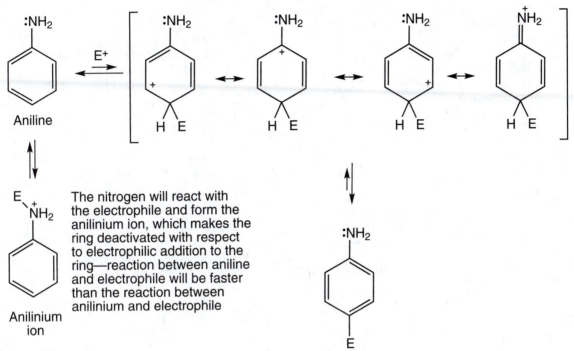

There is a fourth resonance form for this intermediate!

The nitrogen will react with the electrophile and form the anilinium ion, which makes the ring deactivated with respect to electrophilic addition to the ring—reaction between aniline and electrophile will be faster than the reaction between anilinium and electrophile

Problem 15.29 Substitution in the ortho position, like substitution in the para position, requires the close opposition of two positive charges. This charge opposition is strongly destabilizing, and ortho substitution cannot compete with the less destabilizing substitution at the meta position.

Ortho substitution of the trimethylanilinium ion

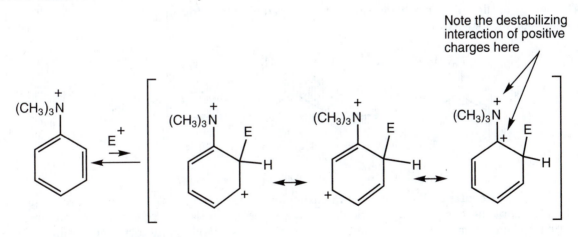

Note the destabilizing interaction of positive charges here

Problem 15.30

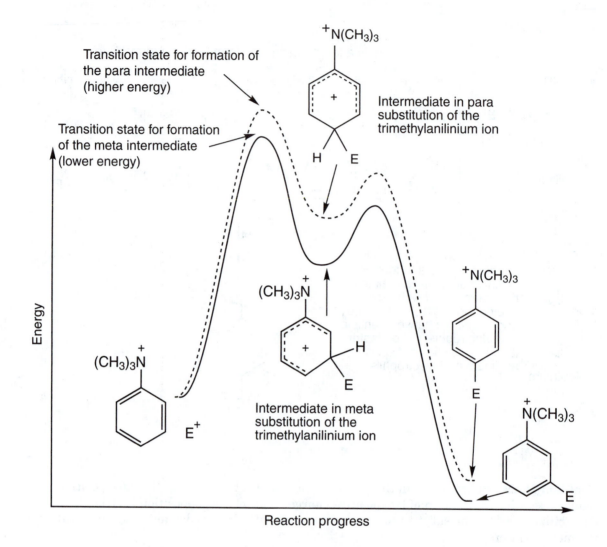

The energy of the nonaromatic intermediate involved in meta addition is lowered as a result of three resonance structures.

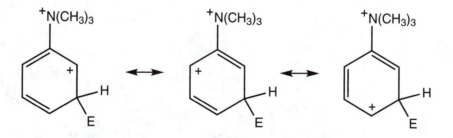

The nonaromatic intermediate involved in para addition is stabilized as a result of three resonance structures, but there is less stabilization in this case because the carbon bearing the ammonium group also bears some of the charge of the carbocation. It is destabilizing to have the same charge on adjacent atoms.

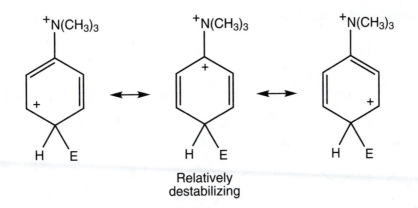

Relatively
destabilizing

Problem 15.31   Like the carbon–oxygen double bond in ethyl benzoate (p. 733), the carbon–nitrogen triple bond in benzonitrile is strongly polarized δ^+ on carbon, δ^- on nitrogen, thus placing a partial positive charge adjacent to the ring. Substitution in the ortho or para position, again as in ethyl benzoate, opposes two like charges and is destabilizing. Substitution in the meta position is favored energetically.

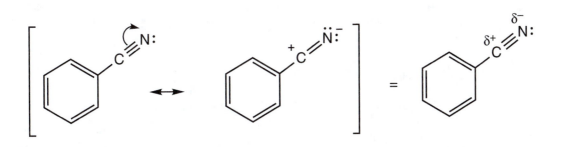

Problem 15.32   As is the case for aniline itself, ortho or para substitution of acetamidobenzene leads to a cyclohexadienyl cation intermediate in which the nitrogen can share the positive charge. In meta substitution, the nitrogen doesn't help, and further substitution will be mainly in the ortho and para positions. Only the sterically favored para substitution is shown in the following figure.

Para substitution of aniline

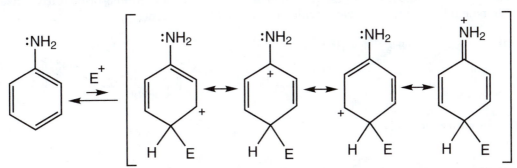

(continued)

Problem 15.32 *(continued)*

Para substitution of acetamidobenzene

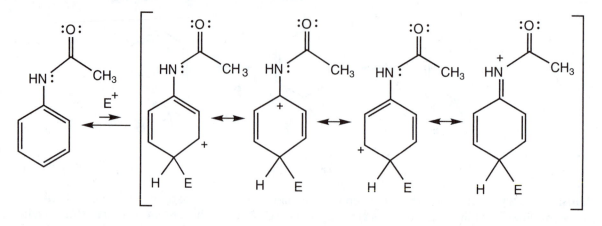

However, in the acyl compound, there is a polarized carbon–oxygen double bond in which the carbon bears a partial positive charge. Accordingly, the resonance form in which the adjacent nitrogen is positive will contribute less than in the intermediate formed from aniline. The acyl compound will be less strongly activating than the free amine.

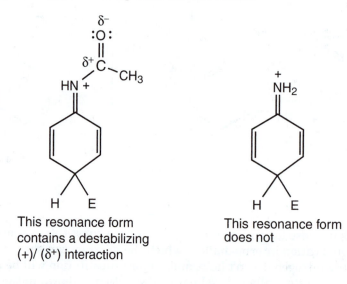

This resonance form contains a destabilizing (+)/ (δ^+) interaction

This resonance form does not

Problem 15.34

(a) We should do a retrosynthesis first. 3-Bromoaniline can be made from 3-bromonitrobenzene, which can be made from nitrobenzene, which can come from benzene.

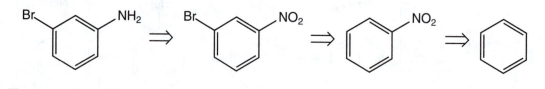

The forward reactions are

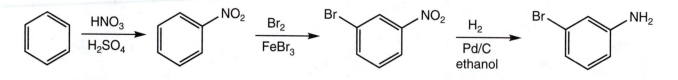

(b) The 3-nitroacetophenone has two meta-directing groups on the ring. We might consider adding the acyl group to nitrobenzene. But the Friedel–Crafts reaction on deactivated rings doesn't work well. So it makes more sense to add the nitro group to acetophenone.

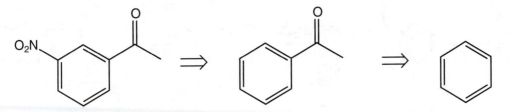

The forward reactions are

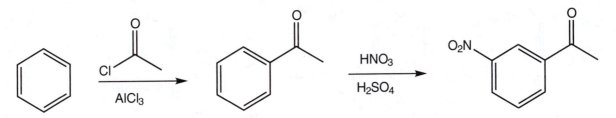

(c) The retrosynthetic analysis tells us that we can use the product from part (b) to make 3-ethyl-nitrobenzene.

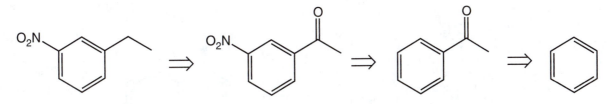

The answer going forward is

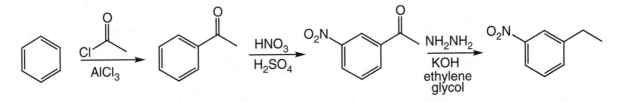

(d) The retrosynthetic analysis for 4-nitropropylbenzene is shown. We know that nitration of propylbenzene will give a mixture of ortho and para substitution, but the para compound should be the major product because of steric interactions at the ortho position.

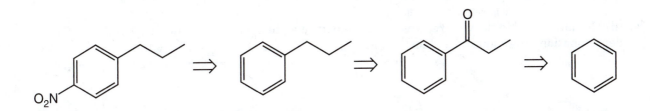

(continued)

The forward process is

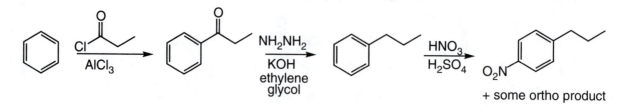

+ some ortho product

Problem 15.36

(a)

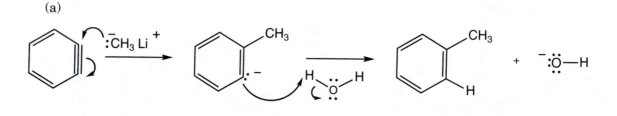

(b and c) These two processes are Diels–Alder reactions. Note that the stereochemical relationships of the diene in (b) are maintained in the product. The reaction must be a concerted one (Chapter 13, p. 615).

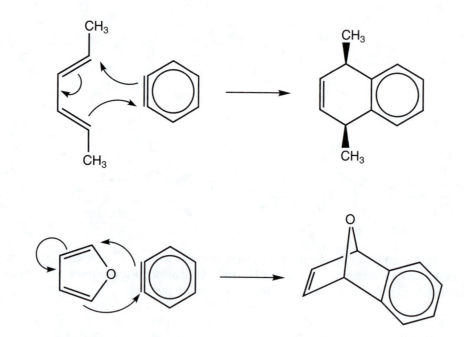

(d) Methanol adds to the very reactive benzyne to give a phenyl anion. Protonation and deprotonation afford anisole and methanol.

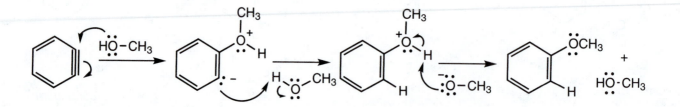

(e) This reaction is a simple dimerization of benzyne.

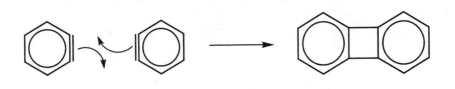

Problem 15.37 Benzene depends for its great stability on perfect overlap of the six 2p orbitals in the ring. When the ring is distorted from planarity, orbital overlap decreases between the p orbitals at the "prow" and "stern" positions and those at the sides of the boat. This loss of overlap is strongly destabilizing and, as a result, the boat benzene would be more reactive than benzene.

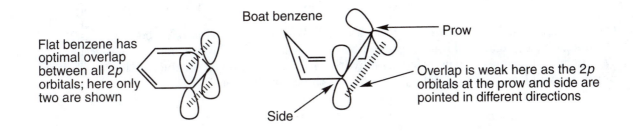

Problem 15.38 Another possible Diels–Alder reaction makes new bonds at the bridgehead carbons of the strained paracyclophane. Although the product has nice symmetry and looks appealing, it is an impossible structure. The bridgehead carbons are now sp^3 hybridized. The seven-carbon bridge will have too much strain to span the ring. You may need to make a model to see the trouble.

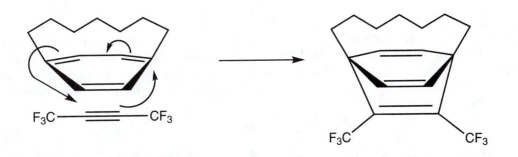

Problem 15.40 The reaction of an R—F with SbF_5 results in the fluorine transferring from the carbon to the antimony making R^+ and $^-SbF_6$. It isn't so much that the fluorine prefers to bond to antimony, it is that SbF_5 is such a strong electrophile. This reaction is very similar to the R—Cl reaction with $AlCl_3$. The SbF_5 is even more electrophilic than $AlCl_3$.

Problem 15.41

(a) The "obvious" ion that should be formed is the 1-butyl cation. As the ^{13}C NMR spectrum of that species must show four signals, something else must be happening. Moreover, it isn't hard to see that the product that shows only two signals must be the relatively stable *tert*-butyl cation. Perhaps the 1-butyl cation is rearranging to the much more stable *tert*-butyl cation.

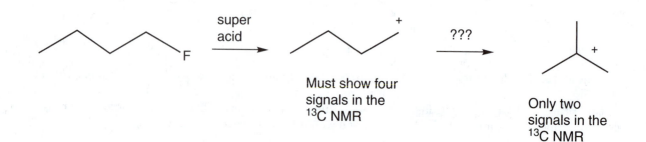

Must show four
signals in the
^{13}C NMR

Only two
signals in the
^{13}C NMR

(b) The vexing problem is that primary carbocation. We have seen primary carbocations as mechanistic stop signs, and it is at least a little worrisome to have to use one in this reaction. We can eliminate one of them by letting a hydride migrate (p. 457) as the fluoride leaves. That migration will lead to the secondary carbocation, the 2-butyl cation.

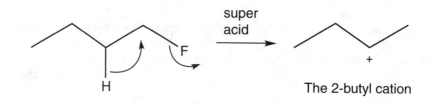

The 2-butyl cation

But now what? It is by no means easy to go from this ion to the *tert*-butyl cation without invoking a primary carbocation (try it). So that's the difficulty. In practice, researchers in the field think that the ions involved are not simple "open" ions, as we have been drawing, but more complex, "bridged" structures. For much more on bridged ions, you can either wait until the end of the course or peek at Chapter 24.

Problem 15.42 The only way you know to construct the necessary six-membered ring is the Diels–Alder reaction, so let's start with that.

s-cis Form of
1,3-butadiene

Now the problem is to elaborate that single double bond into three double bonds. That task is probably hard right now. We can use Chapter 12 chemistry to do it. Allylic halogenation will place a leaving group in the molecule, and an E2 elimination will give us one more double bond. Note the strategy: Use the existing double bond to put in a leaving group, and use the leaving group to make a new double bond.

Repeat the process to add the third double bond. In fact, the last elimination might be so easy as to happen without the added strong base.

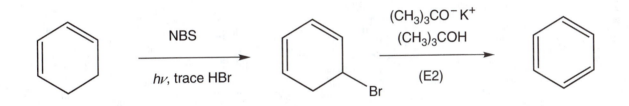

Here are some improvements you might try. In the original Diels–Alder reaction, we could have placed a leaving group on the diene or dienophile (or both), thus decreasing the number of steps necessary.

Problem 15.43 We have to make a bond between the CH_2 of chorismate and one of the ring positions, so let's do that. At the same time, a C—O bond breaks and a C=O is formed. The figure shows you those arrows and rewrites the molecule without moving any atoms. That is very good practice—don't try to leap immediately to the complex product. Push the arrows, rewrite, and only then unravel the molecule. You will make many fewer mistakes this way.

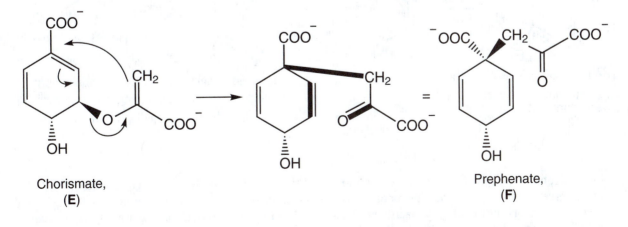

Chorismate, **(E)**

Prephenate, **(F)**

Problem 15.44   The problem is that "extra" OH in tyrosine. In the scheme we have so far, we need that OH in prephenate to act as a leaving group. Tyrosine retains an OH, and so nature must find a way to keep it in.

Additional Problem Answers

Problem 15.45   The key to this problem is to recognize that the carbon–fluorine bond is very polar. Fluorine is very electronegative, and the carbon of the CF_3 group has a partial positive charge.

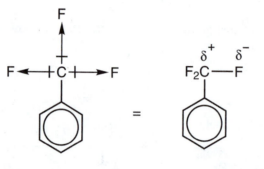

Substitution at either the ortho or para position leads to an opposition of positive charges on adjacent atoms in the cyclohexadienyl cation intermediates (and, most important, in the transition states leading to these intermediates). Substitution in the meta position does not induce this strongly destabilizing opposition of like charges and is therefore favored.

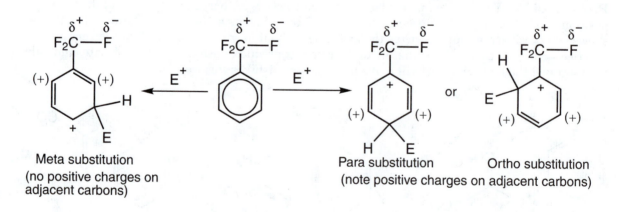

As for nitrobenzene (p. 731) and trialkylammonium ions (p. 730), *any* further substitution of trifluoromethylbenzene will be slow because there is charge–charge opposition in all possible intermediates. Trifluoromethylbenzene will be substituted more slowly than benzene itself because the trifluoromethyl group deactivates the ring.

Problem 15.46 The first three parts are easy. The substituents and the unsubstituted positions to which those substituents direct are shown with little arrows. In each case, both substituents direct further substitution to the same position. In (c), the position between the two methyl groups will be disfavored by steric effects.

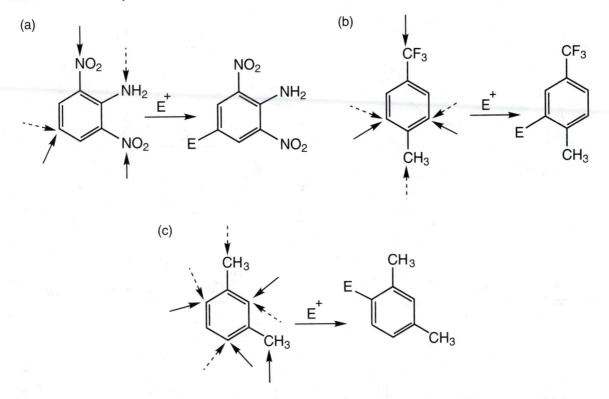

In (d), the two substituents direct further substitution differently. It will be the more powerfully ortho/para-directing amino group that directs future substitution.

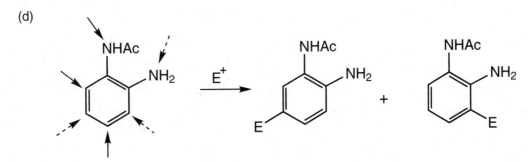

In (e), both nitro groups direct to the same position, although further substitution is sure to be *very* slow. In (f), all positions are equivalent and only one product from single addition is possible.

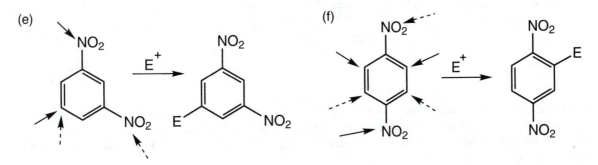

(continued)

Problem 15.46 *(continued)*

In (g), there are two ortho/para-directing groups. The stronger of these, OH, will "win" and further substitution will be ortho to the OH. In (h), all three groups direct to only one open position.

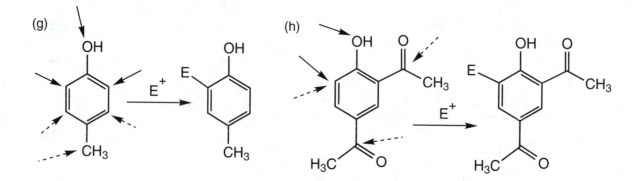

Problem 15.47   Analyze this problem in the usual way. Look at the intermediates formed from ortho, meta, and para substitution.

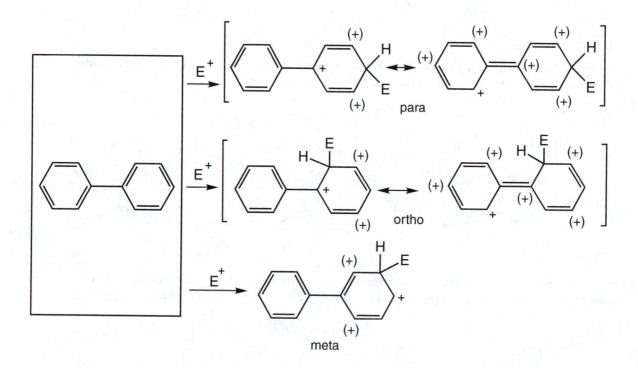

Ortho and para substitution leads to resonance-stabilized intermediates in which the second ring helps stabilize the positive charge, whereas meta substitution does not. Substitution of biphenyl will be in the ortho and para positions, with the ortho position destabilized by steric effects.

Problem 15.48

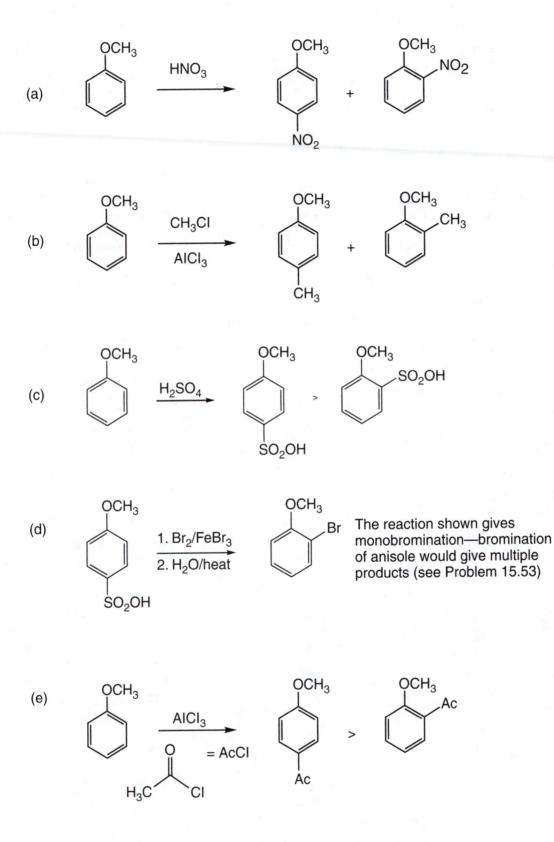

Problem 15.49

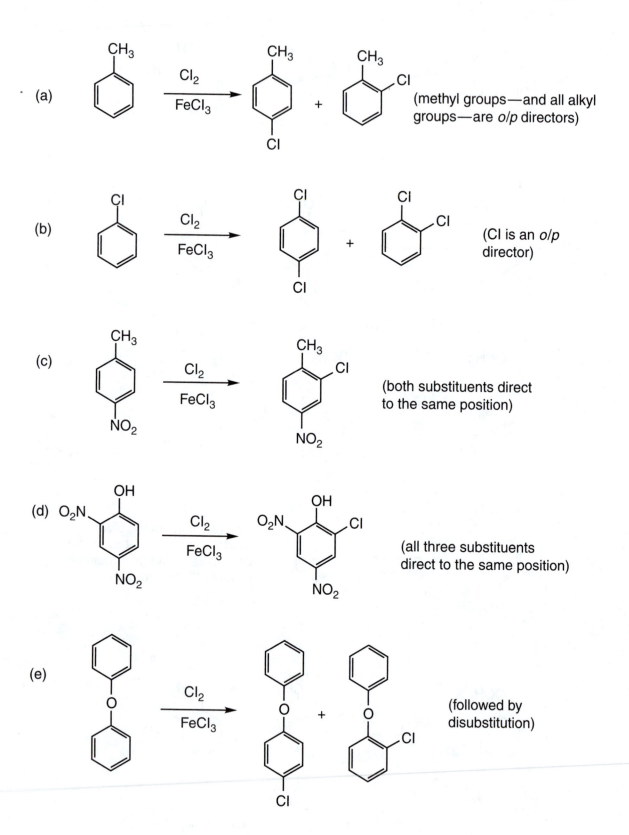

Problem 15.50 Both chlorines are deactivating, so substitution will go in the other ring. The secondary amine is a more powerful ortho/para director than the methylene, and substitution will be ortho or para to it. The ortho position is sterically hindered, so substitution will be para, as shown.

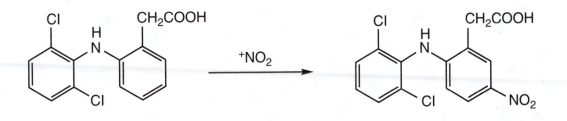

Problem 15.51 In the presence of water, heating the diazonium ion results in formation of the phenyl carbocation, which is captured by water. The product is phenol.

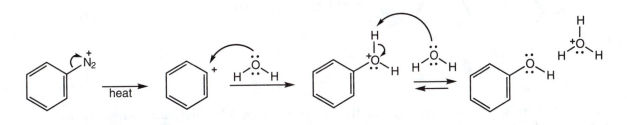

The normal decomposition of the aryldiazonium ion is a radical process. It presumably occurs via a dimer of the diazonium species. The radical abstracts the hydrogen from ethanol forming benzene and a radical on C(1) of ethanol. The ethanol radical can do many reactions (see Chapter 12), but the major product is the dimer (2,3-butanediol).

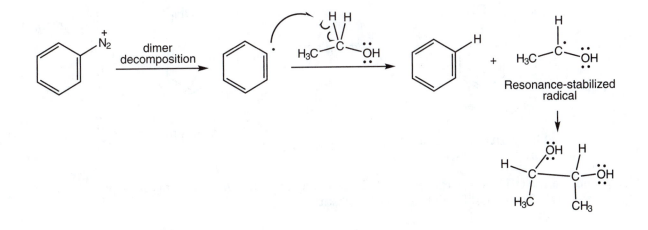

Problem 15.52   Azo dyes are colored because of the extended conjugation and the resulting small HOMO–LUMO gap.

In principle, there are two possible ways to make this compound.

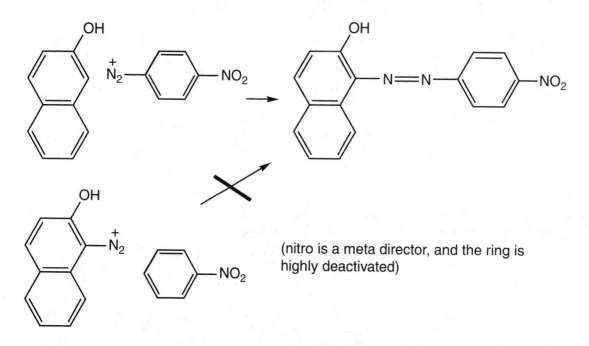

(nitro is a meta director, and the ring is highly deactivated)

In practice, there are two reasons why only the first route is possible. First, nitro is a meta-directing group. Second, it strongly deactivates the ring to which it is attached. By contrast, the OH is activating and an ortho/para-directing group. Note that para substitution is blocked in this case.

Problem 15.53   In strong acid, phenol is essentially completely protonated. The oxonium ion, like the anilinium ion (p. 730), is a meta director.

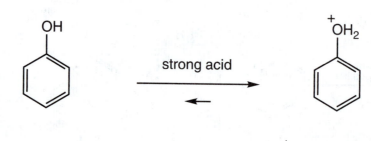

strong acid

OH is an o/p director

$^+OH_2$ is a m director

In phenoxide, the OH has been converted into O⁻, an even more strongly activating, ortho/para-directing group than OH. Thus it is difficult to avoid oversubstitution.

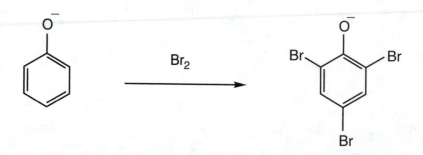

Br₂

Problem 15.54

(a) The retrosynthetic analysis helps us see that we can put the nitro group on the para position of *tert*-butylbenzene. We know how to alkylate the benzene.

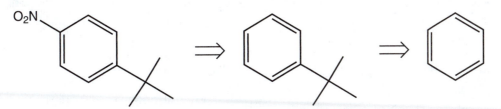

The forward reactions are shown. We don't expect any significant amount of ortho product in the nitration step. The *tert*-butyl group is effectively blocking the ortho position.

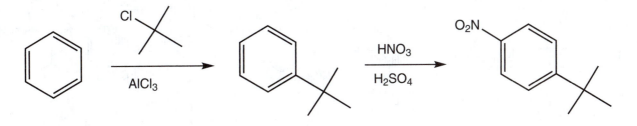

(b) There are many ways to make 3-aminopyridine. One method starts with 3-nitropyridine. We can nitrate pyridine in the 3-position.

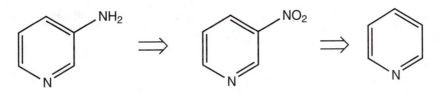

The synthesis is shown. The nitration is a slow process because the pyridine is less reactive than benzene and pyridine will be protonated under the reaction conditions, which will make the ring even less reactive toward electrophilic substitution.

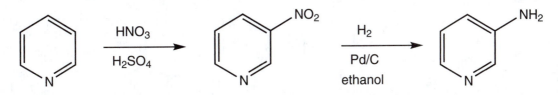

(c) The first thing to figure out is how we can force the bromine and cyclopentyl groups to be ortho to each other. We need to have a group that can be replaced by hydrogen after it has helped direct the ortho substitution. We could use benzenesulfonic acid (p. 703). We also know that the nitro group can be reduced to the amine and converted into the diazonium ion, which can be replaced by hydrogen. The nitro group could either be para to the cyclopentyl group or para to the bromine. We can't use the Friedel–Crafts reaction to alkylate the 4-bromonitrobenzene because the ring is deactivated. But we can brominate 4-cyclopentylnitrobenzene. Nitration of cyclopentylbenzene results in predominant nitration in the para position.

(continued)

Problem 15.54 *(continued)*

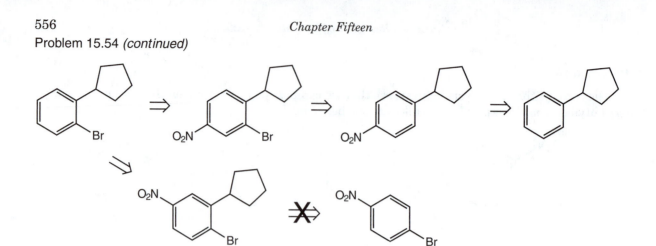

The forward reactions are

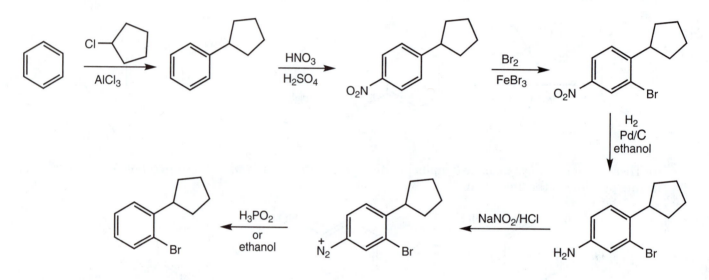

(d) We can make 4-chlorobenzoic acid from the oxidation of an alkyl side chain. It doesn't matter which alkyl group we put on the ring to be oxidized (as long as it isn't the *tert*-butyl group). The isopropyl group works well because we want it to be large enough to block the addition of chlorine to the ortho position. We can chlorinate isopropylbenzene (2-phenylpropane), which can be obtained from benzene.

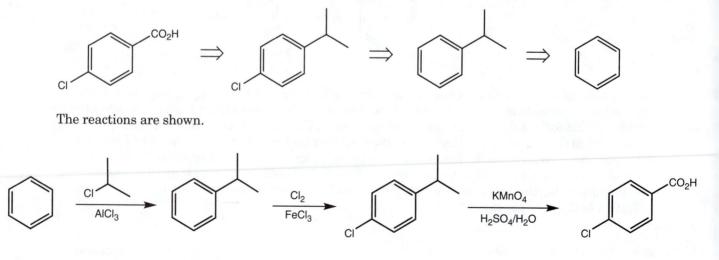

The reactions are shown.

Problem 15.55

(a) The retrosynthesis for 3-bromophenol is shown. We know the OH group likely comes from a diazonium ion, which must ultimately come from a nitro group. We can make nitrobenzene from benzene.

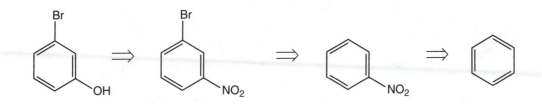

The forward reactions are

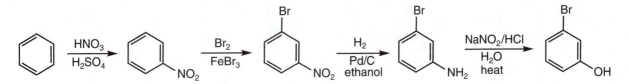

(b) The retrosynthesis is shown.

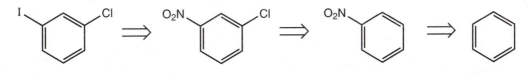

The synthesis is

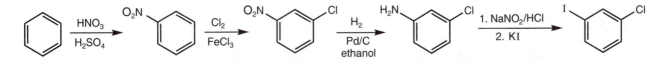

(c) The retrosynthesis takes us back to 4-fluoronitrobenzene, which can be made from 4-nitroaniline. The nitration of aniline requires acetanilide, which will ensure that we only get para substitution. You could use the sulfonic acid blocking/directing group instead of the nitro group.

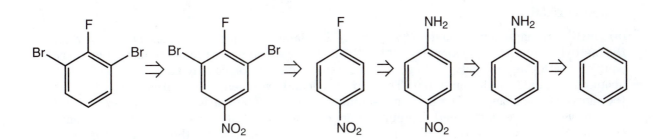

(continued)

Problem 15.55 *(continued)*

The forward reactions are shown. Aniline is acylated on the nitrogen so that the nitration will be selective for the para position. The acyl group can be removed with KOH in water.

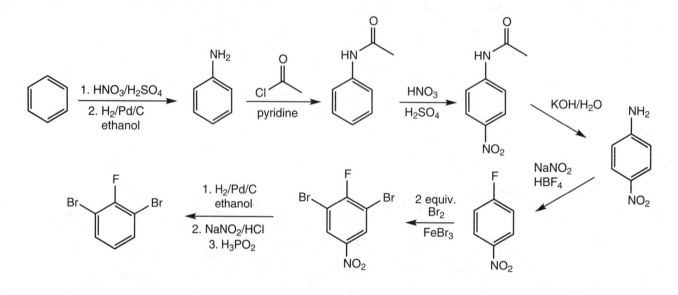

(d) The retrosynthesis for 4-isopropylbenzonitrile is shown. Nitration of isopropylbenzene will be mostly para, because the ortho position is sterically hindered.

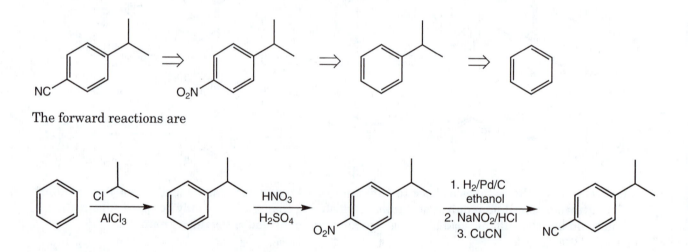

The forward reactions are

Problem 15.56

(a) The first thing to notice about *ortho*-xylene is the plane of symmetry. Because of the symmetry, the methyl groups are equivalent (H_a). Also, the hydrogens that are ortho to the methyl groups are equivalent (H_b), as are the H_c hydrogens. The methyl group hydrogens are benzylic and should be about δ 2.3 ppm (Table 9.5 on p. 400). There are no vicinal hydrogens that will couple with H_a, so the signal will be a singlet. The H_b signal should be around δ 7.3 ppm, which is the chemical shift for benzene hydrogens. The methyl groups do not have a significant impact on the environment of the aromatic ring because the methyl groups do not strongly donate or withdraw electrons from the ring by induction or resonance. The H_b signal would be a doublet because each H_b "sees" one vicinal H_c hydrogen. The H_c hydrogens would also have a chemical shift around δ 7.3 ppm. The H_c signal would be a doublet because each H_c "sees" one vicinal H_b hydrogen. Although the H_c hydrogens are vicinal to each other, equivalent hydrogens do not split each other. As you can see, the H_b and H_c signals will be difficult to distinguish in the NMR spectrum of *ortho*-xylene.

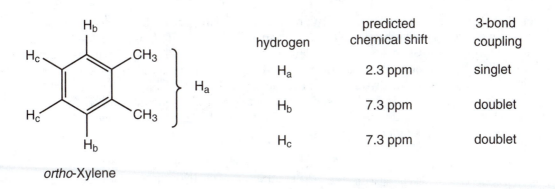

hydrogen	predicted chemical shift	3-bond coupling
H_a	2.3 ppm	singlet
H_b	7.3 ppm	doublet
H_c	7.3 ppm	doublet

ortho-Xylene

There will be non-first-order coupling in this molecule, but that is a detail that you can learn about in an advanced organic chemistry course.

(b) The molecule 1,3-dicyanobenzene is also symmetrical. We have only three types of hydrogens. The H_a hydrogen between the two cyano groups should be the furthest downfield. The cyano groups are electron withdrawing by resonance, which will deshield the hydrogens that are ortho and para to the cyano group. The amount of deshielding is not quantifiable, but a typical region for strongly deshielded aromatic hydrogen is about δ 8.0 ppm. The position ortho to a withdrawing group is usually a little further downfield than the position para to the same withdrawing group. The H_a hydrogen is ortho to two nitriles, so we will predict δ 8.0 ppm. The H_a hydrogen does not have any vicinal hydrogens with which to couple. So our analysis based on 3-bond coupling tells us H_a will be a singlet. However, there will be long-range coupling to the two H_b hydrogens, which will make the H_a signal a triplet. But that coupling is small (1–2 Hz) and likely to be observed only with a very high field NMR. In fact, there is likely long-range coupling between H_a and H_c. That coupling constant is even smaller (0–1 Hz), and it will require an even more trained eye to observe it. Predicting the H_a hydrogen as a singlet is acceptable for our level of NMR analysis in this course. Including all long-range coupling would tell us it should be a doublet of triplets.

hydrogen	predicted chemical shift	3-bond coupling	with long-range coupling
H_a	8.0 ppm	singlet	doublet of triplets
H_b	7.8 ppm	doublet	doublet of doublets
H_c	7.5 ppm	triplet	doublet of triplets

1,3-Dicyanobenzene

The H_b signal will be shifted downfield because of the inductive effect of the nitriles. Each H_b is ortho to one nitrile and para to the other nitrile. So this chemical shift might be a little less than that of H_a. We can predict a shift of δ ~7.8 ppm. The coupling for H_b will be a doublet because of H_c. If we include the long-range coupling, then this signal would be a doublet of doublets because H_b is coupled with H_c and has a small coupling with H_a. The H_c chemical shift can be predicted to be δ ~7.5 ppm, which means that the H_c hydrogen is shifted slightly downfield (remember benzene hydrogens are at δ 7.3 ppm). The H_c hydrogen is meta to the nitrile groups, so they do not feel any of the resonance withdrawal. But there is likely to be some inductive withdrawal by both nitriles. The coupling for H_c will be a triplet as a result of the two H_b hydrogens that are vicinal. If we include the long-range coupling to H_a, then H_c would be a doublet of triplets.

(c) Notice the symmetry in 4-methoxybenzaldehyde; there will be four signals in its ^1H NMR spectrum. The methoxy hydrogens (H_a) will be about δ ~3.7 ppm because the CH_3—O shift would be δ 3.2 ppm and there is an additional δ$^+$ charge on the oxygen as a result of the resonance structure shown. This resonance effect will add an additional δ 0.5 ppm to H_a. There are no vicinal

(continued)

Problem 15.56 *(continued)*

hydrogens, so H_a will be a singlet. The H_b hydrogens on the ring are ortho to the methoxy group, which donates electrons by resonance. That donation will result in H_b being shielded, which means they will appear upfield from δ 7.3 ppm (benzene hydrogens). Such shielded hydrogens are usually at δ 6.8 ppm. The H_b hydrogens have one vicinal hydrogen and therefore will be a doublet. The H_c hydrogens are ortho to a withdrawing group: the aldehyde. We typically see such deshielded hydrogens at δ 7.8 ppm. Each H_c hydrogen has one hydrogen that is vicinal and will appear as a doublet. The last hydrogen is the aldehyde hydrogen. It will be significantly downfield, showing up at δ ~9.5 ppm. It will be a singlet.

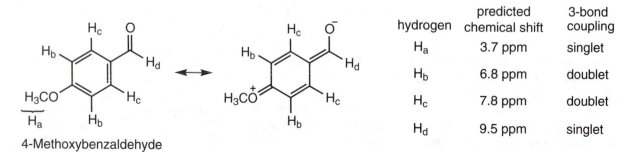

hydrogen	predicted chemical shift	3-bond coupling
H_a	3.7 ppm	singlet
H_b	6.8 ppm	doublet
H_c	7.8 ppm	doublet
H_d	9.5 ppm	singlet

4-Methoxybenzaldehyde

There will be non-first-order coupling in this molecule also, a detail to learn about in an advanced organic chemistry course.

(d) There are six different signals for this lactone. The H_a signal will be furthest downfield. It is ortho to an electron-withdrawing group, so we expect it to be at δ 7.8 ppm. It will be a doublet because of the vicinal H_b hydrogen. If we are able to see all of the long-range coupling, then H_a could be a doublet of a doublet of doublets. Let's just say it's a doublet. We predict the H_b hydrogen will be at δ 7.4 ppm because it is meta to the withdrawing group, so only the induction of the carbonyl will have an effect on the H_b chemical shift. The coupling for H_b will essentially be a triplet because H_b will have nearly the same coupling constant for H_a and H_c. If there is long-range coupling to H_d, then the signal will appear as a doublet of triplets. The chemical shift for H_c will be about δ 7.6 ppm because it is para to an electron-withdrawing group. Resonance will deshield the para position, and that results in a downfield shift. The coupling for H_c will be the same as for H_b. It will look like a triplet unless we can see the long-range coupling, in which case it will be more like a doublet of triplets. The H_d hydrogen will be about the same chemical shift as the H_b hydrogen, about δ 7.4 ppm, because of induction of the carbonyl. The coupling will be a doublet due to H_c. Long-range coupling could look like a doublet of a doublet of doublets. The H_e hydrogens will appear at a chemical shift of δ 3.0 ppm. This shift is because of H_e being benzylic (δ 2.5 ppm) and once removed from the electronegative oxygen (add δ 0.5 ppm). The coupling will be a triplet because there are two vicinal H_f hydrogens. The chemical shift for the H_f hydrogens will be δ 3.8 ppm because they are directly attached to the electronegative oxygen (δ 3.5 ppm) and they are once removed from the carbonyl (add about δ 0.3 ppm). These hydrogens will also be a triplet as a result of the two vicinal hydrogens (H_e).

hydrogen	predicted chemical shift	3-bond coupling	with long-range coupling
H_a	7.8 ppm	doublet	doublet of doublet of doublets
H_b	7.4 ppm	triplet	doublet of triplets
H_c	7.6 ppm	triplet	doublet of triplets
H_d	7.4 ppm	doublet	doublet of doublet of doublets
H_e	3.0 ppm	triplet	triplet
H_f	3.8 ppm	triplet	triplet

Problem 15.57 This kind of extreme high field signal is diagnostic for the ring currents present in aromatic compounds. In a cyclophane of this kind, one or more hydrogens will lie directly, or nearly directly, over the center of the ring. At this point, the induced magnetic field (B_i) will be especially strong and will act to shield the hydrogens effectively. Accordingly, an especially strong B_0 will have to be applied to bring these hydrogens into resonance, and an especially high field signal is observed.

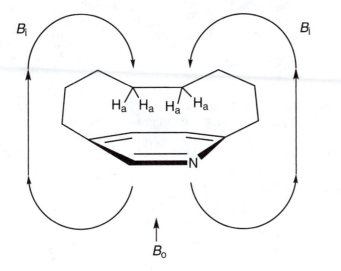

The H_a hydrogens "feel" $B_0 - B_i$ and will be shifted upfield, typically appearing to the right of TMS (negative δ value)

Problem 15.58

(a) The retrosynthesis of 3-deuterioethylbenzene forces us to think about the order of substitution. There is no good way to put a deuterium at the meta position of ethyl benzene. So we need to think of another molecule from which we can obtain the product. We know we can do a reduction of the ketone to get the final product. That's a good precursor because now we have a meta-directing group on the ring. We could get the deuterium from either a diazonium ion (use D_3PO_2 rather than H_3PO_2) or from a Grignard (using D_2O). The Grignard, you will remember, can be made from the bromide. But wait . . . there is a ketone on this molecule. We can't make a Grignard because it will react with the carbonyl on another molecule, as we will see in Chapter 16. So we use the diazonium pathway. That will work.

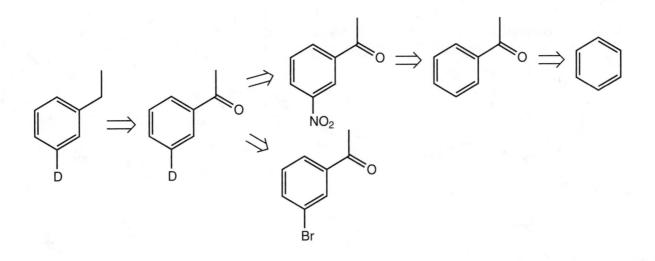

(continued)

Problem 15.58 *(continued)*

The forward steps are shown. The only potential problem is reduction of the nitro group in the presence of the ketone (a Chapter 16 reaction). But nitro groups are easier to reduce with hydrogenation than are ketones, so this route should work.

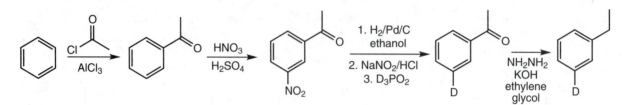

(b) The retrosynthesis of 4-propoxynitrobenzene might take you in several directions. But you know that we can nitrate the propoxybenzene. And putting the nitro group on at this point will maximize the para substitution because of steric interactions at the ortho position. One of the only ways for getting an oxygen on the ring is through the diazonium ion.

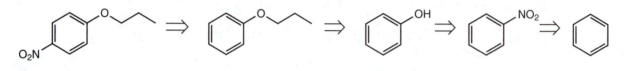

The forward reactions are shown. Remember that the best way to make an ether is usually the Williamson ether synthesis—an S_N2 reaction between an alkoxide and an alkyl halide.

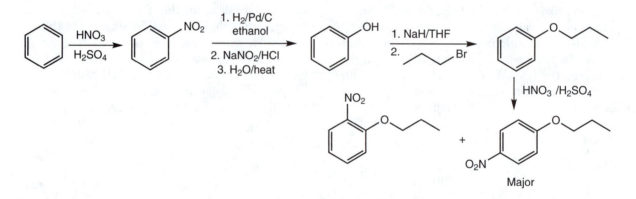

(c) The alkoxy group comes from a phenol. The phenol comes from a nitro group. We can't do a Friedel–Crafts acylation on a nitrobenzene, but we can nitrate propiophenone.

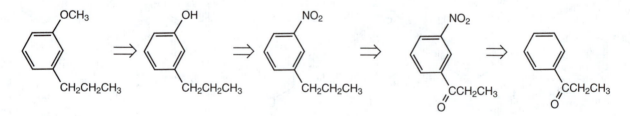

The forward reactions are

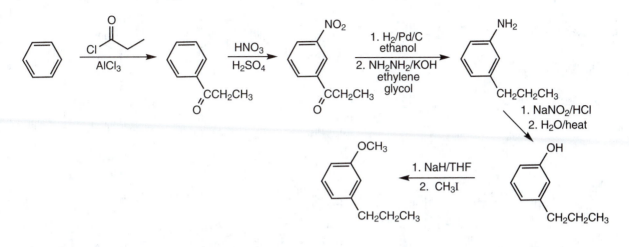

(d) The amino group can be added in an S_NAr fashion. This molecule is a logical precursor because of the ortho/para nitro groups. The chlorodinitrobenzene can come from the 4-chloronitrobenzene. And we can nitrate chlorobenzene.

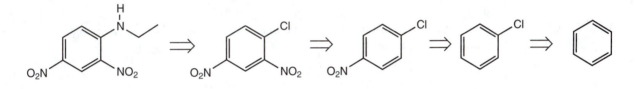

The forward reactions are shown. The nitration of chlorobenzene will give both ortho and para products, but both can be used to nitrate a second time to give a good yield of the 1-chloro-2,4-dinitrobenzene.

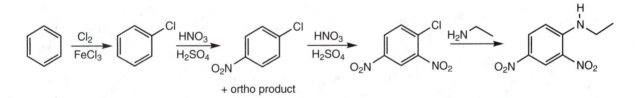

Problem 15.59 Up to compound **D**, this reaction sequence is called the Haworth reaction. The initial step is an intermolecular Friedel–Crafts acylation. Aluminum chloride complexes one carbonyl group, and the anhydride opens to give the acylium ion (see the figure in Problem 15.11 for an analogous reaction). The product of the acylation reaction is the *p*-ketocarboxylic acid **A**. Clemmensen reduction of **A** gives the carboxylic acid **B**. Conversion of **B** into the acid chloride **C**, followed by an intramolecular Friedel–Crafts acylation, yields **D**, even though the new substituent appears meta to the ortho/para-directing methoxy group. That's the only position that can be reached in this intramolecular reaction. The chain is not long enough to reach the other free position on the benzene ring. Finally, reduction of the ketone by a second Clemmensen reduction gives the product, **E** (7-methoxytetralin).

(continued)

Problem 15.59 *(continued)*

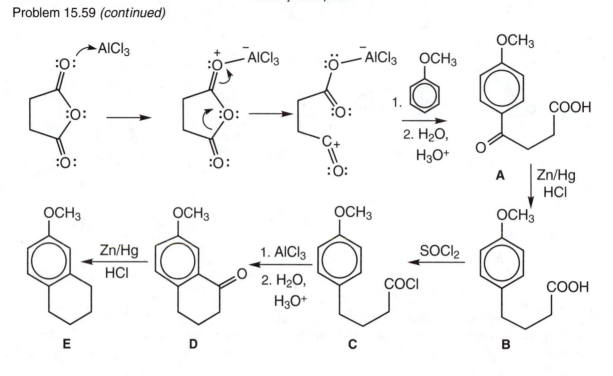

Problem 15.60 Protonation of carbon monoxide leads to an acylium ion. Standard electrophilic aromatic substitution then gives the product.

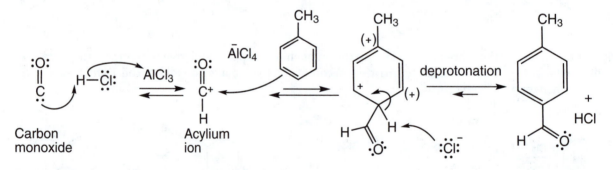

Problem 15.61 The first step is a straightforward Friedel–Crafts acylation to give acetophenone, **A**. Clemmensen reduction gives ethylbenzene, **B**, a molecule that could also be made directly by Friedel–Crafts alkylation.

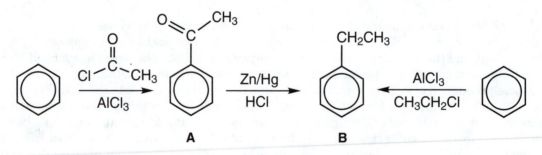

Ethylbenzene can be photochlorinated exclusively in the benzylic position to give **C**. The E2 reaction gives vinylbenzene, better known as styrene, **D**. Ozonolysis with a reductive workup gives benzaldehyde.

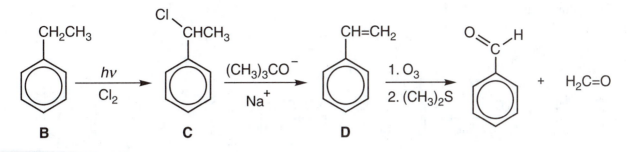

Problem 15.62 It will be far easier to form the resonance-stabilized phenoxide (pK_a of a phenol is ~10) than it will be to form the simple, unstabilized alkoxide (pK_a of ethyl alcohol is 15.9).

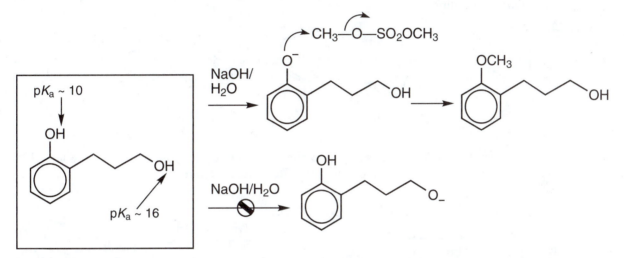

Problem 15.63 At least compound **A** should be easy. A straightforward Chichibabin reaction, right from Figure 15.97, makes 2-aminopyridine. Now you have to recall what nitrous acid does with aromatic amines. It produces the diazonium salts, in this case **B**. Treatment with CuBr leads to 2-bromopyridine in a Sandmeyer reaction.

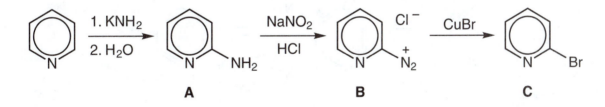

Problem 15.64

(a) The acylated aniline gives para substitution as the major product in reaction with electrophiles.

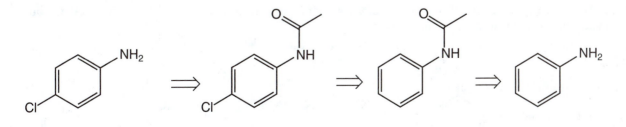

(continued)

Problem 15.64 *(continued)*

The solution is

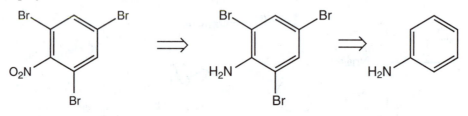

(b) The nitro group can come from the amine via oxidation using a peroxy acid (see pp. 739–740). We can polybrominate aniline.

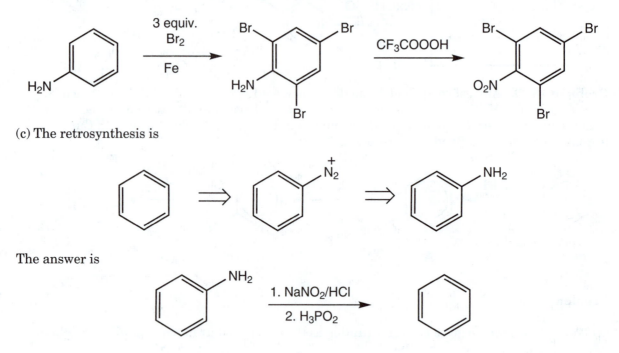

The solution is shown. We don't need a catalyst for the first bromination, but each bromine added deactivates the ring, and it won't hurt to have a little Fe in the reaction.

(c) The retrosynthesis is

The answer is

(d) Aniline cannot be used in Friedel–Crafts reactions because the amine will gum up the AlCl₃, but acylated aniline (acetanilide) can be used.

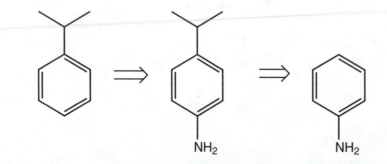

The solution is

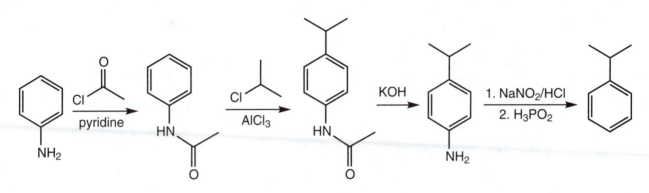

(e) The nitrobenzene will sulfonate in the meta position. We can oxidize aniline to obtain nitro-benzene (see p. 740).

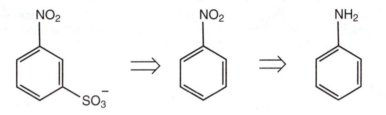

The solution is

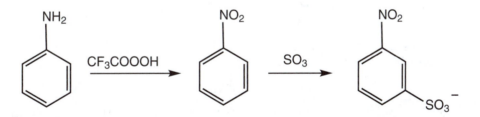

Problem 15.65 This problem presents a pair of "How do we get from here to there?" problems. Neither is particularly hard, and you will probably be able to do them without severe difficulty. Each of these reactions involves protonation of the carbonyl group to generate a carbocation in proximity to a benzene ring. The carbocation is certainly an E^+ reagent and reacts with the ring in a typical electrophilic aromatic substitution to create a new ring. It is well worth your time right now to get into the habit of analyzing this kind of problem before you begin to write structures and push arrows. Analysis is always useful, and when we progress to more difficult problems it will be essential for success. Much of what follows may seem obvious, but do not be insulted. Stating the obvious is often useful—it focuses the mind on the objective at hand. You may eventually come to the point at which you do this kind of analysis automatically, but for now making yourself look at the problem in an analytical way may take some effort. We promise you that it is well worth it.

The first obvious, but important, point is that a ring must be closed in this reaction. There are two rings in the product, and only one in the starting material. Obvious? You bet. Did you think of this when you first looked at the problem? We bet not. The thought that a ring must be closed now leads to the second point: We know exactly where the ring must be closed.

(continued)

Starting material (one ring)   Product (two rings)

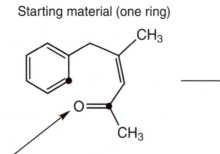

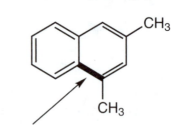

This oxygen is lost in the reaction; a bond must
somehow be formed between the two dotted carbons

Bond that must be closed somehow

It is also clear that an oxygen atom must be lost in this reaction. It is present in the starting
material but absent in the product.

So now we are able to see quite specific goals in this problem. A ring must be closed between the
two dotted carbons, and an oxygen atom must eventually be lost. How many ways do we know to
form a bond to a benzene ring? Not many, that's for certain. This chapter has been primarily
devoted to the reaction of E⁺ reagents with benzene rings. Surely, that is the first thing to think of
when considering ways to close the second ring by forming a new bond to benzene.

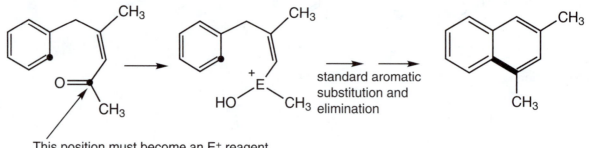

This position must become an E⁺ reagent
to close the second ring

How to form the E⁺ reagent? Simple: Protonate the oxygen. Addition to the ring generates the
usual cyclohexadienyl cation intermediate. Note again the convention by which we show the
positions sharing the positive charge with (+).

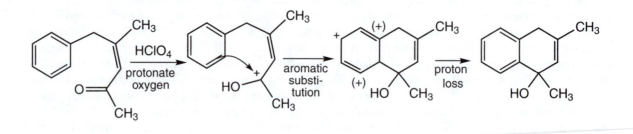

Now the alcohol is protonated, and elimination reactions (E1 as shown or E2) lead to the final
product.

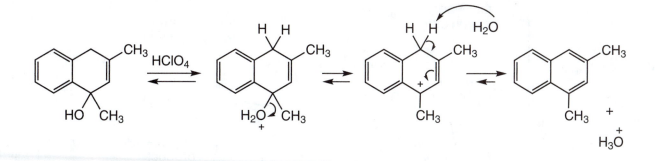

The second example in this problem can be analyzed in a very similar fashion. Do this. Find the position at which the new ring must be formed, and find a way to accomplish this step, using the first example as a model. The following drawings outline the process.

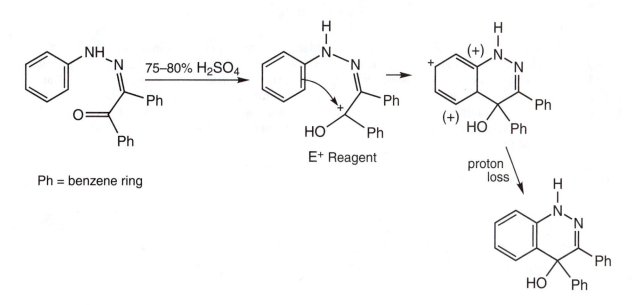

Ph = benzene ring

Now here is the loss of water (elimination) phase.

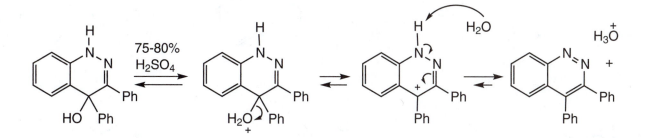

Problem 15.66 These reactions should ring a bell—the higher-energy product is formed preferentially at low temperature, and the lower-energy product is formed at higher temperature. This problem is another example of kinetic versus thermodynamic control. Recall our discussion of the additions of Cl_2 and HBr to 1,3-butadiene (Chapter 13, p. 604) and the temperature effect on exo/endo selectivity in Diels–Alder reactions (Problem 13.54, p. 637). Also, recall that aromatic sulfonation is a reversible reaction, especially at high temperature (Section 15.3b, p. 703). The energy versus reaction progress diagram tells almost all.

(continued)

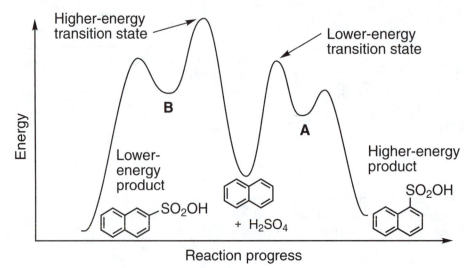

Now we have two questions. First, why is naphthalene-1-sulfonic acid the kinetically preferred (first formed) product, and second, why is naphthalene-2-sulfonic acid the thermodynamic product? As the first question concerns kinetics, we must look at the transition states for the two reactions. The transition states for electrophilic aromatic substitution reactions resemble the cationic intermediates for these endothermic reactions. Recall the Hammond postulate (Chapter 8, p. 351 and Problem 15.26). Therefore, an examination of these intermediates, **A** and **B**, in the previous diagram should prove useful.

For addition at position 1,

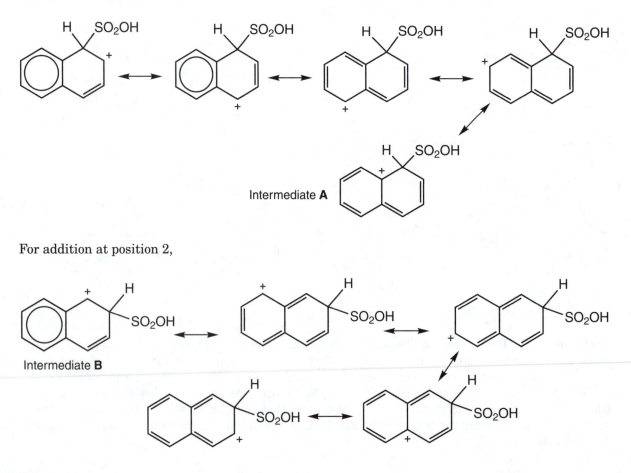

For addition at position 2,

Two of the resonance forms for intermediate **A,** the cation produced from addition to position-1, retain an intact benzene ring (note circles). By contrast, addition to position-2 involves an intermediate (**B**) in which only one form retains an intact benzene ring. Resonance contributors in which a benzene ring is intact are more important contributors (more stable) than the other forms. Accordingly, the intermediate from addition at position-1, *and the transition state leading to it*, are lower in energy than the comparable intermediate, *and the transition state leading to it*, for substitution at the 2-position.

The reason for the increased thermodynamic stability of naphthalene-2-sulfonic acid over naphthalene-1-sulfonic acid is not as clear. If you guessed steric effects, you were right. In the 1-isomer, the sulfonic acid group has an unfavorable steric interaction (called a *peri* interaction) with the hydrogen at the 8-position. The substituents at the 1- and 8-positions are parallel to each other and apparently interact more strongly than other, ortho-substituted groups.

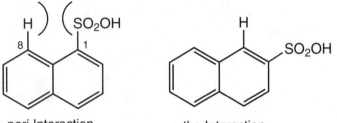

peri Interaction   ortho Interaction

Accordingly, the product of substitution in the 2-position is the more stable compound and will be favored at equilibrium (thermodynamic control).

Problem 15.67 The only difficult thing about this problem is that there is an intermediate alcohol that is transformed into the chloride. Protonation of formaldehyde ($H_2C{=}O$) leads to an E^+ reagent that undergoes electrophilic aromatic substitution at the 1-position of naphthalene to give the alcohol **A**.

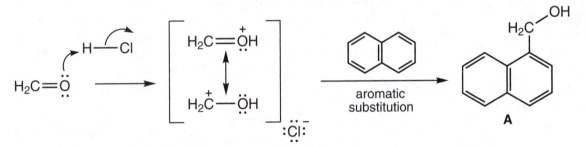

Now, don't forget earlier chemistry. The alcohol is protonated in acid (converting the poor leaving group OH into the good leaving group OH_2). Displacement by chloride gives the final product.

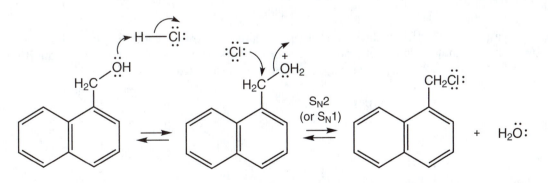

Problem 15.68   The NMR data tell us that the aromatic ring is still intact. There are four hydrogens on the aromatic ring (hydrogens at δ 7.18 and 6.77 ppm), and it is a para-disubstituted ring (the aromatic hydrogens are two doublets). The phenolic hydrogen is still present (δ 5.08 ppm that shifts with D_2O). There are no benzylic hydrogens (δ ~2.3 ppm)! So it must be a quaternary carbon attached to the ring. There is a two-hydrogen quartet at δ 1.58 ppm. That signal must be from a CH_2 that is vicinal to a CH_3, and there is a CH_3 at δ 0.7 ppm that is vicinal to a CH_2. Thus, we have a quaternary carbon with an ethyl group on it. The only remaining signal is the 6H singlet at δ 1.24 ppm, which must be two equivalent methyl groups (because there are six hydrogens) attached to a carbon that has no hydrogens (because it is a singlet). The alkyl group is— $C(CH_3)_2CH_2CH_3$, and the unknown chloride could be 2-chloro-2-methylbutane. It could be a number of other chlorides as the Friedel–Crafts alkylation is accompanied by rearrangement. For example, 1-chloro-2-methylbutane or 2-chloro-3-methylbutane would give the same product.

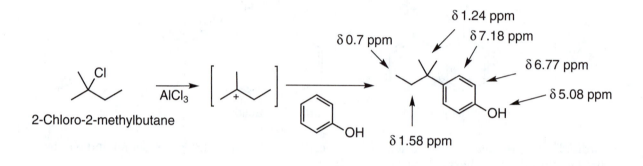

Problem 15.69   This problem is much easier than it looks. It is difficult only for psychological reasons. In this wonderful reaction, brilliantly conceived by a former student of MJ's, Professor L. T. Scott (b. 1944), then at the University of Nevada, Reno, now at Boston College, the ring at the upper right seems to switch its orientation. And that is exactly what happens. The process is triggered by formation of a carbocation through protonation of the naphthalene part of the molecule.

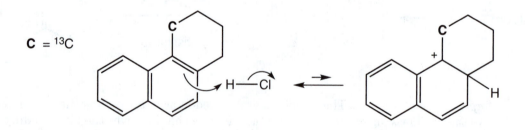

Now, what can happen to this carbocation? It can, of course, rearomatize through proton loss, a reversal of the original protonation, and this must be a very easy reaction. However, in addition, a carbon–carbon bond might rearrange to give an intermediate spiro carbocation.

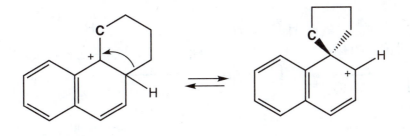

Notice that the spiro carbocation is symmetrical in the sense that there are two equivalent carbon-carbon bonds that can migrate back. If the same bond moves back as originally rearranged, we of course see no change. However, if the other, equivalent carbon–carbon bond moves back, we get a new carbocation. Deprotonation gives the product.

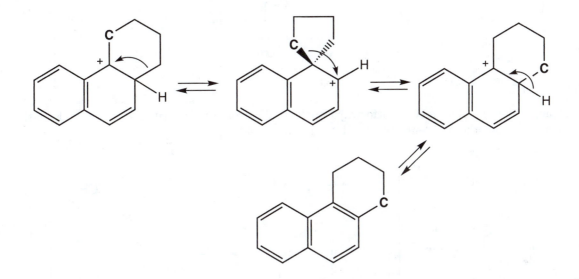

Problem 15.70 Compound **A** is formed through a straightforward Friedel–Crafts alkylation reaction.

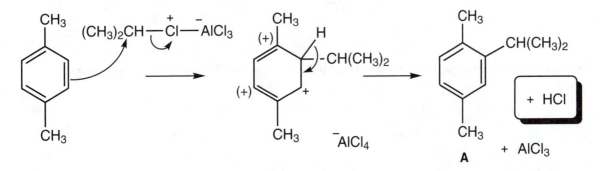

What can HCl do? The hint tells you to focus on this product. If the product **A** is protonated, we can generate an intermediate in which an alkyl shift moves the relatively large alkyl groups apart. That's the key to why this rearrangement happens—the relief of steric strain when the thermodynamically more stable 1,3,5-trisubstituted benzene is formed from a less stable 1,2,4-trialkylbenzene.

(continued)

Problem 15.70 *(continued)*

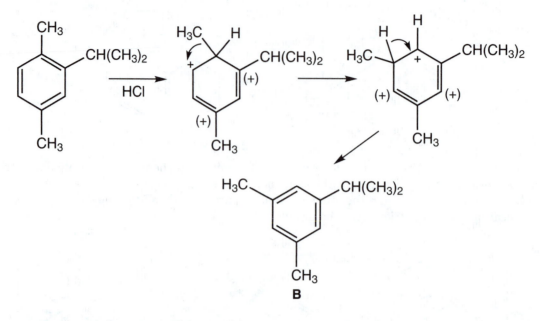

B

Problem 15.71

(a) This problem involves a straightforward nucleophilic aromatic substitution reaction. Like the nitro group, the carbonyl group can stabilize an anion.

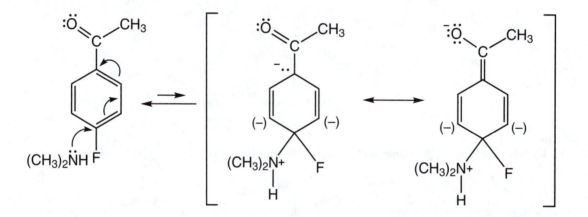

Deprotonation of nitrogen and rearomatization through loss of fluoride complete the reaction.

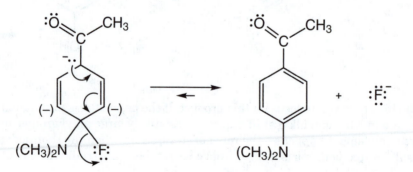

(b) Phthaloyl peroxide decomposes to benzyne when irradiated. Benzyne undergoes a Diels–Alder reaction with the diene to give the product.

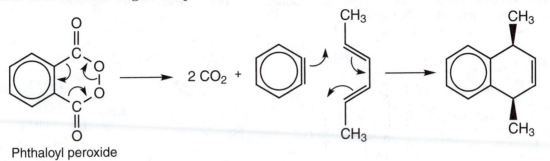

Phthaloyl peroxide

Note the stereochemistry of the product in the addition of benzyne to (*E,E*)-2,4-hexadiene. The cis stereochemistry shows that the addition takes place in a single step. There can be no intermediates capable of allowing rotation around a single bond.

Problem 15.72 Endothermic addition of methoxide gives an intermediate cyclohexadienyl anion. The question is whether the first step or the second step is rate determining. The initial addition to the ring disrupts the aromatic system and is likely to be the slow step in the reaction.

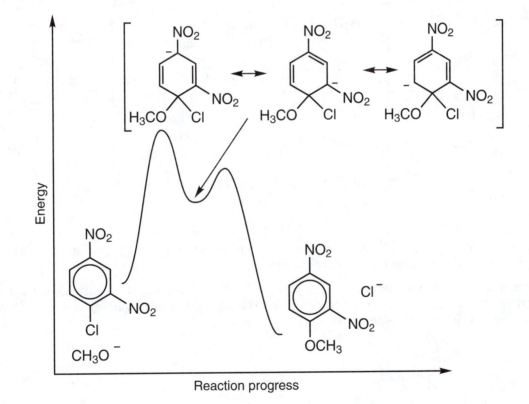

Problem 15.73 The only possible leaving group is chloride.

(a) The three products are the two possible monosubstitution products and the one possible disubstitution product, **A**, **B**, and **C**.

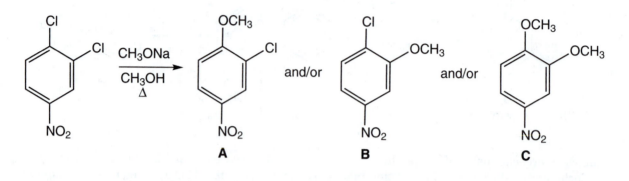

(b) The only product that can be formed in this reaction is **A**, 2-chloro-4-nitroanisole. If we examine the mechanism for this nucleophilic aromatic substitution reaction, we can see why.

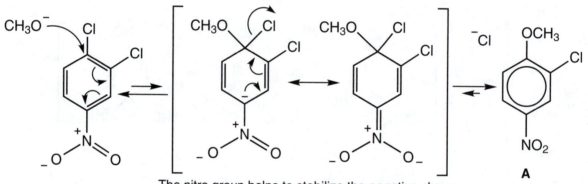

The nitro group helps to stabilize the negative charge

Nucleophilic addition at the position para to the nitro group affords an intermediate in which the nitro group can help stabilize the negative charge.

However, addition of methoxide at the other position bearing a chlorine, the position meta to the nitro group, does not give an intermediate in which the nitro group is stabilizing. Accordingly, no substitution at this position is observed.

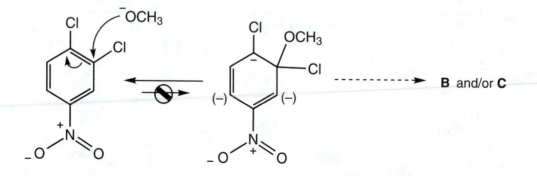

Problem 15.74

(a) It helps to draw a good Lewis structure for benzenediazonium-2-carboxylate. Then the arrow formalism becomes easy to draw.

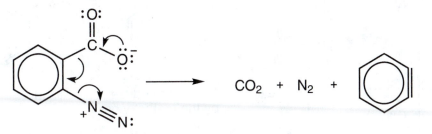

(b) Triptycene is formed through a Diels–Alder reaction across the 9,10-positions of anthracene (see Problem 14.62, p. 691).

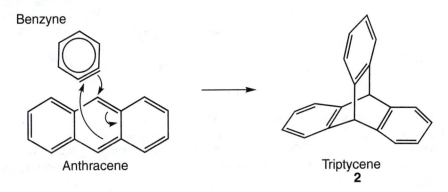

Problem 15.75   In the first reaction, amide ion can add in two ways to the unsymmetrical benzyne intermediate to give the two observed products.

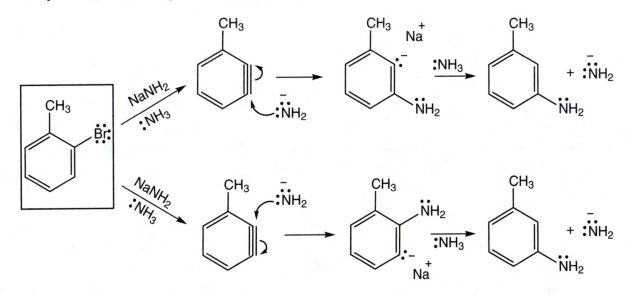

In the second reaction, the same two modes of reaction are theoretically possible, but the presence of a methoxy group strongly influences the situation. Formation of anion **A** is greatly favored over formation of anion **B**, because in **A**, the adjacent methoxy group inductively stabilizes the negative charge and coordinates with the counterion associated with anion. This effect directs the reaction toward the observed product.

(continued)

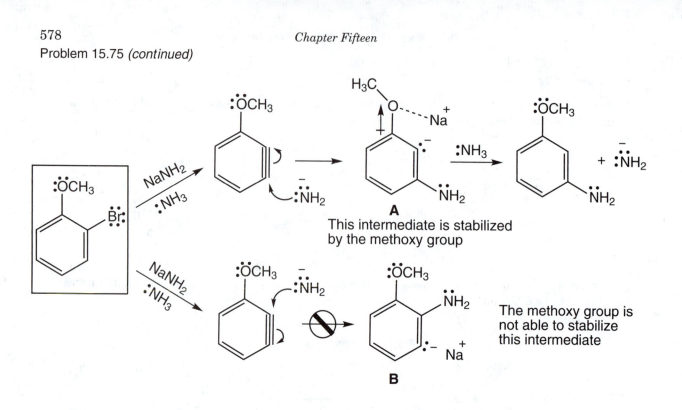

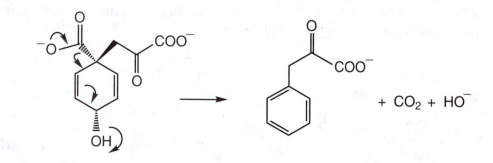

Problem 15.76 In this case, elimination of hydroxide, normally a poor leaving group, has many advantages. First, it is irreversible, as carbon dioxide is lost. Second, the reaction is intramolecular, usually an energetic advantage, and finally, it generates an aromatic compound. An acid catalyst would make the reaction easier by protonating or phosphorylating the OH oxygen to make it a better leaving group.

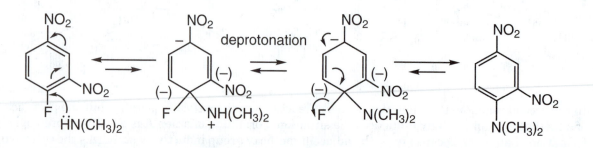

Problem 15.77 Here is the mechanism. The initial ionic product is colored because of the extensive conjugation revealed by the resonance forms shown.

The slow step in this reaction, the rate-determining step, is the initial endothermic addition of the amine. Accordingly, nothing that happens after the addition matters to the rate. Fluorine is more electronegative than chlorine, withdraws electrons better, and thus makes the addition reaction less endothermic and faster.

Problem 15.78 Here are the resonance forms:

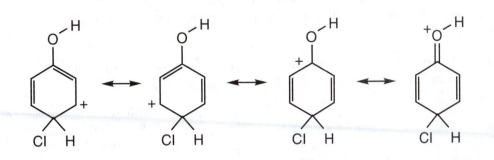

The calculations indicate that there is less cationic character on the oxygen than on any of the ring carbons. Oxygen is more electronegative than carbon, and therefore the resonance form with positive oxygen contributes less to the overall structure than the forms with positive carbon. Keep in mind, however, that the resonance structure with the positive oxygen is the only one that has all atoms with a full octet.

Problem 15.79 The carbon of the acylium ion is *sp* hybridized. You can tell that it is *sp* hybridized because the acylium ion is linear. The weak base that is deprotonating the intermediate is the chloride ion. This deprotonation is a rapid process because it restores aromaticity to the ring. The equilibrium lies far to the right. Therefore the hydrogen on the ring must be more acidic than HCl (pK_a of -8, see back endpaper); its pK_a must be < -8. Calculations suggest the pK_a is close to -20!

Problem 15.80 It looks like the largest LUMO character is at the carbon on the ring that is bonded to the chlorine. The nucleophile attacks at this position. The ammonium ion probably needs to be deprotonated before chloride leaves because otherwise it is a good leaving group. The nitrogen of the amine in the final product is clearly sp^2 hybridized. We can tell that it is sp^2 because the amino group is coplanar with the ring.

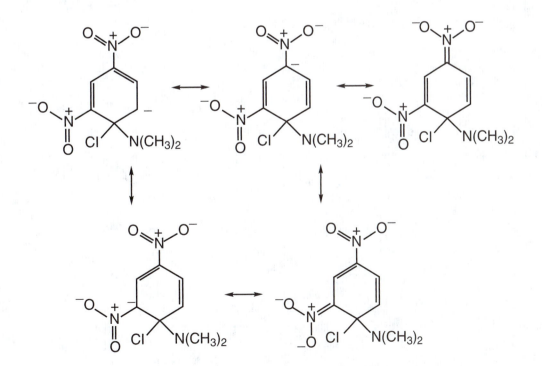

Problem 15.81   The most acidic hydrogens in the reaction mixture are the N—H hydrogens of the ammonia solvent. The second most acidic hydrogens must be the hydrogens ortho to the bromine on the aromatic ring. These hydrogens are the ones with which the strong base ($^-NH_2$) reacts. We expect the strong base to deprotonate the ammonia hydrogens, but that will only regenerate the strong base ($^-NH_2$).

The double bonds are left out of the benzene depiction because the resonance in the ring blurs the double bond nature. There really are no double bonds.

The carbons of the "triple bond" in benzyne (the second intermediate in the animation) share the cation equally. This sharing can be seen in the LUMO animation. The $^-NH_2$ could attack either carbon. The reason both carbons can share the cationic nature is because of resonance between the two carbons. These carbons also share the anion. See the following resonance structures. Notice that even a diradical species is possible.

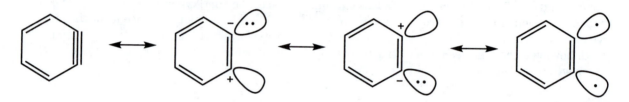

Carbonyl Chemistry 1: Addition Reactions

16

Problem 16.1   There are three different bonds in formaldehyde, the carbon–hydrogen bonds and the σ and π bonds between carbon and oxygen (Fig. 16.2). The carbon–hydrogen bonds are constructed through $1s/sp^2$ overlap, the carbon–oxygen σ bond by $sp^2(C)/sp^2(O)$ overlap, and the carbon–oxygen π bond from $2p(C)/2p(O)$ overlap.

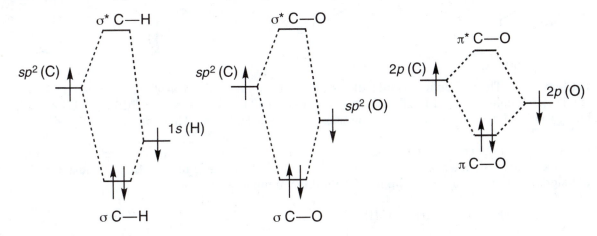

Problem 16.2   The only difference between this model and the one in Figure 16.2 is that the oxygen is hybridized sp. The carbon is still sp^2.

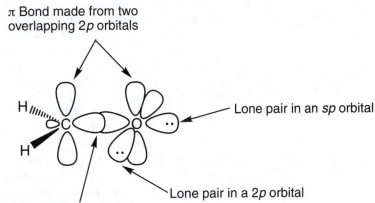

Problem 16.3 Just as the π bond "leans" toward the more electronegative oxygen atom, so will the *sp²/sp²* σ bond. Similarly, as π* leans toward carbon, so will σ*. The σ bond is made of more than 50% of the oxygen orbital and the σ* bond of more than 50% of the carbon orbital.

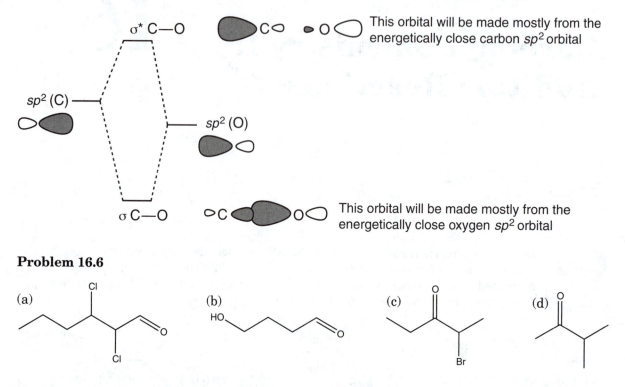

Problem 16.6

(a)   (b)   (c)   (d)

Problem 16.7 The molecular formula allows us to determine the degrees of unsaturation (Ω, DBE, or SODAR, all defined on p. 134). There are 7 carbons and 12 hydrogens, so $\Omega = [7(2) + 2 - 12]/2 = 2$. Thus, we either have two double bonds, one triple bond, one double bond and one ring, or two rings.

The IR band at 1710 cm⁻¹ tells us that we have a carbonyl. There is only one oxygen, so it must be either a ketone or an aldehyde. A quick glance at the NMR data tells us that there is no aldehyde (that signal would be around δ 9.0 ppm). The unknown must be a ketone.

Set up a table for interpreting the NMR data.

δ	Number of Hydrogens	Coupling	Interpretation
4.8	1 H	s (singlet)	The only way a signal can be this far downfield is for the hydrogen to be vinylic.
4.7	1 H	s (singlet)	This signal must be for another vinylic hydrogen, the hydrogen must be on the same carbon or the vinyl signals would be doublets; we must have an alkene.
2.6	2 H	t (triplet)	This signal is for a CH₂ (based on the number of hydrogens) and it is vicinal to a CH₂ (based on the coupling) and next to the carbonyl (based on the δ).
2.3	2 H	t (tripet)	This signal is for a CH₂ (based on the number of hydrogens) and it is vicinal to a CH₂ (based on the coupling) and next to the alkene (based on the δ).
2.2	3 H	singlet	This signal is for a CH₃ (based on the number of hydrogens) and it is vicinal to no hydrogens (based on the coupling) and next to the carbonyl (based on the δ).
1.7	3 H	singlet	This signal is for a CH₃ (based on the number of hydrogens) and it is vicinal to no hydrogens (based on the coupling); it is allylic (based on the δ).

The molecule must be 5-methyl-5-hexen-2-one.

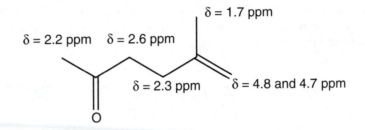

Problem 16.9   Addition to the oxygen end of the carbonyl or to the alkene would transfer the negative charge to carbon and form a weak O—O bond. It is energetically favorable to keep the negative charge on the relatively electronegative oxygen of the hydroxide ion.

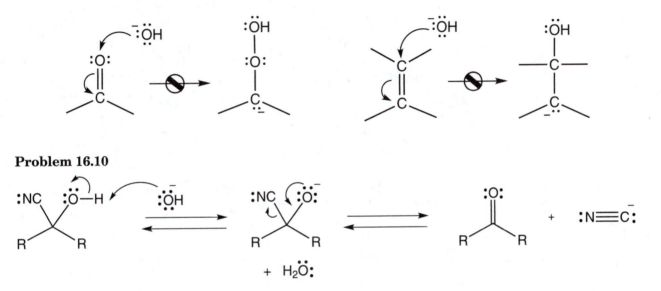

Problem 16.10

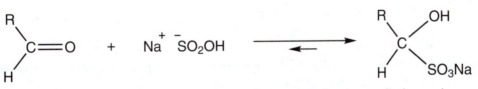

Problem 16.11   Addition reactions are generally less successful with ketones than with aldehydes. Ketones are more substituted than aldehydes and therefore more stable. This stability results in the carbonyl form being relatively favored at equilibrium compared to the addition product.

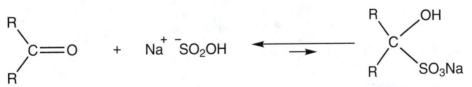

For the less stable aldehydes, it is the addition product that is usually favored

For the more stable ketones, it is the carbonyl compound that is usually favored

Problem 16.12 In the first step, the aldehyde oxygen is protonated. The nucleophilic hydroxyl oxygen adds to the Lewis acid carbonyl group. In the uncatalyzed addition, this generates an ionic structure. Deprotonation of the oxonium ion leads to the final product. These protonations and deprotonations can be accomplished by any acids or bases present. The order of these reactions is not certain, and they are not shown in the figure.

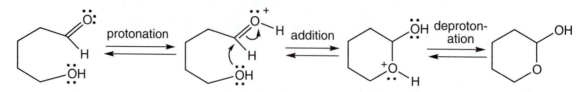

Problem 16.13 D-Glucose can close to the cyclic six-membered hemiacetal with the OH at the indicated carbon (see arrow) either in the axial position or the equatorial position.

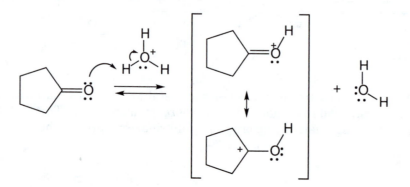

Equatorial OH on C(1)   Axial OH on C(1)

Problem 16.14 As in any acid-catalyzed acetal formation, the first step is protonation of the carbonyl group to give a resonance-stabilized intermediate.

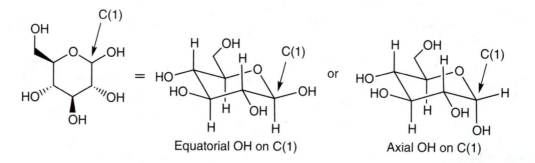

Ethylene glycol is an alcohol and adds to the strongly Lewis acidic protonated carbonyl. Deprotonation and reprotonation give **A**.

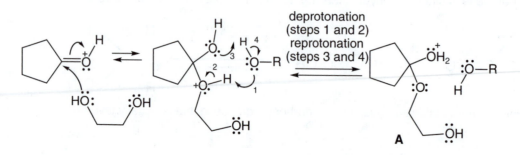

Intermediate **A** loses water to give the resonance-stabilized **B**. Now, in the critical step, the second OH group adds in intramolecular fashion to give **C**. A final deprotonation leads to the acetal.

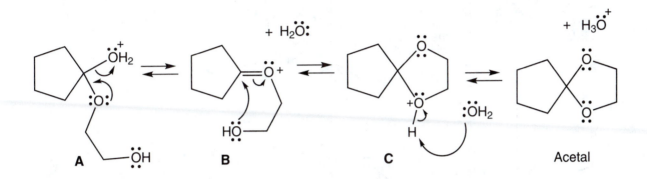

Problem 16.15 Argggh. A trick question. Simply run the mechanism in Problem 16.14 backward!

Problem 16.17 Hydrate formation is reversible. In labeled water, a small amount of hydrate **A** is formed. When **A** reforms acetone, the label will be incorporated half of the time. If enough labeled water is present, eventually all the acetone will acquire the label.

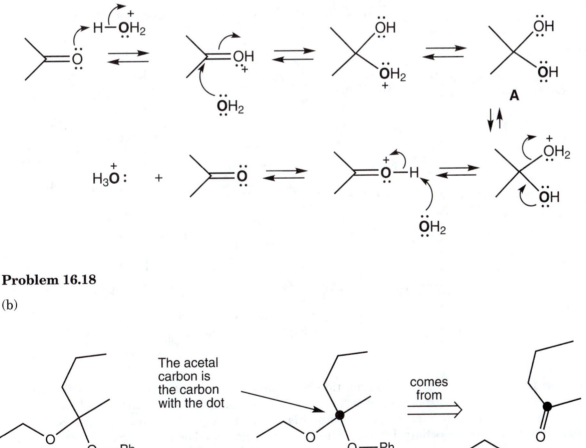

Problem 16.18

(b)

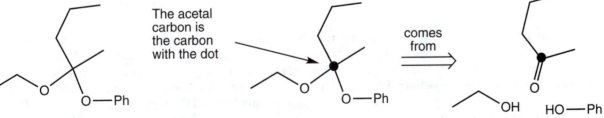

The acetal carbon is the carbon with the dot

comes from

(continued)

Problem 16.18 *(continued)*

(c)

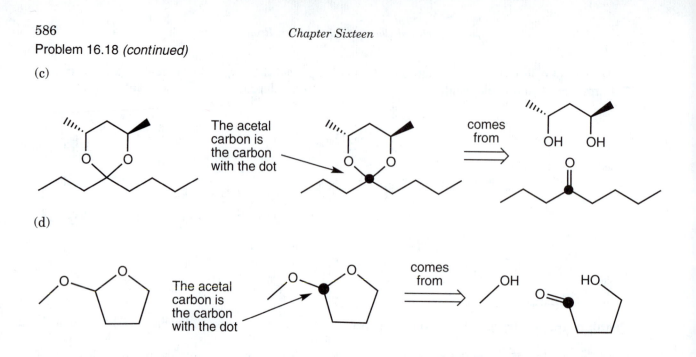

(d)

Problem 16.19 The bases involved are RO⁻, RHN⁻, and HO⁻, the conjugate bases of ROH, RNH$_2$, and H$_2$O. Notice that the second step in the middle reaction is very unfavorable. It is unlikely that an alkoxide will deprotonate an amine.

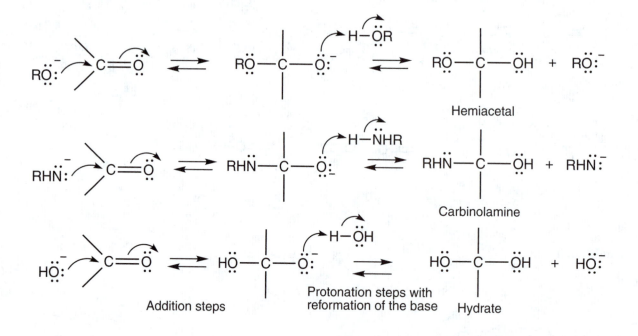

Problem 16.20 In the acid-catalyzed reaction between a ketone and a primary amine, the first step is protonation of the carbonyl oxygen. The nucleophilic amine then adds to the carbonyl carbon to give the tetrahedral intermediate **A**. Deprotonation of the nitrogen gives intermediate **B**, the neutral tetrahedral intermediate. Intermediate **B** can be protonated on the oxygen to give tetrahedral intermediate **C**, which can lose water to give cations **D** and **D′**. Deprotonation of **D** and **D′** leads to the imine product. Each step in the reaction is reversible. The equilibrium will be in favor of the imine formation if water is removed. If water is added, the ketone and primary amine will be favored. Notice that (*E*) and (*Z*) isomers of the imine will be formed.

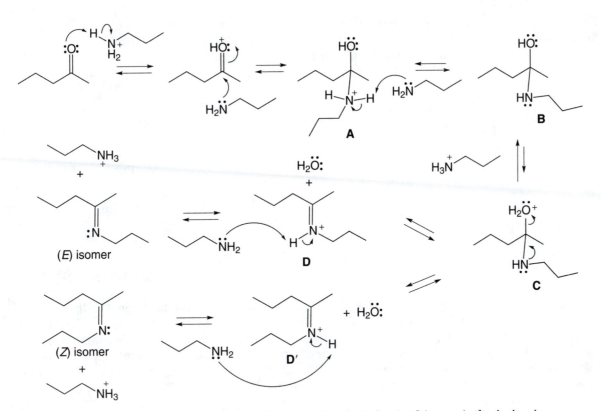

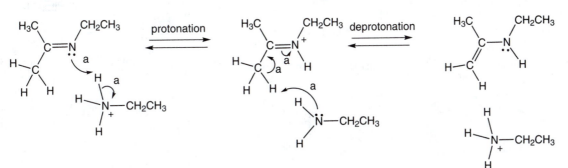

Problem 16.21 In the imine to enamine direction (left to right, in this case), the imine is protonated on nitrogen and then deprotonated on the carbon. The mechanism arrows labeled "a" show this pathway. In the enamine to imine direction (right to left), the enamine protonates on the carbon and deprotonates on the nitrogen. The mechanism arrows for enamine to imine are labeled "b." Thermodynamics will determine where this equilibrium settles out.

Mechanism shown for imine to enamine

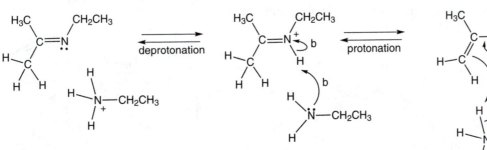

Problem 16.22 Let's work backward. Given the previous paragraph in the text, certainly the final step is the quenching of an organometallic reagent with D_2O. The organolithium or Grignard reagent is made from the bromide and either lithium or magnesium/ether.

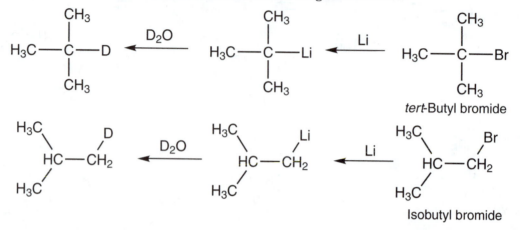

tert-Butyl bromide

Isobutyl bromide

Now, the only remaining question is how to convert the alkene into the bromide. For this you must remember material in Chapter 10. Simple addition of HBr to 2-methylpropene gives *tert*-butyl bromide (Chapter 3, p. 141, Chapter 10, p. 451). In order to make 1-bromo-2-methylpropane, a hydroboration/oxidation sequence is necessary to produce 2-methyl-1-propanol (p. 466), which can then be converted into the bromide with HBr. Alternatively, 2-methylpropene can be brominated in the anti-Markovnikov sense with HBr/ROOR (Chapter 12, p. 559).

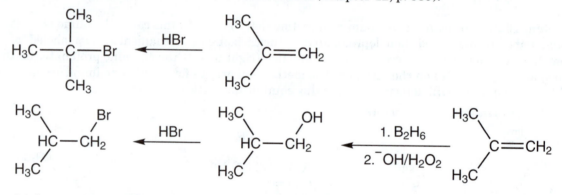

The third part of the problem is similar in its final steps:

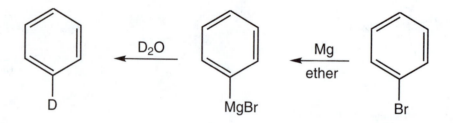

Bromobenzene is made from nitrobenzene by a sequence of conversions encountered in Chapter 15. The last step is a Sandmeyer reaction (Chapter 15, p. 716).

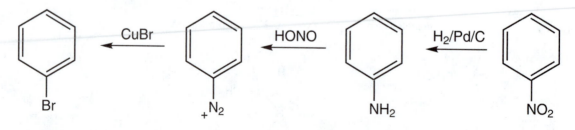

Problem 16.23 Round up the usual suspects! Good leaving groups form stable species, often anions, when they depart. In this case, the negative charge is stabilized by no fewer than three oxygens, and so tosylate is an excellent leaving group.

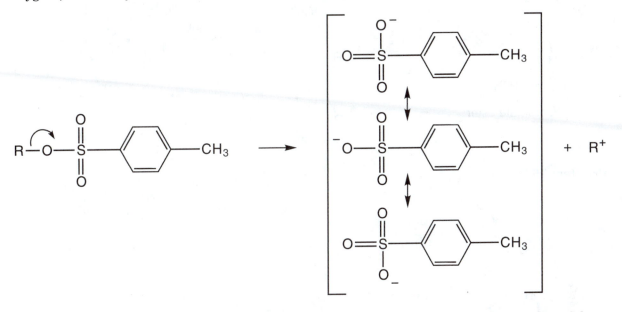

Wait...pause...Is that right? We hope at least some people are wondering why the stability of the final product, the tosylate, affects the ease of its formation. Shouldn't it be the energy of the transition state leading to the product? Yes, indeed it should be. But in the transition state the negative charge will be developing and delocalization will still be important in stabilizing it.

Problem 16.24 One conceptually simple test would be to measure the stereochemistry of the reaction. If the mechanism involves an S_N2 displacement, the reaction must proceed with complete inversion of configuration.

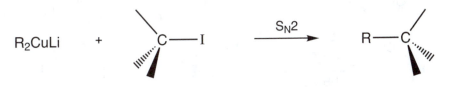

When this experiment is performed, it is found that the stereochemical outcome is *not* pure inversion, but a mixture of inversion, retention, and/or racemization, depending on the nature of the R group and the leaving group. The mechanism of this reaction is more complicated than a simple S_N2 displacement.

Problem 16.25 The reaction mechanism exactly parallels that for primary alcohols. The first step is formation of a chromate ester through addition of the alcohol to CrO_3, followed by protonation and deprotonation steps. The reactions effect the change of leaving group necessary to the success of the following elimination reaction.

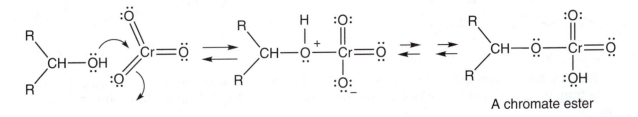

A chromate ester

(continued)

Problem 16.25 *(continued)*

An elimination using any base in the system (pyridine or the alcohol itself are possibilities) generates the ketone. An intramolecular E2 is also possible.

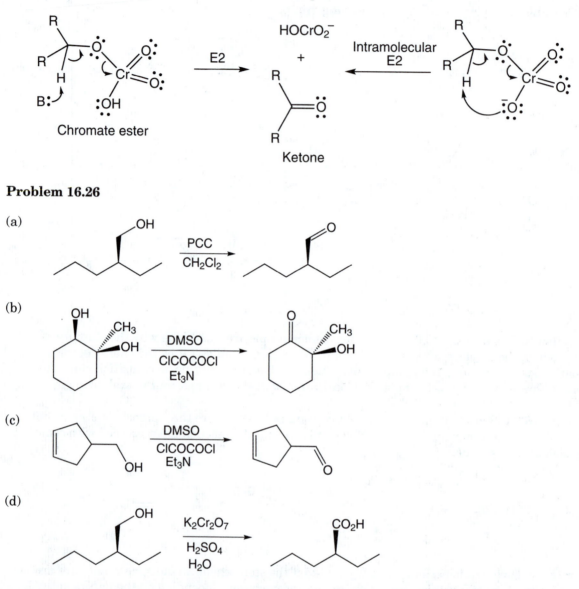

Problem 16.26

(a)

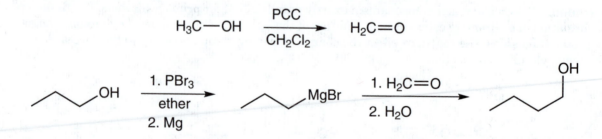

(b)

(c)

(d)

Problem 16.27 (a) Here we need only extend the chain of an alcohol by one carbon. There is a standard technique, and here it is. Formaldehyde is used as the one-carbon fragment.

(b) Here we have a symmetrical secondary alcohol. It can be dissected in only one way. It must come from propyllithium [made in part (a)] and butanal, which can be made from the product of part (a).

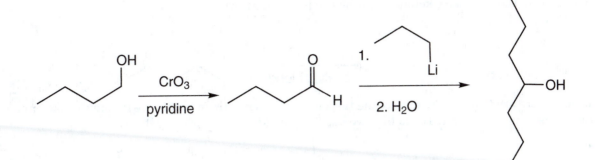

(c) This secondary alcohol is unsymmetrical and can be dissected in two ways:

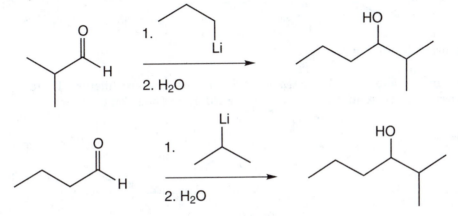

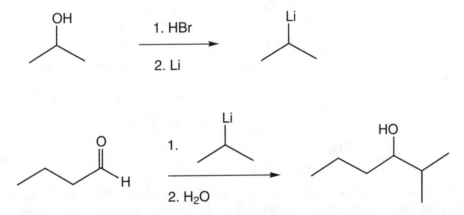

As we just made butanal in the last part, let's use the second version (the first will, of course, work just fine, too). We also need isopropyllithium, which can be made from isopropyl alcohol.

Now we pass on to the three kinds of tertiary alcohols, R_3COH, $R_2R'COH$, and $R''R'RCOH$. If you can make these three, you can make *any* tertiary alcohol.

(d) For R_3COH, there is only one way to do the dissection: RLi must react with $R_2C{=}O$.

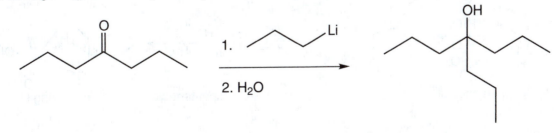

(continued)

Problem 16.27 *(continued)*

We can make the necessary ketone from the product of part (b).

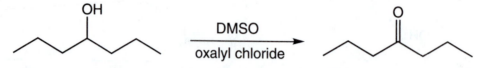

(e) Now there are two ways to dissect the target alcohol (write them both!). We'll take the easier one and add isopropyllithium to the ketone we used in part (d).

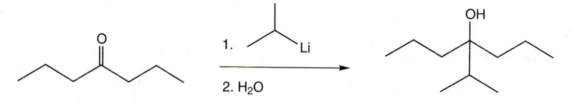

We made isopropyllithium in part (c).

(f) Finally, we must make an alcohol in which all three "R" groups are different. There are three possible dissections, all essentially equally easy or hard. Here's one. You might work out the other two:

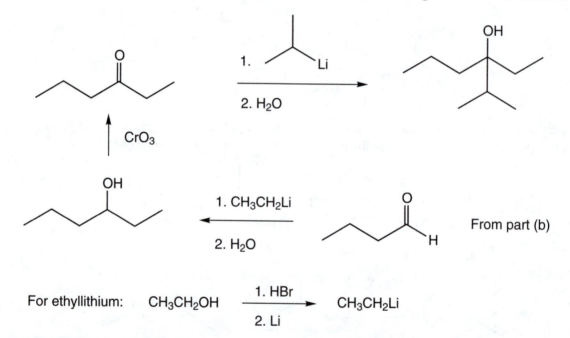

For ethyllithium: CH₃CH₂OH →(1. HBr / 2. Li)→ CH₃CH₂Li

Problem 16.28 To do the periodate cleavage of vicinal diols, it is necessary to form the cyclic ester of periodic acid (Fig. 16.75). The cyclic ester is much easier to form with *cis*-1,2-cyclopentanediol than it is with *trans*-1,2-cyclopentanediol. There is too much ring strain in the cyclic ester that would need to be formed from the trans diol.

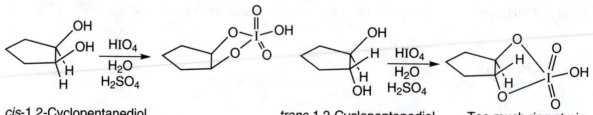

cis-1,2-Cyclopentanediol   *trans*-1,2-Cyclopentanediol   Too much ring strain

Problem 16.30 The first three parts are examples of grind-it-out small alcohol syntheses. In each case we need to use butyllithium, so let's make that reagent first.

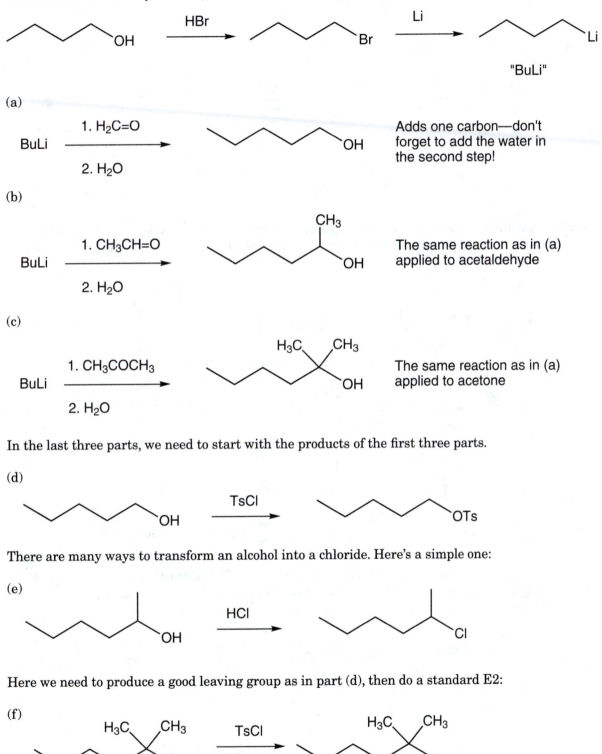

(a)

BuLi

1. $H_2C=O$

2. H_2O

Adds one carbon—don't forget to add the water in the second step!

(b)

BuLi

1. $CH_3CH=O$

2. H_2O

The same reaction as in (a) applied to acetaldehyde

(c)

BuLi

1. CH_3COCH_3

2. H_2O

The same reaction as in (a) applied to acetone

In the last three parts, we need to start with the products of the first three parts.

(d)

TsCl

There are many ways to transform an alcohol into a chloride. Here's a simple one:

(e)

HCl

Here we need to produce a good leaving group as in part (d), then do a standard E2:

(f)

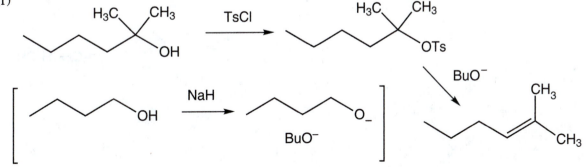

TsCl

NaH

BuO⁻

BuO⁻

Problem 16.31   This reaction is nothing more than an E1 elimination (Chapter 8, p. 332). The oxygen atom is protonated and water is lost to give the tertiary carbocation, **A**.

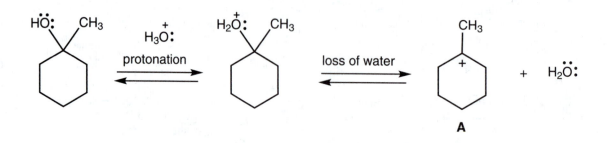

There are two different protons that can be lost in the second step. The one leading to formation of the more stable, trisubstituted double bond will be preferred to the one leading to the less stable, disubstituted double bond.

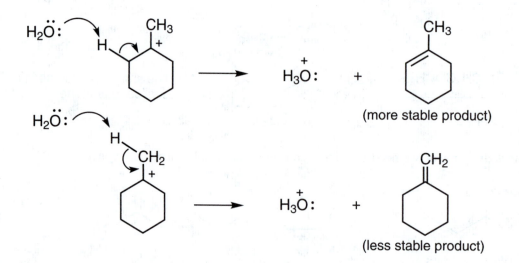

Problem 16.32   Part (a) is a standard Wittig synthesis. The allowed starting materials include the necessary cyclic ketone and all the reagents required to make the Wittig reagent:

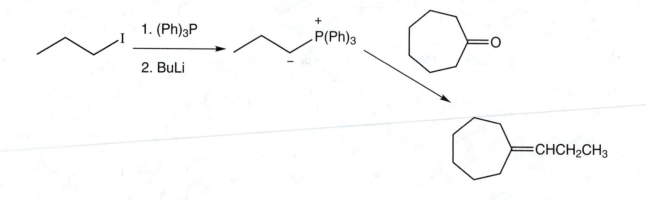

(b) In this part, a Wittig reaction must be followed by a hydrogenation:

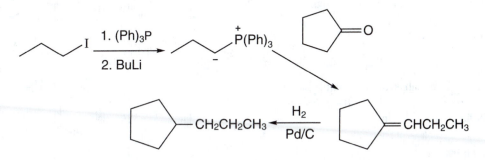

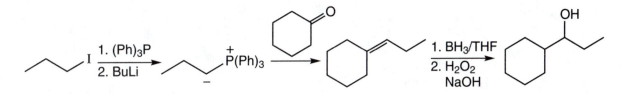

(c) To form the alcohol on the less substituted carbon, the Wittig reaction must be followed by a hydroboration/oxidation reaction.

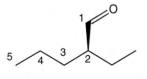

Additional Problem Answers

Problem 16.34

(a) This molecule is (*S*)-2-ethylpentanal.

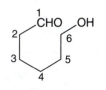

(b) This molecule is 6-hydroxyhexanal.

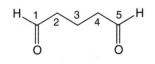

(c) This molecule is pentanedial.

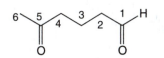

(d) This molecule is 5-oxohexanal.

(continued)

Problem 16.34 *(continued)*

(e) This molecule is 4,4,4-tribromobutanal.

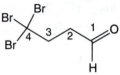

(f) This molecule is 2-hydroxybenzaldehyde
(or *o*-hydroxybenzaldehyde). Its common
name is salicylaldehyde.

Problem 16.35

(a)

(c)

(b)

(d)

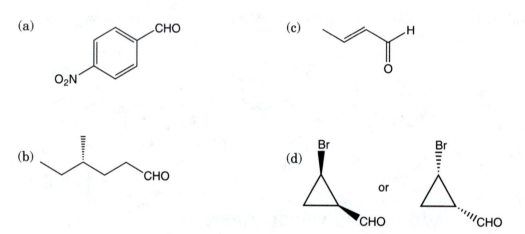

or

Problem 16.36

(a)

(d)

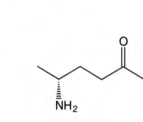

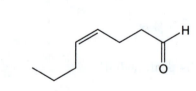

(b)

(e)

(c)

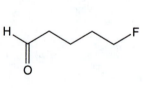

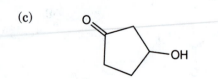

Problem 16.37

(a) 4,4-dimethyl-2-pentanone

(b) 6-hepten-2-one

(c) (2R,4S)-2-chloro-4-methylhexanal

(d) 5-methoxy-3-hexanone

(e) 4-bromocyclohexanone

Problem 16.38

(a) This molecule is (2R,5S)-2,5-dimethylcyclopentanone. We might also name it (2S,5R)-2,5-dimethylcyclopentanone, but in such cases of equivalent groups, the (R) designation gets a lower number than the (S), presumably because (R) comes before (S) in the alphabet.

(b) This molecule is 3-bromo-1-phenyl-1-butanone.

(c) This molecule is 7-amino-4-hydroxy-3-octanone.

Problem 16.39

(a)

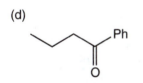

(b)

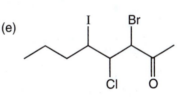

(c)

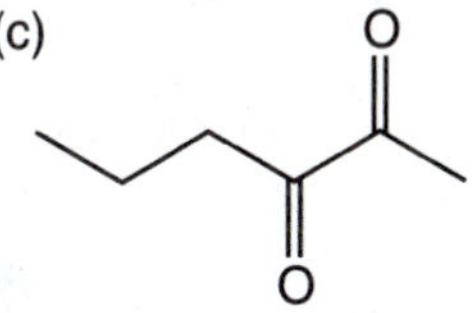

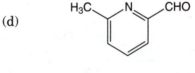

Note that (a) and (b) are the same!

(d)

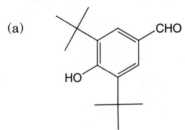

(e)

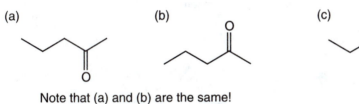

Problem 16.40

(a)

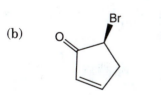

(d)

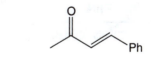

(b)

(e)

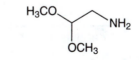

(c)

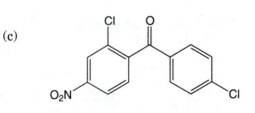

Problem 16.41

(a) Cyclopentanones should absorb in the IR at about 1750 cm⁻¹. Acyclic ketones should absorb at about 1720 cm⁻¹. In fact, cyclopentanone itself absorbs at 1751 cm⁻¹ and 4-penten-2-one at 1723 cm⁻¹.

(b) Cyclobutanones absorb about 40 cm⁻¹ higher than the less strained cyclopentanones. 3-Methyl-cyclobutanone absorbs at 1789 cm⁻¹. See Table 16.2 (p. 774) for the carbonyl stretching frequencies of the cyclic ketones.

(c) The appearance of a downfield signal for the carbonyl carbon will be a dead giveaway. Cyclohex-anone absorbs at δ 207.9 ppm, and no signal in the spectrum of diallyl ether will be close to that position. Moreover, the ^{13}C NMR spectrum of cyclohexanone must show signals for four different carbons, whereas the spectrum for diallyl ether will show only three.

(d) There are many ways to make the distinction. One way that definitely won't work is to use the strong IR band for the carbonyl stretch. Small acyclic aldehydes and ketones are too close together for that kind of distinction. These two molecules absorb at 1715 cm⁻¹ and 1719 cm⁻¹, for example. However, you should be able to pick out the two bands for the C—H stretch in the aldehyde spec-trum. These typically appear at about 2850 cm⁻¹ and 2750 cm⁻¹, and this aldehyde shows them at 2840 cm⁻¹ and 2732 cm⁻¹. The ^{1}H NMR spectra should also be definitive. First of all, the aldehyde should show a distinctive downfield signal at about δ 9–10 ppm. This one appears at δ 9.77 ppm. The detailed spectra will also be different, and these are summarized.

Problem 16.42

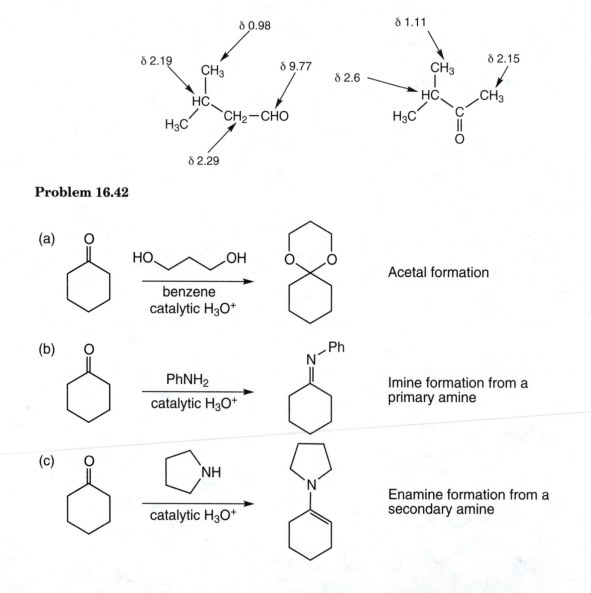

(a) Acetal formation

(b) Imine formation from a primary amine

(c) Enamine formation from a secondary amine

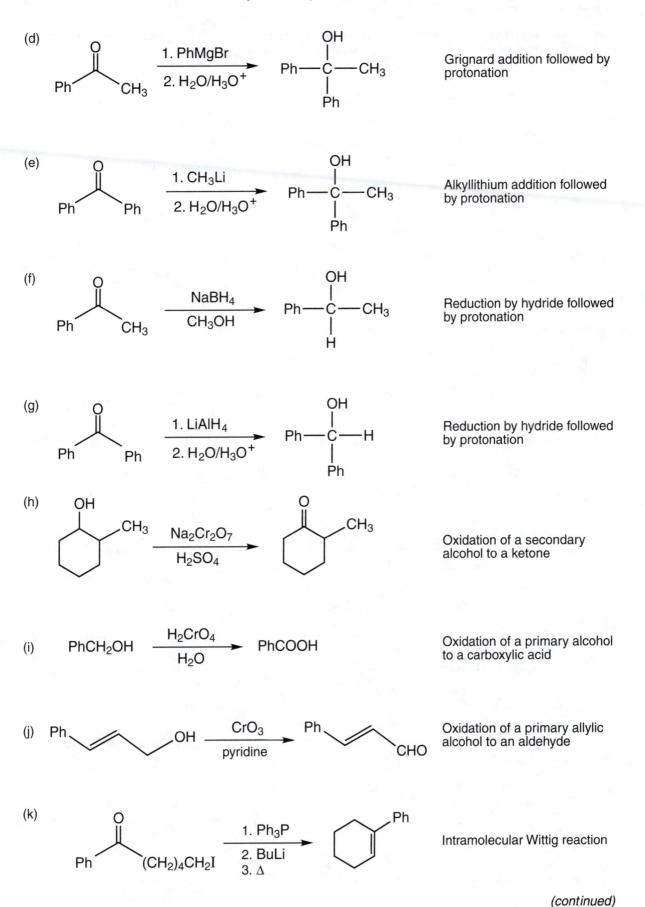

(d)

1. PhMgBr
2. H₂O/H₃O⁺

Grignard addition followed by protonation

(e)

1. CH₃Li
2. H₂O/H₃O⁺

Alkyllithium addition followed by protonation

(f)

NaBH₄
CH₃OH

Reduction by hydride followed by protonation

(g)

1. LiAlH₄
2. H₂O/H₃O⁺

Reduction by hydride followed by protonation

(h)

Na₂Cr₂O₇
H₂SO₄

Oxidation of a secondary alcohol to a ketone

(i)   PhCH₂OH

H₂CrO₄
H₂O

PhCOOH

Oxidation of a primary alcohol to a carboxylic acid

(j)

CrO₃
pyridine

Oxidation of a primary allylic alcohol to an aldehyde

(k)

1. Ph₃P
2. BuLi
3. Δ

Intramolecular Wittig reaction

(continued)

Problem 16.42 *(continued)*

The answer to part (k) on the previous page is the only tricky one. It is hard because it involves a multistep intramolecular reaction. The following intermediates are involved:

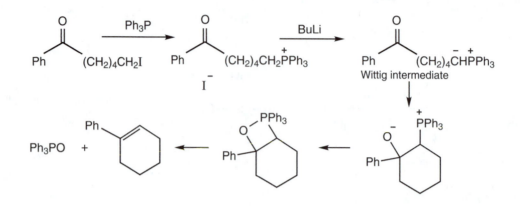

Problem 16.43

(a)

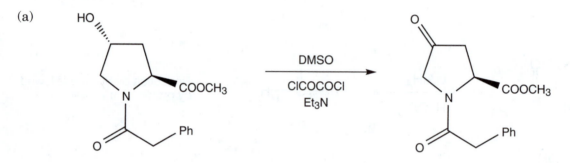

(b) In the first step, the R—MgBr adds to the aldehyde. That reaction forms an alkoxide, which is quenched in the second step by the 5% HCl to give a secondary alcohol. The third step is a Swern oxidation of the secondary alcohol to give the ketone.

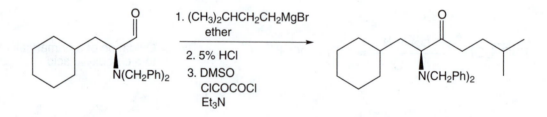

(c) The first two steps produce the primary alcohol. The PCC (pyridinium chlorochromate) oxidation gives the aldehyde shown.

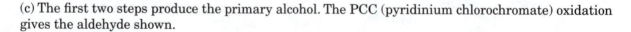

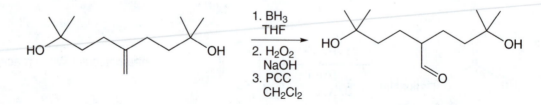

Problem 16.44

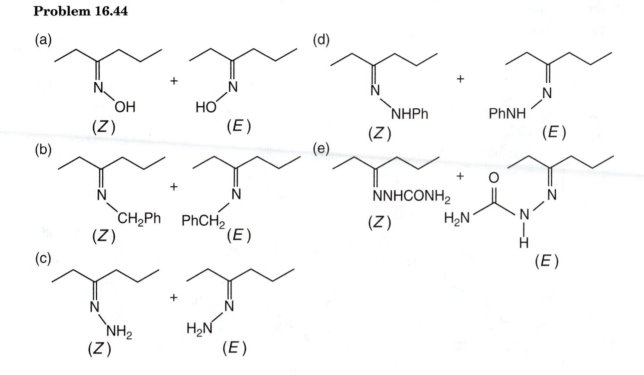

(a)

N–OH (Z) + HO–N (E)

(b)

N–CH₂Ph (Z) + PhCH₂–N (E)

(c)

N–NH₂ (Z) + H₂N–N (E)

(d)

N–NHPh (Z) + PhNH–N (E)

(e)

NNHCONH₂ (Z) + H₂N(C=O)N(H)–N (E)

Problem 16.45 We might argue that the transition state for protonation on oxygen places a partial positive charge on carbon, whereas the transition state for protonation on carbon places the positive charge on oxygen.

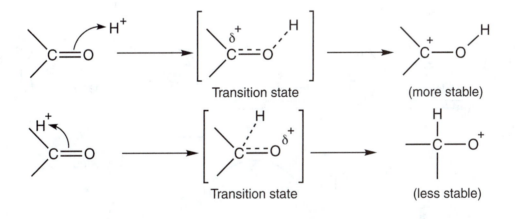

Transition state   (more stable)

Transition state   (less stable)

It is far better energetically to place the positive charge on the relatively electropositive carbon than on the relatively electronegative oxygen. Protonation will be favored on oxygen.

Problem 16.46

(a) This problem is just a reverse acetal formation.

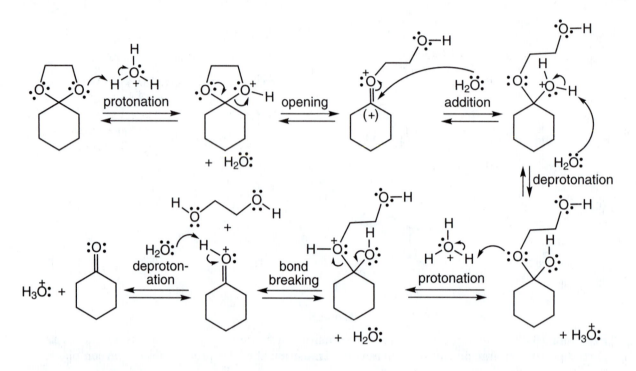

(b) You are asked here to hydrolyze an iminium ion. The analogy is to carbonyl chemistry, with the iminium ion playing the part of a protonated carbonyl.

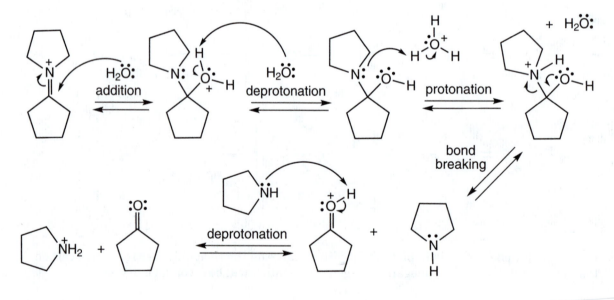

(c) This reaction involves the conversion of a hemiacetal into an acetal. The steps are essentially those of part (a).

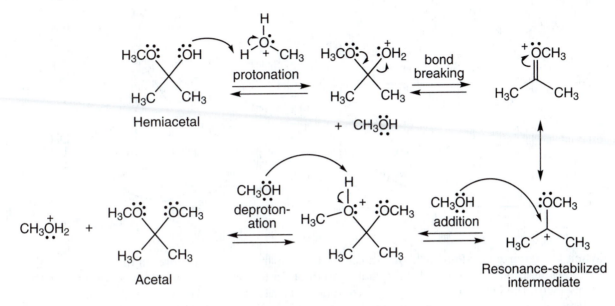

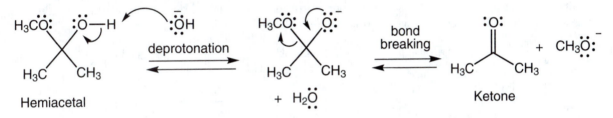

(d) In base, a hemiacetal cannot go on to the full acetal. The only possible reaction here is reversal to the carbonyl compound, generally an exothermic process.

(e) Full acetals are stable in base. There is no reaction here.

(f) This reaction is a version of imine formation in which a carbonyl compound reacts with a primary amine. In this case, the imine has the special name of "oxime."

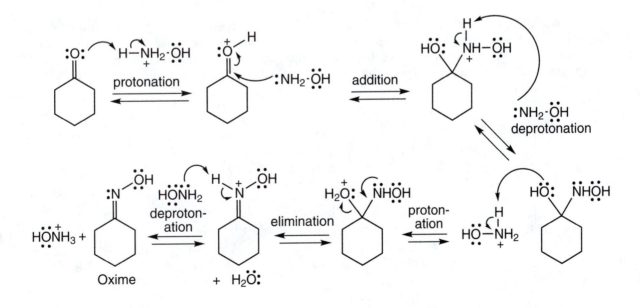

Problem 16.47   There are two possible directions from which the carbonyl group can be reduced. The product is a mixture of diastereomeric alcohols.

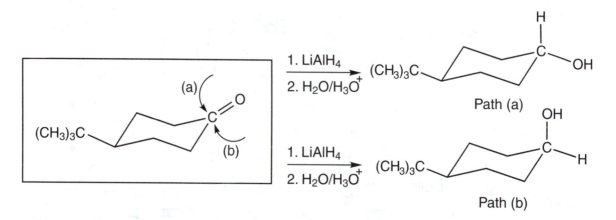

When the environment on each side of the carbonyl group is comparable, as in this ketone, the more stable product is normally the major product. In this case, the more stable product is the alcohol with the OH equatorial. In more complicated metal hydride reductions, the stereochemical outcome is more difficult to rationalize. Both the ease of approach of the reducing agent (steric approach control) and the aluminum coordination to the ketone oxygen are important factors.

Problem 16.48   The difficulty is that carbonyl groups are attacked by lithium aluminum hydride to give alkoxides. This reaction will surely compete with the desired reduction.

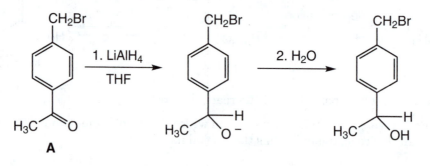

Problem 16.49   We need to protect that carbonyl group from reduction by lithium aluminum hydride, and then we need a way of regenerating the carbonyl group once the bromide is reduced. This classic problem is solved by converting the carbonyl into an acetal, doing the reduction, and then regenerating the carbonyl. The technique is described on page 792. Here is how it would work in this situation.

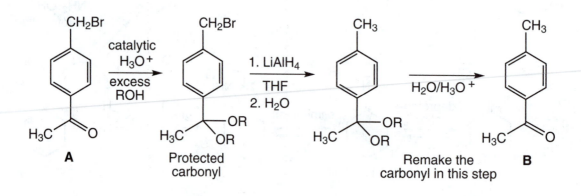

Problem 16.50 The mechanism that accounts for the observed product has a hydride (H⁻) from LiAlH₄ attack C(1) of the epoxide, as shown below. The stereogenic C(2) carbon would retain the (R) configuration. The C(1) stereogenic atom would lose its stereochemistry because it will have two hydrogens after the hydride addition. The alkoxide would be quenched by the 5% HCl to form (2R)-1-methoxy-2-pentanol.

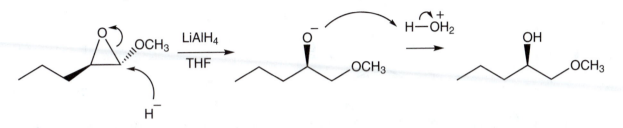

An alternative product would come from the hydride adding to C(2) of the epoxide. This pathway would yield 1-pentanol, as shown below. The initial alkoxide would "kick out" the methoxy group, producing pentanal. In the presence of LiAlH₄, the aldehyde would be reduced to the alkoxide and quenched by 5% HCl to give 1-pentanol.

Alternative pathway:

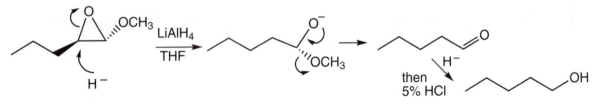

A likely reason that the reaction follows the pathway of hydride attack at the C(1) carbon is that the methoxy oxygen electrons are feeding into the epoxide, weakening the bond between the C(1) carbon and the epoxide oxygen. In terms of resonance, this can be shown as the following (note that no atoms are moving, only electrons):

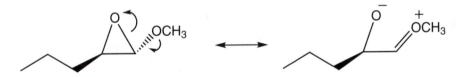

This resonance structure makes the C(1) carbon a much more electrophilic carbon than the C(2) carbon.

Problem 16.51 In the first step, the Grignard reagent is formed. Next, an intramolecular addition takes place to give alkoxide **A**. Protonation gives the product.

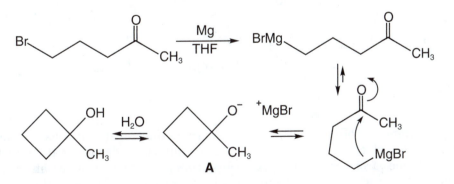

Problem 16.52   Most of these parts require you to transform isopropyl alcohol into a molecule that can incorporate the label in the desired way.

(a) For example, transformation of isopropyl alcohol into 2-bromopropane allows the formation of the alkyllithium reagent or Grignard reagent. Reaction with D_2O completes this part of the question.

(b) Oxidation of isopropyl alcohol to acetone allows reduction in deuterated medium to give 2,2-dideuteriopropane:

(c) For this part, we need to make propene. We can use 2-bromopropane, made in part (a), and do an elimination reaction. "Hydrogenation" using D_2 makes the desired product.

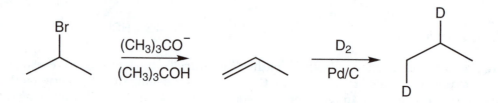

(d) Reduction of the acetone, made in part (b), with $NaBD_4$ or $LiAlD_4$, followed by hydrolysis, will do the job:

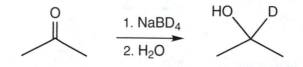

(e) Two propane backbones need to be sewn together here. A nice way is to allow 2-propyllithium part (a) to react with acetone part (b). The resulting alcohol can be transformed as in part (a) to the deuterated material.

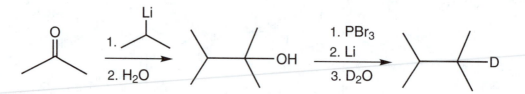

Problem 16.53   Here is another problem requiring sequential transformations of a simple starting material. In each case, at least one carbon atom must be added to the three-carbon starting material. One carbon can be added by transforming propyl alcohol into an organometallic reagent, allowing it to react with formaldehyde to give a new alcohol. Manipulation of this alcohol (butyl alcohol) can lead to all the desired products.

(a) Reaction of butyl alcohol with PBr₃ (or many other reagents) will give the desired bromide.

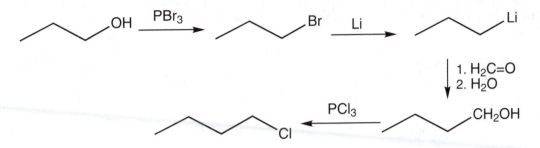

(b) Butyl bromide can also be used as starting material for 1-cyanobutane.

(c) Butyl bromide can be made into the organolithium reagent and treated with water to make butane.

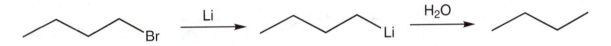

(d) This slightly more complex transformation requires that we find a way to combine a three-carbon fragment with a four-carbon fragment. One way to do this is to allow propyllithium part (a) to react with oxidized butyl alcohol, butanal.

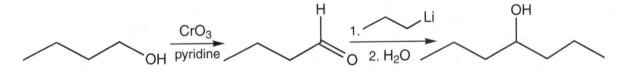

Problem 16.54 This problem demands a series of relatively complex alcohol syntheses. The last two parts require the conversion of an intermediate alcohol into a new material. In questions of this complexity, it is definitely worthwhile to work backward, starting with the desired product and asking, What is the immediate precursor of this molecule? (Retrosynthetic analysis.) In this way, we can always work back to a set of four-carbon alcohols. It is very likely that there will be more than one way to make each target. Notice that in this problem we make use of the special retro-synthetic arrow that points from the desired product to the immediate reactants needed to make it.

(a) In this problem, the target secondary alcohol is potentially available from two different sets of precursors, both involving addition of a Grignard reagent to an aldehyde.

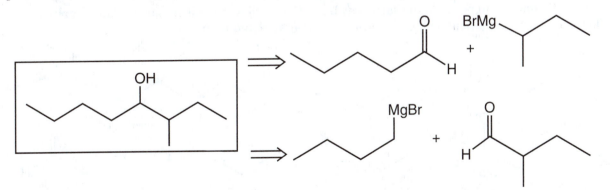

(continued)

Problem 16.54 *(continued)*

Although both of these routes are feasible, we will concentrate on the first possibility. You should work out the other synthesis. The required Grignard reagent for the first route is available from *sec*-butyl alcohol. The reaction partner, pentanal, can be made from butyl magnesium bromide and formaldehyde.

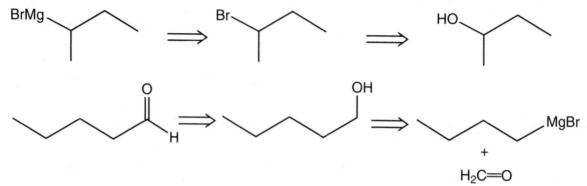

Butyl magnesium bromide is available from butyl alcohol, and formaldehyde can be made through oxidation of methyl alcohol. An outline of the full synthesis follows.

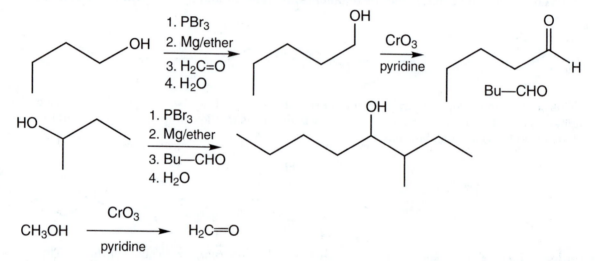

(b) This tertiary alcohol "deconstructs" in only one way. The immediate precursors must be dipropyl ketone and propyl magnesium bromide (propyl Grignard).

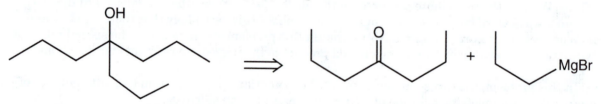

Propyl Grignard can be made the same way butyl Grignard was made in part (a), and dipropyl ketone comes from 4-heptanol, itself formed from butanal and propyl Grignard.

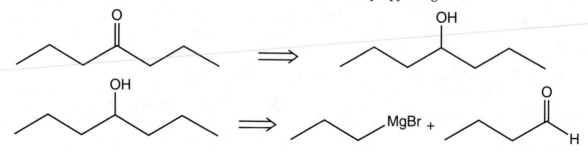

Butanal can be made from oxidation of butyl alcohol, so the synthesis looks as follows:

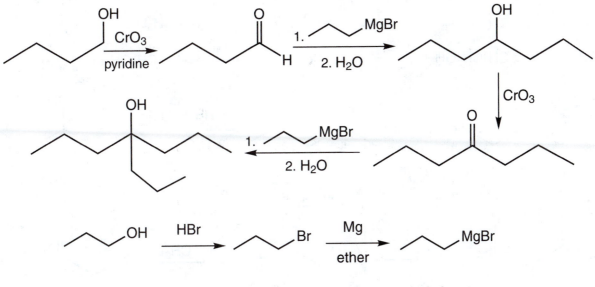

(c) The target tertiary bromide is available from the corresponding alcohol.

The alcohol, in turn, is available from three different sets of precursors, all combinations of organometallic reagents and ketones:

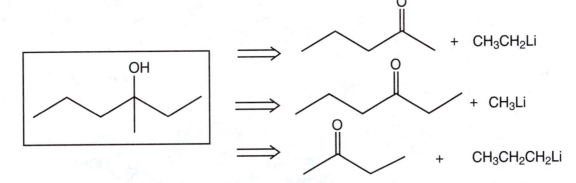

We will follow the second path, but you might work out the other two for practice. The required ketone, 3-hexanone, is available from butanal and ethyllithium. Butanal can be made from butyl alcohol, and ethyllithium and methyllithium derive, ultimately, from ethyl alcohol and methyl alcohol.

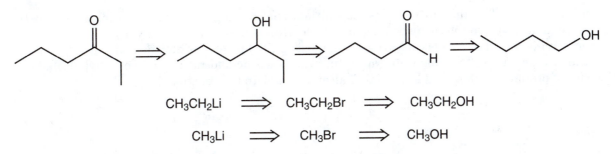

(continued)

Problem 16.54 *(continued)*

So, here is the synthesis:

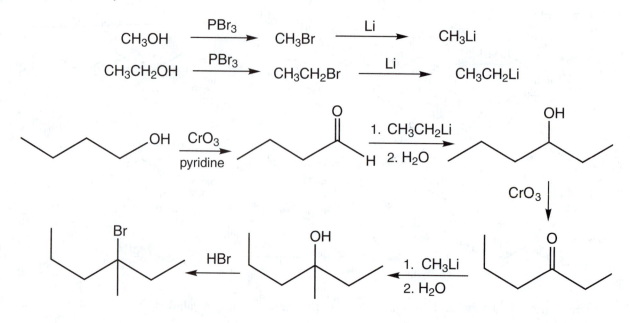

(d) This last problem is easy. The target molecule comes from dipropyl ketone, a molecule we made in (b).

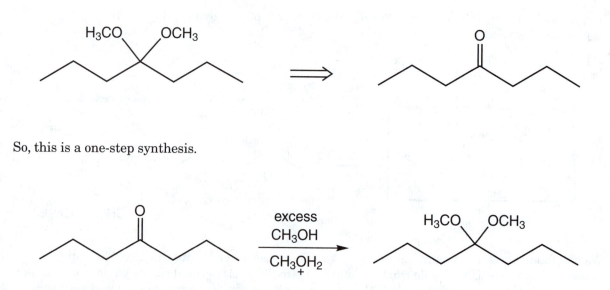

So, this is a one-step synthesis.

Problem 16.55   The difficulty here is recognition. The starting material is an acetal and must reverse in acid to give the corresponding aldehyde. However, the standard hydrolysis can be short-circuited in this case because there is a nucleophile lurking within the molecule in the form of the nitrogen atom. This nitrogen can capture the Lewis acid formed by loss of ethyl alcohol. A series of proton transfers and loss of ethyl alcohol complete the reaction.

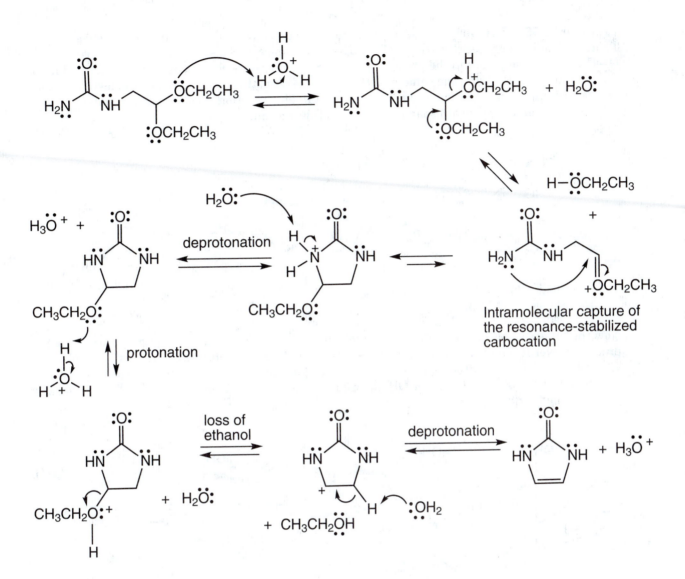

The process shown above looks complicated, but it is really not so bad. One critical insight is necessary, and the rest is nuts and bolts protonation–deprotonation and addition–elimination611 chemistry. How to see the crucial ring-forming step? Analyze before you begin. Look at the product and compare it to the starting material. What needs to be accomplished? In this case, it is very likely that the series N—CO—N in the five-membered ring product (arrows) is formed from the same N—CO—N series in the starting material. If this is so, it becomes easy to see where the connection must be made. Then the task is to see how to do the connecting.

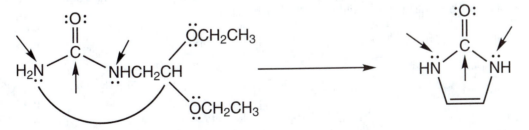

If this N—CO—N sequence becomes the N—CO—N sequence in the product (very likely), the points of ring formation are identified. That makes the problem much easier.

Problem 16.56   Here we have a seven-membered acetal converted into a five-membered acetal. This problem is tricky because the O—C(CH₃)₂—O sequence in the starting material is *not* the O—C(CH₃)₂—O sequence in the product. The seven-membered acetal opens, and an intermediate is captured by the OH within the same molecule. In this sense, this problem is like the last one. Start by working out the mechanism for opening the acetal. This reaction involves only protonation of a ring O and opening to give a resonance-stabilized carbocation.

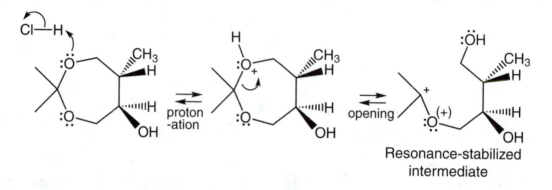

Now the "upper right" of the product molecule is in place. That's the kind of thing that makes a problem solver feel he or she is on the right track. If parts of the molecule begin to fall into place as you work through a potential answer, that's a good sign. In this case, we need only close up the five-membered ring using the other OH, deprotonate, and we are done. It helps greatly to redraw the molecule first.

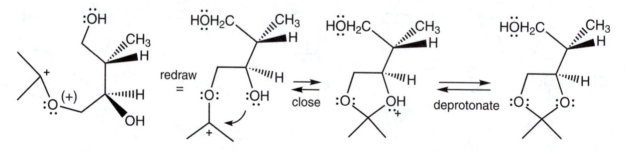

Why does the reaction go? There is more strain in a seven-membered ring than in a five-membered ring.

Problem 16.57   Here is yet another problem involving an acetal. This problem becomes much easier if we do a little analysis before we start. The question to ask is, What must be the precursor to the final acetal? This is a "retrosynthetic" question. Acetals are produced from carbonyl compounds and alcohols under acidic conditions. The alcohol is there as one starting material, and the reaction conditions are surely acidic, so this seems a promising line of thought. The final acetal could be formed from the starting diol and the aldehyde shown (**A**).

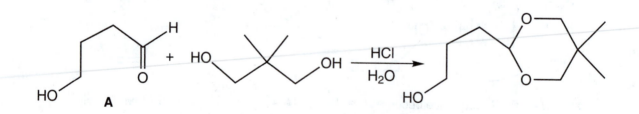

Now the problem is to find a way to make aldehyde **A** from the five-membered ring in the problem. That's not so hard. Protonation to give the resonance-stabilized carbocation is followed by addition of water to give a hemiacetal. Opening of the hemiacetal gives exactly the aldehyde we need.

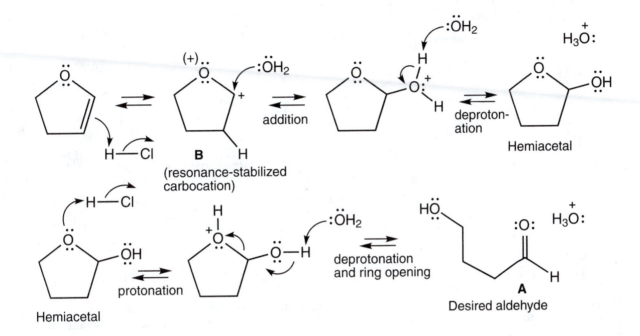

B
(resonance-stabilized carbocation)

addition

deprotonation

Hemiacetal

Hemiacetal

protonation

deprotonation and ring opening

A
Desired aldehyde

Now, straightforward acetal formation makes the product (mechanism not shown).

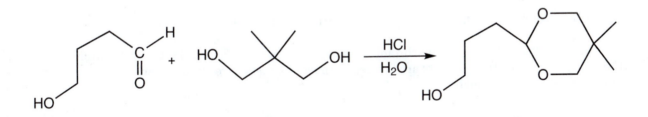

There is another, more imaginative (and probably correct) version of the mechanism. Suppose the resonance-stabilized cation **B** is captured, not by water, but by the alcohol? Now a sequence of protonations, deprotonations, and ring openings and closings leads directly to the product.

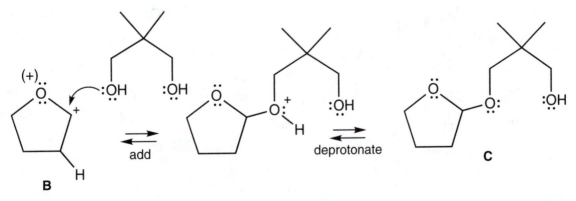

B

add

deprotonate

C

(continued)

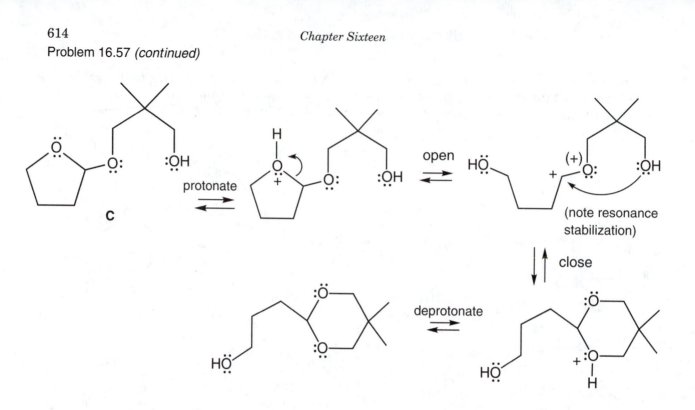

C

open

(note resonance
stabilization)

close

deprotonate

How is one to choose between these two reasonable mechanisms? The answer is that we don't really have to. Both are probably taking place, and the mix will depend on reaction conditions and, especially, on the relative concentrations of water and the diol.

Problem 16.58 Were a Si—C bond to be broken, a bare, unstabilized carbanion would be typically formed. Such species are simply too high in energy to be viable intermediates. By contrast, when the Si—O bond is broken, a relatively stable alkoxide is produced.

HF is used to etch glass. The Si—F bond is even stronger than the Si—O bond.

Problem 16.59 The heavy metal chromium is highly toxic when released into the environment. These days, other oxidizing agents are favored because the costs of recapturing the used chromium oxidizing agents are too high.

Problem 16.60 The NADH nitrogen in the six-membered ring has a lone pair of electrons that can "kick out" one of the hydrogens on C(4) of the ring. The hydrogen would leave with its electrons, so it would be a hydride. The driving force for this reaction is the aromatization of the ring. NADH is nonaromatic and NAD⁺ is aromatic.

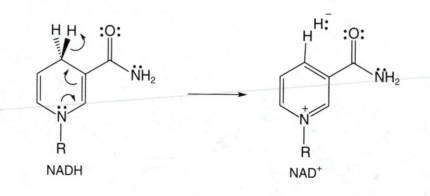

NADH   NAD⁺

Problem 16.61 The formula shows that two nitrogens are incorporated into this product. Like hemiacetals, carbinolamines can react further, in this case with amines to give diamino compounds called aminals. These do not generally predominate at equilibrium, but they are partners in the interconverting mixture of compounds that eventually leads to the imine, the thermodynamically favored structure.

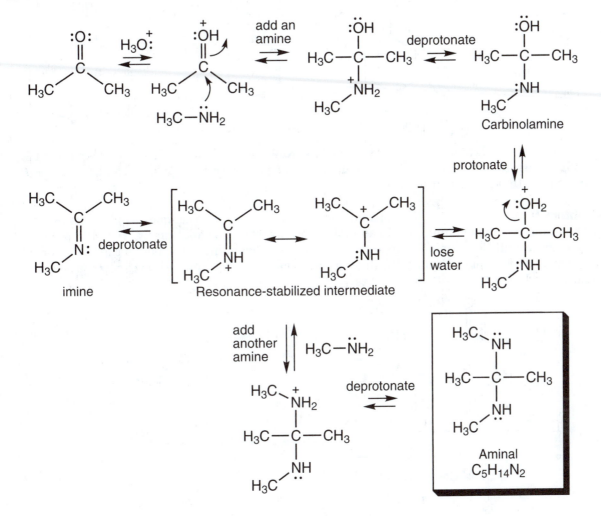

Problem 16.62 There are many different ways to do this problem. Perhaps the best attack is first to recognize that, as compound **A** is aromatic, it can contain only one other carbon. Why? Well, the reaction is a Wittig reaction, and the Wittig reagent used in this case adds two carbons. The products are C_9 compounds, so there must be seven carbons in **A**; six make up the ring, and there is one left over. The Wittig reaction converts C=O into C=C, so **A** must have the following partial structure.

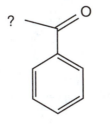

A
(partial structure)

(continued)

The Wittig reaction adds $CH_3CH=$, and as the formulae of **B** and **C** are C_9H_{10}, the "?" must be H. So **B** and **C** must be

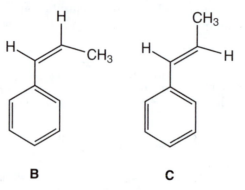

That's fine because **B** and **C** fit the ozonolysis data. Now which is which? There is a discussion of coupling constants on pp. 411–412. The smaller *J* belongs to the cis compound **B** and the larger *J* belongs to **C**, the trans compound.

Problem 16.63 The reaction of propanal and morpholine sets up a complex equilibrium in which the enamine, a carbinolamine, and an aminal are all involved. The analysis by ¹H NMR spectroscopy shows that little enamine is present. Nonetheless, as enamine is used up in the Diels–Alder addition, more is formed. Eventually, all of the components in the equilibrium can be converted into enamine and then to Diels–Alder adducts.

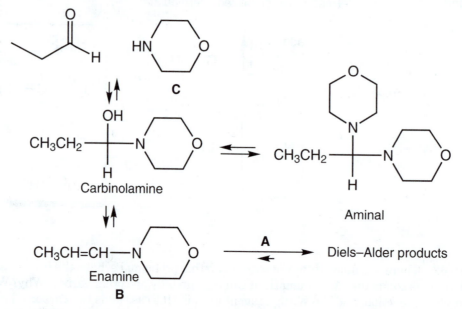

Problem 16.64 The anion formed by loss of a proton is resonance stabilized, and thus that "α" hydrogen is relatively easy to remove.

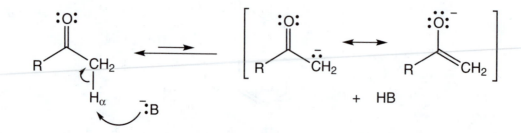

Problem 16.65   You have been prompted by the previous problem to think about the formation of what are called "enolate" anions. Here we have an "α" hydrogen and a base, so let's first form the enolate.

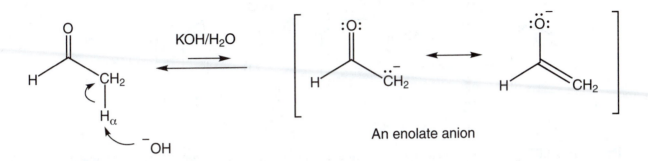

An enolate anion

The problem reminds us that all nucleophiles add to the C=O group. Not only that, but the product of this reaction clearly contains parts of two acetaldehyde molecules. So, let's add the enolate to the C=O of acetaldehyde. We have sewn two acetaldehydes together in just the right way. A straightforward protonation gives the product.

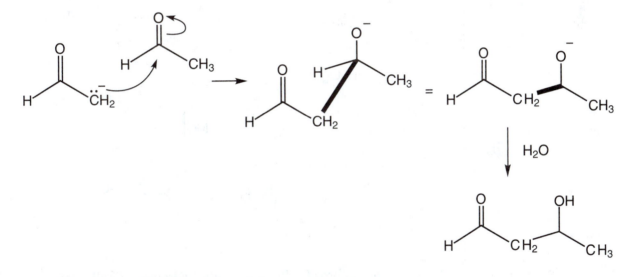

Congratulations, you have just invented the aldol reaction.

Problem 16.66   The first equivalent of methyl Grignard will surely deprotonate the carboxylic acid.

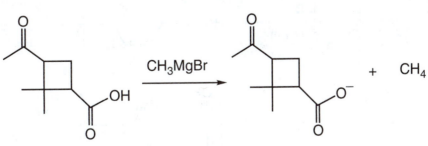

(continued)

Problem 16.66 *(continued)*

Now the second equivalent of Grignard will add to the more reactive ketone carbonyl, not to the already negatively charged carboxylate. Acidification will result in protonation of both oxygens.

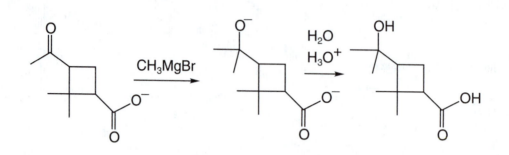

Problem 16.67   This reaction sequence is an example of a synthesis that requires the use of a protecting group. If the Grignard reagent of *p*-bromoacetophenone were prepared directly, it would add to the carbonyl group of another molecule of starting material. To avoid this complication, the carbonyl group of *p*-bromoacetophenone is first protected as a cyclic acetal to give **A**. The Grignard reagent **B** can then be prepared safely.

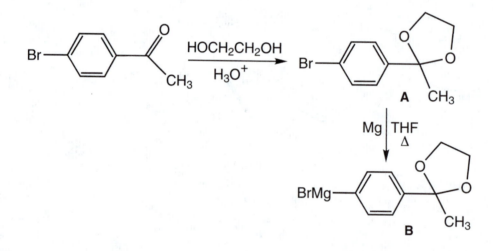

Now reaction with *p*-methoxybenzophenone affords **C** after mild acidic hydrolysis. Finally, the carbonyl group can be regenerated upon more vigorous acidic hydrolysis to yield ketone **D**.

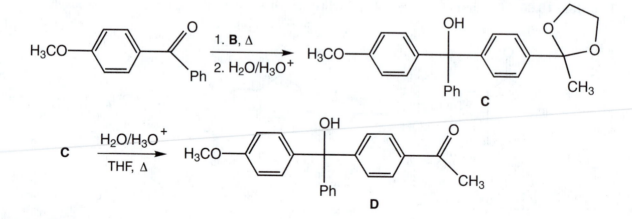

The structure of **D** is supported by the spectral data. The parent ion in the mass spectrum at $m/z = 332$ is consistent with a molecular formula of $C_{22}H_{20}O_3$. The presence of an O—H stretch at 3455 cm^{-1} and a conjugated carbonyl band at 1655 cm^{-1} is also consistent with the proposed structure. The ^1H and ^{13}C NMR data are also consonant with the assignment. Pertinent assignments are indicated below with the ^{13}C NMR chemical shifts in parentheses.

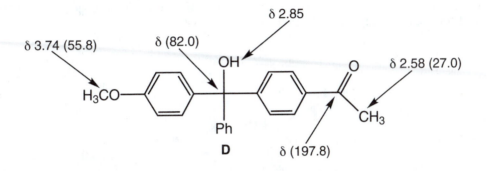

Problem 16.68 This reaction sequence is an example of the Wittig reaction, followed by hydrolysis of the enol ether (**C**), to give 2-ethylbutanal (**D**). This sequence represents a convenient method for the following conversion: R—CO—R′ → RR′CH—CHO.

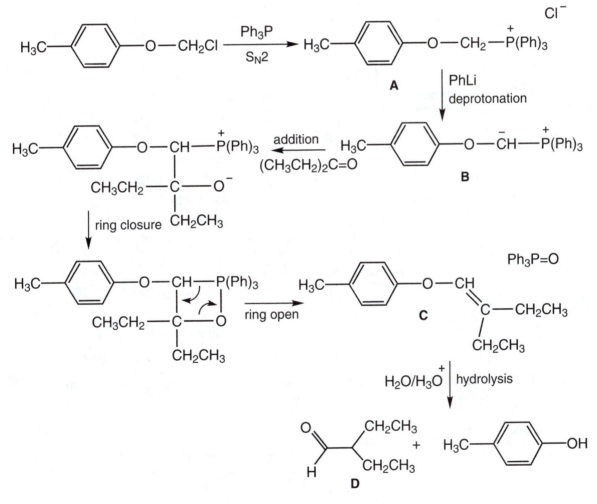

(continued)

Problem 16.68 *(continued)*

The spectral data for **D** are consistent with the proposed structure. The IR spectrum displays the characteristic aldehyde C—H stretching doublet at 2817 and 2717 cm⁻¹, as well as the intense C=O stretch at 1730 cm⁻¹. The presence of the low-field doublet at δ 9.51 ppm in the ¹H NMR spectrum is also indicative of an aldehyde C—H. Note the small coupling to the adjacent methine hydrogen. The methyl groups appear as a triplet centered at δ 0.92, whereas the methylene and methine hydrogens appear as an overlapping multiplet at δ 1.2–2.3 ppm. The ¹³C NMR spectrum is also consistent with the proposed structure. Assignments are shown. Note especially the low-field chemical shift for the aldehyde carbonyl carbon.

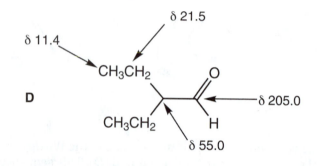

Problem 16.69 The mechanism is just a series of acid-catalyzed eliminations of water, with one crucial conversion of an enol into an aldehyde.

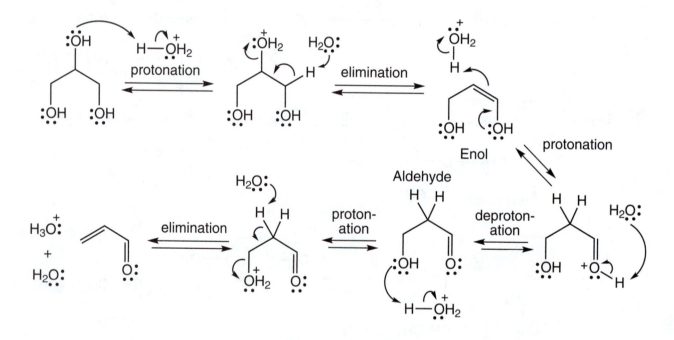

Problem 16.70 Sodium hydride forms the alkoxide by deprotonating the alcohol, and S_N2 displacement of chloride introduces the protecting group.

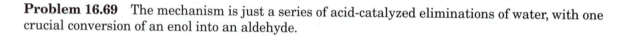

As the MOM-protected alcohol is an acetal, it is not surprising that it is labile in aqueous acid. Treatment with acid regenerates the original alcohol. Protonation of the oxygen atom, followed by S_N1 loss of alcohol, produces the alcohol and a resonance-stabilized carbocation.

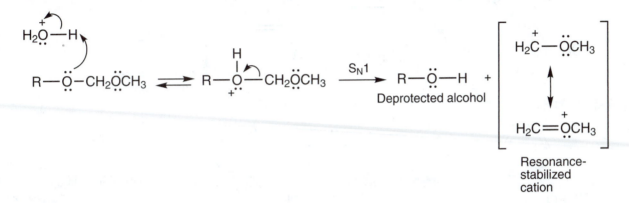

Problem 16.71 The conversion of thiols into disulfides is an oxidation, and its reverse is a reduction. Atmospheric oxygen could do the oxidation, and NADH is a biological agent capable of the reduction.

Problem 16.72 It is not necessary to supply the exact reagents used in the actual synthesis, as long as you give the same type of reagent that will accomplish the purpose. The first step in this synthesis is the reduction of a ketone to an alcohol. Although this reduction could be accomplished by $NaBH_4$, as shown, the actual reducing agent used was aluminum triisopropoxide, the critical ingredient in the Meerwein–Ponndorf–Verley–Oppenauer reaction, to be discussed in Chapter 19.

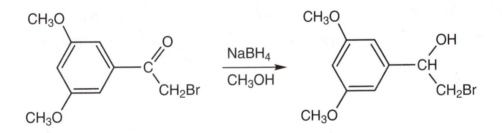

The second step in this reaction sequence involves the formation of an epoxide from a bromohydrin, a reaction we have now seen several times. The base forms a small amount of the alkoxide, which displaces bromide in an intramolecular S_N2 reaction.

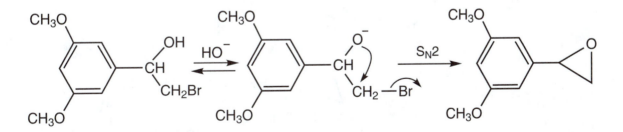

The epoxide is then reopened by isopropylamine in an intermolecular S_N2 reaction that takes place at the less substituted end of the epoxide. Amines are good Brønsted bases and are often conveniently isolated as the salts of mineral acids. In this case, the hydrochloride salt was isolated when the amine was treated with HCl.

(continued)

Problem 16.72 *(continued)*

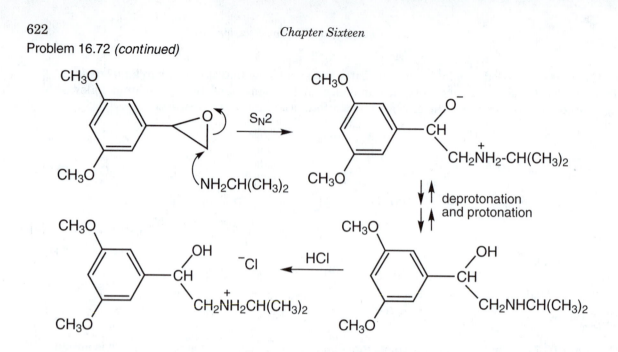

Finally, the methyl protecting groups were removed by treatment with HBr. Be sure you understand the mechanism of this ether-cleavage reaction (p. 298).

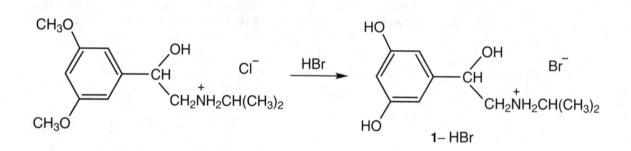

Problem 16.73 From Section 16.16 (p. 814), we know that sulfides such as **1** can be oxidized to sulfoxides or sulfones, depending on the type and amount of oxidizing agent used. The only trick here is to appreciate that sulfoxides are pyramidal. Accordingly, diastereomeric (exo and endo) sulfoxides **2** and **3** are produced upon mild oxidation of **1**. Further oxidation gives the same sulfone, **4**, in each case.

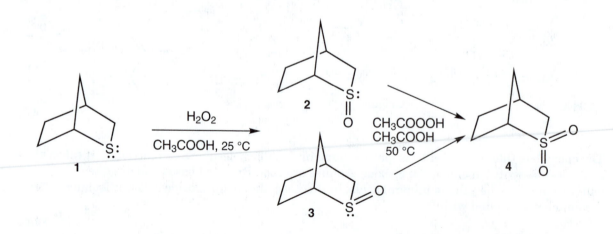

Problem 16.74 This item should be a simple road-map problem, as there are only straightforward steps. The epoxide is first opened by hydride to give the alkoxide, then protonated by water to give **A**. Treatment with PCl₃ gives the corresponding chloride, **B**. Lithium generates the organolithium reagent, and this adds to acetone to give the alcohol **C**, after protonation of the alkoxide. Acid-catalyzed elimination in the Saytzeff (more substituted double bond is formed) sense produces the final product.

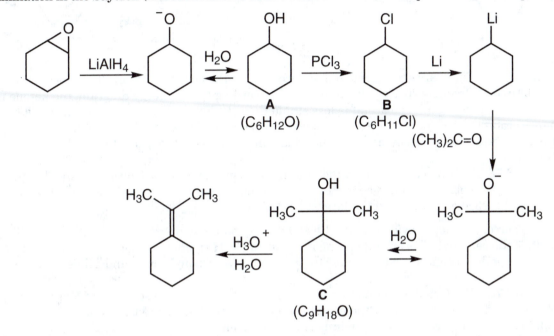

Problem 16.75 This road-map problem is slightly more challenging than the last one. The first step is easy enough; bromine forms a bromonium ion that is opened in water to give the bromo alcohol, **A**. Treatment with base gives a molecule that does not contain bromine. So, we clearly must find some way to get rid of a bromine. The alkoxide is formed and undergoes an intramolecular S_N2 displacement of bromide to give the epoxide, **B**. Compound **B** opens in acid to give diol **C**. 1,2-Diols are cleaved by periodic acid to give a pair of carbonyl compounds. Notice that you could have used that fact to reconstruct **C** from the two carbonyl compounds shown at the end of the problem. That's an important technique. If you get stuck, work a road-map problem from any point of information, especially the end. For example, if you couldn't remember the transformation from **B** to **C**, you could have worked backward to **C** from the structures of the two aldehydes.

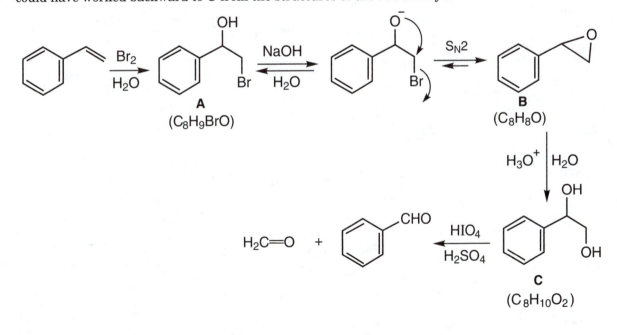

Problem 16.76

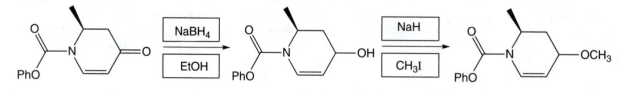

Problem 16.77 There are six intermediates, according to the energy diagram. The most stable intermediate is the neutral species, the hemiacetal. According to the energy diagram, the ketone is more stable than the acetal. The acetal is less stable than the ketone because the double bond has resonance stabilization, and the acetal has some steric crowding. One can drive the ketone (or aldehyde) to the acetal by removing water or using a huge excess of the alcohol.

Problem 16.78 This reaction can be difficult to understand completely. One of the complications is that the acidic conditions should be protonating the amine, not the carbonyl oxygen of acetone. The carbonyl carbon of acetone has the largest LUMO density.

Yes, the hyperconjugation is evident in the LUMO density on the hydrogens that line up with the π system of the carbonyl. This orbital mixing is the filled C—H sigma bond donating electrons into the C—O π antibond. This effect is what makes the α proton of a carbonyl compound acidic.

Problem 16.79 The Grignard animation is complicated by the need to solvate the magnesium species. It is further complicated by the need to show the Mg—R bond and the transfer of the R anion to the carbonyl carbon. The R minus is what we show when we push the electrons on a chalkboard, but there is actually no free carbanion in the reaction. The Wittig reaction starts at a high energy because the very unstable methyl carbanion is part of the reaction.

Carboxylic Acids

17

Problem 17.2 The only tricky part is making sure you figure out the absolute configurations of the two stereogenic carbons in the penultimate example, part (f). Notice that the "cis" designation, necessary when we are talking about the racemic material, becomes unnecessary if we specify the absolute configurations, as the (*R*) and (*S*) designations require the cis stereochemistry in this molecule.

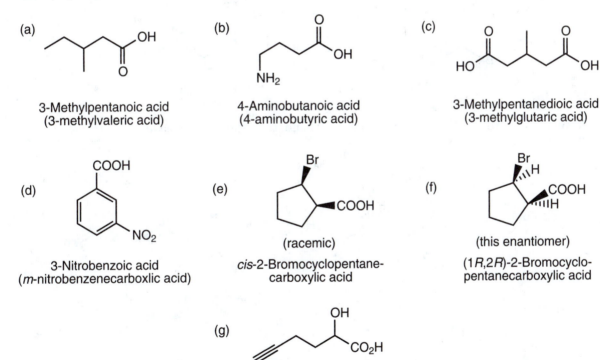

(a)

3-Methylpentanoic acid
(3-methylvaleric acid)

(b)

4-Aminobutanoic acid
(4-aminobutyric acid)

(c)

3-Methylpentanedioic acid
(3-methylglutaric acid)

(d)

3-Nitrobenzoic acid
(*m*-nitrobenzenecarboxlic acid)

(e)

(racemic)
cis-2-Bromocyclopentane-
carboxylic acid

(f)

(this enantiomer)
(1*R*,2*R*)-2-Bromocyclo-
pentanecarboxylic acid

(g)

2-Hydroxy-5-hexynoic acid

Problem 17.3   The s-cis form can hydrogen bond with a water molecule in a comfortable six-member ring structure. All of the bond lengths are just right for this conformation, and the water molecule can take advantage of the acidic hydrogen of the carboxylic acid at the same time as the basic oxygen of the carbonyl. A second reason for favoring the s-cis form is intramolecular hydrogen bonding. Although the bond distance and angle are nonideal, there is likely *some* stabilization for the s-cis compared to the s-trans conformation.

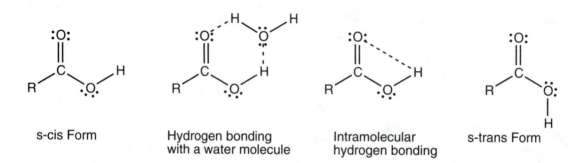

Problem 17.4   We can draw a carboxylic acid that has a keto group (IUPAC would call it an oxo substituent) on the β carbon (IUPAC would call it carbon number 3). The s-cis and s-trans forms are drawn, and there is no obvious reason for the s-trans to be favored.

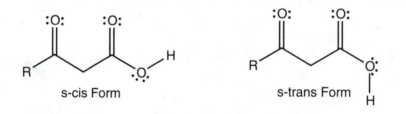

But if we allow for rotation around the C(1)—C(2) bond, then the preference for the s-trans conformation becomes apparent.

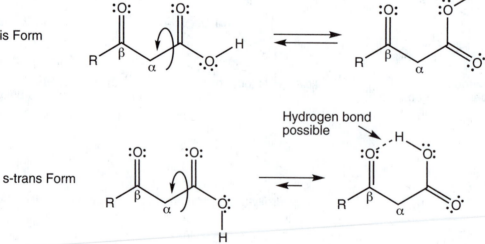

Problem 17.8 It is the same old story. Although the rate of a reaction is determined not by the energy of the product, but by the energy of the transition state leading to the product, very often the factors influencing the energy of the product will also be at work in determining the energy of the transition state. This parallelism is especially true for endothermic reactions in which the transition state will resemble the product. In this case, the transition state for protonation of the carbonyl oxygen will be stabilized by delocalization of the developing positive charge.

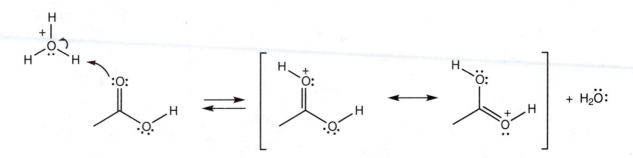

Protonation of the carbonyl oxygen leads to a delocalized intermediate

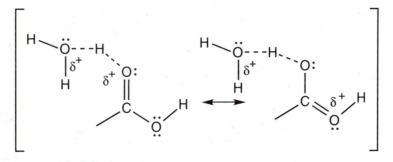

The delocalized transition state for protonation

Problem 17.9

(a) This part only requires you to recall the oxidation of alcohols to carboxylic acids.

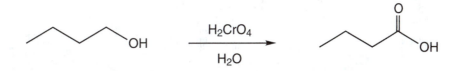

(b) In this part, we have to extend the chain by one carbon. One standard technique uses carbon dioxide as the source of that new carbon. This process arrives directly at the proper oxidation state. One could also imagine making the five-carbon alcohol first (how?) and oxidizing as in part (a).

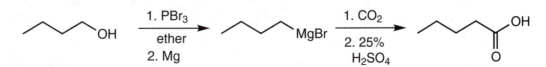

(continued)

Problem 17.9 *(continued)*

(c) Here we have to make the carbon backbone one shorter. The easiest way to do this task is to convert 1-butanol into 1-butene and ozonize with an oxidative workup. But how to make 1-butene? We could imagine converting the alcohol into a good leaving group and doing an E2 elimination. But the best base for an E2 reaction is potassium *tert*-butoxide, which we don't have in our bag of tricks. Perhaps we should simply dehydrate the alcohol. Using a dehydration to obtain 1-butene is not ideal because the product mixture will include (*E*) and (*Z*)-2-butene, but it will be better than trying to make *tert*-butyl alcohol from 1-butanol. Ozonolysis follows.

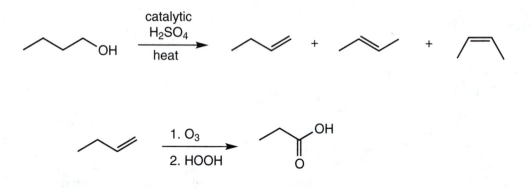

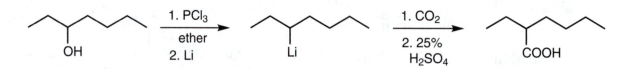

(d) Grind it out and everything's easy. The acid can be made from the corresponding alcohol as in part (b).

Now, we just have to make the complex secondary alcohol, a task made easy by all those alcohol syntheses in Chapter 16.

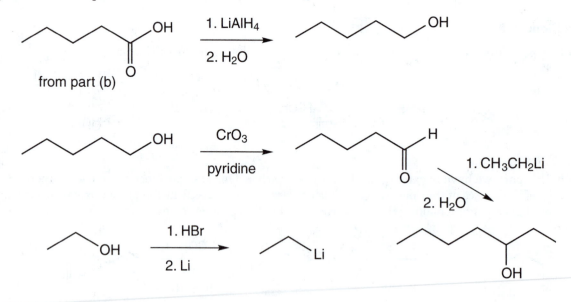

Problem 17.10 The mechanism involves a tetrahedral intermediate **A** that contains three equivalent oxygens, one ^{18}O, the others ^{16}O.

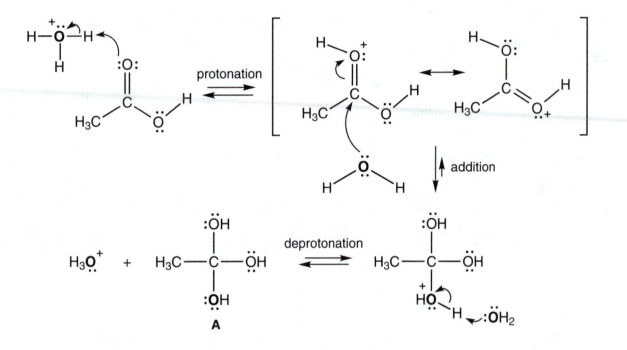

Intermediate **A** can revert to acetic acid to give either carbonyl-labeled acetic acid or hydroxyl-labeled acetic acid. The following drawing only sketches the mechanism (it is, of course, just the reverse of the steps outlined in the first figure of this problem).

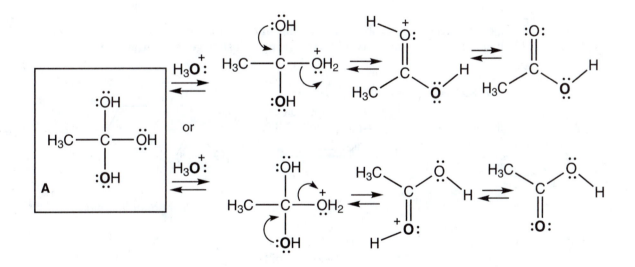

Problem 17.11

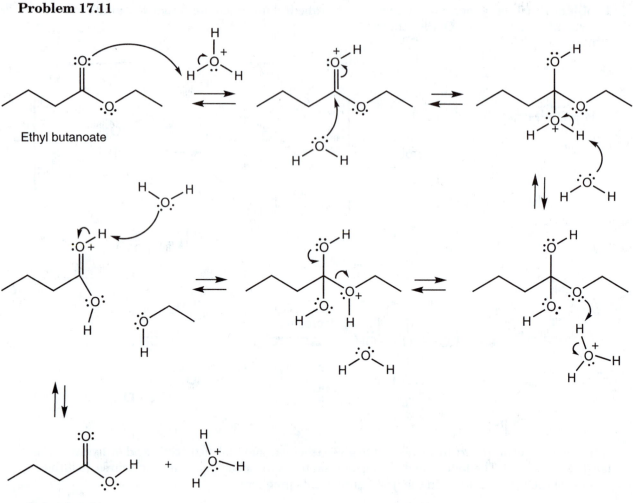

Ethyl butanoate

Butanoic acid

Problem 17.12 These reactions are both addition–elimination processes. In each case, the alcohol adds to the double bond to oxygen, and a good leaving group is lost to generate the ester:

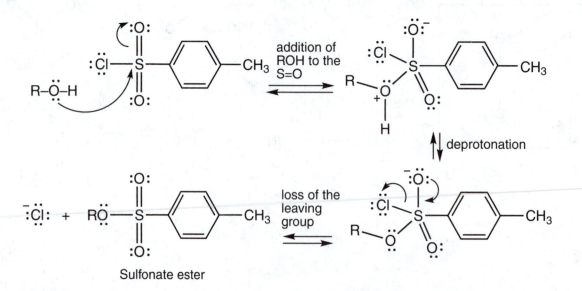

addition of ROH to the S=O

deprotonation

loss of the leaving group

Sulfonate ester

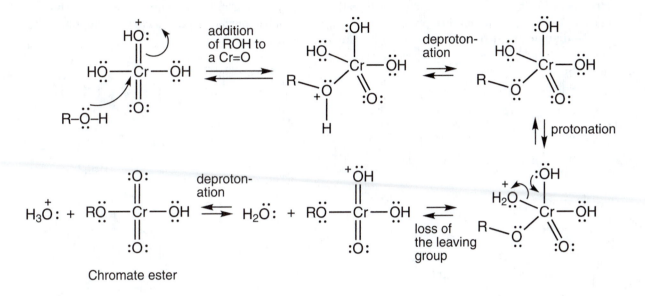

Chromate ester

Problem 17.14 The critical insight is to recognize the product as a cyclic ester, a lactone. This reaction is just an intramolecular Fischer esterification. The steps are identical to those of Figure 17.24, except that they all occur within the same molecule.

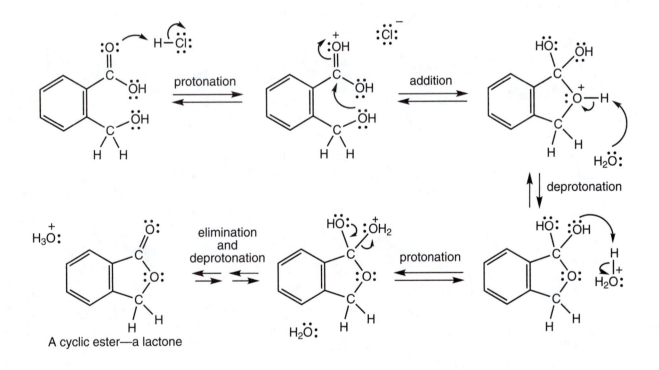

A cyclic ester—a lactone

Problem 17.15   In base, the carboxylate salt will be formed first. Once iodine reacts with the carbon–carbon double bond to produce the iodonium ion, the perfectly poised carboxylate opens it (S_N2—displacement from the rear) to give the lactone, the "neutral product."

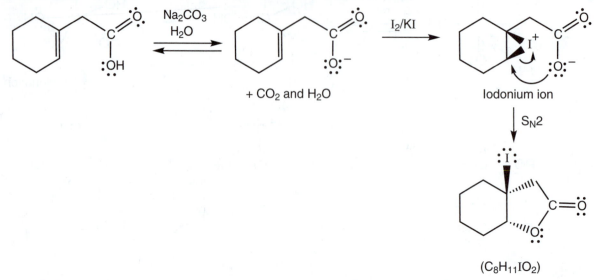

Problem 17.16   Both of these steps are standard addition–elimination reactions. In the first step, the DCC-activated acid (or the protonated acid) is attacked by the carboxylic acid. Loss of the good leaving group leads to the anhydride. Anion **A** is first protonated and then converted into dicyclohexylurea, DCU.

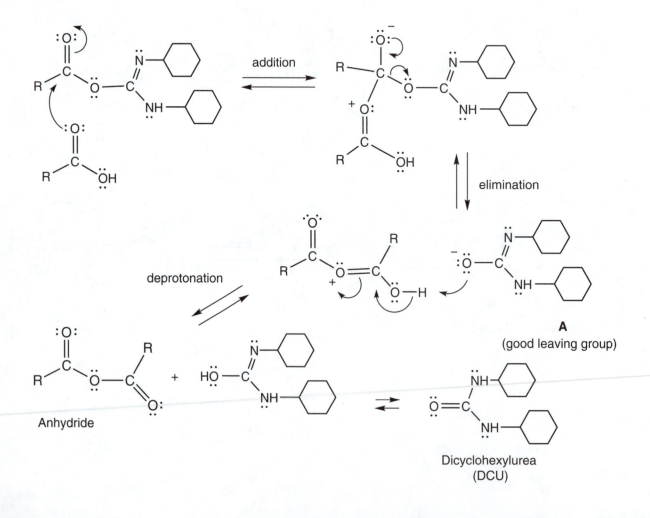

In the second step, the amine adds to the anhydride, and after deprotonation on nitrogen and reprotonation on oxygen, a carboxylic acid is lost in the elimination step that completes the reaction.

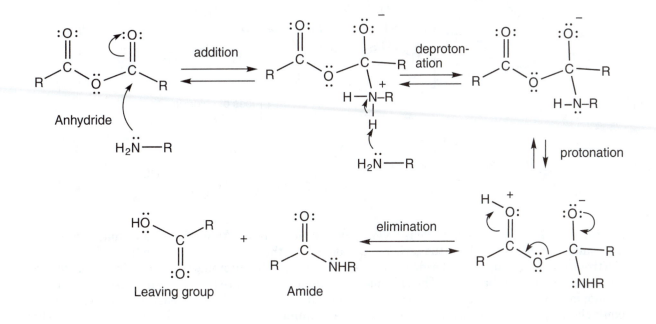

Problem 17.17 Do you get the impression that the addition–elimination mechanism is important? You're right, and here comes yet another example. The most tempting thing is just to use the hydroxyl oxygen as the nucleophile, form intermediate **A**, lose chloride, deprotonate, and be done with it.

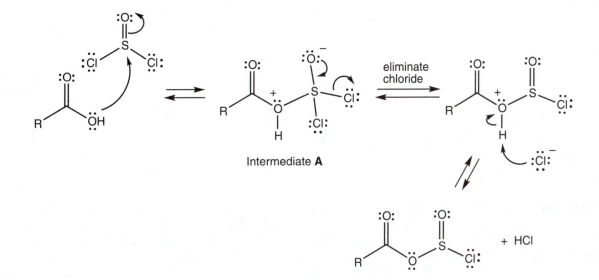

However, there is an error. Addition using the carbonyl oxygen, not the hydroxyl oxygen, leads to a more stable, resonance-stabilized intermediate in which both oxygens and the carbon share the positive charge. The product is the same as in the mechanism shown first, but the mechanistic details are slightly different.

(continued)

Problem 17.17 *(continued)*

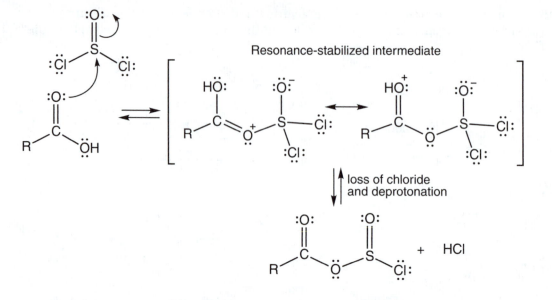

Resonance-stabilized intermediate

loss of chloride and deprotonation

+ HCl

Problem 17.18   The acid chloride is a very reactive electrophile. Ammonia (or a primary amine or a secondary amine) will add at the carbonyl carbon to give the tetrahedral intermediate. The alkoxide electrons will eliminate the chloride very rapidly to produce the protonated amide. The chloride is a better leaving group than NH_3. The chloride could then deprotonate the amide nitrogen, but the ammonia is more basic. Because there are two roles for the ammonia, this reaction requires two equivalents of NH_3 (or primary amine or secondary amine). *Note:* It is possible for the nitrogen to be deprotonated from the first tetrahedral intermediate before the chloride is eliminated.

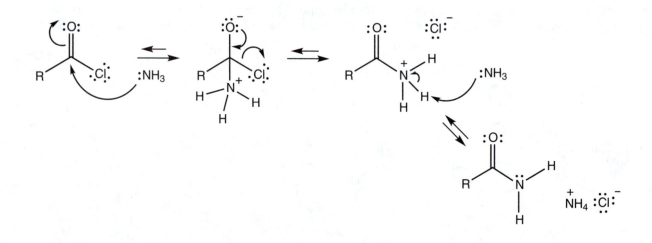

Problem 17.19   Once again, this problem involves a series of addition–elimination steps. Phosgene reacts with an acid in the first addition–elimination sequence to give intermediate **A** and a chloride ion. The chloride ion adds to **A** to give **B**, from which carbon dioxide and chloride can be irreversibly lost. This sequence is the second addition–elimination. A final deprotonation leads to the acid chloride and hydrochloric acid. Many variations on this mechanism are reasonable. For

example, one might write first an equilibrium in which phosgene deprotonates the acid to give a carboxylate and a protonated carbonyl. The carboxylate could add to the protonated carbonyl to give an **A**-like intermediate. The mechanism(s) of reaction with oxalyl chloride is (are) similar.

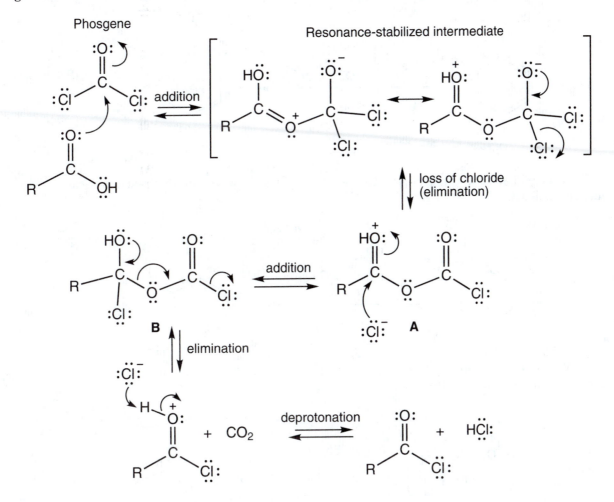

Problem 17.20

(a) The methoxy group is an ortho/para-directing group. The acylation will give two products. The para product is likely favored due to steric interactions involved in the ortho addition.

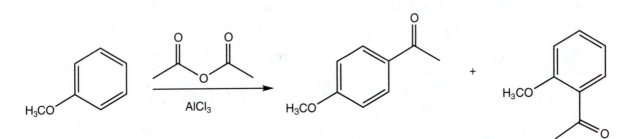

(continued)

Problem 17.20 *(continued)*

(b) Both groups on *p*-methoxytoluene are ortho/para directing. The methoxy group will control the selectivity because it is a stronger directing group.

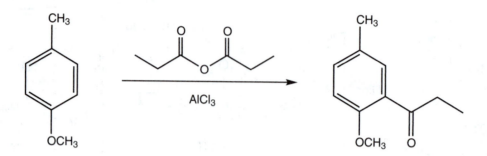

(c) Benzene will add the acyl group once. There is no polysubstitution in this reaction because the acyl group deactivates the ring.

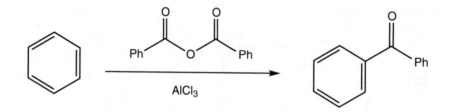

(d) The aromatic furan molecule will add electrophiles to the C(2) position of the furan (p. 721). Some C(3) addition might also be obtained.

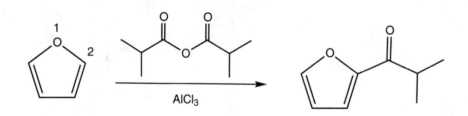

Problem 17.21 There are many ways. The proximate discussion in the text and, especially, Figure 17.50 might lead you to the following short synthesis:

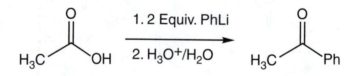

But there are other ways that are just as good. Chapters 14 and 15 might lead you to

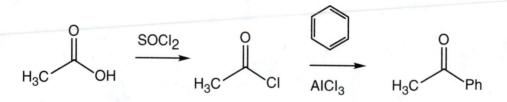

Problem 17.22 In each case carbon dioxide is lost on heating.

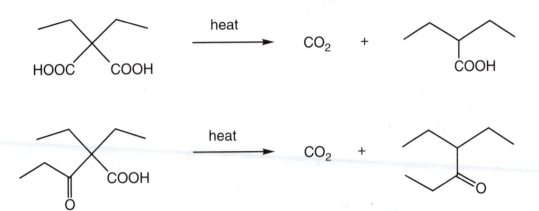

Problem 17.23 Nature has come to rely on carboxylic acids for several reasons. One reason is that the carboxylic acid is able to provide a hydrogen for hydrogen bonding and it is able to receive hydrogen bonds. Another reason for nature using carboxylic acids is that they are often soluble in water (nature's solvent). The carboxylic acid is typically deprotonated at physiologic pH, and the conjugate base of the carboxylic acid is even more soluble in water. A third reason that carboxylic acids are biologically important is that they are the most oxidized organic functional group (other than CO_2). This level of oxidation reflects the oxidizing conditions that are found in nature.

Problem 17.24 Nature has a mechanism for clipping apart long-chain molecules. It is likely that the same enzymes found in bacteria that cleave the two-carbon units of the fatty acid (p. 866) can cleave two-carbon units of the unbranched detergent. The ultimate result would be several acetyl groups and a sulfonic acid derivative. The acetyl groups would be feedstock for the bacteria and lead eventually to CO_2. The literature suggests that the sulfonic acid group does not significantly change the enzymatic pathway. Studies of isotopically labeled detergents have shown that the bacteria that decompose the detergent effectively attack the long-chain portion of the man-made material and produce isotopically labeled CO_2.

Additional Problem Answers

Problem 17.25

(a) This molecule is (*E*)-5-methyl-3-hexenoic acid.

(b) This molecule is (*S*)-2-amino-3-methylbutanoic acid. Its common name is valine.

(c) This molecule is *cis*-3-hydroxycyclobutanecarboxylic acid.

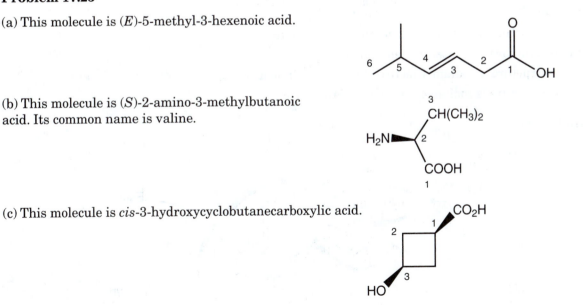

(continued)

Problem 17.25 *(continued)*

(d) This molecule is 3-hydroxybenzoic acid.
It is also correct to call it *m*-hydroxybenzoic acid.

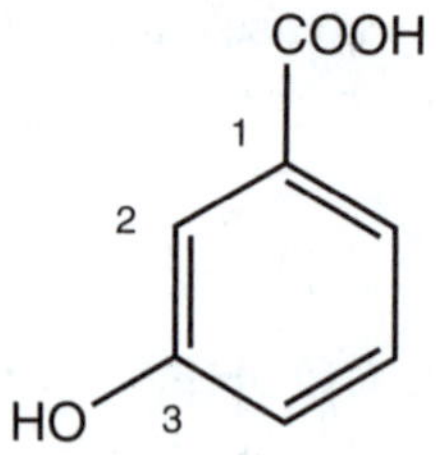

(e) This molecule is 3-phenylbutanoic acid.

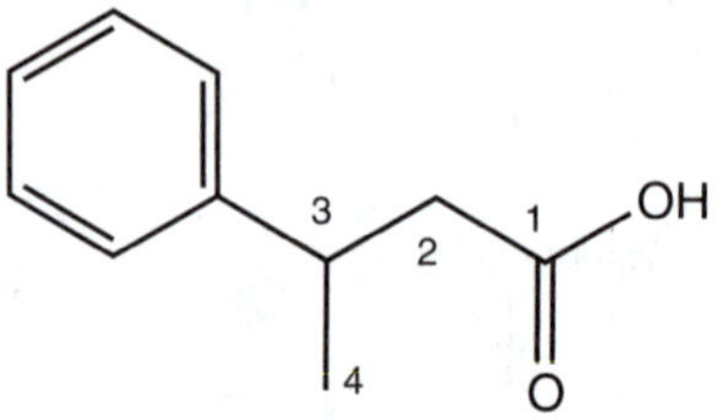

(f) This molecule is 4-methyl-3-phenyl-4-pentenoic acid.

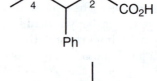

(g) This molecule is 4,4,4-tribromo-3-methylbutanoic acid.

(h) This molecule is 3-amino-2-hydroxypropanoic acid.

Problem 17.26

(a) This structure is lithium (*E*)-5-methyl-3-hexenoate.
(b) This structure is potassium heptanoate.
(c) This structure is sodium 4-methyl-3-phenylpentanoate.
(d) This structure is sodium 4-methylbenzoate.

Problem 17.27

(a)

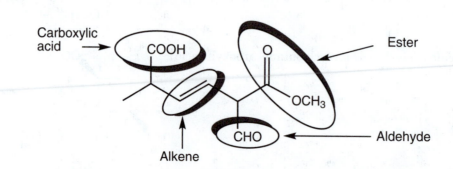

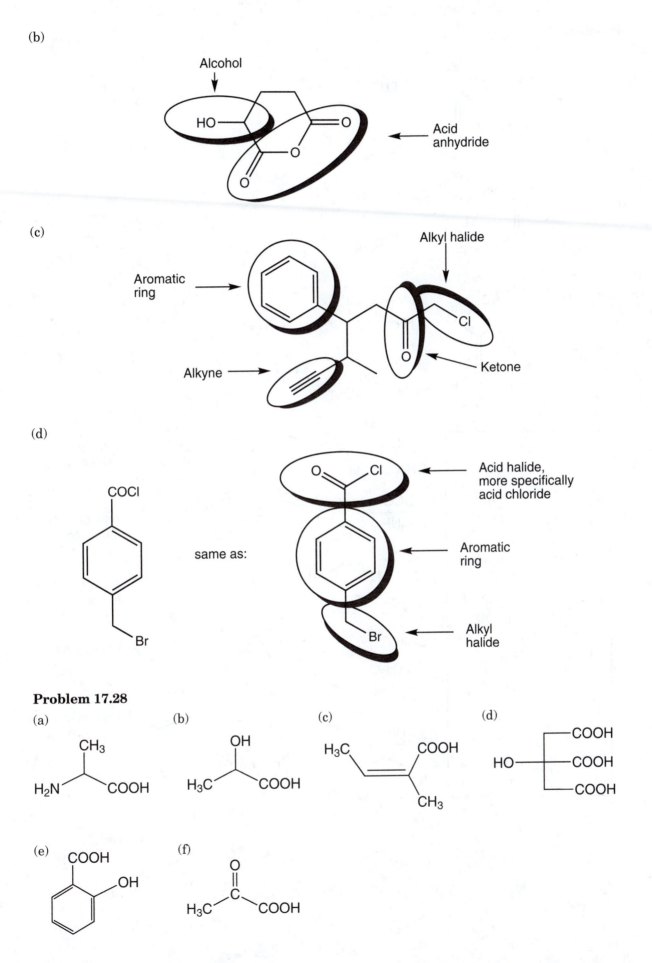

(b)

Alcohol

Acid anhydride

(c)

Alkyl halide

Aromatic ring

Alkyne

Ketone

(d)

Acid halide, more specifically acid chloride

Aromatic ring

Alkyl halide

same as:

Problem 17.28

(a)

(b)

(c)

(d)

(e)

(f)

Problem 17.29

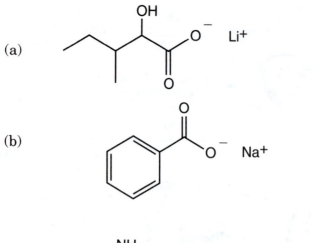

(a)

(d)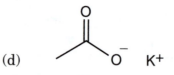

(b)

(e)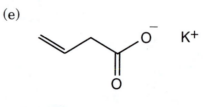

(c)

Problem 17.30

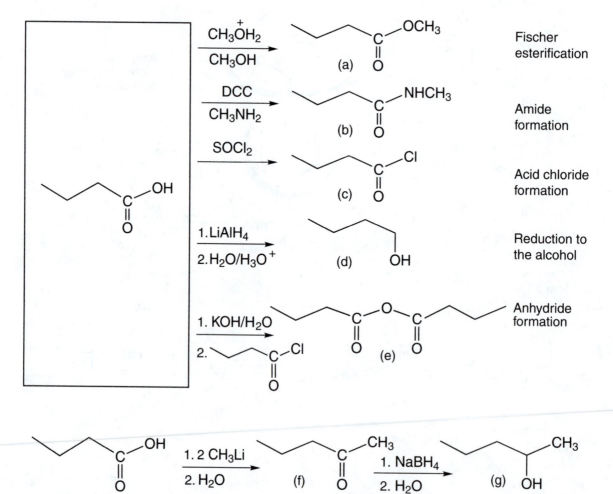

Problem 17.31

(a) This reaction is an example of a standard Fischer esterification. The bromide is not displaced under acidic conditions as there is no nucleophile strong enough.

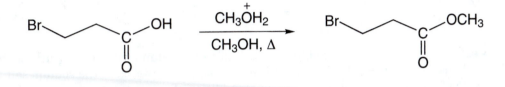

(b) Treatment of a carboxylic acid with DCC results in the formation of an anhydride.

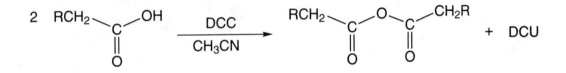

(c) This reaction is a classic example of the saponification of a glyceride (a fat) to afford a fatty acid and glycerol.

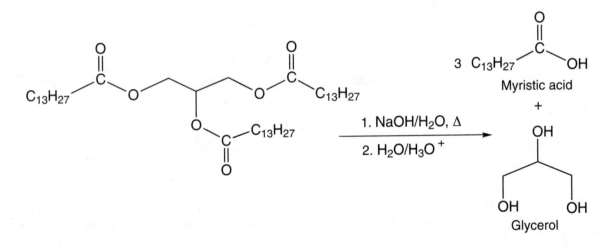

(d) In the presence of an amine, the anhydride (formed from the carboxylic acid and DCC) yields an amide.

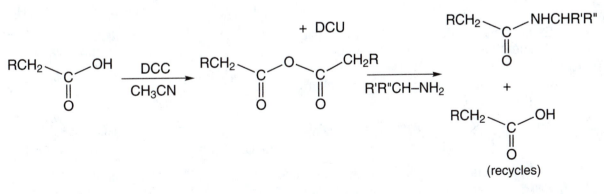

(continued)

(e) Carboxylic acids react with bases such as NaH to form carboxylate anions. Acidification of the salt merely regenerates the carboxylic acid.

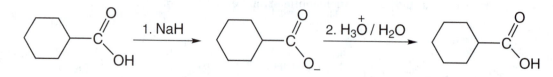

(f) This process is a bit more complicated. It is a variation of the standard ketone synthesis from a carboxylic acid and two equivalents of an alkyllithium reagent. In this case, the sodium salt of the carboxylic acid is formed upon treatment with NaH. Addition of one equivalent of organolithium reagent, followed by hydrolysis, gives the ketone.

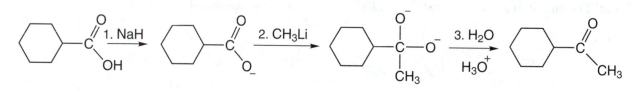

(g) This reaction involves acid chloride formation followed by a Friedel–Crafts acylation.

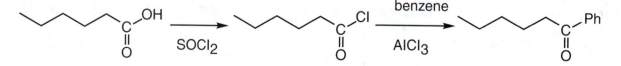

(h) This reaction is a classic example of the carboxylation of a Grignard reagent.

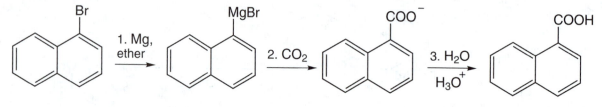

Problem 17.32   This problem is an exercise in making the same compound, **A = C = F**, several times in slightly different ways. Here are the structures and kinds of reactions involved. Look them up in the chapter if you are uncertain.

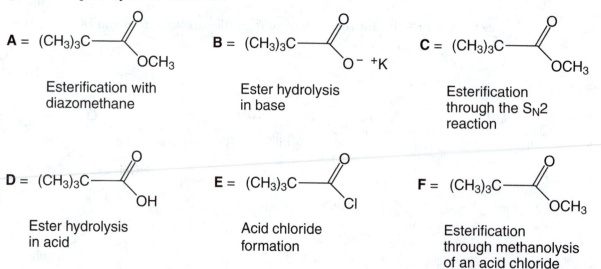

A = (CH₃)₃C—C(=O)—OCH₃

Esterification with diazomethane

B = (CH₃)₃C—C(=O)—O⁻ ⁺K

Ester hydrolysis in base

C = (CH₃)₃C—C(=O)—OCH₃

Esterification through the S$_N$2 reaction

D = (CH₃)₃C—C(=O)—OH

Ester hydrolysis in acid

E = (CH₃)₃C—C(=O)—Cl

Acid chloride formation

F = (CH₃)₃C—C(=O)—OCH₃

Esterification through methanolysis of an acid chloride

Problem 17.33 Here are the structures and the kinds of reactions involved.

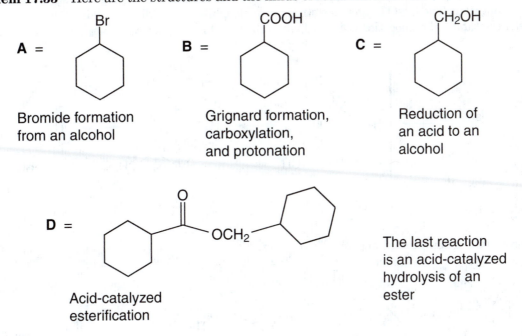

A =

Bromide formation
from an alcohol

B =

Grignard formation,
carboxylation,
and protonation

C =

Reduction of
an acid to an
alcohol

D =

Acid-catalyzed
esterification

The last reaction
is an acid-catalyzed
hydrolysis of an
ester

Problem 17.34 Here are the structures and kinds of reactions involved.

A =

Acid chloride
formation

B =

α-Halogenation
of an acid chloride

C =

Hydrolysis of an
acid chloride

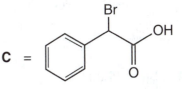

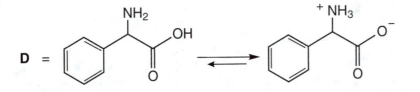

D =

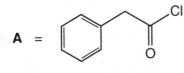 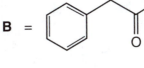

Displacement of an
α-halogen (S$_N$2) by a
good nucleophile

Problem 17.35 The first step is surely going to be protonation of the alkene to give the tertiary
carbocation.

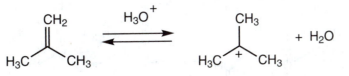

(continued)

Problem 17.35 *(continued)*

Now this carbocation adds to a molecule of acetic acid. Be careful: It is tempting to use the OH oxygen, but that's not right. The C=O oxygen is a stronger nucleophile, and addition will be there. Why? A resonance-stabilized carbocation is formed when the C=O reacts, but not when the OH is used.

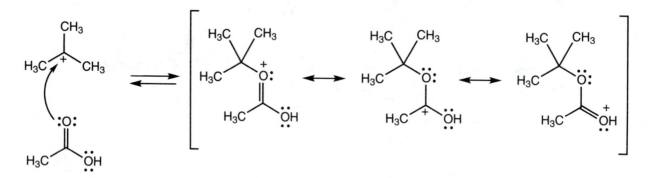

A straightforward deprotonation gives the product.

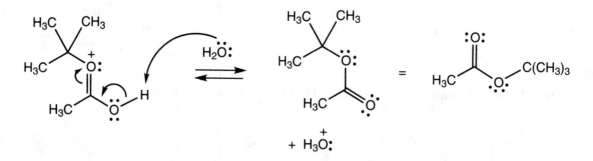

Problem 17.36

(a) The reason the nitroacetic acid is more acidic is the inductive effect of the nitro group (note the positive charge on N). This induction will stabilize the conjugate base, making the acid easier to deprotonate.

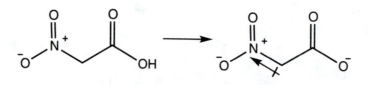

The conjugate base is stabilized, therefore the acid is more acidic

(b) The reason 2-oxopropanoic acid is more acidic is also induction. The carbonyl group has a partially positive carbon, which is withdrawing by induction. That effect stabilizes the conjugate base, making the acid easier to deprotonate.

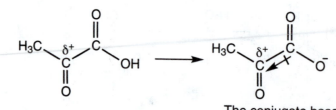

The conjugate base is stabilized, therefore the acid is more acidic

(c) The cis diacid is stabilized by intramolecular hydrogen bonding, thus significantly stabilizing the conjugate base, making it easier to deprotonate.

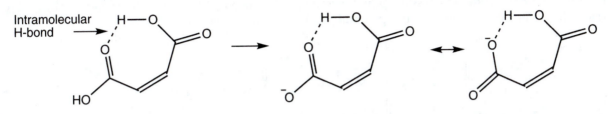

Intramolecular H-bond

The conjugate base is stabilized by the available intramolecular proton, therefore the initial acid is more acidic

(d) The relative acidity in this series of molecules is a result of resonance and induction. The methoxy group destabilizes the conjugate base by resonance, as shown, so *p*-methoxybenzoic acid is the least acidic of these three. Benzoic acid has no substituent affecting it. The *p*-nitrobenzoic acid is the most acidic of the three because the nitro group is electron-withdrawing by resonance and induction, which stabilizes the conjugate base, making the acid easier to deprotonate.

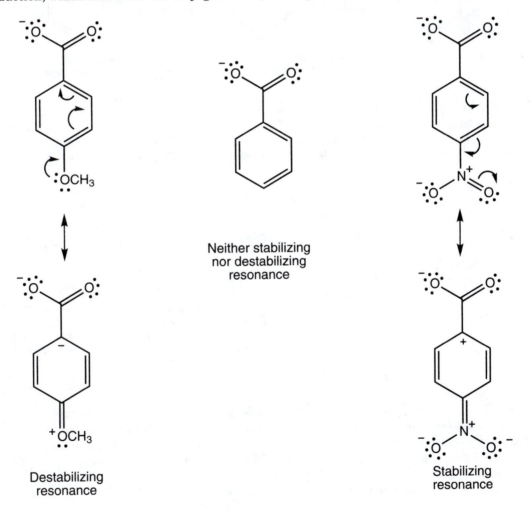

Destabilizing resonance

Neither stabilizing nor destabilizing resonance

Stabilizing resonance

(continued)

Problem 17.36 *(continued)*

(e) The dicarboxylates that will be formed in each case are either on opposite sides of the alkene or on the same side of the alkene. It is more favorable for the two negative charges to be as remote from each other as possible. This structural effect is why the second hydrogen of the trans diacid is more acidic than the second hydrogen of the cis diacid.

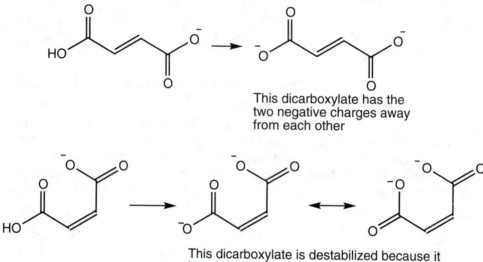

This dicarboxylate has the
two negative charges away
from each other

This dicarboxylate is destabilized because it
has the two negative charges next to each other

(f) The thioacid is more acidic because of the size of the sulfur. The conjugate base of the thioacid is better able to take on the negative charge because of its size. In addition, the S—H bond is weaker than the O—H bond because of the size of the sulfur. It has poor orbital overlap with the hydrogen. Both of these factors contribute to the increased acidity of the thioacid.

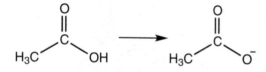

Sulfur is a larger atom
and therefore makes a
weaker bond to the
hydrogen

Sulfur is a larger atom
and is therefore better
able to stabilize the
negative charge

Problem 17.37 Although *N,N'*-carbonyldiimidazole (CDI) is a derivative of phosgene, it is more convenient to use because it is a nontoxic solid. CDI behaves similarly to phosgene in its reactions with carboxylic acids (see Problem 17.19). Both reagents react through addition–elimination sequences, but the by-product of the reaction of CDI is the relatively inert molecule imidazole, rather than HCl, as it is for reaction with phosgene. In the final elimination step, the imidazole may well be protonated, perhaps in an intramolecular fashion.

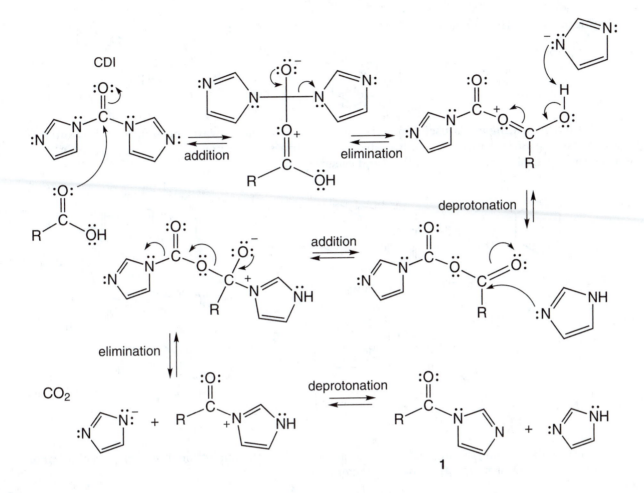

Imidazolides (**1**) look like amides but react like acid chlorides. Why? There is minimal amide resonance because the lone pair on N is required for aromaticity. The nucleophilic nitrogen of the amine adds to the carbonyl group, and an excellent leaving group (why?) is then lost in the elimination phase of this addition–elimination process.

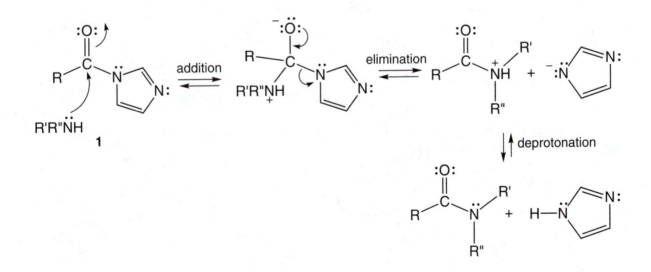

Problem 17.38   Here's our target.

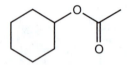

DCC is a dehydrating agent useful in activating carboxylic acids toward the addition–elimination mechanism (p. 854). A retrosynthetic analysis would direct us to esterification of acetic acid. All we need to do is treat acetic acid with DCC and cyclohexanol.

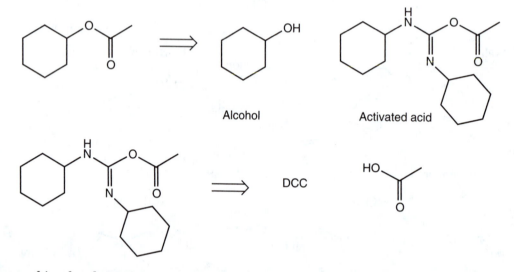

Alcohol   Activated acid

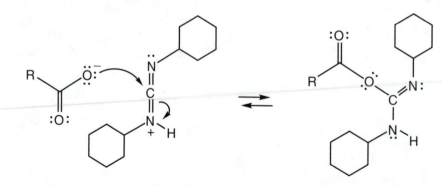

DCC

Cyclohexanol is a legal starting material, and acetic acid can be made from the oxidation of ethanol with H_2CrO_4/H_2O, or many other oxidizing agents.

Problem 17.39   The first step is proton transfer from the acid to DCC (p. 855).

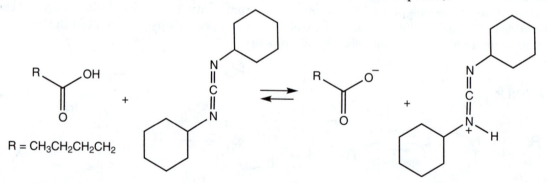

$R = CH_3CH_2CH_2CH_2$

Addition of the carboxylate to the new Lewis acid gives us the first adduct. We have just played the game of "change the leaving group."

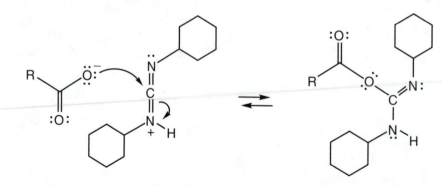

Addition–elimination makes our ester.

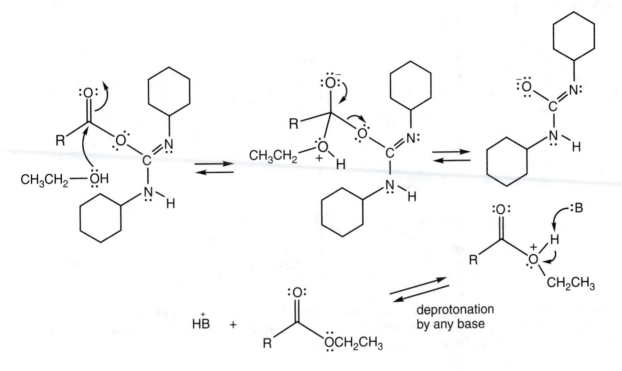

Problem 17.40 The first step in this process is formation of a "mixed anhydride" through reaction of the carboxylic acid with acetic anhydride. Addition of the carboxylic acid to the anhydride carbonyl, followed by loss of acetate and deprotonation, gives the mixed anhydride intermediate.

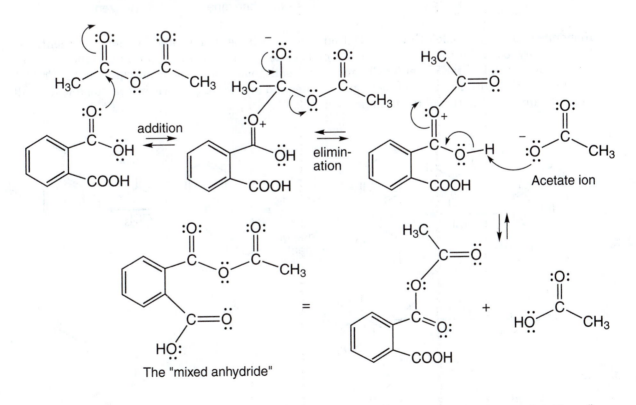

(continued)

Problem 17.40 *(continued)*

The mixed anhydride now undergoes an intramolecular reaction very similar to the intermolecular reaction just shown.

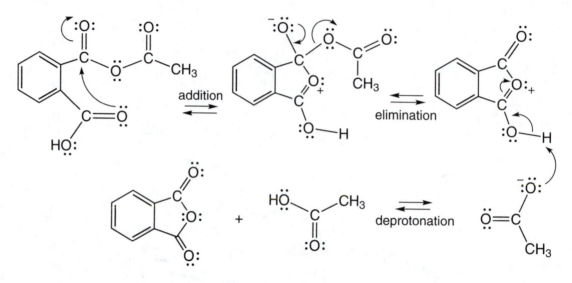

Problem 17.41 First you should do a retrosynthetic analysis for the aryl ketone. You know that you could add the acyl group to the isobutylbenzene. Isobutylbenzene could come from benzene by a Friedel–Crafts reaction. So now let's do the forward reaction.

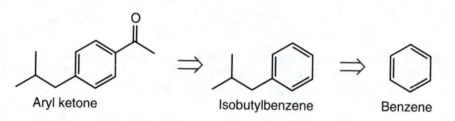

Aryl ketone   Isobutylbenzene   Benzene

Remember that the Friedel–Crafts alkylation reaction will be complicated by rearrangement. So the best way to synthesize the isobutylbenzene is to do Friedel–Crafts acylation and then reduce the carbonyl with Wolff–Kishner (or Clemmensen) conditions. Isobutylbenzene can undergo Friedel–Crafts acylation with acetyl chloride to give the desired aryl ketone. Steric interactions will result in predominant formation of the para product.

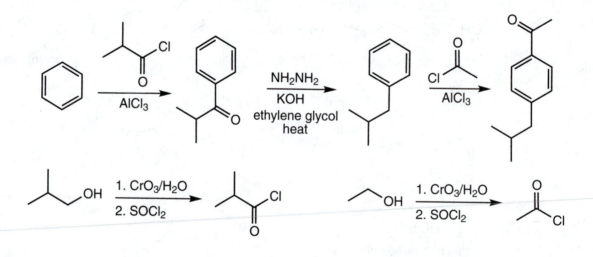

There are many ways to make ibuprofen from the aryl ketone. One method to make a carboxylic acid is to use the Grignard reaction. That process requires a bromide, and we know how to make a bromide from an alcohol, which could come from the ketone.

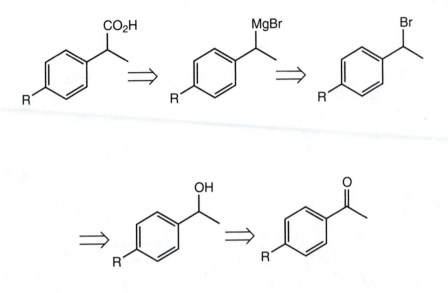

The forward process that follows from this retrosynthetic analysis is (1) reduce the ketone with $LiAlH_4$ (or $NaBH_4$), (2) convert the alcohol into a bromide with PBr_3, (3) make the Grignard with Mg, and (4) add CO_2 to make, after acidification, the carboxylic acid. There would be no stereo-control of the first step (the reduction), so the final product would be a racemic mixture.

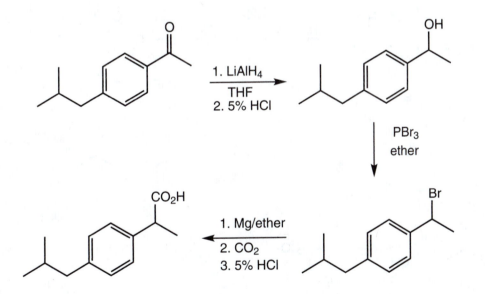

Problem 17.42 The Vilsmeier reagent is a strong Lewis acid and reacts with the carboxylic acid in an addition reaction. Deprotonation and loss of chloride lead to intermediate **A**. Addition of chloride gives an intermediate **B**, which eliminates DMF to give the product.

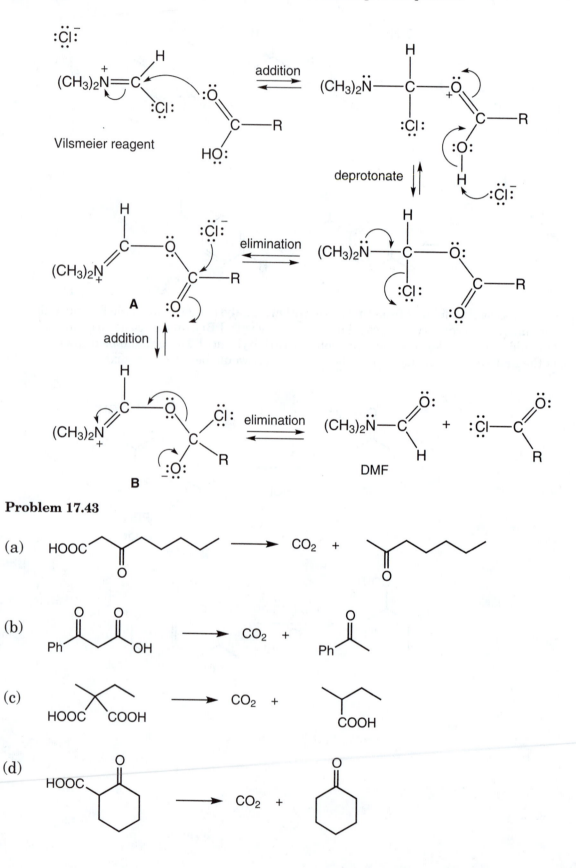

Problem 17.43

(a) HOOC ⟶ CO_2 +

(b) Ph ⟶ CO_2 + Ph

(c) HOOC COOH ⟶ CO_2 + COOH

(d) HOOC ⟶ CO_2 +

Problem 17.44   Nothing tricky here—all those OH groups become chlorines.

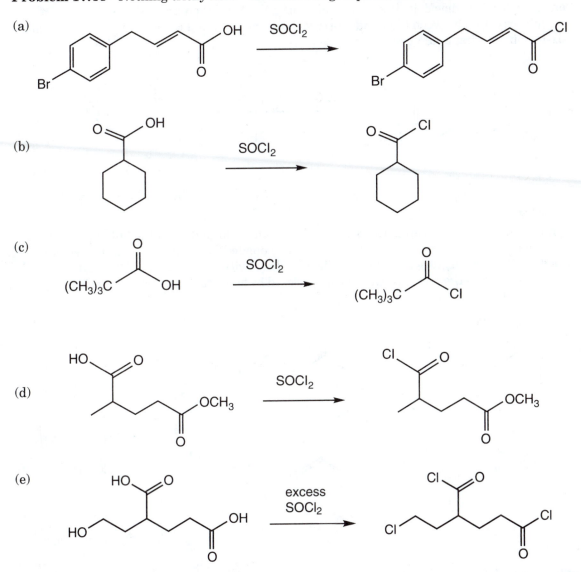

(a)

SOCl₂

(b)

SOCl₂

(c)

SOCl₂

(d)

SOCl₂

(e)

excess
SOCl₂

Problem 17.45   If we allow the amine to react with a carboxylic acid, the first thing that will happen is the basic amine will deprotonate the carboxylic acid. So it might be tempting to try the reaction of pentanedioic acid with benzylamine in dilute conditions so that the diacid will react twice with the amine, but we would need to heat the tar out of it in order to drive the reaction to the imide.

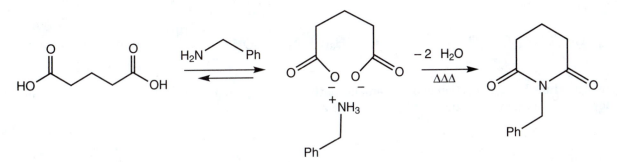

H₂N⁀Ph

− 2 H₂O
ΔΔΔ

(continued)

Problem 17.45 *(continued)*

Alternatively, we could activate the carboxylic acid (p. 855) with DCC. This reaction would form the amide on one side of the diacid, giving us an amido acid. We could then add DCC again, and the intramolecular reaction between the amide nitrogen and the activated carboxylic acid to give closure to the six-membered ring would be favored.

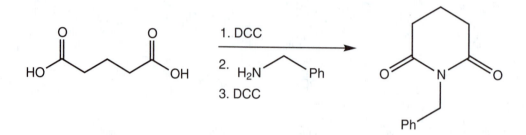

But you might notice that activating the diacid would result in formation of the anhydride. This is a good example of "if nature gives you a lemon, make lemonade." The anhydride can be used to make the desired molecule. Reaction of the cyclic anhydride with benzylamine will result in the same amido acid as does the process above. It can be closed by activating the carboxylic acid using DCC.

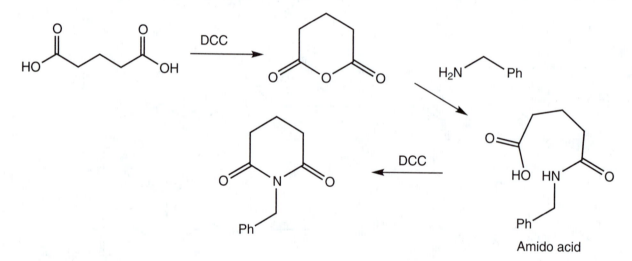

Amido acid

Problem 17.46

(a) This conversion requires the formation of a bromide containing the same number of carbons as the starting carboxylic acid. Acids can be converted into alcohols through reduction with lithium aluminum hydride. Alcohols can be made into bromides in many ways, including reaction with PBr₃. The combination of these two steps gives the first product.

$$CH_3(CH_2)_{10}COOH \xrightarrow[\text{2. } H_2O]{\text{1. LiAlH}_4} CH_3(CH_2)_{10}CH_2OH \xrightarrow{PBr_3} CH_3(CH_2)_{11}Br$$

(b) The Hunsdiecker reaction is a convenient method for preparing an alkyl bromide containing one fewer carbon atom than the starting carboxylic acid.

$$CH_3(CH_2)_{10}COOH \xrightarrow[\text{2. AgNO}_3]{\text{1. KOH}} CH_3(CH_2)_{10}COOAg \xrightarrow[\text{CCl}_4, \, \Delta]{Br_2} CH_3(CH_2)_{10}Br$$

(c) The third part of this problem requires the addition of a single carbon, which can easily be done through formation of the Grignard reagent, reaction with carbon dioxide, and neutralization with dilute acid. The starting material was made in part (a).

$$CH_3(CH_2)_{11}Br \xrightarrow[\substack{2.\ CO_2 \\ 3.\ H_2O/H_3O^+}]{1.\ Mg/ether} CH_3(CH_2)_{11}COOH$$

Problem 17.47 This reaction is an example of the Kolbe electrolysis with a few added twists. First of all, it is a "mixed" Kolbe in which two different carboxylates are employed. Second, in one of the carboxylates, there is a carbon–carbon double bond appropriately positioned to trap the initially generated alkyl radical **A** to give the cyclized radical **C**. It is also worth noting that this process is most efficient when hexanoic acid is used in excess. In this way, radical **B** reacts preferentially with radical **C** rather than with itself to give a dimer. When a fourfold excess of hexanoic acid was used, the product tetrahydrofuran was obtained in a 52% yield.

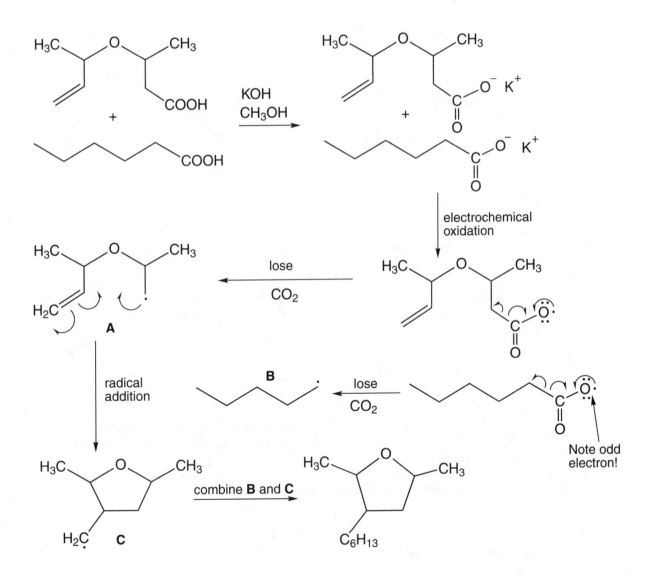

Problem 17.48   The reaction of the (E) ester is a simple acid-catalyzed reverse acetal formation. This is an important review problem. If formation of the diol seemed mysterious, BE CERTAIN to go back to Section 16.9 and review acetal formation.

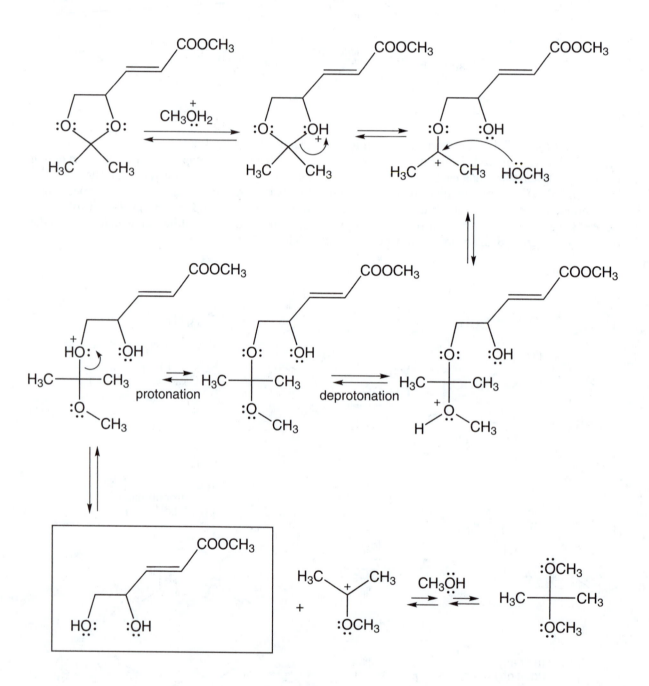

In the (Z) isomer, there are more options. The ester is in position to react with the product diol. This reaction is really only a Fischer esterification carried out in an intramolecular way. The steps leading to the diol are exactly the same as in the (E) isomer. Now new things happen as a hydroxyl group is within range of the ester group. Protonation is followed by addition of the hydroxyl to give a five-membered ring. Proton transfers, loss of methyl alcohol, and deprotonation give the lactone product.

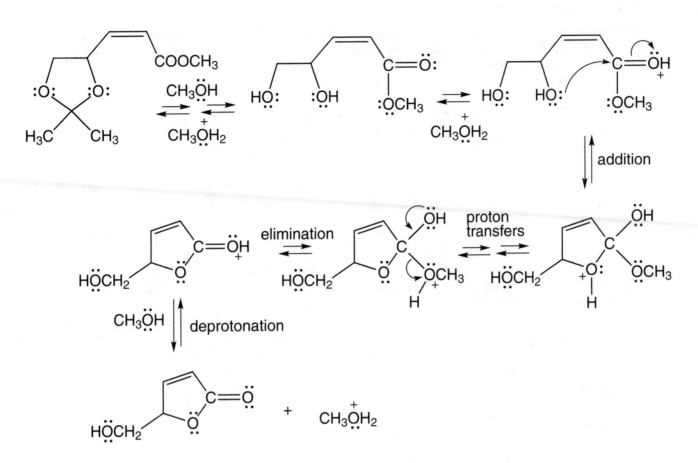

Problem 17.49

(a) This one is simple. A Kolbe electrolysis will do the trick.

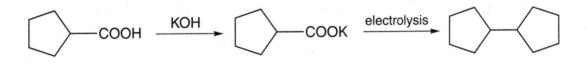

(b) Here we need an acid chloride and an amine. Addition–elimination will give the desired product.

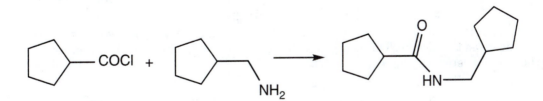

(continued)

Problem 17.49 *(continued)*

The acid chloride can be made directly from cyclopentanecarboxylic acid, and the desired amine by reduction of the acid, followed by conversion of the OH into a good leaving group such as chloride, and treatment with an excess of NH_3.

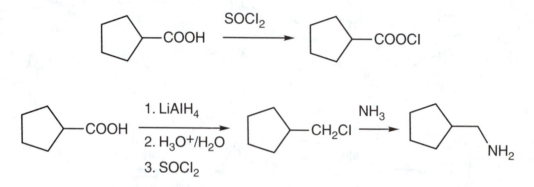

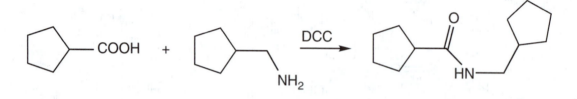

Alternatively, we could use cyclopentanecarboxylic acid and the amine directly, using DCC to facilitate coupling (p. 854).

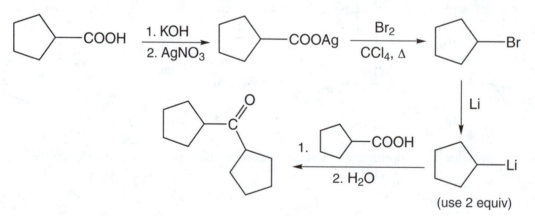

(c) In this part, you need to do two things with your starting material. First, use the Hunsdiecker reaction to make bromocyclopentane. This compound is converted into the corresponding organolithium reagent. Two equivalents of this reagent will react with the starting acid to make dicyclopentylketone.

(d) Here, too, a number of things must be done. Reduction of the starting acid with lithium aluminum hydride gives the alcohol. This compound can be used in a Fischer esterification reaction to make the desired ester.

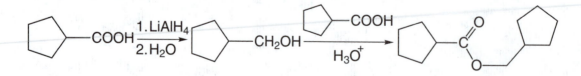

Problem 17.50   The IR spectrum of compound **3** (as well as its appearance in this chapter) suggests the presence of a carboxylic acid. The critical bands are the broad intense OH stretch at 3330–2500 cm^{-1} and the carbonyl stretch at 1696 cm^{-1}. The presence of the carboxylic acid is further supported by the presence of a 1H singlet at δ 11.6 ppm in the ^{1}H NMR spectrum. In addition, the ^{1}H NMR spectrum of **3** implies the presence of three adjacent methylene groups as indicated by the quintet at δ 1.95 ppm and the two triplets at δ 2.34 and 2.59 ppm, each integrating for two hydrogens. The 3H singlet at δ 3.74 ppm suggests the presence of some sort of deshielded methyl group. Finally, the set of doublets at δ 6.75 and 7.05 ppm with J = 8 Hz is most consistent with a para-disubstituted benzene ring.

An atom inventory of the suspected structural fragments indicates a mass of 178 g/mol. As the mass spectrum of **3** shows a parent ion at m/z = 194, we are missing a mass of 16, most easily accommodated by an oxygen atom. Thus, the molecular formula is $C_{11}H_{14}O_3$.

	Mass	Fragment(s)
	45	COOH
	42	$(CH_2)_3$
	15	CH_3
	76	C_6H_4
Total	178	$C_{11}H_{14}O_2$ (plus one more O to accommodate the mass spectrum) = $C_{11}H_{14}O_3$

It also appears reasonable that the oxygen atom is attached to the methyl group as in OCH_3, because of the low-field position of the signal for the methyl singlet (δ 3.74 ppm).

There are only two reasonable structures that can be assembled from these structural pieces: 4-(4-methoxyphenyl)butanoic acid (**3**) and 4-(3-methoxypropyl)benzoic acid (**3a**).

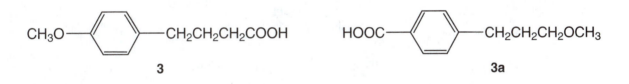

Possibility **3a** can be eliminated from consideration because the methylene hydrogens adjacent to the methoxy group would be expected to appear at δ 3.5 ppm, which is downfield of the observed δ 2.34 or 2.59 ppm.

With a structure for **3** in hand, it is now possible to speculate about the structures of **2** and estragole (**1**) itself. From the method of preparation of **3**, it appears reasonable that compound **2** is 4-(3-bromopropyl)anisole.

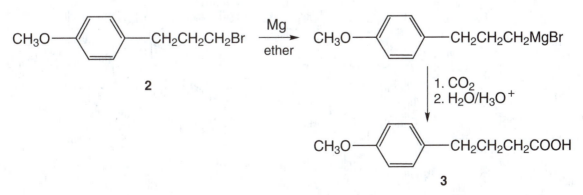

(continued)

Problem 17.50 *(continued)*

Finally, compound **2** appears to be the product of a peroxide-induced anti-Markovnikov addition of HBr to an alkene. That alkene, estragole, must be 4-allylanisole.

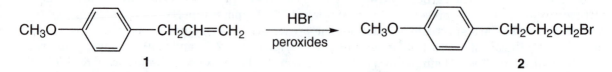

Problem 17.51   The reaction of cyclopentadiene and maleic anhydride should be familiar stuff. This example of the Diels–Alder reaction gives **1**, *endo*-norbornenedicarboxylic acid anhydride.

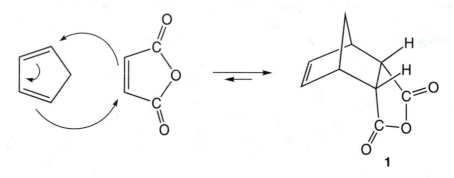

The spectral data for compound **1** are in accord with the proposed structure. The weak molecular ion at $m/z = 164$ is consistent with a molecular formula of $C_9H_8O_3$. The two carbonyl stretches at 1854 and 1774 cm^{-1} are indicative of a five-membered cyclic anhydride. The ^{1}H NMR spectral assignments are shown.

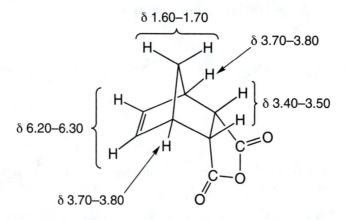

It must be admitted that, while the endo cycloadduct **1** is formed kinetically, this level of analysis does not rule out the exo isomer.

Hydrolysis of anhydride **1** gives the dicarboxylic acid **2**, as shown below:

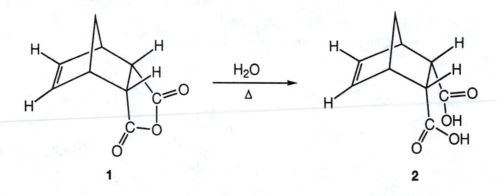

Again, the spectral data for diacid **2** are consistent. Note, in particular, the broad, intense O—H stretch in the IR, as well as the appearance of a single C=O stretch at lower frequency than that for the anhydride. In addition, the appearance of a 2H singlet at δ 11.35 ppm is consistent with the OH hydrogens of a dicarboxylic acid. Note that this is given in the problem as 1H: Integrals are relative, and you must adjust the ratios given to match the molecular formula.

So far, so good. Now comes the hardest part of this problem, deducing the structure of **3**. First, we know that compound **3** is isomeric with **2** ($C_9H_{10}O_4$). Second, the IR spectrum of **3** still shows the presence of a carboxylic acid (broad hydroxyl band at 3450–2500 cm^{-1} and the carbonyl stretch at 1690 cm^{-1}). However, there is also a higher-frequency carbonyl band at 1770 cm^{-1} so far unexplained. Finally, the ^1H NMR spectrum of **3** is much more complicated than that of either **1** or **2**. Clearly, compound **3** is less symmetric than its precursors. Notice that there is only <u>one</u> carboxylic acid group remaining in **3**, as indicated by the <u>1</u>H singlet in the NMR at δ 12.4 ppm. There is a new 1H signal in the NMR at δ 4.75–4.85 ppm, and the 2H signal for the hydrogens attached to the double bond of **2** (δ 5.6–5.7 ppm) is gone.

It appears that one carboxyl group and the carbon–carbon double bond in **2** have reacted to form **3**. Now comes the mechanistic analysis. What can compound **2** do in concentrated sulfuric acid? How about protonation of the double bond to give carbocation **A**?

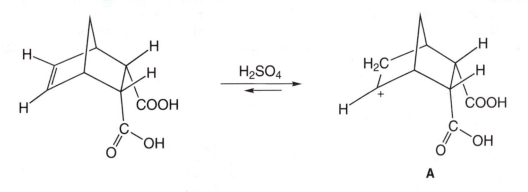

As carbocation **A** inherits endo stereochemistry from its ancestors **1** and **2**, intramolecular capture of the carbocation by a carbonyl oxygen is possible. Deprotonation then leads to the lactone **3**.

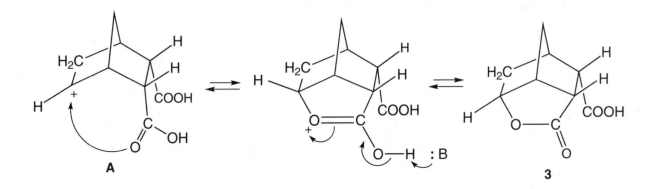

(continued)

Problem 17.51 *(continued)*

Lactone **3** is much less symmetrical than either anhydride **1** or diacid **2**, accounting for the far more complex NMR spectrum. The 1770 cm^{-1} band in the IR is appropriate for a five-membered lactone carbonyl group. Note the absence of hydrogens attached to double bonds in **3** and the presence of a single, low-field hydrogen, **H$_a$**, deshielded by the adjacent oxygen.

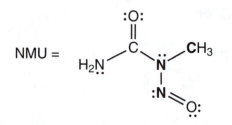

δ 4.75–4.85 (m)

3

Problem 17.52   In this problem, there is some serious electron and hydrogen pushing, this time in base. One important part of this tough problem is figuring out where the heavy atoms of diazomethane (C—N—N) come from. The methyl group and the attached two nitrogen atoms of NMU (shown in boldface type) seem likely candidates. The problem now is how to get rid of the "extra" atoms.

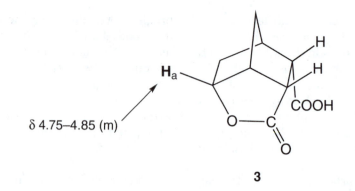

Addition of hydroxide to the carbonyl group is followed by loss of CH$_3$NNO$^-$, a resonance-stabilized anion, and therefore a decent leaving group. Protonation, loss of hydroxide, and deprotonation complete the reaction.

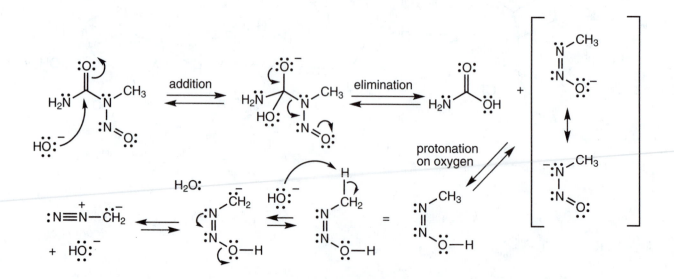

Problem 17.53 The carbonyl oxygen is the nucleophilic site because protonation on that oxygen gives the more stable intermediate. Protonation is a reversible process, so one can argue that the thermodynamic stability will control the outcome. The carbonyl-protonated intermediate is more stable than the protonated alcohol because the positive charge is shared between both oxygens and the carbon (see the figure). The LUMO of the first intermediate shows that the two oxygens are equally participating in the positive charge. It is interesting to note that the carbonyl carbon carries most of the positive charge in the LUMO.

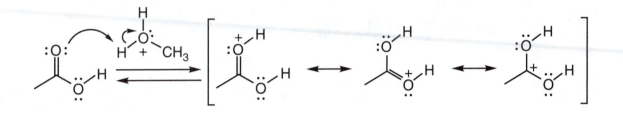

Problem 17.54 The SOCl$_2$ is pyramidal and SO$_2$ is bent. Both structures are telling us that there is a lone pair of electrons on the sulfur. The reaction diagram indicates the reaction is exothermic because the SOCl$_2$ is so reactive and the SO$_2$ is so stable. The sum of the energies for *all* the reactants is higher than the energies for *all* the products. The difference in energy between the SOCl$_2$ and the SO$_2$ is more than the difference in energy between the RCO$_2$H and the RCOCl.

Problem 17.55 There are two intermediates in this reaction. The first probably has a very short lifetime because the tetrahedral intermediate is able to expel either the OH group (return to starting material) or the OR group. Either process will regain the more stable carbonyl group. It is the symmetry of the elimination steps that makes the process appear rapid. The bond between the carbonyl carbon and the OH group is forming at the same rate that the bond between the carbonyl carbon and the OR group is breaking.

Derivatives of Carboxylic Acids: Acyl Compounds

18

This chapter continues the exploration of the addition–elimination mechanism begun in Chapter 17. This process dominates the chemistry of the acyl compounds explored here. There are some quite taxing problems in this section. Do try to analyze what needs to be done before you start. It is much easier that way.

Problem 18.1

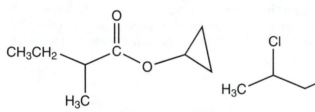

Cyclopropyl 2-methylbutanoate
(Cyclopropyl 2-methylbutyrate)

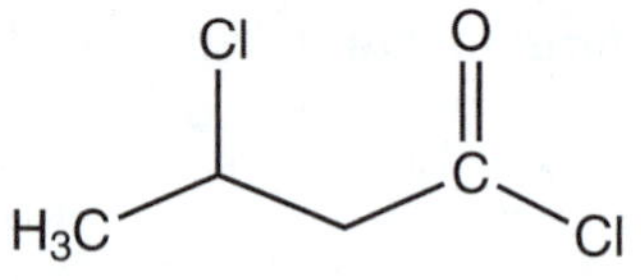

3-Chlorobutanoyl chloride

Benzoic propanoic anhydride
(Benzoic propionic anhydride)

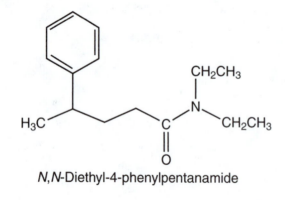

N,N-Diethyl-4-phenylpentanamide

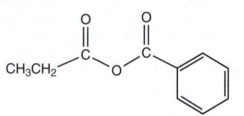

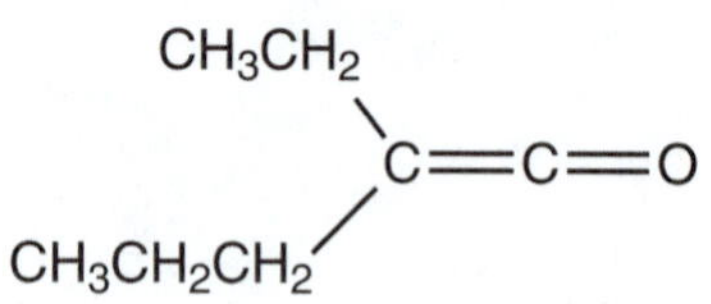

Ethylpropylketene

Problem 18.2 *N,N*-Dimethylformamide is stabilized by resonance, like any ester or amide in which heteroatoms bearing lone pairs of electrons are adjacent to the carbon–oxygen double bond. There is double-bond character to the carbon–nitrogen bond.

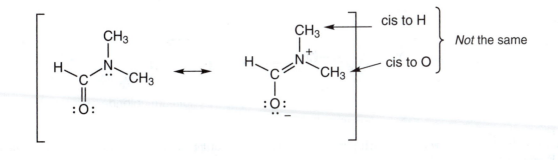

There must be a barrier to rotation about the carbon–nitrogen bond, *and the two methyl groups are different*! At room temperature there is not enough energy to overcome the barrier to rotation, and the two different methyl groups appear at different positions in the ^{1}H NMR spectrum. At higher temperature there is enough energy to overcome the barrier to rotation, and the methyl groups become equivalent. The figure shows both resonance forms of the amide. The carbon–nitrogen bond is not a full double bond, but it does have partial double-bond character.

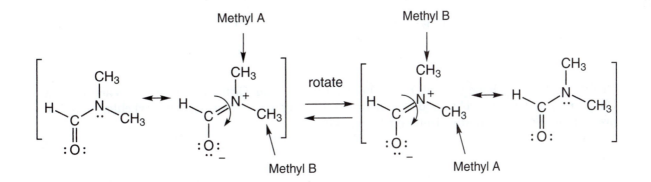

Problem 18.3 Your nasal passages are moist (or at least they should be). When an inhaled acid chloride encounters that moisture, a hydrolysis reaction takes place and HCl is formed. That is what is detected, usually quite unpleasantly.

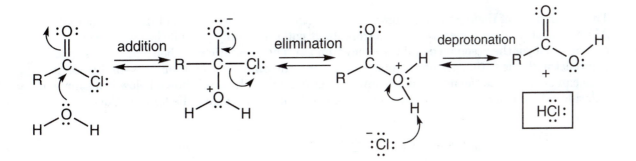

(continued)

Problem 18.3 *(continued)*

The first specific example of Figure 18.21 has acetyl chloride reacting with trimethylaniline. When an amine reacts with an acid chloride, HCl is a by-product. Building up a concentration of HCl can be a problem for organic molecules. We can use pyridine as a solvent to remove HCl from the reaction. Pyridine is weakly basic and reacts with the HCl to make pyridinium chloride, as shown below.

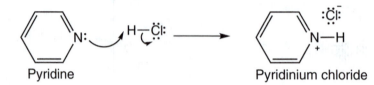

As a result of the HCl reaction with pyridine, we can write out both products of the specific reaction in Figure 18.21.

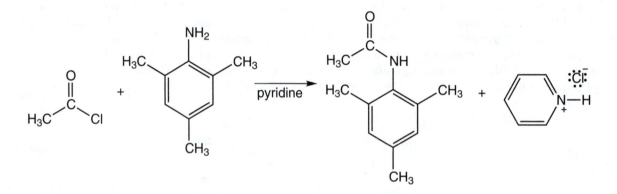

Problem 18.4   We hope you will remember that the Friedel–Crafts acylation is a reaction from Chapter 15. The product obtained from benzene reacting with 2-methylpropanoyl chloride in the presence of AlCl$_3$ is 2-methyl-1-phenyl-1-propanone.

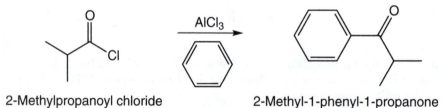

The mechanism of the reaction is shown below. The first step is formation of the acylium ion. Aluminum chloride is a very strong electrophile. It will essentially pull off the chloride of an acid chloride to form the resonance-stabilized acylium ion. The aromatic ring, benzene in this case, then can react with the cation giving the nonaromatic intermediate that is resonance stabilized. Any base can deprotonate the nonaromatic intermediate. The equilibrium shown below is a source of chloride, which can deprotonate the nonaromatic intermediate to give the final product.

$$:\overset{..}{\underset{..}{Cl}}{-}\bar{A}lCl_3 \quad \rightleftharpoons \quad :\overset{..}{\underset{..}{Cl}}{:}^{-} \quad + \quad AlCl_3$$

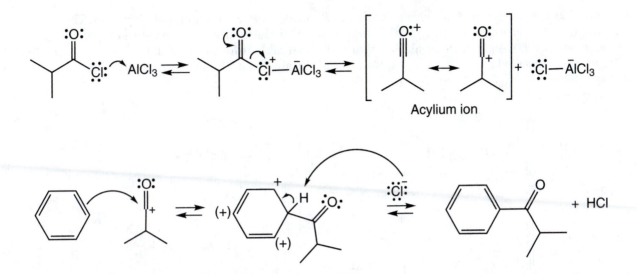

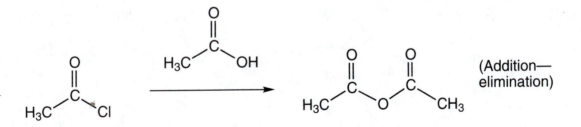

Problem 18.5 The first three parts of this problem require only very simple reactions of the starting acid chloride with nucleophiles:

(a)

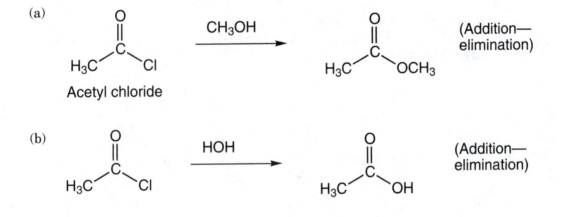

(a)

$$H_3C-C(=O)-Cl \xrightarrow{CH_3OH} H_3C-C(=O)-OCH_3 \quad \text{(Addition—elimination)}$$

Acetyl chloride

(b)

$$H_3C-C(=O)-Cl \xrightarrow{HOH} H_3C-C(=O)-OH \quad \text{(Addition—elimination)}$$

(c) Use acetic acid, made in part (b), as the nucleophile to make the anhydride:

$$H_3C-C(=O)-Cl + H_3C-C(=O)-OH \longrightarrow H_3C-C(=O)-O-C(=O)-CH_3 \quad \text{(Addition—elimination)}$$

(d) Remember the synthesis of ketones from carboxylic acids in Chapter 17? Now is the time to use it. Another possible answer is suggested in part (e).

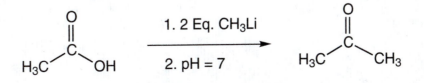

$$H_3C-C(=O)-OH \xrightarrow[\text{2. pH = 7}]{\text{1. 2 Eq. CH}_3\text{Li}} H_3C-C(=O)-CH_3$$

(continued)

Problem 18.5 *(continued)*

(e) We have to make a tertiary alcohol with two R groups of one kind and one of another. The one thing we can't do is to allow acetyl chloride to react with methyllithium, then with ethyllithium (why not? p. 892). A grind-it-out alcohol synthesis is possible from acetic acid, but far quicker is the following suggested route that takes advantage of the single addition to acid chlorides of organo-cuprates:

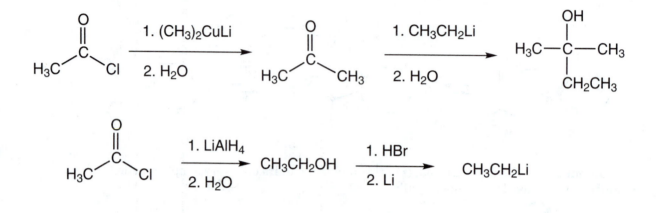

Problem 18.6 This problem is somewhat complicated, although no individual step is very hard. The first steps involve opening the anhydride in a typical addition–elimination process, with the nitrogen of aniline acting as the nucleophile.

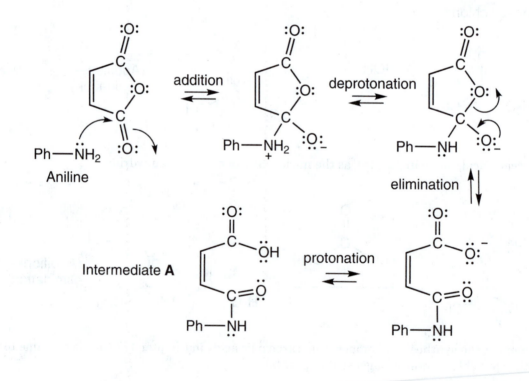

The reaction sometimes goes on to product without the aid of acetic anhydride, but it is greatly facilitated by the ability of the anhydride to create a good leaving group out of the poor leaving group, OH (a repeating theme if ever there were one).

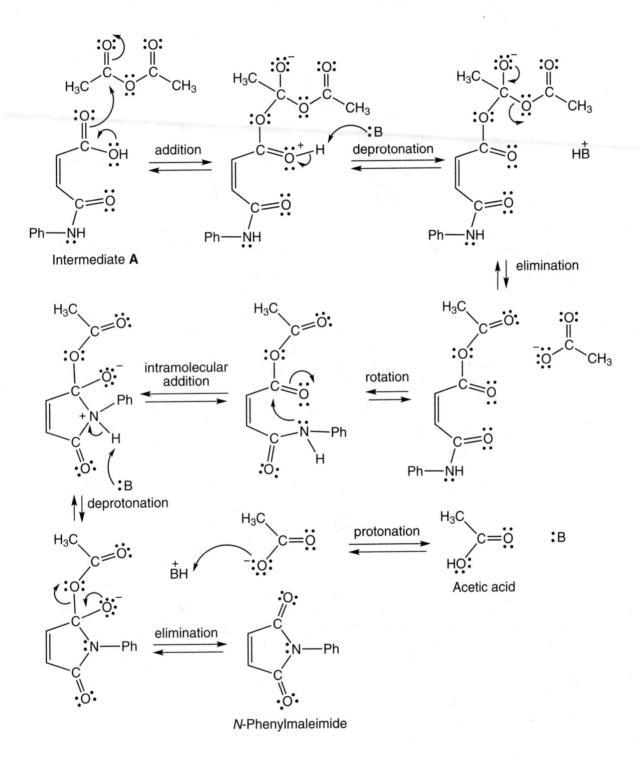

Problem 18.7 The mechanism is the usual one. There is complex formation and generation of an E⁺ reagent, here the acylium ion, followed by addition to benzene to give a resonance-stabilized cyclohexadienyl cation. Deprotonation by any base in the system leads to rearomatization and formation of the substituted benzene.

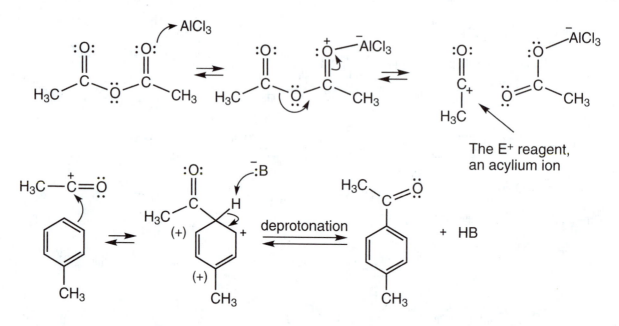

The E⁺ reagent, an acylium ion

deprotonation

+ HB

Problem 18.8 The hydrolysis of ethyl butanoate gives butanoic acid and ethanol. The mechanism for the acid-catalyzed hydrolysis is shown here. There are five intermediates. Only the first and last intermediates have important resonance structures.

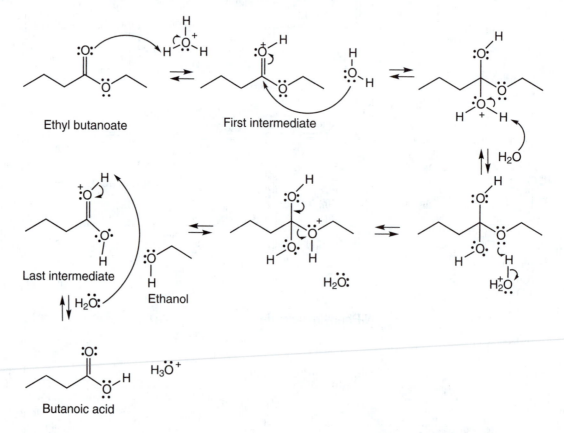

Ethyl butanoate

First intermediate

H_2O

Last intermediate

$H_2\ddot{O}$:

Ethanol

$H_2\ddot{O}$:

Butanoic acid

$H_3\ddot{O}^+$

The resonance structures for the first and last intermediates are drawn below.

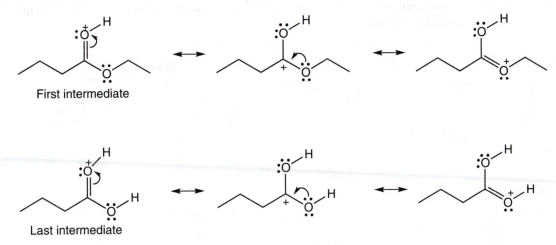

First intermediate

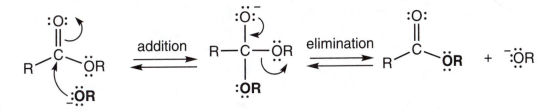

Last intermediate

Problem 18.9   The base-induced reaction involves the usual addition–elimination process. Alkoxide adds reversibly, and loss of the original alkoxide leads to the new ester.

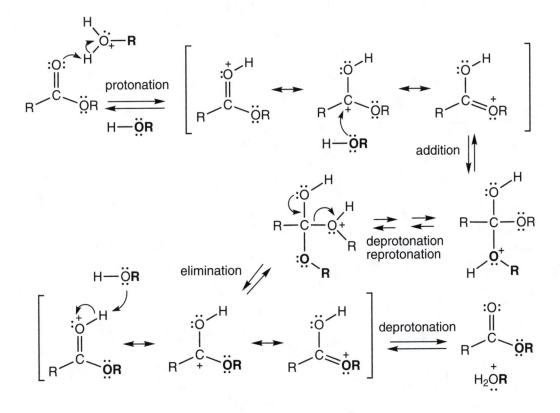

In acid, protonation leads to a resonance-stabilized intermediate to which alcohol adds. Deprotonation and reprotonation lead to an intermediate that can lose alcohol and generate the new ester.

Problem 18.12   This problem involves a garden-variety addition–elimination process. Ammonia adds to the ester, a second molecule of ammonia deprotonates the ammonium of the tetrahedral intermediate, and methoxide is eliminated.

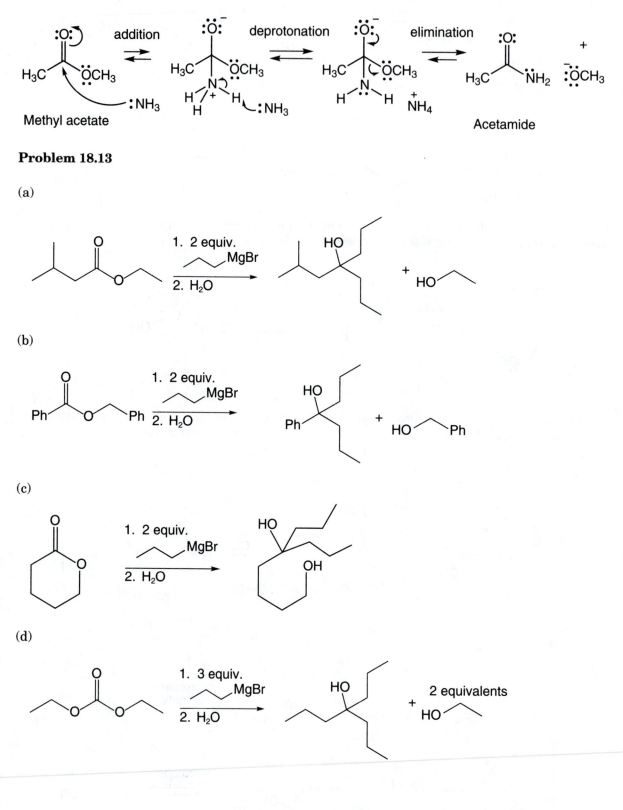

Problem 18.13

(a)

(b)

(c)

(d)

Problem 18.14 One reason is simple. Diisobutylaluminum hydride is much bigger than lithium aluminum hydride and addition reactions are naturally slower. It's a case of steric inhibition of reactivity.

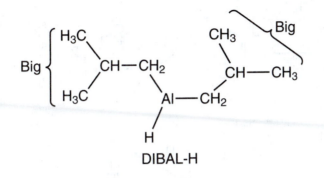

DIBAL-H

Problem 18.15 The first molecule of hydroxide adds to the carbonyl group just as shown in Figure 18.38. Loss of amide ion is slow, and a second molecule of hydroxide removes the OH proton to give a dianion. The dianion is able to eliminate amide ion.

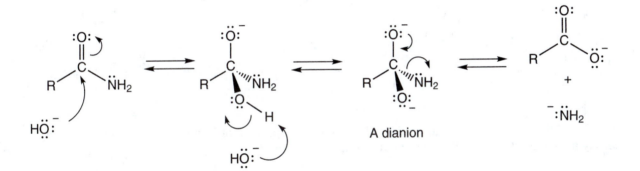

A dianion

Problem 18.17 Pentanoic acid and aniline (aminobenzene). In acid, aniline will be protonated to give the anilinium ion.

Problem 18.18 For an *sp*-hybridized nitrogen, the lone-pair electrons are in an *sp* orbital with 50% *s* character. They are held more tightly than electrons in an *sp*² orbital (33% *s* character) and will therefore be less available for reaction. Electrons in an *sp*³ hybrid orbital (25% *s* character) are even less tightly held and will be more easily used in reactions with Lewis acids.

Problem 18.19

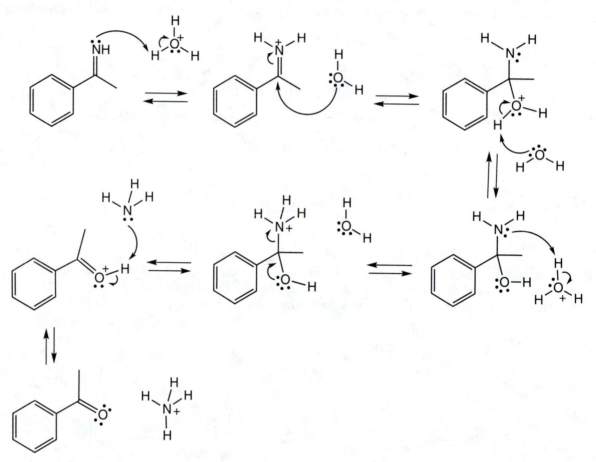

Problem 18.20   Here is the mechanism. Just follow the pattern of Figure 18.51. Addition of the peroxy acid is followed by rearrangement and a final proton transfer.

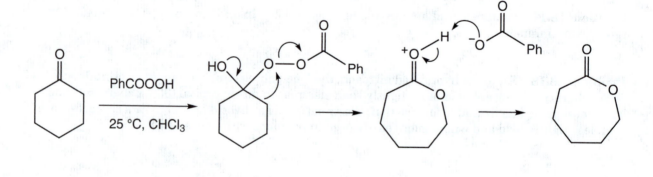

Problem 18.21 The migrating group can never leave the framework along which it is migrating. If any kind of free species, be it radical, cation, or anion, were formed, racemization would result.

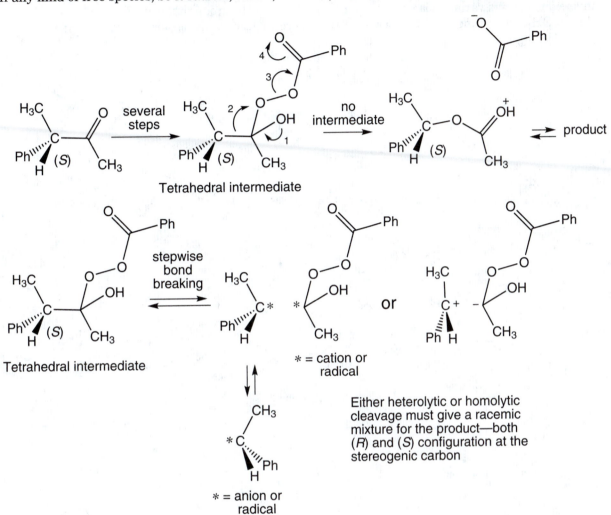

Tetrahedral intermediate

Tetrahedral intermediate

* = cation or radical

* = anion or radical

Either heterolytic or homolytic cleavage must give a racemic mixture for the product—both (*R*) and (*S*) configuration at the stereogenic carbon

Problem 18.22 Just follow the mechanism for the acyclic version that is shown in Figure 18.54.

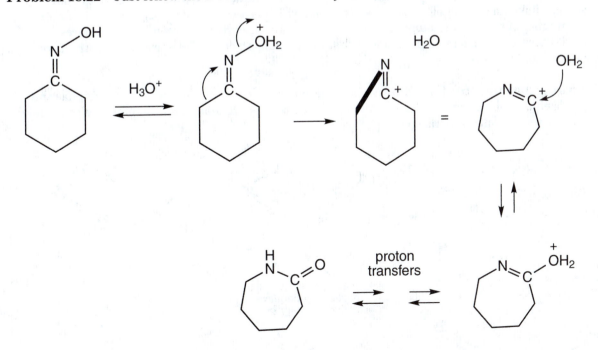

H₂O

proton transfers

Problem 18.23   Look closely at the rearrangement step. It is, in fact, nothing more than an intra-molecular S$_N$2 displacement of water by the migrating pair of electrons. All S$_N$2 reactions go with inversion, and this variation is no exception. So it is always the group that is anti to the leaving group, the old OH in the oxime itself, that moves. This preference leads to the observed stereospecificity.

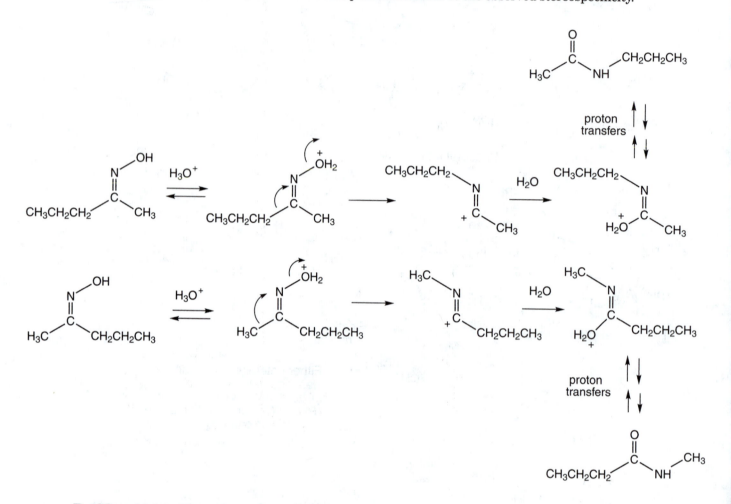

Problem 18.24   There are more than three possible resonance structures, but there are only three that have an octet of electrons on each atom, which is a way of saying that there is a maximum number of bonds (rule number 1, p. 28). If you have drawn other resonance structures, check to see if there are atoms in your structure that are lacking an octet of electrons. Such structures are minor contributors.

Comparing the three structures drawn below, we can see that the middle representation has the negative charge on the oxygen. That is the most electronegative atom, and therefore we can safely assume it would contribute most.

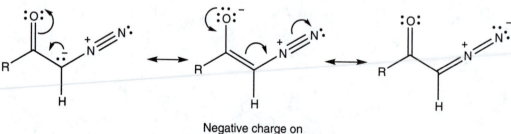

Negative charge on
most electronegative
atom

Problem 18.27

(a) In this part, we need to extend the chain of a carboxylic acid by one carbon. That's the Arndt–Eistert synthesis.

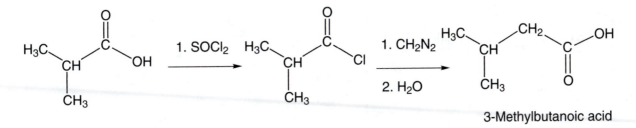

3-Methylbutanoic acid

(b) This part is easy—just oxidize the alcohol and proceed as in part (a).

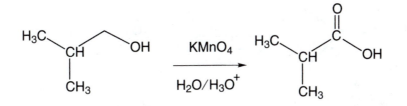

(c) Here we need only convert the primary iodide into the alcohol of part (b) and proceed as in parts (a) and (b).

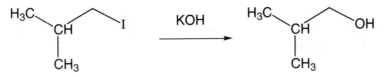

(d) Hydroboration gives us the alcohol of part (b). Now proceed as in parts (a) and (b).

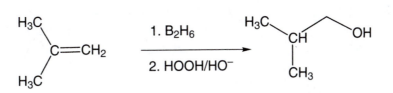

Problem 18.28 If nitrene chemistry is akin to carbene chemistry, there should be addition reactions to alkenes and carbon–hydrogen insertion reactions with alkanes, and there are, as the following two reactions show.

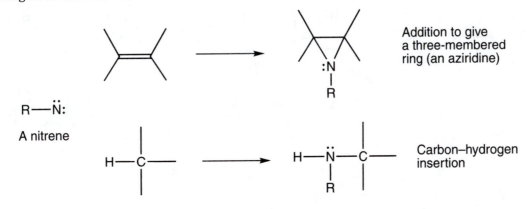

Addition to give a three-membered ring (an aziridine)

A nitrene

Carbon–hydrogen insertion

Problem 18.29   Addition to the carbon–oxygen double bond takes place in the usual way to give a resonance-stabilized intermediate. Deprotonation and protonation steps give the carbamate ester.

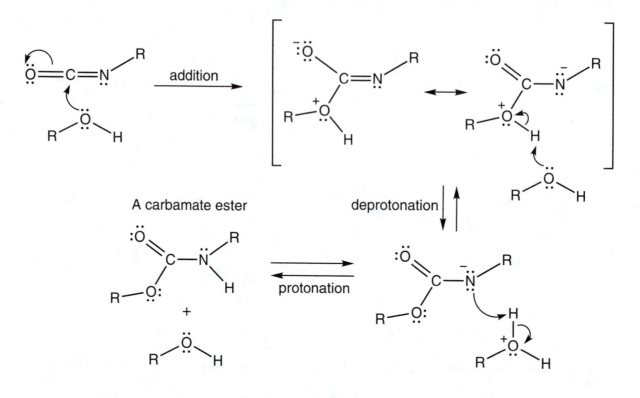

A carbamate ester

Problem 18.30   The reaction begins exactly as in Problem 18.29 with the substitution of water for the alcohol. This addition leads to a carbamic acid. Carbamic acids are not isolable, but decarboxylate to give CO_2 and an amine.

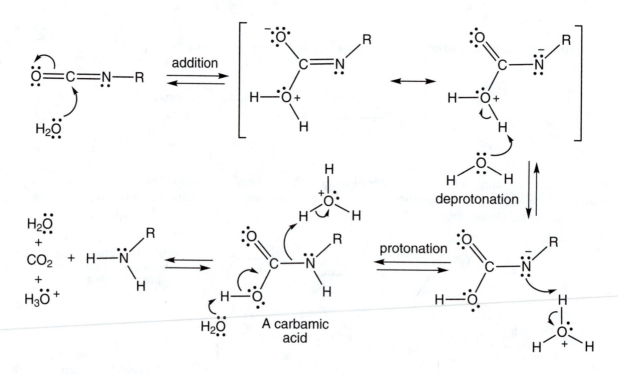

A carbamic acid

Additional Problem Answers

Problem 18.31

(a) This molecule is benzyl 4-bromobenzoate. The benzyl term is technically not systematic, but IUPAC would recognize it. The systematic name is phenylmethyl 4-bromobenzoate.

(b) This molecule is propyl 2-methylpropanoate.

(c) This molecule is ethyl 4-methyl-3-oxopentanoate.

(d) This molecule is methyl (*Z*)-3-pentenoate.

(e) This molecule is methyl 4-methyl-5-oxopentanoate. It is not methyl 4-formylpentanoate because we choose the longest carbon chain containing the most functional groups.

(f) This molecule is dimethyl butanedioate. The common name is dimethyl succinate.

Problem 18.32   We should always do a retrosynthetic analysis first. The anhydride can be made from an acid chloride and a carboxylic acid. The acid chloride comes from the corresponding acid and the carboxylic acids can be made from ethyl propanoate.

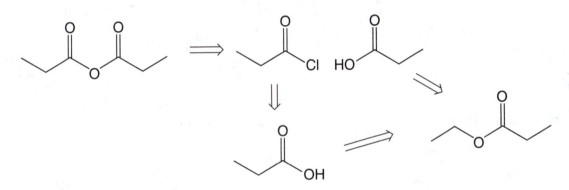

The forward process is straightforward. Hydrolysis of the ethyl propanoate gives propanoic acid. Some of the propanoic acid can be converted into propanoyl chloride using thionyl chloride. Then the reaction between propanoyl chloride and propanoic acid gives the anhydride.

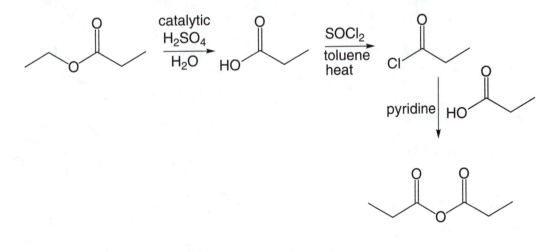

Problem 18.33 These two problems are straightforward addition–elimination sequences. In these cases, an amide and an acyl azide are the products.

(a)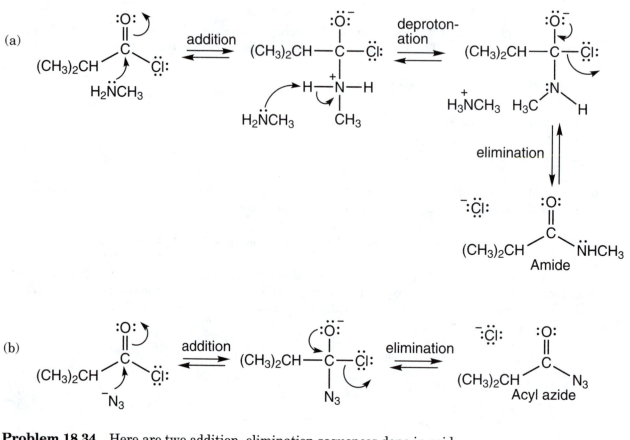

(b)

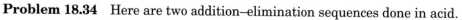

Problem 18.34 Here are two addition–elimination sequences done in acid.

(a)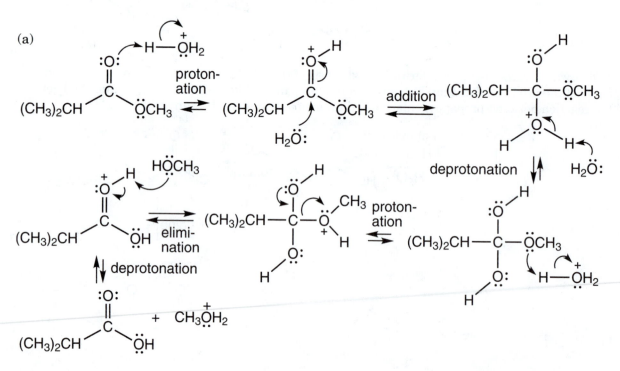

(b)

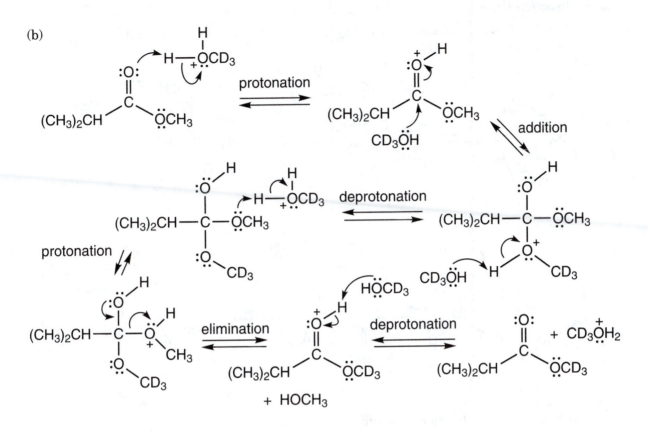

Problem 18.35 Here is another series of protonations, deprotonations, additions, and eliminations. Be careful about the details.

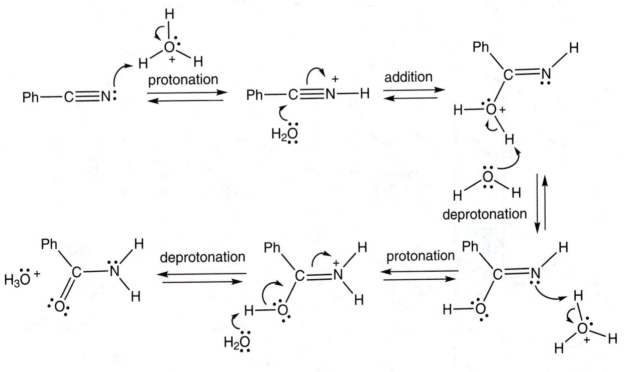

(continued)

Problem 18.35 *(continued)*

The amide is not stable in acid and is further hydrolyzed to the carboxylic acid (p. 902).

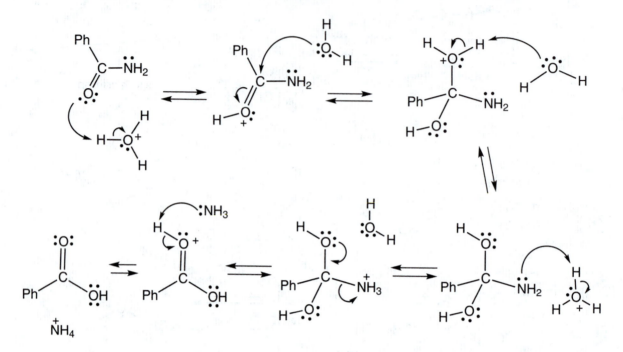

Problem 18.36   Just Fischer esterification:

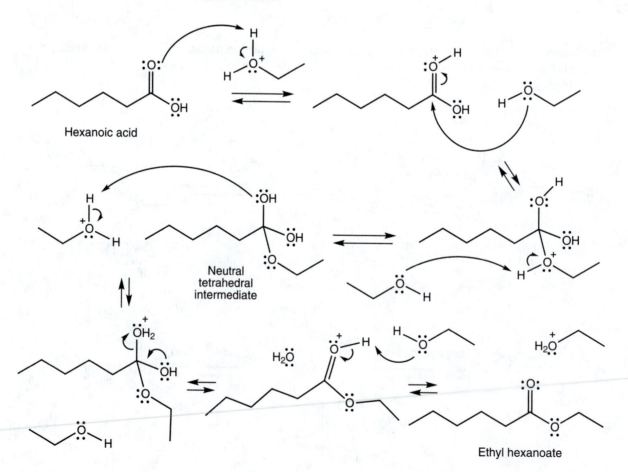

Problem 18.37 Just two steps: oxidation followed by Fischer esterification.

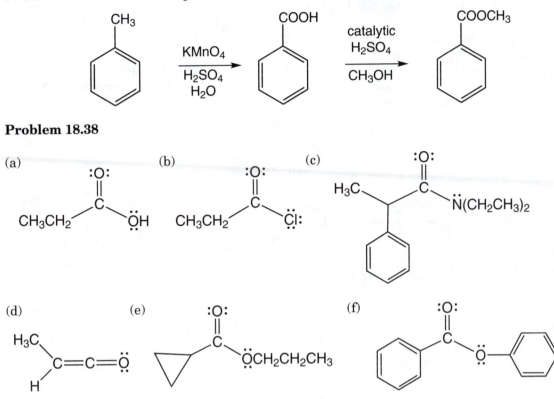

Problem 18.38

(a) (b) (c)

(d) (e) (f)

Problem 18.39

(a) Methyl cyclohexanecarboxylate

(b) Ethyl ethanoate (ethyl acetate is it's common name)

(c) Cyclohexyl 3-methylbutanoate

(d) (*S*)-3-Bromo-4-methylpentanoyl chloride

(e) (*R*)-4-Methylhexanamide

(f) 2-Chloro-3,*N*-diethylpentanamide

(g) 3-Nitrobenzamide (*m*-nitrobenzamide)

(h) 2-Amino-*N*-methylbutanamide

Problem 18.40 First, convert the alcohol into a leaving group. A tosylate will work well. Then do an S_N2 reaction using excess aniline and the primary tosylate. The aniline anion would undergo S_N2 reaction more effectively than aniline.

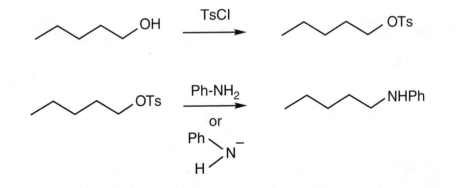

Problem 18.41   We need to make a tertiary *N*-oxide here and then do a Cope elimination (p. 358, Problem 8.18). So, we need to alkylate that primary amine starting material first.

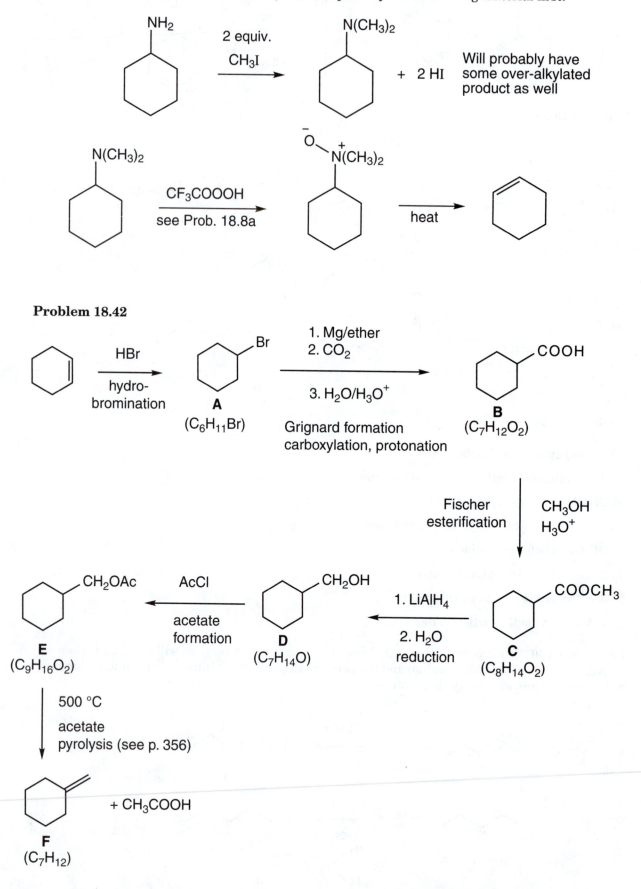

Problem 18.42

Problem 18.43

(a) Consider the resonance descriptions of acetone and acetaldehyde shown.

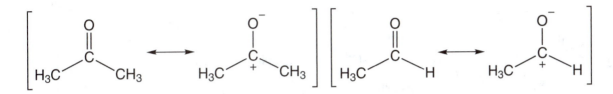

The "extra" methyl group in acetone will stabilize the polar resonance form in acetone relative to that in acetaldehyde. The polar form contributes more to the resonance hybrid in acetone than it does in acetaldehyde. As this form has only a single bond between carbon and oxygen, the carbonyl bond will be weaker in acetone than in acetaldehyde, and therefore the stretching frequency is lower (1719 vs. 1733 cm^{-1}). But wait. Doesn't the "extra" methyl group in acetone also stabilize the carbon–oxygen double bond relative to that in acetaldehyde? Sure, but these hyperconjugative effects are especially strong for charged species, and it is the effect of the methyl group on the polar form that dominates.

(b) Look again at the contributing resonance forms.

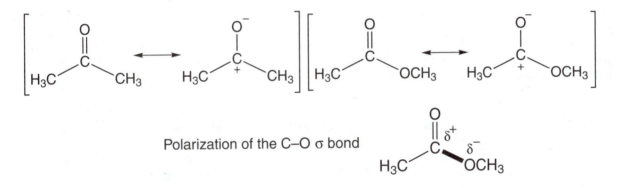

Polarization of the C–O σ bond

As the figure indicates, now the polarity of the carbon–oxygen σ bond makes the polar form in the ester especially unfavorable. The polar form will contribute less to the resonance hybrid than it does in acetone, and the carbonyl group will be relatively more "double," and therefore stronger. Accordingly, it appears at higher frequency than in acetone (1750 vs. 1719 cm^{-1}).

But wait again. Doesn't the ester oxygen affect the situation in another way? Isn't there another resonance form in which there is a single bond in the original carbonyl group? Shouldn't this make the carbon–oxygen double bond weaker, and the absorption at lower frequency?

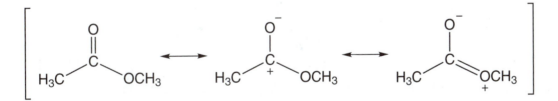

Yes, but the inductive effect mentioned earlier must dominate. Here is where we are so perilously close to circular reasoning. The statement "The inductive effect must dominate the resonance effect, and therefore the absorption appears at higher frequency" is valid, but, it surely is clear that in the absense of a measured spectrum, it would be difficult to predict the relative positions of the absorptions.

(continued)

Problem 18.43 *(continued)*

(c) Once again, start by looking at the contributing forms.

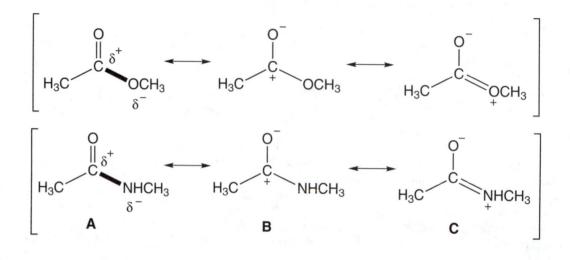

The problem is to explain why in the amide, the dipole shown in form **A** doesn't result, as it does for the ester, in a high-frequency absorption. Once again, one can rationalize. First of all, the dipole in form **A** does not have as big an effect in the amide as it does in the ester. Nitrogen is less electronegative than oxygen. Second, form **C** will be more important for the amide than it is for the ester, as a positive charge on nitrogen (less electronegative) is more stable than it is on oxygen (more electronegative). In the amide, the combination of a reduced dipole and increased contribution of form **C** leads to a weaker carbon–oxygen double bond and a lower-frequency absorption.

Problem 18.44   Yes. The greater the positive charge on the carbon of the carbonyl group, the lower the position of the carbon in the ^{13}C NMR spectrum. In the ester, the dipole in the carbon–oxygen σ bond reduces the contribution of the dipolar resonance form. The carbon is less positive than that in acetone and appears far upfield (δ 169 vs. 205 ppm) of the carbonyl carbon in acetone.

Problem 18.45

(a)

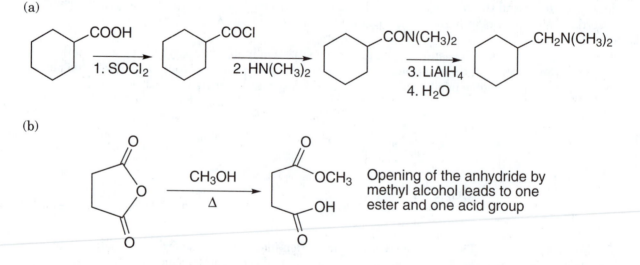

(b)

Opening of the anhydride by methyl alcohol leads to one ester and one acid group

This part includes a base-catalyzed transesterification. The cage is hiding a simple reaction.

(c)

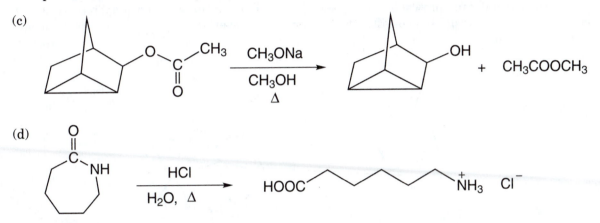

(d)

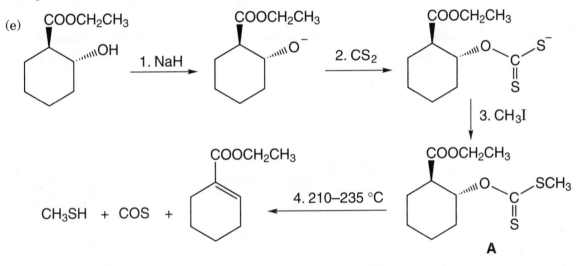

The cyclic amide is opened by water to give the carboxylic acid and the amine salt. In this case, both products are in the same molecule.

(e)

The xanthate ester **A** is first prepared, and then undergoes a thermal elimination reaction to give the conjugated ester, which is more stable than the unconjugated ester. (See Problem 8.17, p. 358.)

Problem 18.46   Decarboxylation leads first to an enol, which then equilibrates with the generally more stable keto form. In this case, the enol that would be formed is shown below.

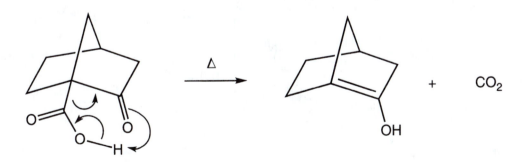

(continued)

Problem 18.46 *(continued)*

The putative product contains a bridgehead double bond, and that is bad news in energy terms. Although the paper doesn't protest as we write such a structure, a three-dimensional picture shows that the orbitals making up that so-called double bond do not overlap at all well. There really is no double bond possible.

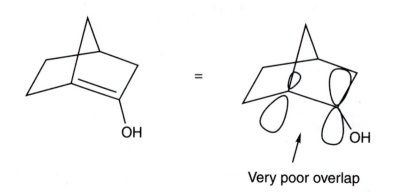

Very poor overlap

Problem 18.47 The conversion of aldoximes to nitriles is formally a dehydration reaction. This reaction is yet another example of the conversion of a poor leaving group (hydroxyl) into a better one (acetic acid).

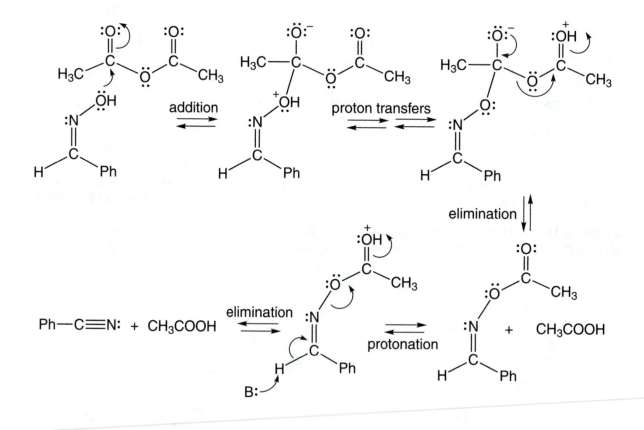

Oximes (R_2C=N—OH) are formed from the reaction of aldehydes (and ketones) with hydroxylamine.

Problem 18.48 This process is named the Ritter reaction. The first step in the sequence is the formation of the relatively stable *tert*-butyl cation.

$$(CH_3)_3COH \quad \xrightleftharpoons[]{H_2SO_4} \quad (CH_3)_3C\overset{+}{O}H_2 \quad \rightleftharpoons \quad (CH_3)_3\overset{+}{C} \quad + \quad H_2O$$

The *tert*-butyl cation is then captured by the nucleophilic nitrile to give the resonance-stabilized cation **A**.

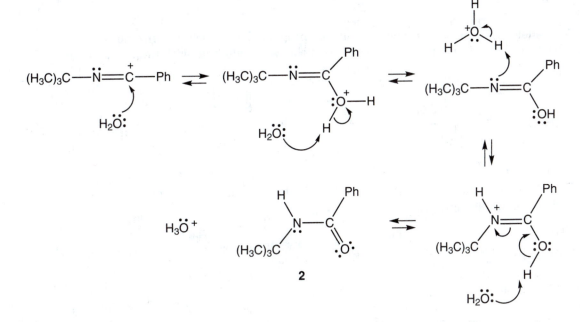

Addition of water to **A**, followed by a series of proton shifts, gives the observed product.

2

Problem 18.49

(a) Tertiary alcohols in which all three R groups are the same are easily made from the reaction of a carbonate with a Grignard reagent.

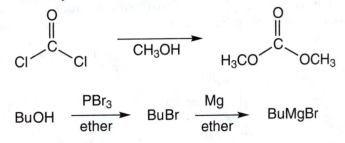

Dimethyl carbonate

The carbonate comes from reaction of phosgene and methyl alcohol, and the Grignard reagent can be made from treatment of butyl bromide with Mg.

(continued)

Problem 18.49 *(continued)*

(b) The reaction of propanoic acid with two equivalents of butyllithium, followed by protonation, will do the trick. Butyllithium comes from butyl bromide [part (a)] and lithium.

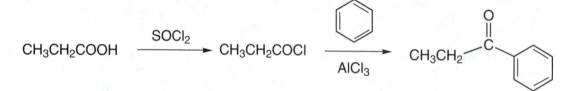

(c) Reaction of propanoic acid with ethyl alcohol and an acid catalyst (Fischer esterification) leads to ethyl propanoate. Reaction with butyllithium, followed by protonation, completes the synthesis.

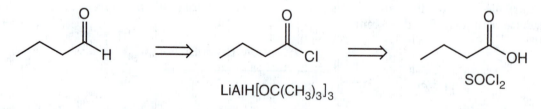

(d) A Friedel–Crafts acylation reaction between propanoyl chloride and benzene will produce the product. The acid chloride is made directly from the carboxylic acid with thionyl chloride.

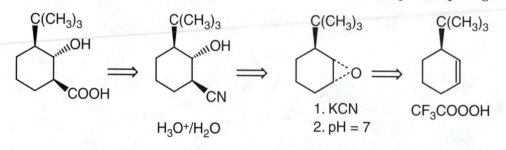

(e) Dehydration of propanamide will make the nitrile. The amide can be made from the acid chloride [part (d)] and ammonia.

$$CH_3CH_2COCl \xrightarrow{NH_3} CH_3CH_2CONH_2 \xrightarrow[\Delta]{P_2O_5} CH_3CH_2CN$$

Problem 18.50

(a) Here we just need to make the acid chloride and reduce it. Rosenmund reduction (look it up!) would also work.

(b) The key to this problem is remembering that cyanide can be transformed into a carboxylic acid through hydrolysis (when writing this version of the Study Guide, MJ didn't and had to ask SF, so don't be unhappy if this one didn't come easily). The vicinal cyanohydrin can be made through the opening of an epoxide. That *tert*-butyl group is crucial, though, because it both directs epoxide formation to the side away from it and shields the adjacent carbon in the epoxide opening.

(c) Regiospecific hydroboration followed by intramolecular transesterification does it.

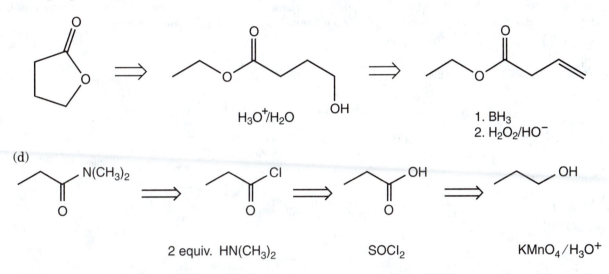

H_3O^+/H_2O

1. BH_3
2. H_2O_2/HO^-

(d)

N(CH$_3$)$_2$ Cl   OH   OH

2 equiv. HN(CH$_3$)$_2$ SOCl$_2$ KMnO$_4$/H$_3$O$^+$

Problem 18.51   This problem only requires that you remember the Beckmann rearrangement for part (a) and the Baeyer–Villiger reaction for part (b) (pp. 909 and 907).

(a)

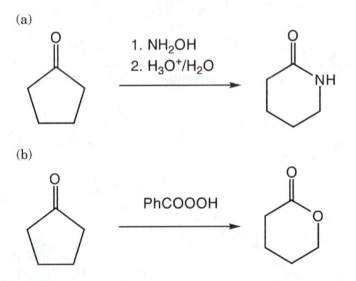

1. NH$_2$OH
2. H$_3$O$^+$/H$_2$O

NH

(b)

PhCOOOH

Problem 18.52   The first step is surely protonation of the carbonyl (not the OR!). Bromide then does an S$_N$2 reaction to give one of the products, benzyl bromide, and an unstable carbamic acid (p. 917). Proton transfers and decarboxylation conclude the process.

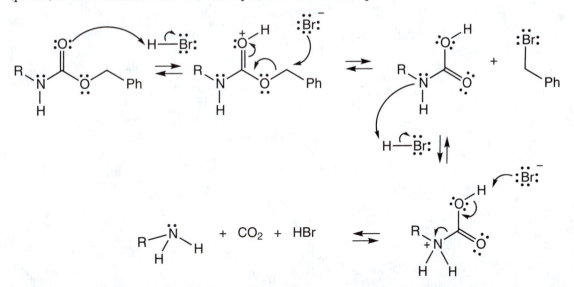

Problem 18.53 The hint tells you how to start; the nucleophile must be the oxygen of DMF. The DMF and oxalyl chloride react first in an addition–elimination process to give **A**, a very strong Lewis acid. Addition of chloride, followed by another elimination, gives the Vilsmeier reagent:

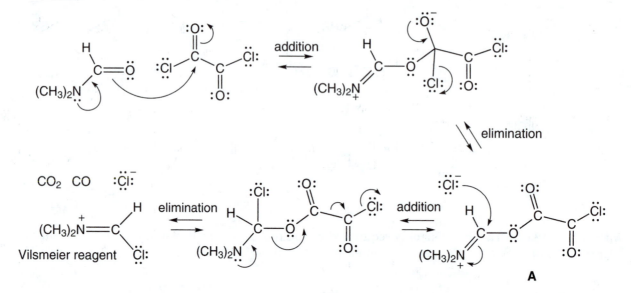

Problem 18.54 Reaction of ethyl butanoate (butyrate) and DIBAL-H initially affords addition product **A**. Hydrolysis at −70 °C ultimately gives butanal (butyraldehyde). The success of this aldehyde synthesis is undoubtedly the result of the stability of the addition product **A** at −70 °C.

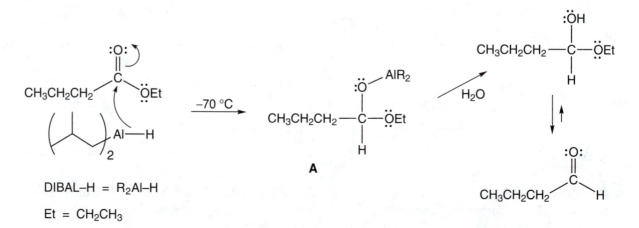

However, when the temperature of the reaction mixture is allowed to rise, compound **A** decomposes to butanal and ethoxydiisobutylaluminum (EtOAlR$_2$).

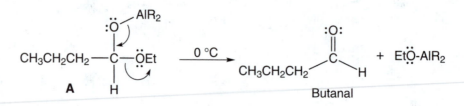

The liberated butanal can now react with compound **A** to give intermediate **B** that can decompose to ethyl butanoate (starting material) and butoxydiisobutylaluminum (**C**) that is hydrolyzed to butyl alcohol.

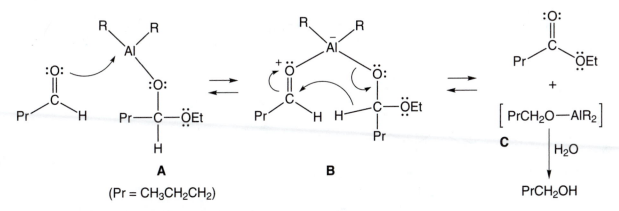

(Pr = CH₃CH₂CH₂)

Problem 18.55 In the first step, the dienoic acid is deprotonated by the amine base. Reaction with ethyl chloroformate through an addition–elimination process gives **A**.

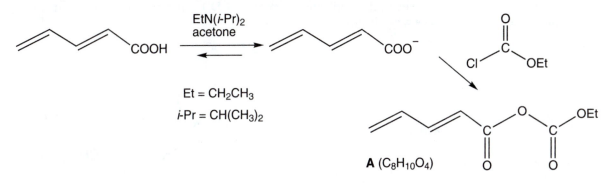

$Et = CH_2CH_3$

$i\text{-Pr} = CH(CH_3)_2$

A ($C_8H_{10}O_4$)

In Section 18.13b, we saw that acyl azides are available from the reaction of acid chlorides and azide ion. Here a similar reaction occurs, with **A** playing the part of an acid chloride. Once again, the mechanism is addition–elimination.

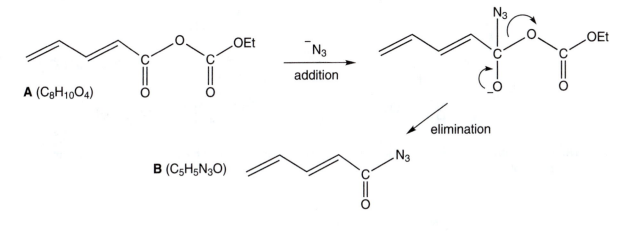

(continued)

Problem 18.55 *(continued)*

A Curtius rearrangement of the acyl azide **B** leads to an isocyanate that is captured by the solvent *tert*-butyl alcohol to produce carbamate **C**.

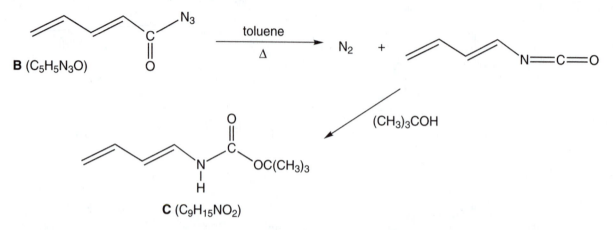

Carbamate **C** then undergoes a Diels–Alder reaction with methyl β-nitroacrylate to give cyclo-adduct **D**. You should be able to rationalize the stereochemistry of **D** through principles first encountered in Chapter 13. However, the regiochemistry (**D** not **D′**) is not obvious from what we know. The structure of **G** shows what it must be.

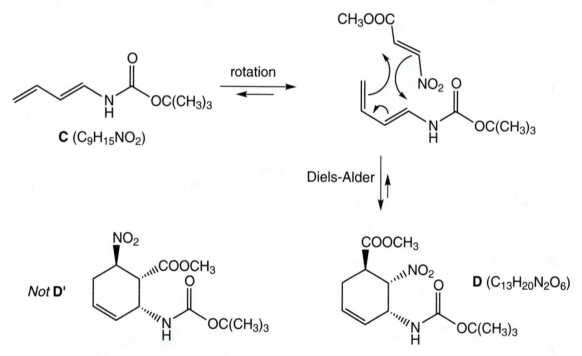

The transformation of **D** to **E** involves loss of HNO_2. There must be a base-induced elimination reaction, probably E1cB.

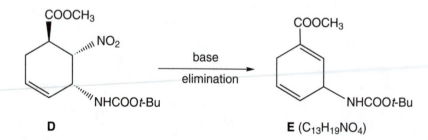

Two hydrolysis steps follow. The ester group of compound **E** is hydrolyzed in base to the carboxylic acid salt that is protonated in the second step to acid **F**. A final reaction in HCl/H$_2$O leads to the product **G**.

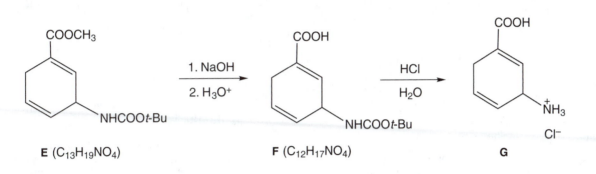

E (C$_{13}$H$_{19}$NO$_4$)   **F** (C$_{12}$H$_{17}$NO$_4$)   **G**

Problem 18.56 From the mass spectrum of **2**, we see a molecular ion of m/z = 137. The ^{13}C NMR spectrum of **2** indicates a minimum of seven different carbons, and the ^{1}H NMR spectrum suggests a minimum of seven hydrogens. The formula C$_7$H$_7$ has a mass of 91 g/mol. So, we need to account for the missing mass of 46 g/mol. The ^{1}H NMR spectrum of **2** shows four aromatic hydrogens between δ 6.5 and 7.7 ppm. The coupling pattern suggests the possibility of an unsymmetrically 1,2-disubstituted benzene in which H$_3$ and H$_6$ appear as doublets and H$_4$ and H$_5$ appear as rough triplets. Remember that no meta or para couplings were observed.

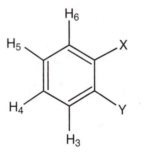

Now, what are the substituents X and Y? The IR spectrum of **2** helps in this regard. First, note the broad band at 3300–2400 cm^{-1} and the intense band at 1665 cm^{-1}. These two bands are consistent with the O—H and C=O stretches of a carboxylic acid. The singlet at δ 169.5 ppm in the ^{13}C NMR spectrum of **2** is also consistent with the carbonyl carbon of a carboxylic acid. If we subtract the mass of two oxygen atoms (2 × 16 = 32) from the missing mass of 46, we still need to account for a mass of 14, which could be the result of a nitrogen atom. Once again, the IR spectrum of **2** helps to confirm this conjecture. The two bands at 3490 and 3380 cm^{-1} are consonant with the N—H stretching doublet expected for a primary amine. Thus, the spectral data suggest that compound **2** is an aminobenzoic acid—an anthranilic acid. The ^{1}H NMR chemical shifts of the aromatic hydrogens are summarized on the next page. Because of rapid exchange reactions, the carboxylic acid hydrogen and the two amine hydrogens appear collectively as a broad signal centered at δ 8.60 ppm. These hydrogens shift as they exchange rapidly in the presence of D$_2$O.

(continued)

Problem 18.56 *(continued)*

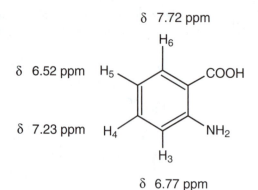

δ 7.72 ppm

δ 6.52 ppm

δ 7.23 ppm

δ 6.77 ppm

The conversion of phthalimide (**1**) into anthranilic acid (**2**) is just a Hofmann rearrangement with the added complication that **1** is an imide rather than an amide. Bromination of nitrogen probably occurs first to give *N*-bromophthalimide (**A**), which is followed by opening of the imide ring. The Hofmann rearrangement then occurs to give isocyanate **B**, which is trapped by hydroxide to give the carbamic acid salt **C**. Salt **C** is not stable and decarboxylates to give the anion **D**. Finally, adding acetic acid into the reaction results in protonation of **D** to give the benzoic acid **2**.

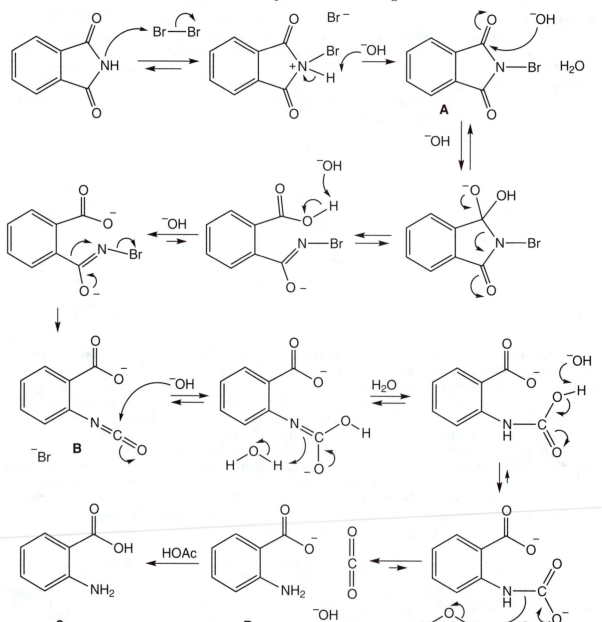

Problem 18.57 There is an obvious similarity between an acid chloride and a sulfonyl chloride. We would expect the reactivity to be similar. An alcohol adds to the carbonyl carbon of an acid chloride. In the same manner we might expect the alcohol to add to the sulfonyl group.

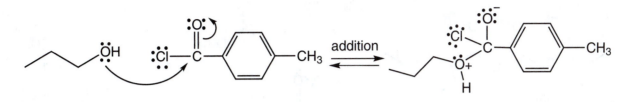

After the addition, the basic solvent (pyridine) will deprotonate the oxygen, which will be followed by elimination of the chloride in the last step. Pyridine can help the reaction along by lowering the transition state to the first intermediate as it starts to pull off the alcohol hydrogen.

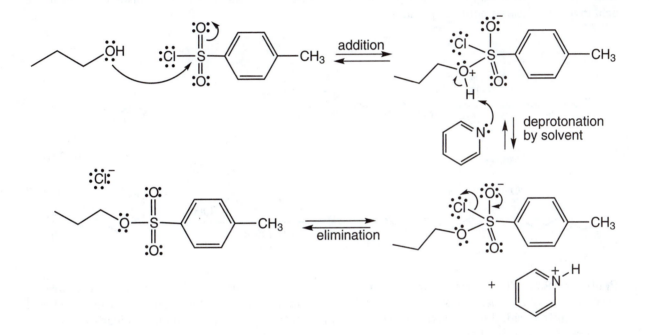

There is another mechanism that we should consider. There is evidence (*Chem. Soc. Rev.*, **1989**, 123) that suggests the alcohol reacts with TsCl in an S_N2 fashion. How might you test for this pathway?

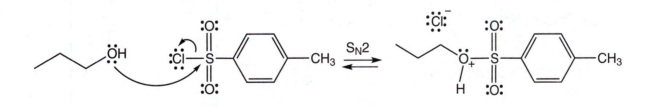

Problem 18.58   The retrosynthetic analysis is similar to others we've done. Anhydrides come from the reaction of the acid chloride and a carboxylic acid. The carboxylic acid can come from a primary alcohol, and the primary alcohol can come from an alkene.

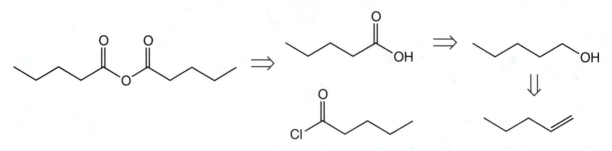

The forward synthesis can now be written easily. The first step is a hydroboration–oxidation reaction of 1-pentene to give 1-pentanol. Chromium oxidation of the primary alcohol gives pentanoic acid, which can be converted into pentanoyl chloride. Reaction between pentanoyl chloride and pentanoic acid gives the desired pentanoic acid anhydride.

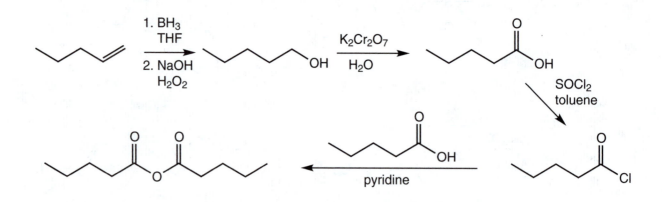

Problem 18.59   The hints tell you how to start: Form the acylium ion. Protonation of the hydroxyl oxygen, followed by loss of water, does the trick. Of course the hydroxyl oxygen is not the more basic position in the acid, the carbonyl O is. However, protonation of the carbonyl oxygen leads nowhere—it is a mechanistic dead end.

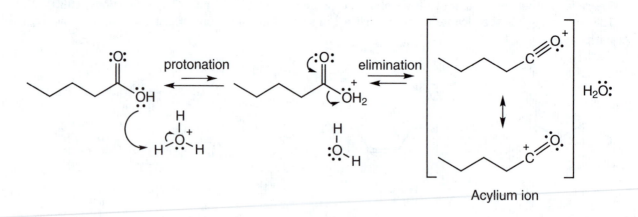

Addition of azide ion to the acylium ion makes acyl azide, an intermediate capable of rearrangement to an isocyanate. As in the related Hofmann rearrangement, the isocyanate is unstable under aqueous conditions and is rapidly hydrolyzed to a carbamic acid that decarboxylates to give the amine. Mechanisms are not shown for the last two steps (see p. 917).

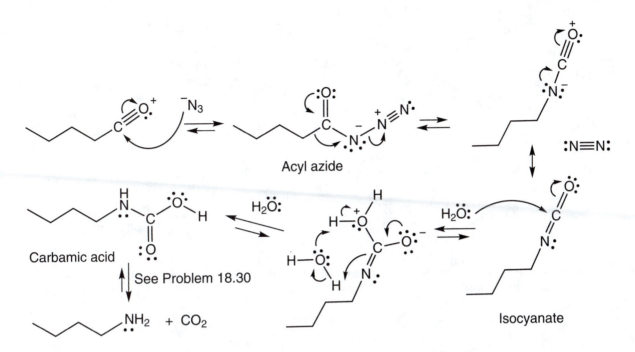

Acyl azide

Carbamic acid

See Problem 18.30

Isocyanate

Problem 18.60 This problem is a straightforward Cope elimination (p. 358, Problems 8.18 and 18.41) preceded by a sequence of reactions designed to make the starting *N*-oxide. Here are the structures.

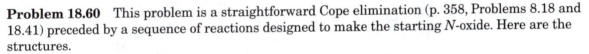

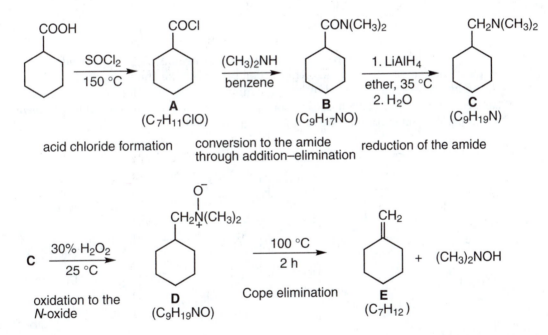

Problem 18.61 A quick check of the molecular formula of compound **3** reveals that it is a 1:1 adduct of **1** and **2**. We are most likely dealing with a Diels–Alder reaction here. However, 1,3-diphen-ylisobenzofuran possesses two different diene systems. Which one will react with the dienophile **2**? Reaction of **2** and the furan diene would lead to **A**, whereas reaction with the cyclohexadiene would give **B**. In the formation of **A**, a benzene ring is generated, but in **B** it is not. We might well expect the path to **A** to be preferred. In addition, **A** has the proper carbon framework to give **4**, but **B** does not.

(continued)

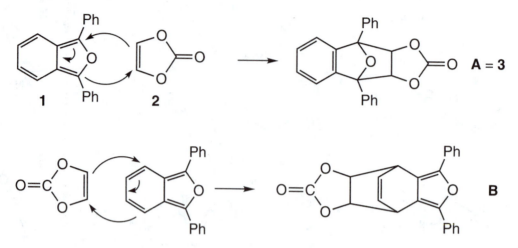

In fact, compound **A** is formed, mainly in its endo form, although small amounts of the exo compound are also isolated.

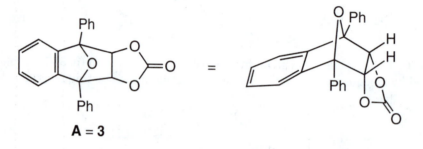

The transformation of cycloadduct **3** to naphthalene **4** involves aromatization and hydrolysis of the carbonate. These reactions can occur in either order, but let's take the aromatization first. Protonation of the bridge oxygen, followed by ring opening, gives carbocation **C**. Deprotonation gives one of the required carbon–carbon double bonds. Now, protonation of the alcohol, followed by dehydration, completes formation of the aromatic compound **D**.

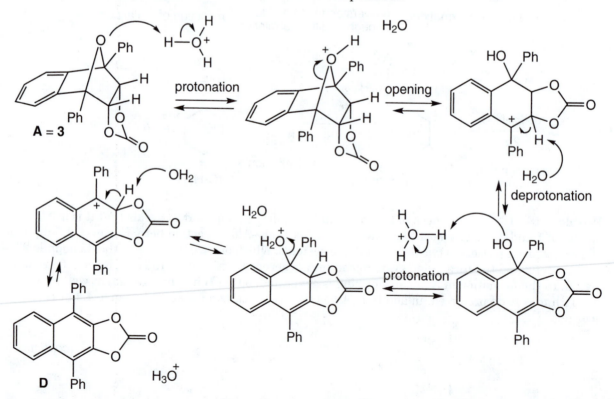

Now hydrolysis of the carbonate, a dazzling sequence of protonations, deprotonations, and additions of water, generates the final product, **4**.

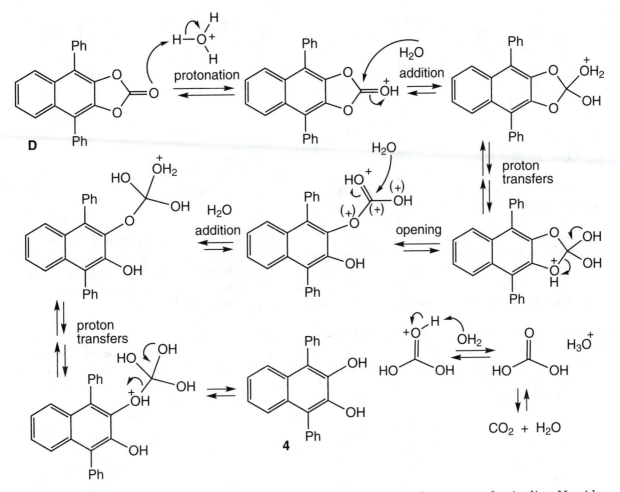

Problem 18.62 The reaction begins with addition of the nucleophilic oxygen of quinoline *N*-oxide to benzoyl chloride. The elimination phase of this addition–elimination process expels chloride and leads to **A**.

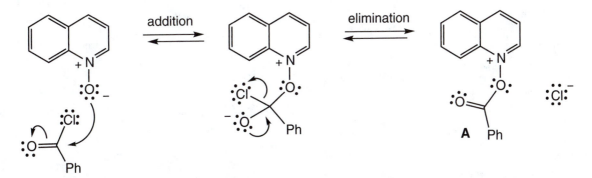

What has been accomplished? It is an old story. An impossibly bad leaving group has been transformed into a good one. In addition, a good Lewis acid lies waiting for cyanide, a good nucleophile. Why does cyanide add to the 4-position and not to the 2-position? That's hard to answer; perhaps steric inhibition by the benzoyl group is to blame. In any event, the structure of the product tells you that it is the 4-position that is the more reactive one. Addition of cyanide and elimination of the good leaving group complete the reaction.

(continued)

Problem 18.62 *(continued)*

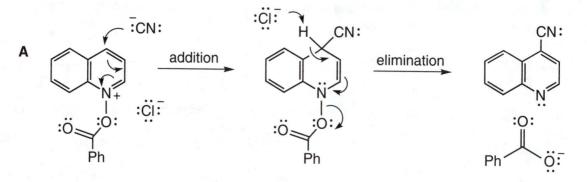

Problem 18.63 The ester hydrolysis, or saponification, is not base catalyzed because the hydroxide group (the base) is consumed during the reaction. It does require a stoichiometric amount of base. That demand means there is a 1:1 ratio between the moles of ester and the moles of hydroxide needed to do the complete hydrolysis.

Problem 18.64 The LUMO of the ester is concentrated mostly on the carbonyl carbon. This position is where we would expect to find the most electrophilic site because it is the most positive atom in the ester. The LUMO density is also slightly on the carbonyl oxygen. The most interesting site of LUMO density is on the two hydrogens on the alpha carbon of the ester. These two hydrogens must be acidic. We will learn more about their acidity in the next chapter.

Problem 18.65 The animation shifts to the second LUMO (the one perpendicular to the C—N axis) because the water orbital mixes with this orbital in order to move to the next intermediate. Yes, this means that there are other options for the first intermediate. In order for the reaction to move forward, it must be a water molecule that attacks the nitrile carbon. But there are other nonproductive, reversible steps that could occur.

Carbonyl Chemistry 2: Reactions at the α Position

19

In this chapter, we reach a point we have been approaching for some time. We now know so much of the grammar and vocabulary of the discipline, and have done so many simple problems (in this analogy, "written sentences"), that we can now go on to do quite complicated problems, that is, to write some of those "paragraphs" we have been warning you about. Now we have a full arsenal at our command: a great deal of structure, most basic reactions, and an ability to determine structure through spectroscopy. We can now deal with problems that resemble the questions chemists deal with in the real world of chemistry. It will be very important to analyze each hard problem. These exercises become much, much easier when you have a sense of where you are going, of what bonds you are trying to make or break. That idea sounds simple, but it is amazing how few people really start problems with that simple question to oneself: What happens in this reaction?

There are still drill problems (some are review) in this chapter, and their solutions should come relatively simply. But don't worry if the hard problems do not come so easily; they are meant to tax you, to demand some hard work and careful thought. Some hard problems will be dealt with best over time. If a problem resists solution, and some will, come back to it after a while; let your subconscious work on it for a time. Most research chemists carry unsolved problems around in their heads, sometimes for years, returning to them now and then. There is nothing wrong with emulating that process, at least with hard, real-world problems. People think at vastly different rates, and it is a rare moment indeed that requires a *rapid* solution of a problem (unfortunately, hour exams tend to be such moments).

Why put such "hard stuff" in an introductory book? That's a good question and deserves a thoughtful answer. Is it possible to do a reasonable job in a course in organic chemistry without having a great deal of success with problems at the level of difficulty of some of the problems in this chapter? You bet. If you can do the bread and butter exercises well, that certainly constitutes a respectable performance. But you should at least be exposed to more complicated material. The fun in this business comes in solving problems, and for some of you it will be the pleasure of working out complex problems that leads you on in chemistry. In this chapter, we also lapse further into schematic drawings. Full Lewis "dot" structures are not always drawn, for instance. Be careful to fill in electron dots where necessary.

Problem 19.3 The following drawing shows exchange through formation of the anion and reprotonation using D_2O to give the deuterioaldehyde. However, the anion formed by removal of the aldehydic hydrogen is *not* resonance stabilized and cannot be formed easily. If the anion is not formed, exchange cannot take place.

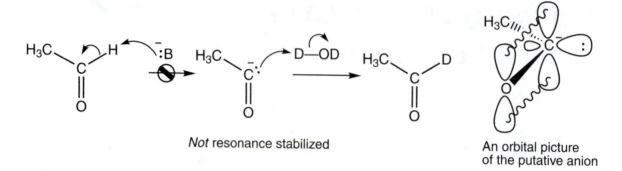

Not resonance stabilized

An orbital picture
of the putative anion

Problem 19.4 This problem sets up an apparent contradiction: α hydrogens exchange in base—but this particular α hydrogen doesn't. Why? The way to start on this question is to remind yourself why α hydrogens exchange—the formation of resonance-stabilized enolates. For example, the two other α hydrogens exchange in the usual way.

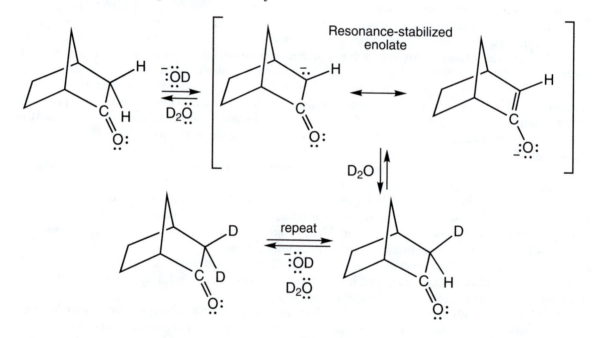

Why might the enolate formed through removal of the bridgehead α hydrogen not be resonance stabilized? It seems on paper that a similar mechanism can be written for exchange of the bridgehead hydrogen, but this notion is an illusion.

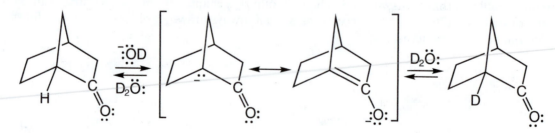

In reality, there is no resonance stabilization. Although it is easy to draw the delocalization on paper, there really is no resonance. The problem is that the orbital at the bridgehead does not overlap with the orbitals making up the π system of the carbon–oxygen double bond. The two-dimensional drawings on the paper are lying to you: There is no resonance stabilization because the non-overlapping orbitals do not allow the electrons to be delocalized. If the anion is not resonance stabilized, it cannot be formed. If the anion is never formed, there can be no exchange.

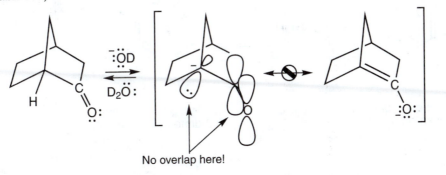

No overlap here!

Problem 19.7 The carbonyl compound appears to be somewhat more stable than the enol form. In doing this ΔH calculation, we are making many (unwarranted) assumptions including, for example, the notion that all carbon–hydrogen and all carbon–carbon bonds are equal in energy. We find the enolization reaction to be somewhat endothermic in the general case.

Ketone		Enol	
Bond	Energy (kcal/mol)	Bond	Energy (kcal/mol)
C=O	178.8	C=C	174.1
2 C—C	202.8	C—C	101.4
6 C—H	606.6	C—O	92.1
	988.2	H—O	104.6
		5 C—H	505.5
			977.7

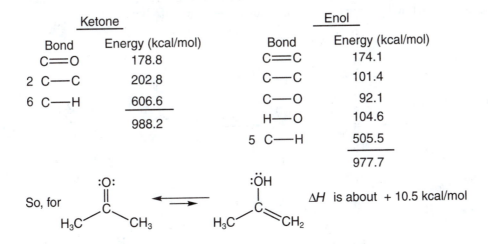

So, for   ΔH is about + 10.5 kcal/mol

Problem 19.8 In each case, a resonance-stabilized intermediate is formed, but the amidate is favored because the charge is partially borne by a relatively electronegative nitrogen rather than a carbon.

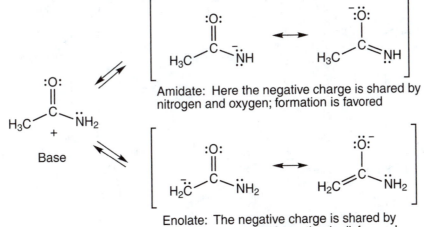

Amidate: Here the negative charge is shared by nitrogen and oxygen; formation is favored

Enolate: The negative charge is shared by carbon and oxygen; formation is disfavored

Problem 19.9

(a) The predicted product for this reaction is the conjugate base of 2,2-dimethylpropanoic acid. This problem involves the iodoform reaction and it converts a methyl ketone into the conjugate base of the corresponding carboxylic acid and iodoform.

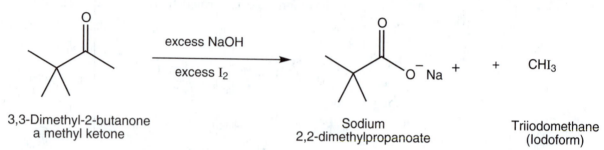

3,3-Dimethyl-2-butanone
a methyl ketone

Sodium
2,2-dimethylpropanoate

Triiodomethane
(Iodoform)

The mechanism for the iodoform reaction is shown here. The methyl group α to the carbonyl has slightly acidic hydrogens that will be deprotonated by hydroxide. The I_2 is an electrophile that will successively replace the hydrogens on the methyl group. Once the α carbon has three iodides, it can be a leaving group. The $^-CI_3$ is a sufficient leaving group because the mildly electronegative iodides stabilize the anion via induction. Loss of $^-CI_3$ from the tetrahedral intermediate generates the carboxylic acid, which will provide a proton to form the conjugate base of the carboxylic acid and iodoform.

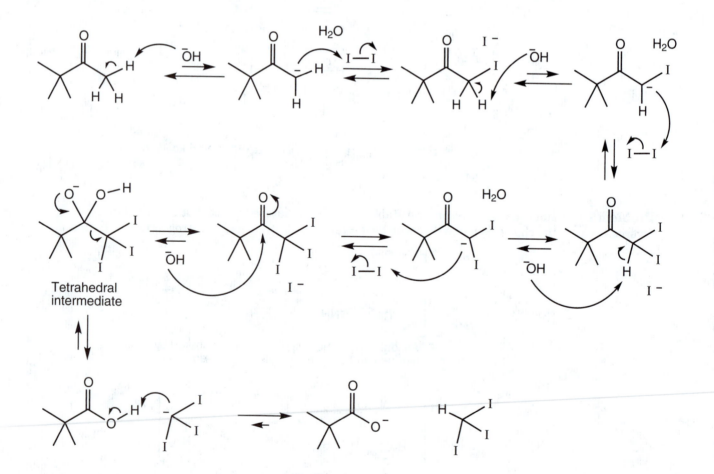

Tetrahedral
intermediate

(b) Although these conditions are the same as for the iodoform reaction, the starting material is not a methyl ketone. In this case, the two α hydrogens are replaced by iodide. It must be that the $^-CI_2CH_3$ is not a good enough leaving group to be kicked out from the tetrahedral intermediate shown in the mechanism of the reaction.

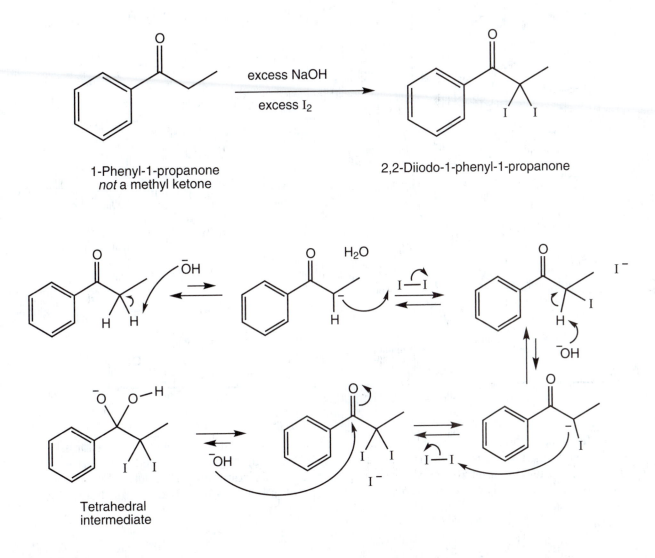

Problem 19.10 This problem recalls a similar situation involving the S_N1 reaction. The rate-determining step of the S_N1 reaction is the slow ionization to give a carbocation. The faster product-determining steps in which nucleophiles capture the carbocation follow. In each of the three seemingly different reactions of this problem, the rate-determining step is the same, enolate formation. Only after the endothermic formation of the enolate do the subsequent, product-determining steps take place.

(continued)

Problem 19.10 *(continued)*

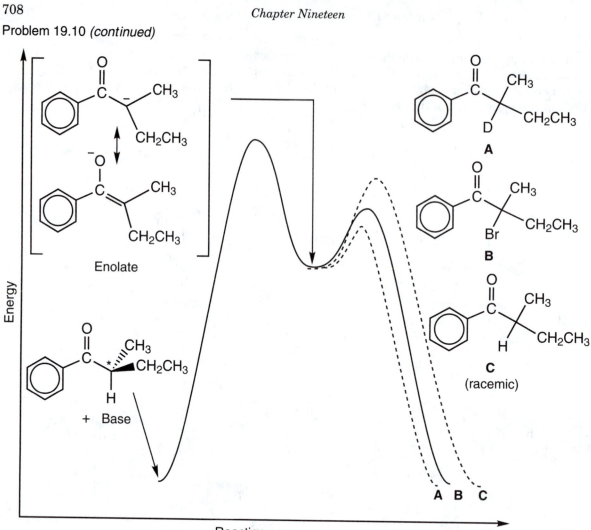

Enolate

+ Base

A

B

C
(racemic)

Energy

Reaction progress

Problem 19.11   More addition–elimination reactions! The same mechanism will suffice for all three reactions. The nucleophile first adds to the carbonyl group to give a tetrahedral intermediate, and the good leaving group bromide (or chloride in the specific example) is then lost. The reaction is unlikely to be seriously complicated by S_N2 displacement of the α-bromide because alcohol is a poor nucleophile.

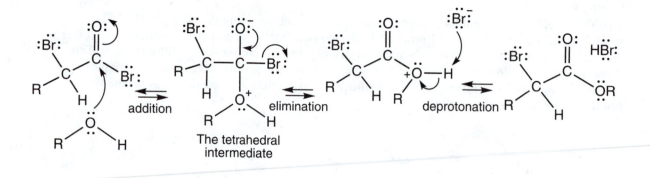

addition   elimination   deprotonation

The tetrahedral
intermediate

Problem 19.12 Lithium diisopropylamide (LDA) is a poor nucleophile because of its steric bulk and because the nitrogen anion is a better base than nucleophile (p. 294). The nitrogen anion is better able to bond with the 1s orbital of an acidic hydrogen than with the π antibond of the carbonyl. A similar nonnucleophilic base is potassium *tert*-butoxide.

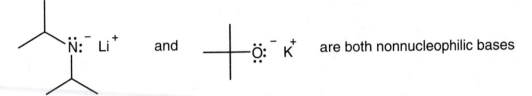

Problem 19.13 The most important resonance form will be the one with both negative charges on the relatively electronegative oxygen atoms. However, the HOMO will have more electron density on carbon, and carbon will be the more nucleophilic site in alkylation reactions.

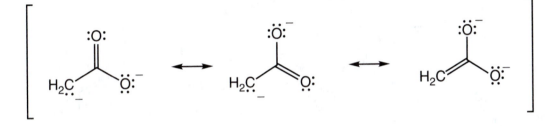

Problem 19.14 In each case, formation of the dianion is followed by alkylation at carbon. A final protonation completes the reaction. In acid, the γ-hydroxy acid would probably form the related lactone. Can you write a mechanism?

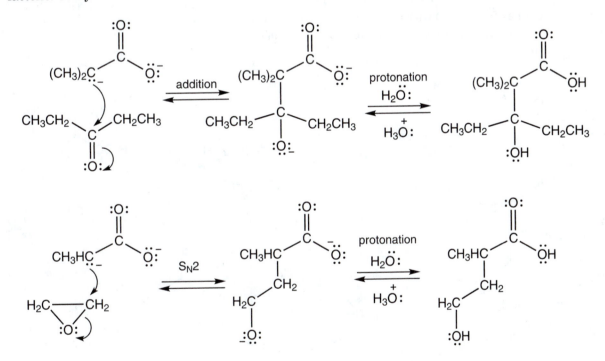

Problem 19.15

(a) Another easy one. Malononitrile is acidic (Table 19.3) because the anion is so well stabilized. Formation of the anion followed by S_N2 displacement on ethyl iodide leads to the once-alkylated product. Repetition can give the dialkylated compound.

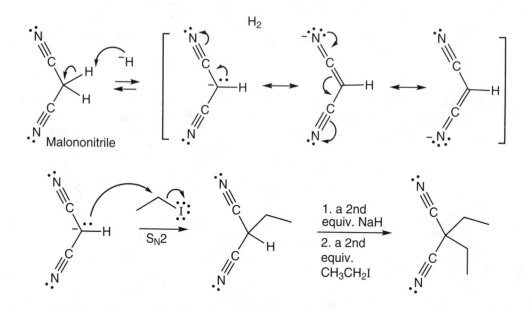

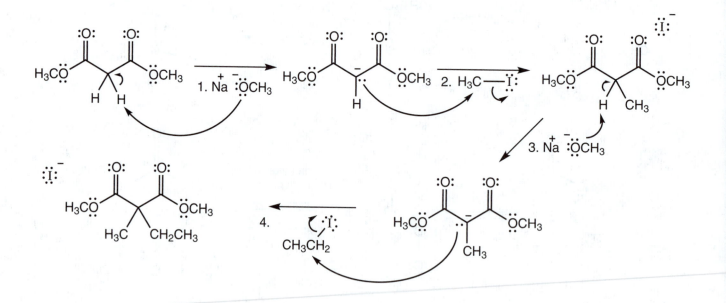

(b) Dimethyl malonate can be alkylated twice. Deprotonation using one equivalent of base gives the stabilized anion, which can react with methyl iodide to give the dimethyl methylmalonate. Deprotonation of the second acidic proton gives the second anion that can be alkylated with ethyl iodide. Two ideas to note: (1) we could have switched the order of alkylation (ethyl iodide first and methyl iodide second); (2) we cannot make the dianion of dimethyl malonate.

Problem 19.16

(a) Ethyl 3-oxobutanoate can be alkylated twice, first with methyl iodide, then with ethyl iodide. Hydrolysis and decarboxylation give the desired product.

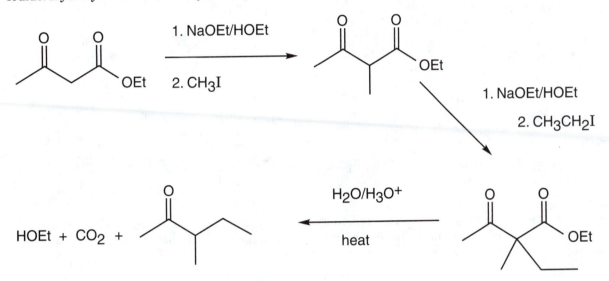

(c) Hydrolysis and decarboxylation of the β-keto ester shown will do the job.

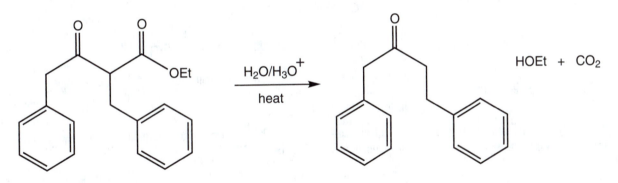

Problem 19.17

(b) Two alkylations are followed by hydrolysis and decarboxylation.

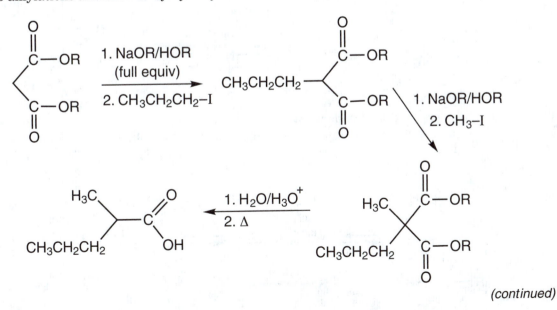

(continued)

Problem 19.17 *(continued)*

(c) This part is very similar to part (b), except that a final esterification of the acid must be included.

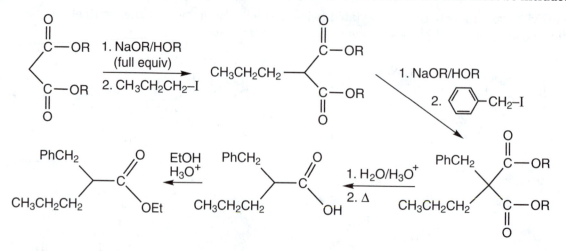

Problem 19.18   In the acid-catalyzed reaction of carbonyl compounds with amines, the first steps produce a protonated carbinolamine, **A**. As long as the nitrogen has at least one attached hydrogen, deprotonation to a carbinolamine can occur.

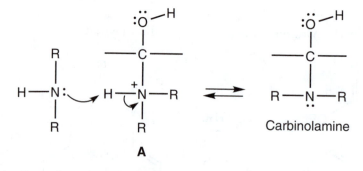

When the amine is tertiary, there is no hydrogen that can be removed from the nitrogen atom of **A**. The only possible reaction is reversal to starting material.

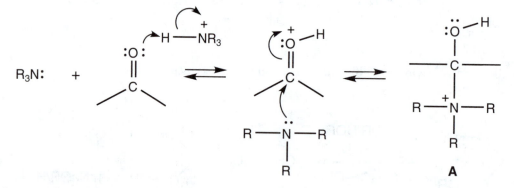

When the amine is primary or secondary, proton loss from **A** *can* occur, and the carbinolamine can be formed. If the amine is primary, protonation of the oxygen can be followed by elimination of water to give an imine.

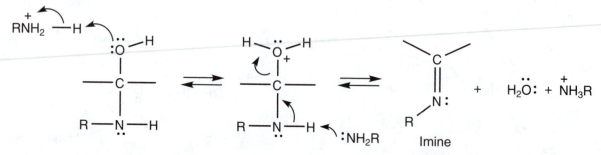

If the amine is secondary, no imine is possible. Instead, water loss leads to an iminium ion. If there is an available hydrogen on carbon, it can be lost to give the enamine.

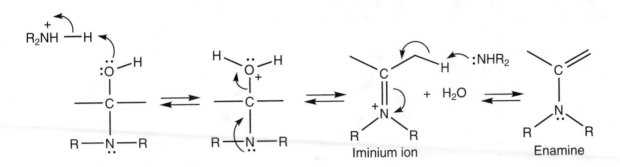

Only secondary amines can give enamines because only secondary amines can give carbinolamines (tertiary amines cannot) but are blocked from imine formation. Primary amines can give imines or enamines, but the imines are usually favored.

Problem 19.19 The reaction between a secondary amine and a ketone (or aldehyde) produces an enamine. Compound **A** is an enamine [both (*E*) and (*Z*) isomers would be formed]. Enamines are sufficiently nucleophilic to react with alkyl halides in an S_N2 fashion. In this case, enamine **A** gives the iminium intermediate **B**, which will react with water. Hydrolysis of an iminium ion occurs by water attacking the iminium carbon and proton transfers to give the tetrahedral intermediate with the protonated nitrogen leaving group. Elimination of the secondary amine regenerates the carbonyl double bond, and deprotonation in the last step produces the alkylated ketone.

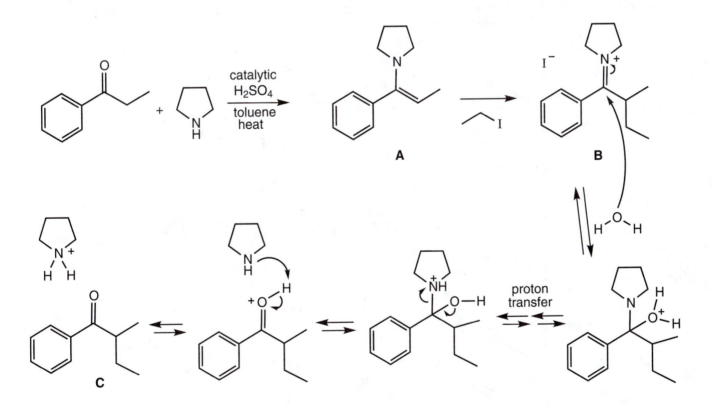

Problem 19.20   The protonated alcohol is approximately 10^8 less acidic (higher pK_a) than the protonated aldehyde. The protonated aldehyde is a much stronger acid. The electrons on the alcohol are in approximately sp^3 orbitals and are held less strongly than the electrons on the aldehyde oxygen, which are in approximately sp^2 orbitals.

lone pairs $\sim sp^2$ lone pairs $\sim sp^3$

So, the OH protonates more easily to give the weaker acid.

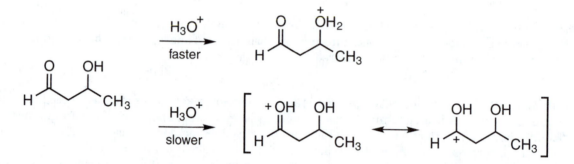

However, despite the relative ease of forming the protonated alcohol, it is often the protonated carbonyl that leads on to product, as you will see in Problem 19.23.

Problem 19.21   In each case, the β-hydroxyketone or β-hydroxyaldehyde can dehydrate to give the α,β-unsaturated carbonyl product.

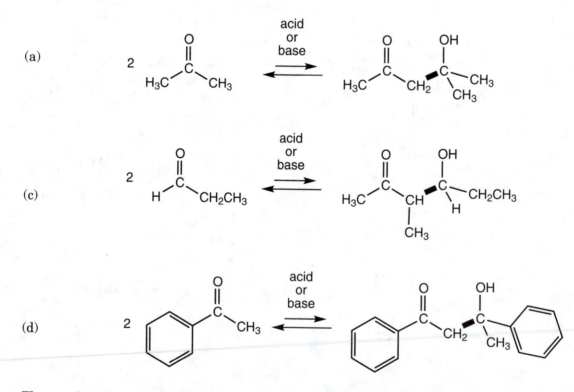

The new bond between the two carbonyl compounds is shown as a bold line. The condensations are completed by elimination of water. Practice seeing these reactions in reverse. Ask yourself what compounds could make these β-hydroxy ketones and aldehydes.

Problem 19.22   This problem is simple. All the products are α,β-unsaturated carbonyl compounds formed by loss of water from the β-hydroxy carbonyl compounds.

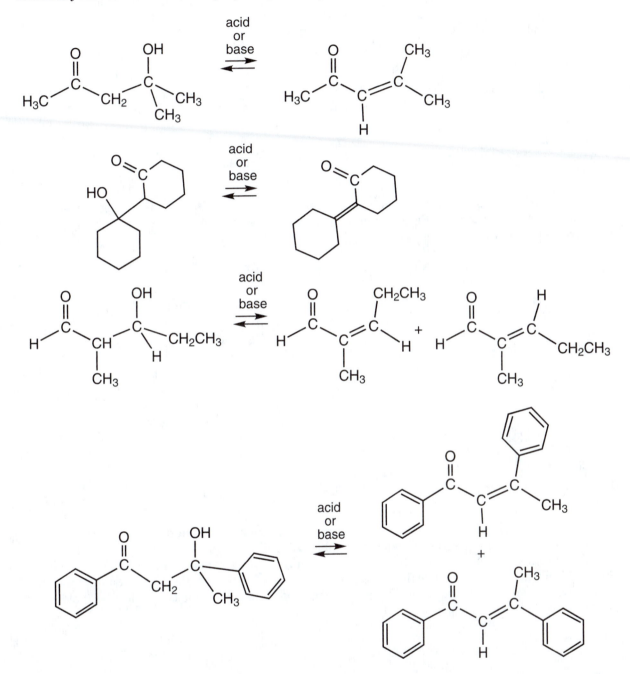

Now ask yourself a harder question: What carbonyl compounds could lead to these β-hydroxy carbonyl compounds?

Problem 19.23   More drill here. As we have often said, it is important to know reactions backward and forward—to be able to write mechanisms in both directions. These mechanisms are exactly the reverse of the "forward" processes. In acid, the first step is protonation of the carbonyl oxygen. This molecule decomposes to an enol and protonated carbonyl compound. Equilibration leads to two molecules of acetone and regenerates the catalyst for the reaction, H₃O⁺.

(continued)

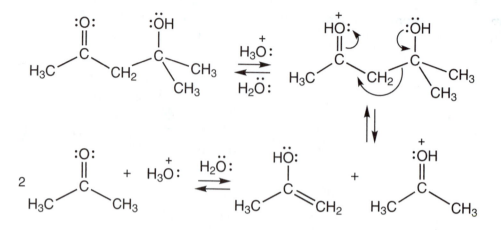

In base, the first step is formation of the alkoxide ion. Now decomposition leads to a molecule of acetone and one of the enolate. Protonation by water gives a second molecule of acetone and regenerates the catalyst, hydroxide ion.

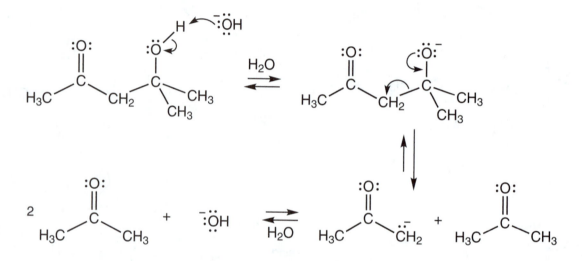

Problem 19.24   This problem provides practice in seeing backward from products to potential starting materials. First just deconstruct the double bond of the α,β-unsaturated carbonyl compound to give the β-hydroxy ketone from which it must have been formed. Then, further deconstruct the β-hydroxy ketone into the two molecules [or two pieces of one molecule in part (c)] from which it must have been constructed by breaking the critical bond attaching the two molecules (boldface bond).

(a)

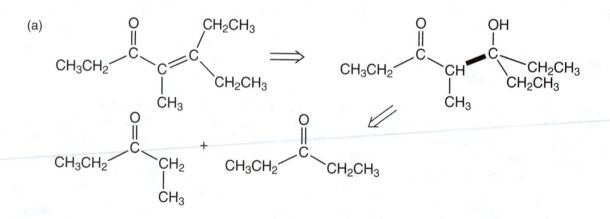

(c)

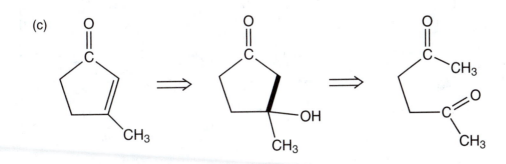

Problem 19.25

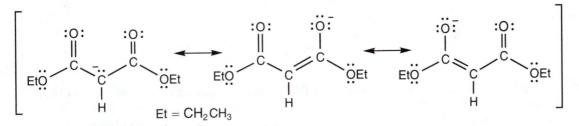

Et = CH₂CH₃

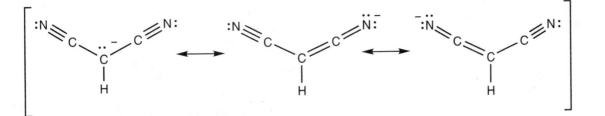

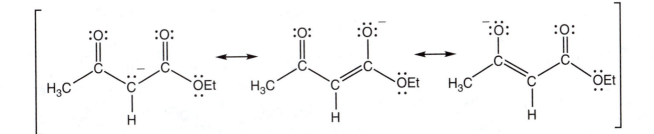

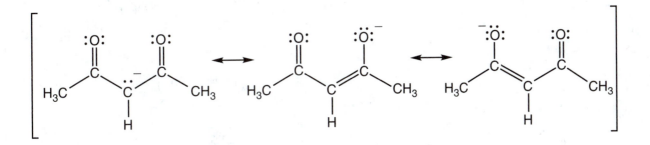

(continued)

Chapter Nineteen

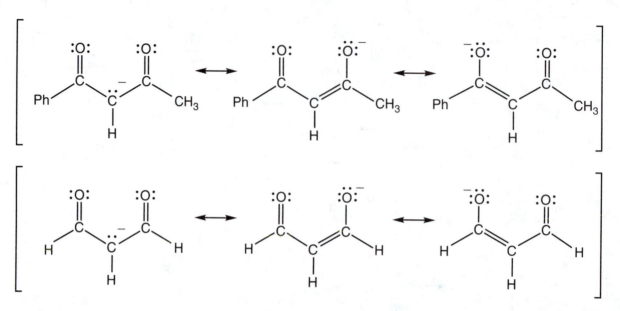

Problem 19.26 Retrosynthetic analysis leads to the following ideas. In each case, various basic catalysts could be used. In particular, NaOEt/HOEt should suffice in each reaction. Dehydration probably will take place under the conditions of the condensation reaction or on subsequent acidification (Et = CH_2CH_3).

(a)

(b)

(c)

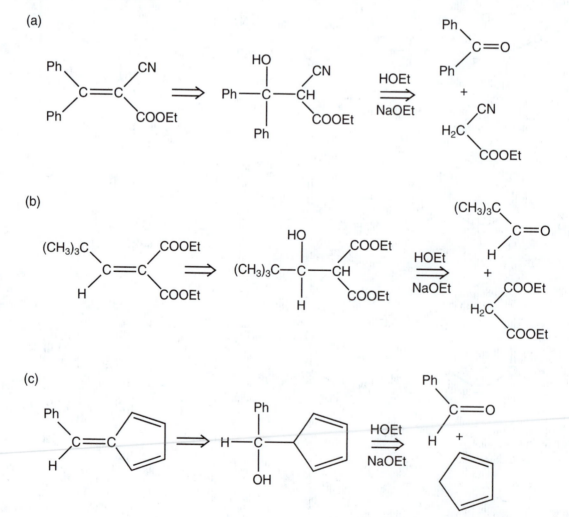

Problem 19.27 In acid, methyl phenyl ketone (acetophenone) will be in equilibrium with its enol form, and the other ketone will be protonated, thus forming a strong Lewis acid. Addition of the modest nucleophile, the enol, to the strong Lewis acid leads to intermediate **A**, in a standard, acid-catalyzed Michael reaction. Subsequently, **A** is deprotonated and ketonized to give the product. If the formation of the ketone from the enol is either surprising or difficult, *please* go over the mechanism.

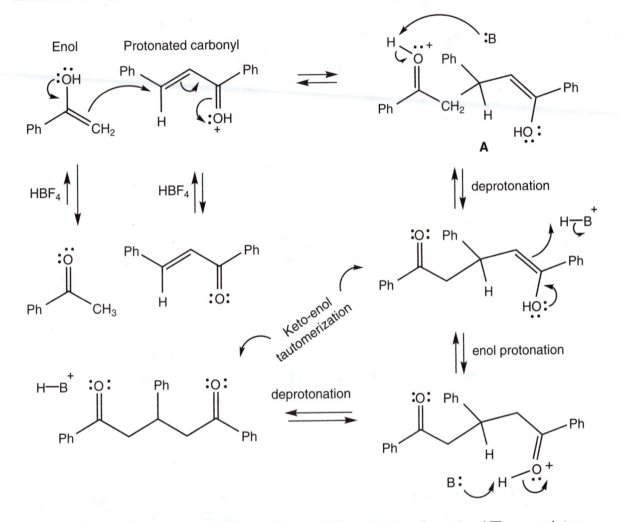

Problem 19.28 This problem involves still more drill on aldol condensations! These are intra-molecular reactions and are a little harder to do than their intermolecular versions. Nothing fundamental has changed, however. In base, the first step is formation of an enolate.

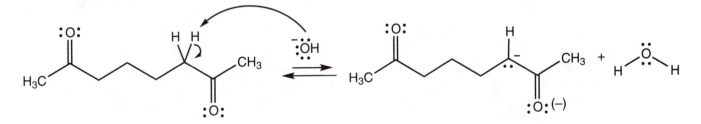

Now the enolate could add to another molecule of the starting dione in a standard, base-catalyzed aldol condensation. But, a Lewis acid carbonyl compound lurks within the same molecule, well within reach of the anion. Addition leads to a β-hydroxy ketone that can dehydrate to product.

(continued)

Problem 19.28 *(continued)*

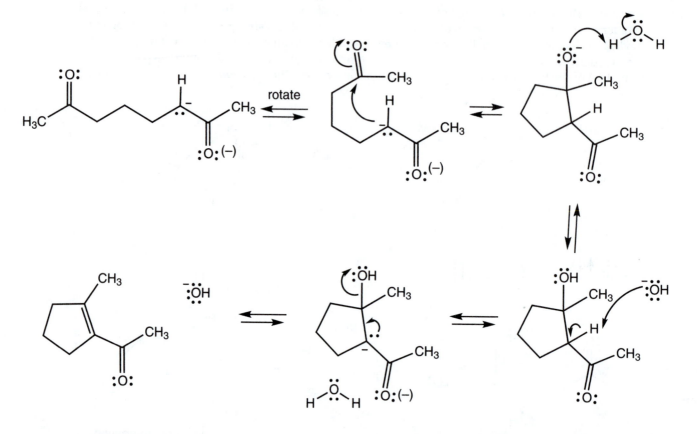

In acid, the first steps are formation of one enol and protonation of one of the carbonyl groups.

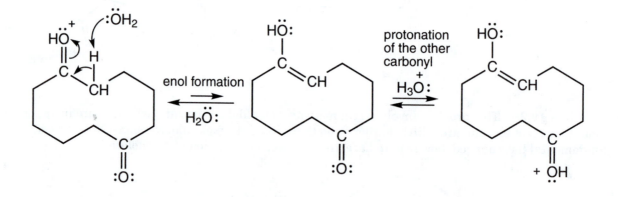

Now the enol, a modest nucleophile, adds to the protonated carbonyl group, a strong Lewis acid. As in the base-catalyzed reaction, intramolecular reaction is more favorable than intermolecular reaction. A sequence of deprotonation, protonation, and elimination steps completes the reaction and regenerates the catalyst, H_3O^+.

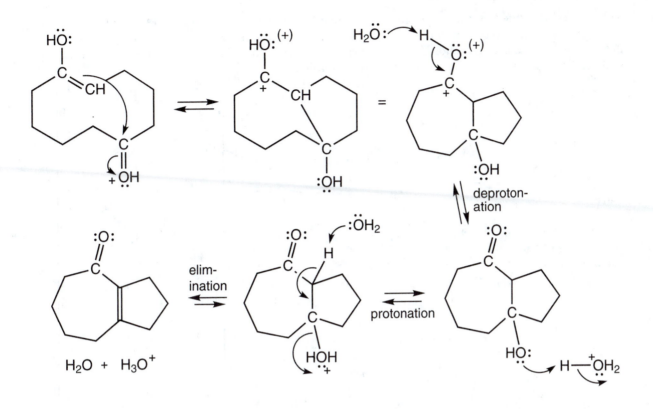

Problem 19.29 More drill, this time on the reverse intramolecular aldol condensation. The mechanistic steps of Problem 19.28 are repeated in the reverse order, starting at the β-hydroxy ketone stage. This problem is very similar to Problem 19.23, except that this time the reaction is intramolecular. In acid, protonation of the carbonyl is followed by the reverse aldol step in which the ring is opened. Deprotonation leads to an enol that regenerates the diketone. If formation of the ketone from the enol in acid is the least bit obscure, by all means write out the detailed mechanism.

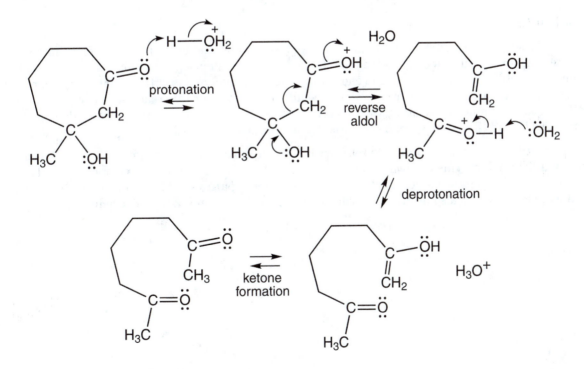

(continued)

Problem 19.29 *(continued)*

In base, the alkoxide is formed and the ring opened to give an enolate. The enolate is protonated at carbon to give the diketone and regenerate hydroxide ion catalyst. Protonation may well occur faster at oxygen to give the enol. The enol, of course, will equilibrate with the diketone, and the diketone will be very strongly favored thermodynamically.

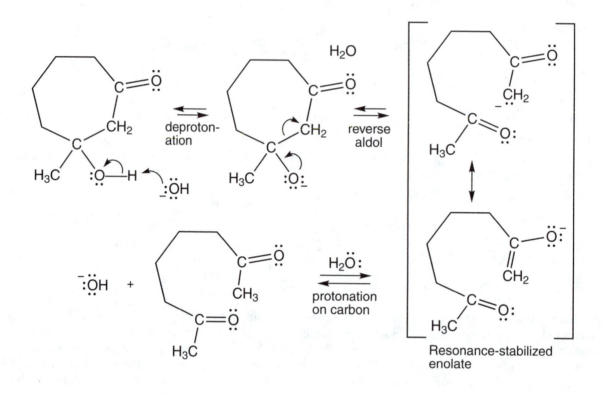

Resonance-stabilized enolate

Problem 19.30

(a) The 1-phenyl-1-propanone (propiophenone) can be deprotonated by hydroxide to form an enolate. It is a reversible acid–base reaction. Benzaldehyde has no acidic hydrogens, so it can only be an electrophile in this reaction. The enolate of propiophenone reacts with benzaldehdye to make the crossed aldol product. Dehydration would be difficult to stop for this reaction because the dehydrated product is much more stable as a result of extensive conjugation. There are two dehydrated products [(E) and (Z) isomers] from the crossed aldol product.

The propiophenone enolate could also react with another molecule of propiophenone to form the normal aldol product. This pathway would not be as fast as the reaction between the enolate and benzaldehyde. The crossed aldol reaction will be the favored pathway. If the propiophenone aldol product *is* formed, it is also prone to dehydrate and form the (E) and (Z) enone products.

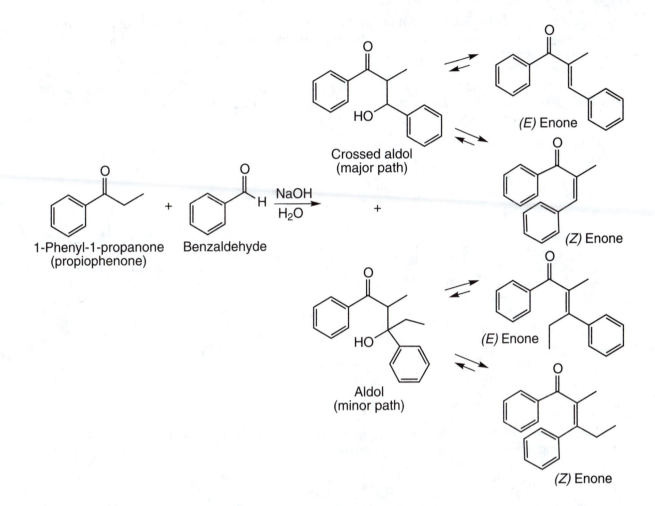

(b) The 4-heptanone can form an enolate and react with benzaldehyde. We will also obtain some aldol product from the 4-heptanone reacting with itself. The crossed aldol product is likely to dehydrate to give the (*E*) and (*Z*) enones.

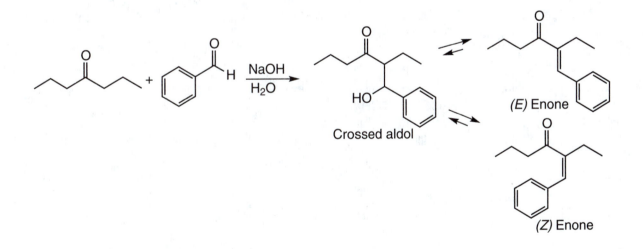

(continued)

Problem 19.30 *(continued)*

(c) Both of the aldehydes, butanal and propanal, can form enolates in the presence of hydroxide. There are two possible crossed aldol products. Crossed aldol product **A** comes from the butanal enolate reacting with propanal. Crossed aldol product **B** comes from the propanal enolate reacting with butanal. We will also observe aldol products **A** (butanal enolate reacting with butanal) and **B** (propanal enolate reacting with propanal). With heat, these β-hydroxy aldehydes will dehydrate to give the α,β-unsaturated aldehydes. As no heat is indicated, and as there is no extended conjugation in the system, we predict the products shown.

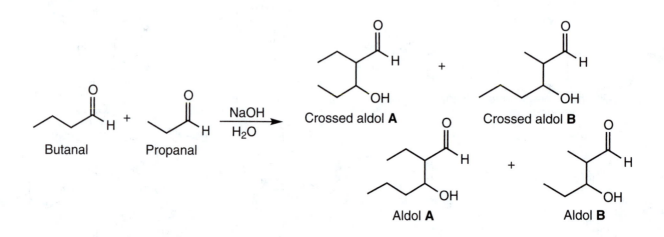

Problem 19.31 Review again. Esters are more strongly stabilized by resonance than are aldehydes and ketones, and thus are less reactive.

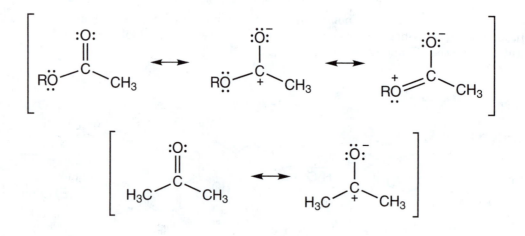

Problem 19.32   This reaction is just like every other Claisen condensation. Remember that the thermodynamically unfavored β-keto ester is formed at the end in the absence of base and cannot revert to the more stable starting material—a pair of esters.

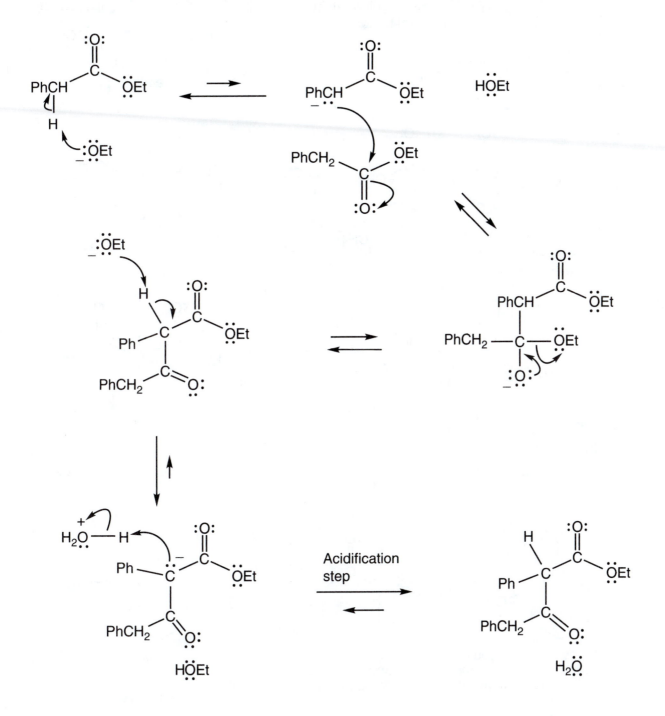

Problem 19.34   Here is a detailed mechanism for one product.

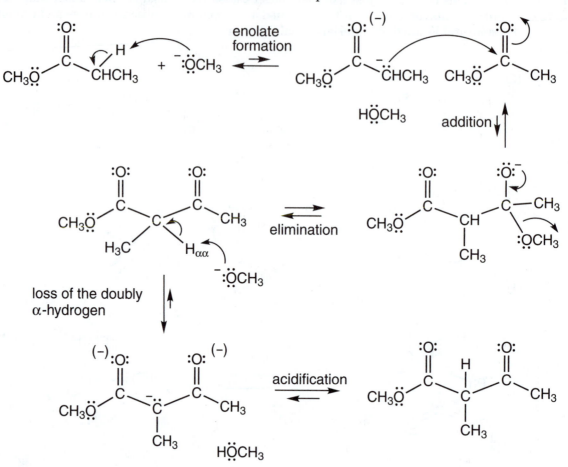

loss of the doubly
α-hydrogen

Problem 19.35

(a) Acid hydrolysis generates a β-keto acid that, like almost all other β-keto acids, is prone to decarboxylation.

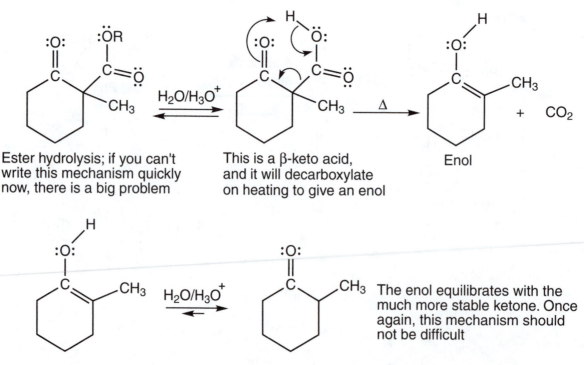

Ester hydrolysis; if you can't
write this mechanism quickly
now, there is a big problem

This is a β-keto acid,
and it will decarboxylate
on heating to give an enol

Enol

The enol equilibrates with the
much more stable ketone. Once
again, this mechanism should
not be difficult

In base, a reverse Claisen takes place—note the lack of doubly α hydrogens in the starting material.

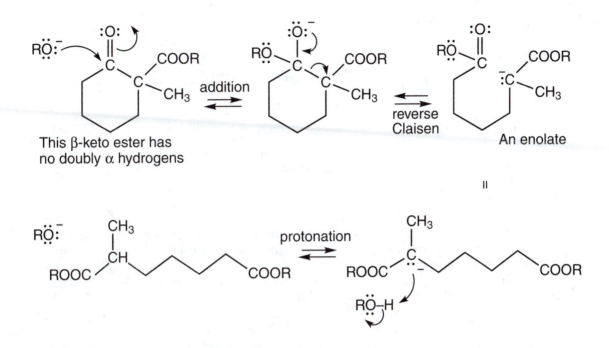

This β-keto ester has
no doubly α hydrogens

An enolate

(b) Here we are faced with a common problem: how to choose among several possible modes of reaction. At this point it's hard to make this choice, and you may just have to explore as many possibilities as you can. However, we can always try to use the Wisdom of the Ages, obey Magid's Second Rule (Chapter 19, p. 997), and try the Michael reaction first, using the enolate formed from dimethyl malonate.

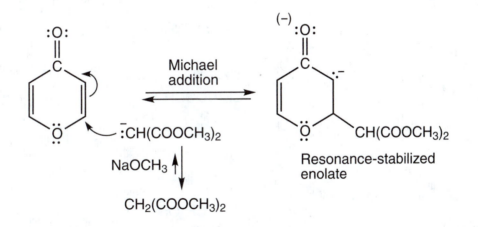

The addition generates an enolate, so the Michael addition to the carbon–carbon double bond is a reasonable reaction. Now what's possible? The reaction can always reverse, but that path only leads back to the starting molecules. Perhaps an examination of the first arrow leading along the reversal path will give you an idea. It is also possible, as the carbon–carbon double bond reforms, to open the ring through generation of a new enolate.

(continued)

Problem 19.35 *(continued)*

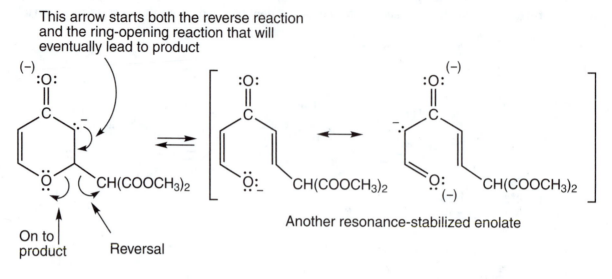

This arrow starts both the reverse reaction and the ring-opening reaction that will eventually lead to product

(−)

On to product

Reversal

Another resonance-stabilized enolate

One of the hardest things that you must learn is to analyze the problem before you start to push arrows and electrons around. In this case, the critical thing to see is that the original ring *must* open. There must be a way to get the ether oxygen out of the ring, and ring opening is the only possible way. As the product contains an all-carbon ring, there must also be a way to close the ring up again. Problem 19.35(b) is an "open–close" problem, one of *many*.

Protonation of the enolate at carbon gives an aldehyde. Notice that the ring opening has generated a triply α hydrogen that can easily be removed in base to give a new enolate. This enolate sits poised, ready to add to the new aldehyde. So the ring opening has generated the two species necessary to close the all-carbon ring, the nucleophilic enolate, and the Lewis acid aldehyde carbonyl. Protonation of the alkoxide formed through addition gives an alcohol.

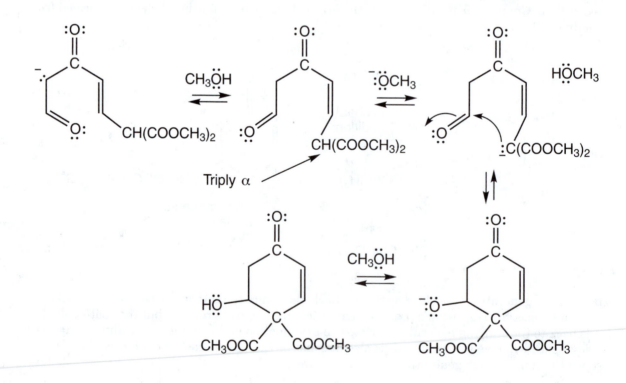

Triply α

Finally, an elimination reaction generates the product diene. It is true that hydroxide is a poor leaving group, but the anion, an enolate, is easy to form, and this E1cB reaction can take place.

This α-hydrogen can be removed in base to give the enolate

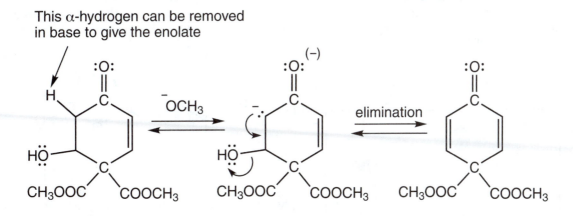

In the second part of this problem, an ester group is removed and an aromatic compound is formed. One can already see *why* this reaction happens; it must be driven by the great stabilization of the aromatic product. Again, the tough part is seeing how to start. An analysis of the problem tells us that an ester is lost, so that should focus attention on reaction with an ester group. Addition of alkoxide to the ester gives a tetrahedral intermediate (**A**). This species can either revert to regenerate the starting ester (hardly a productive pathway) or lose dimethyl carbonate to generate the aromatic system. This route surely seems a good way to try. A simple protonation completes the reaction.

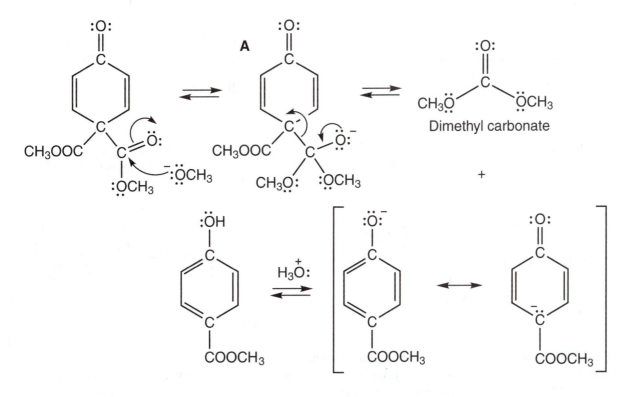

(continued)

Problem 19.35 *(continued)*

(c) This problem is easy but it *looks* hard because the starting material and products have such different structures. That can make a problem difficult, especially at this stage. We'll try to solve this problem by exploring all the possible reactions of the starting material in the hopes that one will reveal an obvious path to the product. There are two different α positions that can lead to two enolates, and two different carbonyl groups that could participate in addition reactions.

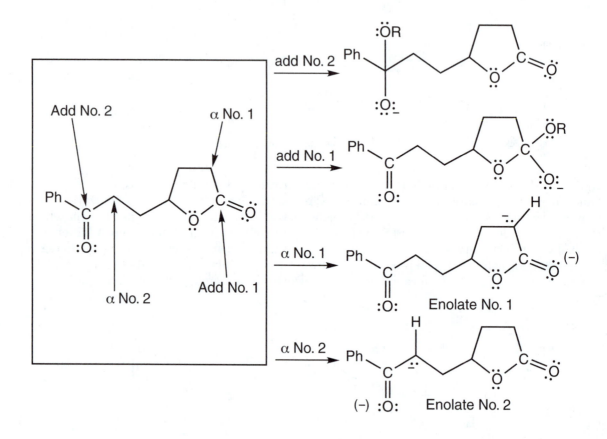

Enolate No. 2 can open directly to the product by an intramolecular S_N2 reaction! The carboxylate anion acts as leaving group. So here's a good solution: two simple steps, enolate formation and S_N2 ring opening.

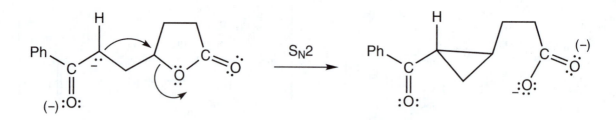

Problem 19.36 This beautiful problem is tantalizing because it involves serious misdirection. It has an "obvious" answer that is wrong. By far the hardest part of this very hard problem is resisting the temptation to take the short, wrong route. The reaction is called the Stobbe condensation.

First of all, it is a member of the class of ester plus ketone reactions we discussed in Section 19.8a. As benzaldehyde has no α hydrogens, the answer must start with removal of the α hydrogen of the succinic ester.

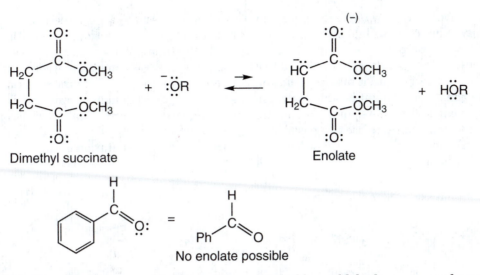

Dimethyl succinate   Enolate

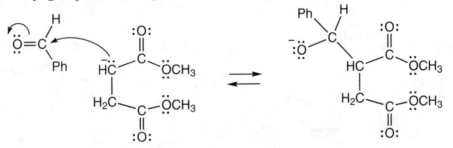

No enolate possible

The end product clearly is a combination of the ester and benzaldehyde, so we need waste no time with self-condensation of the ester but turn our attention immediately to a reaction between the two partners. The new bond is simple to find, and easily made through attack by the enolate on the aldehyde carbonyl group. So far, the problem seems simple, and it is.

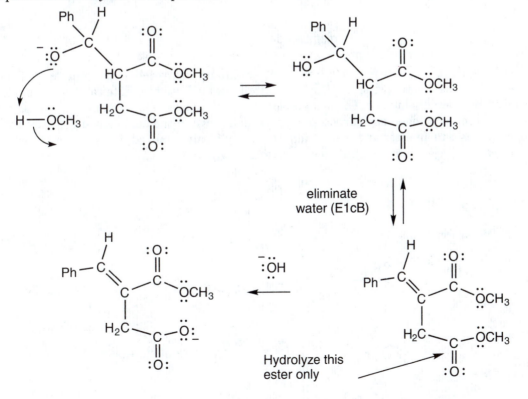

We are almost there, or so it seems. Now temptation enters. How easy it is to protonate the newly formed alkoxide, lose water in an elimination, and then hydrolyze one ester to give the disarmingly simple monocarboxylate in the product.

eliminate
water (E1cB)

Hydrolyze this
ester only

(continued)

Problem 19.36 *(continued)*

But wait—why hydrolyze only *that* ester? Why not the other one? And how does the hydrolysis work? The only hydroxide available is that produced in the elimination reaction, and it is hard to believe that it could compete effectively in attacking one of the esters. We see no good answers to these questions. Unless you do, you are forced to conclude that there is something wrong with this answer, no matter how painful it may be to do so. It is necessary to backtrack to the alkoxide and ask what other reaction it might be able to do. There's not much else possible beyond attack of the alkoxide on one of the esters, an intramolecular, base-induced transesterification reaction. There are two esters and therefore two possible reaction paths, (a) and (b). Don't worry that (b) seems awkwardly long. That's an artifact of the two-dimensional representation of these three-dimensional molecules.

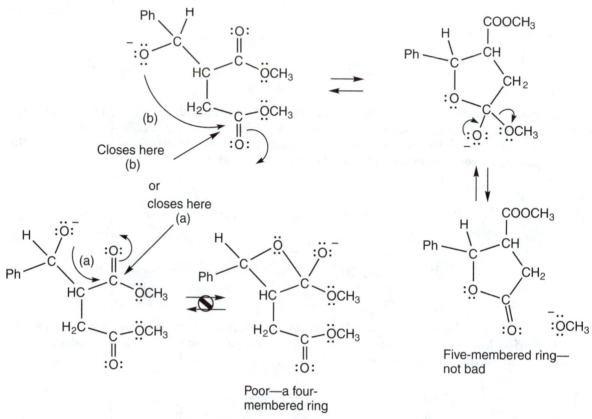

Path (a) produces a strained four-membered ring and should be disfavored, so concentrate first on the five-membered ring-forming reaction, (b). The initially formed intermediate can lose alkoxide to generate a five-membered γ-lactone. The last step of the reaction is an elimination. The breaking bonds are the carbon–hydrogen and carbon–oxygen bonds shown in the next drawing. Notice that the hydrogen lost is α to an ester and is therefore relatively easily removed. The leaving group is a rather stable carboxylate anion. The final result is exactly what we want.

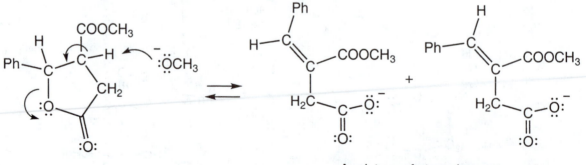

A mixture of stereoisomers

This mechanism *demands* the formation of the CH₂COO⁻ group found in the product. It is the specificity of this mechanism that gives us confidence in it. Rather than present further puzzles— Why does hydrolysis of one ester occur, but not of the other? Why any hydrolysis at all?—this mechanism requires that just the carboxylate found in the product be formed and no other. It is also comforting to know that γ-lactones can sometimes be isolated in the Stobbe condensation.

Problem 19.37 This problem insists that we do the aldol first, so there is no doubt about where to start. Notice that an acidic hydrogen has been generated after dehydration. This is the difficult part of the problem—seeing that a resonance-stabilized carbanion can be generated after the initial aldol/dehydration sequence.

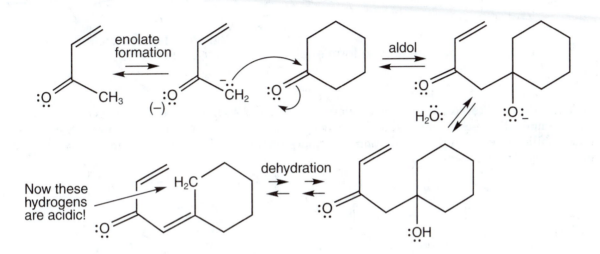

Once the new enolate is formed, a Michael reaction can take place to close the second ring. Protonation at carbon gives the product.

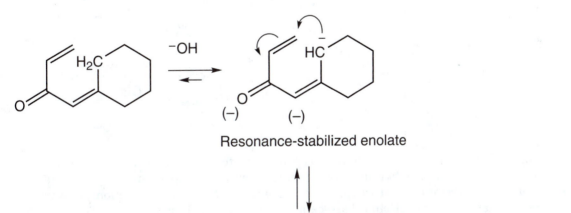

Resonance-stabilized enolate

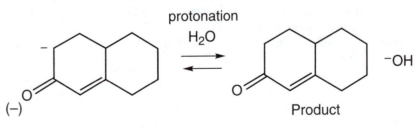

protonation
H₂O

Product

Problem 19.38

(a) The first step is a straightforward base-catalyzed enolization. If this sequence is not obvious, be certain to go back and review keto–enol equilibrations. Once the enolate is formed in base, reformation of the ketone can take place in two ways. If the proton is added from the bottom side, starting material is regenerated. If, however, the proton is added from the top side, the product shown in the problem is formed.

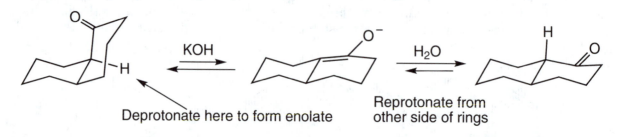

Deprotonate here to form enolate

Reprotonate from other side of rings

(b) The red-dotted carbon is part of a carbonyl group, and the green-dotted carbon is α to a carbonyl. Let's make the enolate and add to the carbonyl to achieve goal number one. Now we have a nucleophilic oxyanion in perfect position to displace bromide and achieve goal number two. With a map and goals it is an easy problem.

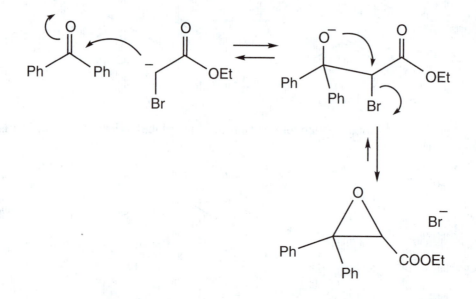

(c) This problem clearly asks you to find a way to open a ring; to break a carbon–carbon bond. We are used to seeing things in the other direction—to finding ways to make carbon–carbon bonds. Success in this problem requires that we see the aldol condensation in both directions, forward and reverse. Here we have an intramolecular base-catalyzed reverse aldol condensation. The key is to recognize the starting material as a β-hydroxy ketone. *All* β-hydroxy ketones are potentially the products of aldol condensations. In this case, the first step is formation of alkoxide **A**. The reverse aldol opens the four-membered ring and forms the resonance-stabilized enolate **B**. Addition of a proton at carbon yields the diketone product.

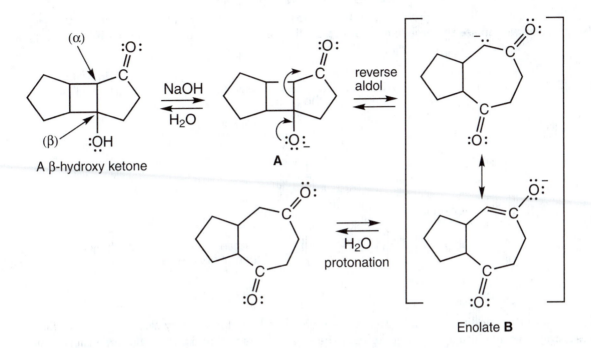

Enolate **B**

(d) The recognition point in this problem is the formation of a cyclohexenone. Cyclohexenones can be made by the combination of Michael and aldol condensations known as the Robinson annulation. In this case, two enolates could be formed: one from methyl vinyl ketone and one from cyclohexanecarboxaldehyde. We'll follow Magid's second rule and use the second enolate to do a Michael addition to methyl vinyl ketone. Protonation and a second enolate formation lead to **A**. The second step in the Robinson annulation is an aldol condensation, which in this case gives **B**. A sequence of protonation, formation of a third enolate, and elimination give the product.

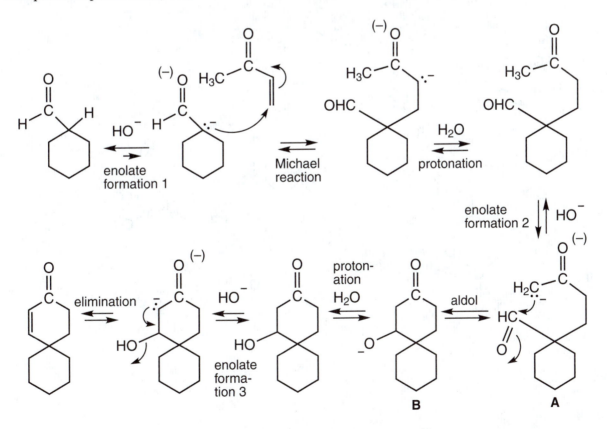

(continued)

Problem 19.38 *(continued)*

(e) Now we come to tougher problems, and analysis before you begin will become ever more important. In this case, we can see the probable final resting place of benzaldehyde within the product.

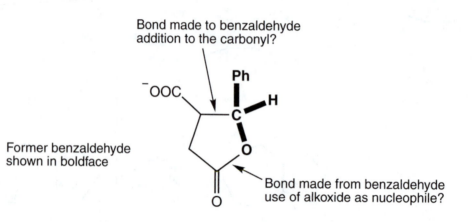

It seems that an addition reaction to benzaldehyde has been followed by incorporation of the resulting alkoxide into a new ring. At least the start of this problem is easy because there is only one possible enolate. Addition to the carbon–oxygen double bond of benzaldehyde leads to **A**, and accomplishes one of our goals, the formation of the bond to benzaldehyde. The hard part of this problem is the next step. The nucleophilic alkoxide ion adds to one carbonyl group to give **B**. How do we choose between the two different carbonyl groups? Addition to the other carbonyl would generate **C**, a strained four-membered ring, and this reaction will be slower than the one shown. Opening of **B** gives the product.

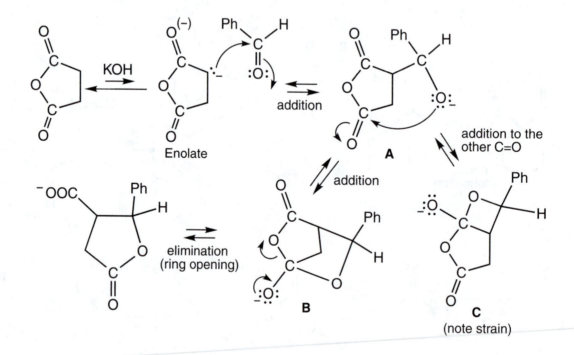

(f) This one is really hard. Many things can happen when the starting material reacts with hydroxide, but we will first try the Michael reaction (Magid's second rule) to give **A**. The hint tells us that a 10-membered ring is involved, so the ring junction must be broken. Let's do a reverse Michael reaction to give **B**. Protonation of enolate **B**, followed by ketonization of the enol gives **C**, a molecule with two "doubly α" hydrogens. Removal of one of them by hydroxide leads to **D**. In **D**, an intramolecular aldol reaction to give **E** is possible. Compound **E** also contains a "doubly α" hydrogen, and elimination of hydroxide leads to the final product.

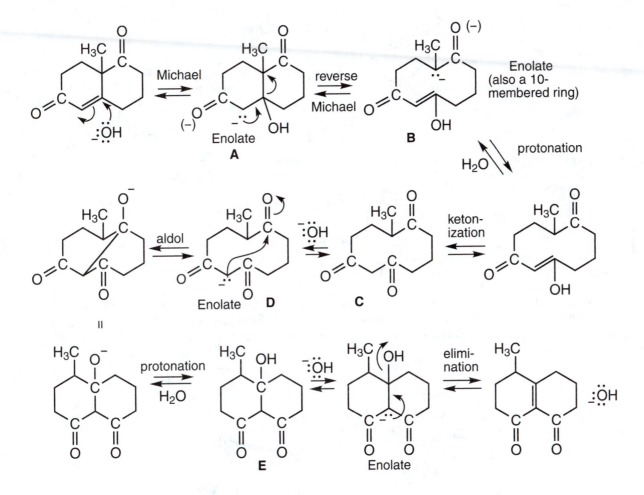

Problem 19.39 Sulfur has empty 3d orbitals that may help stabilize an adjacent negative charge.

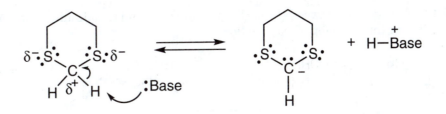

Problem 19.41   In the hydride shift step of the reaction, the carbon–hydrogen bond of one aldehyde is breaking, and the new hydrogen–carbon bond is being made. The dashed lines show the partially made and partially broken bonds.

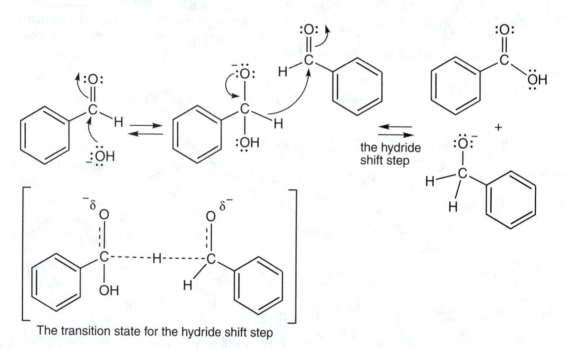

The transition state for the hydride shift step

Problem 19.42   Aluminum sits right below boron in the periodic table. Trisubstituted boron compounds should be familiar. Boron is hybridized sp^2 and there is an empty $2p$ orbital on boron. The situation for aluminum is the same, except that it is a $3p$, not a $2p$, empty orbital.

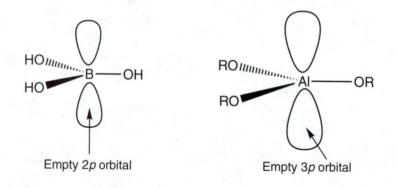

Problem 19.43   A "complex" between the carbonyl compound and the aluminum alkoxide is formed, just as it is in the first step of the reaction.

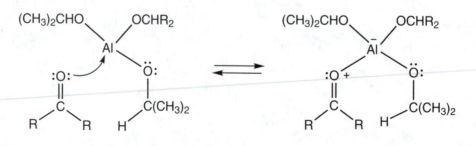

Next, the hydride migrates, reducing the old carbonyl and oxidizing one of the alkoxide groups. This transfer generates a molecule of acetone and $(R_2CHO)_2AlOCH(CH_3)_2$.

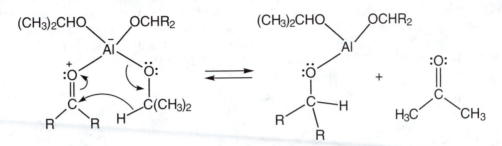

Problem 19.44 This alkoxide, $(R_2CH{-}O)_3Al$, has an empty $3p$ orbital on the aluminum (see Problem 19.42) and is a good Lewis acid. The nucleophile H_2O first adds to the Lewis acid forming a tetrahedral intermediate. This intermediate can reverse to regenerate $(R_2CH{-}O)_3Al$ and water or can lose $R_2CH{-}OH$ and produce $HOAl(OCHR_2)_2$. Repetition of this process two times eventually leads to the final products, $Al(OH)_3$ and three molecules of $R_2CH{-}OH$. We have shown the proton transfers occurring in one step using a water molecule. The process may be a deprotonation, followed by a protonation, followed by an elimination each time.

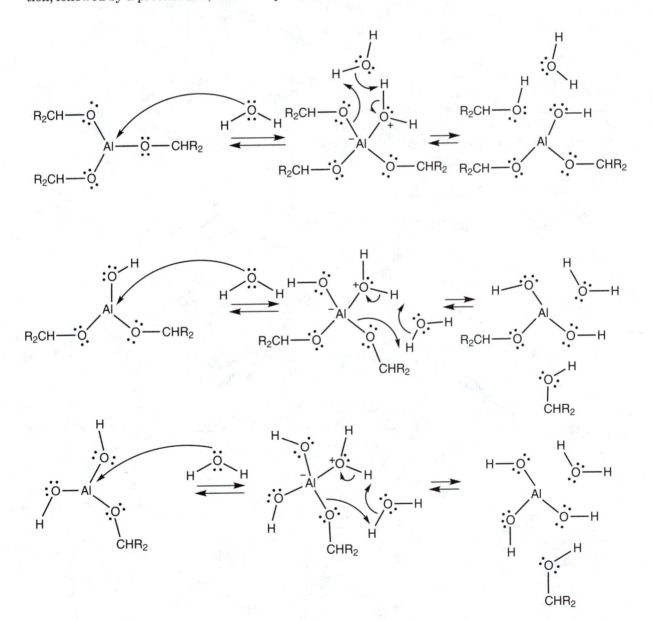

Problem 19.45

(a) Notice first the intramolecular redox reaction. One carbonyl has been reduced to the alcohol stage, and the other has been oxidized to the carboxylic acid stage. Reaction (a) is an intramolecular Cannizzaro reaction. Addition of hydroxide is to the more reactive carbonyl group of the molecule, the aldehyde, to give adduct **A**. Hydride transfer takes place within the molecule to generate **B**. Protonation of the alkoxide and deprotonation of the acid lead to the product.

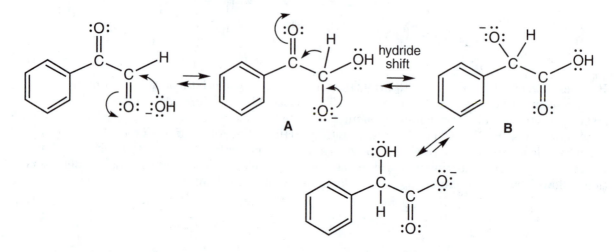

(b) The second reaction occurs in acid, and resembles the pinacol rearrangement. The structure of the product gives the clue to what happens—a way must be found to transfer the deuterium from one side of the molecule to the other. The first steps are protonation of one hydroxyl group and loss of water to generate the tertiary carbocation, **C**. This carbocation is a strong Lewis acid, and the deuteride (D$^-$) shifts to give the resonance-stabilized cation **D**. Deprotonation by water gives the product.

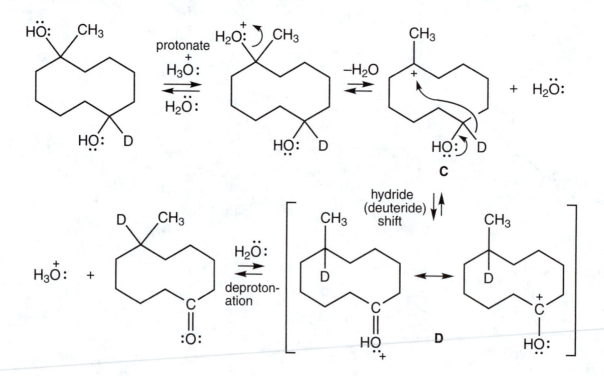

(c) This part advises the construction of a good three-dimensional drawing, so let's start with that. The drawing includes the hydrogen adjacent to the hydroxyl group, drawn in what should be a suggestive way.

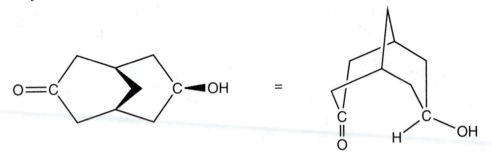

Of course, the four hydrogens adjacent to the carbonyl group will exchange in deuterated base. (If this is the least bit obscure, be certain you can draw the mechanism for this garden-variety α-exchange!) In strong base, another reaction will certainly occur, the loss of the hydroxyl proton to give the alkoxide.

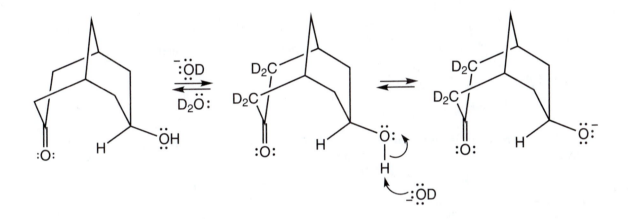

Now an intramolecular Cannizzaro reaction occurs as the hydride is transferred to the strategically placed carbonyl group on the other side of the molecule. The carbonyl group is reduced and the alkoxide is oxidized. The newly formed alkoxide is deuterated by D_2O.

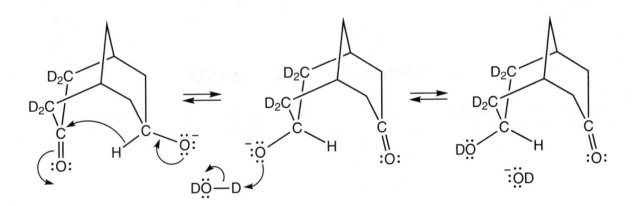

(continued)

Problem 19.45 *(continued)*

The new carbonyl activates the adjacent hydrogens that exchange to give the final molecule in which eight carbon–hydrogen bonds have been exchanged for deuterium.

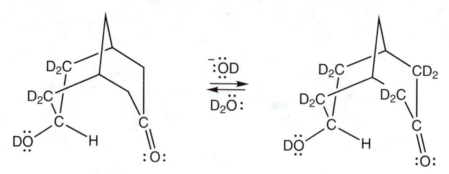

In the other stereoisomer no hydride shift is possible, as the hydrogen points to the outside of the molecule and not toward the acceptor carbon–oxygen double bond.

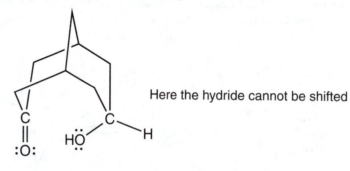

Here the hydride cannot be shifted

The only possible reaction is exchange of the four hydrogens adjacent to the carbonyl, and exchange of the hydroxyl hydrogen.

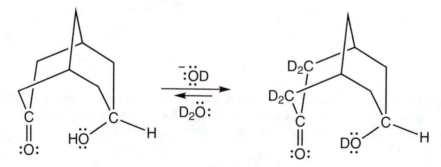

Additional Problem Answers

Problem 19.46   The enol is resonance stabilized and the carbon–oxygen single bond has double bond character. Accordingly, it is shorter than a normal carbon–oxygen bond.

Problem 19.47

(a) This problem involves a straightforward keto–enol equilibration. In base, the active ingredient is the enolate, whereas in acid it is the protonated carbonyl. Both intermediates are resonance stabilized.

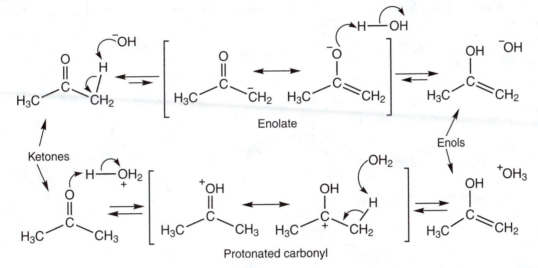

(b) This imine–enamine equilibrium is just the nitrogen counterpart of the keto–enol equilibrium in part (a).

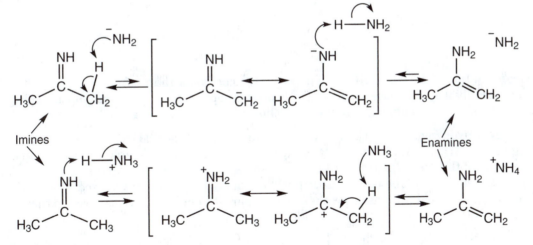

(c) Hydrate formation can be both acid and base catalyzed.

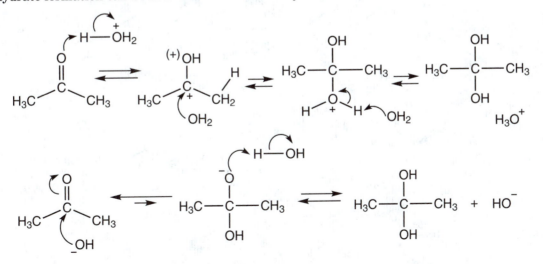

Problem 19.48

(a) Removal of the methylene hydrogen leads to an enolate that is stabilized by delocalization by both the carbonyl group and the aromatic ring. It is surely more stable than the other enolate in which only delocalization by the carbonyl group is possible. Enolate **A** also contains the more substituted double bond and hence is more stable than enolate **B**, which contains the less substituted double bond.

A, The more stable enolate

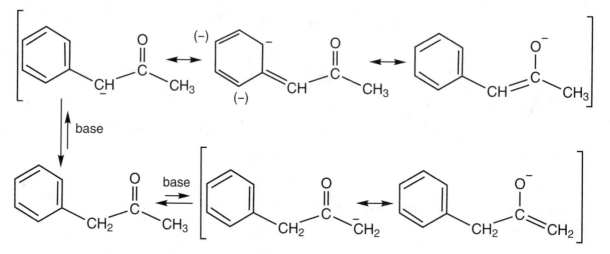

B, The less stable enolate

However, formation of the less stable enolate is likely to be faster than formation of the more stable enolate, especially if a large base such as LDA is used. Approach to the relatively open methyl group is sterically easier than is approach to the more "buried" methylene group.

(b) There are α hydrogens on C(1) and on C(3) of 3-methyl-2-butanone. Deprotonation of the α hydrogen on C(3) gives **A**, the thermodynamic enolate. It is the more stable enolate because the alkene portion of the enolate is tetrasubstituted compared to the disubstituted alkene of enolate **B**. Formation of enolate **B** will be faster, particularly if a bulky base like LDA is used. There are three α hydrogens on C(1), and it is easier for the base to approach one of them compared to the methine hydrogen on C(3).

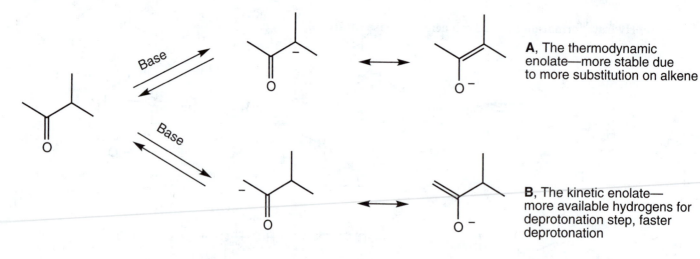

A, The thermodynamic enolate—more stable due to more substitution on alkene

B, The kinetic enolate—more available hydrogens for deprotonation step, faster deprotonation

(c) Enolate **A** is the thermodynamic enolate because extended resonance makes it more stable. Deprotonation of one of the α hydrogens on the methylene group is likely to be faster because these hydrogens are more accessible.

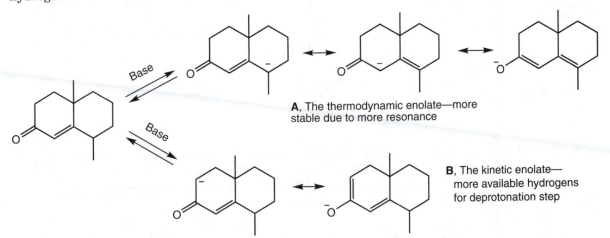

A, The thermodynamic enolate—more stable due to more resonance

B, The kinetic enolate— more available hydrogens for deprotonation step

Note that deprotonation of the other α carbon is not an option. The vinyl anion that would be formed would not be resonance stabilized by the adjacent carbonyl group. Structure **C** would not contribute, so there would be no resonance and the anion would be very unstable.

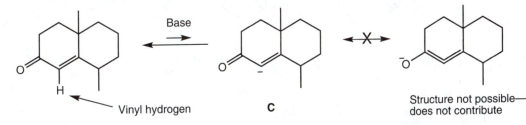

Vinyl hydrogen

C

Structure not possible— does not contribute

Problem 19.49 Parts (a)–(c) and (e) should be straightforward. The positions α to the carbonyl group can be exchanged through the usual formation of the enolate. The products are shown in their exchanged forms.

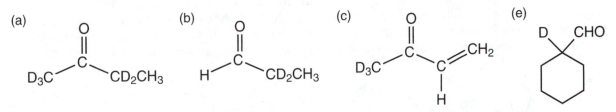

(a)

(b)

(c)

(e)

Part (d) is more troublesome. To get this one completely right you have to recognize two things, and one is tricky. First, the presence of the carbon–carbon double bond in conjugation with the carbonyl group allows the γ position to exchange. The figure shows one γ hydrogen exchanging through enolate **A**.

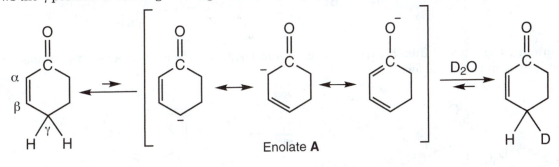

Enolate **A**

(continued)

Of course, the other γ hydrogen and the usual α hydrogens also exchange. So, one might think that the final result will be as shown for part (d).

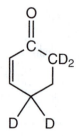

But there is more to it. What if enolate **A** adds deuterium at the α position? Deuterium is now incorporated into the allylic α position of **A** to give **B**. Now either H or D can be removed in base. Removal of D simply reverses the reaction, but removal of H leads to an enolate, **C**, in which deuterium is incorporated at one of the vinyl positions.

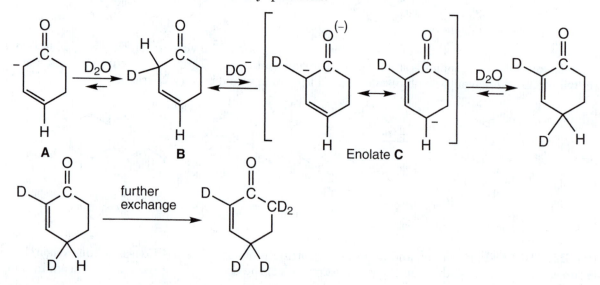

Problem 19.50 Enolate formation demands orbital overlap. Stabilization by an adjacent carbonyl group requires that the π orbitals of the carbon–oxygen double bond overlap with an orbital on an adjacent α carbon. The problem with the hydrogen in question is that, although formally α, there is no overlap of the appropriate orbitals. Accordingly, there can be no enolate formation and no exchange.

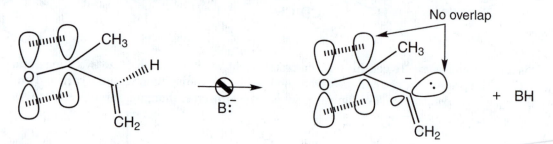

Problem 19.51 Recall from Figure 19.21 that β-dicarbonyl compounds such as 2,4-pentanedione exist largely in their enol forms.

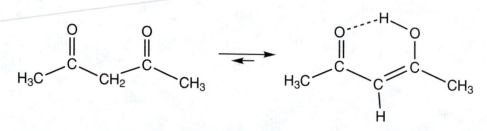

In this case, both the β-diketone and enol form are observed in the ^{13}C NMR spectrum. For the β-diketone form, we would expect to see three signals: one signal for the methyl carbons (δ 24.3 or 30.2 ppm), one signal for the methylene carbon (δ 58.2 ppm), and one signal for the carbonyl carbons (δ 191.4 or 201.9 ppm). Perhaps what is not as easy to see is that the enol form would also display only three different carbons. The trick is that there are two rapidly equilibrating enol forms.

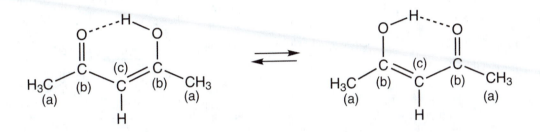

Methyl carbons (a) are equivalent, as are the ketone/enol carbons (b). The actual peak assignments are shown.

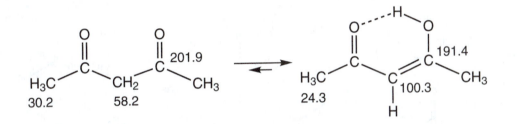

Problem 19.52 These transformations should all be easy to remember. Part (a) is acetal formation, part (b) is reduction to a methylene group (two methods are shown), part (c) is imine (hydrazone) formation, and part (d) is reduction to the alcohol. The only difficulty might come from part (e), which requires you to recall the Wittig reaction from Chapter 16 (p. 816).

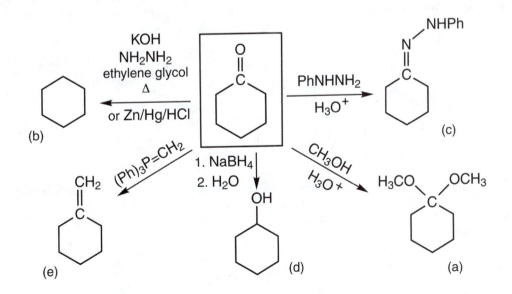

Problem 19.53

(a) Deprotonation of cyclohexanone using LDA at low temperature will form the desired enolate. Because deprotonation with LDA is essentially irreversible, there will be no complications from an aldol reaction. Alkylation of the enolate with methyl iodide will form 2-methylcyclohexanone. Deprotonation a second time using LDA will favor the kinetic enolate. Alkylation with methyl iodide should give 2,6-dimethylcyclohexanone as the major product.

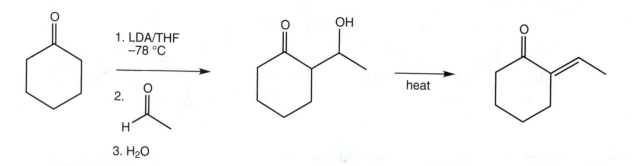

Kinetic enolate

(b) The desired compound is an enone. That means it can be made from an aldol reaction. To avoid having the cyclohexanone enolate react with another molecule of cyclohexanone, we can use LDA as the base, driving all of the cyclohexanone to the enolate. Reaction with acetaldehyde, then adding water and warming the reaction would give the desired product.

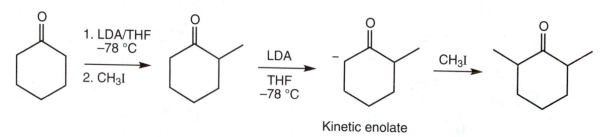

(c) There are many conditions that could be used to make 2-bromocyclohexanone from cyclohexanone. We can brominate a ketone in acidic or basic conditions, or even just by adding bromine to the ketone without acid or base. On paper they are all acceptable. But in the lab, we can be most certain that we get monobromination and no aldol chemistry by using strong base (LDA) irreversibly to make the enolate and then by adding bromine.

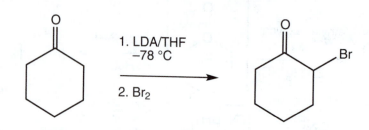

(d) Selectively placing a single deuterium on the C(2) of cyclohexanone involves the same challenge as making 2-bromocyclohexanone. There are many ways to perform this task on paper, but in the lab the most controlled reaction is using LDA to make the enolate and then adding D_2O.

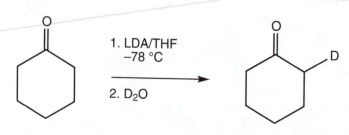

Problem 19.54 This problem shows two isomerizations.

(a) The first of these involves enolate formation and reprotonation. In this case, there are two carbons available for reprotonation. One leads back to starting material, but the other accomplishes the isomerization shown.

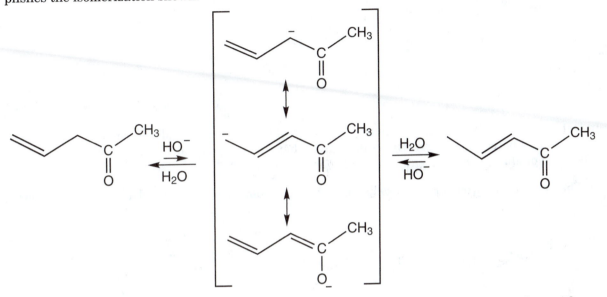

(b) In the second example, a planar enol is produced. Protonation of the enol can occur from either side and give the two products.

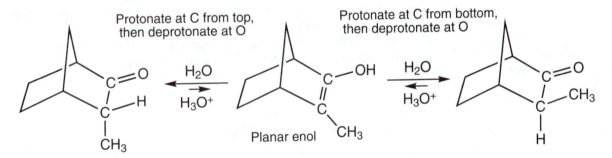

Protonate at C from top, then deprotonate at O

Planar enol

Protonate at C from bottom, then deprotonate at O

Problem 19.55 This problem provides more practice in seeing simple aldol condensations in synthetic terms.

(a)

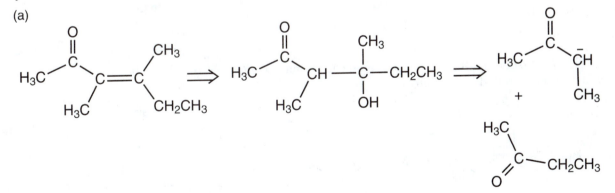

As the retrosynthetic analysis shown above indicates, in principle, the target molecule could be prepared from two molecules of 2-butanone through an aldol reaction and dehydration. However, in practice, a mixture of products would be the likely result. First, the proposed dehydration would probably give an (*E/Z*) mixture of α,β-unsaturated ketones. Second, as 2-butanone has two

(continued)

Problem 19.55 *(continued)*

different sets of α hydrogens, two enolates are possible. Ultimately, the other enolate would lead to an (*E*/*Z*) mixture of another α,β-unsaturated ketone, **1** (write a mechanism for its formation).

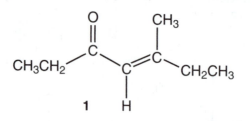

1

In fact, compound **1** appears to be the major product of the base-catalyzed self-condensation of 2-butanone, whereas the target (*E*)-3,4-dimethyl-3-hexen-2-one is the favored (although not exclusive) product of the acid-catalyzed self-condensation.

(b) The enone comes from a β-hydroxy ketone, which can come from the crossed aldol reaction shown.

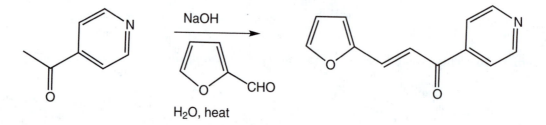

This process would be a one-step synthesis with the desired product as the major isomer formed. The reaction will work well because there is only one enolate that can be formed (α-hydrogen to the ketone deprotonated) and the enolate will undergo crossed aldol reaction with the aldehyde rather than add to another ketone, because aldehydes are more reactive electrophiles than are ketones. The crossed aldol molecule will dehydrate to give the very stable conjugated product. Heat might not even be necessary. The more stable (*E*) isomer will be the major product because the dehydration step is reversible.

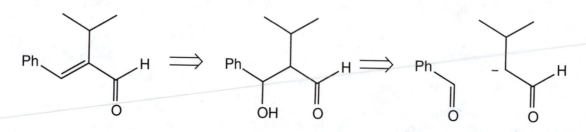

NaOH

H_2O, heat

(c) The (*E*)-2-isopropyl-3-phenyl-2-propenal can be made from the crossed aldol reaction between 3-methylbutanal and benzaldehye.

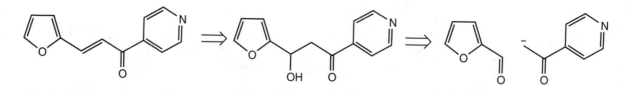
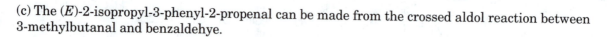

This synthesis will not be as selective as the previous reactions because 3-methylbutanal could undergo aldol addition with itself. To improve the yield of the desired material, we would add an excess of benzaldehyde for the crossed aldol reaction. The dehydration of the β-hydroxy aldehyde will probably give as much (Z) isomer as (E). Both look sterically hindered.

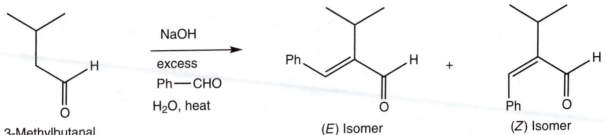

3-Methylbutanal　　　　　　　　　　　　　　　(E) Isomer　　　　　　(Z) Isomer

(d) The formation of 4-ethyl-3-methyl-3-hexen-2-one could come from the crossed aldol reaction between 2-butanone and 3-pentanone.

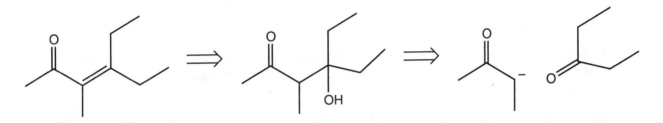

This reaction will be messy. The desired crossed aldol reaction will produce a product, but the reaction will definitely give other products. If we use excess 3-pentanone, then the enolate of 2-butanone will primarily react with the desired 3-pentanone. But there would be an increase of the enolate of 3-pentanone reacting with 3-pentanone. The products of this aldol reaction would dehydrate to give 5-ethyl-4-methyl-4-hepten-3-one. So the reaction will work best with an equal amount of 2-butanone and 3-pentanone. That means that there would also be the product from the enolate of 3-pentanone reacting with 2-butanone and the product of enolate of 2-butanone reacting with 2-butanone. These two undesired aldol reactions would dehydrate to give a mixture of (E) and (Z) isomers of 4,5-dimethyl-4-hepten-3-one and (E) and (Z) isomers 3,4-dimethyl-3-hexen-2-one, respectively. Messy.

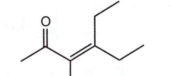

4-Ethyl-3-methyl-3-hexen-2-one

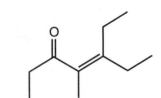

5-Ethyl-4-methyl-4-hepten-3-one

2-Butanone　　+　　3-Pentanone

NaOH
H₂O
heat

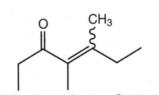

4,5-Dimethyl-4-hepten-3-one

3,4-Dimethyl-3-hexen-2-one

Problem 19.56   This is the iodoform reaction, a way to test for methyl ketones. Compound **1** must be 3,3-dimethyl-2-butanone.

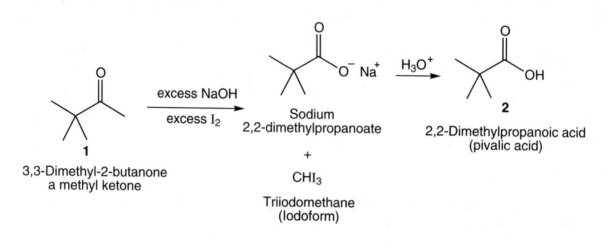

excess NaOH
excess I₂

Sodium
2,2-dimethylpropanoate

+

CHI₃

Triiodomethane
(Iodoform)

2,2-Dimethylpropanoic acid
(pivalic acid)

1

3,3-Dimethyl-2-butanone
a methyl ketone

You have written the mechanism of this reaction. See Problem 19.9(a), which has you analyze the same molecule. Based on the mechanism we wrote, the reaction requires three equivalents of I₂ and four equivalents of hydroxide.

Problem 19.57   As soon as you see LDA, look for an α hydrogen that can be removed by this strong base. Both possible enolates are formed from LDA and molecule **1**, but removal of the methylene hydrogen is far easier (faster) than removal of the methine hydrogen (slower), even though this leads to the less stable enolate, **A**. It is easier to approach the relatively unhindered methylene position than the more encumbered methine group. As enolate formation by LDA is essentially irreversible, the faster formation of **A** ensures that it will lead to the majority of methylated product.

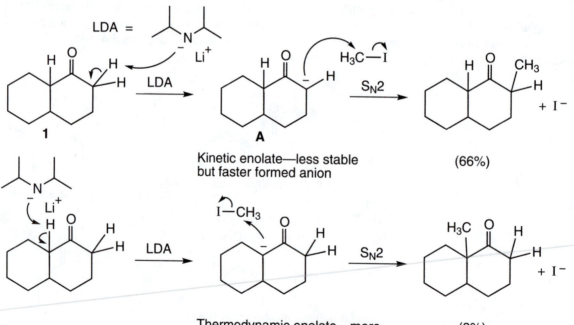

A

Kinetic enolate—less stable
but faster formed anion

(66%)

Thermodynamic enolate—more
stable but more slowly formed anion

(3%)

Problem 19.58 The silicon–oxygen bond is enormously strong (~109 kcal/mol), whereas the silicon–carbon bond is much weaker (about 69 kcal/mol) and this difference drives the reaction toward silicon–oxygen bond formation. So, thermodynamics will clearly favor alkylation on oxygen and formation of the more substituted, more stable double bond. The problem shows that the regiochemistry depends on the base used to form the enolate. As we saw in Problem 19.57, the strong, sterically hindered base LDA irreversibly removes a proton from the position of easier access. In this case, the methylene hydrogen is removed and the less substituted, less stable enolate is formed.

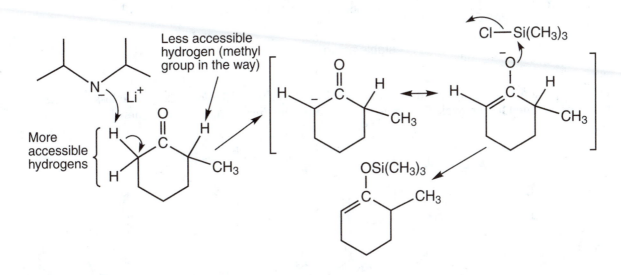

When triethylamine is used as base, enolate formation is reversible. So, even if the more stable enolate is formed more slowly, the reversible nature of the reaction allows it to predominate eventually. Under such conditions, the thermodynamically more stable enolate will lead to most of the silated product.

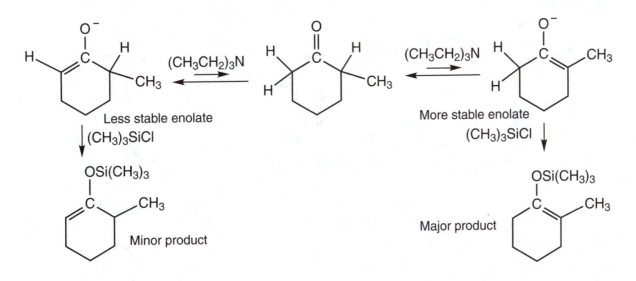

Problem 19.59 Once again, the solution to this problem depends on the irreversible removal of the sterically more accessible hydrogen by LDA. When **1** is deprotonated under equilibrating conditions with potassium *tert*-butoxide in *tert*-butyl alcohol at room temperature, the thermodynamically favored, more substituted enolate dominates and ultimately leads to **2**.

(continued)

Chapter Nineteen

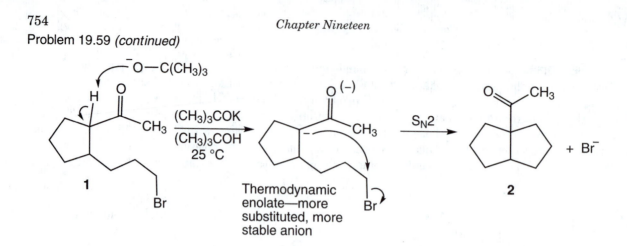

However, when **1** is deprotonated under kinetic conditions with LDA at −72 °C, the less-substituted enolate is formed by removal of the more easily accessible hydrogen and subsequently leads to **3**.

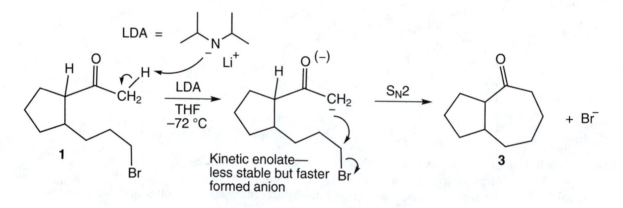

Problem 19.60 This reaction is an example of a "double" crossed aldol condensation (Claisen–Schmidt condensation). Note that only ketone **1** has α hydrogens; in fact, it has two sets of α hydrogens. Also, the diketone **2** is more reactive than an ordinary ketone (Why?). The reaction proceeds first to intermediate **A**, and then through dehydration to the α,β-unsaturated ketone **B**.

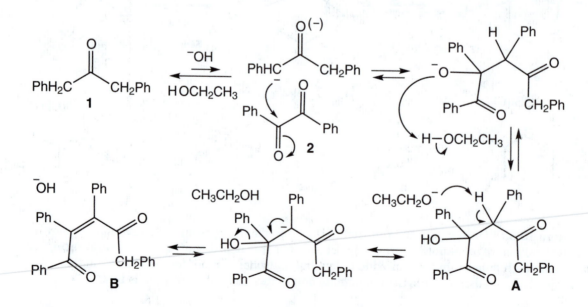

Now the process simply is repeated, this time in an intramolecular sense to give **3**. (It is also possible that the second aldol condensation occurs before the first dehydration from **A** to **B**.)

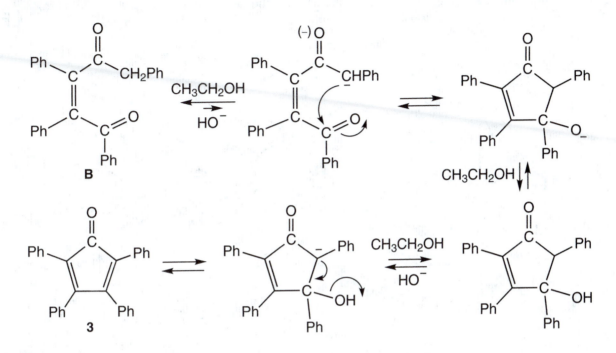

Problem 19.61 Here we have a relatively straightforward "combination" problem. A little analysis shows that we need to find a way to combine two molecules of dimedone with one aldehyde.

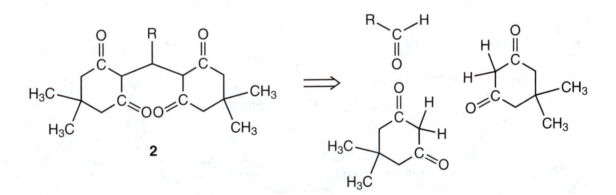

(continued)

Problem 19.61 *(continued)*

The dimedone derivatives, **2**, are formed by a combination of a Knoevenagel condensation (to give **A**) and a Michael reaction. Enolate formation is followed by addition to the aldehyde and dehydration to give the α,β-unsaturated compound **A**. Michael addition of another enolate leads, after protonation, to **2**. (Piperidine could also react with the aldehyde to make the iminium ion, which would make the aldol step faster.)

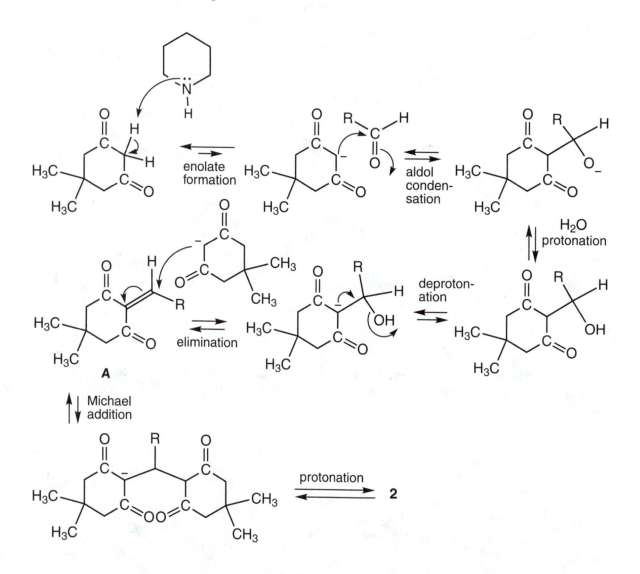

The conversion of **2** into **3** is reminiscent of an acid-catalyzed intramolecular aldol condensation, except that, because of geometrical constraints in this system, it is the enol oxygen, not carbon, that acts as the nucleophile and adds to the protonated carbonyl.

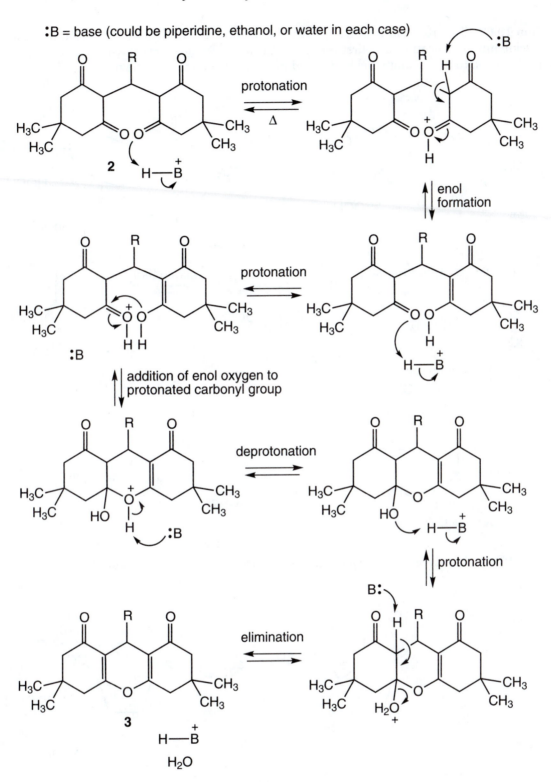

Problem 19.62   This reaction is a straightforward intramolecular Claisen condensation, known as a Dieckmann condensation. The first product is the anion formed by removing the doubly α proton. This enolate is protonated by acetic acid in a second step to give the product.

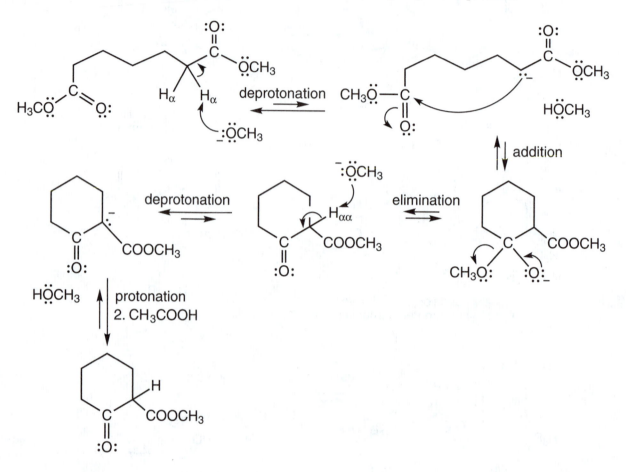

Problem 19.63   In this process, the steps of the forward Dieckmann of Problem 19.62 are reversed.

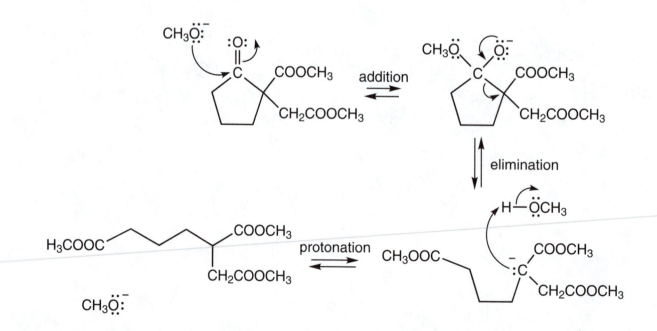

Problem 19.64

(a)

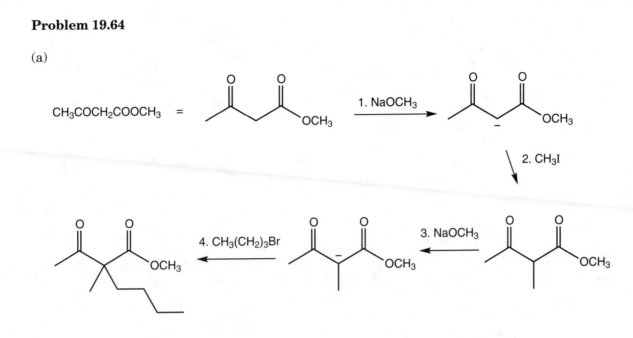

$CH_3COCH_2COOCH_3$ =

1. NaOCH$_3$

2. CH$_3$I

3. NaOCH$_3$

4. CH$_3$(CH$_2$)$_3$Br

(b) This reaction would involve a dianion. One equivalent of ethyl iodide would react with the more basic site, the carbon that had the less acidic hydrogen. One equivalent of ethyl iodide would give the product shown.

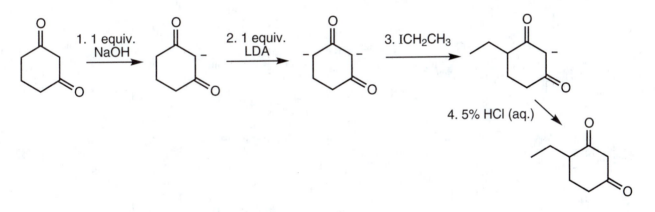

1. 1 equiv. NaOH

2. 1 equiv. LDA

3. ICH$_2$CH$_3$

4. 5% HCl (aq.)

Problem 19.65   The enolate is first formed and then adds to the Lewis acid carbonyl group.

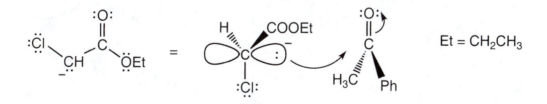

Et = CH$_2$CH$_3$

But this is not the only way addition can take place. There are two possible orientations for addition, which lead to two diastereomeric alkoxides.

(continued)

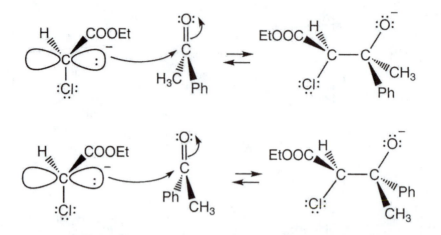

In the final step of the Darzens condensation, the alkoxides undergo an intramolecular S$_N$2 reaction in which the alkoxide displaces chloride (from the rear) to give the two epoxides.

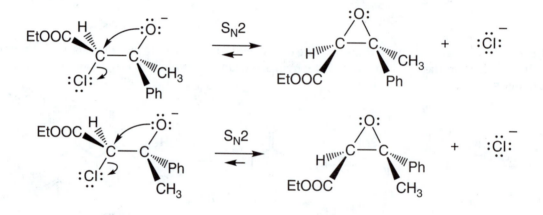

Problem 19.66   This problem involves formation of a thioacetal and three of its subsequent reactions: alkylation, hydrolysis to a carbonyl compound, and reduction.

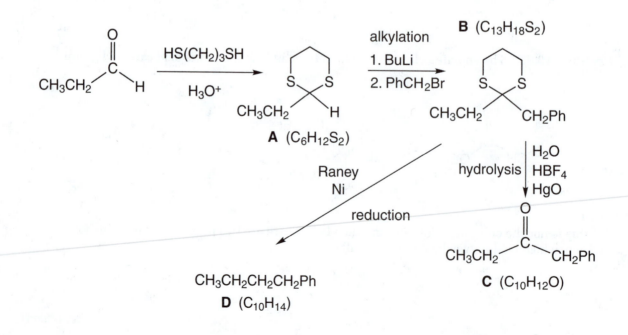

Problem 19.67   Formation of **A** is nothing more than a Fischer esterification.

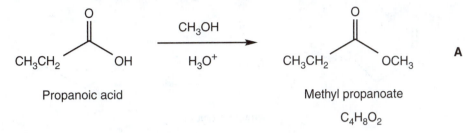

Propanoic acid

Methyl propanoate

$C_4H_8O_2$

The transformation of **A** into **B** is more complex. A Claisen condensation is followed by careful acidification.

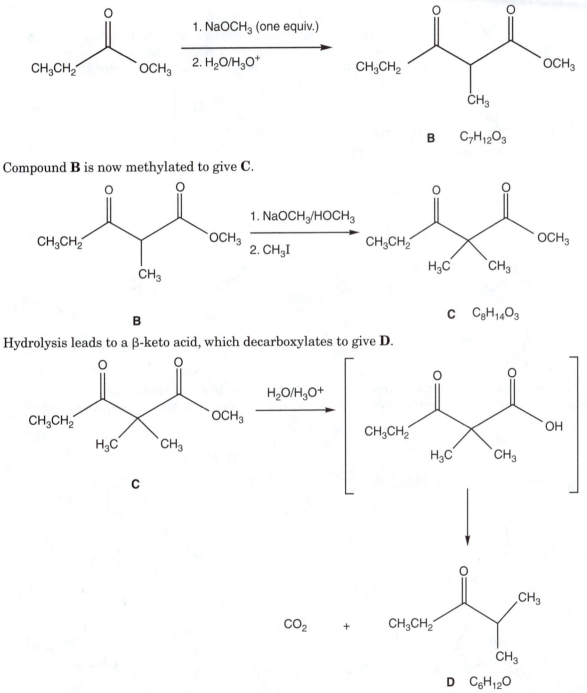

1. NaOCH$_3$ (one equiv.)

2. H$_2$O/H$_3$O$^+$

B $C_7H_{12}O_3$

Compound **B** is now methylated to give **C**.

1. NaOCH$_3$/HOCH$_3$

2. CH$_3$I

B

C $C_8H_{14}O_3$

Hydrolysis leads to a β-keto acid, which decarboxylates to give **D**.

H$_2$O/H$_3$O$^+$

C

CO$_2$ +

D $C_6H_{12}O$

(continued)

Problem 19.67 *(continued)*

Finally, **D** is reduced in a Wolff–Kishner reaction (p. 712) to give 2-methylpentane.

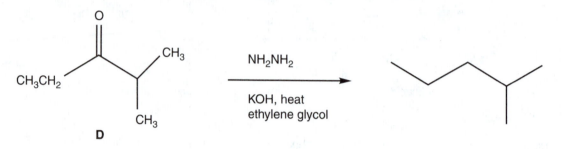

Problem 19.68

(a) This problem involves nothing more than the standard sequence of anion formation, addition, and dehydration. The only trick is seeing that a nitro group can nicely stabilize an adjacent anion. That effect appeared earlier when we described nucleophilic aromatic substitution (Chapter 15, p. 741).

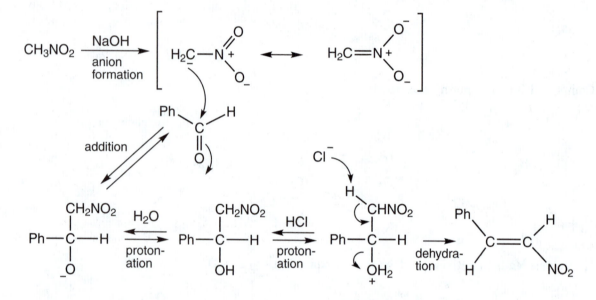

(b) Here the strong base is able to remove a benzylic hydrogen to give an anion that is stabilized by the adjacent benzene ring. Intramolecular addition and dehydration finish the sequence.

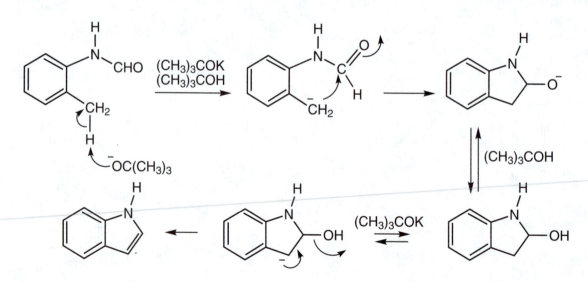

Problem 19.69

(a) We must find a way to combine two molecules of **2** with one of **1**. In this problem, the "standard" sequence of enolate formation, addition, and elimination leads to a molecule that can act as an acceptor in a Michael reaction (see Problem 19.61 for a similar sequence). This reaction is a tandem Knoevenagel–Michael. Notice that both the enolates in this sequence result from removal of an especially acidic "doubly α" hydrogen.

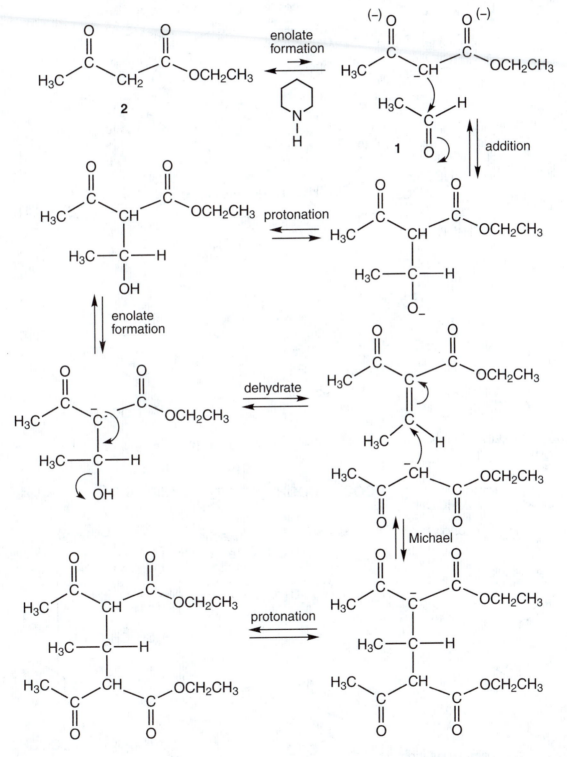

Problem 19.69 (continued)

(b) Here condensation occurs under reducing conditions (note the H₂/Pd), and the α,β-unsaturated carbonyl compound is hydrogenated as it is formed. Start with the formation of the better enolate by removal of the most acidic hydrogen. Then add to the carbonyl group of propionaldehyde and dehydrate. Hydrogenation gives the final product. In both reactions, the piperidine could be acting as an organocatalyst, rather than as a simple base. Piperidine can react with the aldehyde to form a more reactive electrophile, the iminium ion.

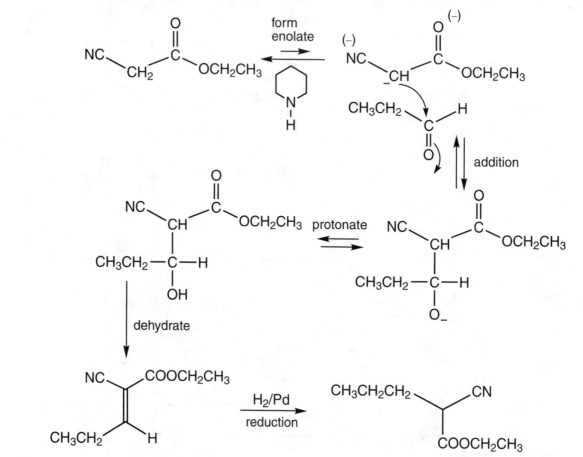

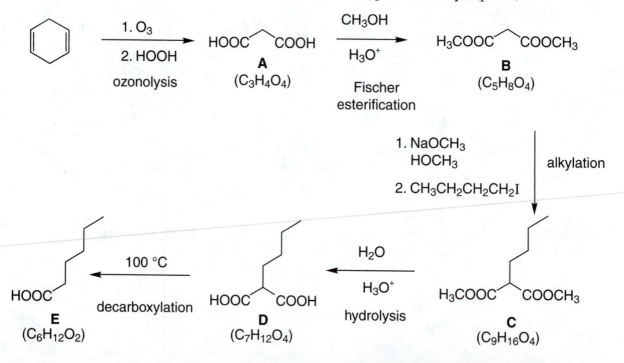

Problem 19.70 This sequence is a malonic ester synthesis, pure and simple (p. 957).

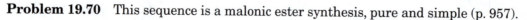

Problem 19.71   This reaction is nothing more than an intramolecular base-catalyzed aldol condensation. However, it is complicated because, as you undoubtedly noticed after a little analysis, there are four possible enolates [actually more if you consider (*E/Z*) stereochemistry] that can be formed from **1**. In fact, all four are probably formed. The simpler but closely related compound **3** exchanges nine hydrogens for deuterium when treated with D_2O/DO^-, for example, so all four enolates of **3** must be generated.

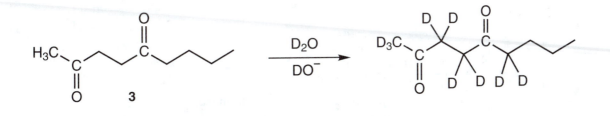

Of the four possible enolates that can be formed from **1**, two can only form three-membered rings through intramolecular additions to carbon–oxygen double bonds, and these will surely be disfavored thermodynamically. (You should draw these two enolates and the three-membered rings they form.) The two remaining possible enolates can each lead to an energetically more favorable five-membered ring. However, it is the more substituted, and thus more stable, enolate that leads to the observed *cis*-jasmone (**2**).

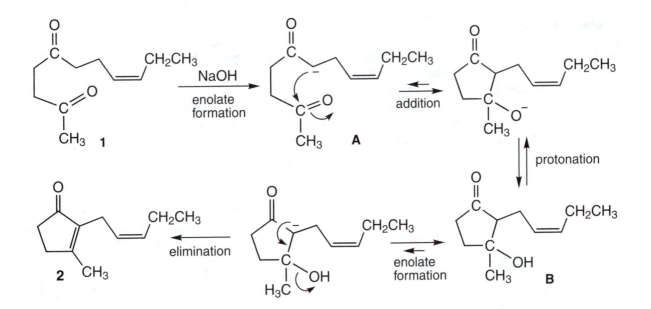

The other cyclopentenone could arise from the less stable, less substituted enolate in the following way. The steps are equivalent to those in the previous figure. Enolate formation, intramolecular addition to the carbonyl group, and dehydration would give **4**.

(continued)

Problem 19.71 *(continued)*

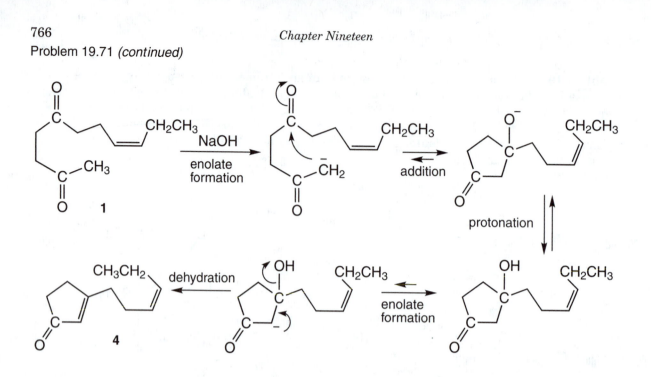

Why is *cis*-jasmone (**2**) formed in this base-catalyzed aldol process? Two immediate possibilities come to mind. Perhaps the tetrasubstituted enone **2** is thermodynamically more stable than the trisubstituted enone **4**, and under the reaction conditions the production of **4** is reversible through a retro-aldol reaction (write a mechanism). Alternately, *cis*-jasmone may be the kinetically favored product. The situation is far from resolved. Notice that we could have worked this problem nicely backward. Enone **2** must have come from dehydration of **B**, and **B** is the inevitable result of aldol condensation of enolate **A**.

Problem 19.72

(a) Notice that the desired product is a ketone (3-ethyl-5-hexen-2-one). Because we are going to use ethyl acetoacetate as our starting material, we know that we need to do a decarboxylation at some point. The standard synthesis using acetoacetate calls for the decarboxylation as the last step. Therefore, in the retrosynthesis the precursor to our product is the β-keto acid, which must have come from the dialkylated ethyl acetoacetate. We know that the alkyl groups can be put on sequentially starting with ethyl acetoacetate.

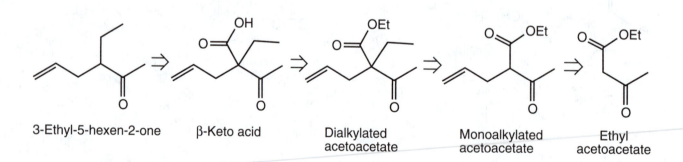

3-Ethyl-5-hexen-2-one   β-Keto acid   Dialkylated acetoacetate   Monoalkylated acetoacetate   Ethyl acetoacetate

In the forward direction with the reagents we have

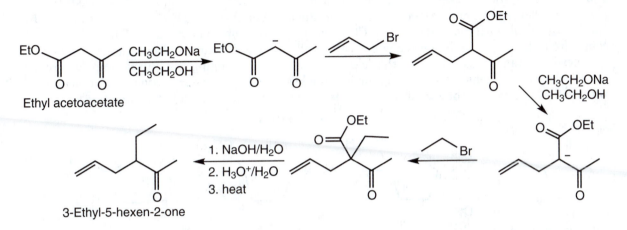

3-Ethyl-5-hexen-2-one

(b) The desired product is 2-methyldecanoic acid. The starting material is diethyl malonate. That means we must do a decarboxylation. As we saw in part (a), the standard procedure is to do the decarboxylation last. That means that the immediate precursor to our target molecule is the disubstituted malonic acid, which could come from the disubstituted malonate diester. The disubstituted compound can be formed from the monosubstituted malonate, which can be made from diethyl malonate.

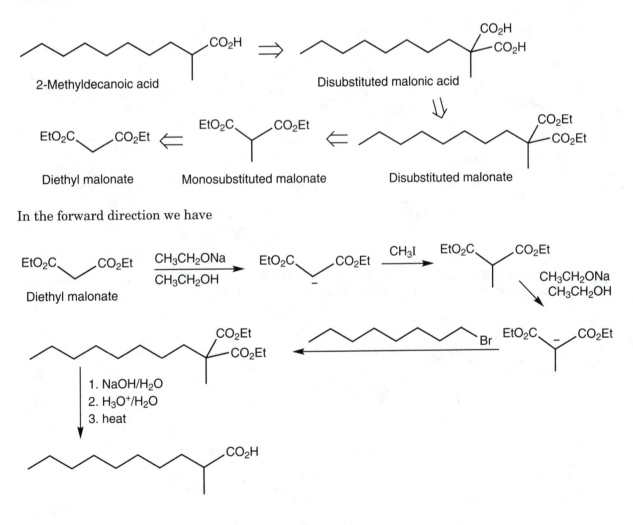

2-Methyldecanoic acid

Disubstituted malonic acid

Diethyl malonate   Monosubstituted malonate   Disubstituted malonate

In the forward direction we have

Diethyl malonate

1. NaOH/H$_2$O
2. H$_3$O$^+$/H$_2$O
3. heat

(continued)

Problem 19.72 *(continued)*

(c) We must have a decarboxylation in this synthesis. Let's assume it is in the last step (that's the standard procedure). We need to have a β-keto acid for the decarboxylation and there are two places the acid group could be—either on C(2) of the benzyl-substituted cyclopentanone or on C(5). But the molecule with the acid group on C(5) doesn't have an obvious precursor. The C(2) acid comes from the corresponding ester, which can be alkylated easily coming from the ester substituted cyclopentanone. This retrosynthetic step is the most challenging one. But perhaps you recall the Dieckmann condensation, which is the intramolecular Claisen condensation. That reaction would allow us to make the substituted cyclopentanone from diethyl 1,6-hexanedioate.

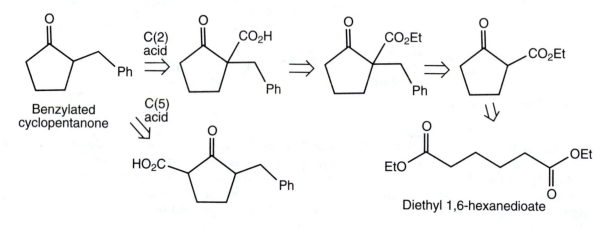

The forward reactions are

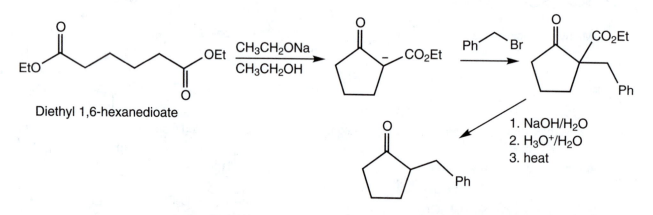

(d) This synthesis will require the formation of the five-membered ring. Therefore an intramolecular reaction must occur. We use can diethyl malonate and a dihalide to make the desired ring.

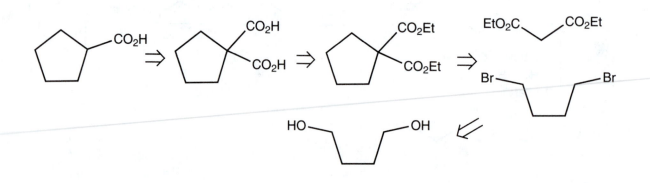

The forward reactions are

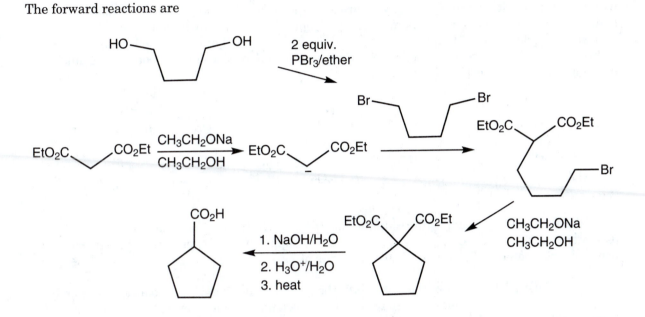

Problem 19.73 The parent masses (**1** = 146 g/mol; **2** = 234 g/mol) allow us to see that compound **1** is a 1:1 combination of acetone and benzaldehyde (minus water), and compound **2** incorporated two molecules of benzaldehyde with one of acetone (again, minus two waters). Both **1** and **2** are formed by crossed aldol condensations (Claisen–Schmidt reactions). Notice that acetone has α hydrogens whereas benzaldehyde does not. The presence of an excess of acetone should favor "single" aldol reaction/dehydration.

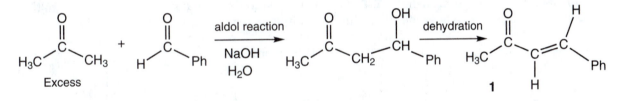

With an excess of benzaldehyde, a "double" aldol reaction–dehydration should be favored.

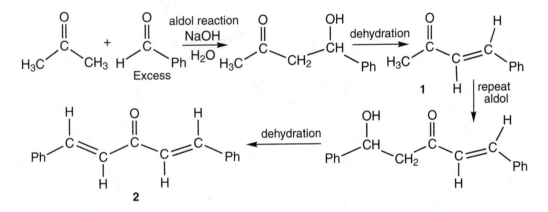

The spectral data for compounds **1** and **2** support these conjectures. The mass spectrum for compound **1** is consistent with the proposed α,β-unsaturated ketone, 4-phenyl-3-buten-2-one ($C_{10}H_{10}O$, MW = 146 g/mol). The IR spectrum of compound **1** shows a strong carbonyl stretch at 1667 cm^{-1}, consistent with a conjugated ketone. Finally, the ^{1}H NMR spectrum is also consonant

(continued)

Problem 19.73 *(continued)*

with the proposed structure. The methyl hydrogens appear at δ 2.38 ppm, and the aromatic hydrogens appear as a 5H multiplet at δ 7.30–7.66 ppm. The olefinic hydrogen resonances, which appear as doublets at δ 6.71 and 7.54 ppm, are quite informative. First, note the low field position of one of the hydrogens. This signal is the β-vinyl hydrogen and is deshielded both by the adjacent benzene ring and the carbonyl resonance:

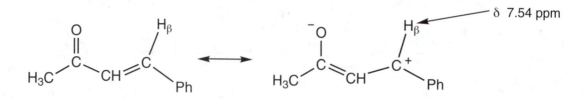

Note also that the magnitude of the coupling constant, J = 16 Hz, is appropriate for trans coupling (Chapter 9, p. 411). The stereochemical assignment is supported by the presence of a =C—H bending (wag) band at 973 cm^{-1} in the IR spectrum of **1**.

The spectral data for **2** are also consistent with the proposed "double" aldol product, 1,5-diphenyl-1,4-pentadien-3-one. The molecular ion in the mass spectrum is appropriate for a molecular formula of $C_{17}H_{14}O$ (MW = 234 g/mol). The IR spectrum of **2** displays a strong carbonyl stretch at 1651 cm^{-1}, once again consistent with a conjugated carbonyl group. Notice the absence of a methyl group signal in the ^{1}H NMR spectrum of **2**. The trans stereochemistry of compound **2** is supported by the magnitude of the coupling constant for the vinyl hydrogens (16 Hz) and the characteristic =C—H bending at 984 cm^{-1} in the IR spectrum of **2**.

Problem 19.74   This reaction is the Michael-like addition of an enamine to an alkene activated by a nitro group to give intermediate **A**. The nitro group stabilizes an adjacent negative charge more strongly than does a benzene ring, which explains the regiochemistry of the addition.

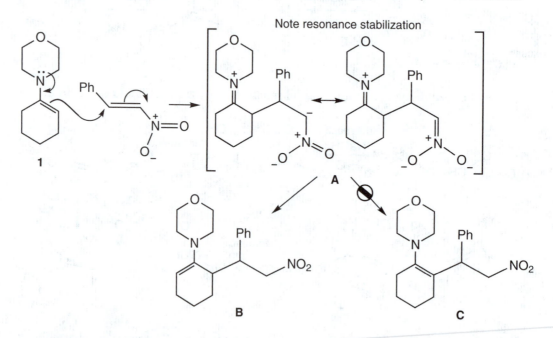

Deprotonation and reprotonation lead to a new enamine, **B**. There is experimental evidence that the structure is as shown, but you may well wonder why deprotonation does not form the other possible enamine, **C**. The answer isn't known, but it may involve nothing more complicated than the relative ease of access to a methylene group compared to that for the more congested methine hydrogen. Acid-catalyzed hydrolysis leads to the observed nitro ketone, **3**.

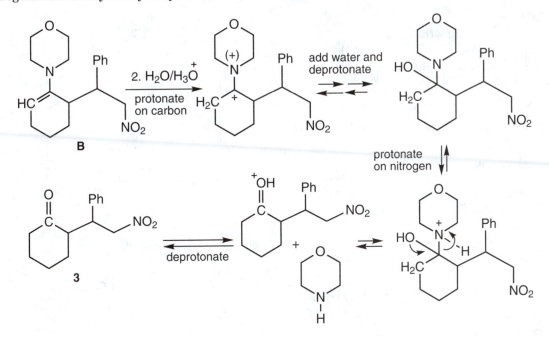

Problem 19.75 These are both double Michael problems. Part (a) is nothing more than a pair of straightforward addition reactions. Part (b) incorporates the complication of an intramolecular displacement after the second Michael.

(a)

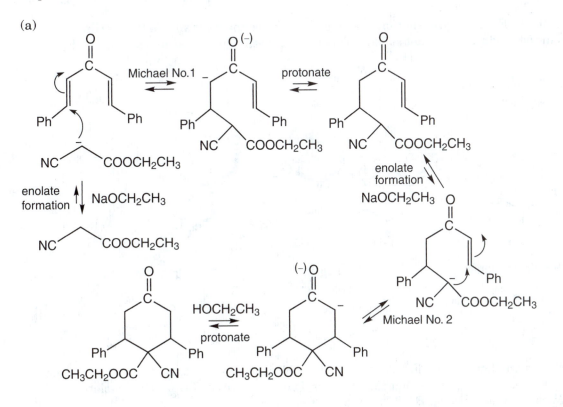

(continued)

Problem 19.75 *(continued)*

(b)

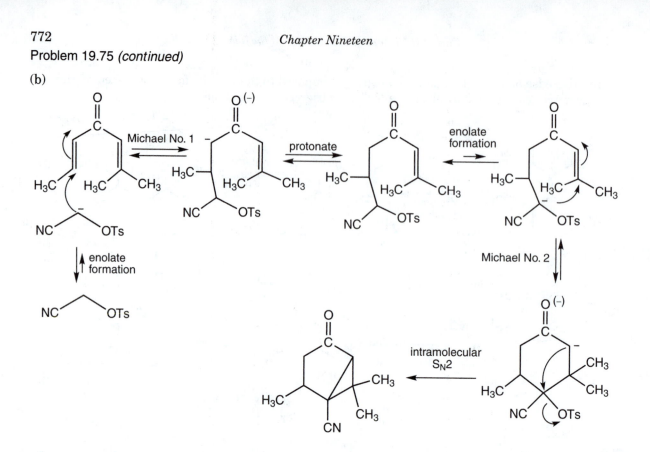

Problem 19.76 First analyze what must be done. Try to see the remnants of the starting materials in the product. That's not so hard if we use the methyl groups as markers (R = CH₂CH₃).

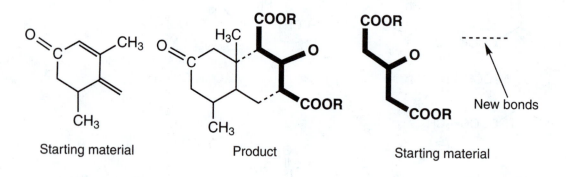

| Starting material | Product | Starting material |

This reaction is another double Michael, but the two additions must be done in the correct order. If you try the wrong one first, the second cannot occur. There are two choices once an enolate has been formed, "Michael **A**" or "Michael **B**."

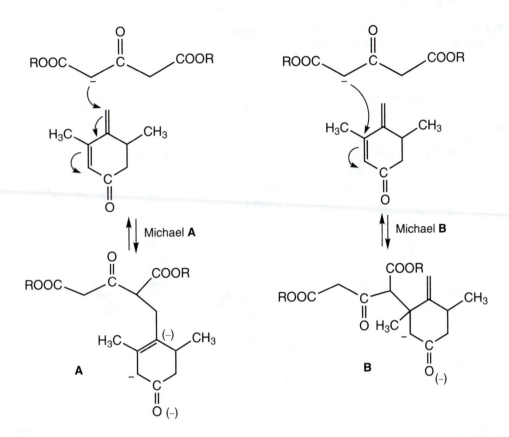

Intermediate **A** can protonate to give a new α,β-unsaturated ketone, **C**, but **B** cannot. If it is not clear how the protonation of **A** occurs, by all means carefully draw out all the resonance forms for **A**. Enolate formation and a second Michael addition, followed by protonation, give the product.

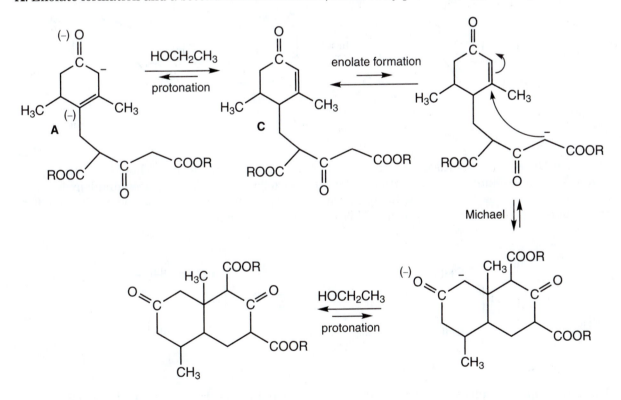

Problem 19.77   On the surface, this is a simple alkoxide formation, followed by alkylation. This reaction is a Williamson ether synthesis.

$$R-OH \xrightarrow{\text{NaH}} R-O^- \xrightarrow[\text{(S}_N2)]{H_3C-I} R-O-CH_3 \;+\; I^-$$

However, there is the problem of the seemingly strange stereoisomerization that takes place. Alcohol **1** is a β-keto alcohol, which means that it can, at least in principle, be made through an aldol condensation. We can find that aldol condensation in many ways, but one good method is to write the reverse aldol condensation starting from **1**.

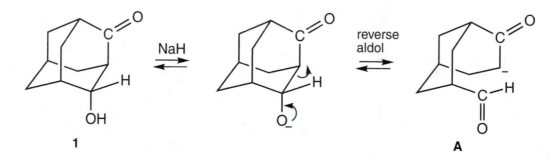

What can the enolate **A** do? Of course it can simply reverse to give **1**, that's merely the forward aldol. But there are two ways this can happen! Opened intermediate **A** can undergo carbon–carbon bond rotations to give **A′**, and aldol condensation of **A′** gives the isomerized alkoxide, **B**. Methylation gives the product.

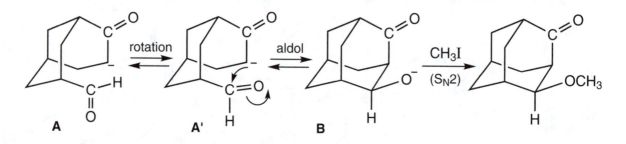

Problem 19.78   This synthesis of tropinone (**3**), aptly called the Robinson–Schöpf reaction, involves a double Mannich condensation, followed by two decarboxylations. Succindialdehyde (**1**) and methylamine react to yield the bis-iminium salt **A**. We won't write the detailed mechanism for this simple transformation, but you should be certain you can do it easily.

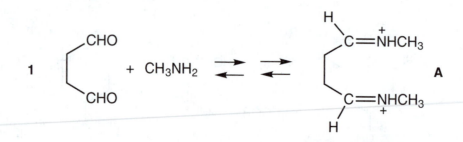

The first Mannich condensation now occurs between one iminium salt and the enol form of **2**. The amine **B** is the product. Now intramolecular cyclization takes place to give a new iminium salt **C**.

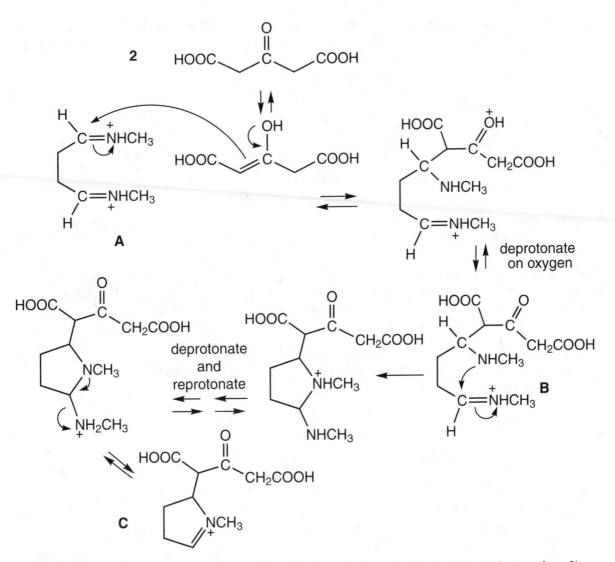

Now a second, intramolecular Mannich reaction takes place to give **D**. As **D** is a β-ketocarboxylic acid, two decarboxylations readily occur to give tropinone (**3**).

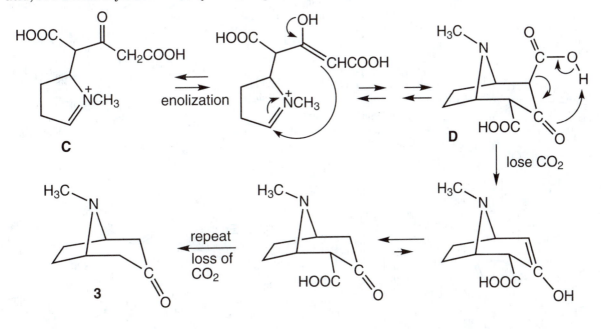

Problem 19.79   Do some analysis first. A new ring must be made. It is likely that the oxygen atom in the ring is the oxygen of the phenol starting material. The other atoms in the ring probably come from the starting materials as shown.

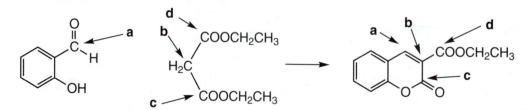

The new bonds to be made are between the oxygen and carbon **c**, and carbons **a** and **b**. We are now directed to find ways to make these bonds. That's a great advantage. Now we can start this problem with confidence.

The most acidic carbon-bound hydrogens are the "doubly α" hydrogens of malonic ester. The base piperidine removes one of these hydrogens, and the resulting enolate adds to the aldehyde to give **A**. (Piperidine could react with the aldehyde to form the more reactive iminium ion, rather than acting as a simple base.) One of the new bonds, carbon **a** to carbon **b**, has been made. The alkoxide is protonated and water eliminated to give **B**. At this point, we must think about closing the new ring, incorporating the oxygen atom. The quite acidic phenolic hydrogen can be removed to give an alkoxide poised perfectly to do an intramolecular transesterification of one of the ethyl esters and make the oxygen–carbon **c** bond. This closes the ring and leads to the product.

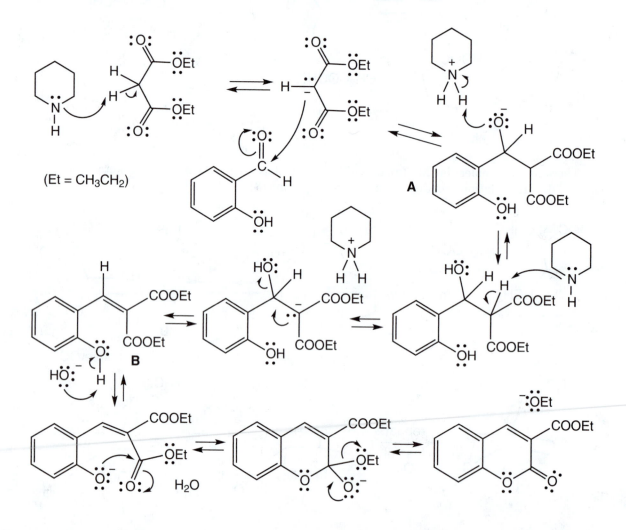

Problem 19.80 Treatment of diethyl malonate with base gives the enolate. In Section 11.4c, we saw that anions can open epoxides in S_N2 reactions. Here, there are two possible modes of opening. Although it was originally thought that only the less hindered route was followed, it is now clear that both openings occur to give a pair of alkoxides. The molecular formulas of **A** and **B** show that ethanol has been lost, and so it is clear that intramolecular transesterifications have occurred through the usual addition–elimination reactions.

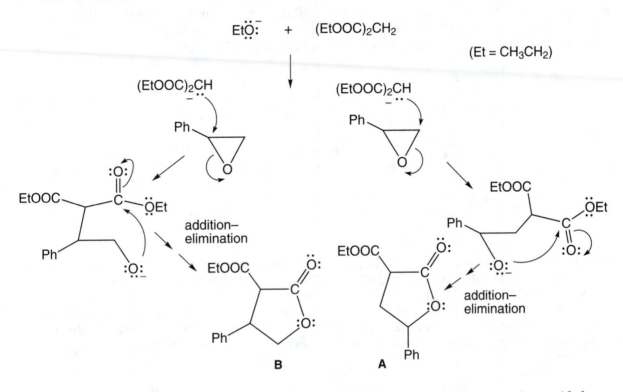

Saponification (treatment with base, followed by acidification) transforms the esters into acids by straightforward ester hydrolyses. All acids β to carbonyl groups are prone to decarboxylation (acetoacetic acids and malonic acids are examples in this chapter), and **C** and **D** lose CO_2 on heating to give lactones **E** and **F**.

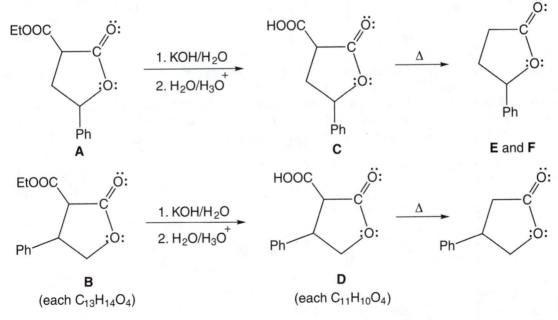

(continued)

Problem 19.80 *(continued)*

Now we are left to decide which lactone is **E** and which is **F**. Both **E** and **F** show a molecular ion in the mass spectrum with $m/z = 162$, consistent with the proposed structures. In addition, the IR spectra of both **E** and **F** exhibit an intense band at about 1780 cm^{-1}, consonant with a carbonyl stretch of a γ-lactone. The crucial spectral distinctions occur in the ^{1}H and ^{13}C NMR spectra. In particular, **E**, γ-phenyl-γ-butyrolactone, shows a single low-field hydrogen (H$_x$) at δ 5.45–5.55 ppm in the ^{1}H NMR spectrum. This hydrogen is deshielded by both the adjacent phenyl ring and the adjacent oxygen atom. There is no corresponding hydrogen in **F**. The carbon to which H$_x$ is attached in **E** appears as a doublet at δ 81.2 ppm in the ^{13}C NMR spectrum. In **F**, β-phenyl-γ-butyrolactone, two hydrogens (H$_x$ and H$_{x'}$) on the carbon adjacent to the other oxygen appear at δ 4.28 and 4.67 ppm, and the carbon to which they are attached is a triplet at δ 73.8 ppm.

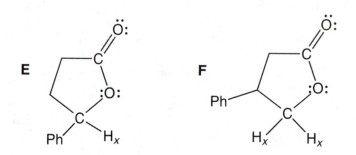

Problems 19.81   The difficulty with the mechanism shown in the Problem (and in more than one place in the chemical literature) is the increase in strain incurred as the four-membered ring is closed to give the bicyclo[2.1.0]pentane system. One's intuitive feeling is that there must be a better way! The anion shown, **A**, is the most stable enolate possible in this system, so this does seem a good way to start. The hint tells you to think about reducing strain. Opening the three-membered ring surely relieves strain and in this case generates a new enolate, **B**.

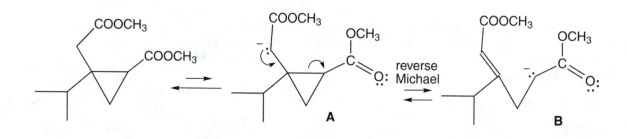

Here comes the key step. The problem is to find a way to close a five-membered ring so as to create the product we know is formed. This cannot be done from **B** directly. So, either there is a way for **B** to rearrange, or **B** is a dead end. Notice that a proton shift, or more likely a protonation–deprotonation sequence, generates **C**, a resonance-stabilized enolate even more stable than **B**. This anion can close to product, although a final double bond isomerization is also necessary (**D** to **E**).

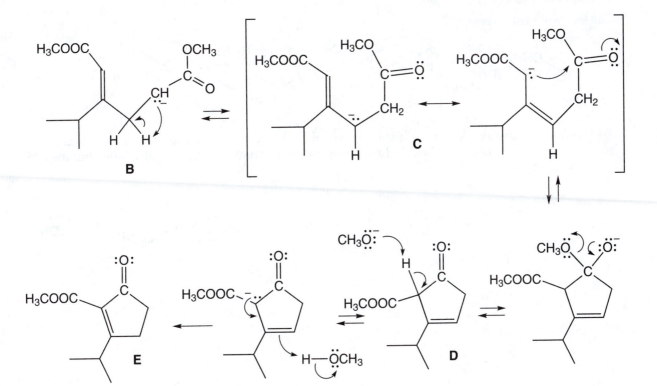

So, how is one to know? That's a tough question. For some time we have offered 10 hour-test points for a workable way to distinguish the two mechanisms. Although there have been some good suggestions, no one has ever won the full 10 points. One of the main difficulties is that isotopic labels seem ineffective in this case, and other, more intrusive labels such as methyl groups have the potential of changing the molecule so that the mechanism changes. You try.

Problem 19.82

(a) This ester (butyl butanoate) can be made from reaction of butanoic acid and 1-butanol. The butanoic acid can be made from 1-butanol.

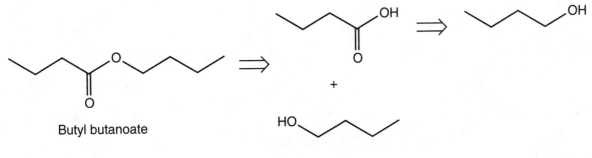

Butyl butanoate

(continued)

Problem 19.82 *(continued)*

The forward synthesis is

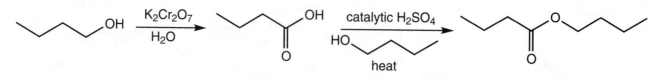

(b) The easiest way to make 2-ethylhexanoic acid, at least on paper, is to alkylate the dianion of butanoic acid with 1-bromobutane. Butanoic acid can come from 1-butanol, and 1-bromobutane can come from 1-butanol.

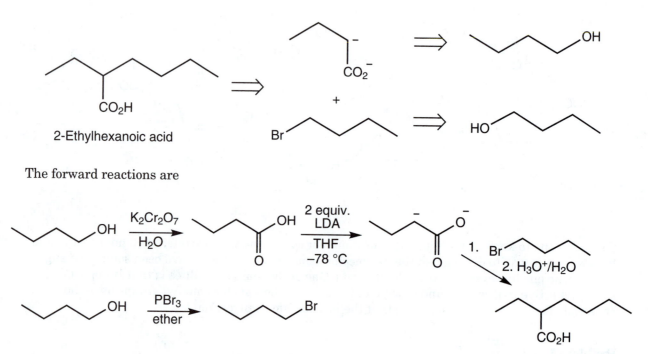

2-Ethylhexanoic acid

The forward reactions are

(c) There are a number of ways to make a primary amine. For example, it could come from reduction of an amide, Hofmann rearrangement of an amide, reduction of a nitrile, or S_N2 reaction of an alkyl halide. You might notice that the amide route fits nicely with the 2-ethylhexanoic acid we synthesized in (b).

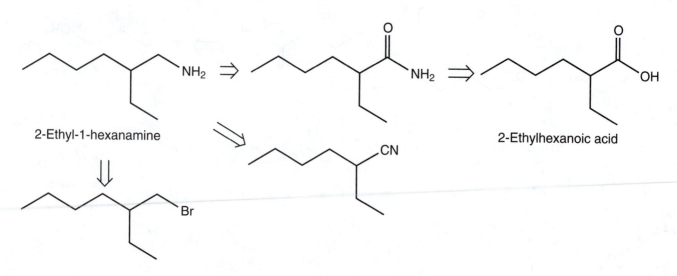

2-Ethyl-1-hexanamine   2-Ethylhexanoic acid

The forward path is

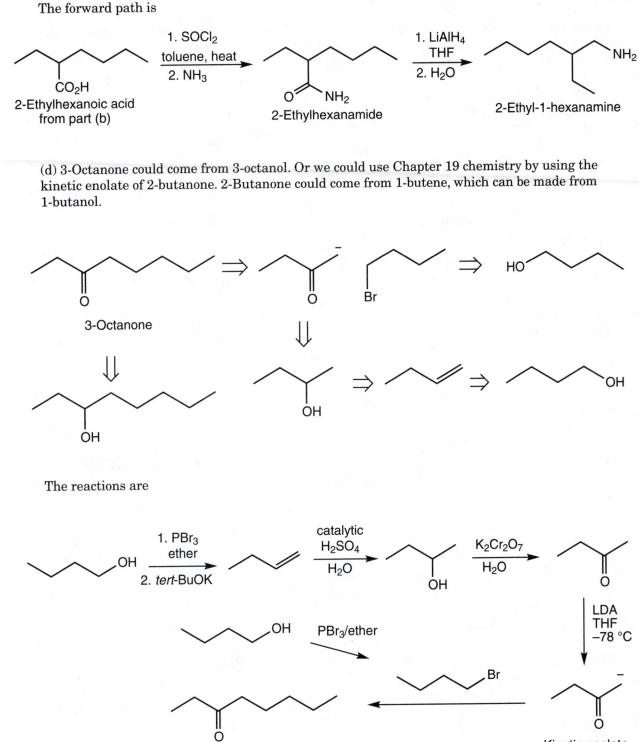

(d) 3-Octanone could come from 3-octanol. Or we could use Chapter 19 chemistry by using the kinetic enolate of 2-butanone. 2-Butanone could come from 1-butene, which can be made from 1-butanol.

The reactions are

Problem 19.83

(a) The first step is a mixed Claisen condensation. Note that only one partner in the condensation has an α hydrogen and can form an enolate. As usual in the Claisen condensation, the initial product is a salt. Neutralization gives the β-keto ester.

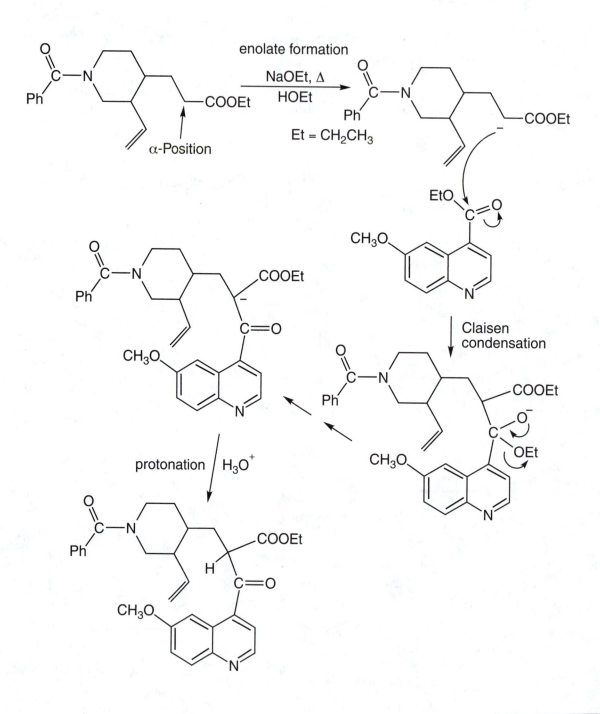

(b) First, the α position is brominated.

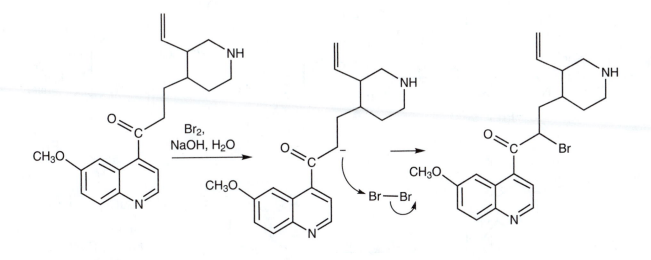

Now, treatment with base generates the amide ion. Intramolecular displacement of the α-bromide produces the bicyclic compound. The arrow looks hideously long, but this is an artifact of the two-dimensional representation of these three-dimensional molecules.

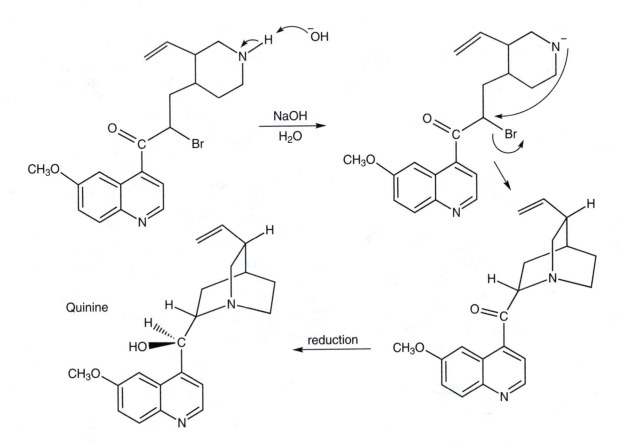

Problem 19.84

(a)

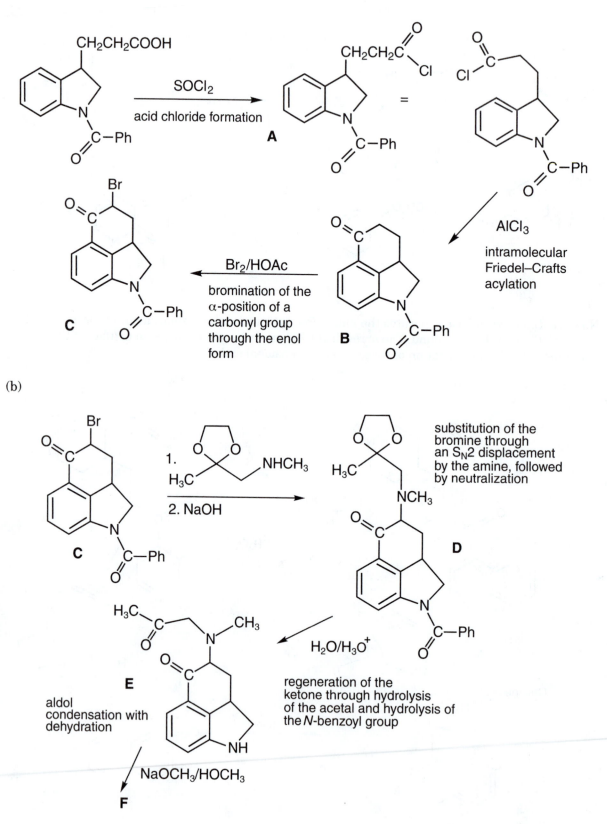

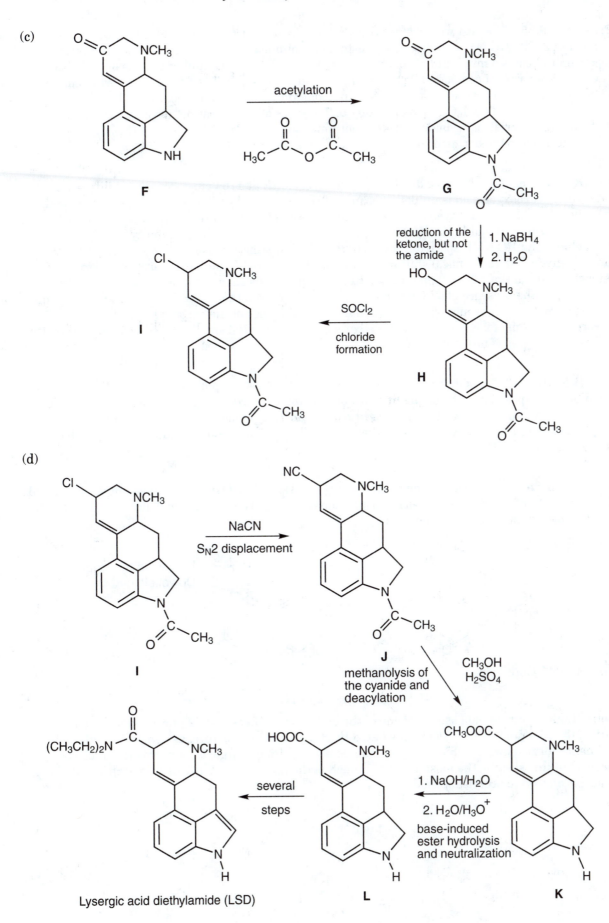

(c)

F

acetylation

G

reduction of the ketone, but not the amide
1. NaBH₄
2. H₂O

H

SOCl₂
chloride formation

I

(d)

I

NaCN
S_N2 displacement

J

methanolysis of the cyanide and deacylation

CH₃OH
H₂SO₄

K

1. NaOH/H₂O
2. H₂O/H₃O⁺
base-induced ester hydrolysis and neutralization

L

several steps

Lysergic acid diethylamide (LSD)

Problem 19.85   The best way to start this problem is to deduce the molecular formula of compound **2**. The ^{13}C NMR spectrum suggests a minimum of nine carbons, and the ^{1}H NMR spectrum indicates a minimum of six hydrogens. The formula C_9H_6 implies a mass of 114 g/mol. The mass spectrum of compound **2** shows a molecular ion of $m/z = 146$, so we are short a mass of 32 g/mol. This deficit can be most easily accommodated by two oxygen atoms, and this produces a molecular formula of $C_9H_6O_2$, corresponding to 7 degrees of unsaturation. Given the structure of the starting material, salicylaldehyde, it is not unreasonable to attribute four of the seven degrees of unsaturation to the benzene ring.

The IR spectrum of compound **2** displays a carbonyl stretch at 1704 cm^{-1}. The possibility of an aldehyde or carboxylic acid is unlikely because of the absence of the corroborating C—H or O—H stretches. In addition, aldehydes and carboxylic acids can be eliminated by the absence of the expected low-field signal in the ^{1}H NMR spectrum of **2**. This leaves the possibility of either an ester or a ketone. The ^{13}C NMR spectrum serves to make this distinction. The carbonyl carbon of ketones (and aldehydes) has a chemical shift in the 190–220 ppm range, whereas the carbonyl carbon of esters appears in the 160–190 ppm range. In this case, the absence of a singlet in the 190–220 range makes the presence of a ketone most unlikely. Moreover, there is a signal at 160.6 ppm, quite consistent with the presence of an ester. Finally, the frequency of the carbonyl stretch in the infrared spectrum of **2** is appropriate for a *conjugated* ester (1704 cm^{-1}). The ester group accounts for the fifth degree of unsaturation.

The ^{1}H NMR spectrum of **2** shows two 2H multiplets at δ 7.2–7.6 ppm, consistent with a disubstituted benzene (possibly ortho disubstituted, considering the structure of the starting salicylaldehyde). Further, there are two 1H doublets at δ 6.42 and 7.72 ppm, coupled to each other. This coupling constant, 9 Hz, suggests a cis alkene. The alkene double bond accounts for the sixth degree of unsaturation.

All the atoms are now accounted for ($C_6H_4 + CO_2 + C_2H_2 = C_9H_6O_2$), but we have not yet identified the last degree of unsaturation, which must be the result of a ring. All the evidence suggests that **2** is the compound known as coumarin:

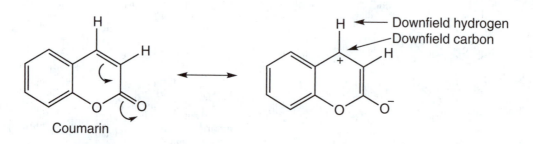

Coumarin

Notice that the downfield position of one of the olefinic hydrogens (δ 7.72 ppm) and one olefinic carbon (δ 143.4 ppm) is nicely accommodated by the coumarin structure, as demonstrated by the resonance form shown in which the β-carbon bears a positive charge. Now, how is this molecule formed? In this system, the most acidic hydrogen is the phenolic proton. The first step is O-acylation to give acetylsalicylaldehyde, **3**.

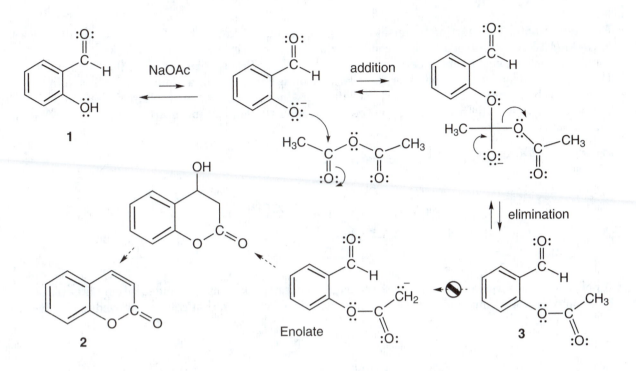

The most economical route to **2** would seem to be formation of the enolate, intramolecular addition to the aldehyde group and dehydration, as outlined with dotted arrows in the figure. However, this idea cannot be correct, as the problem tells us that **3** does not go efficiently to **2** in the absence of acetic anhydride. Acetic anhydride must somehow be involved in the **3** to **2** conversion. If you get stuck at this point, don't worry. It has been proposed that a so-called Perkin reaction now takes place to give **4**. Addition–eliminations lead to **5** (use the indicated oxygen of **4**), and an elimination of acetate gives **2**.

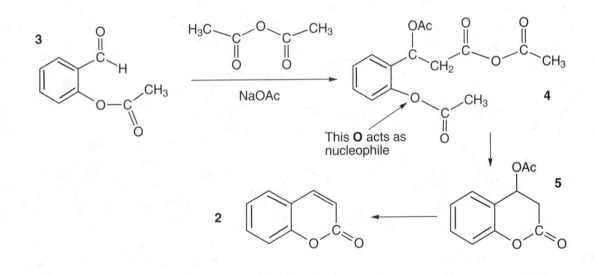

Problem 19.86   The starting material for this problem is a β-keto ester, the potential product (as are *all* β-keto esters) of a Claisen condensation (here in its cyclic version, a Dieckmann condensation). However, remember (or, as the hint says, see Problem 19.35a) that the Claisen and Dieckmann condensations depend for their success on the presence of a "doubly α" hydrogen in the product so that a salt can be formed. This Claisen product does not have such a hydrogen, and so the reverse Claisen, the reversion to the thermodynamically more stable starting materials, is inevitable.

The beginning: a reverse Dieckmann

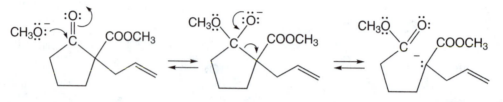

A new enolate is formed through proton transfers and a new Dieckmann condensation takes place to give compound **A**. Compound **A** does contain a "doubly α" hydrogen that can be removed to give **B**.

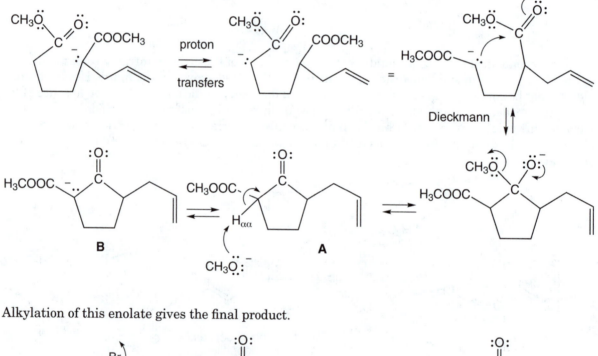

Alkylation of this enolate gives the final product.

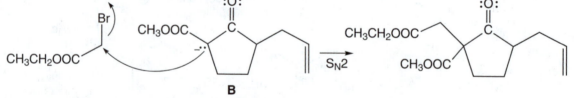

We could probably have done this problem backward. It is quite clear that the final product comes from an alkylation of enolate **B**. How can **B** arise? Only through the Dieckmann condensation shown. Now the problem is to find a way to make the critical enolate **B**. Somehow the ring must open. That's a critical insight. Now you are directed to find a way to open the ring, and that makes starting the problem much easier.

Problem 19.87 Clearly, a ring is to be closed. A quick analysis should show where the starting pieces are in the product (bold bonds), and where the new bonds must be formed. Now think about reactions that let you make those new bonds.

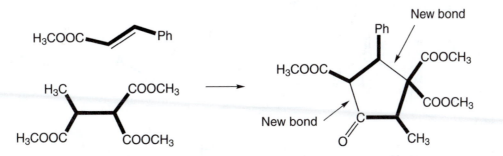

This problem is an example of a "tandem" reaction. In this case, a Michael addition is followed by a Dieckmann condensation. Removal of a "doubly α" hydrogen from the starting material by methoxide gives enolate **A**. This enolate then adds in Michael fashion to methyl cinnamate to give a new enolate, **B**, and to make one of the required new bonds. Intermediate **B** is poised to close the five-membered ring in a Dieckmann condensation to give **C** and make the second new bond. A final protonation leads to the observed product.

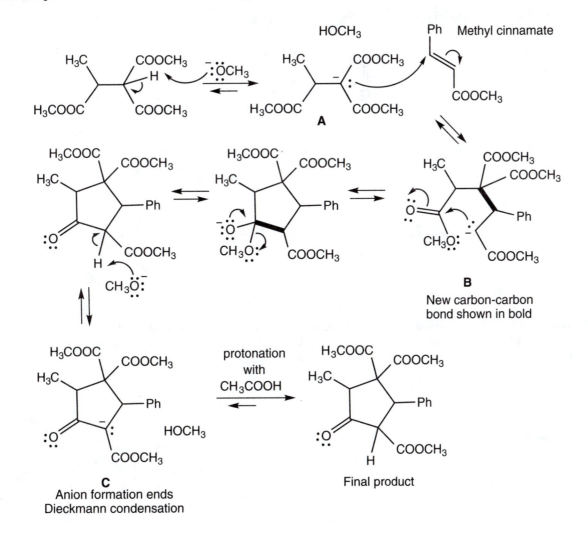

Problem 19.88   Reduction of one equivalent of carbonyl compound by DIBAL-H is simple to explain. The DIBAL-H acts as any hydride reducing reagent (LiAlH$_4$ and NaBH$_4$ are other examples) and delivers a hydride to the Lewis acidic carbon of the carbonyl group.

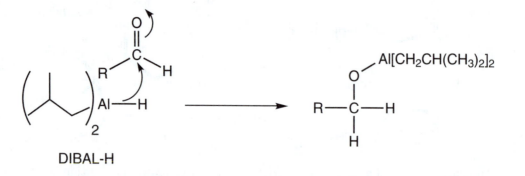

The second and third equivalents of hydride are available from the isobutyl groups, in a manner reminiscent of the Meerwein–Ponndorf–Verley–Oppenauer equilibration (p. 1002).

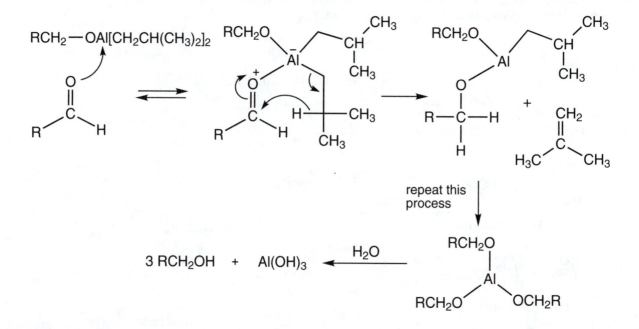

Problem 19.89   This problem starts with an easy reaction. Cyclopentyl methyl ketone (**1**) and pyrrolidine (**2**) must react to give an enamine. In this case, there are two enamines possible, **A** and **A′**. Be sure you can write mechanisms for their formations.

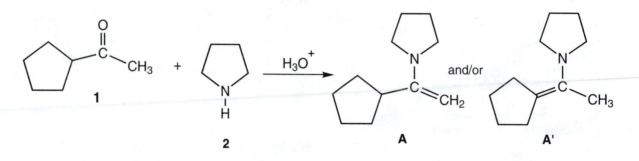

In the original literature, this enamine was represented as the less-substituted isomer **A**. Alkylation of the enamine with methyl α-(1-bromomethyl)acrylate (**3**) affords compound **B**. The two enamines should give different alkylation products, **B** and **B′**. Which is formed? Recall that mild hydrolysis of **B** yields compound **E** for which we have spectral data. The ^{1}H NMR spectral data are most compatible with the ketone derived from the less substituted enamine. For example, structure **E′** would be expected to exhibit a 3H singlet at δ 2.0–2.5 ppm for the ketone methyl group. Such a signal is clearly not present.

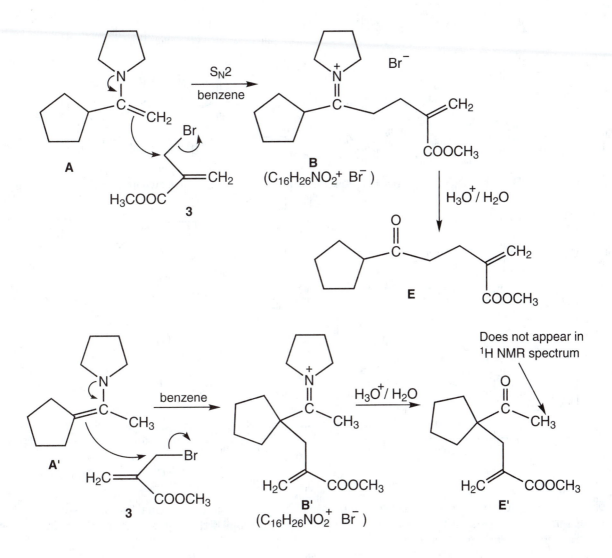

So, apparently the less substituted enamine **A** selectively undergoes alkylation with **3** to give salt **B**. Even if enamine **A** is the minor component of a mixture of enamines **A** and **A′**, it appears to be the more reactive isomer, presumably for steric reasons. Remember that as long as the two possible enamines are in equilibrium, the equilibrium will shift to replace the minor component as it is used up in the alkylation reaction. It is also worth commenting on the alkylation reaction itself, as the methyl α-(1-bromomethyl)acrylate (**3**) is such an unusual electrophile. The alkylation reaction could proceed by three different mechanistic pathways: (1) an S_N2 displacement as shown, (2) an S_N2' displacement, and (3) a Michael reaction followed by expulsion of bromide. The last possibility is illustrated next. Unfortunately, it is not known what the correct mechanistic path(s) is (are).

(continued)

Problem 19.89 *(continued)*

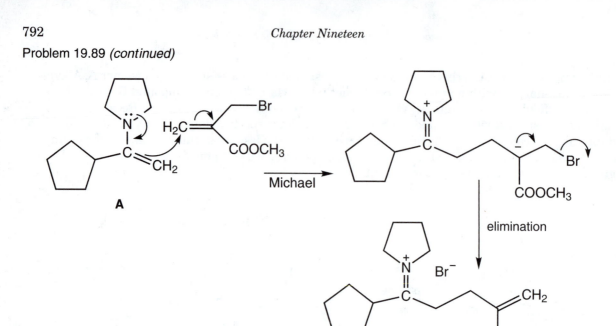

A   **B**

Treatment of **B** with triethylamine generates a new enamine; once again, two enamines, **C** and **C′**, are possible.

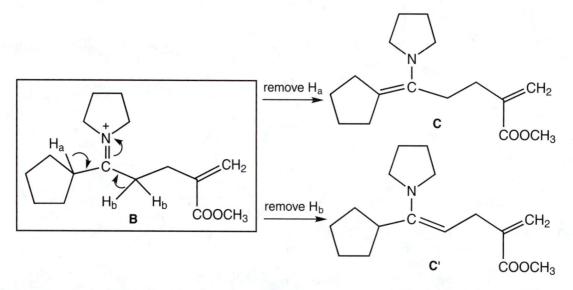

Even if an equilibrating mixture of the two possible enamines **C** and **C′** forms, the more substituted enamine **C** leads ultimately to the observed spirodecanone **4** through an intramolecular Michael addition and mild hydrolysis.

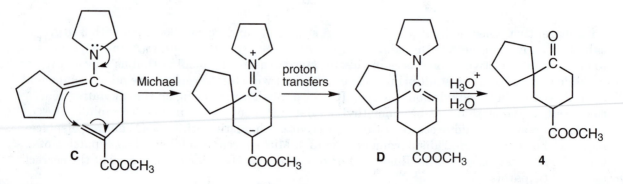

Note that the less substituted enamine **C'** could also undergo an intramolecular Michael addition. However, in this case, a thermodynamically unfavorable four-membered ring would result.

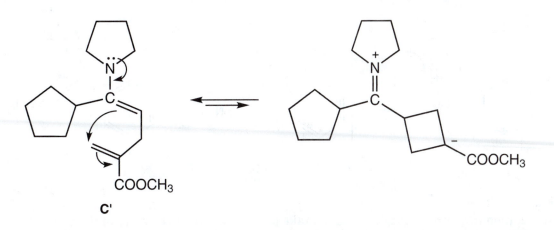

C'

Although there are many mechanistic possibilities in this synthetic sequence, the process is still quite efficient as the spirodecanone **4** was obtained in an overall yield of 78% from **A** and **3**.

Problem 19.90 Let's take those hints and avoid the terrible step of removing the aldehydic hydrogen.

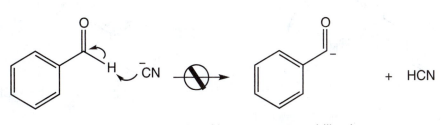

Not resonance stabilized

Instead, let's think about what any nucleophile is likely to do in the presence of a carbonyl group—add to the carbon atom. Protonation gives the cyanohydrin.

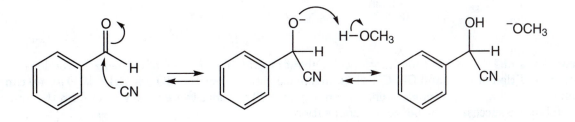

(continued)

Problem 19.90 *(continued)*

Now the fundamental nature of the aldehydic hydrogen is changed—it is now acidic because removal leads to an anion that is resonance stabilized by the cyano group.

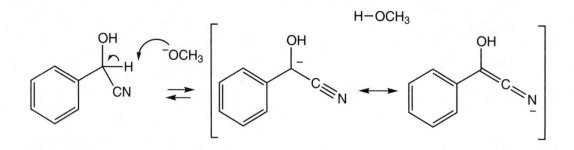

Next, addition to another benzaldehyde can take place to sew the two halves of the molecule together. Loss of the catalyst, cyanide, and protonation gives benzoin. Be sure you can push the arrows for those changes. This reaction provides a wonderful, challenging problem, filled with misdirection. If you got it, feel very good.

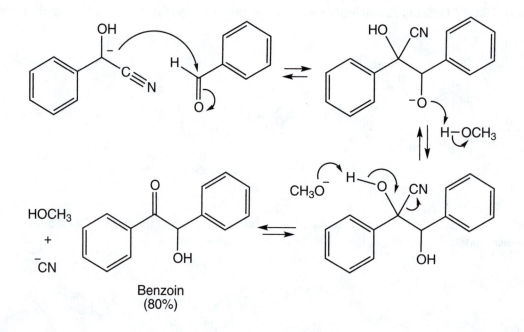

Benzoin
(80%)

Problem 19.91 There is something special about cyanide's promotion of the benzoin condensation. Addition of HO⁻ and $CH_3CH_2O^-$ takes place, but neither the HO nor CH_3CH_2O group can stabilize an adjacent pair of electrons as can a cyanide. Therefore there is no way to form the anion crucial to the success of the benzoin condensation.

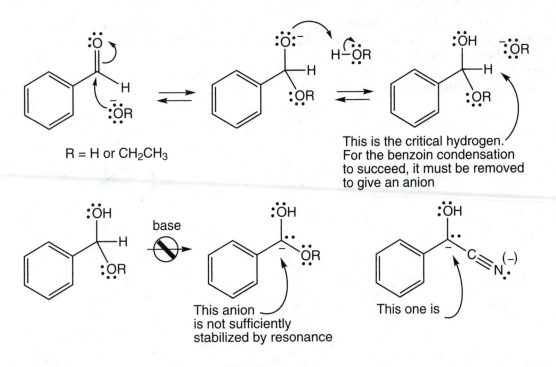

R = H or CH₂CH₃

This is the critical hydrogen. For the benzoin condensation to succeed, it must be removed to give an anion

base

This anion is not sufficiently stabilized by resonance

This one is

Problem 19.92 Like cyanide, this catalyst will stabilize an adjacent anion. The first step in the condensation reaction is addition. Now the critical hydrogen becomes acidic, as an adjacent anion can be stabilized through delocalization after this nucleophile adds.

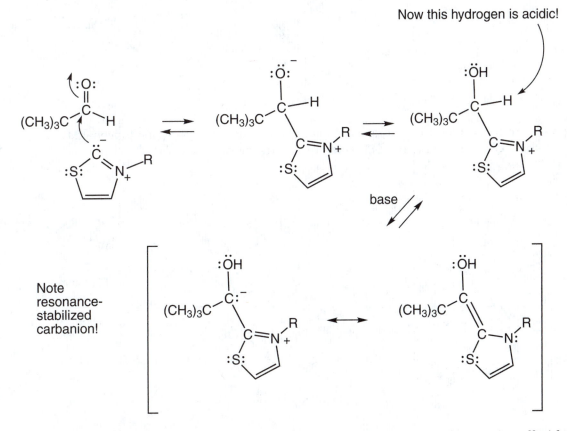

Now this hydrogen is acidic!

base

Note resonance-stabilized carbanion!

This drawing shows how this catalyst can mimic cyanide. You should finish the reaction off with the remaining steps of the benzoin-like condensation. How is the product formed?

Problem 19.93   Recall from Problem 19.92 that deprotonated thiazolium ions can catalyze benzoin-type condensations. That is what is happening here, with the added complication of a decarboxylation. We first deprotonate thiamine pyrophosphate (TPP) with an enzymatic base. The resulting anion adds to the α-keto group of pyruvate **1** to give intermediate **A**. Decarboxylation then occurs to give the resonance-stabilized intermediate **B**.

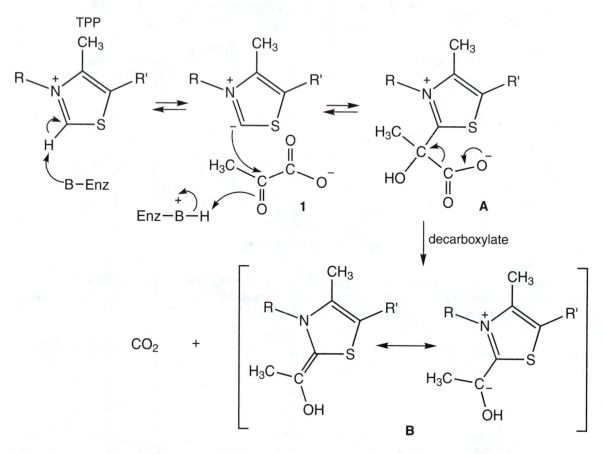

Intermediate **B** now adds to the α-keto group of a second molecule of pyruvate to yield **C**. Intermediate **C** can eliminate the TPP anion to give the product, acetolactate (**2**).

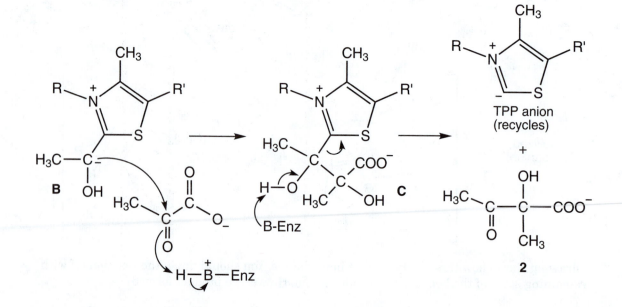

Problem 19.94 Note first that this is a redox reaction. One carbonyl group has been reduced and the other oxidized. This redox process should make you consider a hydride shift, but there is one very substantial problem: There is no hydrogen to be shifted! It is not only hydride that can shift. Here there is a migration of a phenyl group, a benzene ring, with its pair of electrons. In the trade, this reaction is called the benzilic acid rearrangement. There really is only one way to start. There are no α hydrogens, so no enolates can be formed. Hydroxide must add to one of the two equivalent carbonyl groups. In Problem 19.45a, a hydride migrated in what was essentially an intramolecular Cannizzaro reaction. Here there is no hydride, and it is a benzene ring that plays the role of the migrating group. This reaction generates an alkoxide, and proton transfer to make the far more stable carboxylate anion must be very fast. A final acidification generates the free acid.

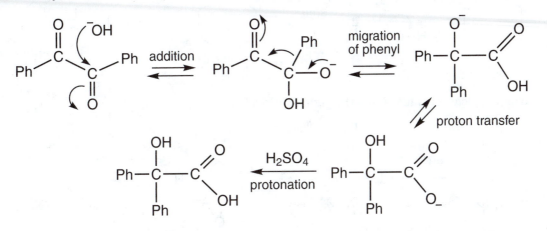

The key to this problem is the recognition that a benzene ring has moved. That much is obvious from a glance at the starting material (two separated benzene rings) and product (two benzene rings on one carbon). Although that much *is* obvious, it is nonetheless important to go through the analysis that makes you articulate the thought that a ring must move. Then you can set out to accomplish that task, asking at every point in the problem, Is there a way now to move the ring?

Problem 19.95 The first step in this condensation reaction is an acid-catalyzed Michael addition of the enol of methyl ketone **2** to the protonated α,β-unsaturated ketone **1**. This reaction ultimately leads to 1,5-diketone **A**.

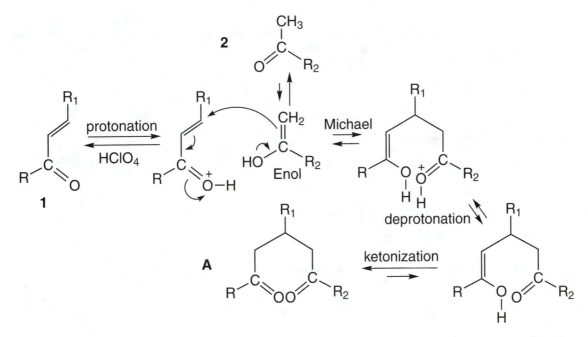

(continued)

Problem 19.95 *(continued)*

We now need to convert **A** into pyrilium salt **3**. Two things need to be accomplished in this transformation. One is obvious, but the other is more subtle. The obvious step is a cyclization with
dehydration—we clearly must close a ring somehow if we are to make **3**. The other step is an
oxidation. Unfortunately, we need to accomplish the subtle step before we can do the easy one. The
oxidation can best be accomplished by a hydride transfer from **A** to the conjugate acid of **1**. This
transfer gives carbocation **B**, the key to the cyclization step, and ketone **5**.

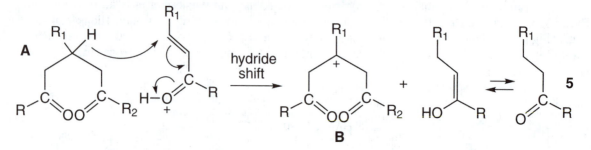

The enol of carbocation **B** leads to **3** through a cyclodehydration.

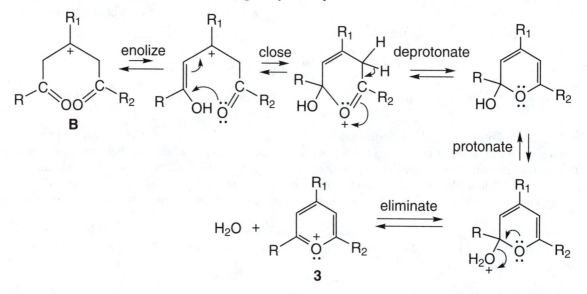

Now the pyrilium salt **3** is easily converted into pyridine **4**, as shown below, in a classic "open–close"
process. Ammonia first adds to C(2), and then the ring-opening, ring-closing sequence takes place.
The final steps are a dehydration and deprotonation.

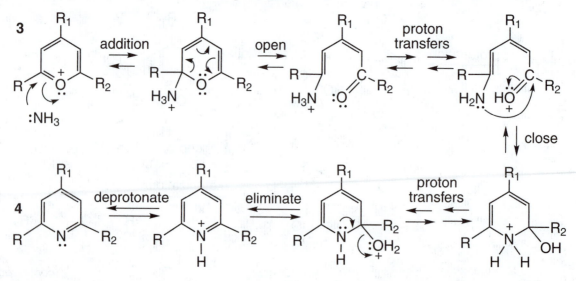

Problem 19.96　The dehydration of a β-hydroxy aldehyde to give the unsaturated α, β-unsaturated aldehyde will occur on heating under acidic conditions. The HOMO of the enolate shows that there is more anionic character on the carbon of the enolate than on oxygen. Because the HOMO is the reacting orbital, it is this nature of the HOMO that promotes reaction at the carbon rather than the oxygen of the enolate.

Problem 19.97　This reaction has a cyclic six-membered transition state. The transition state is most likely in a chair conformation, which has a significant impact on the stereochemical outcome of such aldol reactions. The transition state is called a Zimmerman–Traxler transition state based on the work by Howard E. Zimmerman (1926–2012) and his undergraduate student Margaret Traxler. The larger groups will orient themselves so that they are equatorial in this chair-shaped transition state.

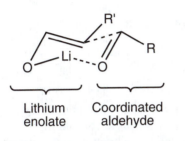

Chair-like structure for the transition state of the lithium-coordinated aldol reaction, called Zimmerman–Traxler transition state

Lithium enolate　　Coordinated aldehyde

Problem 19.98　There is not a large visible difference between the density of the LUMO on the β carbon and the density of the LUMO on the carbonyl carbon. Based on this observation, it is difficult to say whether a nucleophile would react at the β carbon or the carbonyl carbon. Because the reaction is usually reversible, we can argue that the thermodynamic stability controls whether the nucleophile adds to the β carbon (1,4-) or the carbonyl carbon (1,2-). The faster process (1,2- or 1,4-addition) is probably controlled by the location of the nucleophile. So the Grignard reaction is likely to occur in a 1,2-fashion when it is nicely coordinated next to the carbonyl carbon by the interaction of the magnesium with the carbonyl oxygen. This situation would be similar for hydride addition using $LiAlH_4$. Coordination can also be used to explain the selective 1,4-addition of cuprate complexes.

Problem 19.99　For a hydrogen to be removed by base, the C—H bond must line up with the π bond of the carbonyl group. The aldol condensation begins when the aligned hydrogen is removed by base.

Carbohydrates

20

Even though there are few new reactions in this chapter, sugar chemistry can certainly be vexing. The profuse functionality makes for many possible choices in most reactions, and the stereochemical complexity is certainly serious. There is even a new stereochemical convention to master, the Fischer projection. Still, if you keep your wits about you, go slowly, and always keep in mind that what you are doing is applying old knowledge in a new setting, even sugar problems become easy. Always pay close attention to stereochemistry, however. Here, the Additional Problems section (Section 20.8) picks up with some building block problems before going on to more detailed, challenging exercises.

Problem 20.2 *Remember*: All vertical bonds are heading away from you and all horizontal bonds are coming toward you. The process of turning these Fischer projections into three-dimensional drawings begins with a transformation that puts in the wedges. Then, translate this picture into a "sawhorse form," and finally draw this sawhorse in the staggered, energy-minimum form.

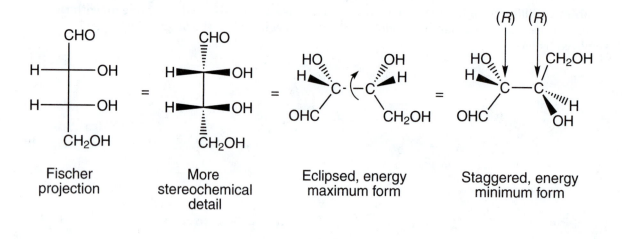

Fischer projection = More stereochemical detail = Eclipsed, energy maximum form = Staggered, energy minimum form

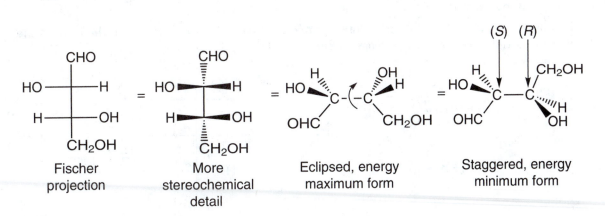

| Fischer projection | More stereochemical detail | Eclipsed, energy maximum form | Staggered, energy minimum form |

Problem 20.3 The convention for drawing Fischer projections is to draw the carbon chain verti-
cally with the most oxidized carbon (usually an aldehyde) at the top. Then each carbon is shown as
an intersection of a horizontal line with the vertical line. When the typical sugar is drawn this way,
those carbons that have an OH group on the right and an H on the left have the (R) configuration. If
the OH group is on the left and the H is on the right, then it is an (S) configuration. There are many
ways to go from a wedge/dash drawing that has 3D perspective to a Fischer projection. Perhaps the
most reliable method is to figure out the (R/S) configuration at each stereogenic carbon on the
perspective drawing and represent it correctly on the Fischer projection.

(a) Notice first that the carbon on the left is an aldehyde carbon. It is C(1). There are five carbons.
The C(2) carbon has the (R) configuration, C(3) is (S), and C(4) is (R).

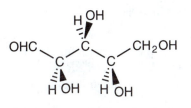

Original perspective structure

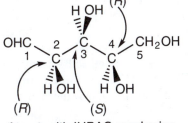

Atoms with IUPAC numbering, (R/S) determined

Translates to:

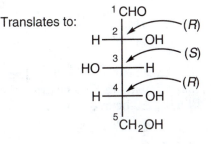

Fischer projection with atoms numbered and (R/S) matching original structure

(b) The aldehyde carbon is on the right. It is C(1). For this structure, C(2) has the (S) configuration,
C(3) is (S), C(4) is (R), and C(5) is (R).

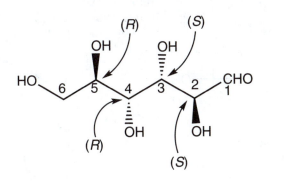

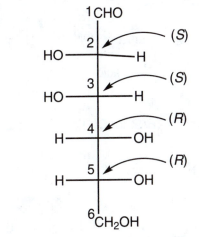

(continued)

Problem 20.3 *(continued)*

(c) This molecule is (2*R*,3*R*,4*R*)-2,3,4,5-tetrahydroxypentanal. Each OH will be on the right of the Fischer projection because each stereogenic carbon is (*R*).

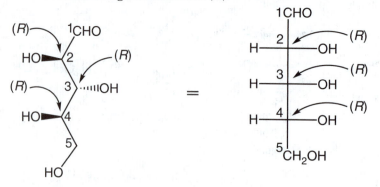

(d) This molecule is (2*S*,3*R*,4*R*)-2-amino-3,4,5-trihydroxypentanal.

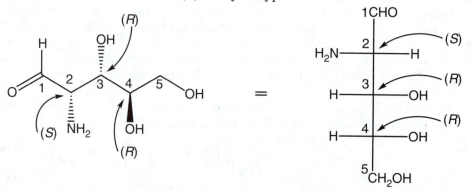

(e) Sialic acid is not a typical sugar. So we will need to be careful about the assumptions for (*R/S*) assignments in Fischer projection. The IUPAC name for this molecule is (4*R*,5*R*,6*R*,7*R*,8*R*)-4,5,6,7,8,9-hexahydroxy-2-oxononanoic acid. The (*R/S*) determination for C(4) is shown below.

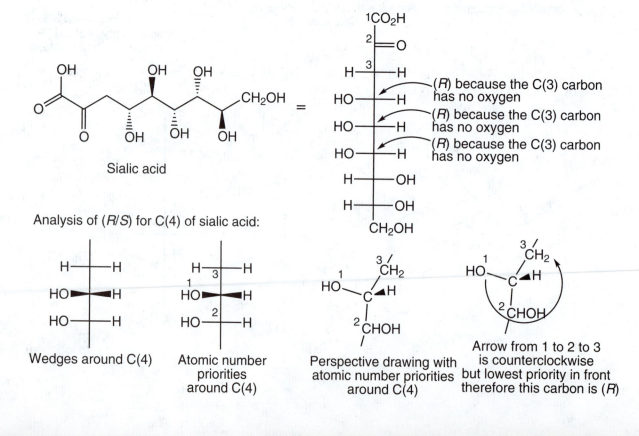

Problem 20.4 This problem is easy as long as you take a systematic approach. The question asks for L sugars so the "bottom" OH, in other words, the OH on the configurational carbon, must be on the left. Start by putting all three OH groups on one side and then work out the three possible ways to have two OH groups on one side and one on the other (without making the molecule a D sugar).

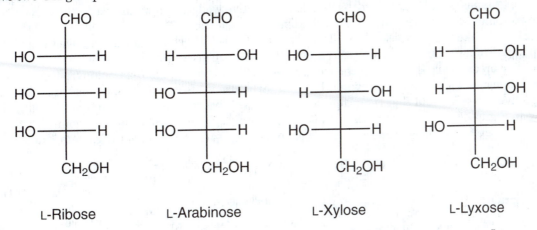

L-Ribose L-Arabinose L-Xylose L-Lyxose

Problem 20.5 There are five stereogenic carbons in an aldoheptose, so there will be $2^5 = 32$ possible stereoisomers, 16 in the D series and 16 mirror-image L isomers.

Problem 20.6 D-Glucose is (2R,3S,4R,5R)-2,3,4,5,6-pentahydroxyhexanal. We can draw the line structure with all of the hydroxyl groups on hexanal, then figure out the correct use of wedges or dashes at each stereogenic carbon to get the three-dimensional perspective drawing with the correct configurations at each carbon.

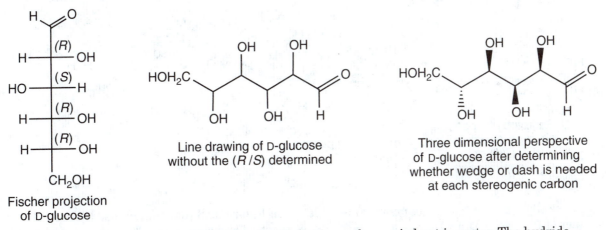

Fischer projection of D-glucose

Line drawing of D-glucose without the (R/S) determined

Three dimensional perspective of D-glucose after determining whether wedge or dash is needed at each stereogenic carbon

The reaction of the aldehyde with sodium borohydride can be carried out in water. The hydride source attacks the carbonyl carbon to give the alkoxide. The alkoxide can deprotonate water to give D-glucitol.

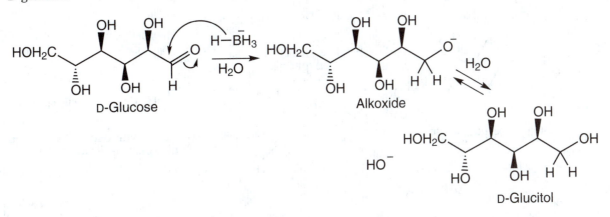

D-Glucose Alkoxide D-Glucitol

Problem 20.7 Drawing D-galactopyranose in three-dimensional perspective is a good challenge. Start with the Fischer projection as shown in Figure 20.8 and then add wedges to the horizontal bonds. You can add the dashes to the groups at the top of the projection and the bottom (adding more dashes will likely only add confusion). Now rotate the perspective drawing 90° clockwise, so that the aldehyde group is on the right going back. The perspective wedges can be shown on the carbon framework more clearly now, so you can visualize looking at the C(3)—C(4) bond in front of the molecule. To make the pyranose we need to rotate the C(4)—C(5) bond in order to have the C(5) OH group in position to add to the aldehyde carbon. Ring closure gives the pyranose with the OH group on C(1) either up or down (shown with the squiggly line). That structure is D-galactopyranose drawn in three-dimensional perspective. Drawing the chair structure is also a way of showing three-dimensional perspective. Keep in mind that D-glucopyranose has every substituent in the equatorial position. Notice in Figure 20.8 that D-galactose is a C(4) epimer of D-glucose. We could have gone directly to the chair structure for D-glucopyranose and just changed the C(4) position to get D-galactopyranose.

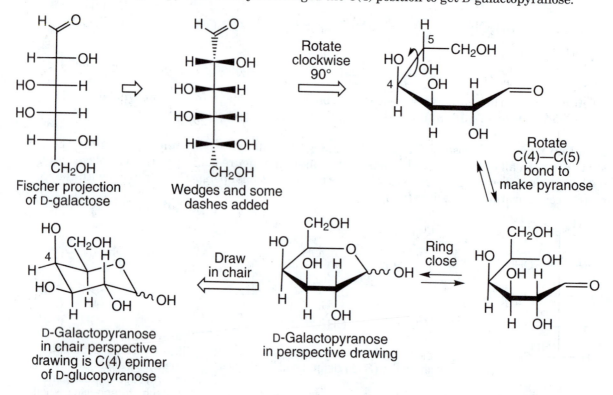

Fischer projection of D-galactose → Wedges and some dashes added → Rotate clockwise 90° → Rotate C(4)—C(5) bond to make pyranose → D-Galactopyranose in perspective drawing → Ring close → Draw in chair → D-Galactopyranose in chair perspective drawing is C(4) epimer of D-glucopyranose

Let's use that approach for D-mannopyranose. We see in Figure 20.8 that D-mannose is the C(2) epimer of D-glucose. Draw D-glucose in the chair perspective and change the C(2) OH from equatorial position to axial position.

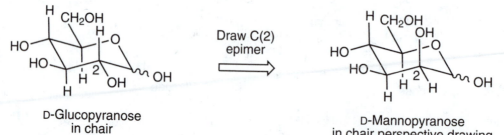

D-Glucopyranose in chair → Draw C(2) epimer → D-Mannopyranose in chair perspective drawing

Now we have a twist. We are asked to draw L-gulopyranose. The important thing to remember about L sugars is that they are the enantiomers of the D sugars. So let's not make this problem any harder than it needs to be. Draw D-gulopyranose and then draw its enantiomer. D-Gulose is an epimer of D-glucose at C(3) and C(4).

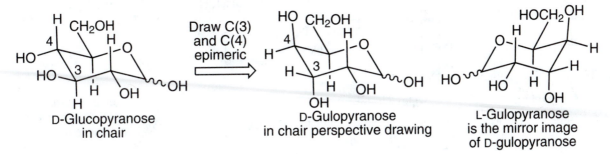

Draw C(3) and C(4) epimeric

D-Glucopyranose in chair

D-Gulopyranose in chair perspective drawing

L-Gulopyranose is the mirror image of D-gulopyranose

Problem 20.8   The numbers show the group priorities.

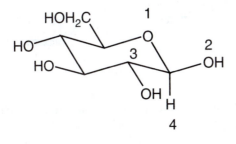

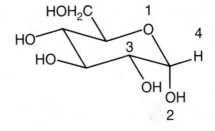

β Anomer is (*R*)

α Anomer is (*S*)

Problem 20.9

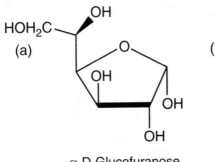

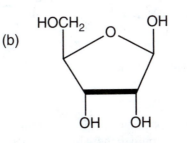

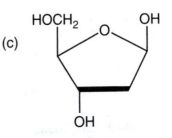

α-D-Glucofuranose

β-D-Ribofuranose

β-D-2-Deoxyribofuranose

Problem 20.11   Just follow the procedure outlined in the text.

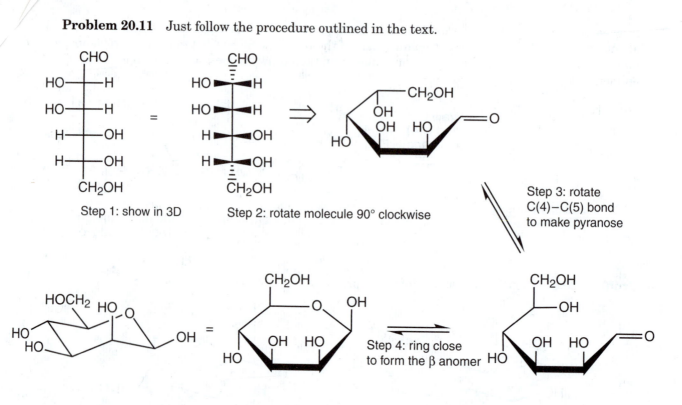

The problem tells you to draw the β anomer, so the OH at the "squiggled" position is equatorial. As you can see, D-mannose is the C(2) epimer of D-glucose.

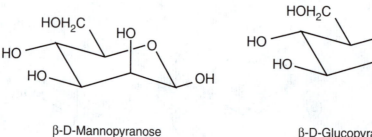

Problem 20.12   The Kiliani–Fischer synthesis begins with the reaction between an aldose and NaCN that gives two diastereomeric cyanohydrins. The cyano group for each is then reduced to the corresponding imines by hydrogenation with a poisoned Pd-catalyst. The imines, without being isolated, undergo hydrolysis in water to give the chain-extended sugars that are epimeric at C(2). L-Ribose gives L-allose and L-altrose.

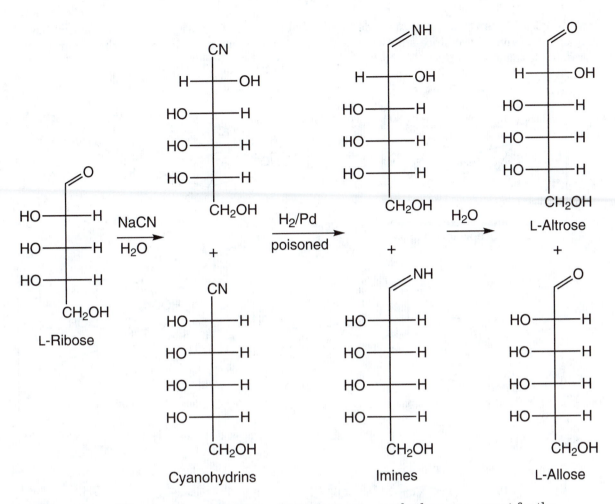

Problem 20.13 The two sugars, D-gulose and D-idose, are exactly the same, except for the stereochemistry at C(2). They can be made through the Kiliani–Fischer synthesis from the aldopentose one carbon shorter, D-xylose.

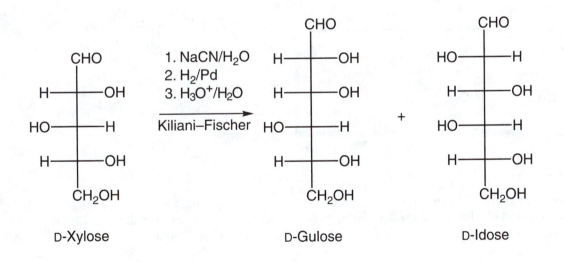

Problem 20.14   An ether oxygen of one hemiacetal pyranose form of glucose is protonated. The ring opens to give a resonance-stabilized cation. Reclosure can regenerate either the original pyranose or the other anomer.

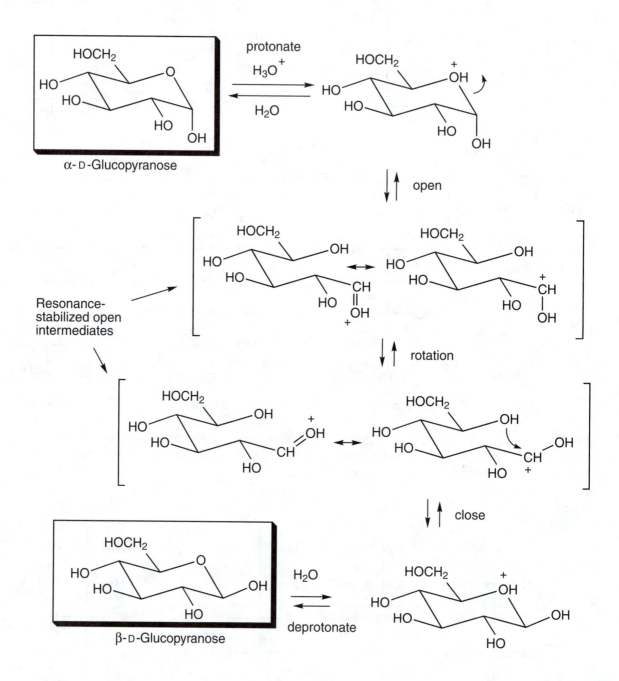

Problem 20.15   The furanose form contains a five-membered ring, and thus it is the OH group attached to C(5) that is involved in the ring [remember that the carbonyl group is at C(2) in fructose]. In the six-membered pyranose form, it is the OH group at C(6) that participates in ring formation.

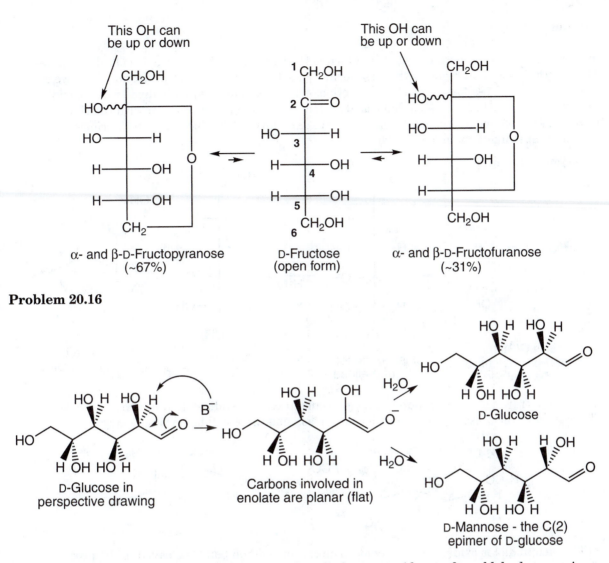

This OH can be up or down

CH$_2$OH

α- and β-D-Fructopyranose
(~67%)

^{1}CH$_2$OH
^{2}C=O

D-Fructose
(open form)

This OH can be up or down

CH$_2$OH

α- and β-D-Fructofuranose
(~31%)

Problem 20.16

D-Glucose in
perspective drawing

Carbons involved in
enolate are planar (flat)

D-Glucose

D-Mannose - the C(2)
epimer of D-glucose

Problem 20.18 Remember that aldohexoses show little or no evidence of an aldehyde group in their IR and NMR spectra. The reason is that there is little of the open aldehyde present; most of the molecule is tied up as the stable hemiacetal. A similar phenomenon occurs with aldonic acids. There is an intramolecular reaction between one of the OH groups and the acid to form a cyclic ester, a lactone. Here is an example.

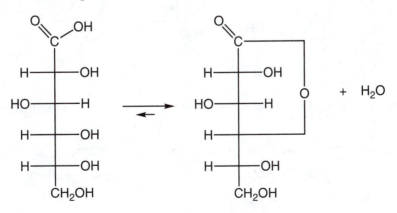

Open form of D-gluconic acid

One possible cyclic ester
made using the OH at C(4)

+ H$_2$O

Problem 20.19

(a) First, note that the pyranose that is shown is the C(3) epimer of glucose. It is D-allopyranose. The Br$_2$/H$_2$O oxidation requires the open form. Draw D-allose in Fischer projection. Now we can draw the oxidized D-allose in Fischer projection. An aldose that has its aldehyde carbon oxidized to the carboxylic acid is called an aldonic acid. Aldonic acids are in equilibrium with the ring-closed lactones. The five-membered ring lactones are usually favored. We have shown the ring-closed product that is likely the major product.

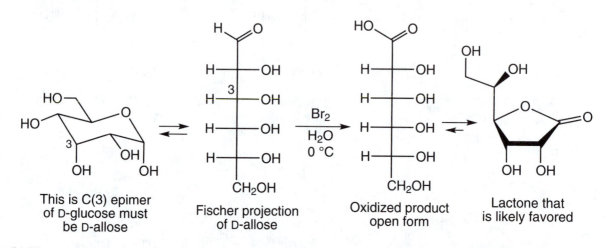

This is C(3) epimer of D-glucose must be D-allose

Fischer projection of D-allose

Oxidized product open form

Lactone that is likely favored

(b) There is only one alcohol group in this molecule. It can be oxidized to the carbonyl by PCC. There is no water present in this reaction, so we assume there will be no ring opening.

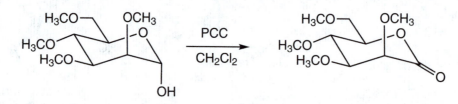

(c) The vicinal diol is oxidized by periodate. The carbon–carbon bond is cleaved in the process. There is only one vicinal diol in this molecule, so the product is the dialdehyde shown. The stereochemistries for the stereogenic carbons at C(1), C(4), and C(5) are retained in the process.

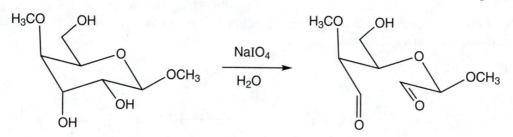

Problem 20.20   The first step is loss of bromide to give a resonance-stabilized intermediate.

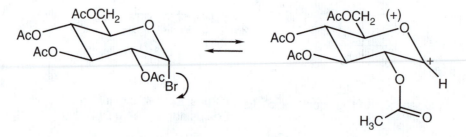

Now look at that acetate carbonyl in just the right position to close a ring to make a second resonance-stabilized intermediate.

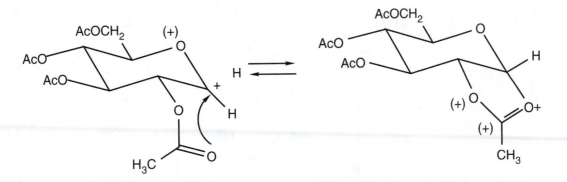

Now inversion through S_N2 opening by alcohol, followed by deprotonation, must give the observed product.

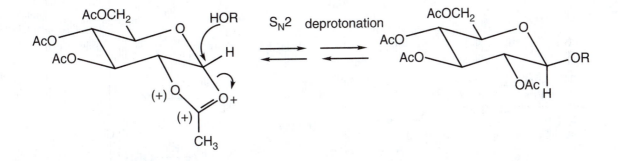

Problem 20.23 This problem seems really hard, but it is actually not so tough. The only real problem is to deal with the attachment point, and that turns out to take care of itself in most cases.

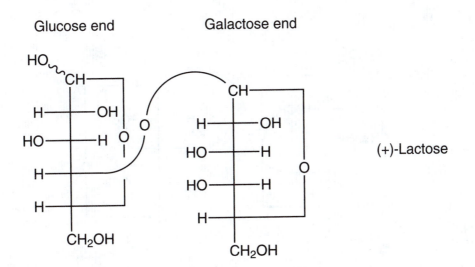

Problem 20.24 Two glucose molecules are attached in this disaccharide. Methylation with dimethyl sulfate, followed by acid hydrolysis, pinpoints the position at which the two sugars were attached. This position appears as a bare OH group, in this case at C(4).

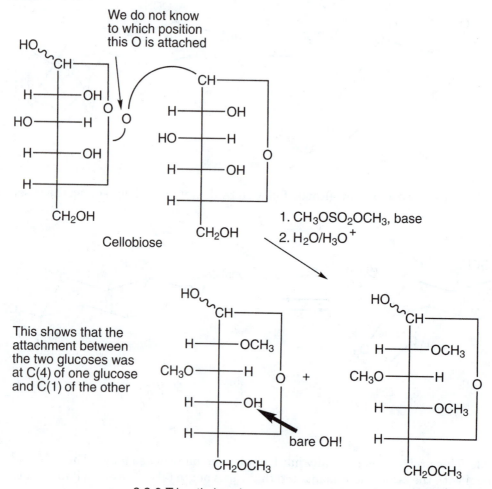

2,3,6-Trimethyl-D-glucopyranose   2,3,4,6-Tetramethyl-D-glucopyranose

Octamethylcellobiose must be

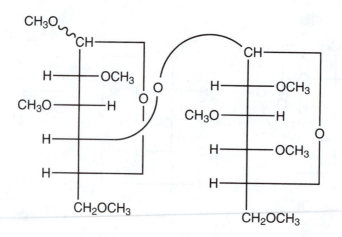

As cellobiose is cleaved by lactase, the two sugars are attached in β-fashion. Here is a three-dimensional drawing.

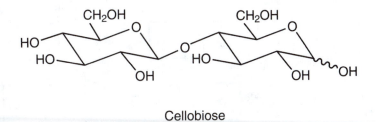

Cellobiose

As you can see, there are two stereoisomers of cellobiose—the "squiggled" OH can be up or down.

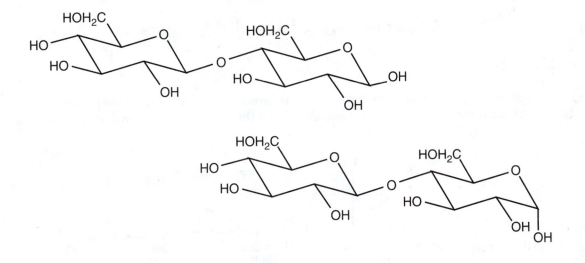

Problem 20.25

(a) An easy way to do this problem is to start with D-galactose and then modify the structure until you get to L-fucose.

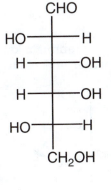

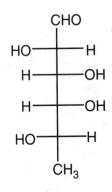

D-Galactose   L-Galactose   6-Deoxy-L-galactose = L-fucose

(continued)

(b) The boxes represent the sugars of the structures on pp. 1068 and 1069.

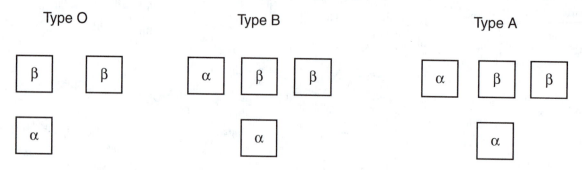

Type O   Type B   Type A

(c) To be a reducing sugar, a structure must be a hemiacetal, which will have some amount of the open, carbonyl-containing form. Thus, it cannot be connected to another sugar through the OH on C(1). As these sugars are all acetals, they are all nonreducing sugars.

Answers to Additional Problems

Problem 20.26

(a) There are many possible answers, but all must be seven-carbon, C_7, sugars containing an aldehyde. As the problem specifies a D sugar, the OH on the stereogenic carbon furthest from the aldehyde must be on the right in Fischer projection.

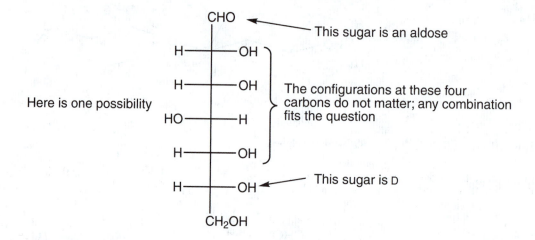

(b) This answer must show a five-carbon sugar containing a ketone. The OH group adjacent to the lower primary alcohol must be on the left in Fischer projection.

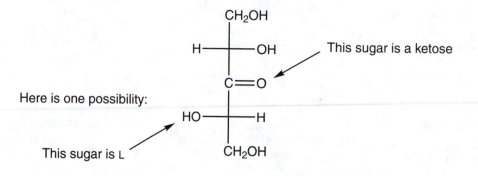

(c) There are only two D-aldotetroses: D-threose and D-erythrose. Only D-erythrose will give a meso diacid (*meso*-tartaric acid) on oxidation with nitric acid.

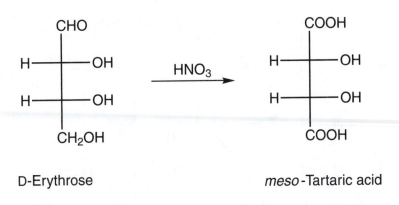

D-Erythrose   *meso*-Tartaric acid

(d) The problem tells us that the sugar galactose is in its pyranose form, so we must draw it as a six-membered ring. The sugar is present as a methyl glycoside, so the anomeric OH (the one newly created in the six-membered ring) is methylated. The α anomer is specified, and that means that the OCH$_3$ must be axial (down) in the chair representation.

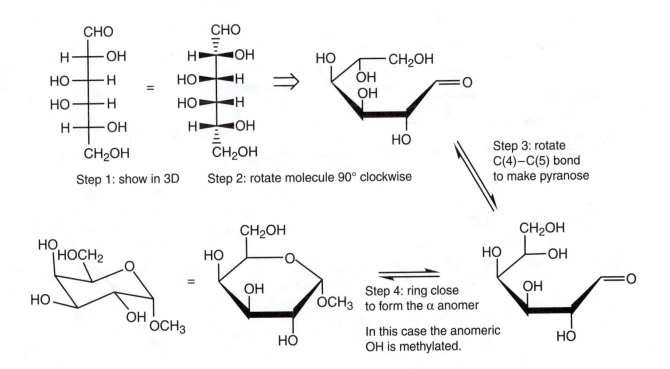

(continued)

(e) D-Allose is the aldohexose with all secondary OH groups on the right in Fischer projection. It is present as the osazone, and that means that both the aldehydic carbon and C(2) have been converted into phenylhydrazones.

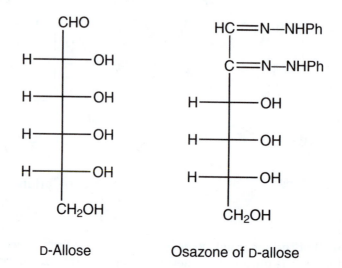

D-Allose   Osazone of D-allose

(f) In this part, the aldopentose ribose is present as a five-membered ring. The anomeric position is β, so the OH is up in the conventional Haworth drawing of a D sugar and is phenylated.

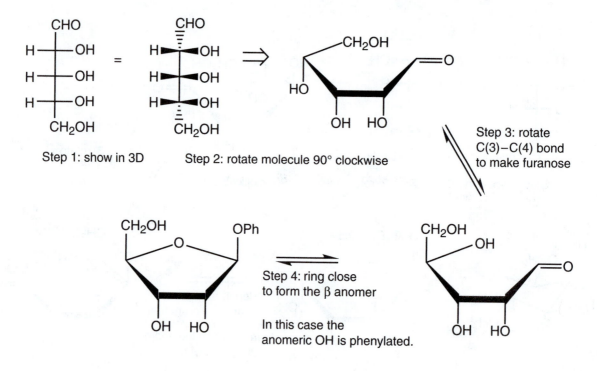

Step 1: show in 3D   Step 2: rotate molecule 90° clockwise

Step 3: rotate
C(3)–C(4) bond
to make furanose

Step 4: ring close
to form the β anomer

In this case the
anomeric OH is phenylated.

Problem 20.27 First of all, there will be four pyranose forms, as both the α- and β-forms can exist as mixtures of two chairs. Follow the procedure in the chapter (or any other method that you devise) to draw the molecules. First, draw the Fischer projections.

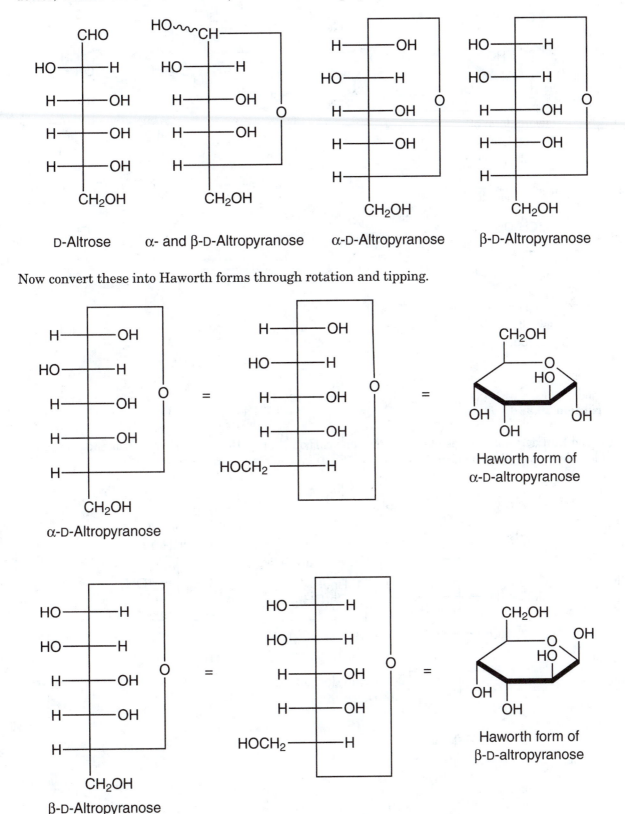

Now convert these into Haworth forms through rotation and tipping.

(continued)

Problem 20.27 *(continued)*

Now let the Haworth forms relax to pairs of equilibrating chair forms.

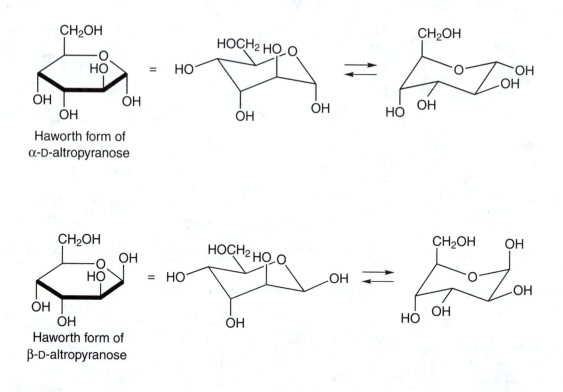

Haworth form of
α-D-altropyranose

Haworth form of
β-D-altropyranose

Problem 20.28

(a) The Ruff degradation results in a sugar one carbon shorter than the original. It is the original aldehyde that is clipped off in the procedure. So, D-gulose results in D-xylose.

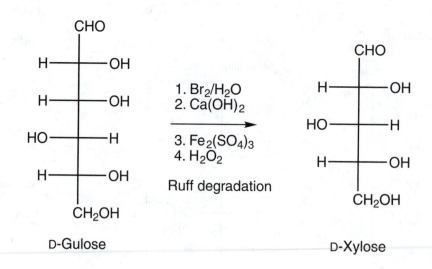

D-Gulose   D-Xylose

(b) The Kiliani–Fischer synthesis lengthens the chain of a sugar and creates both possible isomers at the new stereogenic carbon. So, in this case both D-galactose and D-talose must result.

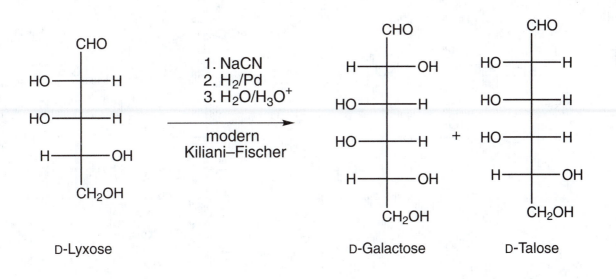

Problem 20.29 Nitric acid oxidizes both the aldehyde and the primary alcohol to carboxylic acids. The ends of the sugar are no longer differentiated in an aldaric acid.

(a) There is one other sugar, **A** (written with the aldehyde at the bottom and the primary alcohol at the top), that will give the same aldaric acid as D-talose. Now flip **A** over into proper Fischer form with the aldehyde group at the top, and we can see that this sugar is D-altrose.

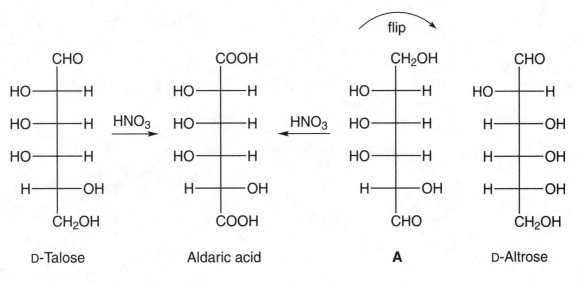

(continued)

Problem 20.29 *(continued)*

(b) Here we use the same technique as in part (a). In this case, the answer is L-xylose, the enantiomer of the original sugar.

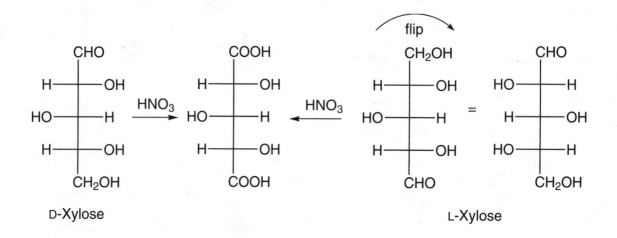

D-Xylose

L-Xylose

(c) Oxidation of D-idose leads to **B**. If we flip idose, we get a "new" sugar that also would give **B**. But, this is not really a new sugar; it is D-idose all over again. There is no other sugar that can give **B**, only D-idose.

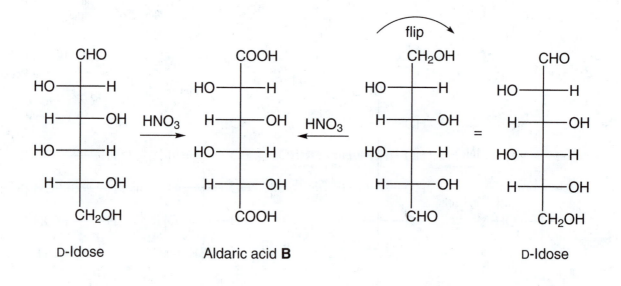

D-Idose   Aldaric acid **B**   D-Idose

Problem 20.30 Osazone formation destroys stereochemistry at C(2) through phenyhdrazone formation, as well as transforming the aldehyde at C(1) into a phenylhydrazone. Hence, the other sugar must be the same as L-talose in every respect, except at C(2). It is the C(2) epimer of L-talose, L-galactose.

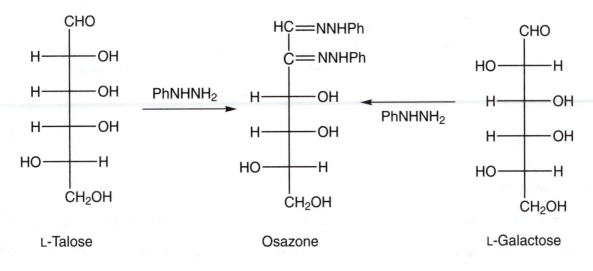

Problem 20.31

(a) Reduction of the aldehyde group to an alcohol. (b) Oxidation of only the aldehyde to the aldonic acid.

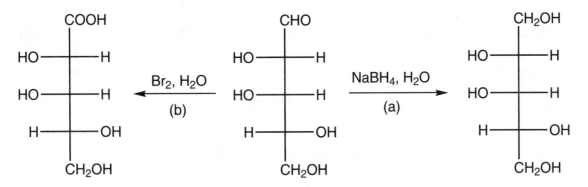

(c) Oxidation of both the aldehyde and primary alcohol to give the aldaric acid.

(d) Methylation of all free OH groups. The OH involved in the furanose ring cannot be methylated.

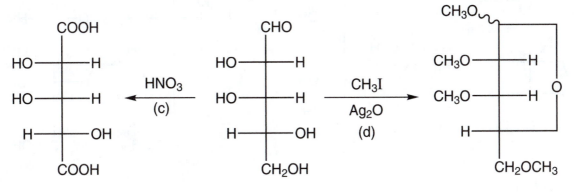

(continued)

Problem 20.31 *(continued)*

(e) In acid, only the OH on the anomeric carbon is methylated.

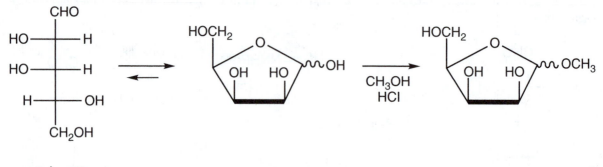

D-Lyxose

Problem 20.32

(a) Acetylation will occur at every free OH, leaving only the oxygen involved in the ring untouched.

(b) Under these conditions, only the hemiacetal OH will be converted into OCH_3.

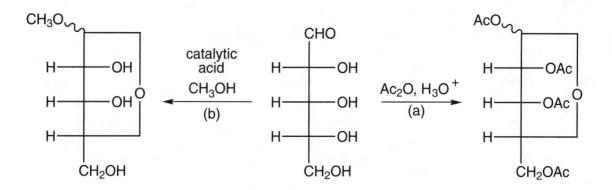

(c) Osazone formation; C(1) and C(2) will be turned into phenylhydrazones.

(d) This reaction is the Kiliani–Fischer synthesis. Two new sugars, one carbon longer, and epimeric at the new C(2), will be formed.

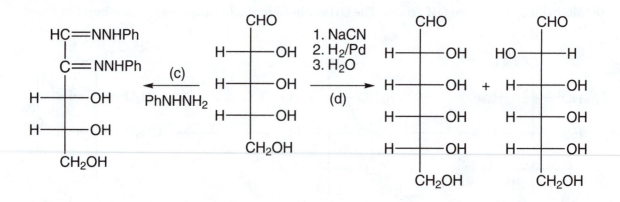

(e) The Ruff degradation leads to the sugar one carbon shorter. In this case, the product is D-erythrose.

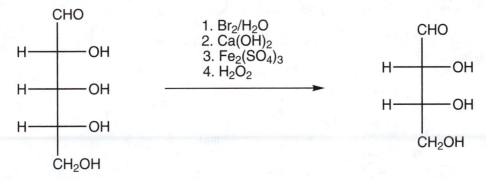

Problem 20.33 Let's just sequentially modify β-D-glucopyranose until we come to 2-amino-2-deoxy-β-D-glucopyranose.

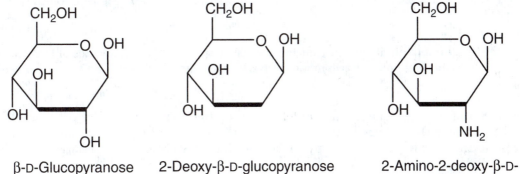

β-D-Glucopyranose   2-Deoxy-β-D-glucopyranose   2-Amino-2-deoxy-β-D-glucopyranose

As a chair form:

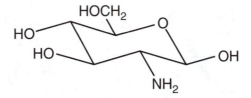

Problem 20.34 It's easy. Just take the ^{13}C NMR spectra. The diester from D-galactose will show four signals, whereas that from D-talose will show eight. There are other similar answers.

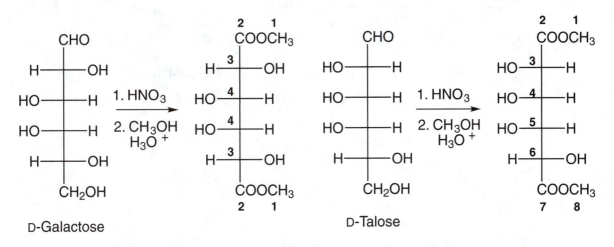

D-Galactose   D-Talose

Problem 20.35   Table 20.2 shows that at equilibrium D-allose exists 16% in the α-pyranose form and 76% in the β-pyranose form. Thus D-allose is 76/92 = 82.6% in the β-pyranose form and 16/92 = 17.4% in the α-pyranose form. The equilibrium constant is 82.6/17.4 = 4.75.

If $\Delta G = -2.3\,RT \log K,$ and recalling that at 25°C, 2.3 RT is just about 1.36, then $\Delta G = -1.36 \log 4.75 = -0.92$ kcal/mol.

Problem 20.36

Let X = fraction α, Y = fraction β.
X(112) + Y(18.7) = 52.7 and
X + Y = 1, Y = 1 − X

So, 112X + 18.7 − 18.7X = 52.7,
93.3X = 34
X = 0.364, or 36.4%; Y = 0.636, or 63.6%. Table 20.2 shows that this answer is right on the nose.

Problem 20.37   First of all, it is clear that maltose is composed of two molecules of D-glucose. The problem is only to unravel how the two monosaccharides are attached and then to make a good drawing.

Bromine in water oxidizes only the free aldehyde group to the acid. The aldehyde of one glucose, bound up in the linkage to the other monosaccharide, will not be affected. So, using Fischer projections for the two D-glucopyranoses, we have a very preliminary structure for maltose, and MBA.

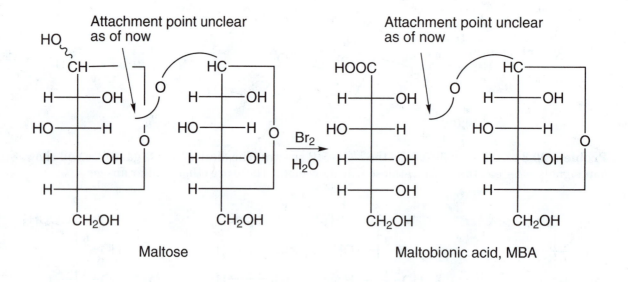

The strategy used is to uncover the attachment point by methylating all free hydroxyl groups in MBA. Hydrolysis will reveal the attachment point as an unmethylated hydroxyl. So,

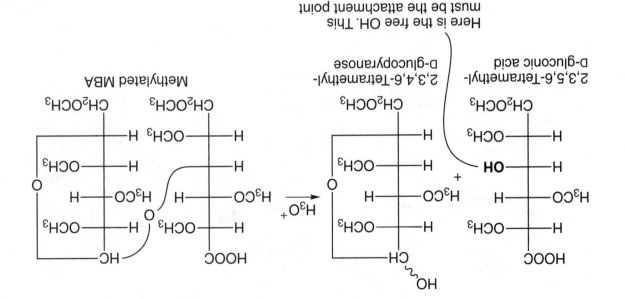

Here is the free OH. This must be the attachment point

Now we know the structure of methylated MBA, and, by implication, MBA itself and maltose.

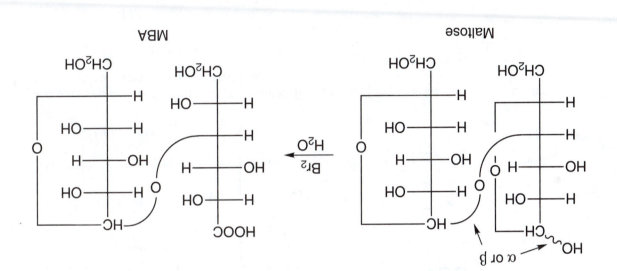

What remains are only the stereochemical details and the translation into three dimensions. The enzyme maltase cleaves only α linkages, so the attachment from C(4) of the left-hand glucose to C(1) of the right-hand glucose is α. Of course, the OH at C(1) of the left-hand glucose can be either α or β.

(continued)

Problem 20.37 *(continued)*

To make the drawing, start with the right-hand glucose, and let Glu stand for the other glucose for the moment.

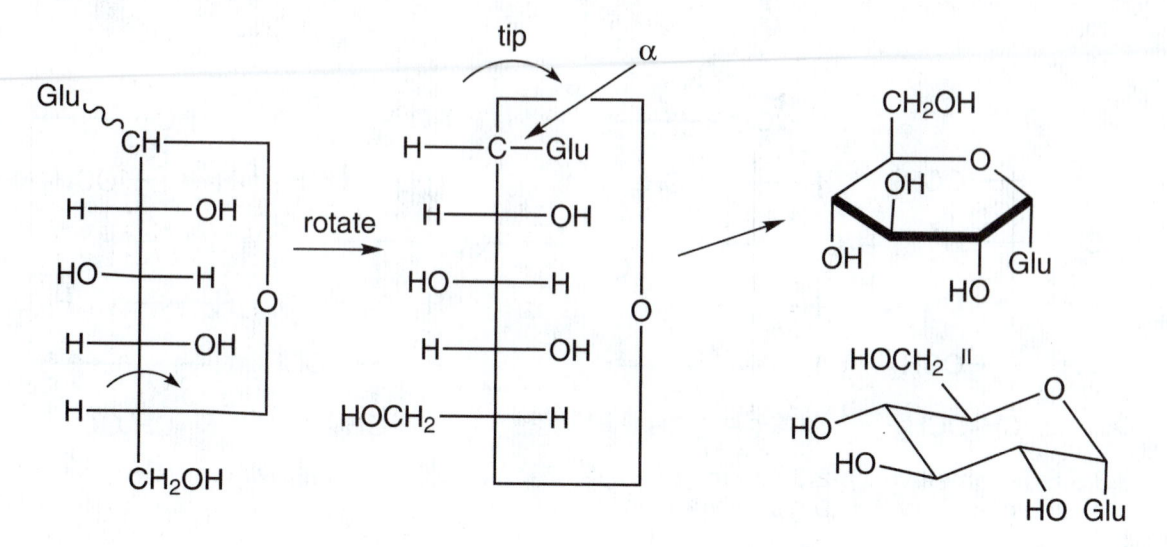

Now repeat the procedure for the other glucose.

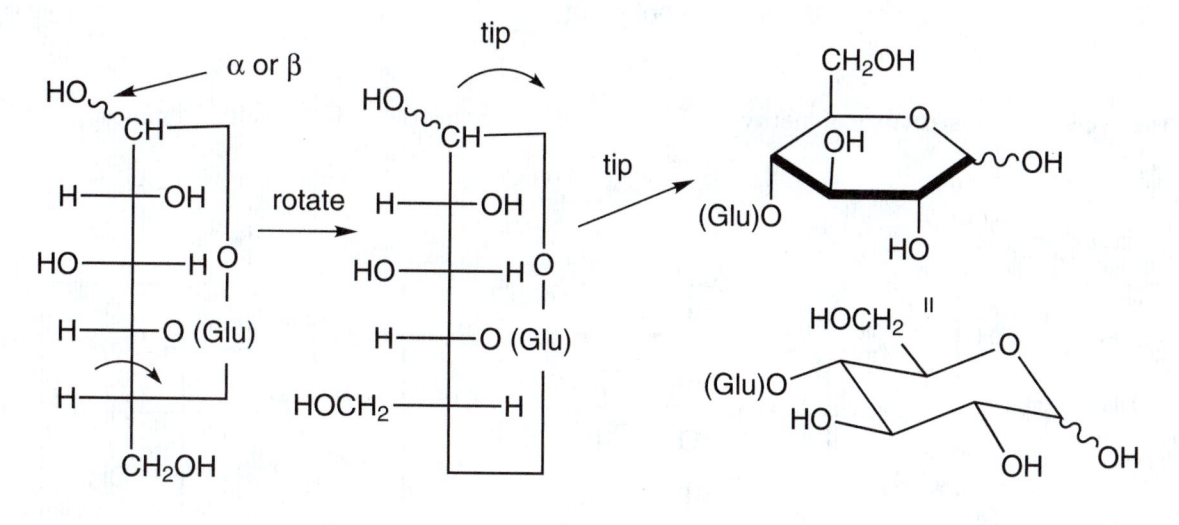

Now it only remains to combine the two three-dimensional forms for the monosaccharides, being careful to attach them in α-fashion.

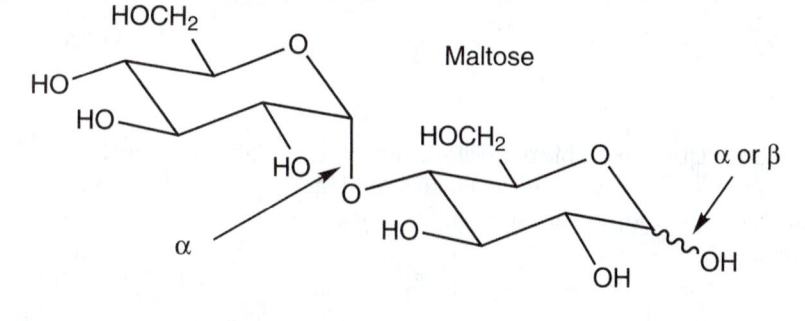

Problem 20.38 The question specifies a nonreducing disaccharide. Thus, there can be no amount of free aldehyde present. The two sugars must be attached at C(1) as parts of a full acetal. Otherwise, one sugar would always have some amount of free aldehyde present, and the disaccharide would be a reducing sugar.

First, let's draw D-altrose and D-allose in their pyranose forms.

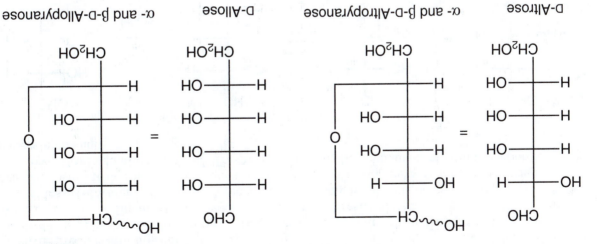

| D-Altrose | α- and β-D-Altropyranose | D-Allose | α- and β-D-Allopyranose |

These must be connected by an acetal linkage of the two C(1) positions. The squiggly bonds still mean that we are not specifying the stereochemistry of the linkage (α or β).

Here is a schematic view, followed by the chair forms ββ, αβ, βα, and αα.

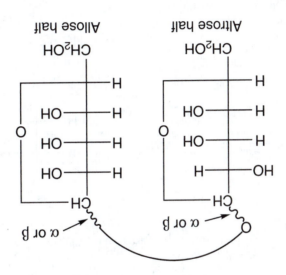

Altrose half   Allose half

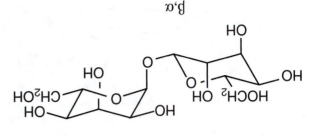

β,β   β,α

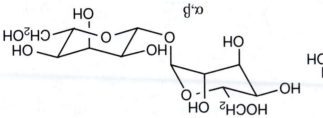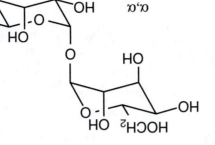

α,β   α,α

Problem 20.39   Let's think a bit about strategy and what must happen in this reaction. An aldehyde group is still present in the product, so, at least for a start, it seems best to leave the aldehyde alone in the starting material. In contrast, the primary alcohol, CH_2OH, has vanished. We need to close a ring and to remove some OH groups. Presumably, that removal will take the form of losses of water in elimination reactions generating the double bonds in the product. There are many ways to do this reaction. One good way is to set up ring formation by converting the primary OH into another aldehyde. This reaction is done through an elimination reaction to generate an enol.

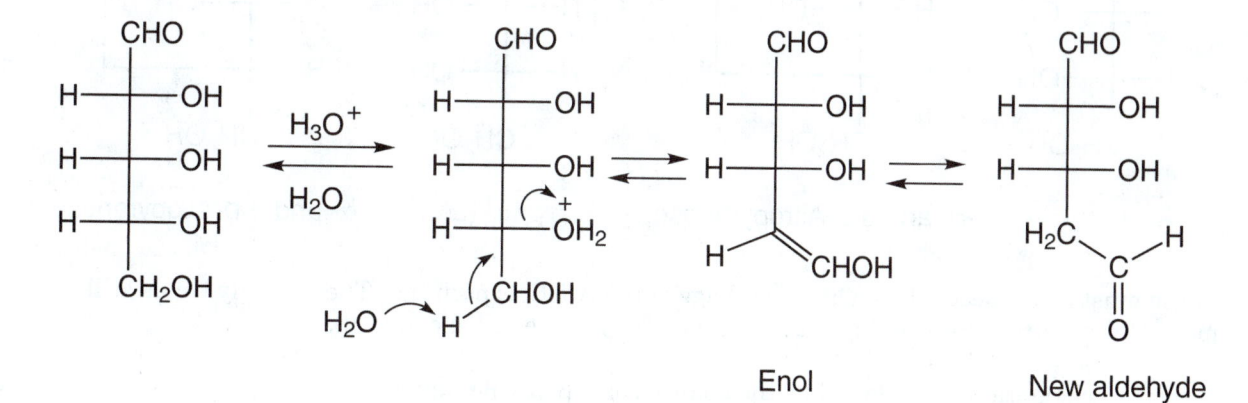

Now we need to close a five-membered ring. The ring size dictates which oxygen atom to use in what is nothing more than an acid-catalyzed hemiacetal formation.

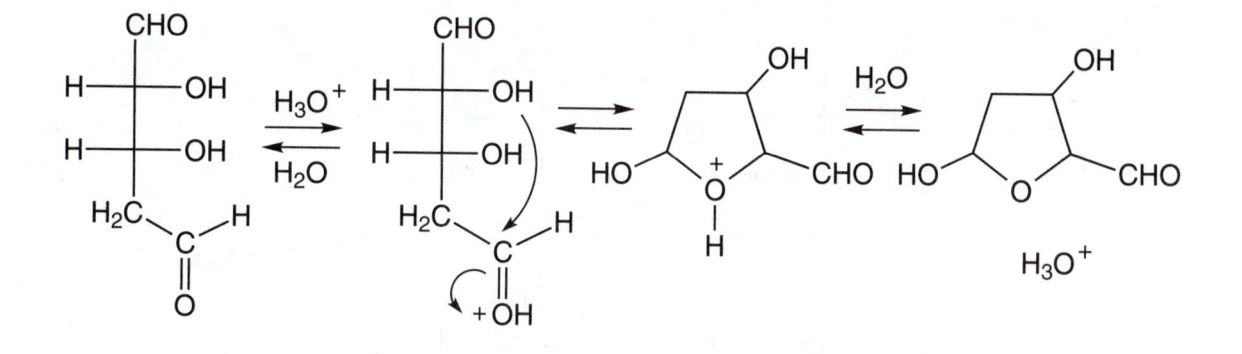

Now we need only protonate an OH and eliminate water twice to get furfural. In the figure, only one E1 reaction is shown. An E2 process is also reasonable. A second elimination gives the product.

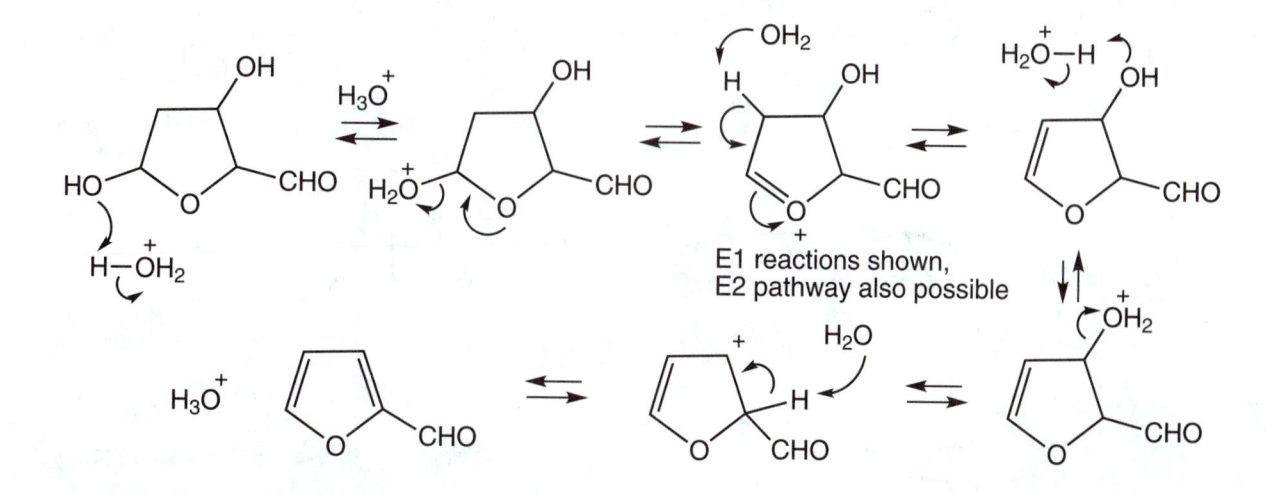

Problem 20.40 In base, not only will the enolate shown in Figures 20.25 and 20.26 form, but alkoxides will be generated as well. In one of these, a hydride shift (intramolecular Cannizzaro reaction) leads to fructose. Reversal of the hydride shift can occur in two ways to regenerate D-glucose or make the isomerized molecule, D-mannose.

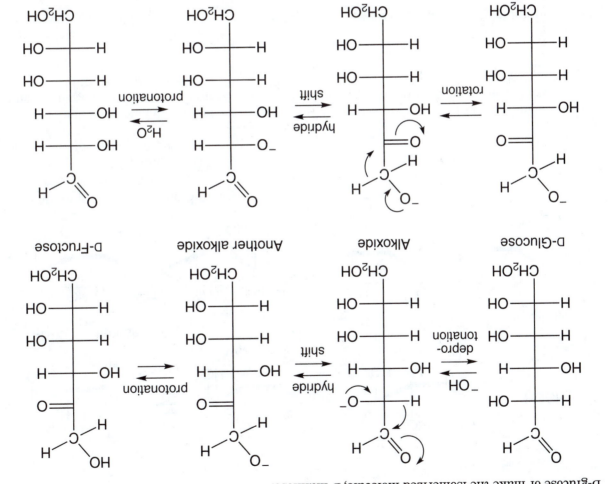

Problem 20.41

(a) First of all, a xylopyranose must be a six-membered ring, and there is only one possible α-pyranose for D-xylose. Here it is in Haworth form.

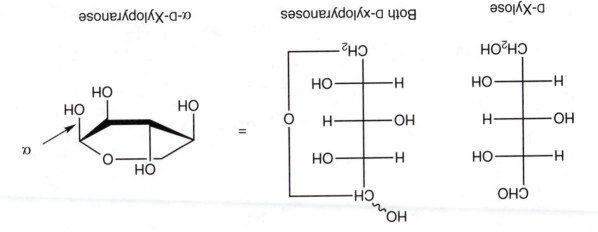

(continued)

Problem 20.41 *(continued)*

The mechanism for methyl glycoside formation goes through a carbocation to which methyl alcohol can add in two ways to give the major products, α- and β-methyl D-xylopyranoside.

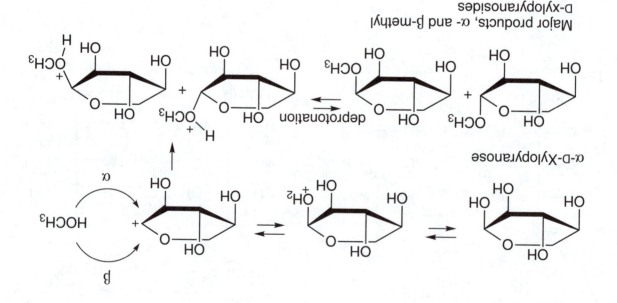

Major products, α- and β-methyl
D-xylopyranosides

(b) There will be a small amount of the open pentose form in equilibrium with the six-membered ring pyranose form. The open form can close to an equilibrium mixture of pyranose and five-membered furanose forms.

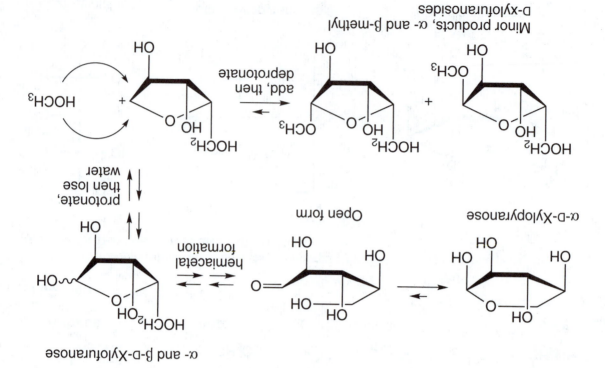

α- and β-D-Xylofuranose

Minor products, α- and β-methyl
D-xylofuranosides

Problem 20.42 Carbocations are not likely intermediates under these basic conditions, so please don't protonate and form a carbocation as in Problem 20.41. The pyranose form of glucose is in equilibrium with the open form, and this compound can react with ammonia to give an imine. In this answer, we have used Haworth forms for simplicity.

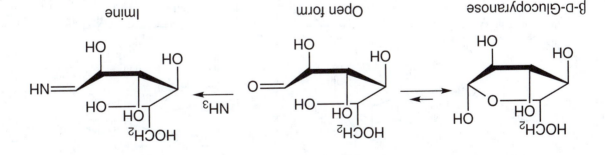

Now reclosure can occur to give the amino sugar. The β-form is shown.

Problem 20.43 This problem is very similar to Problem 20.42. The pyranose equilibrates with a small amount of the open form, which reacts with ethyl mercaptan to give the thioacetal (see p. 999).

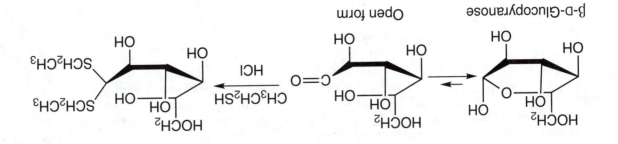

Problem 20.44 First, draw D-glucose in its β-pyranose form and then flip the six-membered ring. Of course this flipped form is less stable than the unflipped form, but some of it will be present at equilibrium. Now protonation, followed by an S$_N$2 displacement reaction, can take place to make **1**.

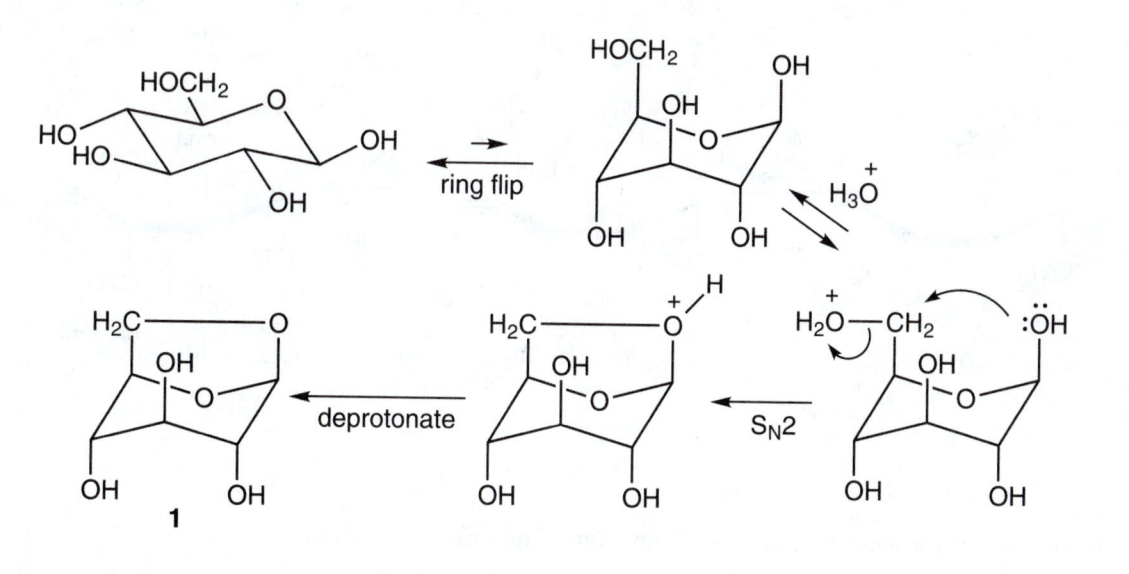

Another mechanism that might account for the formation of compound **1** involves intramolecular attack of the anomeric carbon as shown.

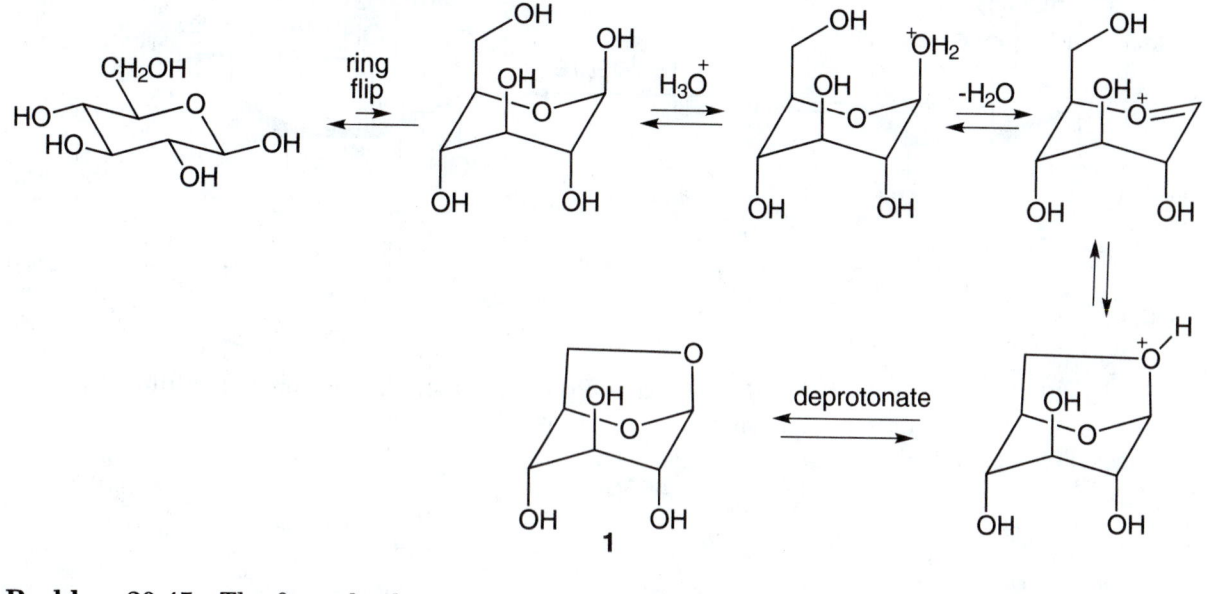

Problem 20.45 The formula shows you that parts of two molecules of acetone have become incorporated into the new molecule (six carbons have been added to the sugar). What is the acid-catalyzed reaction between a diol and a carbonyl compound? Acetal formation (see Chapter 16). In this case, a double acetal is formed. As we know that the primary alcohol group is not involved (the question tells us this), there are only two possible 1,2-diols from which to make the acetals.

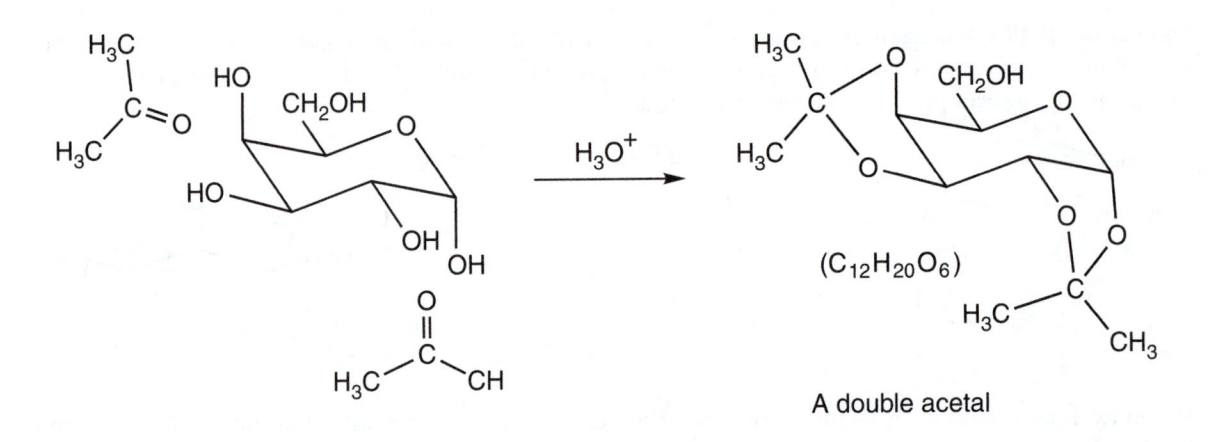

A double acetal

Problem 20.46 All OH groups but the primary alcohol are protected as parts of acetals, which leaves the primary alcohol free to react, and this is where all the action is. First, a tosylate (**A**) is formed (remember that OTs is an excellent leaving group, Chapter 7). Displacement by iodide gives **B**, and reduction leads to a methyl group, as in **C**. Finally, treatment with a catalytic amount of acid removes the acetals and frees the four OH groups. The result is **D**.

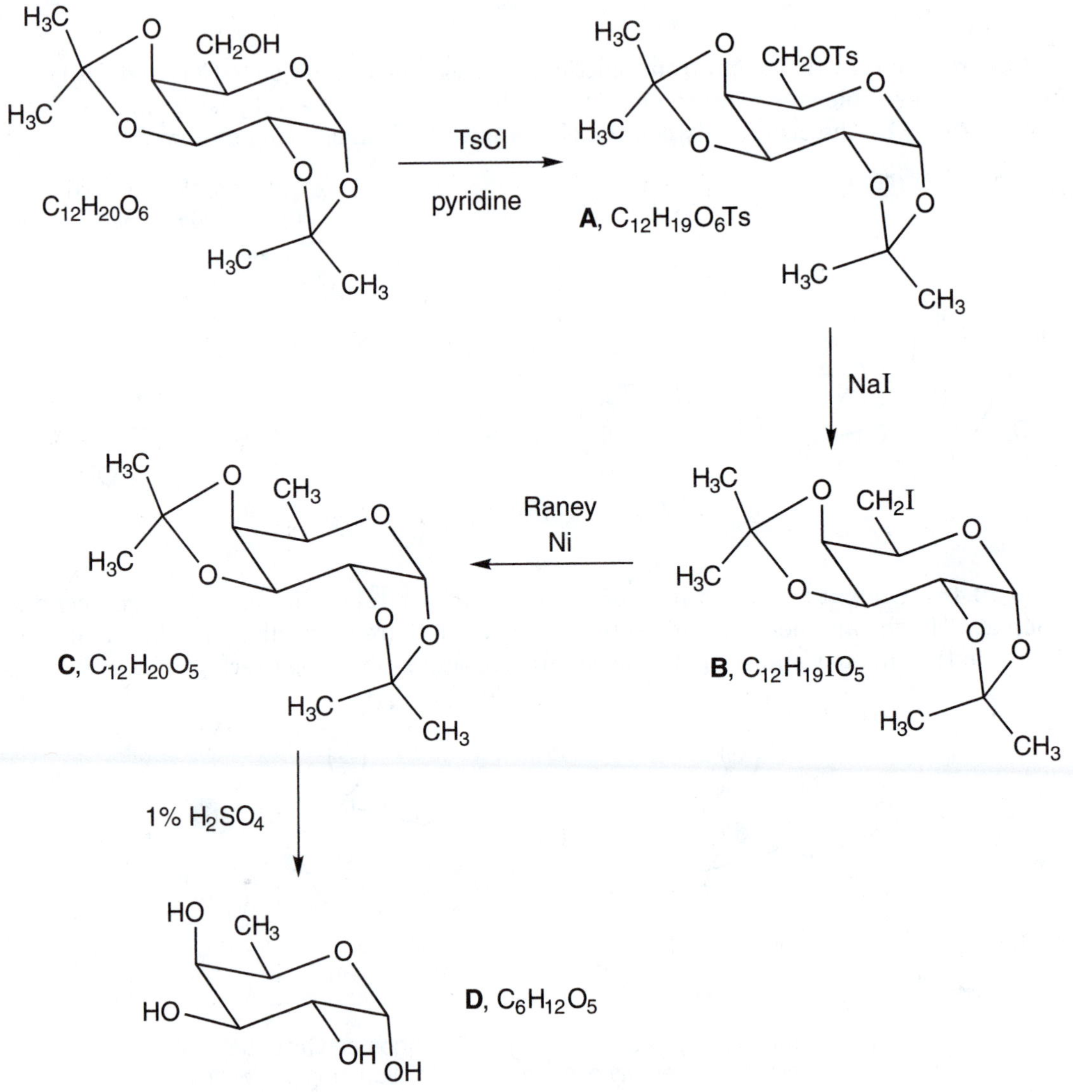

Problem 20.47   Reaction between D-glucopyranose and benzaldehyde gives a chair conformation for the acetal that is formed at the hydroxyl groups on C(4) and C(6). The phenyl group of the acetal is able to occupy the equatorial position.

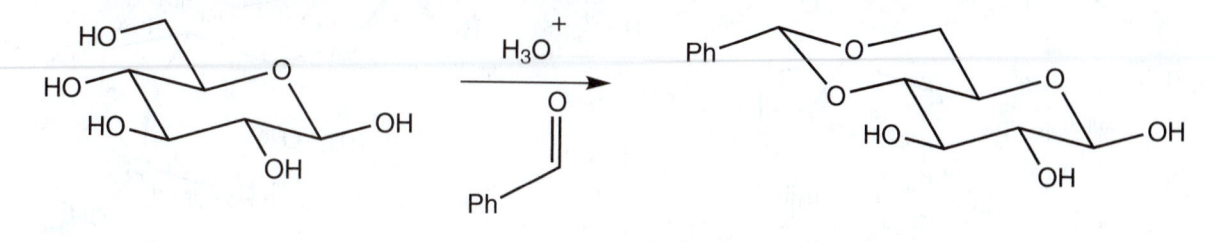

When acetone reacts with D-glucopyranose, the major product is formation of the acetal between C(3) and C(4).

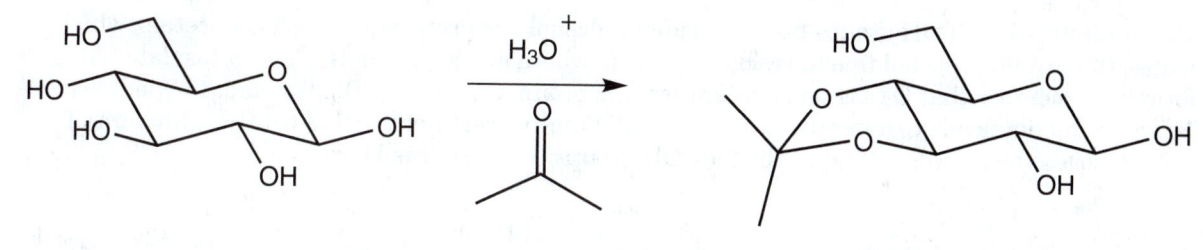

Perhaps the reason that the acetal from reaction with acetone doesn't go to C(4) and C(6) is because of the inevitable steric interaction of a methyl group in the axial position. The literature does not address why the acetal between the C(2) and C(3) alcohols is not observed.

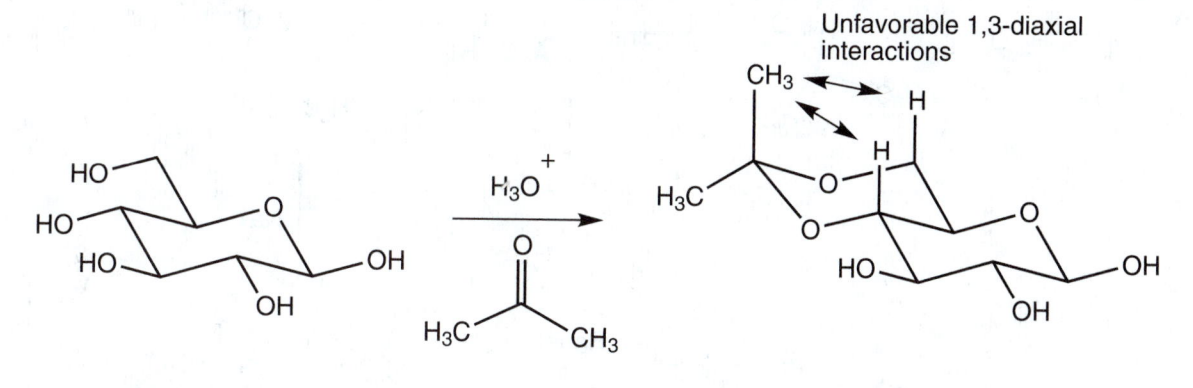

Problem 20.48   Look at the orientation of the empty σ* orbital of the carbon–oxygen bond to the OH group at C(1), the anomeric carbon. If the OH is axial, there is excellent overlap with one of the 2p orbitals on the ring oxygen. If the OH is equatorial, that stabilizing overlap is lost.

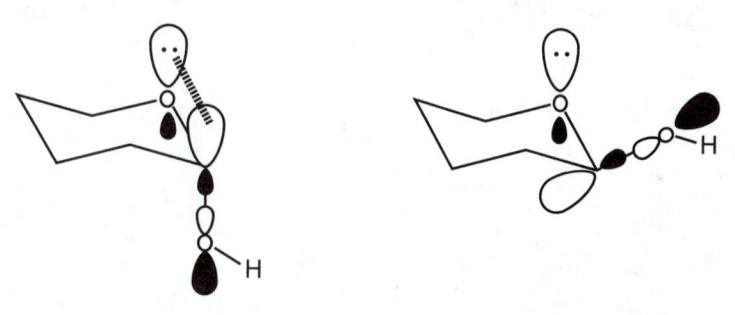

α Anomer, OH axial
excellent filled-empty overlap

β Anomer, OH equatorial
no filled-empty overlap

Problem 21.9 An indole is less basic than an amine because the nitrogen lone pair of electrons of indole is involved in an aromatic ring. These electrons are not available for reaction. In fact, when indole does react with acid, it isn't the nitrogen that gets protonated, because protonating the carbon gives a resonance-stabilized intermediate.

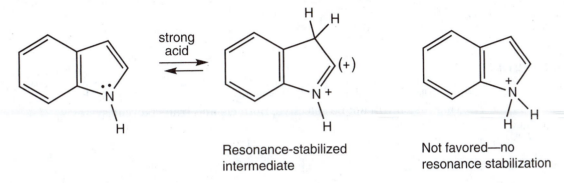

Resonance-stabilized intermediate

Not favored—no resonance stabilization

Problem 21.11 Azide is a terrific nucleophile. The tosylate is a great leaving group. The first step is an S_N2 reaction, which will occur with inversion. The second step is a reduction of the azide to give the amine product.

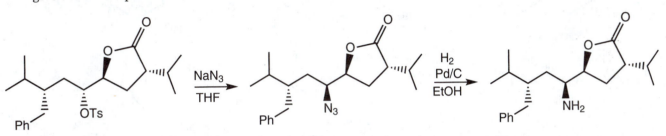

Problem 21.12 The fact that most of the metal hydrides ($LiAlH_4$ in particular) react with water means that these reagents are not compatible with biological conditions. There may be a rare case where a metal hydride is involved in an amide reduction, but it would need to be protected from water some way.

Problem 21.13 Do a retrosynthetic analysis first. The 2-methyl-1-propanamine could come from 2-methylpropanamide or from 2-methylpropanenitrile or from 1-bromo-2-methylpropane. These reagents can come from the other precursors. The route from the amide will require adding CO_2 to a Grignard. We'll assume carbon dioxide is not organic. The route using the nitrile will require adding cyanide anion to a secondary halide (some E2 will compete with the desired S_N2). The reaction that goes through the alkene might come from a Wittig reaction. The Wittig reagent can come from methyl iodide.

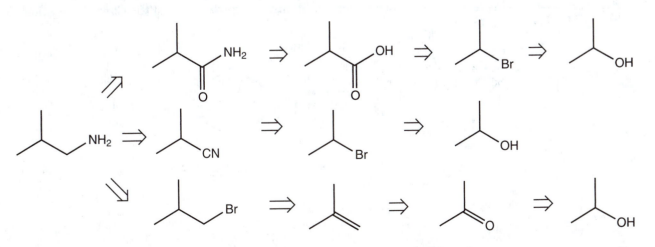

(continued)

Problem 21.13 *(continued)*

The amide route looks best. Here are the forward steps.

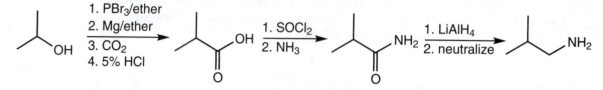

Problem 21.14

(a) The ammonia will do a Michael addition on both enones (one at a time) to give the six-membered ring. The ring fusion will not be selective. Both cis and trans bicyclic systems were formed in about a 1:1 ratio.

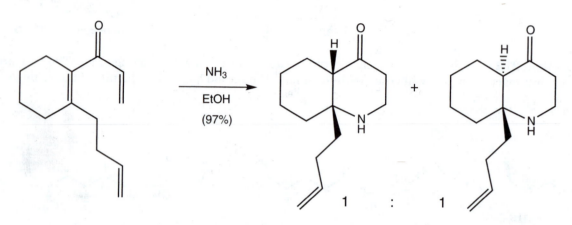

(b) The amine is the best nucleophile, and the acid chloride is the most electrophilic site.

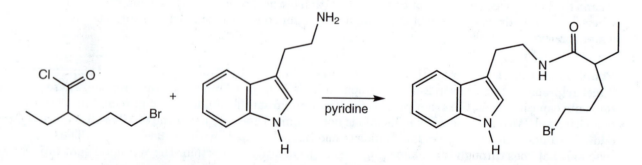

Problem 21.15   This reaction is an intramolecular elimination process called the Cope elimination. (See p. 358, Problem 8.18.)

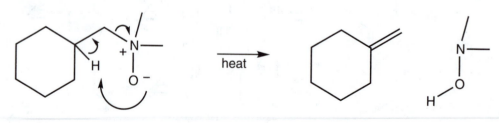

Problem 21.16 The aryldiazonium ion is more stable than the alkyldiazonium ion because the bond between the carbon and the nitrogen is stronger. It is a stronger bond because there is *p* orbital overlap between the aromatic ring and the nitrogen triple bond. The connection is more than just a single bond.

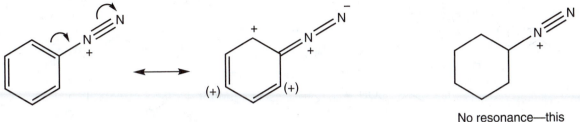 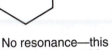

Resonance—this ion is relatively stable

No resonance—this ion is very unstable

The diazonium group is a great leaving group because it leaves irreversibly as N_2, which is a very stable molecule and a gas.

Answers to Additional Problems

Problem 21.18 There are many ways to synthesize each of these molecules. We are showing one route for each. Unfortunately, it is impossible to show all of the ways. It is certainly possible that you will have an equally acceptable synthesis, maybe even a better route.

(a) A secondary amine is most easily obtained from a secondary amide. An amide can be formed from an acid chloride and an amine. The acid chloride comes from the carboxylic acid, which is available from the primary alcohol. The amine is made from an alkyl halide, which is made from the corresponding alcohol.

The forward reactions are

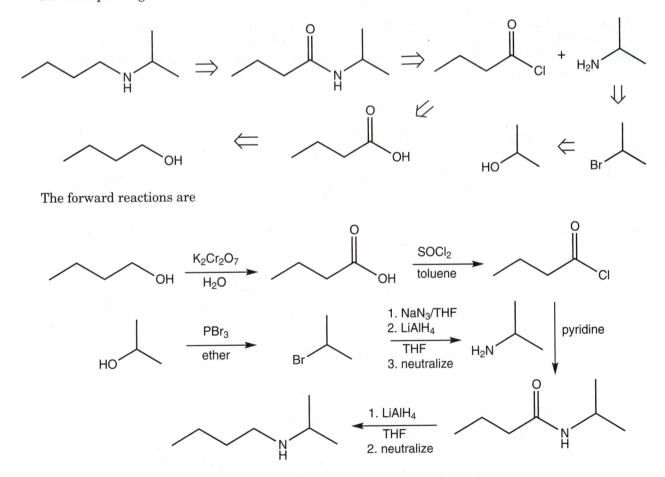

(continued)

Problem 21.18 *(continued)*

(b) The *sec*-pentylacetamide can be made from 2-pentanamine and acetyl chloride. The amine comes from a halide, which comes from an alcohol. This alcohol contains five carbons, so we have to resort to Chapter 16 chemistry to make it. The Grignard reaction between acetaldehyde and propylmagnesium bromide will do the trick. Both can be made from the corresponding alcohols.

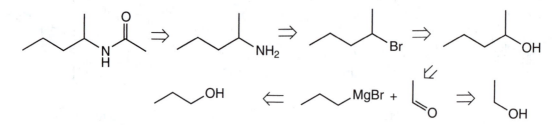

The forward reactions are shown:

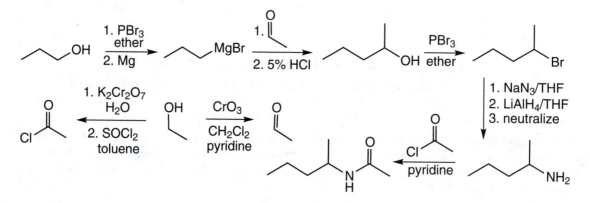

(c) The five-membered ring can be made from the reaction of an amine with a dihalide. The intramolecular double S$_N$2 reaction is favored for this size ring. Running the reaction in dilute conditions will help avoid the reaction of one end of the dibromide with ethanamine and the other end with a different ethanamine.

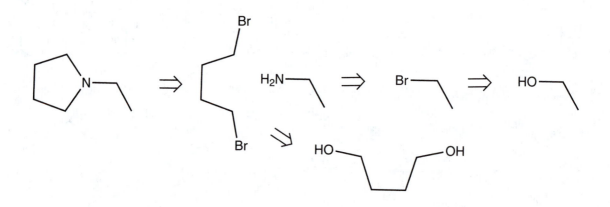

The forward reactions are shown here:

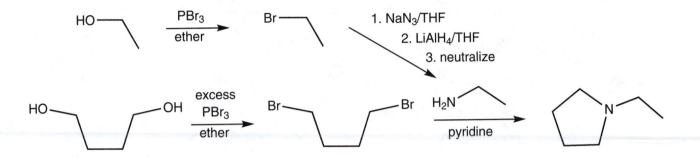

(d) This molecule is 1-(*N*-propylamino)-2-propanol. It can be obtained from an amine reacting with propylene oxide. The epoxide will be opened by the amine attacking at the less substituted carbon. That should be the major product anyway. The epoxide comes from propene (Chapter 11 chemistry), which can be obtained from 2-propanol.

The forward reactions are

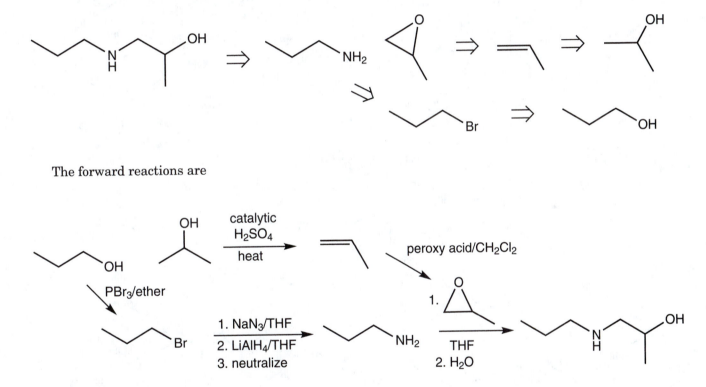

Problem 21.19 The challenge for this problem is to figure out which carbon is the nucleophile and which carbon is the electrophile. We hope you can spot the C—C bonds that need to be made. Remember from Chapter 15 that an aromatic ring with several electron-donating groups (such as OH) is a reasonable nucleophile. The ring with the oxonium ion is electrophilic. So we can imagine that the electrons would come from one anthocyanin to the para position of the oxonium ion ring of another anthocyanin. Then we need to reduce the double bond of the pyran ring. That requires an H⁺ (it is shown as an intramolecular deprotonation, but that's just a guess) and then an H⁻. Now we're ready to add another nucleophile to the oxonium ion ring. This time we'll add a malvidin. Then we can transfer the acidic proton the same way we did before. Addition of an H⁻ gives us the observed product. Why would the hydride addition occur at the middle pyran ring rather than at the malvidin oxonium ion? Who knows? Maybe it is another mechanism altogether. Can you come up with one?

(continued)

Problem 21.19 *(continued)*

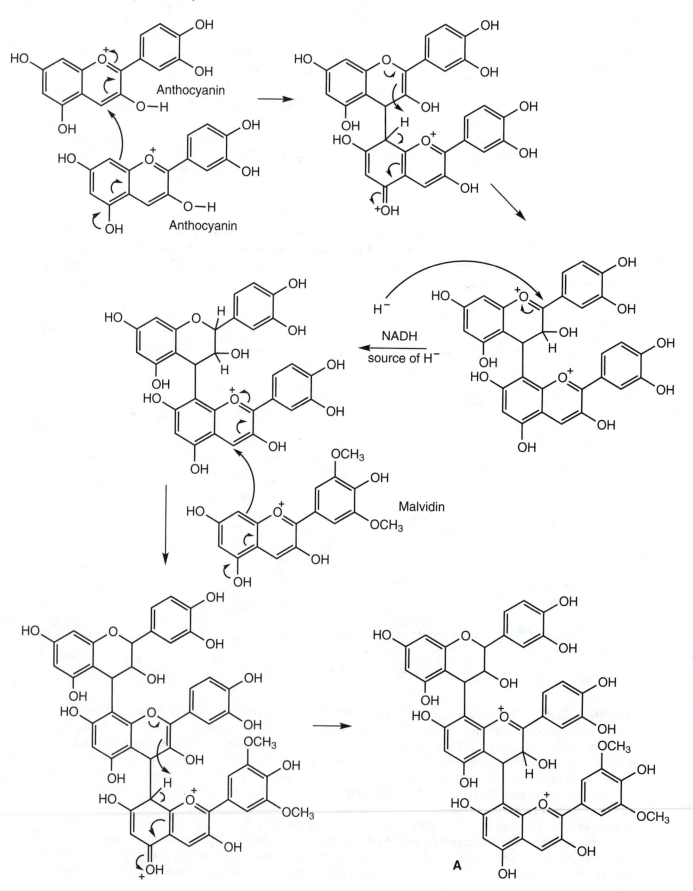

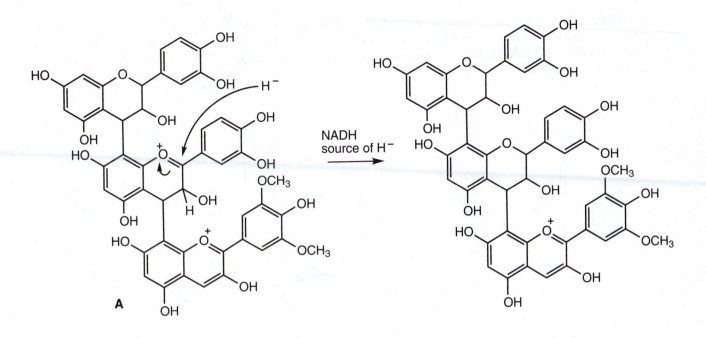

Problem 21.20 The amine can do an intramolecular S_N2 reaction to give the aziridinium ion. Hydroxide will open the three-membered ring by attacking one of the carbons of the three-membered ring. We would predict that the attack would favor the more substituted carbon, as it is a positive-charged aziridinium intermediate. In this case, the bond between the nitrogen and the more substituted carbon is not weak enough (the carbon isn't substituted enough) to show a preference for nucleophilic attack of the more substituted carbon. You might need to number the carbons to keep track of them in this reaction.

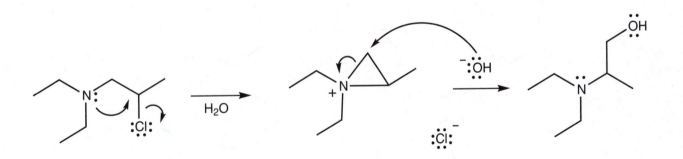

Problem 21.21 These reactions are all reductive aminations. According to the literature, there is a pH requirement for this reaction. If the reaction is too acidic, then the aldehyde or ketone is reduced. If the reaction is too basic, then the intermediate imine isn't reduced. The ideal pH is 6–7 (*Synthesis*, **1975**, 135).

(a)

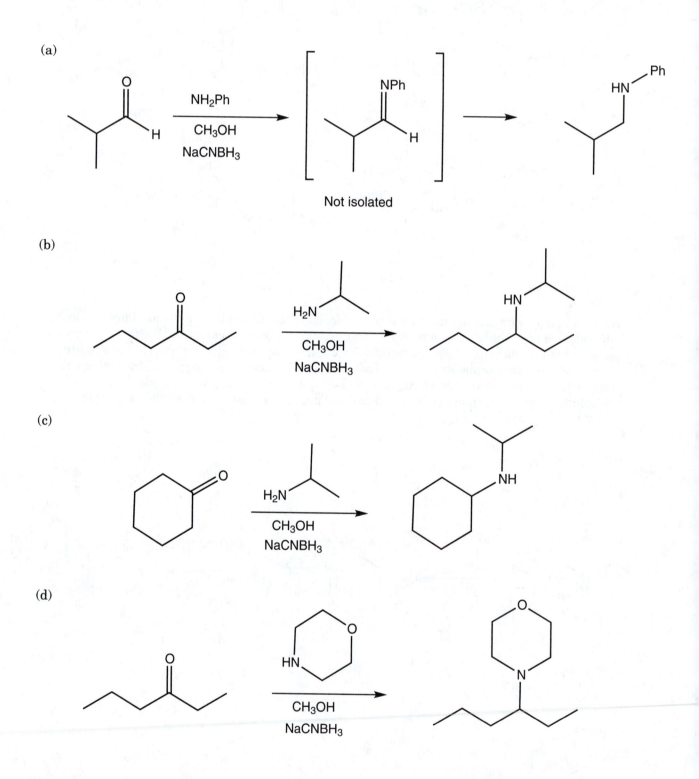

Not isolated

(b)

(c)

(d)

Problem 21.22 This reaction is a double reductive amination. The first reaction is between one of the aldehydes and the amine. That process gives the intermediate shown. Then the amine can react intramolecularly with the aldehyde to form the six-membered ring.

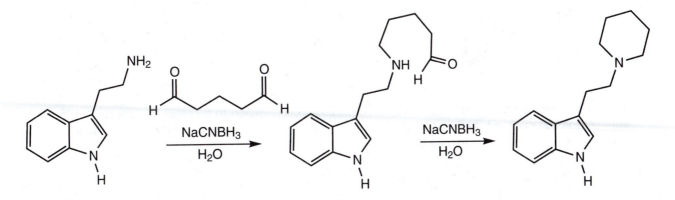

Problem 21.23 This reaction is an intramolecular example of imine formation. There is considerable driving force for this reaction because the ring that is formed in the reaction is aromatic.

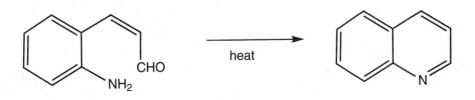

Problem 21.24 The primary amine is nucleophilic, and the acetylenic diester is an excellent Michael acceptor. Michael addition gives the intermediate allenic enolate. The proton transfer shown in the mechanism is an intramolecular transfer, but it needn't be. The proton will be removed from the nitrogen, and the enolate will pick up a proton to return to the ester.

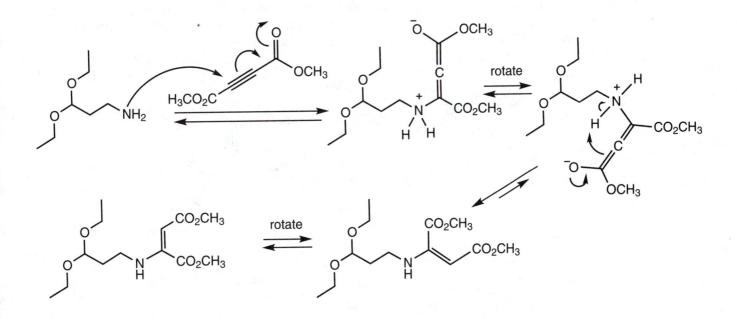

Problem 21.25   There are many ways to add a formyl group to an enol (or enolate). The methyl formate (or any formate ester) will react to produce the β-keto aldehyde.

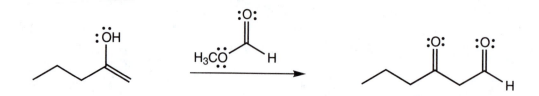

The mechanism for the reaction is a standard nucleophilic addition to an ester. In this case, it is the enol that is the nucleophile. The tetrahedral intermediate is not a final product; it will go to a carbonyl under these reaction conditions.

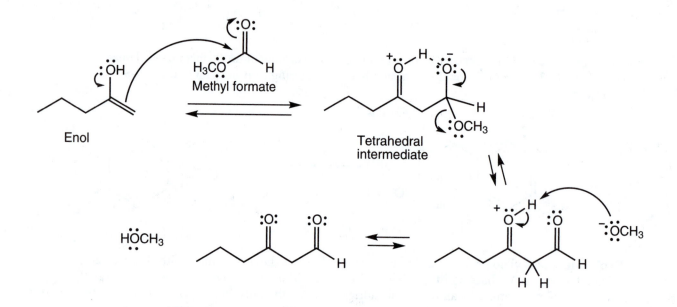

Problem 21.26

(a) Azides, alkenes, and nitriles can all be reduced by hydrogenation using palladium on carbon (Pd/C). Aromatic rings can also be hydrogenated using this catalyst, but only with high pressure or under forcing conditions.

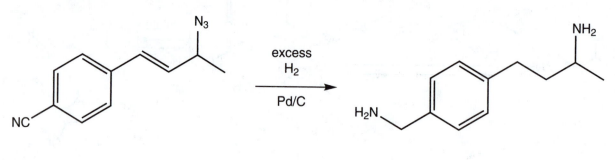

(b) The hemiacetal is in equilibrium with the aldehyde. The aldehyde will react with the amine to make the imine. The imine will be able to reclose to form the *N*-glycoside.

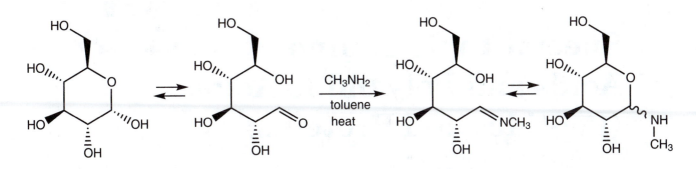

(c) This reaction is messy. The primary amine is converted into the diazonium ion with HONO. The heat will result in carbocation chemistry, and this molecule can do a number of rearrangements. We can't know which would be the major product, but we can see that the 4-methyl-1,3-pentadiene is the most stable predicted product. The most likely products are listed below.

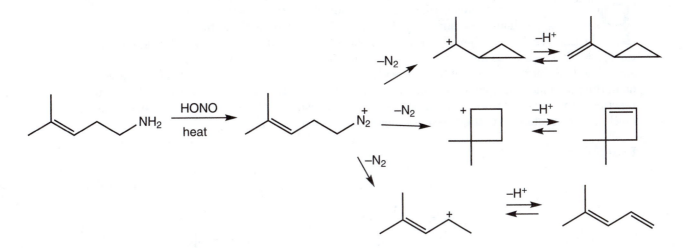

(d) This reaction is the Claisen condensation reaction (from Chapter 19). The enolate of the ester (a nucleophile) is formed in the presence of the ester (an electrophile). The two will react to give the β-keto ester.

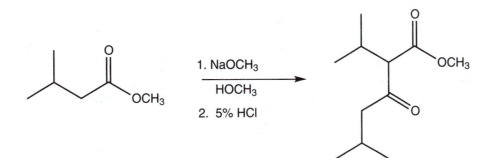

Special Topic: Amino Acids and Polyamino Acids (Peptides and Proteins)

22

Polyamino acids are complex; there can be no doubt of that. Structural matters become more severe as questions of secondary and tertiary structure arise. Nonetheless, the microchemistry of the amino acids and their oligomers and polymers can still be understood through normal reactions, usually Lewis acids and Lewis bases reacting in familiar ways. "Round up the usual suspects!" says Claude Raines at the end of the movie *Casablanca,* and he's exactly right. That's what you should do in the problems in this chapter. Don't be put off by the polyfunctionality; forge ahead with the usual reactions, and they will work out. This section starts, as usual, with relatively simple structural and mechanistic questions and then becomes more complicated.

Problem 22.1

Valine

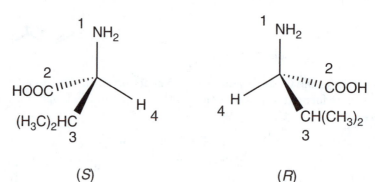

Aspartic acid

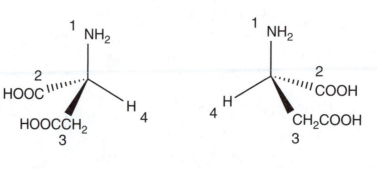

Problem 22.2

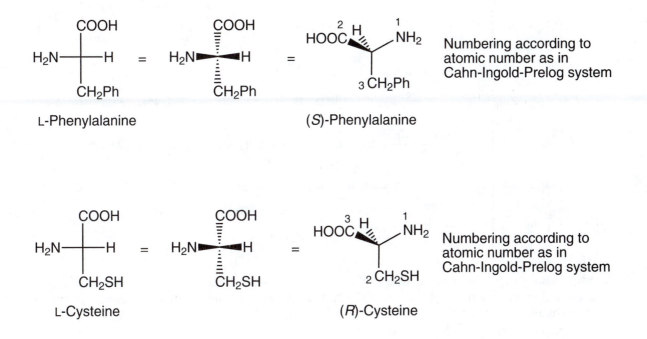

Problem 22.3 This is an example of the effect that induction (I of ISHARE) can have on acidity. The $^+NH_3$ group is strongly electron withdrawing inductively, and will stabilize a neighboring carboxylate anion, which increases the acidity of the acid and lowers the pK_a. The small "dipole arrow" shows the polarization of the electrons in the carbon–nitrogen σ bond.

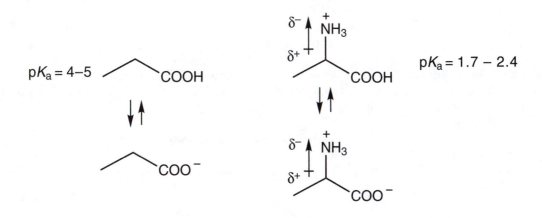

Problem 22.4   The isoelectric point pH is the average of the two pK_a values.

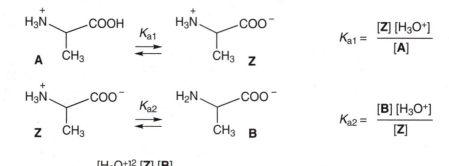

so, $[K_{a1}][K_{a2}] = \dfrac{[H_3O^+]^2\,[Z]\,[B]}{[A]\,[Z]}$ or, at the isoelectric point where by definition, $[A] = [B]$

$$[K_{a1}][K_{a2}] = [H_3O^+]^2$$

and, taking the negative log of each side: $pK_{a1} + pK_{a2} = -\log[H_3O^+]^2 = 2\,pH$

The pH at the isoelectric point is the average of the two pK_a values:

$$\frac{pK_{a1} + pK_{a2}}{2} = (2.3 + 9.7)/2 = 6.0$$

Problem 22.5   Arginine at pH of 12 would be mostly the neutral structure in the sequence shown here. The pH values for the protonation of the carboxylic acid, the amine, and the side chain can be found in the fifth, sixth, and seventh columns of Table 22.1.

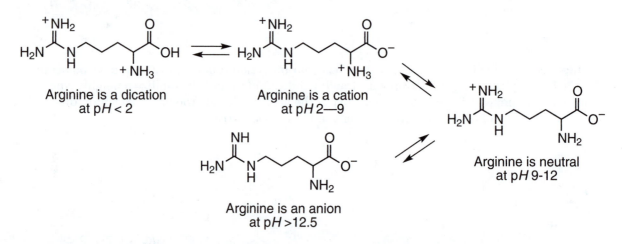

The arginine side chain has three nitrogens. We can compare protonation of each in order to see which is preferred. Protonation of the imine nitrogen (the sp^2 hybridized N) is favored because the resulting cation is resonance stabilized.

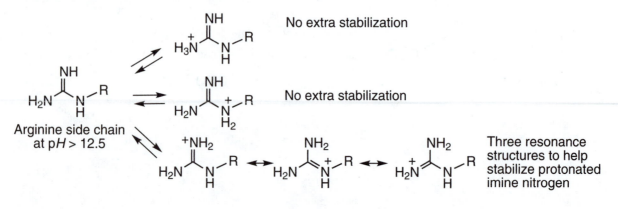

Problem 22.6 Let's look at the two possibilities:

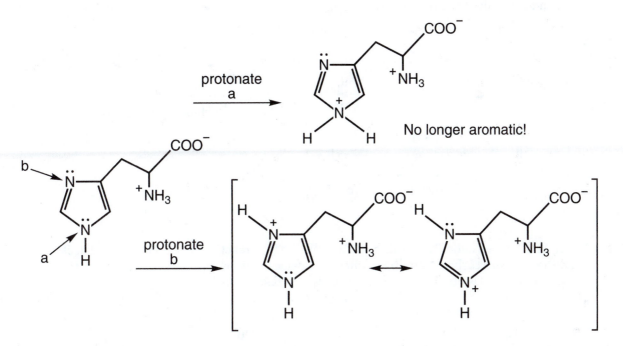

Protonation of nitrogen a interrupts the conjugation in the ring. No longer is the molecule aromatic. By contrast, protonation on nitrogen b leads to an aromatic, resonance-stabilized cation and will be greatly preferred.

Problem 22.7 All acids with an acidic α hydrogen can be brominated in the α position with PBr_3 or the equivalent, a mixture of phosphorus and bromine (the Hell–Volhard–Zelinsky reaction, Chapter 19, p. 947), so the last step is the same in each example. The first acid can be made by oxidation of the corresponding alcohol.

(a) This part of the problem is simple, as we are allowed to start with the four-carbon alcohol butanol.

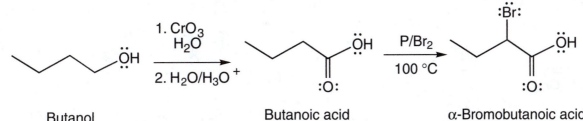

(continued)

Problem 22.7 *(continued)*

(b) This part is a bit harder because it is necessary to build up to a five-carbon acid: straightforward Grignard chemistry does the trick (p. 845).

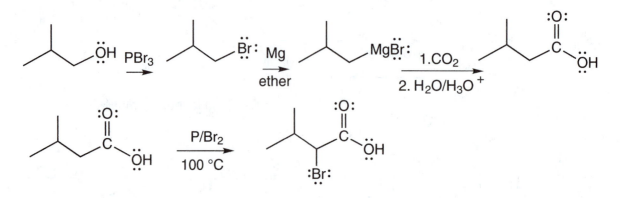

Problem 22.8   A real oldie. The transition state for the S_N2 reaction will benefit from delocalization, and the adjacent carbonyl group provides that delocalization. If the transition state (the top of the activation barrier) is stabilized, the reaction will go faster.

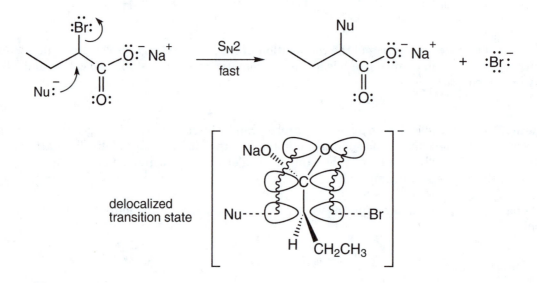

Problem 22.9   The first step is, as advertised in the figure, a simple S_N2 displacement of bromide by the phthalimide anion.

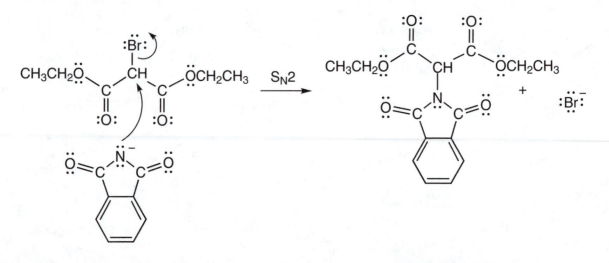

Now, an alkylation reaction takes place as the doubly α proton is removed in base to give an anion that can be used to displace chloride from the alkylating agent.

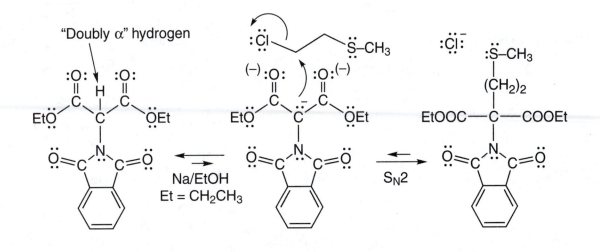

In base, four base-induced hydrolyses take place. Hydroxide hydrolyzes the esters to carboxylate anions and at the same time removes phthalic acid as the dianion, phthalate. In the figure, only one of these four essentially identical steps is shown in detail.

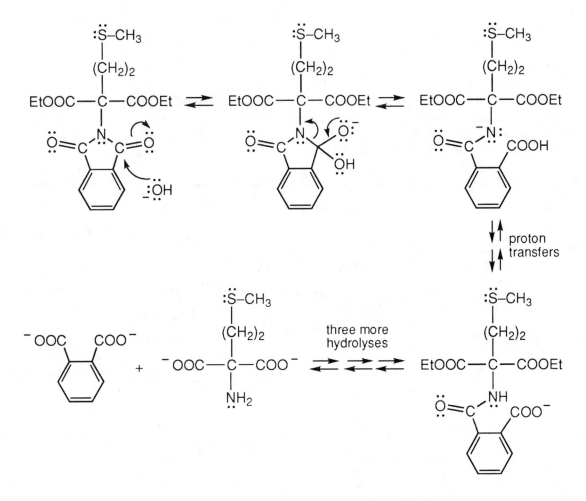

(continued)

Problem 22.9 *(continued)*

The acid neutralizes the anions, forming phthalic acid and a substituted malonic acid.

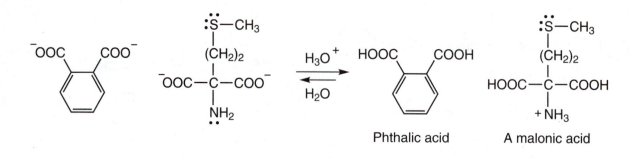

Gentle heating decarboxylates the malonic acid (Chapter 19, p. 958) to give the product.

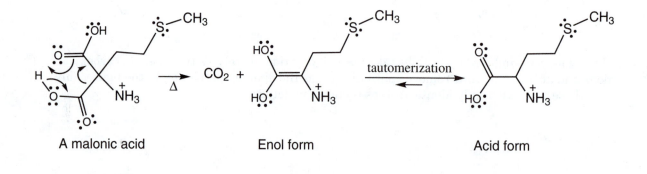

A malonic acid	Enol form	Acid form

Problem 22.11   It is the ability to rotate the plane of plane-polarized light that differentiates the two. One enantiomer will rotate the plane of plane-polarized light to the right, the other by the same amount to the left (Chapter 4, p. 160). Enantiomers can be differentiated in a chiral environment and plane-polarized light is chiral light.

Problem 22.12   Both reactions are examples of straightforward addition–elimination processes.

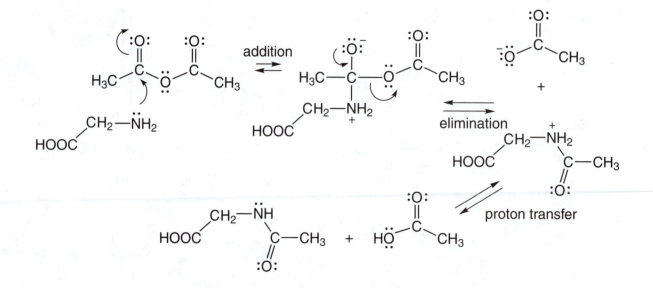

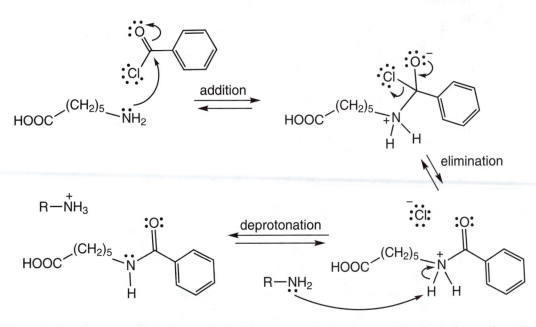

Problem 22.13   It is true that the amino acids are mostly in their zwitterionic forms, but there is always enough of the free acid and free amine groups present so that typical acid and amine chemistry is possible. The amino acids are not *exclusively* in their zwitterionic forms.

Problem 22.14   Vicinal (1,2) diones are destabilized by the repulsions of the adjacent dipoles in which like charges are opposed. In the trione, there are two 1,2-dipolar repulsions. Hydration at the central carbonyl removes both 1,2-interactions, whereas hydration at the side carbonyl leaves one 1,2-repulsion intact.

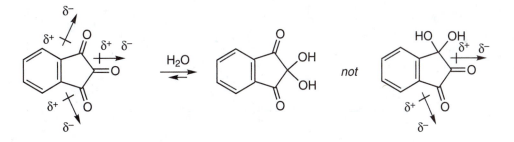

Problem 22.15   The structures are shown with stereochemistry, assuming L-amino acids.

(a) Pro·Gly·Tyr (P·G·Y)

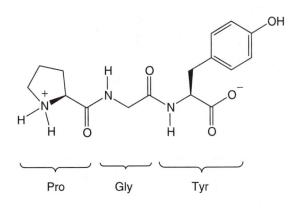

(continued)

Problem 22.15 *(continued)*

(b) K·F·C (Lys·Phe·Cys) (or Kentucky Fried Chicken?)

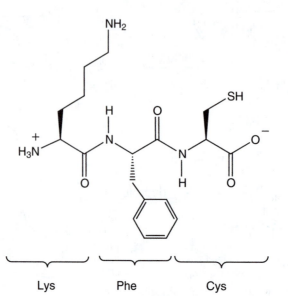

Lys	Phe	Cys

(c) D·H·C (Asp·His·Cys)

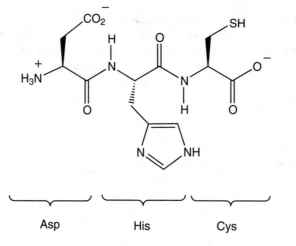

Asp	His	Cys

(d) S·N·Q (Ser·Asn·Gln)

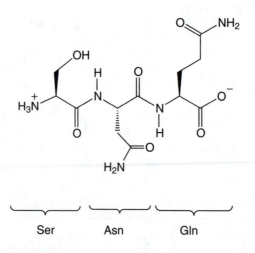

Ser	Asn	Gln

(e) E·T·P (Glu·Thr·Pro)

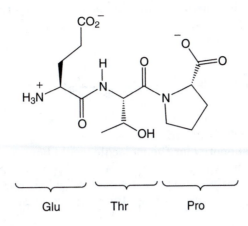

Glu   Thr   Pro

Problem 22.16   (a) Phe·Lys·Trp   (b) Asp·Met·Ile

Problem 22.18   It's nucleophilic aromatic substitution. Let's abbreviate the tripeptide of Figure 22.32 as H_2N—R. Addition to the benzene ring at the same position as the F (ipso attack) leads to an ion in which both nitro groups help stabilize the negative charge. Without the nitro groups, there is no such stabilization, and the reaction fails. Loss of fluoride completes the reaction as the aromatic benzene ring is regenerated.

(Many other resonance forms are possible)

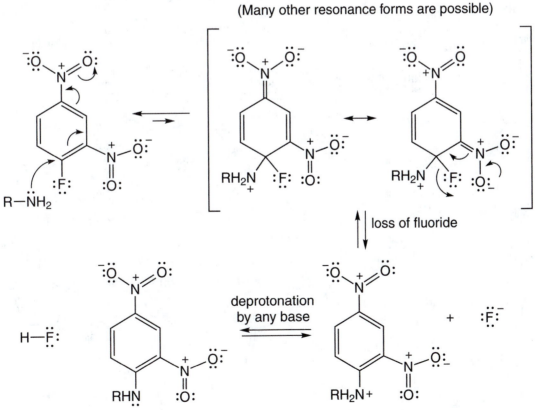

Problem 22.19 The hybridization for the N and C in isothiocyanate is *sp*. We can see that the central carbon is using two *p* orbitals in bonding. The resonance structures shown below help us see that the nitrogen is also using two *p* orbitals. Atoms that are hybridized, but retain two *p* orbitals for π bonds, are *sp* hybridized.

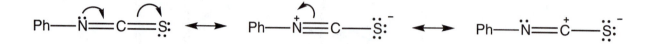

It is difficult to know the sulfur hybridization. The S is using one *p* orbital to make the π bond to C. It is likely that the lowest energy for the molecule would have one of the sulfur lone pairs of electrons be in an orbital that mixes with the NC π bond. That would have to be a *p* orbital. It would mean that S has two *p* orbitals and thus it would be *sp* hybridized also. The additional resonance structure shown below helps us understand the potential stabilization.

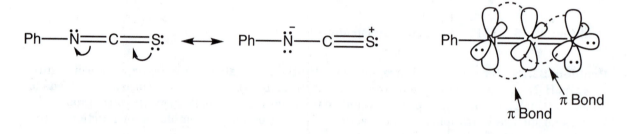

Problem 22.20 The critical, if obvious, realization is that this one must be an "open–close" problem. After all, the sulfur is in the ring in the starting material and out of the ring in the product. We have to find a way to open the ring by breaking a carbon–sulfur bond. Here's a possibility.

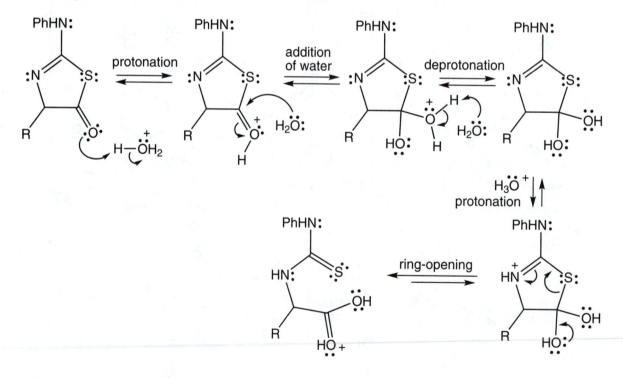

The ring is reclosed through amide formation, and a series of protonation–deprotonation steps finishes the reaction.

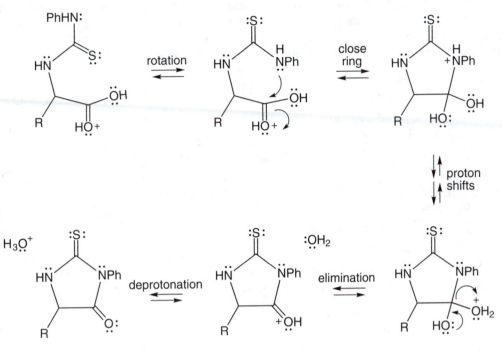

Problem 22.21 The acid hydrolysis gives the constituent amino acids of bradykinin. The key observation is that the only amino acid present in bradykinin after which chymotrypsin cleaves is Phe.

There are several ways in which the pieces could be fit together.

(a) The pentapeptide could start the sequence. If so, there are two possibilities:

1. (Arg·Pro·Pro·Gly·Phe) (Arg) (Ser·Pro·Phe)

2. (Arg·Pro·Pro·Gly·Phe) (Ser·Pro·Phe) (Arg)

The pentapeptide could be in the center, which gives two more possibilities:

3. (Arg) (Arg·Pro·Pro·Gly·Phe) (Ser·Pro·Phe)

4. (Ser·Pro·Phe) (Arg· Pro·Pro·Gly·Phe) (Arg)

Finally, the pentapeptide could end the sequence. Again there are two possibilities:

5. (Ser·Pro·Phe) (Arg) (Arg·Pro·Pro·Gly·Phe)

6. (Arg) (Ser·Pro·Phe) (Arg·Pro·Pro·Gly·Phe)

But the only amino acid present after which chymotrypsin cleaves is Phe. So, only possibilities 2 and 4 remain. Only these two would give the three pieces found.

However, only possibility 2 has the same amino acid at the amino and carboxy terminus, and so it must be the structure of bradykinin.

Problem 22.22   The question asks us to draw out the "steps" of the deprotection of a Cbz-protected amine. The mechanism is not well known, so you won't be expected to show the electron pushing. Let's take the Cbz-protected ethyl ester of valine as an example. We deprotect by hydrogenation, which cleaves the benzyl bond. It doesn't alter the amide or the ester functional groups. The intermediate product is a carbamic acid, which is unstable. The carbamic acid will decarboxylate to give the deprotected amine (in this case the ethyl ester of valine).

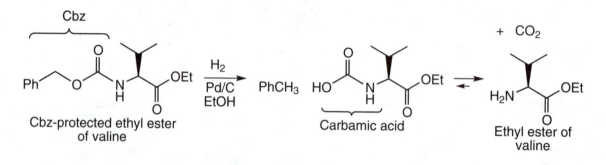

Problem 22.23   If the polymer is based on benzene rings, the chloromethyl groups are benzylic chlorides. A cyclohexane ring-based polymer would have simple primary chlorides:

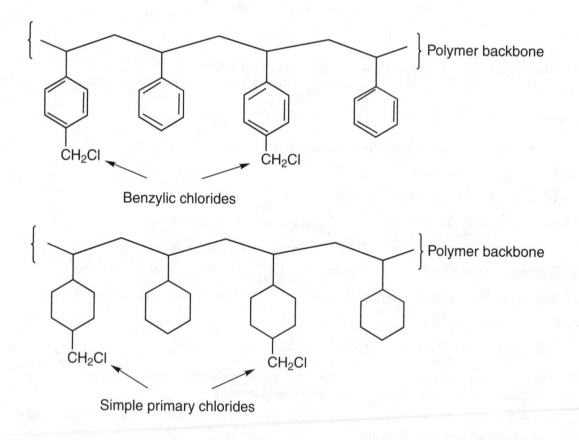

Attachment of the amino acid is through S_N2 displacement of chloride. These S_N2 reactions are much faster at the benzylic position than at an unconjugated primary position.

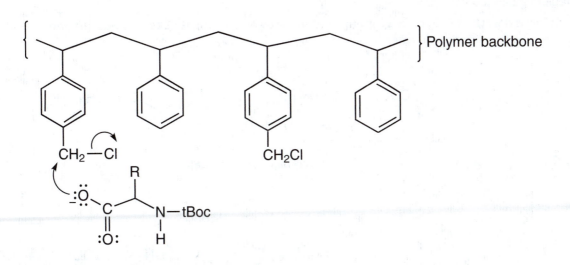

Polymer backbone

Problem 22.24   First of all, what's Leu·Ala? It is just the reverse of the example used in the chapter, Ala·Leu.

Leu Ala

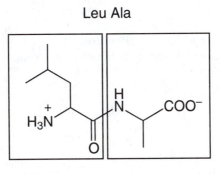

Leu portion   Ala portion

To do this synthesis efficiently, we need to block reaction at the amino end of leucine and the carboxy end of alanine. We also need to activate the carboxy group of leucine.

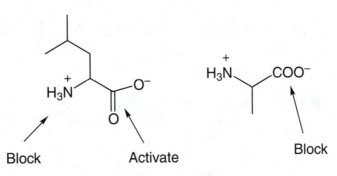

Block   Activate   Block

So, first we will protect the amino group of leucine by using either tBoc or Cbz—we'll show tBoc.

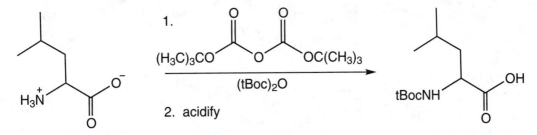

1.

$(H_3C)_3CO$ $OC(CH_3)_3$

(tBoc)$_2$O

2. acidify

(continued)

Problem 22.24 *(continued)*

Next, we'll deactivate the carboxy group of alanine by esterifying it. A tertiary amine can be added to form the free amino acid ester.

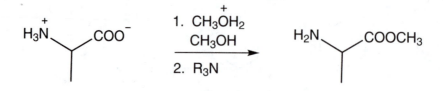

Now, we need to do the condensation itself. We'll use DCC to mediate it.

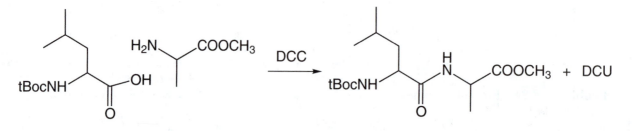

Finally, we need to deprotect—to remove the tBoc from the leucine end and to convert the ester at the alanine end back into the carboxylic acid. Mild acid accomplishes the first goal, and base hydrolysis the second. We're finished.

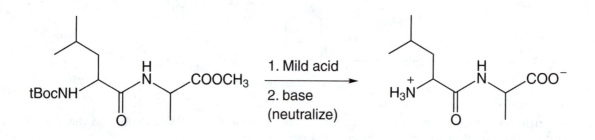

Answers to Additional Problems

Problem 22.26

(a) Protonation on oxygen leads to a resonance-stabilized cation, whereas protonation on nitrogen does not.

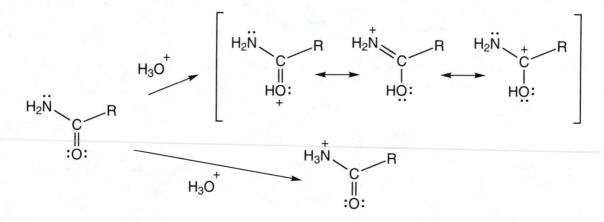

(b) The resonance stabilization in amides makes them more stable, relative to their conjugate acids, than amines. This stability makes them less reactive—weaker bases.

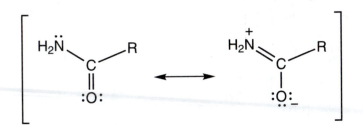

Problem 22.27   We will assume that these are the L-amino acids and that the pH is 7.

(a) Ala·Ser·Cys·Arg is named alanylserinylcysteinylarginine.

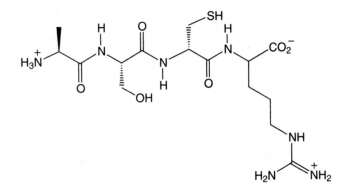

(b) Met·Phe·Pro is named methionylphenylalanylproline.

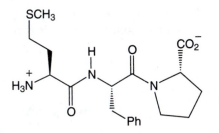

(c) Val·Asp·His is named valinylaspartylhistidine.

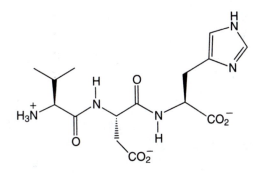

(continued)

Problem 22.27 *(continued)*

(d) W·I·L·L (Trp·Ile·Leu·Leu) is named tryptophanylisoleucinylleucinylleucine.

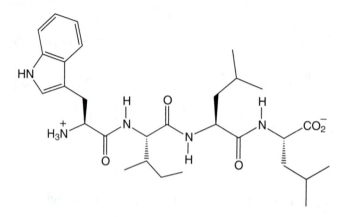

(e) R·A·K (Arg·Ala·Lys) is arginylalanyllysine.

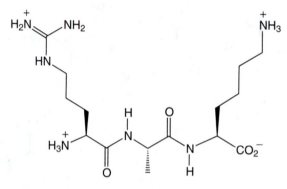

Problem 22.28   (a) Thr·Leu, threonylleucine   (b) Lys·Tyr, lysinyltyrosine

Problem 22.29   The numbers in the following drawing show the priorities. Remember, if the 1 → 2 → 3 arrow runs clockwise, the compound is (*R*); if it runs counterclockwise, it is (*S*).

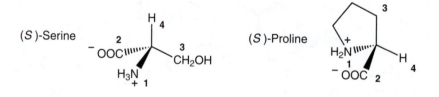

Problem 22.30 L-Leucine and L-proline are drawn in Fischer projection. Proline, like almost all amino acids, has one stereogenic carbon.

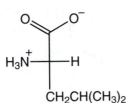

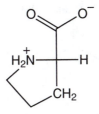

L-Leucine same as (S)-leucine   L-Proline same as (S)-proline

Isoleucine has two stereogenic carbons.

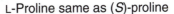
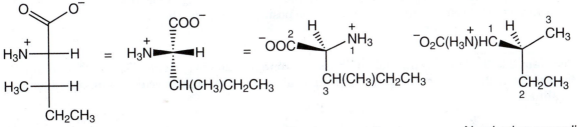

L-Isoleucine same as (S)-isoleucine. There are two stereogenic carbons, both are (S) configuration in nature

Numbering according to Cahn-Ingold-Prelog showing (S) configuration at C(2)

Numbering according to Cahn-Ingold-Prelog showing (S) configuration at C(3)

Problem 22.31 The isoelectric points are His = 7.6, Arg = 10.8, and Phe = 5.9. Above the isoelectric point, the amino acids will be net negatively charged. Below the isoelectric point, they will be net positively charged. The structures will be

At pH 3,

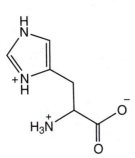

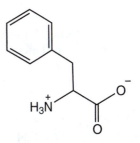

His (isoelectric point 7.6)   Arg (isoelectric point 10.8)   Phe (isoelectric point 5.9)

At pH 12,

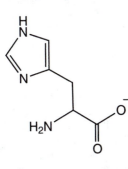

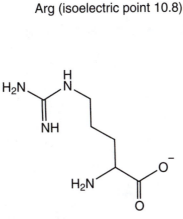

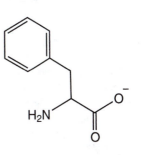

Problem 22.32   At pH 7, the structures will be

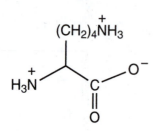

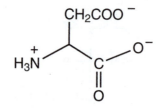

Lys (isoelectric point 9.7)   Asp (isoelectric point 2.9)

So Lys, net positively charged, will migrate to the negatively charged electrode, the cathode; and Asp, net negatively charged, will migrate to the positive electrode, the anode.

Problem 22.33   The L-amino acid with the (R) configuration is cysteine. For the other amino acids in Table 22.1, the second-priority group is always the carboxylic acid, as illustrated for L-valine.

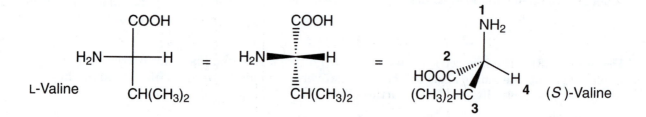

However, for cysteine, the second-priority group is the CH_2SH as sulfur has a higher atomic number than oxygen. Thus, L-cysteine has the (R) configuration.

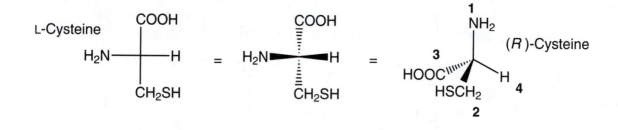

Problem 22.34 By analogy to the reaction of primary amino acids with ninhydrin (see Fig. 22.18), the free amine form of the secondary amino acid proline, a small amount of which is present at equilibrium, reacts with indan-1,2,3-trione to yield, after dehydration, **A**. Intermediate **A** then undergoes a decarboxylation to produce compound **1**. Unlike the related imine of Figure 22.18, compound **1** appears to be stable under normal reaction conditions.

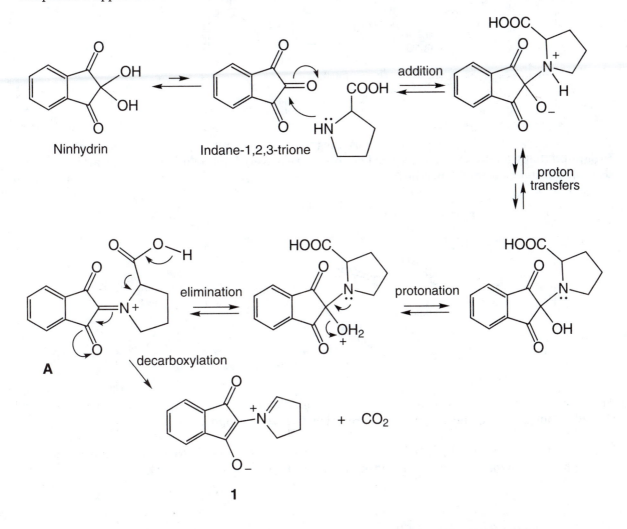

Problem 22.35 This problem involves the use of 2,4-dinitrofluorobenzene to label the amino end terminus of the peptide through a process of nucleophilic aromatic substitution.

(a) Of the common amino acids, only Phe brings with it five aromatic hydrogens. Together with the three remaining ring hydrogens of the 2,4-dinitrobenzene group, this makes eight. The amino terminus of this peptide must be phenylalanine, Phe.

(b) The region δ 3.5 ppm is appropriate for a hydrogen adjacent to an ether or alcohol oxygen atom. Only threonine would show an eight-line signal at this position. The amino end of this peptide must be threonine, Thr. The actual NMR signal for the C(3) hydrogen in threonine appears as a quintet at δ 4.2 ppm.

(continued)

Problem 22.35 *(continued)*

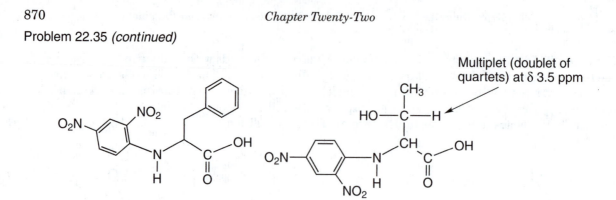

Eight "aromatic" hydrogens
(from Phe)

From Thr

Problem 22.36   The Edman reaction determines the amino terminus. The only phenylthiohydantoin that would show *only* a methyl doublet at relatively high field is **A**, and the amino terminus of this tripeptide must be Ala.

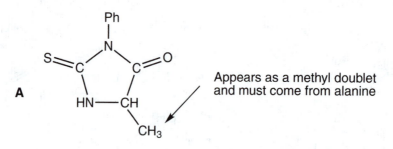

A

Appears as a methyl doublet
and must come from alanine

The other two amino acids are Met and Cys. If the middle amino acid were Met, BrCN would cleave the tripeptide after the Met. As there is no reaction, the structure must be Ala·Cys·Met.

Problem 22.37   First of all, the amino terminus must be one of the Val residues, as the structure of the phenylthiohydantoin produced in the Edman procedure is **A**.

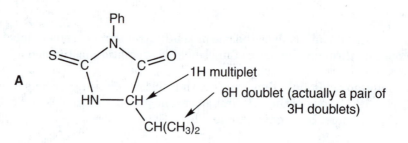

A

1H multiplet

6H doublet (actually a pair of
3H doublets)

Second, trypsin cleaves after Arg or Lys. Given that the polypeptide must start with Val, this observation means that there are only four possibilities. The arrows show the trypsin cleavage points.

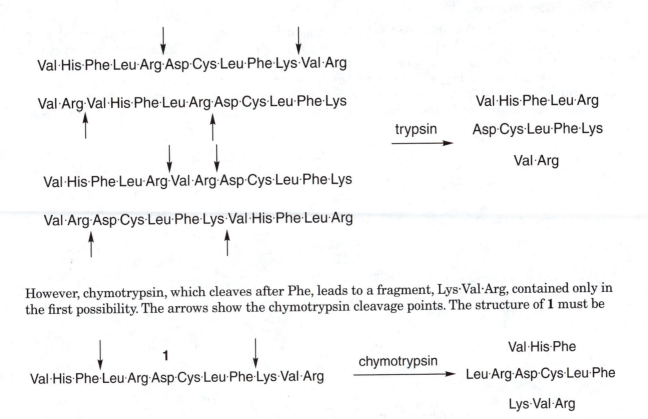

However, chymotrypsin, which cleaves after Phe, leads to a fragment, Lys·Val·Arg, contained only in the first possibility. The arrows show the chymotrypsin cleavage points. The structure of **1** must be

Val·His·Phe·Leu·Arg·Asp·Cys·Leu·Phe·Lys·Val·Arg $\xrightarrow{\text{chymotrypsin}}$

Val·His·Phe

Leu·Arg·Asp·Cys·Leu·Phe

Lys·Val·Arg

Problem 22.38

(a) This problem is just another example of the standard addition–elimination mechanism. Note that the amino group, a better nucleophile than the carboxylate anion, adds to the dicarbonate **1**. It is also worth observing that half of the dicarbonate is wasted, as CO_2 and *tert*-butyl alcohol are lost in the process. The actual product of the reaction is the tBoc carboxylate. The free tBoc carboxylic acid is liberated by careful neutralization with acid.

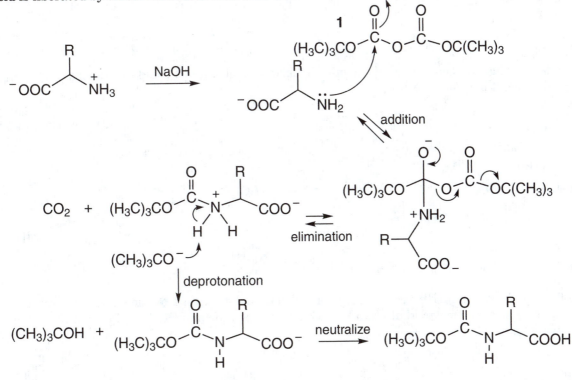

(continued)

Problem 22.38 (continued)

(b) The removal of the tBoc group occurs by an uncommon cleavage mechanism, which operates for *tert*-butyl, and other *tert*-alkyl, esters. The first step is protonation of the "ester" carbonyl oxygen. The protonated ester then undergoes a straightforward E1 ionization to give a carbamic acid and a tertiary carbocation. The key to this mechanism is the formation of the relatively stable tertiary carbocation. The carbocation is deprotonated to form 2-methylpropene, and the carbamic acid decarboxylates to give the trifluoroacetate salt of the amine. The free amine is then liberated upon reaction with triethylamine.

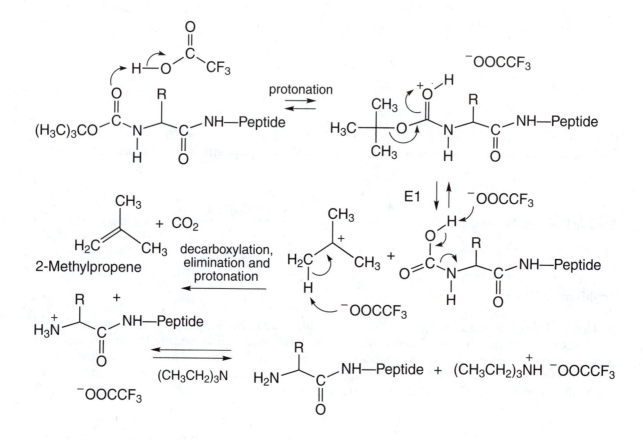

Problem 22.39   It is possible that the initially formed *O*-acylurea **1** could function as the acylating agent in some cases.

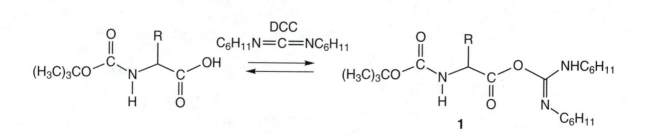

Addition of the new amino ester to the right-hand carbonyl group of **1** by the standard addition–elimination mechanism would ultimately produce the dipeptide and DCU, as shown in the following figure.

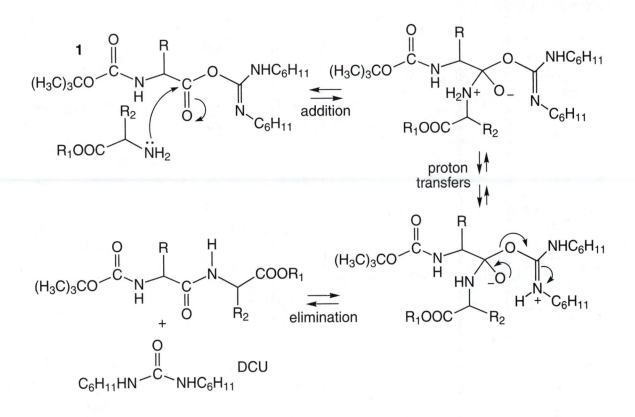

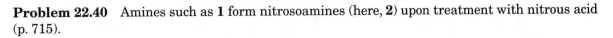

Problem 22.40 Amines such as **1** form nitrosoamines (here, **2**) upon treatment with nitrous acid (p. 715).

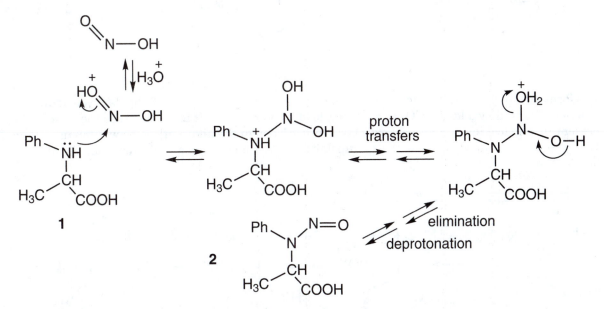

As we saw previously, carboxylic acids react with acetic anhydride to give mixed anhydrides. In this case, the mixed anhydride **A** undergoes a cyclization reaction involving the nitroso oxygen atom to give, after deprotonation, the sydnone **3**.

(continued)

Problem 22.40 *(continued)*

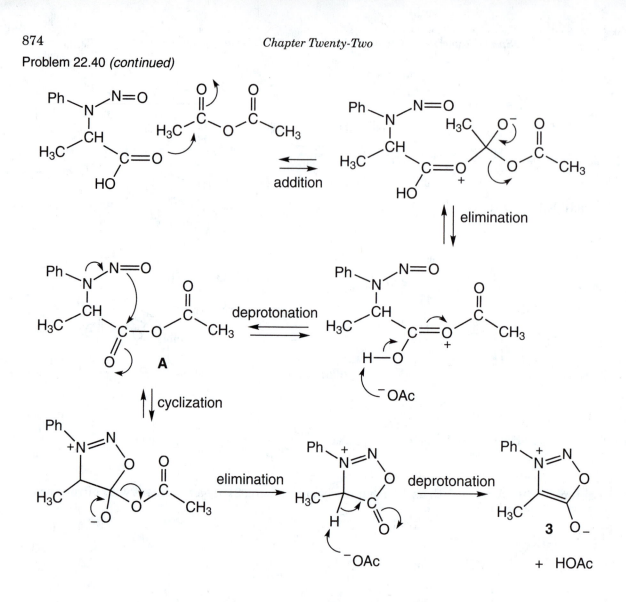

The formation of pyrazole **4** from the reaction of **3** and DMAD involves a 1,3-dipolar cycloaddition reaction, followed by a loss of CO₂ in a retrocycloaddition reaction. It is easier to see the 1,3-dipole present in **3** by looking at another Lewis structure. Although the addition can be written using any valid resonance form, we will use the 1,3-dipolar form. You might "push the arrows" for the other forms.

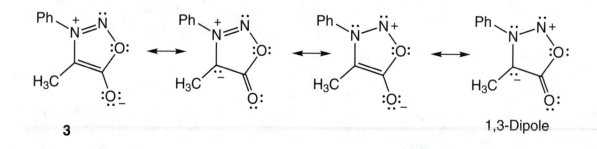

3   **1,3-Dipole**

Here's the rest of the mechanism.

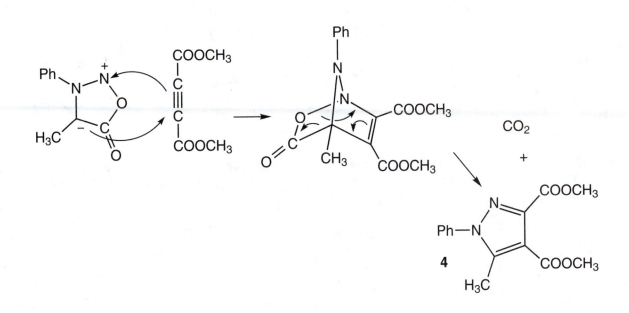

Problem 22.41 The mechanism for the formation of *N*-acetylglycine in the answer to Problem 22.12 is fine as far as it goes. In the presence of excess acetic anhydride, however, the carboxylic acid reacts further with acetic anhydride to form the mixed anhydride **A**. The amide oxygen of **A** is now poised to displace acetate (addition–elimination once again) to give, after deprotonation, the azlactone **1**. Thus, *N*-acetylglycine undergoes a "cyclodehydration" reaction to give azlactone **1**.

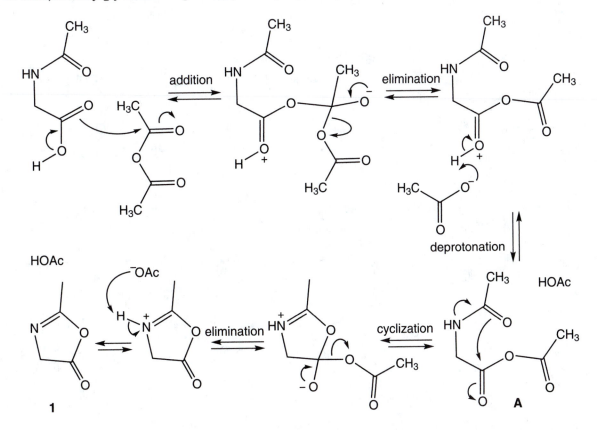

(continued)

Problem 22.41 *(continued)*

Hydrolysis of azlactone **1** to *N*-acetylglycine is essentially the reverse of this process.

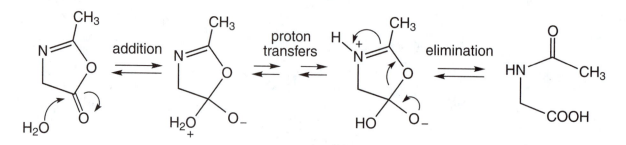

Problem 22.42   As we saw in Problem 22.41, reaction of glycine with acetic anhydride (Ac_2O) first gives *N*-acetylglycine. This product then reacts further with acetic anhydride to yield azlactone **1**.

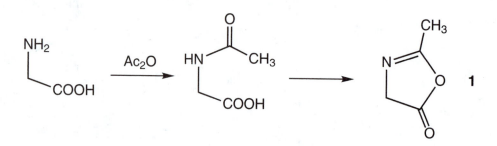

Azlactone **1** can readily be deprotonated at C(4) by sodium acetate to give the resonance-stabilized anion **A**. Anion **A** then adds to the carbonyl group of benzaldehyde to give, after dehydration, the benzylidene azlactone **2**. This reaction sequence is known as the Erlenmeyer azlactone synthesis. Note the formation of the resonance-stabilized intermediate **B**.

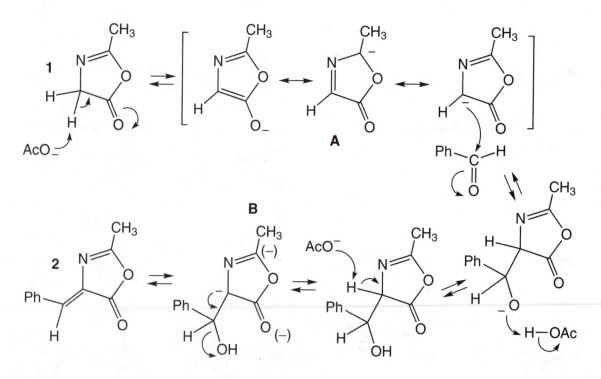

Problem 22.43 The first step in both of these reactions is the *N*-acetylation of **1** to form *N*-acetyl-*N*-methylalanine (**2**). *N*-Acylamino acid **2** then reacts with acetic anhydride to form the mixed anhydride **A**. We won't write the mechanisms for these reactions, as they were written in the preceding problems, but this omission provides you with an opportunity to practice.

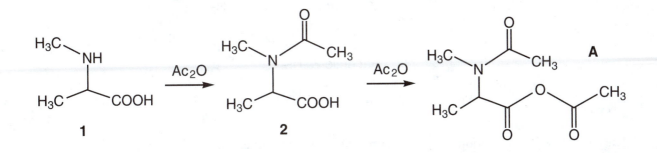

The amide oxygen of **A** is poised to displace acetate in an addition–elimination sequence to yield oxazolium salt **B**. Salt **B** cannot form an azlactone because there is no hydrogen on nitrogen to lose. However, loss of the hydrogen at C(4) affords the compound **C**, commonly known as a münchnone. Münchnones were first investigated by Professor Rolf Huisgen and his collaborators at the University of Munich, hence the name.

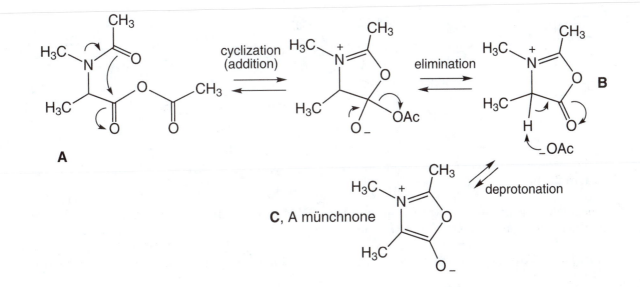

Münchnone **C** is the common intermediate that is responsible for both the racemization of **2** and the formation of pyrrole **3**. Note that the hydrogen at the stereogenic carbon C(4) has been lost. When **C** undergoes hydrolysis to form *N*-acetyl-*N*-methylalanine (**2**), this hydrogen can be readded to either face of the achiral münchnone **C**.

(continued)

Problem 22.43 *(continued)*

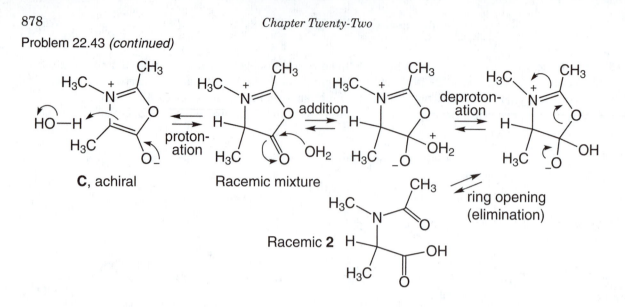

C, achiral   **Racemic mixture**

Racemic 2

The ease of cyclodehydration and, hence, racemization (via münchnones or azlactones) upon carboxy activation with reagents such as DCC explains why simple *N*-acylamino acids are not employed in peptide synthesis. *N*-Alkoxycarbonylamino acids (that is, tBoc- or Cbz-protected amino acids) do not undergo cyclodehydration reactions as easily.

Münchnone **C** is also a 1,3-dipole, similar to the sydnone of Problem 22.40, and can be trapped in a 1,3-dipolar addition reaction by DMAD. The primary cycloadduct **D** easily loses CO_2 to give the observed pyrrole **3**. This reaction is aptly known as the Huisgen pyrrole synthesis, and constitutes a pyrrole synthesis of broad scope.

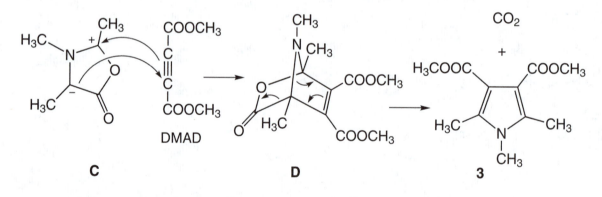

C   DMAD   **D**   **3**

Problem 22.44

(a) The facile base pairing of A with T (or with U in RNA) and G with C means that the mRNA sequence must be A U G C C C A A A U A G.

(b) Just look up the meanings in Table 22.2 (p. 1145). This message translates to Start, add Pro, add Lys, Stop.

Special Topic: Reactions Controlled by Orbital Symmetry

23

Problem 23.2 There are always two conrotatory and two disrotatory modes. The two conrotatory and two disrotatory modes will always give the same bonding or antibonding interaction at the newly formed bond. Here are the "missing" two modes from Figure 23.14.

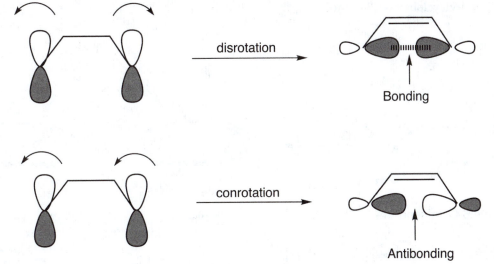

Problem 23.3 The two possible disrotatory modes lead to the same isomer in which there is one cis and one trans double bond. Be certain that you see that these planar molecules are the same.

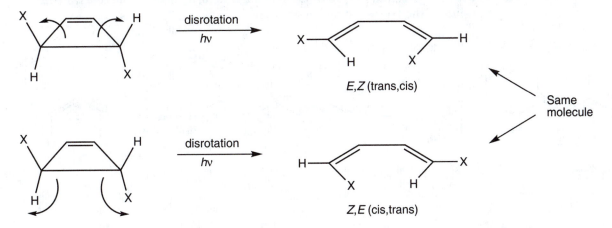

Problem 23.4 The counterclockwise conrotation gives *trans*-5,6-dimethylcyclohexa-1,3-diene. Compare this product to the structure obtained from clockwise conrotation from Figure 23.19 and see that they are enantiomers.

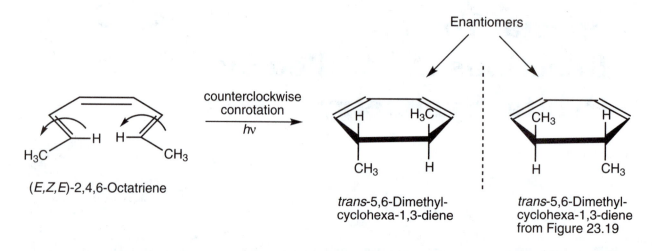

Enantiomers

trans-5,6-Dimethyl-
cyclohexa-1,3-diene

trans-5,6-Dimethyl-
cyclohexa-1,3-diene
from Figure 23.19

(*E,Z,E*)-2,4,6-Octatriene

The disrotatory process in Figure 23.19 has the methyl groups both going up. If we draw the disrotatory process with both methyl groups going down, we obtain the same meso product (*cis*-5,6-dimethylcyclohexa-1,3-diene).

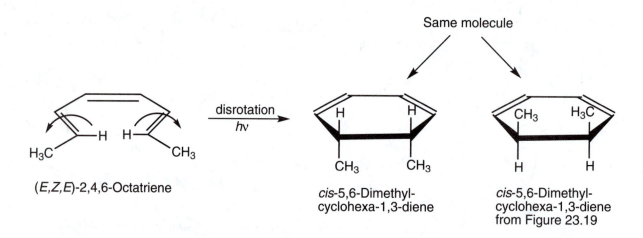

Same molecule

(*E,Z,E*)-2,4,6-Octatriene

cis-5,6-Dimethyl-
cyclohexa-1,3-diene

cis-5,6-Dimethyl-
cyclohexa-1,3-diene
from Figure 23.19

Problem 23.5 Draw out the (*E,E,E*)-2,4,6-octatriene. You will see that this molecule has no conformation available for the C(2) and C(7) to be close enough to form a bond. The central trans double bond keeps the ends of the triene too far apart.

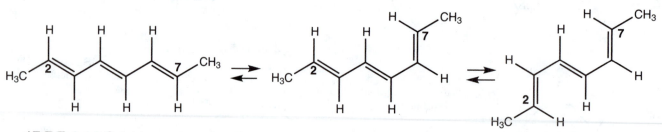

(*E,E,E*)-2,4,6-Octatriene

Problem 23.6 For the six-electron electrocyclic ring closures of the (*Z,Z,E*) and the (*Z,Z,Z*)-2,4,6-octatrienes, the thermal reaction is disrotatory and the photochemical reaction is conrotatory. We have shown one disrotatory closure for each. The thermal reaction of the (*Z,Z,E*) isomer gives a racemic mixture of the *trans*-5,6-dimethylcyclohexa-1,3-diene. We have shown the disrotatory process of the top of the *p* orbitals on C(2) and C(7) going in to form the new σ bond in the trans product. The disrotatory process of the bottom of the *p* orbitals on C(2) and C(7) going in to form the new σ bond gives the enantiomer of the trans product we have shown. The photochemical reaction of the (*Z,Z,E*) isomer gives the cis product regardless of the clockwise or counterclockwise nature of the process.

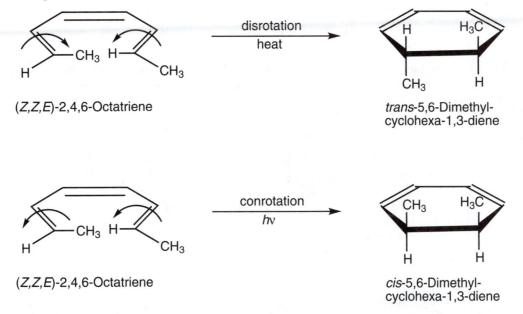

(*Z,Z,E*)-2,4,6-Octatriene → disrotation / heat → *trans*-5,6-Dimethyl-cyclohexa-1,3-diene

(*Z,Z,E*)-2,4,6-Octatriene → conrotation / *h*ν → *cis*-5,6-Dimethyl-cyclohexa-1,3-diene

The thermal reaction of the (*Z,Z,Z*)-2,4,6-octatriene gives the *cis*-5,6-dimethylcyclohexa-1,3-diene as a result of the disrotatory ring closure with either the top lobes of the *p* orbitals on C(2) and C(7) making the new σ bond or the bottom lobes of the *p* orbitals making the new σ bond. Note that this process is difficult because of the steric interactions between the two methyl groups in the starting material. The photochemical reaction gives a racemic mixture. The counterclockwise conrotatory process is shown and gives the trans product. The clockwise process will give the enantiomer of the product shown.

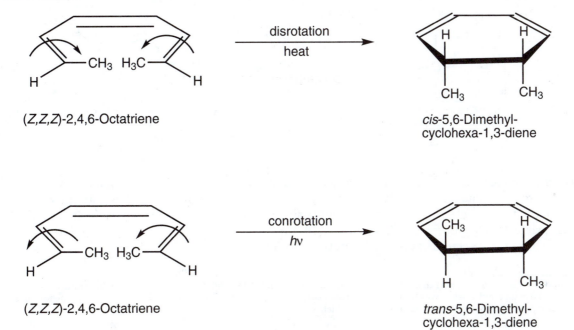

(*Z,Z,Z*)-2,4,6-Octatriene → disrotation / heat → *cis*-5,6-Dimethyl-cyclohexa-1,3-diene

(*Z,Z,Z*)-2,4,6-Octatriene → conrotation / *h*ν → *trans*-5,6-Dimethyl-cyclohexa-1,3-diene

882

Chapter Twenty-Three

Problem 23.7

(a) This molecule incorporates an eight-electron system, a $4n$ number, so we should expect the thermal reaction to take place in a conrotatory manner (Table 23.1). We can verify this idea by looking at the molecular orbitals involved. The HOMO for octatetraene is Φ_4. Here are the four lowest molecular orbitals.

Top views of the molecular orbitals of octatetraene

Φ_4 + + − − + + − −   HOMO
Φ_3 + + o − − o + +   o = node
Φ_2 + + + + − − − −
Φ_1 + + + + + + + +

Now the task is to see how the orbitals must move. The figure shows the ends of the HOMO, Φ_4, as well as one conrotatory motion. If you follow through where the methyl groups must go, you can see why it is the trans stereochemistry that appears in the product.

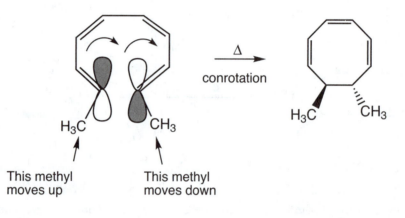

This methyl moves up

This methyl moves down

(b) The rearrangement of Dewar benzene to benzene involves the opening of a cyclobutene to a diene. The thermal motion must be conrotatory, and this conrotation must lead to a trans double bond. Of course, a trans double bond cannot be accommodated in a six-membered ring. There is no simple, disrotatory motion possible. Orbital symmetry will not permit it.

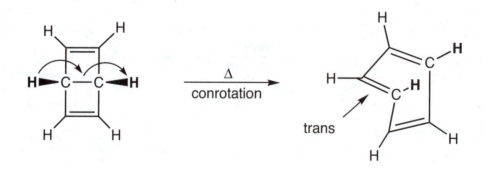

(c) Compound **X** is formed by opening the cyclobutene. A thermal opening of a cyclobutene must be conrotatory, so **X** must incorporate one trans double bond. Compound **X** contains a 10-membered ring, so the trans double bond is not impossibly strained, but it is unstable and rearranges quickly to the product shown in the problem, *trans*-9,10-dihydronaphthalene.* The problem now is why it is the trans stereoisomer that is formed in this second step.

---

*MJ and HLG have argued long and bitterly over the naming of this compound. One of us argues for the traditional name given to the compound when it was first made (by this same author) in 1967. The other claims that the more modern name, 4a,8a-dihydronaphthalene, should be used. Guess which author had the last turn with the proofs?

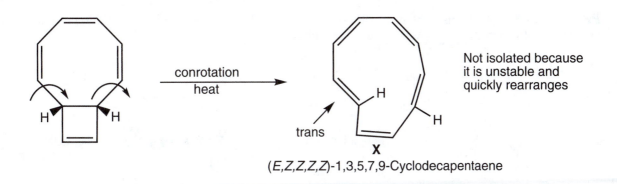

X

(*E,Z,Z,Z,Z*)-1,3,5,7,9-Cyclodecapentaene

Further reaction of **X** involves a hexatriene–cyclohexadiene interconversion and must take place thermally in a disrotatory way (Table 23.1). This rotation leads directly to the *trans*-9,10-dihydronaphthalene product.

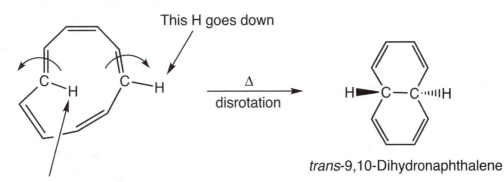

trans-9,10-Dihydronaphthalene

(d) Each reaction involves two hexatriene–cyclohexadiene reactions. The thermal reactions must be disrotatory and the photochemical reactions must be conrotatory (Table 23.1).

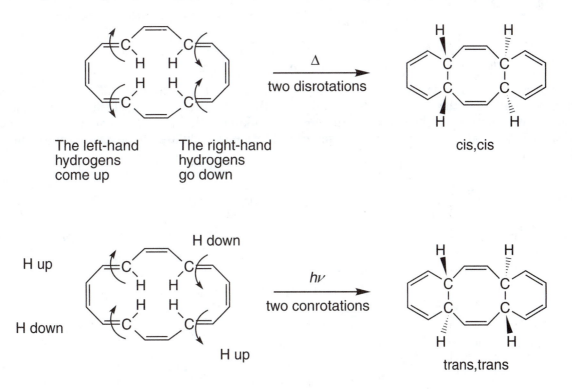

Problem 23.8 In this case, the interactions would be HOMO (ethylene = π^*) – LUMO (butadiene = Φ_3). The photochemical Diels–Alder is still forbidden.

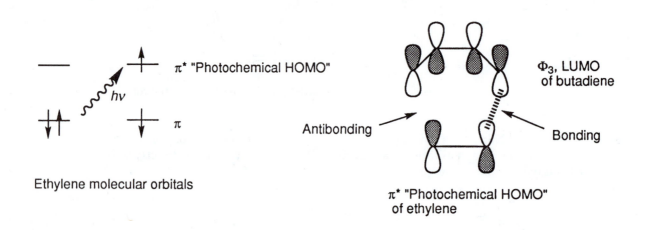

Ethylene molecular orbitals

Problem 23.9 A ΔH calculation can easily be performed.

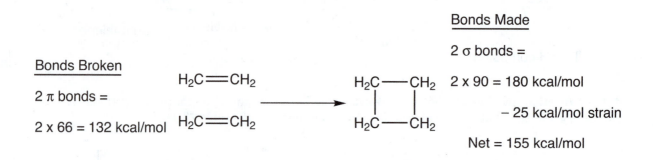

So, the reaction is calculated to be exothermic by approximately 23 kcal/mol (155 – 132 kcal/mol). In this calculation, the carbon–carbon bond strength for ethane was used (Table 7.2, p. 227). More appropriate would be a lower value for the breaking of the central carbon–carbon bond in butane. If this value (~88 kcal/mol) is used, the reaction is calculated to be exothermic by only 19 kcal/mol.

Problem 23.10 As Figure 23.25 shows, the dimerization of two ethylenes is likely to proceed through an intermediate diradical. There will be free rotation about carbon–carbon single bonds in that diradical, and all possible stereoisomers must be formed.

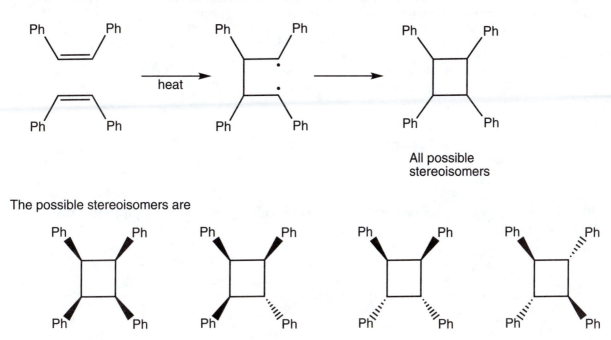

All possible stereoisomers

The possible stereoisomers are

Problem 23.11 The more substituted a double bond, the more stable it is (Chapter 3, p. 119). The isomers on the far left and far right of the figure are the more stable compounds. For example:

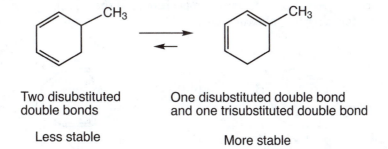

Two disubstituted double bonds

Less stable

One disubstituted double bond and one trisubstituted double bond

More stable

Yes, the [1,5] shift of hydrogen in (Z)-3-methyl-1,3-pentadiene is degenerate.

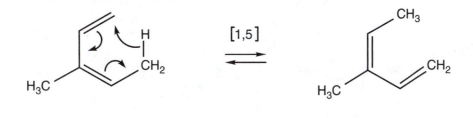

(Z)-3-Methyl-1,3-pentadiene   (Z)-3-Methyl-1,3-pentadiene

Problem 23.13   Reaction (a) is the only degenerate process. It is the only reaction where the starting material and the product are the same.

(a) This problem involves a straightforward [1,9] shift of hydrogen. Don't forget, the shift numbering protocol has nothing to do with the IUPAC naming protocol.

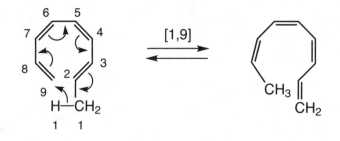

(b) This part is tricky only because of the way the molecule is drawn. It is a [1,3] shift of the pentadienyl group.

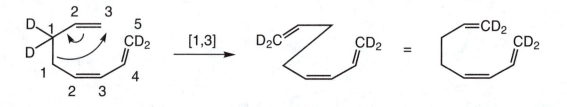

(c) The label lets you follow this through as a [3,5] shift of a pentadienyl group. The bond breaks at position 1 and re-forms at position 5, attaching to position 3 of the allyl fragment. It is the 3- and 5-positions that come together to make the new bond.

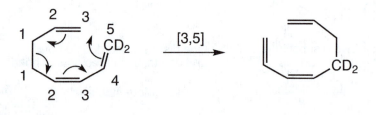

(d) This reaction is a simple [3,3] shift. The 3-position of the starting migrating group becomes attached to the 3-position of the framework over which the migration takes place.

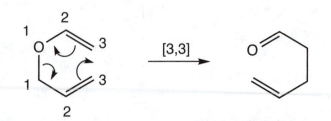

(e) This problem is another hard one. It is a [1,5] shift of the indicated carbon. The R group is there only as a marker so that you can tell that a reaction has taken place.

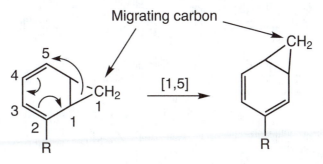

Problem 23.15 Not a chance. The HOMO is Φ_2 of allyl. Therefore, the thermal [1,3] shift requires an antarafacial shift of hydrogen, and the $1s$ orbital is not large enough to span the distance.

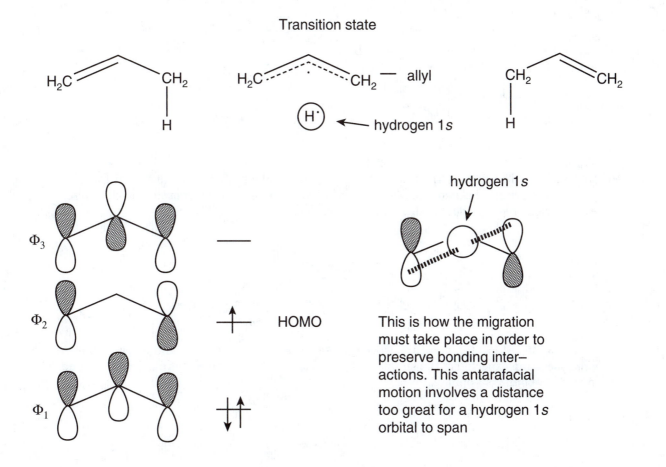

Problem 23.16

(a, b) The key reaction is a [1,5] shift to give the nonaromatic molecule "isoindene." Isoindene can be formed by a [1,5] shift of either hydrogen or deuterium, and both reactions are shown in the figure. A rapid re-formation of the aromatic ring through another [1,5] shift, again of either hydrogen or deuterium, leads to the isomerized products. The difficulty in this problem lies in seeing the necessity for the very endothermic first step. The convention is to count the shift as a [1,5] shift, even though the departure and arrival atoms are attached.

(continued)

Problem 23.16 *(continued)*

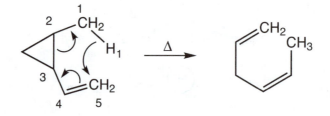

(a)   D [1,5]   H   H [1,5]

Isoindenes

(b)   H [1,5]   H   H [1,5]

(c) Three-membered rings can often play the parts of double bonds. This reaction shows just such a case as the [1,5] shift of hydrogen takes place with opening of the cyclopropane. The reaction is quite analogous to a [1,5] shift in a 1,3-pentadiene.

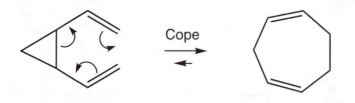

Problem 23.18   In this case, the Cope rearrangement involves a large amount of strain relief. The three-membered ring is opened in a disrotatory fashion to give the cycloheptadiene. Of course, the relatively strain-free cycloheptadiene will be strongly favored at equilibrium.

Cope

Problem 23.19   At low temperature, the Cope rearrangement is stopped, and one sees the ^{1}H NMR spectrum expected of the simple, static structure. For example, there are four olefinic hydrogens at about δ 5.8 ppm.

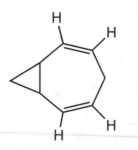

At high temperature, the Cope rearrangement is fast, and a completely averaged spectrum is obtained. There are only two hydrogens that are always olefinic, and these appear at about δ 5.8 ppm (H_a). Four hydrogens are half olefinic and half cyclopropyl, and these appear at about δ 3.7 ppm (H_b).

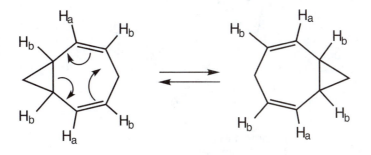

In between these extremes of the frozen-out spectrum and the averaged spectrum, we see the 20 °C intermediate spectrum. We see a spectrum of no particular definition as incomplete averaging of the signals is taking place. Near room temperature we are at neither the "frozen out" nor completely averaged regime, but somewhere in between.

Problem 23.22 Photochemical 2 + 2 reactions are allowed by orbital symmetry, and this reaction is just a retro 2 + 2 addition. The formation of benzene drives the reaction to the right.

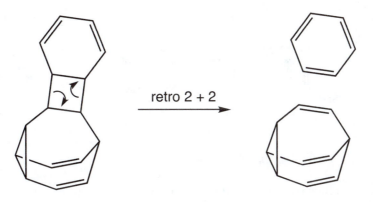

retro 2 + 2

Problem 23.23 This problem shows two ways of disguising the Cope rearrangement.

(a) Whenever you see a complex molecule containing double bonds, think "Cope." This advice is especially useful when the subject of thermal reactions is being discussed, and, of course, that's where we are right now. Count to see if there is a 1,5-diene present.

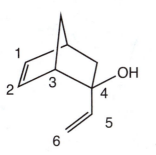

A 1,5-diene, so a Cope is possible

If there is, try the Cope, no matter how odd the system looks. Write the three "Cope" arrows, forming the bond between carbons 1 and 6 and breaking the bond between carbons 3 and 4. After you draw the arrow formalism, be sure to redraw the molecule first without moving anything! Now you can refashion the molecule in a reasonable way. In this case, the product is an enol that is, of course, in equilibrium with the related keto form.

(continued)

Problem 23.23 *(continued)*

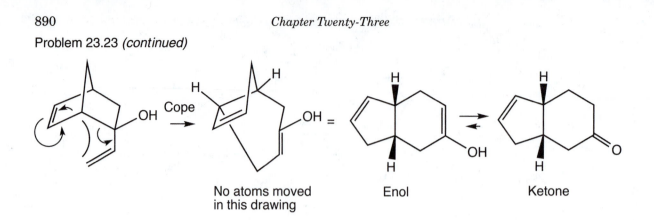

No atoms moved in this drawing   Enol   Ketone

(b) In this example, the aromatic ring helps to hide the 1,5-diene at the heart of the Cope rearrangement. Nonetheless, it is a 1,5-diene, and a Cope-like rearrangement is possible. In this case, it leads to a ketone that is much less stable than its enol form, a phenol. This variation of the Cope is called the Claisen rearrangement. It's the same Claisen of the Claisen condensation (Chapter 19, p. 985) and the Claisen–Schmidt condensation (Chapter 19, p. 981).

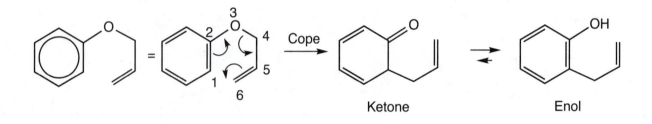

Ketone   Enol

Problem 23.24 Let's follow the suggestions of Section 23.8a. The numbers follow those in the section.

(1) A ring is surely closing, so the reaction must be electrocyclic.

(2) Here is an arrow formalism.

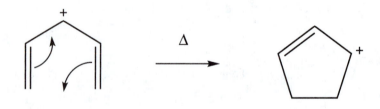

(3, 4, and 5) Here are the MOs for pentadienyl. For the pentadienyl cation, there will be four electrons, and the HOMO will be as shown.

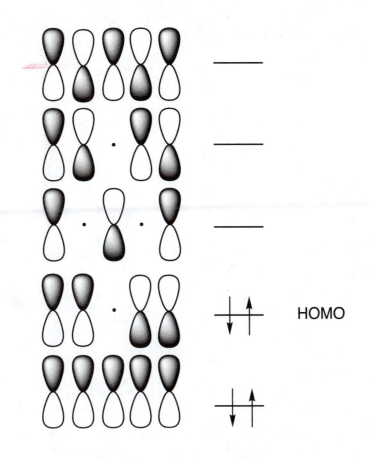

(6) To close the ring, motion must be conrotatory—a bonding interaction must be produced where the new σ bond is.

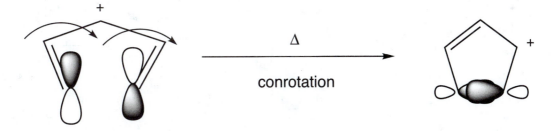

Problem 23.25

(1) A ring is being formed from two fragments. This process must be a cycloaddition reaction. We could just look at Table 23.2, count electrons (six in this case), and decide that the reaction will be successful thermally. But let's analyze instead.

(2) Here is an arrow formalism.

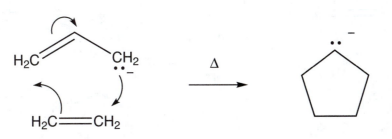

(continued)

Problem 23.25 *(continued)*

(3, 4) Here are the π MOs and the HOMOs and LUMOs.

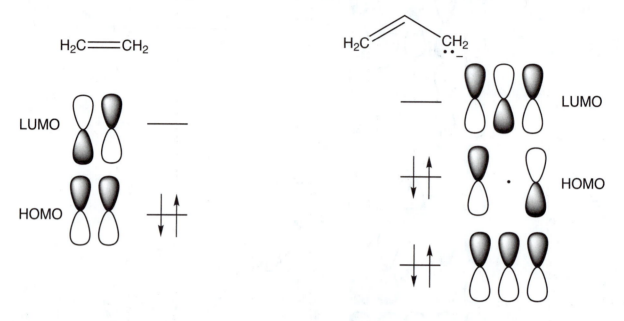

(5) Here is one of the HOMO–LUMO interactions. The other works just as well. As you can see, two new bonding interactions are formed, and the reaction is allowed thermally:

Transition state model

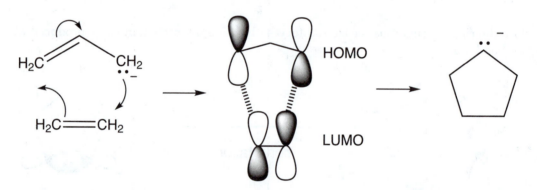

Problem 23.26

(1) A hydrogen migrates from the left-hand side of the molecule to the right-hand side. This process must be a sigmatropic shift.

(2) The arrow formalism shows it best.

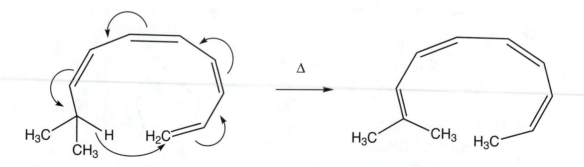

(3) A hydrogen atom is migrating across the π system of a nonatetraenyl radical. Writing out all the MOs of nonatetraenyl is tedious if not, ultimately, difficult. You can shortcut matters substantially if you remember the form of the HOMO of all odd fully conjugated radicals: as seen from the top, + 0 − 0 + 0 − 0 + 0 − etc., etc. We worked all this out in Chapter 10, Problem 10.61.

(4 and 5) Here is a picture that shows the migration of the hydrogen across the HOMO of nonatetraenyl, with the two bonding interactions in the suprafacial migration:

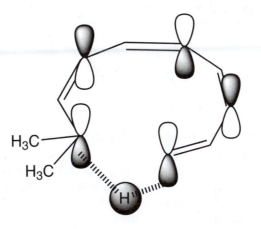

Answers to Additional Problems

Problem 23.27 First of all, stereochemistry is very important. If there is a trans double bond in the heptatriene, there is no way that the ends of the system can come together. The reaction must fail; no [1,7] shift is possible.

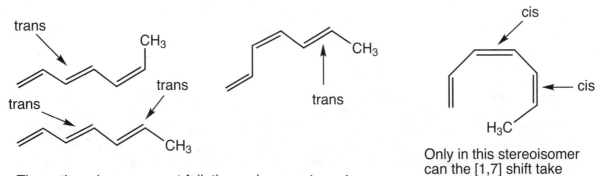

These three isomers must fail; the ends cannot meet

Only in this stereoisomer can the [1,7] shift take place

For heptatrienyl, there are seven π electrons, and therefore the HOMO is Φ_4. For photochemical transformations, the HOMO will be Φ_5. We need only examine the ends of the π system to see what is possible. In this case, the photochemical [1,7] shift must be suprafacial and the thermal [1,7] shift must be antarafacial. In other words, the [1,7] shift looks just like the [1,3] shift.

(continued)

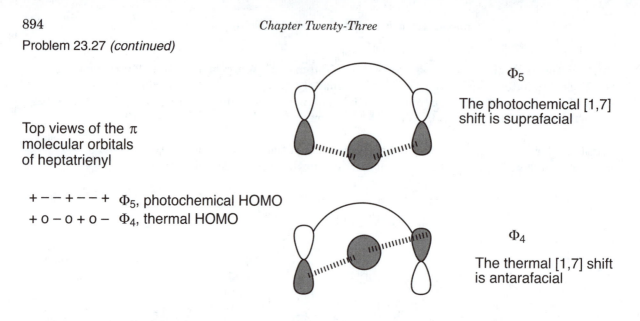

Top views of the π molecular orbitals of heptatrienyl

Φ_5

The photochemical [1,7] shift is suprafacial

Φ_4

The thermal [1,7] shift is antarafacial

+ − − + − − + Φ_5, photochemical HOMO
+ o − o + o − Φ_4, thermal HOMO

Problem 23.28   The conversion of 7-dehydrocholesterol (**1**) to previtamin D_3 (**2**) involves an electrocyclic ring opening of a 1,3-cyclohexadiene to a 1,3,5-hexatriene. The figure shows that this change occurs in a conrotatory fashion. The methyl group moves in a clockwise fashion, as does the hydrogen. Be sure you can see this. This process must be photochemical, as the reaction involves six ($4n + 2$) π electrons, and conrotatory, where six-electron reactions are photochemically, not thermally, allowed. Be certain you can explain why this is so.

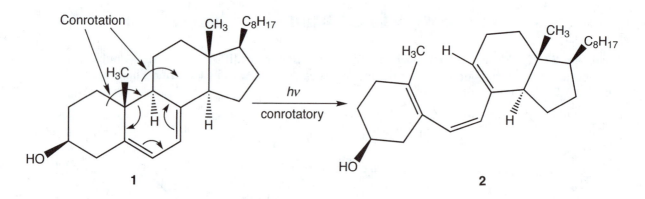

In the next reaction, we get to use the answer to the preceding problem. The transformation of previtamin D_3 (**2**) to vitamin D_3 (**3**) itself involves a [1,7] sigmatropic shift. If this shift occurs under thermal conditions, it must be antarafacial. If it is a photochemical process, it must be suprafacial. Although these two processes cannot be differentiated using these molecules, the antarafacial nature of a similar thermal [1,7] shift of hydrogen was recently demonstrated using a deuterium labeling experiment.

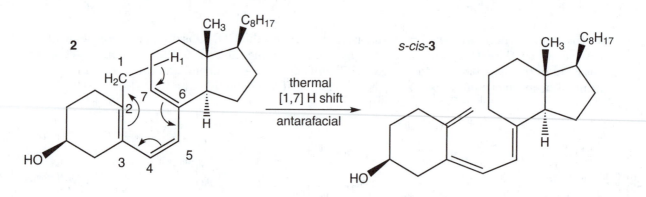

Problem 23.29   The set of arrows in (a) is misleading. The arrow formalism implies a butadiene–cyclobutene interconversion. Under thermal conditions, this process requires a conrotatory motion, and must generate not **2**, but a trans version of **2**, too highly strained to be isolable.

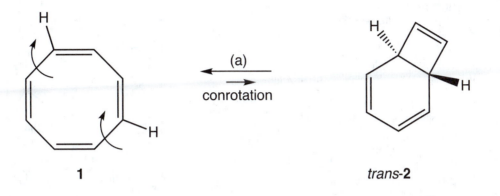

If we look at the reaction in the other direction, we can see that a conrotatory opening of **2** must lead not to **1**, but to a version of **1** in which there is a trans double bond. This molecule is too strained to be formed.

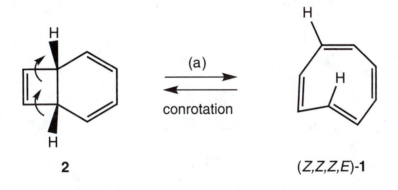

By contrast, arrow formalism (b) shows a thermal, disrotatory interconversion of a 1,3,5-hexatriene and a 1,3-cyclohexadiene. There is no difficulty with stereochemistry or strain here. This arrow formalism is much more appropriate.

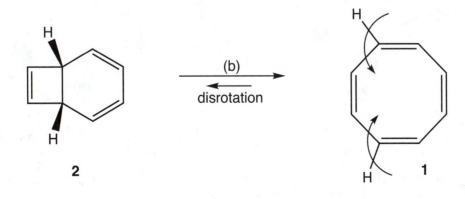

Problem 23.30 The charges disturb some people, but there is no reason to be upset. They become important only when it becomes time to count electrons.

(a) This reaction is an electrocyclic interconversion of the cyclopropyl and allyl cations, and the question to be decided is whether the reaction is conrotatory or disrotatory.

This bond breaks—will it be conrotatory or disrotatory?

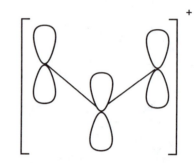

The cyclopropyl cation—note the 2p orbital at one carbon

The allyl cation

The number of electrons involved should tell us the answer without looking any further. Now the positive charge is important because it allows us to count electrons properly. The allyl cation has only two π electrons (destined, on closing, to become the two electrons in the new cyclopropane bond). Two electrons is a $(4n + 2)$ number, $n = 0$, and so we can expect this reaction to be disrotatory. So

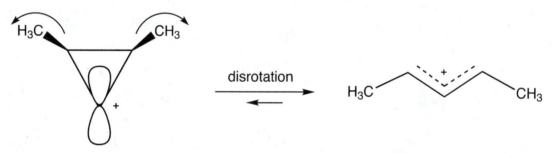

Let's do a better job of analysis. As usual, we will look at the open partner in the reaction, in this case the allyl cation. The HOMO for the cation is Φ_1. In order to make a bonding interaction, rotation must be in the disrotatory sense.

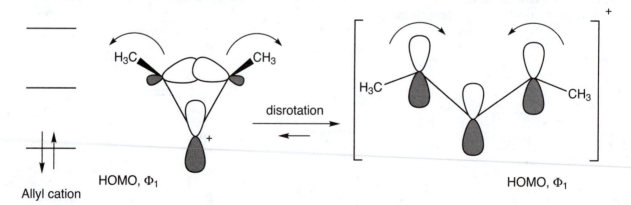

The other possible disrotation leads to the allyl cation with the two methyl groups inside. This process will be disfavored thermodynamically for steric reasons.

(b) For the anion, there are four electrons, and we can expect a thermal, conrotatory process in this 4n system. Here is the orbital analysis.

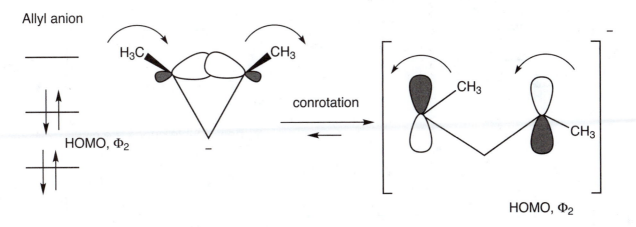

Allyl anion

HOMO, Φ_2

conrotation

HOMO, Φ_2

The other possible conrotation leads to the same product.

Problem 23.31   First of all, here are arrow formalisms for the two possible reactions.

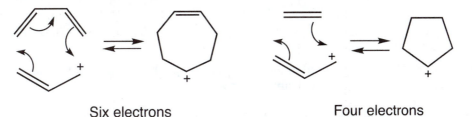

Six electrons   Four electrons

Once again, if we count electrons, we can answer the question with very little analysis. The first process is a six-electron cycloaddition, and, like the Diels–Alder reaction, would be expected to be a thermal process. The second reaction is a four-electron cycloaddition and, like the cycloaddition of a pair of ethylenes, should be a photochemical reaction. Now, let's do a better HOMO–LUMO analysis to see which of the two should be concerted photochemically. We will use the "photochemical HOMO" of allyl and the LUMO of the reaction partner (although an analysis that did just the opposite, and used the "photochemical HOMO" of the reaction partner and the LUMO of the allyl cation, would work out exactly the same way).

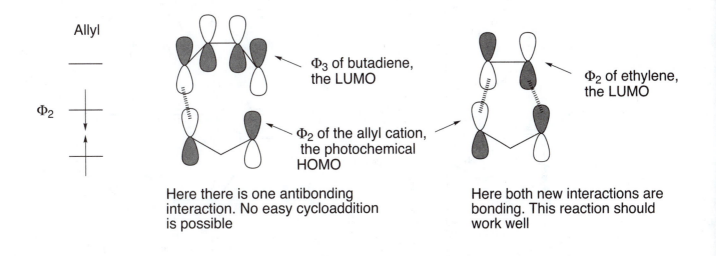

Allyl

Φ_2

Φ_3 of butadiene, the LUMO

Φ_2 of the allyl cation, the photochemical HOMO

Φ_2 of ethylene, the LUMO

Here there is one antibonding interaction. No easy cycloaddition is possible

Here both new interactions are bonding. This reaction should work well

Problem 23.32   The thermal opening of a cyclobutene must be a conrotatory process. Follow through the two reactions and look for a difference.

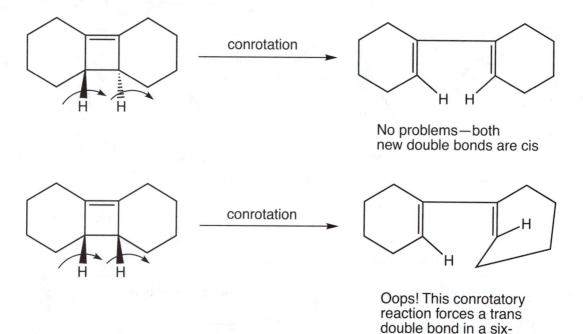

No problems—both
new double bonds are cis

Oops! This conrotatory
reaction forces a trans
double bond in a six-
membered ring. Bad, bad, bad!

It's not hard to find. The cis compound must produce a compound containing a trans double bond in a six-membered ring. This reaction will surely be a slow process.

Problem 23.33   Concerted photochemical 2 + 2 cycloadditions and their reversals are allowed by orbital symmetry, and it is a reverse 2 + 2 cycloaddition that takes place here. As the reactions are concerted, the cis cyclobutene leads to the cis alkene and the trans cyclobutene leads to the trans alkene.

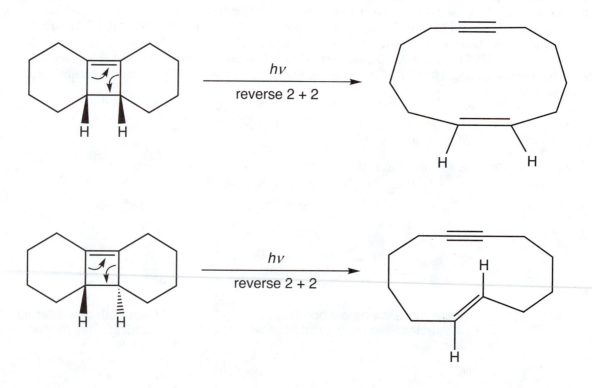

Problem 23.34 There are several clues in the drawing. First of all, whenever you see maleic anhydride, think of the Diels–Alder reaction. Maleic anhydride is a favorite dienophile. Second, this question immediately follows a problem on the opening of cyclobutenes. Finally, all cyclohexenes are potential Diels–Alder products, and you might be able to take the product apart to its component parts. In any event, this problem involves the opening of the starting material to a diene, followed by capture by maleic anhydride. Notice how the last step recovers the aromaticity lost in the original opening. That is why the endocyclic 1,3-diene does not undergo a Diels–Alder addition.

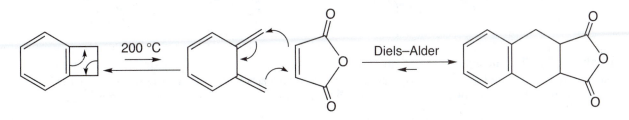

Problem 23.35 Now we have to worry about stereochemistry. The thermal opening of the cyclobutene to the diene will be conrotatory, and in this case that will lead to intermediate **A**, as shown. The concerted, thermal, Diels–Alder reaction of **A** will lead to the compound in which the phenyl groups are cis.

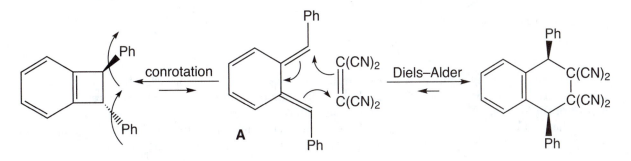

Problem 23.36 Let's do what the problem suggests and work backward from the structure of the product. Let's also be mindful of the presence of maleic anhydride. A Diels–Alder reaction is likely to be involved somehow. So, let's deconstruct the cyclohexene in the product to its constituent parts, the diene **A** and a dienophile (maleic anhydride).

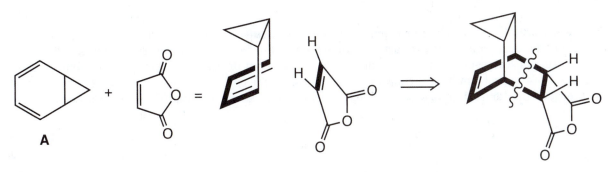

Cyclohexene in bold lines

How can **A** arise from our starting material, cycloheptatriene? We saw many cyclohexadiene–hexatriene rearrangements in this chapter, and this reaction involves another. In this reaction, cyclohexatriene first closes to **A** (called norcaradiene), and **A** is then captured by maleic anhydride

(continued)

Problem 23.36 *(continued)*

to give the product. The closing to **A** is sure to be endothermic because of the strain added by the three-membered ring, but a little **A** is formed, and as it is captured by maleic anhydride, more is produced.

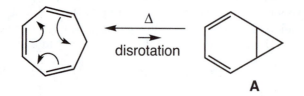

Problem 23.37   The arrow formalism is relatively easy, although it may be difficult to draw the polycyclic product molecule correctly. As usual in such cases, we have first drawn the molecule with only minimal movement of the reaction partners, and then "relaxed" to the product structure.

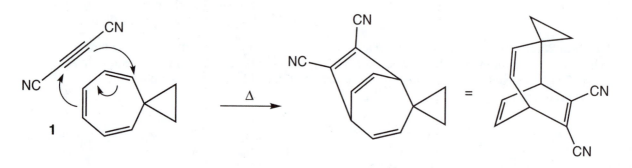

Now the question is why compound **1** reacts in normal Diels–Alder fashion, whereas other cycloheptatrienes do not. As shown in Problem 23.36, they usually first isomerize to a small amount of the "norcaradiene" form, and then react. Look at what happens when **1** isomerizes to a norcaradiene. A compound with two spiro three-membered rings (**A**) is formed. It is this extra strain that prevents **1** from forming enough of the norcaradiene to undergo the "usual" reaction.

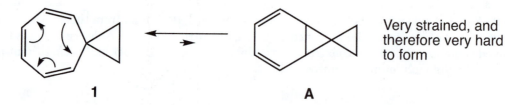

Very strained, and therefore very hard to form

Problem 23.38   In this classic problem, we face two difficulties. First, we need a general mechanism for the formation of the product, and then we need to worry about the stereochemical details. The mention of stereospecificity should make you think about orbital symmetry (not to mention the placement of the problem in this chapter). Protonation of the carbonyl oxygen is the first step (you can't forget Chapter 16, and acid–base chemistry, just because we are in an "orbital" chapter).

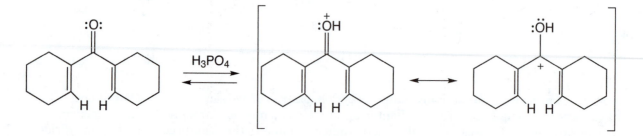

The figure shows two resonance forms for the protonated carbonyl compound, but there are others. In fact, protonation has yielded a pentadienyl cation. Electrocyclic closure of the cation leads to **A**, and deprotonation and a keto–enol equilibration give the product.

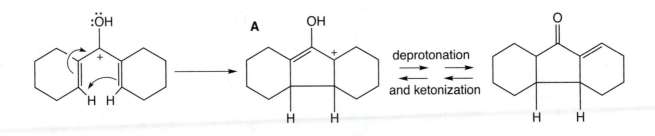

Now, what about the stereochemistry? Let's look at the thermal electrocyclic closure of a pentadienyl cation. The system and top views of its π molecular orbitals are shown in the figure.

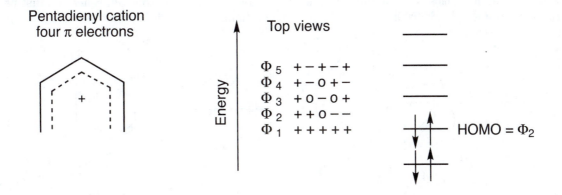

The HOMO is Φ_2 for this four-electron system, and closure must be conrotatory.

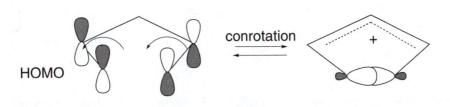

Now look at the more complicated pentadienyl cation in this problem. Conrotatory closure must lead to the observed trans stereochemistry.

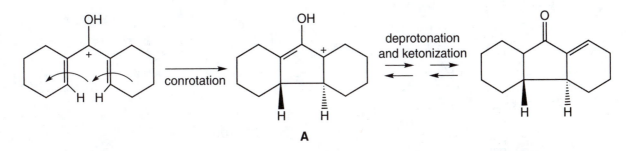

Problem 23.39 This problem is difficult only because there is an intermediate in the reaction and we didn't tell you. There is no way to go directly from starting material to product. Surely, in an orbital symmetry chapter, a cyclobutene must conjure the notion of opening to a butadiene. The thermal opening must be conrotatory, and this leads to intermediate **A**, in which there is one trans double bond.

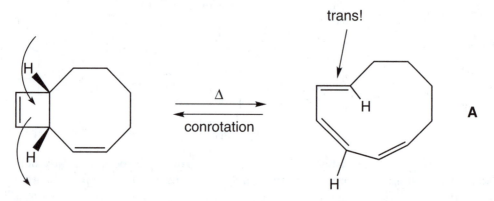

Now **A**, a hexatriene, closes to the final product, a cyclohexadiene. This thermal reaction must be disrotatory, and this leads to the observed trans stereochemistry. The figure first shows the arrow formalism and then the stereochemistry.

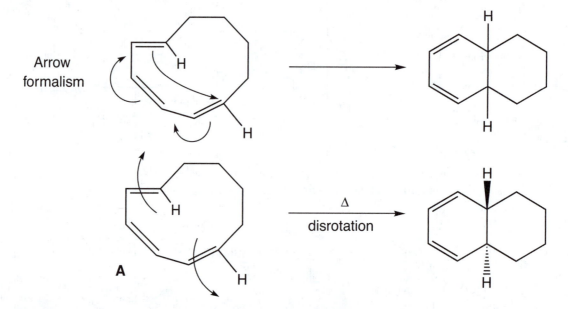

Problem 23.40 Here we have a problem much like Problem 23.30. Despite the nitrogen atom, this is really nothing more than a cyclopropyl anion–allyl anion equilibrium. Like the cyclopropyl anion, this is a four-electron system. The nonbonding pair of electrons on nitrogen and the pair of electrons in the bond are the ones involved.

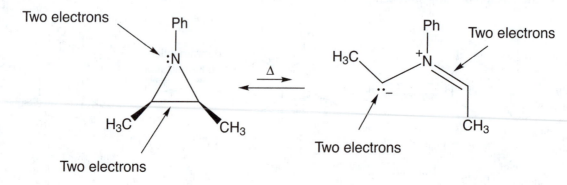

A thermal, four-electron process will be conrotatory (it's the same as the cyclobutene–butadiene four-electron system), and the observed stereochemistry is exactly what conrotation must produce.

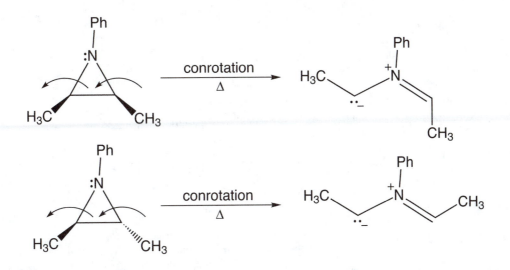

Problem 23.41 Both parts of this problem involve a Diels–Alder reaction that cannot take place until the starting material undergoes another reaction first.

(a) Watch that cyclobutene. There probably has never been a cyclobutene in an orbital symmetry problem that didn't open, and this one is no exception. Once it opens thermally (conrotation, please), the double bond attached to the carbonyl group captures the diene to give the product.

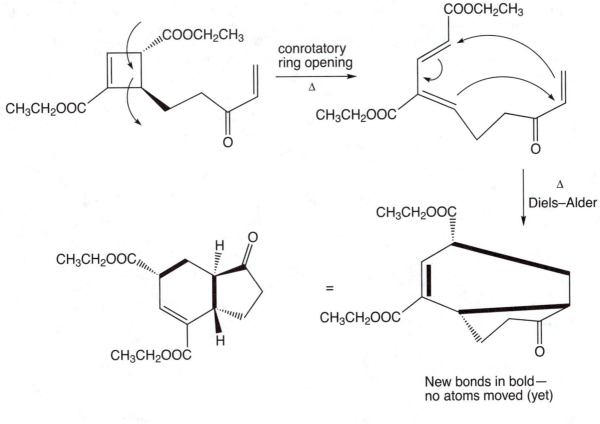

New bonds in bold—
no atoms moved (yet)

(continued)

Problem 23.41 *(continued)*

(b) Here, the first reaction is a disrotatory six-electron thermal closure of a 1,3,5-hexatriene to a 1,3-cyclohexadiene. Once this has happened, the lurking double bond (dienophile) captures the 1,3-diene.

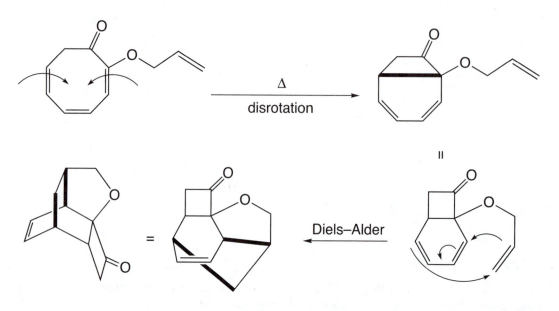

Problem 23.42   The first step is an S_N2 reaction between the basic amine and allyl bromide. The stereochemical preference for the β (top) face of the bicyclic system because of the steric hindrance on that face. There is less difficulty adding to the side of the hydrogen than the side of the ring. The second step is an acid-catalyzed dehydration. The benzylic alcohol is easily dehydrated to give the substituted and very conjugated styrene group. The third step of the reaction occurs at room temperature. It is a Cope-type rearrangement, an aza-Cope. The aza-Cope is a six-electron thermal [3,3] sigmatropic shift, so we expect the new bond to be formed on the same side of the π system as the bond that is breaking. The final step is hydride addition to the iminium ion. There is no stereochemical information. Presumably the hydride addition occurs from either face.

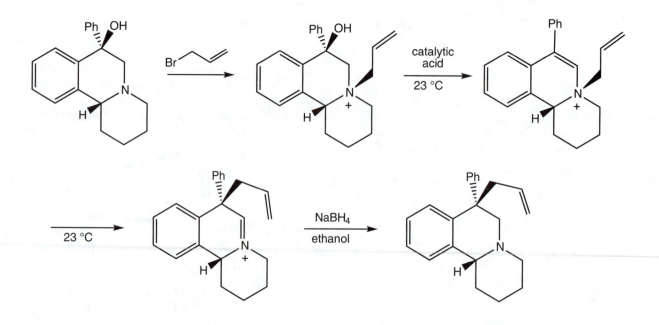

Problem 23.43 This reaction is a [3,3] sigmatropic shift (a Cope rearrangement). It is accelerated by the base deprotonating the allylic alcohol. The alkoxide helps drive the sigmatropic shift, as shown in the mechanism. The numbers next to the mechanism arrows on the alkoxide indicate the direction of the electron flow.

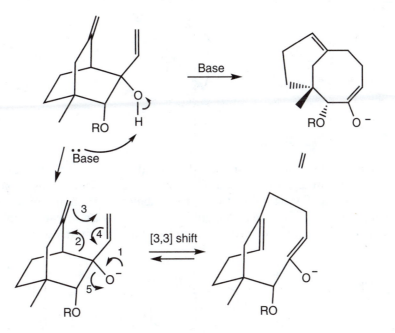

Problem 23.44

(a) The hint says that there are two intermediates in which the three-membered ring is gone, so it surely makes sense to focus on this part of the molecule. Presumably, the cyclopropane is destroyed in a first step and reconstituted later. There are two things to think about whenever you see a cyclopentadiene. The first is a Diels–Alder reaction, but there is no visible dienophile to go with the cyclopentadiene, so this line of inquiry doesn't seem promising. The second is a thermal [1,5] sigmatropic shift. That turns out to be a useful way to think in this case. A [1,5] shift of carbon generates intermediate **A**. There are two things **A** can do. It can go right back to starting material, or it can do another [1,5] shift of hydrogen to produce intermediate **B**. Now a [1,5] shift in **B** regenerates the cyclopropane and makes the product. This problem is hard for two reasons. First, there are two intermediates in between starting material and product, and that level of complexity almost always makes for a difficult problem. Second, there is misdirection here. It looks as if the methyl group is migrating somehow, and this is not the case. It remains steadfast throughout, always attached to the same carbon as the molecule rearranges around it, eventually moving it to another position.

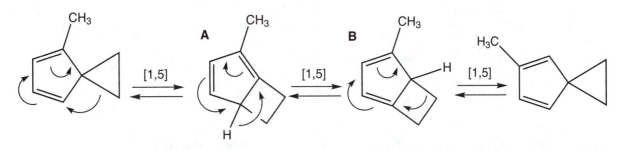

(continued)

Problem 23.44 *(continued)*

(b) This part is hard because it is difficult to visualize the structures involved. What can happen photochemically? The two most common reactions that we know about are 2 + 2 cycloadditions of alkenes and [1,3] shifts. In this case, it is a simple (?) photochemical [1,3] shift of carbon that leads to product.

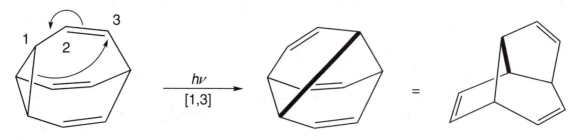

Problem 23.45

(a) There are many possible labeling experiments that could be (and have been) used here, including isotopic labels. Perhaps the simplest label would be a methyl group or groups attached to either C(1) or C(3) of the allyl group. For example, the methyl-labeled allyl phenyl ether **A** would afford only the labeled *o*-allylphenol **B** if the rearrangement occurred by an intramolecular concerted [3,3] shift. Note the so-called "allylic shift" in which the methyl label originally α to the oxygen is found γ to the ring.

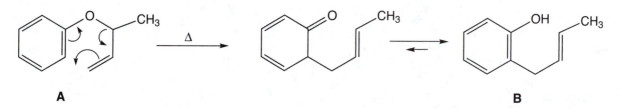

By contrast, if the rearrangement were intermolecular and involved radical intermediates, both phenol **B** and phenol **C** would be expected. The allyl radical can recombine with the dienone radical in two different ways, and conversion of the ketone to the enol forms would give not only **B**, but **C** as well.

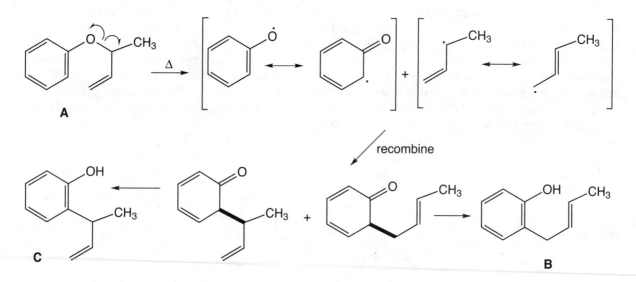

In fact, the experiment reveals only **B**, and the mechanism must be concerted.

(b) The mechanism for the para-Claisen rearrangement begins the same way, with an intramolecular [3,3] shift to give dienone **D**. However, in this case, formation of the aromatic phenol cannot take place because there is no hydrogen α to the carbonyl group. The ortho positions are blocked with methyl groups.

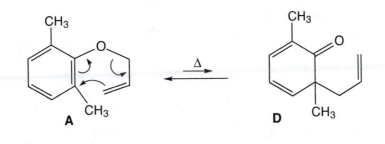

Because aromatization is blocked, compound **D** undergoes another [3,3] shift (a Cope rearrangement) to give dienone **E**. Compound **E** easily rearranges to the aromatic compound.

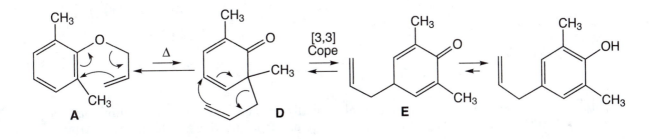

If we use the α-methyl label here, the intramolecular [3,3] shift mechanism predicts only the formation of **F**. An intermolecular mechanism would give a mixture of **F** and **G**. Experimentally, only **F** is observed.

Intramolecular:

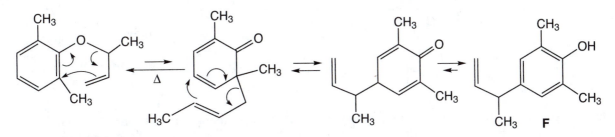

(continued)

Problem 23.45 *(continued)*

Intermolecular:

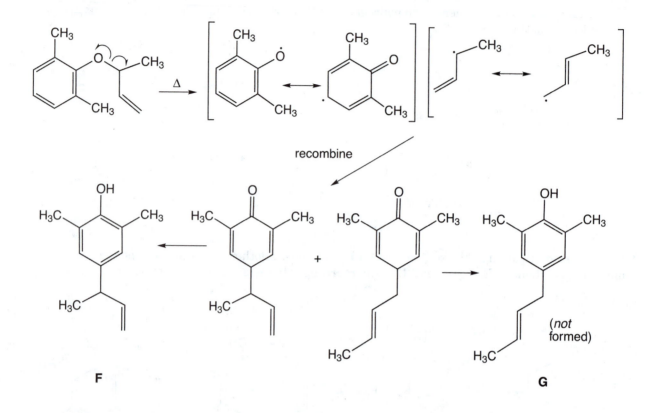

F   G

Problem 23.46 Because the starting material is an allyl aryl ether, you should be thinking Claisen rearrangement (especially considering the two previous questions!). Although it is true that allyl aryl ethers with the para and ortho positions blocked normally don't react, we have a special case here. Begin as before, with a concerted [3,3] shift to give dienone **A**.

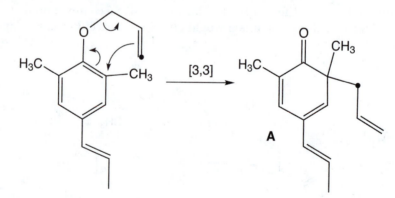

Because there is no hydrogen α to the carbonyl group in **A**, this dienone cannot rearomatize to a phenol. What can it do instead? Often students make the interesting observation that a thermally allowed [1,5] shift of carbon would give dienone **B**, which could give the desired phenol. Although this mechanism is certainly economical, it is also incorrect. The proposed [1,5] shift involves passage over a long distance and cannot occur in an intramolecular fashion. You might make a model to convince yourself that this is so.

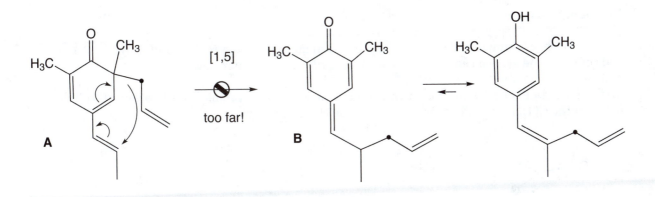

What else can dienone **A** do? How about another [3,3] shift to give dienone **C**, just as we saw in the para-Claisen rearrangement in Problem 23.45? Now dienone **C** can undergo yet another Cope rearrangement to give **B**, the molecule we failed to make earlier.

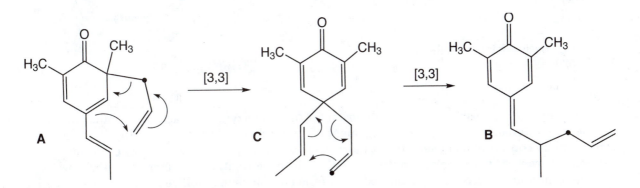

Now, although a [1,5] hydrogen shift, followed by phenol formation, rearomatizes and apparently generates the product, this [1,5] shift passes over far too long a distance to be reasonable. An intermolecular reaction is far more likely. Simply protonate on oxygen and deprotonate on carbon. Where does the acid come from? In practice, there is usually enough acid on the glass surface to accomplish such reactions.

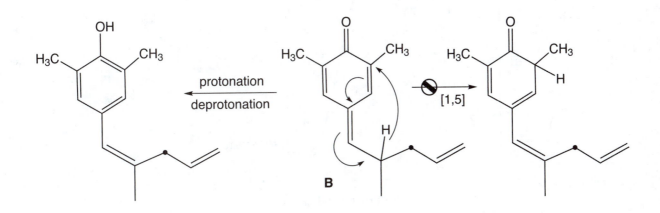

Problem 23.47

(a) This answer is simple and shows clearly in the diagrams in the problem and below. In **2**, the ends of the diene system cannot reach each other, whereas in conformation **1** they can.

(b) Conformational preference for **2** is presumably a result of intramolecular hydrogen bonding between the C(4) hydroxyl hydrogen and the side chain carboxylate.

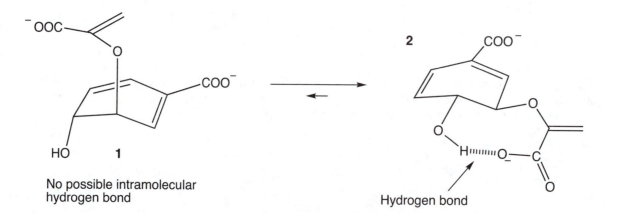

No possible intramolecular
hydrogen bond

Hydrogen bond

(c) The energy versus reaction progress diagram for the nonenzymatic reaction is similar to the one we encountered for the Cope rearrangement of *cis*-1,2-divinylcyclopropane (Fig. 23.59) in that it involves an equilibrium between two conformations, and only the higher-energy conformation can react. However, it is suggested in this case that the conformational isomerism is the slow (rate-determining) step. So, the postulated role of the enzyme is to stabilize the transition state for this conformational isomerization.

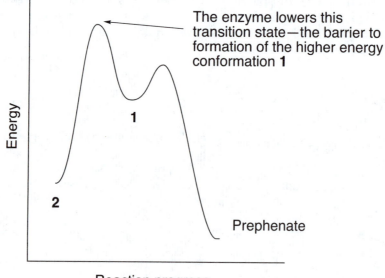

The enzyme lowers this transition state—the barrier to formation of the higher energy conformation **1**

Reaction progress

Unfortunately, this simple picture is not correct. Largely through the efforts of Professor Jeremy Knowles and his co-workers, it has been shown that chorismate mutase selectively binds the 10%–20% of the pseudodiaxial conformer of chorismate (**1**) present in solution. As conformational interconversion is fast, there is no need to postulate an enzyme-catalyzed conformational change. Consequently, the enzyme must stabilize the transition state for the Claisen rearrangement itself.

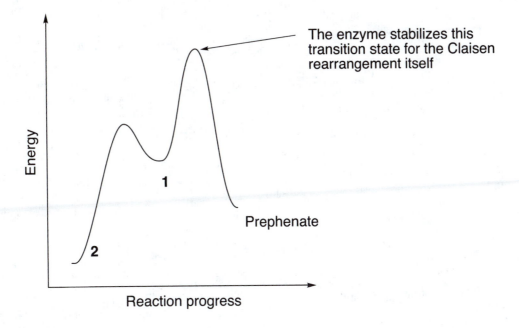

Problem 23.48

(a) As we saw in Chapter 18 (p. 907), a variety of nucleophiles adds to the carbonyl group of ketenes. In this case, the nitrogen nonbonding electrons of the imine **2** add to the carbonyl group of ketene **1** to give the dipolar intermediate **A**. Dipolar intermediate **A** can then undergo cyclization to yield the observed β-lactam **3**.

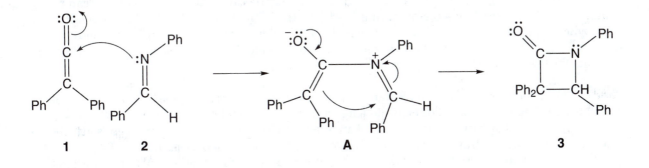

(b) The stereochemistry of these cycloaddition reactions has been rationalized on the basis of a conrotatory ring closure of the dipolar intermediate [**A** above in part (a)] in which steric interactions are minimized. This intermediate is a 4 π electron system, and therefore a conrotatory thermal process is expected (*Remember*: Once again, the cyclobutene-1,3-butadiene interconversion). In this case, conrotatory closure of **B** leads to the observed β-lactam.

(continued)

Problem 23.48 (continued)

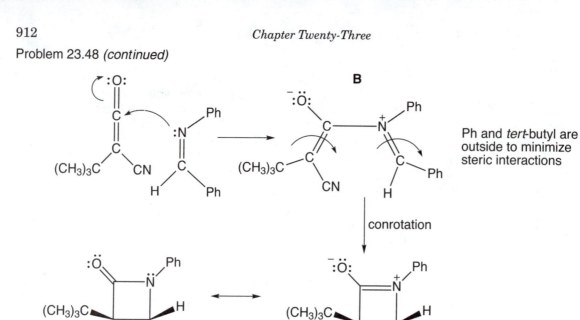

B

Ph and *tert*-butyl are
outside to minimize
steric interactions

conrotation

Problem 23.49   Because compound **2** is an isomer of **1**, it must have a molecular formula of
$C_{11}H_{12}O_2$. There are 6 degrees of unsaturation for this molecular formula. It is not unreasonable to
assume that 4 of the 6 degrees of unsaturation are the result of the presence of the benzene ring,
particularly given the presence of a benzene ring in starting material.

The IR spectrum of compound **2** suggests an ester, as indicated by a strong carbonyl absorption at
1740 cm^{-1}. The presence of an ester is further supported by the signal at δ 170.6 ppm in the
^{13}C NMR spectrum of **2**. This chemical shift is consistent with a carbonyl carbon of an ester or a
carboxylic acid, but not with that of an aldehyde or ketone. The carboxylic acid possibility can be
eliminated by the absence of a corroborating O—H stretch in the IR spectrum and a low-field
carboxylic acid hydrogen in the ^1H NMR spectrum of **2**. The frequency of the stretch in the IR
spectrum suggests that the ester C=O group is not conjugated. Note that the ester carbonyl group
accounts for the fifth degree of unsaturation.

The ^1H NMR spectrum of compound **2** shows the following signals: a methyl singlet at δ 2.08 ppm
(possibly attached to a carbonyl or benzene ring), a methylene doublet at δ 4.71 ppm (attached to
an electronegative substituent such as O and an adjacent CH), two signals for hydrogens attached
to a double bond (accounting for the final degree of unsaturation) at δ 6.27 ppm and 6.63 ppm (the
signal at δ 6.27 is coupled to the methylene signal at δ 4.71 ppm), and finally, a 5H aromatic
multiplet at δ 7.2–7.4 ppm (clearly a monosubstituted benzene ring).

Putting all these structural fragments together gives cinnamyl acetate as the most reasonable
structure for compound **2**. The isomeric **3** can be eliminated from consideration because the methyl
group of **3** would have a much lower chemical shift than δ 2.08 ppm and 20.9 ppm in the ^1H and
^{13}C NMR spectra.

PhCH=CHCH$_2$O—C(=O)CH$_3$

PhCH=CHCH$_2$—C(=O)OCH$_3$

2

3

Two other structures might be considered as well, **4** and **5**, even though there is no obvious mechanistic pathway to them.

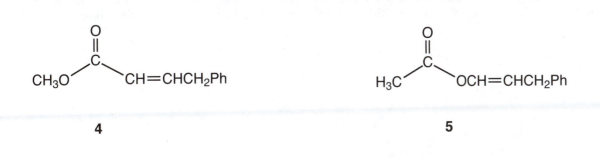

4   **5**

Compound **4** can quickly be rejected on the same grounds that served to eliminate **3**, the absence of a signal in the ¹H NMR spectrum for an OCH₃ group. Compound **5** is tougher. There are hints in both the ¹H and ¹³C NMR spectra that **5** cannot be correct. In compound **5**, a resonance form, **5′**, will contribute. That should result in a relatively high-field absorption for the β carbon and its hydrogen.

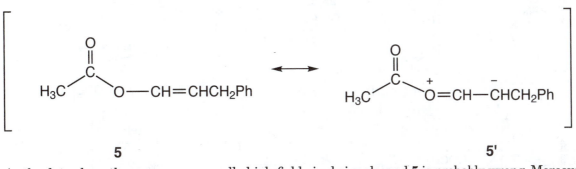

5   **5′**

As the data show, there are no unusually high-field vinyl signals, and **5** is probably wrong. Moreover, there is a straightforward mechanism for producing **2**. Let's do a mechanistic analysis now.

When α-phenylallyl acetate (**1**) and cinnamyl acetate (**2**) are drawn in more suggestive conformations, it is easy to see that this isomerization is just a thermally induced [3,3] shift.

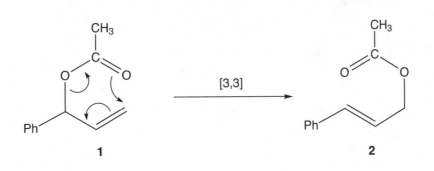

1   **2**

(continued)

Problem 23.49 *(continued)*

What is the stereochemistry of the double bond in cinnamyl acetate (**2**)? Hydrogens that are substituted cis on a double bond have smaller coupling constants (*J* = 6 to 12 Hz) than those substituted trans (*J* = 12 to 18 Hz, p. 411). In the case of compound **2**, the coupling constant is 16 Hz, strongly suggesting a trans double bond. This stereochemical assignment is also supported by the olefinic C—H bending vibration at 960 cm^{-1}, characteristic of trans disubstituted alkenes.

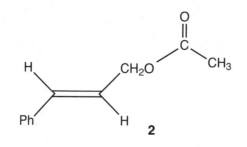

Problem 23.50   A sequence of alternating Diels–Alder and retro-Diels–Alder reactions gets the job done.

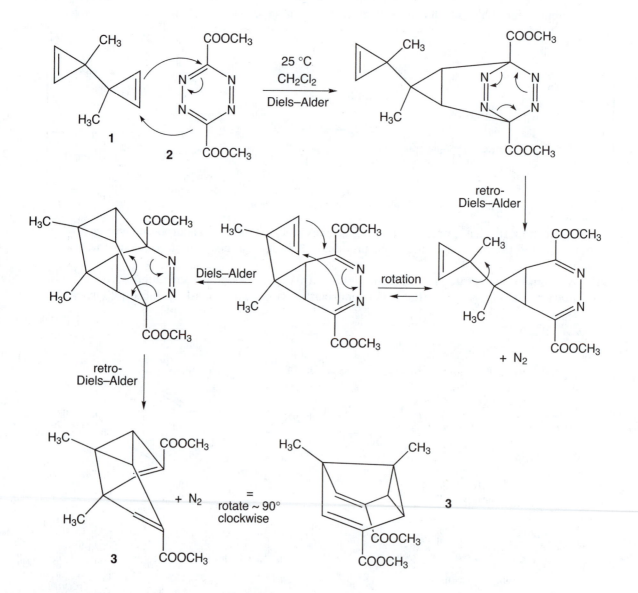

The semibullvalene product **3** undergoes a fast Cope rearrangement that makes the two methyl groups equivalent (δ 1.13 ppm singlet) and the vinyl and cyclopropyl hydrogens equivalent (δ 4.79 ppm singlet, shown as "H" in figure). The methoxy hydrogens (δ 3.73 ppm singlet) are also equivalent. The ^{13}C NMR spectrum can be rationalized in a similar way.

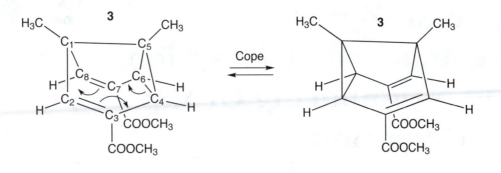

δ 14.9 (q)	Methyls
51.4 (q)	OCH$_3$ groups
60.6 (s)	C(1) and C(5)
93.7 (d)	C(2), C(4), C(6), and C(8)
127.2 (s)	C(3) and C(7)
164.7 (s)	Carbonyl carbon

Problem 23.51 The chairlike transition state should give the (*E,E*) and (*Z,Z*) dienes. The boatlike transition state should give the (*Z,E*) diastereomer.

As it is the chairlike transition state that is preferred, the products formed should be the (*E,E*) and (*Z,Z*) isomers, with the (*E,E*) form preferred for steric reasons.

The chairlike transition state (four-center) is favored:

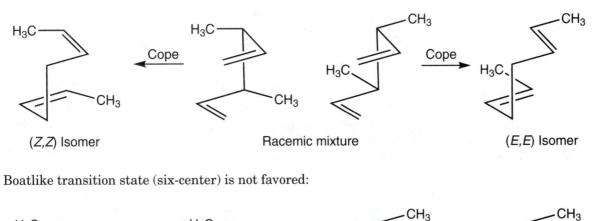

Boatlike transition state (six-center) is not favored:

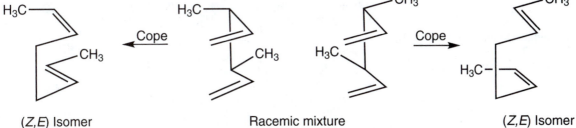

24

Special Topic: Intramolecular Reactions and Neighboring Group Participation

Intramolecular reactions play a very large role in organic chemistry. For example, both product structures and rates of reactions are strongly influenced by the presence of internal nucleophiles. Intramolecular displacements and additions are very common. The following problems show both simple chemistry and more complex reactions in which neighboring group participation, or "anchimeric assistance," appears.

Problem 24.2 This problem is nothing more than a standard, no-frills, hemiacetal formation in an intramolecular setting. Addition of the hydroxyl oxygen atom (a nucleophile) to the carbonyl group within the same molecule is followed by protonation and deprotonation steps. This reaction is usually acid- or base-catalyzed.

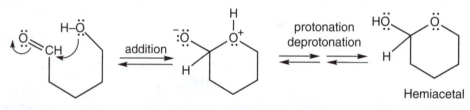

Problem 24.4 Acetic acid can attack at two "side" carbons to give the major products of ring opening. However, there is a third carbon that can participate in the S_N2 reaction with acetic acid, the methyl carbon. This reaction retains the ring and leads to 2-methyltetrahydrofuran.

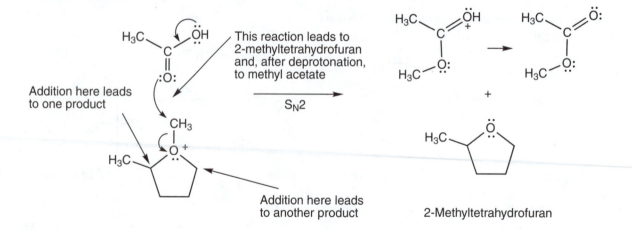

2-Methyltetrahydrofuran

Problem 24.5 Almost any labeling experiment will do. In one, we might use ^{13}C to produce two products in which the ^{13}C appears in a different position.

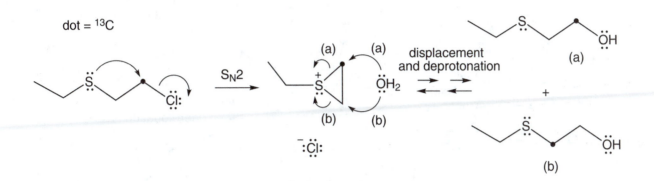

In another, we might use a methyl group to the same effect.

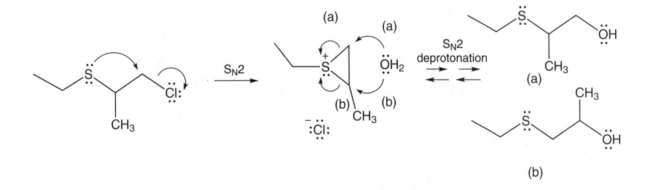

Neighboring group effects would be signaled by the formations of the rearranged products labeled (b) in each case.

Problem 24.6 The substitution reaction that occurs in Koenigs–Knorr reaction should occur via the oxonium ion shown in the reaction sequence. The reaction is run under S_N1 conditions. The intermediate oxonium ion should give a mixture of the α and β glycosides. The selective formation of only the β glycoside suggests that neighboring group assistance is involved. The acetyl group on the C(2) carbon can force the alcohol to attack only from the top face, giving selective formation of the β glycoside.

(continued)

Problem 24.6 *(continued)*

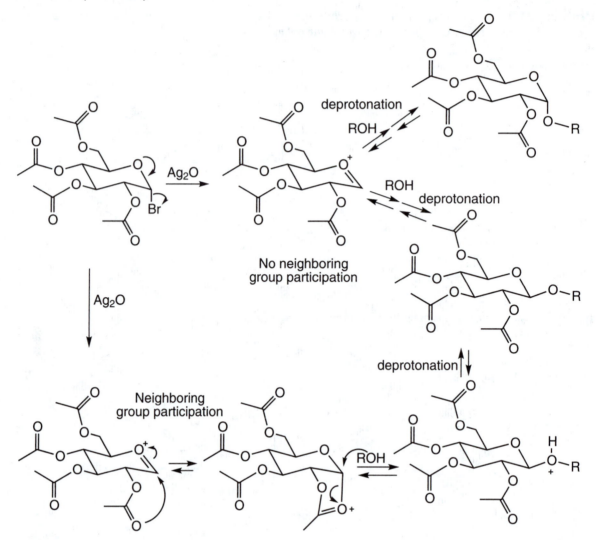

Problem 24.7 The structure of the product shows that a ring bond must be broken. The five-membered ring containing nitrogen can be spotted in the starting material and must be retained in the reaction. Although there is no intramolecular S$_N$2 reaction possible because of the orientation of the leaving group (no frontside S$_N$2 reactions!), there is an intramolecular elimination that can take place. This reaction leads directly to the product.

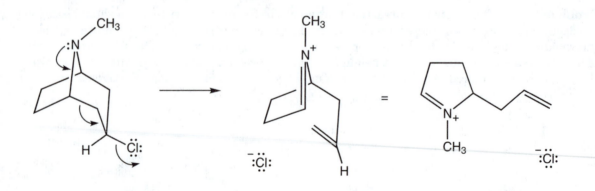

Problem 24.8

(b) The two ester groups remain unchanged in the product, and the three carbons in the chain must become the three carbons of the cyclopropane ring. An easy intramolecular S_N2 displacement by the enolate forms the product in a single step.

Et = CH₃CH₂

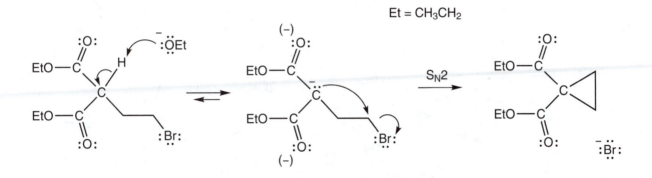

(c) This problem should also be easy. The nitrogen acts as an internal nucleophile, displacing chloride to give the cyclic ammonium ion. Hydroxide ion opens the ion at the less-hindered position to give the product. The usual pattern of an intramolecular S_N2 displacement followed by an intermolecular S_N2 displacement appears.

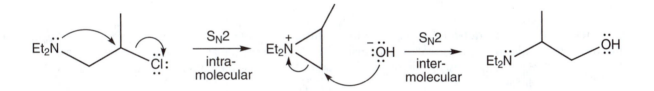

Problem 24.9 If fluorine acts as a neighboring group, a positively charged fluorine atom, a fluoronium ion, would result. Fluorine is very electronegative, and positively charged fluorine is very high in energy. So, such intermediates are only rarely encountered.

Fluoronium ion

Problem 24.10 A really easy one. Simply add bromine to *cis*-2-butene.

Problem 24.11   All you have to do in this problem is follow the pattern of Figure 24.21. If you do, you can see that the other enantiomer would also give the meso dibromide if a conventional mechanism were followed. Protonation, followed by displacement of water by bromide ion, gives the achiral meso compound. The product is shown first in the energy-maximum eclipsed form, then in the energy-minimum staggered arrangement.

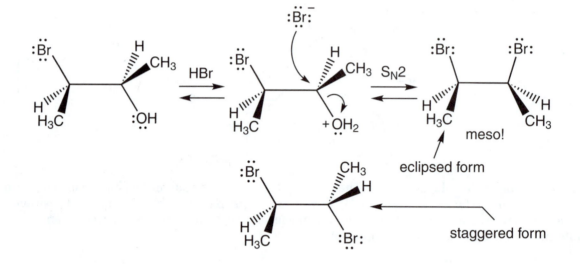

Problem 24.12   Bromination begins with bromonium ion formation. Opening of the bromonium ion by water leads to a bromohydrin. The bromonium ion from *trans*-2-butene can be formed in two ways (addition to the top or bottom of the alkene) to give a pair of enantiomeric bromonium ions.

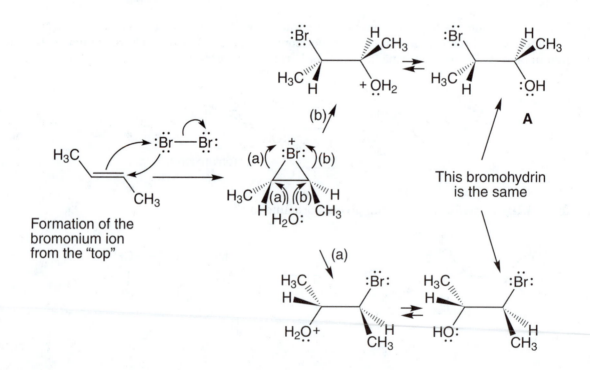

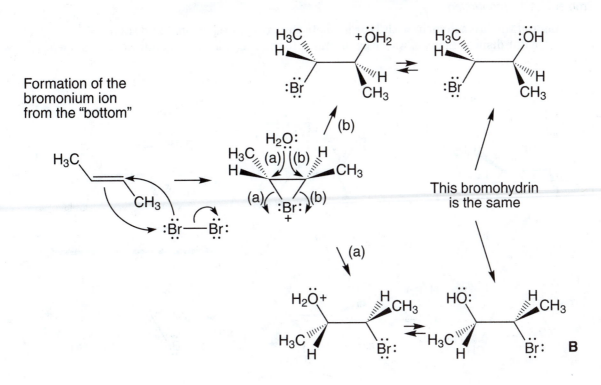

Formation of the bromonium ion from the "bottom"

This bromohydrin is the same

The two possible bromohydrins, **A** and **B**, are enantiomers.

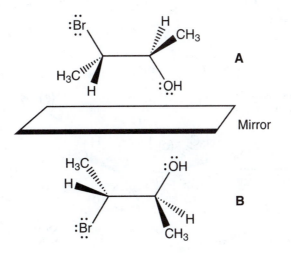

A

Mirror

B

(continued)

Problem 24.12 *(continued)*

Treatment of the bromohydrin with HBr leads to the protonated alcohol and then, by intermolecular displacement of water by bromide, to the racemic pair of enantiomers shown.

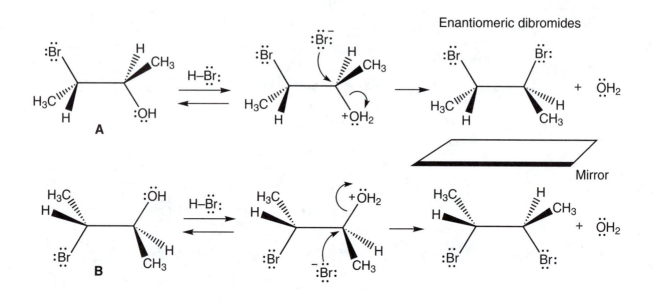

However, that's not what happens. The reaction takes another course, formation of *meso*-2,3-dibromobutane, and a neighboring group must be involved. It is a displacement by intramolecular bromine that takes place on the protonated alcohol.

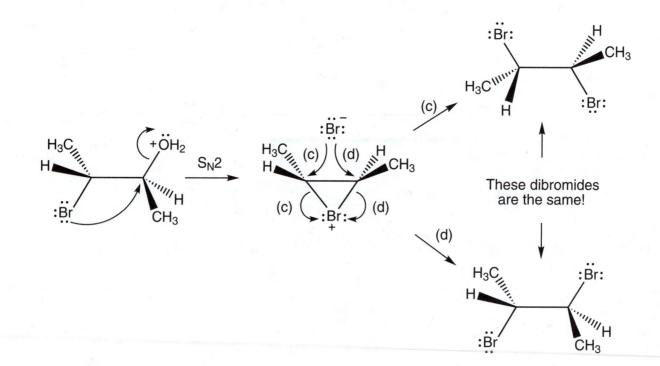

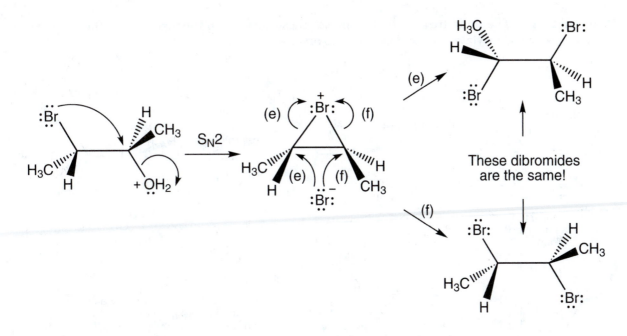

In fact, *all* these dibromides are the same. Each is the meso isomer.

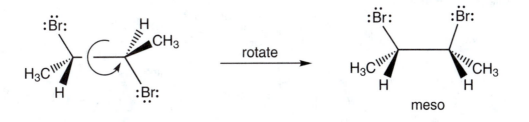

Problem 24.14   They are stereoisomers, but not mirror images. They must be diastereomers.

Problem 24.15   As for any phenonium ion (see Chapter 15 for countless examples of such species formed as intermediates in aromatic substitution reactions), there are three forms in which different carbons share the positive charge.

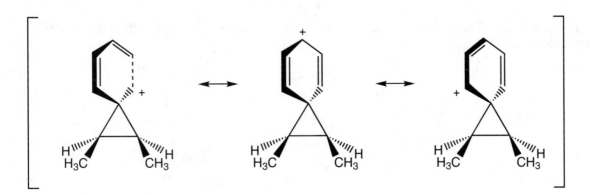

Problem 24.16   The steps are exactly the same as those outlined in the text for the other isomer of 3-phenyl-2-butyl tosylate. Follow Figures 24.26 and 24.28.

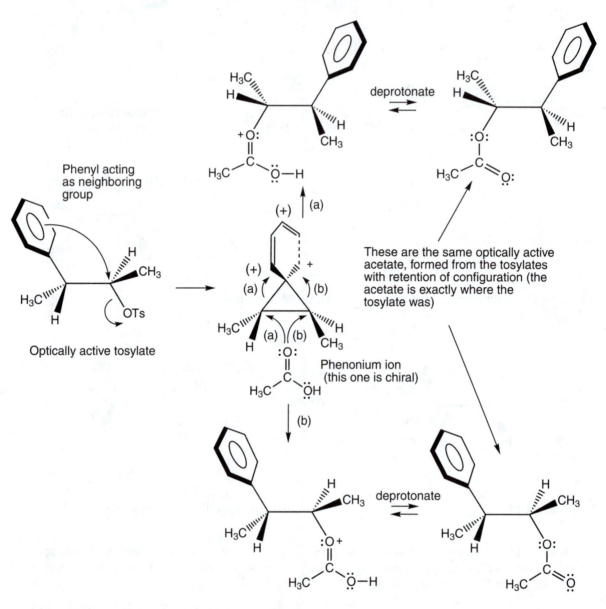

Phenyl acting as neighboring group

Optically active tosylate

deprotonate

These are the same optically active acetate, formed from the tosylates with retention of configuration (the acetate is exactly where the tosylate was)

Phenonium ion (this one is chiral)

deprotonate

A rearranged product

Problem 24.17   The tosylate of *syn*-7-norbornenol can undergo an alkyl shift to displace the tosylate leaving group.

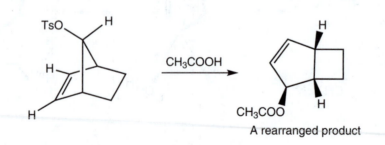

The mechanism for the reaction is shown. The intermediate is drawn as a nonclassical ion with three carbons sharing the positive charge. Attack at the bridgehead carbon forms the bicyclo-[3.2.0]heptene, which leads to the rearranged product.

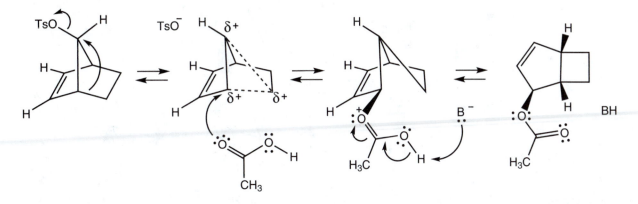

Problem 24.18 Acetic acid could attack either of the carbons bearing the partial positive charge. The result would be formation of the racemic mixture of the tricyclic acetate shown. The mechanism for its formation is shown below. Perhaps the tricyclic product is not observed because there is less carbocation character on the two alkenyl carbons. There is also more ring strain in the tricyclic product and, because the solvolysis reaction is reversible, the higher-energy product would be disfavored at equilibrium.

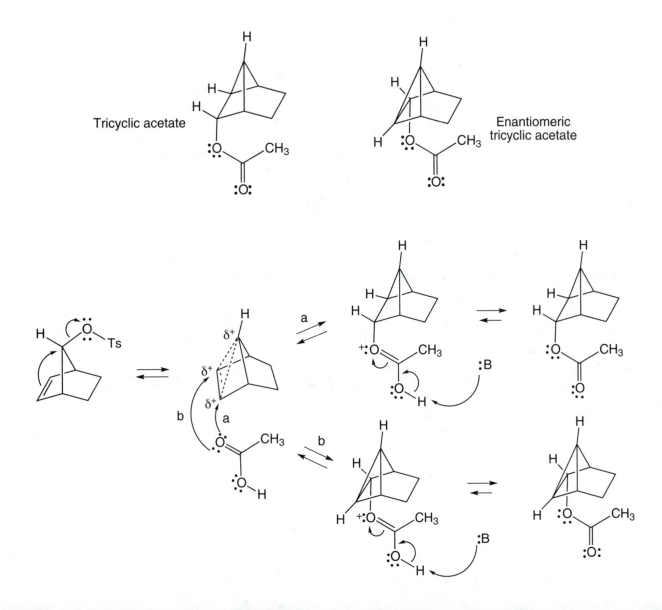

Problem 24.19   Let's combine the π molecular orbitals of ethylene with a carbon 2p orbital. The 2p orbital interacts with π, but there is no interaction between the 2p orbital and π*.

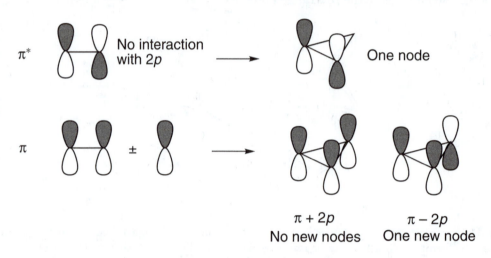

To order these, simply count nodes. These molecular orbital energies could also be generated by using a Frost circle (Chapter 14, p. 654).

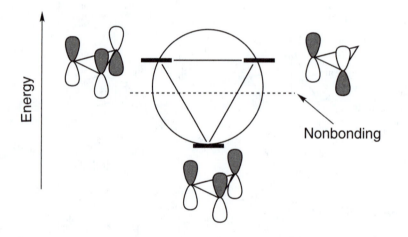

Problem 24.20   The positive charge is shared among the three carbons, as shown.

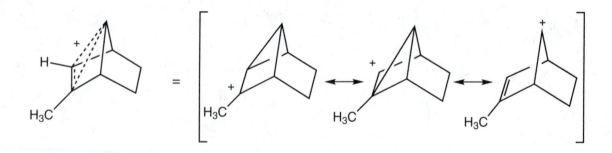

Problem 24.21 There is, by definition, only one barrier in the concerted rearrangement (**A**). In the stepwise process (**B**) there must be two transition states, one for formation of the less stable secondary carbocation and the second for rearrangement (hydride shift) to the more stable tertiary carbocation.

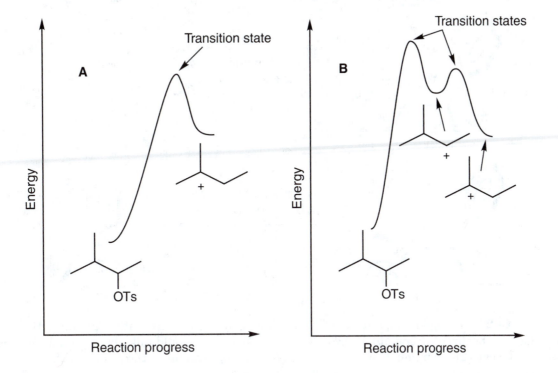

Problem 24.22 There really should not have been any controversy over phenonium ions. They are not substantively different from the intermediate ions involved in aromatic substitution.

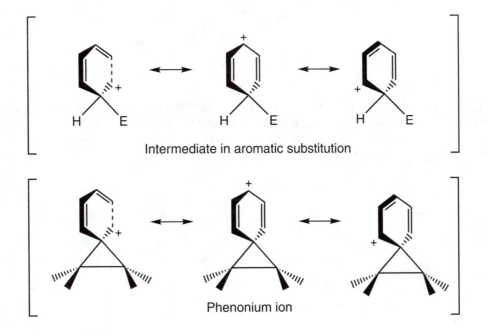

Intermediate in aromatic substitution

Phenonium ion

(continued)

Problem 24.22 *(continued)*

Perhaps π participation to form the phenonium ion was confused with σ participation to form something quite different—a three-center, two-electron bonding system in which the positive charge is not delocalized into the benzene ring.

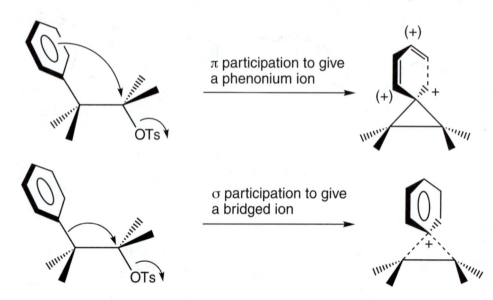

Problem 24.23   The open ion, if it exists, would be like any other carbocation, planar with a central, sp^2-hybridized carbon atom. If you want more detail, we could talk about resonance forms in which the carbon–hydrogen bonds of the methyl group help stabilize the positive charge (hyperconjugation).

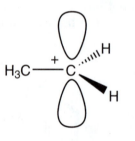

The cyclic ion will have an MO pattern that resembles all triangular arrays of orbitals. The cyclopropenium ion (Problem 24.19) is one example. There will be one bonding molecular orbital and two antibonding molecular orbitals. There are only two electrons in this system, and they will go into the strongly bonding molecular orbital.

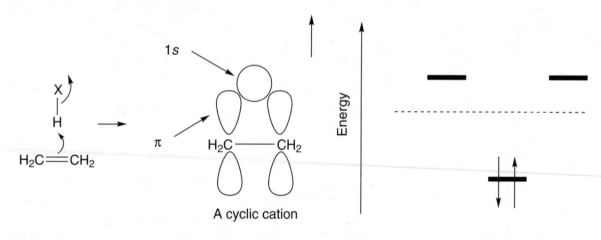

A cyclic cation

So the question is whether this stabilization will "pay" for the angle strain of the three-membered intermediate. That question is not at all easy to solve. Currently, calculations suggest that, at least in the gas phase, the cyclic ion is slightly better. In solution, ions are strongly solvated, and it is not clear whether the same stability order will obtain.

Problem 24.24 The positive charge is shared among the three carbons, as shown.

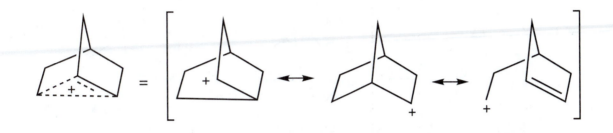

Problem 24.26 This problem is tough. Let's take the simple Coates' cation and do one rearrangement. What's this ion? Label three carbons with dots, and the two three-membered rings with boldface lines, and then redraw. It's the same cation! It's Coates' cation all over again. The rearrangement is degenerate.

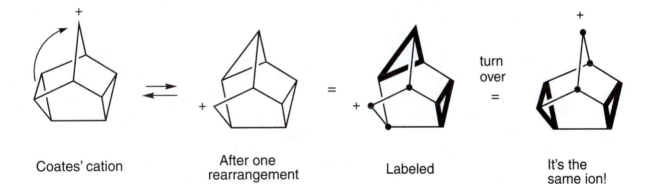

| Coates' cation | After one rearrangement | Labeled | It's the same ion! |

In this system, every rearrangement leads to the same ion. In fact, if you do enough rearrangements, all the carbons become equivalent. Start with a Coates' cation in which we have labeled the four equivalent carbons with dots. One rearrangement leads to an ion in which the "top" bridge and the cyclopropyl position have become involved. We have now shown that all carbons except the bridgehead are equivalent.

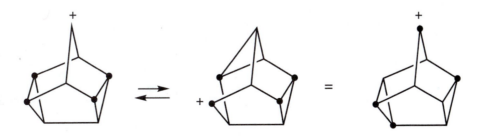

(continued)

Problem 24.26 *(continued)*

What happens to the bridgehead carbons in this rearrangement process? The labeled ion below answers this question. They equilibrate with the cyclopropyl carbons and thus are equivalent to all the others. All nine carbons are indeed the same, as long as rearrangement is rapid. The ^{13}C NMR spectrum would show only a single signal, not the three that are found experimentally.

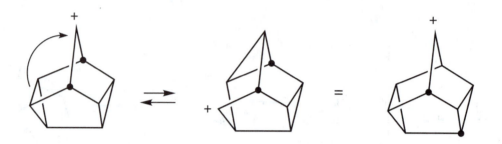

Answers to Additional Problems

Problem 24.27   This reaction involves a straightforward intramolecular displacement of bromide by the nucleophilic amine. As always, it probably helps to draw good Lewis structures for the starting material first. Then attention is focused on the nonbonding pair of electrons on the nitrogen atom that will act as the nucleophile displacing bromide. Deprotonation by hydroxide completes the reaction.

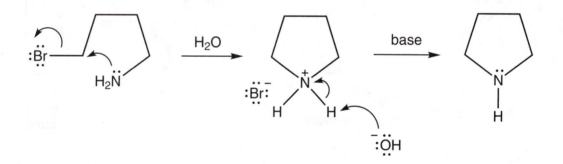

Problem 24.28   Here, too, an intramolecular displacement is followed by deprotonation. In the smaller molecule, intramolecular displacement involves the formation of a highly strained four-membered ring and is slow. In the first example, intramolecular formation of the relatively strain free five-membered ring is much faster than the intermolecular solvolysis reaction on a primary chloride.

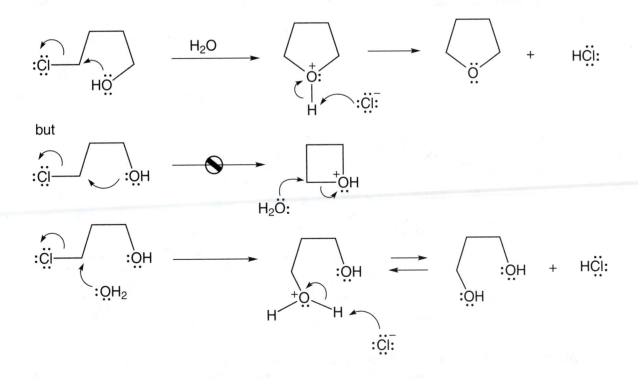

but

Problem 24.29 A straightforward acid-catalyzed hydration of the double bond could lead to the first product shown. But formation of the second product requires a seemingly strange rearrangement—clue number 2 to a neighboring group participation. Surely that bromine, with its three pairs of nonbonding electrons, is playing the role of neighboring group. Let's start the addition reaction and look for an opportunity for the bromine to act. The first step is protonation to give the secondary carbocation.

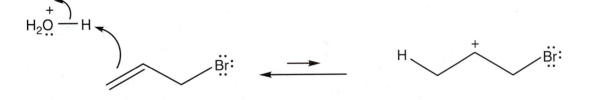

Addition of water to this intermediate can give only the first product, but if bromine acts as nucleophile, we get a bromonium ion that can open in two ways.

(continued)

Problem 24.29 *(continued)*

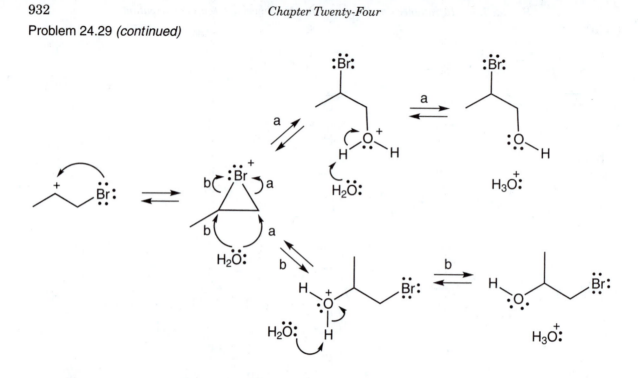

We expect that the 1-bromo-2-propanol from path b would be the major product. We have learned that the bromonium (and mercurinium, chloronium, protonated epoxide, and protonated episulfide) ion is attacked by a nucleophile at the more substituted carbon. This generalization is not always correct. There are cases where the steric factors outweigh the partial positive charge on the three-membered ring. But without further information, it is logical to assume attack at the more substituted carbon of the bromonium ion will be favored.

Problem 24.30 This problem is basically the same as Problem 24.29. Protonate, form the bromonium ion, and open.

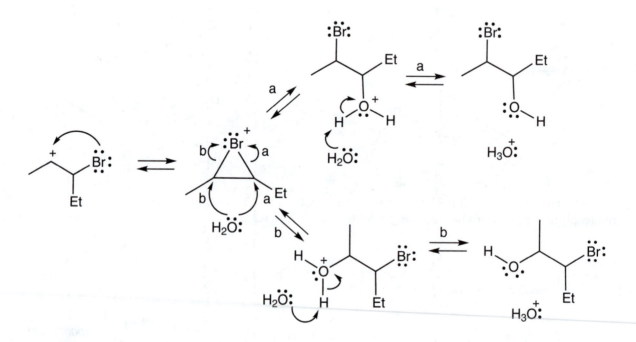

In this case, the bromonium ion has an equal amount of substitution on the two carbons. Therefore, we expect little preference for either product.

Problem 24.31   Here, intramolecular S_N2 displacement of the secondary chloride wins over intermolecular reactions. There is a cyclic ammonium ion (aziridinium ion) intermediate that is opened in a second, intermolecular S_N2 reaction.

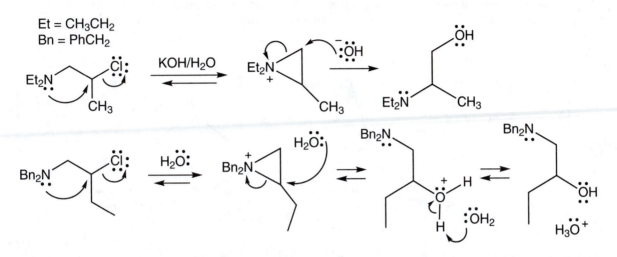

The question becomes why hydroxide opens the aziridinium ion at one position and water at the other. Strong nucleophiles (here, hydroxide) react at the sterically less demanding point, whereas weaker nucleophiles (here, water) break the weaker (more substituted) bond.

Problem 24.32   This problem follows the pattern of Problems 24.29 and 24.30. Protonation is followed by chloronium ion formation. Opening of the chloronium ion by chloride will occur with inversion (it's an S_N2 reaction) at the more substituted position, thus breaking the longer, weaker carbon–chlorine bond. The result is the (1R,2R)-dichloride.

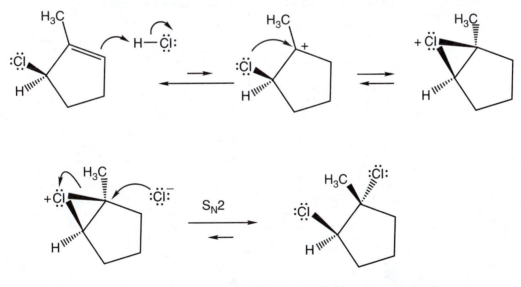

(1R,2R)-1,2-Dichloro-1-methylcyclopentane

The reaction is slow because the electron-withdrawing inductive effect of the chlorine makes protonation to give an adjacent carbocation difficult.

Problem 24.33   Compound **1** can do an intramolecular displacement, whereas neither **2** nor **3** can, which accounts for the increase in rate. However, the data do not allow us to determine whether there is a bridged intermediate or an open intermediate involved in this reaction. The rate increase could come from assistance by the π bond in either a symmetrical fashion to give a delocalized ("bridged") ion or an unsymmetrical fashion giving rapidly interconverting classical carbocations.

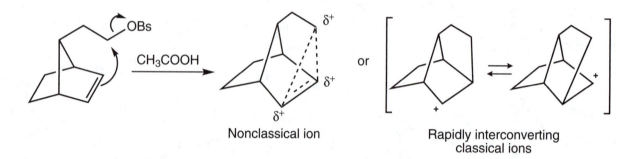

Problem 24.34   The product is formed with retention, and there is a rate difference between **1** and **2** of 10^5 to 10^6. Two clues to neighboring group participation are present. Compound **3** must be formed by neighboring group participation of sulfur. It takes a look at these molecules in three dimensions to see the reason why **1** can undergo an intramolecular displacement and **2** cannot. In **1**, the internal nucleophile, sulfur, can displace the leaving group from the rear (even though this displacement must take place in the less favored, diaxial conformation). In **2**, there is no conformation in which intramolecular S_N2 is possible. A frontside displacement is required, and such reactions do not occur. A relatively slow intermolecular S_N2 can occur. (In the second step of the reaction, it is the water rather than the ethanol that will react as the nucleophile. Water is slightly more nucleophilic than an alcohol, and it is less sterically encumbered.)

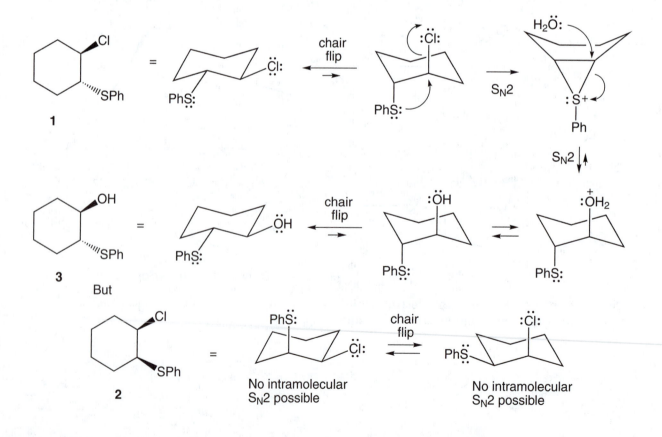

Problem 24.35   Consider the transition states for intramolecular displacement by the two phenyl groups in question.

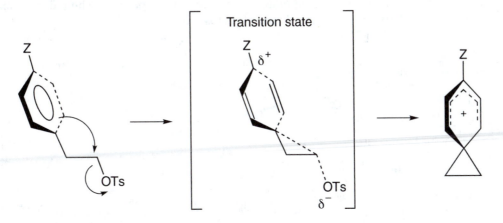

When Z = OCH$_3$, the transition state is stabilized, as the methoxy group helps to delocalize the adjacent partial positive charge. The energy of the transition state is lowered, and the rate of intramolecular displacement is increased. When Z = NO$_2$, the electron-withdrawing nitro group destabilizes the transition state, and the rate of intramolecular displacement is diminished.

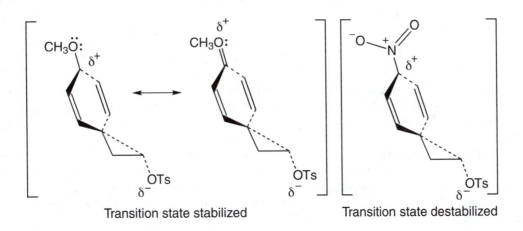

Problem 24.36   The double bond in **1** is not in a position to participate in ionization of the tosylate, but one of the σ bonds is (boldface line). An allylic carbocation (**A**) is formed and then captured by water to produce **2**.

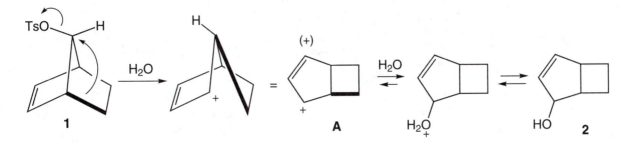

In compound **1**, participation of the σ bond leads to a relatively stable allylic cation. That would not be the case in **3**, and reaction is much slower.

Problem 24.37 Here the clues to neighboring group participation are the "strange rearrange-ment" in which the chlorine and methoxy groups seem to change places, and the mysterious switch in stereochemistry from cis in the starting material to trans in the product. The methoxy oxygen is the neighboring group. A three-dimensional picture helps a lot.

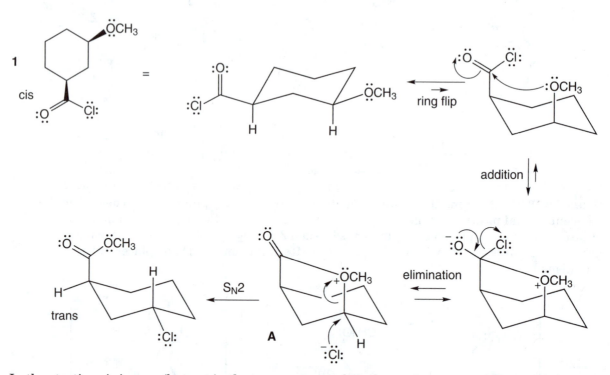

In the starting cis isomer (but not in the trans compound **2**), the methoxy group is in position to undergo an addition–elimination reaction generating bicyclic ion **A**. Opening by chloride (S_N2) from the rear yields the product.

Problem 24.38 Bromine and a double bond add up to bromonium ion formation. This inter-mediate is opened by the nearby acid carbonyl and deprotonated to the product. There are two possible points of attack on the bromonium ion. One, the route actually followed, leads to lactone **A**. A reasonable answer to this problem might take the other approach, opening the bromonium ion to a six-membered ring, and ultimately giving **B**.

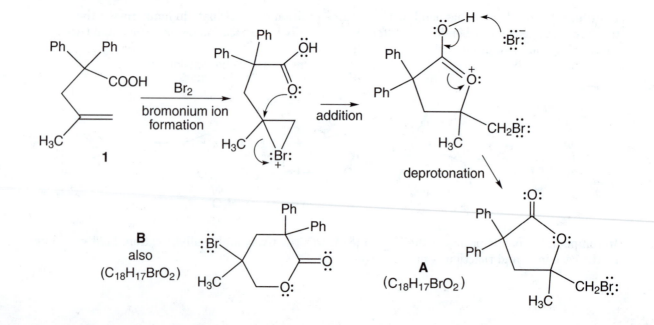

Problem 24.39 This reaction is an example of neighboring group participation by the π system of a benzene ring in which the product is sufficiently stable to be isolable. Potassium *tert*-butoxide is sterically hindered and is unlikely to be effective at intermolecular displacements. There is little strain in the five-membered ring product. Deprotonation of the phenol generates a phenoxide that undergoes a rapid intramolecular displacement of the leaving group (BsO = p-BrC$_6$H$_4$SO$_2$O) at the nearby carbon.

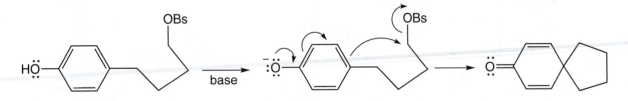

Problem 24.40 Surely the first step in acid will be protonation of the epoxide. Normally, a protonated epoxide would be opened from the backside by water. In this case, an adjacent carbon–carbon σ bond (boldface line) is poised in perfect position to do a faster intramolecular displacement to give delocalized intermediate **A**.

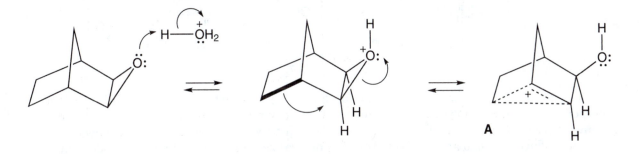

Now let's redraw **A** to show the symmetry, and then let water add at C(1) to give, after deprotonation of the oxonium ion, the product.

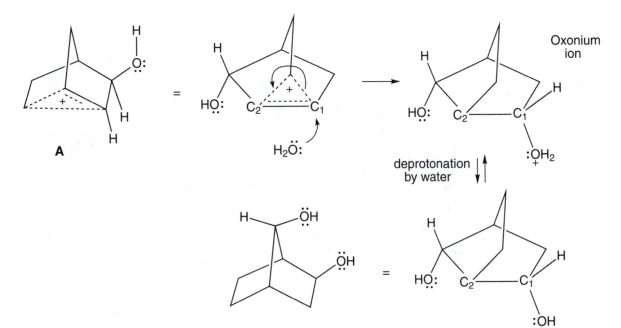

(continued)

Problem 24.40 *(continued)*

Addition of water at the more hindered C(2) would lead not to the product, but to the unobserved diol, **B**.

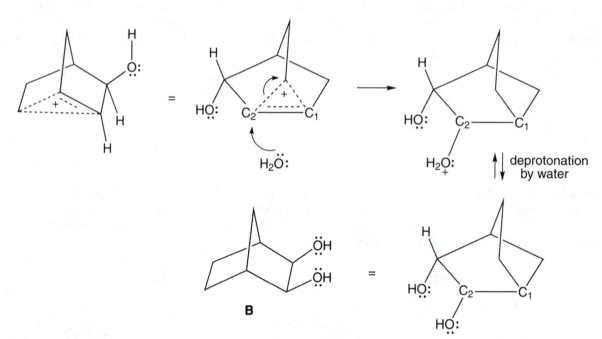

Problem 24.41   As we saw in Problem 24.25, this reaction is just another way of forming the 2-norbornyl cation. Symmetrical displacement of the leaving group by the double bond of the five-membered ring gives the delocalized ion **A**. Opening by acetic acid in an intermolecular S_N2 reaction produces the exo acetate. Note that the delocalized ion for the monomethyl derivative **A** (R = CH$_3$, R' = H) is not symmetrical. From studies of the products formed, it became apparent that acetic acid addition occurred only at the tertiary carbon of the bridged ion.

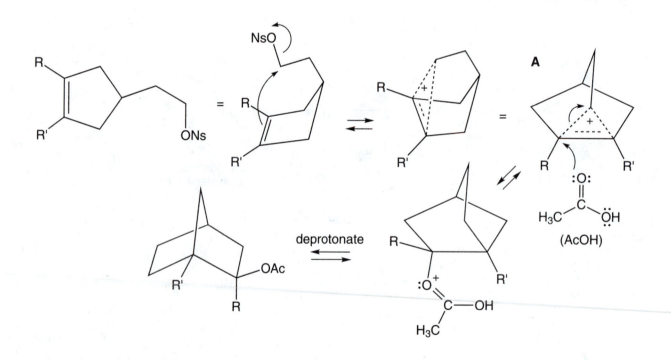

What do the kinetic data tell us about the nature of the carbocation? Although we have drawn them as delocalized, it is possible to view the reaction as proceeding through pairs of equilibrating open cations. As before (Section 24.3b), the rate data help us to differentiate delocalized from classical ions. The monomethyl derivative reacts seven times faster than the parent. This is, of course, compatible with either mechanistic scheme. In the delocalized ion, one of the contributing structures contains a tertiary carbocation. Similarly, for the open ions, one of the equilibrating structures is a tertiary carbocation. However, the two mechanistic schemes make different predictions for the dimethyl compound. In a delocalized carbocation, both methyl groups contribute at the same time. If one methyl group produces a rate enhancement of 7, two must produce $7^2 = 49$. In contrast, if it is the formation of an open, tertiary carbocation that is important, the second methyl group increases this possibility by only a factor of 2. The two methyl groups never exert their influence at the same time. Thus the predicted rate increase would be $7 + 7 = 14$. The experimental rate for the dimethyl derivative is 38.5, much closer to the predicted value for the delocalized ion than to that for a system of equilibrating open cations.

Problem 24.42

(a) This problem is an example of the Hofmann rearrangement using chlorine instead of bromine.

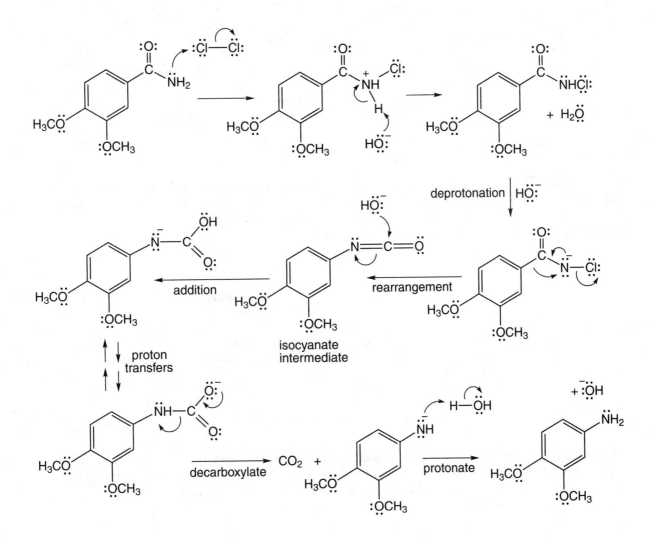

(continued)

Problem 24.42 *(continued)*

(b) This reaction is a Wolff rearrangement in which an initially formed ketene is trapped by methyl alcohol reaction solvent. Begin by drawing a more detailed and suggestive view of the diazo ketone. Loss of nitrogen with bond migration (boldface line) gives the ketene. Capture by methyl alcohol ultimately gives the ester.

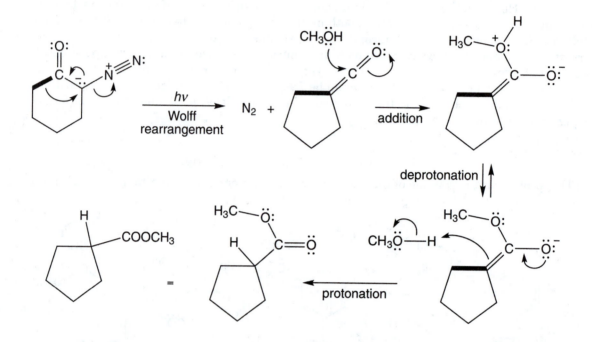

Problem 24.43 The problem here is not so much finding the neighboring group participation, but rather the transformation of the initial product into isobutyraldehyde. It should be obvious now that the rate increase comes from the intramolecular displacement of the leaving group by the oxygen of the methoxy group to give oxonium ion **A**.

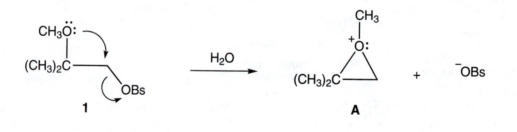

Although **1** can do this, the neopentyl compound **2**, of course, cannot. Now, how to get to the final product from **A**? The main question seems to be how to lose the required one-carbon atom. The most serious candidate for loss would seem to be the methyl group attached to oxygen.

A stepwise mechanism involves opening **A** to give a tertiary carbocation, **B**. A hydride shift generates the resonance-stabilized **C**, and removal of the methyl group through displacement by water (don't just pop it off as $^+CH_3$) gives the product.

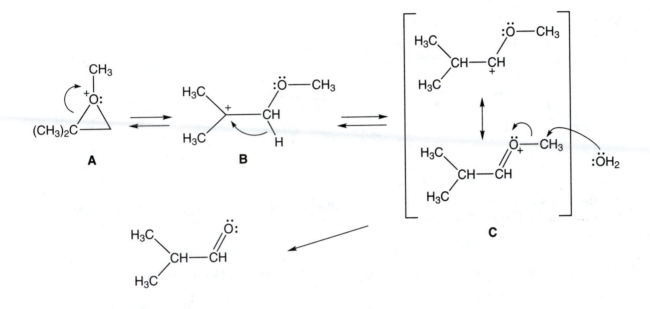

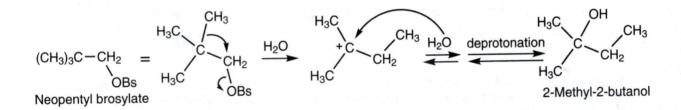

The product that we predict in the hydrolysis of neopentyl brosylate is 2-methyl-2-butanol via a methyl shift.

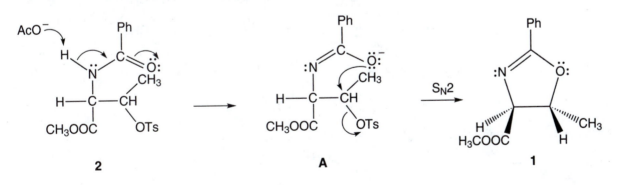

Problem 24.44 In base, removal of the amide hydrogen by acetate takes place to give a resonance-stabilized anion **A**. Displacement of tosylate by the oxygen end of the anion then occurs to give **1**.

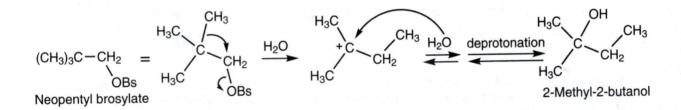

(continued)

Problem 24.44 *(continued)*

We can use the stereochemistry of **1** to deduce the stereochemistry of **2**. Displacement of tosylate by oxygen is an S$_N$2 reaction and must occur with inversion. Thus, we can specify the geometry of **A**, and of **2**.

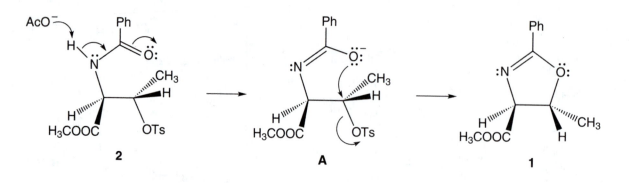

Problem 24.45   In HCl, the first step is surely protonation of the alcohol, converting the poor leaving group $^-$OH into OH$_2$, a good leaving group.

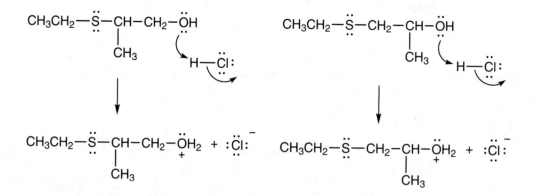

Ordinarily, one would now expect the chloride ion to displace water and produce two different chlorides. But the hint tells you to focus on the sulfur. What can sulfur do? You know that sulfur is a powerful nucleophile, and it is lurking right there within the same molecule as the leaving group, water. Try a pair of intramolecular S$_N$2 displacements.

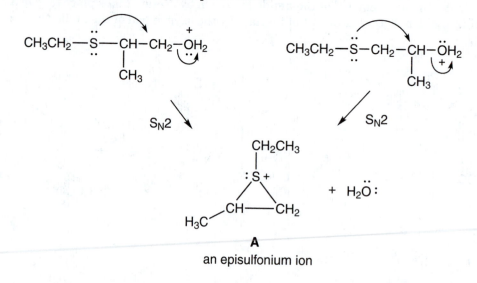

an episulfonium ion

They give the same product, **A**! This molecule is called an episulfonium ion, and it behaves much like a bromonium or oxonium ion. As with bromonium ions, chloride ion opens the episulfonium ion by adding to the more substituted carbon (Chapter 11, p. 494). Of course, as **A** is formed from *each* starting alcohol, the product must be the same in each case.

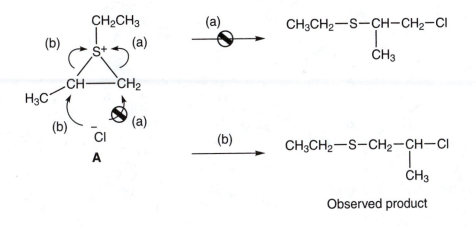

Observed product

Problem 24.46 Once again, the pattern of Problems 24.29, 24.30, and 24.32 emerges. Bromonium ion formation is the first step, but opening by bromide cannot give both products. There has been a "mysterious rearrangement," clue 2 to the operation of a neighboring group. This time the neighboring group is phenyl. The bromonium ion is opened by the neighboring benzene ring to give a phenonium ion. Subsequent opening by bromide in the two possible ways gives the two products.

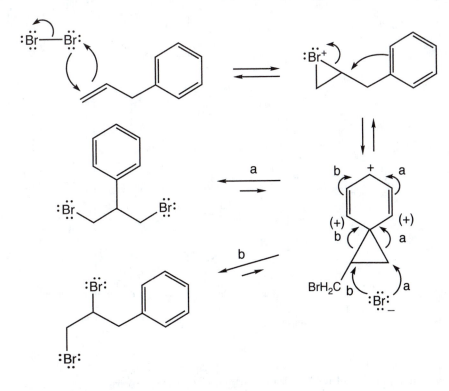

Problem 24.47

(a) Conversion of **1** into **2** could involve the standard alcohol-to-chloride transformation via a chlorosulfite intermediate **A**.

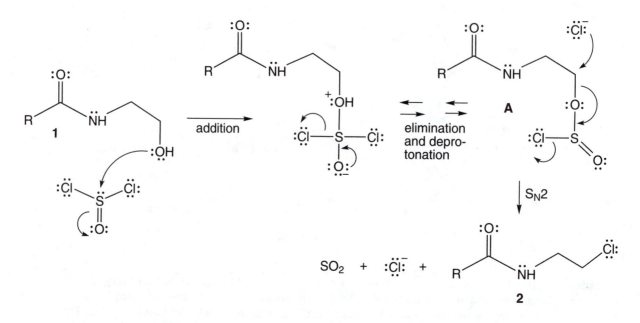

Treatment of **2** with base leads to proton loss to give the resonance-stabilized anion **B**. Intramolecular displacement of chloride leads to **3**.

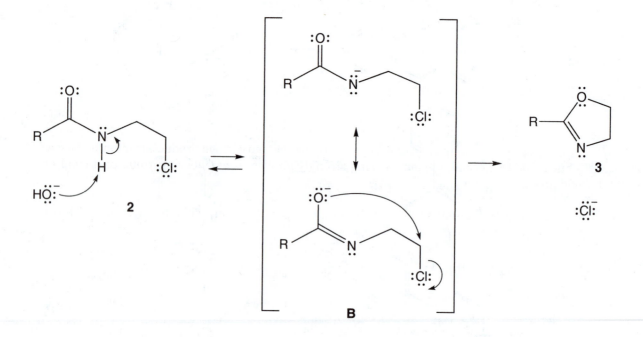

(b) However, when **1** is treated with thionyl chloride at low temperature, compound **4** is isolated rather than **2**. What could compound **4** be? Notice that in chlorosulfite **A** the amide oxygen is poised to displace sulfur dioxide and chloride ion in intramolecular fashion to produce the oxazoline salt, **4**.

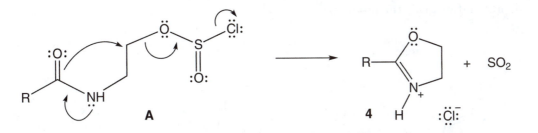

The salt **4** would be water soluble and has the same elemental composition as **2**. Upon heating, oxazoline salt **4** affords **2** by a second intermolecular S$_N$2 reaction.

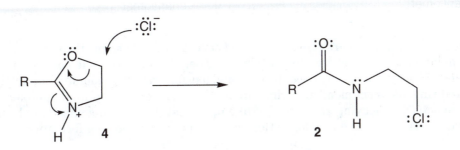

Neutralization of salt **4** with aqueous carbonate, of course, liberates **3**.

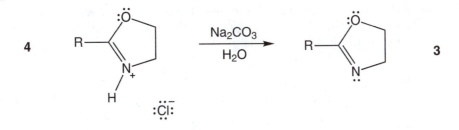

In fact, there was much confusion in the early literature with regard to β-chloroethylamides **2** and oxazoline salts **4**, as they had the same elemental compositions and both afforded oxazoline **3** upon treatment with base.

Problem 24.48   Let's start with two observations and then do the mechanism. First, because the starting materials are all achiral, there can be no optical activity in the product. It must be formed as a racemic mixture. Second, the lactone ring strongly implicates neighboring group participation by the acid group.

The first step in the mechanism is surely iodonium ion formation. Now opening can take place with the carbonyl oxygen acting as nucleophile.

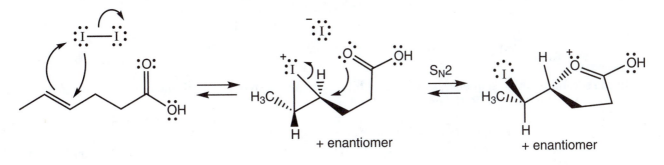

(continued)

Problem 24.48 *(continued)*

A final deprotonation gives the product.

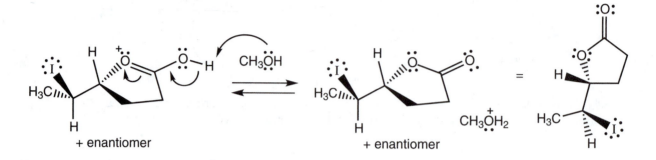

The formation of two enantiomers can be explained several ways. Perhaps the easiest is to note that iodonium ion formation can occur equally from either side of the double bond, each iodonium ion leading to one enantiomer. Why does the carbonyl oxygen open the iodonium ion to give a five-membered ring, not a six-membered ring? In any S_N2 reaction the optimal approach is from the rear of the departing leaving group; a 180° line up is best. That arrangement can be better approximated in five-membered ring formation than in six-membered ring formation.

Problem 24.49 In base, the dicarboxylic acid is first converted into the dicarboxylic acid salt. Bromine then adds to the double bond to give bromonium ion **A**. Addition is from the more accessible, exo face of the molecule. The bromonium ion is now opened from the backside by an intramolecular S_N2 reaction of the perfectly positioned carboxylic acid salt. Neutralization of the carboxylate salt with hydrochloric acid gives the product.

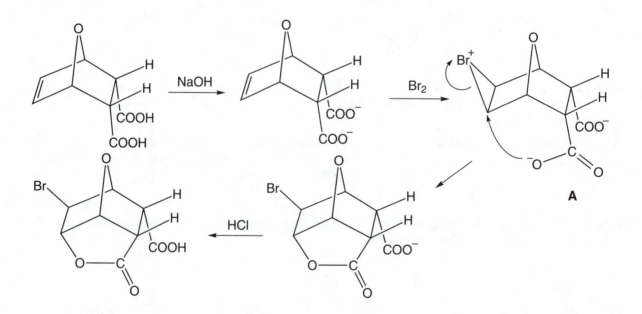

In the second reaction sequence, things are a bit more complicated. Begin by opening the anhydride to give the dicarboxylate **B**, the exo version of **A**.

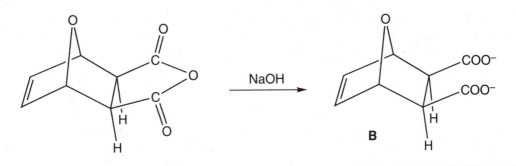

Bromination of the double bond of **B** from the exo face still occurs. This time the product is bromonium ion **C**. However, now neither carboxylate is in a position to open the three-membered ring. Neither can reach.

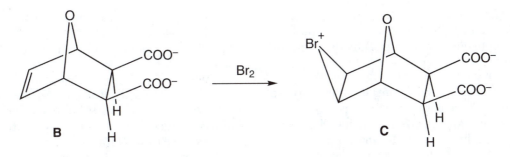

What can happen instead? Note that one of the adjacent carbon–carbon σ bonds is positioned to open the bromonium ion from the backside (boldface line). The resulting carbocation **D** is resonance stabilized by the adjacent oxygen atom, as shown by the (+). Also, one of the carboxylate groups is now poised to capture the carbocation. Acidification of the carboxylate affords the observed product.

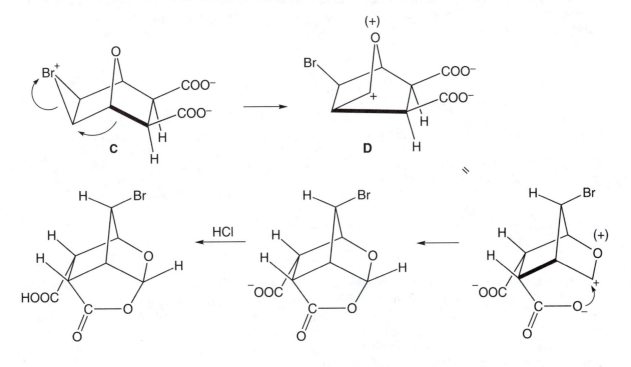

Problem 24.50 It is easiest to rationalize first the ^{13}C NMR spectrum of the 2-norbornyl cation at the lower temperature. Redraw the 2-norbornyl cation so that its symmetry is apparent. In addition, the figure shows the hydrogens at C(1), C(2), and C(6).

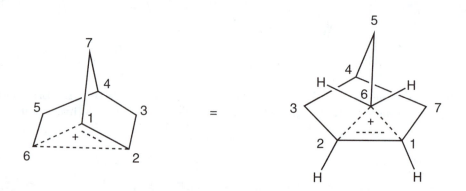

At –159 °C, the structure of the 2-norbornyl cation is static. Accordingly, it is easy to see that C(1) and C(2) are equivalent, as are C(3) and C(7). However, C(4), C(6), and C(5) are each different from all the other carbons. There is a total of five different carbons for this static structure, and the ^{13}C NMR spectrum at –159 °C shows exactly that. Note that the chemical shift of C(1) and C(2) relative to that of C(6) shows that most of the positive charge resides on C(1) and C(2) (more substituted).

At –80 °C, there is a simpler, three-line spectrum in which somehow C(1), C(2), and C(6) have become equivalent, as have C(3), C(5), and C(7). Only C(4) is unique. These equivalencies have been rationalized by postulating a series of fast hydride shifts between the 6-, 1-, and 2-positions. These shifts do not occur at the lower temperature—there is not enough energy. The barrier to these hydride shifts has been determined to be about 6 kcal/mol.

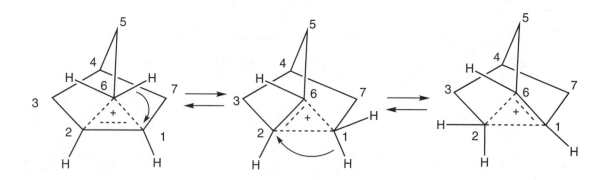

Problem 24.51 Notice the desired aziridine **2** has a molecular formula of $C_{12}H_{17}N$. Accordingly, isomers **3** and **4** have incorporated the additional elements of SO_2. If we examine the chlorosulfite intermediate for this reaction, it is apparent that a five-membered heterocycle **A** with the correct molecular formula is available upon nucleophilic attack of nitrogen on the sulfinyl group to give **B**. (Alternatively, amino alcohol **1** could first react with thionyl chloride via the more nucleophilic amine nitrogen, followed by alcohol cyclization to give **C**.)

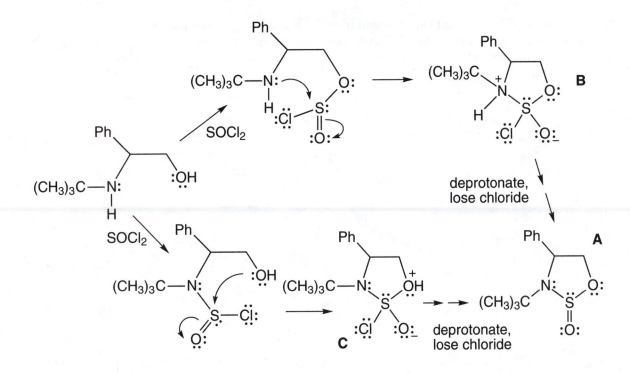

So far, so good. But where is the second compound $C_{12}H_{17}NO_2S$? If you examine heterocycle **A** closely, you will notice that there are two stereogenic atoms. The stereogenic carbon is probably obvious, but the second atom, a stereogenic sulfur, is not so clear. However, **A** is actually a pair of diastereomers, **3** and **4**.

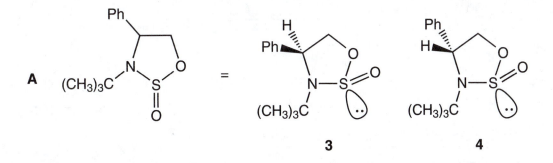

Problem 24.52 The starting material is in a chair conformation. The S_N2 reaction occurs from the less-stable chair structure because the nucleophilic attack must take place at the backside of the carbon–bromine bond. This route is not accessible in the more stable chair structure.

Problem 24.53 The nucleophilic atom in the reaction is the oxygen anion, the alkoxide. The animation doesn't show the HOMO density on the oxygen very clearly, but it must be there. The reaction can only proceed if the oxygen anion reacts as a nucleophile displacing the bromide.

Problem 24.54 The clue that this reaction involves a neighboring group is the fact that there is a rearrangement, classic clue number 2. The neighboring group participation is shown in the last step of the overall reaction.

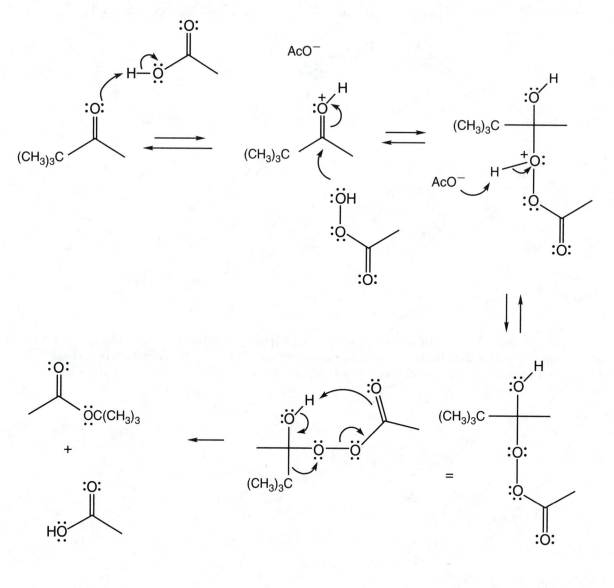